FOURTH EDITION

Basic Electronics

BERNARD GROB
Instructor, Technical Career Institutes, Inc.
(formerly RCA Institutes, Inc.)

Gregg Division
McGraw-Hill Book Company

New York
St. Louis
Dallas
San Francisco
Auckland
Bogotá
Düsseldorf
Johannesburg
London
Madrid
Mexico
Montreal
New Delhi
Panama
Paris
São Paulo
Singapore
Sydney
Tokyo
Toronto

Library of Congress Cataloging in Publication Data

Grob, Bernard.
  Basic electronics.

  Bibliography: p.
  Includes index.
  1. Electronics.   I. Title.
TK7816.G75   1977   537.5   76-45812
ISBN 0-07-024923-7

Other Books by the Author

Applications of Electronics (with Milton S. Kiver)
Basic Television

*In memory of
my father and mother*

BASIC ELECTRONICS

Copyright © 1977, 1971, 1965, 1959 by McGraw-Hill, Inc. All rights reserved. Printed in the United States of America. No part of this publication may be reproduced, stored in a retrieval system, or transmitted, in any form or by any means, electronic, mechanical, photocopying, recording, or otherwise, without the prior written permission of the publisher.

  1 2 3 4 5 6 7 8 9 0   VHVH   8 9 8 7 6 5 4 3

The editors for this book were Gordon Rockmaker and Alice V. Manning, the designer was Marsha Cohen, the art supervisor was George T. Resch, and the production supervisor was Iris A. Levy. It was set in Souvenir by York Graphic Services, Inc.
Printed and bound by Von Hoffman Press, Inc.

Contents

PREFACE		xiii
SURVEY OF ELECTRONICS		1
CHAPTER 1	ELECTRICITY	11
	1-1 Negative and Positive Polarities	11
	1-2 Electrons and Protons in the Atom	12
	1-3 Structure of the Atom	14
	1-4 The Coulomb Unit of Charge	17
	1-5 The Volt Unit of Potential Difference	20
	1-6 Charge in Motion Is Current	22
	1-7 Resistance Is Opposition to Current	27
	1-8 The Closed Circuit	28
	1-9 Direct Current (DC) and Alternating Current (AC)	31
	1-10 Sources of Electricity	33
CHAPTER 2	OHM'S LAW	39
	2-1 The Current $I = V/R$	39
	2-2 The Voltage $V = IR$	41
	2-3 The Resistance $R = V/I$	41
	2-4 Practical Units	42
	2-5 Multiple and Submultiple Units	43
	2-6 The Linear Proportion between V and I	43
	2-7 Power	45
	2-8 Power Dissipation in Resistance	47
	2-9 Electric Shock	49
CHAPTER 3	SERIES CIRCUITS	54
	3-1 Why I Is the Same in All Parts of a Series Circuit	54
	3-2 Total R Equals the Sum of All Series Resistances	56
	3-3 Series IR Voltage Drops	58
	3-4 The Sum of Series IR Drops Equals the Applied V_T	59
	3-5 Polarity of IR Voltage Drops	61
	3-6 Polarities to Chassis Ground	62
	3-7 Total Power in a Series Circuit	63
	3-8 Series-Aiding and Series-Opposing Voltages	64
	3-9 Analyzing Series Circuits	64
	3-10 Effect of an Open Circuit in a Series Path	67

CHAPTER 4 PARALLEL CIRCUITS — 74

- 4-1 The Applied Voltage V_A is the Same across Parallel Branches — 74
- 4-2 Each Branch I Equals V_A/R — 75
- 4-3 The Main-Line I_T Equals the Sum of the Branch Currents — 76
- 4-4 Resistances in Parallel — 78
- 4-5 Conductances in Parallel — 82
- 4-6 Total Power in Parallel Circuits — 83
- 4-7 Analyzing Parallel Circuits — 84
- 4-8 Effect of an Open Branch in Parallel Circuits — 84
- 4-9 Effect of a Short Circuit across Parallel Branches — 85

CHAPTER 5 SERIES-PARALLEL CIRCUITS — 91

- 5-1 Finding R_T for Series-Parallel Resistances — 91
- 5-2 Resistance Strings in Parallel — 92
- 5-3 Resistance Banks in Series — 94
- 5-4 Resistance Banks and Strings in Series-Parallel — 95
- 5-5 Analyzing Series-Parallel Circuits — 97
- 5-6 Wheatstone Bridge — 100
- 5-7 Chassis-Ground Connections — 101
- 5-8 Voltages Measured to Chassis Ground — 102
- 5-9 Opens and Shorts in Series-Parallel Circuits — 104

REVIEW OF CHAPTERS 1 TO 5 — 113

CHAPTER 6 VOLTAGE DIVIDERS AND CURRENT DIVIDERS — 116

- 6-1 Series Voltage Dividers — 116
- 6-2 Current Divider with Two Parallel Resistances — 118
- 6-3 Current Division by Parallel Conductances — 119
- 6-4 Series Voltage Divider with Parallel Load Current — 120
- 6-5 Design of a Loaded Voltage Divider — 122

CHAPTER 7 DIRECT-CURRENT METERS — 126

- 7-1 Moving-Coil Meter — 126
- 7-2 Measurement of Current — 129
- 7-3 Meter Shunts — 131
- 7-4 The Ayrton or Universal Shunt — 133
- 7-5 Voltmeters — 135
- 7-6 Loading Effect of a Voltmeter — 140
- 7-7 Ohmmeters — 142

	7-8	Multimeters	147
	7-9	Digital Meters	150
	7-10	Meter Applications	151
	7-11	Checking Continuity with the Ohmmeter	153

REVIEW OF CHAPTERS 6 AND 7 — 159

CHAPTER 8 KIRCHHOFF'S LAWS — 161

	8-1	Kirchhoff's Current Law	161
	8-2	Kirchhoff's Voltage Law	163
	8-3	Method of Branch Currents	164
	8-4	Node-Voltage Analysis	167
	8-5	Method of Mesh Currents	169

CHAPTER 9 NETWORK THEOREMS — 175

	9-1	Superposition	175
	9-2	Thevenin's Theorem	177
	9-3	Thevenizing a Circuit with Two Voltage Sources	180
	9-4	Thevenizing a Bridge Circuit	182
	9-5	Norton's Theorem	184
	9-6	Thevenin-Norton Conversions	187
	9-7	Conversion of Voltage and Current Sources	189
	9-8	Millman's Theorem	191
	9-9	T and π Networks	193

REVIEW OF CHAPTERS 8 AND 9 — 201

CHAPTER 10 CONDUCTORS AND INSULATORS — 203

	10-1	Function of the Conductor	203
	10-2	Standard Wire Gage Sizes	204
	10-3	Types of Wire Conductors	206
	10-4	Printed Wiring	207
	10-5	Switches	208
	10-6	Fuses	210
	10-7	Pilot Lamps	212
	10-8	Wire Resistance	212
	10-9	Temperature Coefficient of Resistance	214
	10-10	Ion Current in Liquids and Gases	216
	10-11	Electrons and Hole Charges in Semiconductors	218
	10-12	Insulators	219

CHAPTER 11 RESISTORS — 225

	11-1	Resistor Types	225
	11-2	Variable Resistors	227

11-3	Potentiometers and Rheostats	228
11-4	Resistor Color Coding	230
11-5	Power Rating of Resistors	232
11-6	Choosing the Resistor for a Circuit	232
11-7	Series and Parallel Combinations of Resistors	233
11-8	Resistor Troubles	234

CHAPTER 12 BATTERIES 240

12-1	Functions of Batteries	240
12-2	The Voltaic Cell	241
12-3	Carbon-Zinc Dry Cell	243
12-4	Series and Parallel Cells	245
12-5	Lead-Acid Wet Cell	247
12-6	Types of Electromotive Cells	250
12-7	Internal Resistance of a Generator	252
12-8	Matching a Load Resistance to the Generator	257

REVIEW OF CHAPTERS 10 TO 12 263

CHAPTER 13 MAGNETISM 265

13-1	The Magnetic Field	265
13-2	Magnetic Flux ϕ	267
13-3	Flux Density B	268
13-4	Induction by the Magnetic Field	270
13-5	Air Gap of a Magnet	271
13-6	Types of Magnets	272
13-7	Ferrites	275
13-8	Magnetic Shielding	275
13-9	The Hall Effect	276

CHAPTER 14 MAGNETIC UNITS 280

14-1	Ampere-turns (NI)	280
14-2	Field Intensity (H)	281
14-3	Permeability (μ)	282
14-4	B-H Magnetization Curve	284
14-5	Magnetic Hysteresis	285
14-6	Ohm's Law for Magnetic Circuits	287
14-7	Relations between Magnetic Units	288
14-8	Comparison of Magnetic and Electric Fields	289

CHAPTER 15 ELECTROMAGNETIC INDUCTION 294

15-1	Magnetic Field around an Electrical Current	294
15-2	Magnetic Polarity of a Coil	296

15-3	Motor Action between Two Magnetic Fields	298
15-4	Induced Current	300
15-5	Lenz' Law	301
15-6	Generating an Induced Voltage	302
15-7	Faraday's Law of Induced Voltage	303

CHAPTER 16 ALTERNATING VOLTAGE AND CURRENT   310

16-1	Alternating-Voltage Generator	311
16-2	The Sine Wave	314
16-3	Alternating Current	316
16-4	Voltage and Current Values for a Sine Wave	317
16-5	Frequency	319
16-6	Period	321
16-7	Wavelength	322
16-8	Phase Angle	323
16-9	The Time Factor in Frequency and Phase	327
16-10	AC Circuits with Resistance	327
16-11	Nonsinusoidal AC Waveforms	329
16-12	Harmonic Frequencies	331

CHAPTER 17 THE 60-Hz AC POWER LINE   337

17-1	Advantages of 120-V 60-Hz AC Power	337
17-2	Motors and Generators	338
17-3	Fluorescent Lighting	340
17-4	Residential Wiring	341
17-5	Three-Phase Power	344

REVIEW OF CHAPTERS 13 TO 17   348

CHAPTER 18 INDUCTANCE   352

18-1	Induction by Alternating Current	352
18-2	Self-inductance	354
18-3	Self-induced Voltage v_L	356
18-4	How v_L Opposes a Change in Current	356
18-5	Mutual Inductance	358
18-6	Transformers	360
18-7	Core Losses	366
18-8	Types of Cores	367
18-9	Variable Inductance	368
18-10	Inductances in Series or Parallel	369
18-11	Stray Inductance	371
18-12	Energy in Magnetic Field of Inductance	372
18-13	Troubles in Coils	372

CHAPTER 19	INDUCTIVE REACTANCE	379
19-1	How X_L Reduces the Amount of I	379
19-2	$X_L = 2\pi fL$	381
19-3	Series or Parallel Inductive Reactances	384
19-4	Ohm's Law Applied to X_L	385
19-5	Applications of X_L for Different Frequencies	385
19-6	Waveshape of v_L Induced by Sine-Wave Current	386

CHAPTER 20	INDUCTIVE CIRCUITS	394
20-1	Sine-Wave i_L Lags v_L by 90°	394
20-2	X_L and R in Series	395
20-3	Impedance (Z)	398
20-4	X_L and R in Parallel	400
20-5	Q of a Coil	403
20-6	AF and RF Chokes	405
20-7	The General Case of Inductive Voltage	407
20-8	Calculating the L/R Time Constant	408

REVIEW OF CHAPTERS 18 TO 20		415
CHAPTER 21	**CAPACITANCE**	**418**
21-1	How Charge Is Stored in the Dielectric	418
21-2	Charging and Discharging a Capacitor	420
21-3	The Farad Unit of Capacitance	422
21-4	Typical Capacitors	424
21-5	Capacitor Color Coding	429
21-6	Parallel Capacitances	430
21-7	Series Capacitances	430
21-8	Stray Capacitive and Inductive Effects	431
21-9	Energy in Electrostatic Field of Capacitance	434
21-10	Troubles in Capacitors	434

CHAPTER 22	CAPACITIVE REACTANCE	442
22-1	Alternating Current in a Capacitive Circuit	442
22-2	$X_C = 1/(2\pi fC)$	444
22-3	Series or Parallel Capacitive Reactances	446
22-4	Ohm's Law Applied to X_C	447
22-5	Applications of Capacitive Reactance	448
22-6	Sine-Wave Charge and Discharge Current	448

CHAPTER 23	CAPACITIVE CIRCUITS	457
23-1	Sine-Wave v_C Lags i_C by 90°	457
23-2	X_C and R in Series	458
23-3	X_C and R in Parallel	461

23-4	RF and AF Coupling Capacitors	463
23-5	Capacitive Voltage Dividers	464
23-6	The General Case of Capacitive Current i_C	465
23-7	Calculating the RC Time Constant	466

CHAPTER 24 RC AND L/R TIME CONSTANTS 472

24-1	Response of Resistance Alone	472
24-2	L/R Time Constant	473
24-3	High Voltage Produced by Opening RL Circuit	474
24-4	RC Time Constant	476
24-5	RC Charge and Discharge Curves	479
24-6	High Current Produced by Short-circuiting RC Circuit	479
24-7	RC Waveshapes	481
24-8	Long and Short Time Constants	483
24-9	Charge and Discharge with Short RC Time Constant	483
24-10	Long Time Constant for RC Coupling Circuit	484
24-11	Universal Time Constant Graph	486
24-12	Comparison of Reactance and Time Constant	488

REVIEW OF CHAPTERS 21 TO 24 494

CHAPTER 25 ALTERNATING-CURRENT CIRCUITS 498

25-1	AC Circuits with Resistance but No Reactance	498
25-2	Circuits with X_L Alone	499
25-3	Circuits with X_C Alone	501
25-4	Opposite Reactances Cancel	501
25-5	Series Reactance and Resistance	503
25-6	Parallel Reactance and Resistance	505
25-7	Series-Parallel Reactance and Resistance	507
25-8	Real Power	507
25-9	AC Meters	510
25-10	Wattmeters	510
25-11	Summary of Types of Ohms in AC Circuits	511
25-12	Summary of Types of Phasors in AC Circuits	512

CHAPTER 26 COMPLEX NUMBERS FOR AC CIRCUITS 519

26-1	Positive and Negative Numbers	519
26-2	The j Operator	520

	26-3	Definition of a Complex Number	521
	26-4	How Complex Numbers Are Applied to AC Circuits	522
	26-5	Impedance in Complex Form	523
	26-6	Operations with Complex Numbers	524
	26-7	Magnitude and Angle of a Complex Number	526
	26-8	Polar Form of Complex Numbers	527
	26-9	Converting Polar to Rectangular Form	528
	26-10	Complex Numbers in Series AC Circuits	530
	26-11	Complex Numbers in Parallel AC Circuits	532
	26-12	Combining Two Complex Branch Impedances	533
	26-13	Combining Complex Branch Currents	534
	26-14	Parallel Circuit with Three Complex Branches	535

CHAPTER 27 RESONANCE 541

	27-1	The Resonance Effect	541
	27-2	Series Resonance	542
	27-3	Parallel Resonance	546
	27-4	The Resonant Frequency $f_r = 1/(2\pi\sqrt{LC})$	550
	27-5	Q Magnification Factor of Resonant Circuit	552
	27-6	Bandwidth of Resonant Circuit	555
	27-7	Tuning	558
	27-8	Mistuning	560
	27-9	Analysis of Parallel Resonant Circuits	561
	27-10	Damping of Parallel Resonant Circuits	563
	27-11	Choosing L and C for a Resonant Circuit	564

CHAPTER 28 FILTERS 570

	28-1	Examples of Filtering	570
	28-2	Direct Current Combined with Alternating Current	571
	28-3	Transformer Coupling	574
	28-4	Capacitive Coupling	575
	28-5	Bypass Capacitors	577
	28-6	Filter Circuits	580
	28-7	Low-pass Filters	581
	28-8	High-pass Filters	582
	28-9	Resonant Filters	584
	28-10	Interference Filters	586

REVIEW OF CHAPTERS 25 TO 28 593

CHAPTER 29 VACUUM TUBES 597

	29-1	Rectifiers, Amplifiers, and Oscillators	598
	29-2	Construction of Tubes	599

29-3	Diodes	600
29-4	Plate Current	602
29-5	Diode Rectifier Circuit	602
29-6	Triodes	604
29-7	How a Triode Amplifies the Control-Grid Voltage	606
29-8	Triode Characteristics	608
29-9	Tube Parameters	609
29-10	Tetrodes	610
29-11	Pentodes	612
29-12	Tube Types	614
29-13	The Cathode-Ray Tube (CRT)	616
29-14	Troubles in Vacuum Tubes	617

CHAPTER 30 SEMICONDUCTOR DIODES AND TRANSISTORS — 624

30-1	Types of Semiconductor Devices	625
30-2	Characteristics of Semiconductors	626
30-3	N-type and P-type Doping	627
30-4	Current in Semiconductors	629
30-5	The PN Junction	630
30-6	Diode Rectifier Circuits	633
30-7	Transistors	636
30-8	Transistor Amplifier Circuits	639
30-9	The CE Amplifier Circuit	642
30-10	Collector Characteristic Curves	645
30-11	Load-Line Analysis	647
30-12	Bias Stabilization	651
30-13	Field-Effect Transistor (FET)	652
30-14	Silicon Controlled Rectifier (SCR)	655
30-15	Types of Transistors	657
30-16	Special-Purpose Diodes	659
30-17	Transistor Troubles	660

REVIEW OF CHAPTERS 29 AND 30 — 669

CHAPTER 31 INTEGRATED CIRCUITS — 673

31-1	Types of Integrated Circuits	673
31-2	Production of Integrated Circuits	675
31-3	Integrated Components	676
31-4	The Differential Amplifier	679
31-5	Linear IC Applications	681

CHAPTER 32 DIGITAL ELECTRONICS — 685

32-1	Binary Numbers	685
32-2	Binary Arithmetic	688

32-3	Symbolic Logic for Switching Circuits	689
32-4	Truth Tables	691
32-5	Basic Logic Circuits	693
32-6	Diode Gate Circuits	694
32-7	Diode-Transistor Logic (DTL)	695
32-8	Transistor-Transistor Logic (TTL)	698
32-9	Multivibrator (MV) Circuits	700
32-10	Flip-flop Circuits	701
32-11	Counters	704

REVIEW OF CHAPTERS 31 AND 32	711
BIBLIOGRAPHY	713
APPENDIX A ELECTRONIC FREQUENCY SPECTRUM	717
APPENDIX B FCC FREQUENCY ALLOCATIONS FROM 30 kHz TO 300,000 MHz	719
APPENDIX C ALPHABETICAL LISTING OF THE CHEMICAL ELEMENTS	721
APPENDIX D PHYSICS UNITS	724
APPENDIX E TRIGONOMETRIC FUNCTIONS	728
APPENDIX F ELECTRICAL SYMBOLS AND ABBREVIATIONS	732
APPENDIX G COLOR CODES	735
APPENDIX H SOLDERING AND TOOLS	741
APPENDIX I SCHEMATIC SYMBOLS	744
ANSWERS TO SELF-EXAMINATIONS	746
ANSWERS TO ODD-NUMBERED PROBLEMS	751
INDEX	759

Preface

This basic text is for beginning students without any experience in electricity and electronics. The first chapter is on elementary electricity; the last chapters cover integrated circuits and digital electronics. In between the topics progress through series and parallel dc circuits, networks, meters, magnetism, ac circuits with inductance and capacitance, vacuum tubes, and transistors. These fundamentals form the basis for the study of advanced applications, including communications electronics in general and radio and television in particular, industrial electronics, and computers.

For each subject, the basic principles are explained first, followed by typical applications and common troubles. This presentation has proved effective in helping students to learn the fundamentals of electronics with a practical approach that is interesting and useful.

Mathematics is held to a minimum. Some numerical problems require powers of 10 because of the metric units. Trigonometric functions are used to explain the details of ac circuits, where the phase angle is important.

The order of topics follows a typical one-year course in electronics fundamentals that includes dc and ac circuits. However, Kirchhoff's Laws and Network Theorems may be too advanced for some courses. These two chapters can be used for a separate course on network analysis, possibly combined with the details of RC and L/R Time Constants.

Similarly, the three chapters Semiconductor Diodes and Transistors, Integrated Circuits, and Digital Electronics can be used for a short course on semiconductor devices.

Organization. The book is divided into 32 chapters for step-by-step development. For example, individual chapters on Ohm's law, series circuits, and parallel circuits build up to more advanced chapters on series-parallel circuits, voltage dividers, and networks.

The chapters on magnetism and electromagnetic induction lead into the development of sine-wave alternating voltage and current. There are separate chapters on inductance and its ac reactance before these fundamentals are combined for inductive circuits. The same sequential development is used for capacitive circuits. Then all these principles of L and C are combined for ac circuits and resonance with sinusoidal waveforms.

The important details of RC and L/R time constants are reserved for another chapter. Here, the effects on dc transients and nonsinusoidal ac waveforms of inductive and capacitive circuits can be compared.

At the end of the book, the chapters on vacuum tubes, transistors, integrated circuits, and digital principles provide an introduction to electronic circuits with rectifiers, amplifiers, and digital logic functions.

Practical approach. Each chapter has a section at the end to explain common troubles in components or applications of the theory. For instance, the effects of an open circuit and a short circuit are explained in the first five chapters on dc circuits. Typical troubles in resistors, coils, capacitors, tubes, and transistors are described in their respective chapters.

For practical theory, the choke coil is explained as an application of inductive reactance; coupling and bypass capacitors as examples of capacitive reactance are described in detail.

Glossary of technical abbreviations. These are listed at the end of the chapters for Integrated Circuits and Digital Electronics. The expanding field of semiconductor devices has its own terminology, especially alphabetical abbreviations such as MOSFET, LSI, and TTL. It is helpful to have these summarized for convenient reference.

Numerical problems. This fourth edition has additional problems for more work in the numerical calculations, especially for reactances in ac circuits. With the use of calculators, difficult numbers can be used without too much busywork for the students.

Programmed questions. A new feature is Practice Problems at the end of each main topic. These questions should help the students check their understanding of each section immediately after reading the material. The answers are at the end of each chapter.

New material. Entire chapters dealing with voltage dividers, the 60-Hz ac power line, integrated circuits, and digital electronics have been added in order to emphasize these important subjects. As in the third edition, separate chapters are devoted to Kirchhoff's laws, network theorems, and time constants.

Chapter 6, Voltage Dividers and Current Dividers, has the original material from the chapter on series-parallel circuits. However, more space is available to include the practical example of voltage dividers with load currents and the case of current division in proportion to the branch conductances.

Chapter 17, The 60-Hz AC Power Line, includes practical material on motors, generators, residential wiring, fluorescent lighting, and three-way switches. The wye and delta connections for three-phase power are also included. This material is in a separate chapter to allow concentration on the theory of alternating current and voltage in Chap. 16.

Chapter 31, Integrated Circuits, describes the construction, packaging, and circuits for IC units, including amplifier and digital applications. Chapter 32, Digital Electronics, explains binary arithmetic, logic gates, and flip-flop circuits.

The chapter Semiconductor Diodes and Transistors has been completely rewritten and updated to present a thorough description of solid-state devices. Included are NPN and PNP bipolar transistors, the field-effect transistor (FET), the silicon controlled rectifier (SCR), and the unijunction transistor (UJT). The IC chapter is an extension of solid-state electronics, as many of the same principles of semiconductor devices apply to integrated circuits and discrete transistors. The digital chapter is an extension of integrated circuits, as IC units are used for practically all digital circuits.

Chapter 7, Direct-Current Meters, now comes a little earlier in the book, before the material on networks. This way the dc meters can be related more closely to measurements in series and parallel circuits for laboratory work. The chapter includes material on digital meters.

The introduction, Survey of Electronics, now has symbols and units for R, L, and C. The purpose is to help in laboratory work at the start of the course.

The SI standard symbol of V or v for voltage is used throughout the book, eliminating the use of E or e. Also, the SI unit of the siemens is used along with the mho unit for conductance. In addition, the SI system is emphasized in magnetic units.

Learning aids. The entire book is written with shorter sentences, shorter paragraphs, and more subhead titles. With the new two-column format, the text should be easier to read.

The practice problems for each main section and answers at the end of each chapter have the purpose of applying principles of programmed learning. This self-testing is in short units and can be reinforced immediately with correct answers.

Each chapter starts with an introduction that states the objective, followed by a listing of topics. At the end of each chapter a short summary lists the main points to remember. The short-answer questions for self-examination are based on the chapter summary. Summaries for groups of chapters are also given as a review, with additional self-examination questions. This definite structure of ideas in the listing of topics, summaries, and self-testing with review helps the student learn the material.

There are many tables in the text and summaries. The tables are used for a concise listing of important points and to compare similar or opposite characteristics. In many cases, it is easier to understand and remember ideas by comparisons.

Each review summary has a short list of reference books for the topics in those chapters. A more complete bibliography on pages 713 to 716 lists books on mathematics, electronics, semiconductors, and digital electronics. The listing also includes manuals for tubes and transistors, trade publications, and sources for training films.

Answers to all self-examination questions and to odd-numbered problems are given at the end of the book. A solutions manual for all the problems is available to instructors.

Credits. The photographs of components and equipment have been provided by many manufacturers, as noted in each legend. In the text, this fourth edition continues material that has been developed in previous editions with the help of my colleagues Harry G. Rice, Philip Stein, and Gerald P. McGinty.

Finally, it is a pleasure to thank my wife, Ruth, for her excellent work in typing the manuscript.

Bernard Grob

Survey of Electronics

Electronics and radio communications are practical applications of the general principles of electricity. The same electricity produced by a battery for a flashlight can be modified to do any number of jobs, from running a motor or producing heat and light to more advanced uses such as working a computer or providing wireless broadcasting for radio and television.

The word *radio* is an abbreviated form of *radiotelegraph* or *radiotelephone*. In its first form, wireless communication was by radiotelegraph, using short dots and longer dashes as symbols for letters in the Morse code. Now radiotelephone is used more, providing wireless voice communications or broadcasting voice and music programs for entertainment. In general, then, radio is the art of wireless communications.

The word *electronics* derives from the electron, which is a tiny, invisible quantity of electricity present in all materials. In terms of its many uses, electronics can be defined to include all applications of electricity flowing in a vacuum, as in vacuum tubes, in gas or vapor, and in certain solid materials such as transistors. More generally, electronics includes all effects of electricity where the action of individual electrons determines the application. The main electronic devices are transistors and vacuum tubes, shown in Fig. 1.

Radio and electronics are closely related. Sometimes they are even joined in their use. For example, an electronic heating unit generates radio waves that go through the work to produce heat. The heat bonds the solid materials together. Even if the applications are not so close, the principles of radio and electronics are essentially the same. Both are based on the fundamental laws of electricity.

DEVELOPMENT OF ELECTRONICS

Wireless transmission can be taken as starting with the work of Heinrich Hertz, a German physicist. In 1887 he was the first to demonstrate by experiment the process of electromagnetic radiation through space. The distance of transmission was only a few feet. However, it demonstrated radio waves traveling from one place to another without the need for any connecting wires between the transmitting and receiving equipment.

Hertz proved that radio waves, although invisible, travel with the same velocity as light waves. In fact, radio waves and light waves are just two examples of electromagnetic waves, a form of energy that combines the effects of

2 Survey of Electronics

FIGURE 1

Electron devices. (*a*) Vacuum-tube. (*b*) Transistors. (*c*) Silicon-diode rectifiers. (*d*) Integrated circuit (IC) with four transistors on one silicon wafer. Length is $\frac{3}{4}$ in.

electricity and magnetism. Additional examples of electromagnetic waves include heat radiation, x-rays, and cosmic rays, among others, all of which can transmit energy through space without the need for any connecting wires.

The work of Hertz followed earlier experiments on electricity and magnetism. In 1820, a Danish physicist, H. C. Oersted, showed that an electrical current produces magnetic effects. Then, in 1831, a British physicist, Michael Faraday, discovered that a magnet in motion can produce electricity. In 1864, the British physicist James Clerk Maxwell, on the basis of work in electricity and magnetism, predicted the electromagnetic waves demonstrated later by Hertz.

In 1895, Guglielmo Marconi used a long wire antenna and developed a practical radio system for long-distance communication. He succeeded in producing wireless communication across the Atlantic Ocean in 1901.

The rapid advances after that are due largely to the introduction and progress of the vacuum tube. In 1906 Dr. Lee De Forest, with his audion tube that could amplify electric signals, was a leader in this field.

As the design of vacuum tubes advanced, radio broadcasting progressed rapidly. Regularly scheduled programs were broadcast in 1920 by station KDKA in the AM (amplitude modulation) radio band. The commercial FM (frequency modulation) broadcast service for sound programs was started in 1939. Stereo broadcasting in the FM radio band began in 1961.

With regard to television, after discarding previous mechanical systems that used rotating drums or disks, commercial television broadcasting was adopted officially in July 1941, although its popular use did not begin until 1945. Our present color-television system was adopted in 1953.

Now, with the invention of transistors in 1948 at Bell Telephone Laboratories, there are new applications in electronics and radio. The transistor is an application of controlled electron flow in solids such as germanium and silicon. Tubes and transistors both have similar applications for amplification or control purposes. The transistor is smaller, however, and more efficient because there is no heater. See Fig. 1. Solid-state electronics using semiconductors includes not only transistors and diodes but also the integrated circuit (IC) in Fig. 1d. It combines these semiconductor components in one solid chip with the required resistors and capacitors.

Radio Broadcast Services. *Broadcasting* means sending out in all directions. As illustrated in Fig. 2, the transmitter radiates electromagnetic radio waves in all directions by means of its antenna. Receivers can pick up the transmitted radio waves by means of a receiving antenna or aerial. Practically all radio receivers now use transistors instead of vacuum tubes (see Fig. 3).

The carrier is an electromagnetic radio wave that includes the variations of the desired voice or music information, inserted by modulation. This technique of modulating a carrier wave is necessary because the desired information itself is not suitable for wireless transmission. The carrier is chosen for the best radio transmission; the modulation provides the information. For the amplitude modulation (AM) shown in Fig. 2, the amplitude of the carrier wave varies with the modulation. In frequency modulation (FM), the modulating voltage varies the frequency of the carrier wave.

Frequency is an important characteristic of an alternating voltage or current. How many times per second the carrier wave varies through a complete cycle of reversals in polarity is the frequency in cycles per second (cps). The unit is the hertz (Hz), where 1 Hz = 1 cps.

4   Survey of Electronics

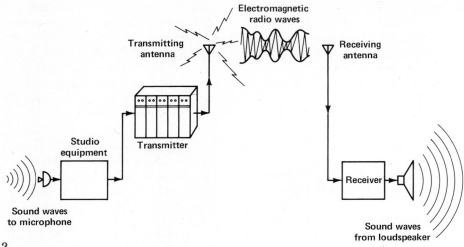

FIGURE 2

Radio broadcasting between the transmitter and receiver. An AM carrier wave is shown.

The transmission distance may be 10 or 5000 miles, depending on the type of radio service. There are many services for different uses, including broadcast radio and television for home entertainment, radio navigation, maritime radio, police radio, amateur radio broadcasting, government radio services, and many others. These are all regulated by the Federal Communications Commission (FCC) in the United States. The FCC assigns the carrier wave to be used by the broadcast station. A few of the more important radio services are listed here, but a more complete list is in Appendix B, with assigned channels.

 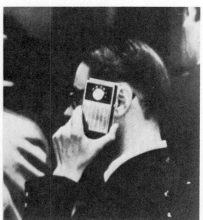

FIGURE 3

The old and the new in radio. (*RCA*)

Standard Broadcast Band. This band is the original system of broadcasting for what we generally call radio, using amplitude modulation in the transmission of the station's assigned carrier wave. The AM radio band is 535 to 1605 kHz. The last digit is not on the dial.

FM Radio Band. This band is 88 to 108 MHz. The FM system reduces static and interference. Also, the FM band is used for broadcasting high-fidelity audio signals. For stereo broadcasting, the left and right audio signals are multiplexed, or combined, on one carrier wave.

Television Broadcasting. Television is just another application of radio broadcasting. Two separate carrier signals are transmitted by the station in its assigned channel. One is the AM picture signal; the other is the FM sound signal. A TV channel is 6 MHz wide to include the picture and sound signals. As an example, channel 4 is 66 to 72 MHz. For color broadcasting, the color signal is multiplexed with the black-and-white signal on the one carrier wave for a picture signal.

Marine Radio. This use is important for ship navigation and safety. In addition to ship-to-ship and ship-to-shore communications, radio is the basis of radar navigation systems.

Aeronautical Radio. In addition to communications, radio is an important part of air navigation. It includes radar, radio compass, radio range, and automatic landing equipment.

Government Radio. There are many radio stations operated by the federal government for civilian and military requirements.

Citizen's Band (CB) Radio. Forty 10-kHz channels from 26.965 to 27.405 MHz are for public use of two-way radio. The CB transceiver includes a transmitter and receiver.

Amateur Radio. This is perhaps the largest noncommercial radio broadcast service. These self-styled "hams" usually build and operate their own transmitters and receivers to call each other in one of the assigned amateur radio bands. The largest organization in this field is the American Radio Relay League (ARRL), Newington, Connecticut.

ELECTRONIC APPLICATIONS

Electronics has its uses in almost all industries for quality control and automation. Just a few examples are given here to indicate its many possibilities. Additional applications are listed in Appendix A.

FIGURE 4

Electronic calculator. (*Texas Instruments*)

Electronic Calculators. See Fig. 4. Using integrated circuits, the calculator adds, subtracts, multiplies, or divides numbers almost instantly. The display is a photoelectric panel that shows the digits in their proper place. The scientific calculator in Fig. 4 can also be used for trigonometric functions, logarithms, and exponents.

Electronic Data Processing (EDP). This application includes computers and automatic equipment to replace manual office routines in filing, sorting, billing, and calculating. Electronic data processing is very common in banks, insurance companies, government agencies, and any office where voluminous records are kept.

Medical. Research in schools and laboratories, diagnosis, treatment, and surgery all use electronic equipment. Examples are the electron microscope, diathermy equipment, and the cardiograph machine.

Electronic Watches. See Fig. 5. This watch uses an integrated circuit to produce timing pulses, instead of using mechanical springs. The digital readout can use the light-emitting diode (LED) or liquid-crystal display (LCD).

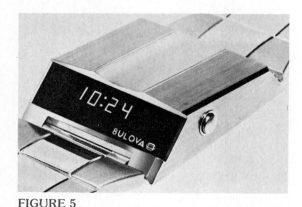

FIGURE 5

Electronic digital watch. (*Bulova Watch Co., Inc.*)

Accuracy of the electronic watch is better than 1 min/year.

Industrial Electronics. These applications include welding, dielectric heating, induction heating, metal detector, smoke detector, moisture control, and computer-controlled machinery. In addition, there are many types of remote-control units, including automatic garage door openers and burglar alarms. Closed-circuit television is often used for surveillance.

Supersonics or Ultrasonics. Electronic equipment also uses sound waves with frequencies above the range of human hearing. Examples are sonar equipment for marine depth equipment, ultrasonic cleaning machines, and remote control units for tuning television receivers to different channels.

Classifications. There are so many applications that they are generally considered in these broad categories:

1. Radio communications. This includes AM radio, FM radio, including stereo, and television broadcasting, including color. Radio can further be subdivided between receivers and broadcast equipment, either at the transmitter or at the studio. High-fidelity audio equipment can be considered a specialized branch of receivers.
2. Electronics. Some of the main subdivisions are computers, industrial control, servomechanisms, testing and recording instruments, and medical electronics. The applications of computers, including EDP, probably form the largest branch of electronics.
3. Electrical power. Generation, distribution, and uses, including dc and ac machinery.

Branches. Specific divisions in radio and electronics are indicated by the following specialized titles for engineers: aeronautical, audio, antennas, communications, computer, engineering management, engineering sales, geophysical, illumination, information theory and coding, magnetics, medical electronics, microwaves, military including guided missiles, nuclear power, packaging and materials, radio astronomy, tubes, semiconductor, space flight including satellites, test equipment, and, finally, ultrasonics. Many of these fields combine the sciences of physics and chemistry.

Job Titles. The types of jobs in each of these fields include management, engineer for research, development, or production, teacher, technician, sales, technical writer, draftsman, service worker, inspector, tester, and wirer. Technicians and service workers are needed for testing, maintenance, and repair on all the types of electronic equipment.

ELECTRONIC COMPONENTS

Considering the many different applications of electronics and radio, we can be a little surprised that there are only five basic types of components for all the different kinds of equipment. Of course, each type has many variations for specific uses. Still, the following is a short list:

1. Electron tubes, including vacuum-tube amplifiers, gas-filled tubes, and the cathode-ray tube (CRT).
2. Transistors. This is probably the most important use of solid-state semiconductors, which includes diodes and integrated circuits.
3. Resistors (Fig. 6).
4. Capacitors, or condensers (Fig. 7).
5. Inductors, or coils (Fig. 8).

Tubes. Types include the two-electrode tubes or diodes, plus triodes, tetrodes, and pentodes.

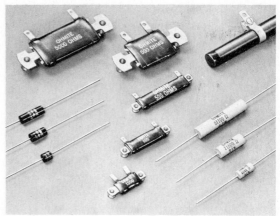

FIGURE 6

Typical resistors. (*Ohmite Mfg. Co.*)

A diode is generally used as a rectifier to change alternating current to direct current. The other tubes are for amplifier circuits, which increase the strength of the input signal.

The cathode-ray tube (CRT) is a vacuum tube with a fluorescent screen. The picture tube for a television receiver is a common example

FIGURE 7

Typical capacitors. (*Cornell-Dubilier Electric Corp.*)

8 Survey of Electronics

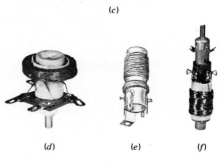

FIGURE 8

Typical inductors. (*Merit Coil and Transformer Corp.*)

of the CRT (Fig. 9). More details of tube types and their schematic symbols are in Chap. 29, Vacuum Tubes.

FIGURE 9

Typical picture tube or CRT.

Semiconductor Devices. Similarly, semiconductor diodes are rectifiers. Transistors correspond to triode tubes for use as amplifiers. Both can control the flow of electric charges between the input and output terminals. Tubes and transistors are used in electronic circuits with resistors, capacitors, and inductors. The transistors and tubes are *active components,* meaning they can amplify or rectify. Resistors, capacitors, and inductors are *passive components*. Integrated circuits combine solid-state transistors and diodes in one IC chip for a complete circuit with the passive components. More details of solid-state devices, with schematic symbols, are in Chaps. 30 to 32.

Resistors. These can be the carbon-composition type or wound with special resistance wire. Their function is to limit the amount of current in a circuit.

The unit of resistance (R) is the ohm (Ω). As an example, $R = 100\ \Omega$ is a common value.

Survey of Electronics 9

TABLE 1. Schematic Symbols for Resistors

TYPE	SYMBOL	NOTES
Fixed	R	Limits current
Variable	R (or) R	Varies current or voltage; volume, contrast, and tone controls

Schematic symbols for R are shown in Table 1. More details are in Chap. 11, Resistors.

Capacitors. A capacitor is constructed as an insulator between two conductor plates. The basic function is to concentrate the electric field of the voltage applied across the insulator or dielectric. As a result, it has the capacity to store electric charge.

With an ac voltage applied, the capacitor can charge and discharge. The practical application of this effect is to use capacitors to pass an ac signal but to block a steady dc voltage. The higher the frequency, the easier it is for the capacitor to couple the ac signal from one amplifier circuit to the next.

The unit of capacitance (C) is the farad (F). However, this is a very large unit requiring metric fractions. As an example, $C = 40 \ \mu\text{F}$ (microfarads), or $40/1,000,000$ F, is a common value for an electrolytic filter capacitor. Schematic symbols for C are shown in Table 2. More details on capacitance are in Chaps. 21 to 23.

Inductors. An inductor is just a coil of wire. The basic function is to concentrate the magnetic field of the current in the coil. An induced voltage is produced when the current with its magnetic field varies.

In the practical application of a *choke,* the inductor can pass a steady direct current better than alternating current. The higher the frequency, the greater is the effect of the choke in reducing the amount of alternating current.

A transformer consists of two or more coil windings in the same magnetic field. The purpose is to increase or decrease the amount of ac voltage coupled between the two windings. Therefore, a transformer can operate only with

TABLE 2. Schematic Symbols for Capacitors

TYPE	SYMBOL	USES
Fixed	C	Stores charge in dielectric; passes ac voltage but blocks dc voltage
Electrolytic	C +	Fixed value with large C but has polarity
Variable	C	Variable capacitor; used for tuning

10 Survey of Electronics

TABLE 3. Schematic Symbols for Inductors or Coils

TYPE	SYMBOL	USES
Air core	(coil symbol, L)	Concentrates magnetic field; for radio frequencies
Iron core	(coil with lines, L)	For 60-Hz and audio frequencies
Variable powdered-iron core	(coil with arrow, L)	Variable inductance; used for tuning
Iron-core transformer	(two coils, L_P / L_S)	Increases or decreases ac voltage from primary L_P to secondary L_S

alternating current. The magnetic field must vary to induce the current that transforms the energy between coils. More details on inductance are in Chaps. 18 to 20.

The symbol for inductance is L, for linkages in the magnetic field. The unit is the henry (H). As an example, $L = 250\ \mu\text{H}$ is a typical value of a choke for radio frequencies. Schematic symbols for L are shown in Table 3.

Electronic Circuits. In conclusion, whether we consider radio communications, electronics, or power machinery, they all depend on the fundamental principles of electricity and magnetism. The applications of electronics are basically electric circuits with transistors or tubes and the required, R, C, and L components. A transistor or tube is needed for amplification of a desired signal.

Electricity

Chapter 1

Electricity is an invisible force that can produce heat, light, motion, and many other physical effects. The force is an attraction or repulsion between electric charges. More specifically, electricity can be explained in terms of electric charge, current, voltage, and resistance. The corresponding electrical units are the coulomb for measuring charge, the ampere for current, the volt for potential difference, and the ohm for resistance. These characteristics can then be applied to electric circuits. The topics explained here are:

1-1 Negative and Positive Polarities
1-2 Electrons and Protons in the Atom
1-3 Structure of the Atom
1-4 The Coulomb Unit of Charge
1-5 The Volt Unit of Potential Difference
1-6 Charge in Motion Is Current
1-7 Resistance Is Opposition to Current
1-8 The Closed Circuit
1-9 Direct Current (DC) and Alternating Current (AC)
1-10 Sources of Electricity

1-1 NEGATIVE AND POSITIVE POLARITIES

We see the effects of electricity in a battery, static charge, lightning, radio, television, and many other applications. What do they all have in common that is electrical in nature? The answer is basic particles of electric charge with opposite polarities. All the materials we know, including solids, liquids, and gases, contain two basic particles of electric charge: the *electron* and the *proton*. An electron is the smallest amount of electric charge having the characteristic called *negative polarity*. The proton is a basic particle with *positive polarity*.

Actually, the negative and positive polarities indicate two opposite characteristics that seem to be fundamental in all physical applications. Just as magnets have north and south poles, electric charges have the opposite polarities labeled negative and positive. The opposing characteristics provide a method of balancing one against the other to explain different physical effects.

It is the arrangement of electrons and protons as basic particles of electricity that determines the electrical characteristics of all substances. As an example, this paper has electrons and protons in it. There is no evidence of electricity, though, because the number of electrons equals the number of protons. In that case the opposite electrical forces cancel, making the paper electrically neutral. The neutral condition means that opposing forces are exactly balanced, without any net effect either way.

When we want to use the electrical forces associated with the negative and positive charges in all matter, work must be done to

FIGURE 1-1

Negative and positive polarities on 1.5-V dry cell.

separate the electrons and protons. Changing the balance of forces produces evidence of electricity. A battery, for instance, can do electrical work because its chemical energy separates electric charges to produce an excess of electrons at its negative terminal and an excess of protons at its positive terminal. With separate and opposite charges at the two terminals, electrical energy can be supplied to a circuit connected to the battery. Figure 1-1 shows a battery with its negative (−) and positive (+) terminals

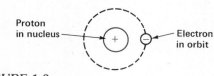

FIGURE 1-2

Electron and proton in hydrogen atom.

marked to emphasize the two opposite polarities.

Practice Problems 1-1
(answers on page 38)
(a) Is the charge of an electron positive or negative?
(b) Is the charge of a proton positive or negative?
(c) Is it true or false that the neutral condition means equal positive and negative charges?

1-2
ELECTRONS AND PROTONS IN THE ATOM

Although there is any number of possible methods by which electrons and protons might be grouped, they assemble in specific combinations that result in a stable arrangement. Each stable combination of electrons and protons makes one particular type of atom. For example, Fig. 1-2 illustrates the electron and proton structure of one atom of the gas hydrogen. This atom consists of a central mass called the *nucleus* and 1 electron outside. The proton in the nucleus makes it the massive and stable part of the atom because a proton is 1840 times heavier than an electron.

In Fig. 1-2, the 1 electron in the hydrogen atom is shown in an orbital ring around the nucleus. In order to account for the electrical stability of the atom, we can consider the electron as spinning around the nucleus, as planets revolve around the sun. Then the electrical force attracting the electrons in toward the nucleus is balanced by the mechanical force outward on the rotating electron. As a result, the electron stays in its orbit around the nucleus.

In an atom that has more electrons and protons than hydrogen, all the protons are in the nucleus, while all the electrons are in one or more outside rings. For example, the carbon atom illustrated in Fig. 1-3a has 6 protons in the nucleus and 6 electrons in two outside rings. The total number of electrons in the outside rings must equal the number of protons in the nucleus in a neutral atom.

The distribution of electrons in the orbital rings determines the atom's electrical stability. Especially important is the number of electrons in the ring farthest from the nucleus. This outermost ring requires 8 electrons for stability, except when there is only one ring, which has a maximum of 2 electrons.

In the carbon atom in Fig. 1-3a, with 6 electrons, there are just 2 electrons in the first ring because 2 is its maximum number. The remaining 4 electrons are in the second ring, which can have a maximum of 8 electrons.

As another example, the copper atom in Fig. 1-3b has only 1 electron in the last ring, which can include 8 electrons. Therefore, the outside ring of the copper atom is less stable than the outside ring of the carbon atom.

When there are many atoms close together in a copper wire, the outermost orbital electrons are not sure which atoms they belong to. They can migrate easily from one atom to another at random. Such electrons that can move freely from one atom to the next are often called *free electrons*. This freedom accounts for the ability of copper to conduct electricity very easily. It is the movement of free electrons that provides electrical current in a metal conductor.

Conductors, Insulators, and Semiconductors. When electrons can move easily from atom to atom in a material, it is a *conductor*. In general, all the metals are good conductors, with silver the best and copper second. Their atomic structure allows free movement of the outermost orbital electrons. Copper wire is generally used for practical conductors because it costs much less than silver. The purpose of using conductors is to allow electrical current to flow with minimum opposition.

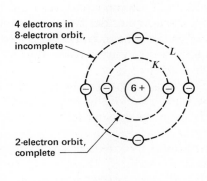
(a)

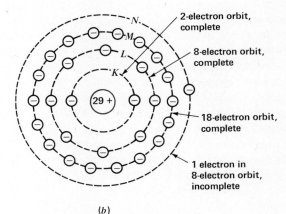
(b)

FIGURE 1-3

Atomic structure showing nucleus and orbital rings. (a) Carbon atom with 6 protons in nucleus and 6 orbital electrons. (b) Copper atom with 29 protons in nucleus and 29 orbital electrons.

A material with atoms in which the electrons tend to stay in their own orbits is an *insulator* because it cannot conduct electricity very easily. However, the insulators are able to hold or store electricity better than the conductors. An insulating material, such as glass, plastic, rubber, paper, air, or mica, is also called a *dielectric*, meaning it can store electric charge.

Insulators can be useful when it is necessary to prevent current flow. In addition, for applications requiring the storage of electric charge, as in capacitors, a dielectric material must be used because a good conductor cannot store any charge.

Carbon can be considered a semiconductor, conducting less than the metal conductors but more than the insulators. In the same group are germanium and silicon, which are commonly used for transistors and other semiconductor components.

Elements. The combinations of electrons and protons forming stable atomic structures result in different kinds of elementary substances having specific characteristics. A few familiar examples are the elements hydrogen, oxygen, carbon, copper, and iron. An *element* is defined as a substance that cannot be decomposed any further by chemical action. The atom is the smallest particle of an element that still has the same characteristics as the element. *Atom* itself is a Greek word meaning a particle too small to be subdivided. As an example of the fact that atoms are too small to be visible, a particle of carbon the size of a pinpoint contains many billions of atoms. The electrons and protons within the atom are even smaller.

Table 1-1 lists some more examples of elements. These are just a few out of a total of 106.* Notice how the elements are grouped.

---
* A more complete listing of the elements, in alphabetical order, is given in Appendix C at the back of the book.

The metals listed across the top row are all good conductors of electricity. Each has an atomic structure with an unstable outside ring that allows many free electrons.

The semiconductors have 4 electrons in the outermost ring. This means they neither gain nor lose electrons but share them with similar atoms. The reason is that 4 is exactly halfway to the stable condition of 8 electrons in the outside ring.

Among the gases, those which are active chemically and electrically have an atomic structure with an incomplete outside ring. The inert gases have a complete outside ring of 8 electrons, which makes them chemically inactive. Remember that 8 electrons in the outside ring is a stable structure.

Molecules and Compounds. A group of two or more atoms forms a molecule. For instance, two atoms of hydrogen (H) form a hydrogen molecule (H_2). When hydrogen unites chemically with oxygen, the result is water (H_2O), which is a compound. A compound, then, consists of two or more elements. The molecule is the smallest unit of a compound, with the same chemical characteristics. We can have molecules for either elements or compounds. However, atoms exist only for the elements.

Practice Problems 1-2
(*answers on page 38*)
(a) Which has more free electrons: metals or insulators?
(b) Which is the best conductor: silver, carbon, or iron?
(c) Which is a semiconductor: copper, silicon, or neon?

1-3
STRUCTURE OF THE ATOM
Although nobody has ever seen an atom, its hypothetical structure fits experimental evi-

TABLE 1-1. Examples of the Chemical Elements

GROUP	ELEMENT	SYMBOL	ATOMIC NUMBER	ELECTRON VALENCE
Metal conductors, in order of conductance	Silver	Ag	47	+1
	Copper	Cu	29	+1*
	Gold	Au	79	+1*
	Aluminum	Al	13	+3
	Iron	Fe	26	+2*
Semiconductors	Carbon	C	6	±4
	Silicon	Si	14	±4
	Germanium	Ge	32	±4
Active gases	Hydrogen	H	1	±1
	Oxygen	O	8	−2
Inert gases	Helium	He	2	0
	Neon	Ne	10	0

*Some metals have more than one valence number in forming chemical compounds. Examples are cuprous or cupric copper, ferrous or ferric iron, and aurous or auric gold.

dence that has been measured very exactly. The size and electric charge of the invisible particles in the atom are indicated by how much they are deflected by known forces. Our present planetary model of the atom was proposed by Niels Bohr in 1913. His contribution was joining the new ideas of a nuclear atom developed by Lord Rutherford (1871–1937) with the quantum theory of radiation developed by Max Planck (1858–1947) and Albert Einstein (1879–1955).

As illustrated in Figs. 1-2 and 1-3, the nucleus contains protons for all the positive charge in the atom. The number of protons in the nucleus is equal to the number of planetary electrons. Thus, the positive and negative charges are balanced, as the proton and electron have equal and opposite charges. The orbits for the planetary electrons are also called *shells* or *energy levels*.

Atomic Number. This gives the number of protons or electrons required in the atom for each element. For the hydrogen atom in Fig. 1-2, the atomic number is 1, which means the nucleus has 1 proton balanced by 1 orbital electron. Similarly, the carbon atom in Fig. 1-3 with atomic number 6 has 6 protons in the nucleus and 6 orbital electrons. Also, the copper atom has 29 protons and 29 electrons because its atomic number is 29. The atomic number is listed for each of the elements in Table 1-1 to indicate the atomic structure.

Orbital Rings. The planetary electrons are in successive shells called K, L, M, N, O, P, and Q at increasing distances outward from the nucleus. Each shell has a maximum number of electrons for stability. As indicated in Table 1-2, these stable shells correspond to the inert gases, like helium and neon.

The K shell, closest to the nucleus, is stable with 2 electrons, corresponding to the atomic structure for the inert gas helium. Once the stable number of electrons has filled a shell, it cannot take any more electrons. The atomic structure with all its shells filled up to the maximum number for stability corresponds to an inert gas.

TABLE 1-2. Shells of Orbital Electrons in the Atom

SHELL	MAXIMUM ELECTRONS	INERT GAS
K	2	Helium
L	8	Neon
M	8 (up to calcium) or 18	Argon
N	8, 18, or 32	Krypton
O	8 or 18	Xenon
P	8 or 18	Radon
Q	8	

Elements with a higher atomic number have more planetary electrons. These are in successive shells, tending to form the structure of the next inert gas in the periodic table.[1] After the K shell has been filled with 2 electrons, the L shell can take up to 8 electrons. Ten electrons filling the K and L shells is the atomic structure for the inert gas neon.

The maximum number of electrons in the remaining shells can be 8, 18, or 32 for different elements, depending on their place in the periodic table. The maximum for an outermost shell, though, is always 8.

To illustrate these rules, we can use the copper atom in Fig. 1-3b as an example. There are 29 protons in the nucleus balanced by 29 planetary electrons. This number of electrons fills the K shell with 2 electrons, corresponding to the helium atom, and the L shell with 8 electrons. The 10 electrons in these two shells correspond to the neon atom, which has an atomic number of 10. The remaining 19 electrons for the copper atom then fill the M shell with 18 electrons and 1 electron in the outermost N shell.

[1] For more details of the periodic table of the elements, developed in 1869 by Dmitri Mendelyeev, refer to a textbook on chemistry or physics, or see "Periodic Chart of the Atoms," Sargent Welch Scientific Co., Skokie, Ill. 60076.

For most elements, we can use the rule that the maximum number of electrons in a filled inner shell equals $2n^2$, where n is the shell number in sequential order outward from the nucleus. Then the maximum number of electrons in the first shell is $2 \times 1 = 2$; for the second shell $2 \times 2^2 = 8$, for the third shell $2 \times 3^2 = 18$, and for the fourth shell $2 \times 4^2 = 32$. These values apply only to an inner shell that is filled with its maximum number of electrons.

Electron Valence. This value is the number of electrons in an incomplete outermost shell. Copper, for instance, has a valence of 1 because there is 1 electron in the last shell, after the inner shells have been completed with their stable number. Similarly, hydrogen has a valence of 1, and carbon has a valence of 4. The number of outer electrons is considered positive valence, as these electrons are in addition to the stable shells.

Except for H and He, the goal of valence is 8 for all the atoms, as each tends to form the stable structure of 8 electrons in the outside ring. For this reason, valence can also be considered as the number of electrons in the outside ring needed to make 8. This value is the negative valence. As examples, the valence of copper can be considered $+1$ or -7; carbon has the valence of ± 4. The inert gases have a valence of 0, as they all have a complete stable outer shell of 8 electrons.

The valence indicates how easily the atom can gain or lose electrons. For instance, atoms with a valence of $+1$ can lose this 1 outside electron, especially to atoms with a valence of $+7$ or -1, which need 1 electron to complete the outside shell with 8 electrons.

Subshells. Although not shown in the drawing here, all the shells except K are divided into

TABLE 1-3. Stable Particles in the Atom

PARTICLE	CHARGE	MASS
Electron, in orbital shells	0.16×10^{-18} C, negative	9.108×10^{-28} g
Proton, in nucleus	0.16×10^{-18} C, positive	1.672×10^{-24} g
Neutron, in nucleus	None	1.675×10^{-24} g

subshells. This subdivision accounts for different types of orbits in the same shell. For instance, electrons in one subshell may have elliptical orbits, while other electrons in the same main shell have circular orbits. The subshells indicate magnetic properties of the atom.

Particles in the Nucleus. A stable nucleus, which is not radioactive, contains protons and neutrons. A neutron is electrically neutral without any net charge. Its mass is almost the same as a proton.

A proton has the positive charge of a hydrogen nucleus. The charge is the same amount as an orbital electron but of opposite polarity. There are no electrons in the nucleus. Table 1-3 lists the charge and mass for these three basic particles in all atoms.

*Practice Problems 1-3
(answers on page 38)*
(a) An element with 14 protons and 14 electrons has what atomic number?
(b) What is the electron valence of an element with atomic number 3?

1-4
THE COULOMB UNIT OF CHARGE

If you rub a hard rubber pen or comb on a sheet of paper, the rubber will attract a corner of the paper if it is free to move easily. The paper and rubber then give evidence of a static electric charge. The work of rubbing resulted in separating electrons and protons to produce a charge of excess electrons on the surface of the rubber and a charge of excess protons on the paper.

Because paper and rubber are dielectric materials, they hold their extra electrons or protons. As a result, the paper and rubber are no longer neutral, but each has an electric charge. The resultant electric charges provide the force of attraction between the rubber and the paper. This mechanical force of attraction or repulsion between charges is the fundamental method by which electricity makes itself evident.

Any charge is an example of *static electricity* because the electrons or protons are not in motion. There are many examples. When you walk across a wool rug, your body becomes charged with an excess of electrons. Similarly, silk, fur, and glass can be rubbed to produce a static charge. This effect is more evident in dry weather, because a moist dielectric does not hold its charge so well. Also, plastic materials can be charged easily, which is why thin, lightweight plastics seem to stick to everything.

The charge of many billions of electrons or protons is necessary for common applications of electricity. Therefore, it is convenient to define a practical unit called the *coulomb* (C) as equal to the charge of 6.25×10^{18} electrons or protons stored in a dielectric (see Fig. 1-4). The analysis of static charges and their forces is called *electrostatics*.

The symbol for electric charge is Q or q, standing for quantity. For instance, a charge of 6.25×10^{18} electrons[1] is stated as $Q = 1$ C. This unit is named after Charles A. Coulomb (1736–1806), a French physicist, who measured the force between charges.

Negative and Positive Polarities. Historically, the negative polarity has been assigned to the static charge produced on rubber, amber, and resinous materials in general. Positive polarity refers to the static charge produced on glass and other vitreous materials. On this basis, the electrons in all atoms are basic particles of negative charge because their polarity is the same as the charge on rubber. Protons have positive charge because the polarity is the same as the charge on glass.

[1] For an explanation of how to use powers of 10, see B. Grob, "Problems Workbook in Technical Mathematics for Basic Electronics," McGraw-Hill Book Company, New York.

Charges of Opposite Polarity Attract. If two small charged bodies of light weight are mounted so that they are free to move easily and are placed close to each other, one can be attracted to the other when the two charges have opposite polarity (Fig. 1-5a). In terms of electrons and protons, they tend to be attracted to each other by the force of attraction between opposite charges. Furthermore, the weight of an electron is only about $\frac{1}{1840}$ the weight of a proton. As a result, the force of attraction tends to make electrons move to protons.

Charges of the Same Polarity Repel. In Fig. 1-5b and c, it is shown that when the two bodies have an equal amount of charge with the same polarity, they repel each other. The two positive charges repel in Fig. 1-5b, while two negative charges of the same value repel each other in Fig. 1-5c.

Polarity of a Charge. An electric charge must have either negative or positive polarity, labeled

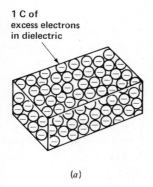

(a)

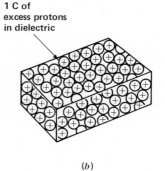

(b)

FIGURE 1-4

The coulomb (C) unit of charge. (a) 6.25×10^{18} excess electrons for negative charge. (b) Same amount of protons for positive charge caused by removing electrons.

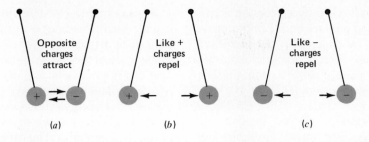

FIGURE 1-5
Force between charges. (*a*) Opposite charges attract. (*b*) and (*c*) charges of the same polarity repel.

$-Q$ or $+Q$, with an excess of either electrons or protons. A neutral condition is considered zero charge. On this basis, consider the following examples, remembering that the electron is the basic particle of charge and the proton has exactly the same amount, although of opposite polarity.

Example 1. A neutral dielectric has added to it 12.5×10^{18} electrons. What is its charge in coulombs?

Answer. This number of electrons is double the charge of 1 C. Therefore, $-Q = 2$ C.

Example 2. A dielectric has a positive charge of 12.5×10^{18} protons. What is its charge in coulombs?

Answer. This is the same amount of charge as in Example 1 but positive. Therefore $+Q = 2$ C.

Example 3. A dielectric with $+Q$ of 2 C has 12.5×10^{18} electrons added. What is its charge then?

Answer. The 2 C of negative charge added by the electrons cancels the 2 C of positive charge, making the dielectric neutral, with zero charge.

Example 4. A neutral dielectric has 12.5×10^{18} electrons removed. What is its charge?

Answer. The 2 C of electron charge removed allows an excess of 12.5×10^{18} protons. Since the proton and electron have exactly the same amount of charge, now the dielectric has a positive charge of $+Q = 2$ C.

Note that we generally consider the electrons moving, rather than the heavier protons. However, a loss of a given number of electrons is equivalent to a gain of the same number of protons.

Charge of an Electron. Fundamentally, the quantity of any charge is measured by its force of attraction or repulsion. The extremely small force of an electron or proton was measured by Millikan[1] in experiments done from 1908 to 1917. Very briefly, the method consisted of measuring the charge on vaporized droplets of oil, by balancing the gravitational force against an electrical force that could be measured very precisely.

[1] Robert A. Millikan (1868–1953), an American physicist. Millikan received the Nobel prize in physics for this oil-drop experiment.

A small drop of oil sprayed from an atomizer becomes charged by friction. Furthermore, the charges can be increased or decreased slightly by radiation. These very small changes in the amount of charge were measured. The three smallest values were 0.16×10^{-18} C, 0.32×10^{-18} C, and 0.48×10^{-18} C. These values are multiples of 0.16. In fact, all the charges measured were multiples of 0.16×10^{-18} C. Therefore, we conclude that 0.16×10^{-18} C is the basic charge from which all other values are derived. This ultimate charge of 0.16×10^{-18} C is the charge of 1 electron or 1 proton. Then

1 electron or $Q_e = 0.16 \times 10^{-18}$ C

The reciprocal of 0.16×10^{-18} gives the number of electrons or protons in 1 C. Then

$1 \text{ C} = 6.25 \times 10^{18}$ electrons

Note that the factor 6.25 equals exactly $\frac{1}{0.16}$ and the factor 10^{18} is the reciprocal of 10^{-18}.

The Electric Field of a Static Charge. The ability of an electric charge to attract or repel another charge is actually a physical force. To help visualize this effect, lines of force are used, as shown in Fig. 1-6. All the lines form the electric field. The lines and the field are imaginary since they cannot be seen. Just as the field of the force of gravity is not visible, however, the resulting physical effects prove the field is there.

Each line of force in Fig. 1-6 is directed outward to indicate repulsion of another charge in the field with the same polarity as Q, either positive or negative. The lines are shorter further away from Q to indicate that the force decreases inversely as the square of the distance. The larger the charge, the greater is the force.

The electric field in the dielectric between two plates with opposite charges is the basis for the ability of a capacitor to store electric charge. More details are explained in Chap. 21, Capacitance. In general, any charged insulator has capacitance. A capacitor unit is constructed in a form to concentrate the electric field.

Practice Problems 1-4
(answers on page 38)
(a) How many electron charges are there in the practical unit of one coulomb?
(b) How much is the charge in coulombs for a surplus of 25×10^{18} electrons?

1-5
THE VOLT UNIT OF POTENTIAL DIFFERENCE

Potential refers to the possibility of doing work. Any charge has the potential to do the work of moving another charge, by either attraction or repulsion. This ability of a charge to do work is its potential. When we consider two unlike charges, they have a difference of potential.

A charge is the result of work done in separating electrons and protons. Because of the separation, there is stress and strain associated with opposite charges, since normally they would be balancing each other to produce a neutral condition. We could consider that the

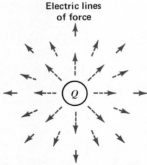

FIGURE 1-6
Electrostatic field around a stationary charge Q.

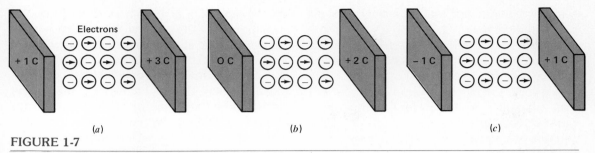

FIGURE 1-7

The work required to move electrons between two charges depends on their difference in potential. This difference is equivalent to 2 C of charge in (a), (b), and (c).

accumulated electrons are drawn tight and are straining themselves to be attracted toward protons in order to return to the neutral condition. Similarly, the work of producing the charge causes a condition of stress in the protons, which are trying to attract electrons and return to the neutral condition. Because of these forces, the charge of electrons or protons has potential, as it is ready to give back the work put into producing the charge. The force between charges is in the electric field.

Potential between Different Charges. When one charge is different from the other, there must be a difference of potential between them. For instance, consider a positive charge of 3 C shown at the right in Fig. 1-7a. The charge has a certain amount of potential, corresponding to the amount of work this charge can do. The work to be done is moving some electrons, as illustrated.

Assume a charge of 1 C can move 3 electrons. Then the charge of +3 C can attract 9 electrons toward the right. However, the charge of +1 C at the opposite side can attract 3 electrons toward the left. The net result, then, is that 6 electrons can be moved toward the right to the more positive charge.

In Fig. 1-7b, one charge is 2 C, while the other charge is neutral with 0 C. For the difference of 2 C, again 2 × 3 or 6 electrons can be attracted to the positive side.

In Fig. 1-7c, the difference is still 2 C between the charges. The +1 C attracts 3 electrons to the side. Also, the −1 C repels 3 electrons. This effect is really the same as attracting 6 electrons.

Therefore, the net number of electrons moved in the direction of the more positive charge depends on the difference of potential between the two charges. This difference corresponds to 2 C for all three cases in Fig. 1-7. Potential difference is often abbreviated P.D.

The only case without any potential difference between charges is where they both have the same polarity and are equal in amount. Then the repelling and attracting forces cancel and no work can be done in moving electrons between the two identical charges.

The Volt Unit of Potential Difference. This unit is named after Alessandro Volta (1754–1827). Fundamentally, the volt is a measure of the work needed to move an electric charge. When 0.7376 ft·lb (foot-pound) of work is required to move 6.25×10^{18} electrons between two points, each with its own charge, the potential difference is 1 V.

Note that 6.25×10^{18} electrons make up one coulomb. Therefore the definition of a volt is for a coulomb of charge.

Also, 0.7376 ft · lb of work is the same as 1 joule (J), which is the practical metric unit of work[1] or energy. Therefore, we can say briefly that *one volt equals one joule of work per coulomb of charge.*

The symbol for potential difference is V for voltage. In fact, the volt unit is used so often, potential difference is often called voltage. Remember, though, that voltage is the potential difference between two points. Two terminals are necessary to measure a potential difference.

Consider the 2.2-V lead-acid cell in Fig. 1-8. Its output of 2.2 V means that this is the amount of potential difference between the two terminals. The cell then is a voltage source, or a source of *electromotive force* (*emf*).

Sometimes the symbol E is used for emf, but the standard symbol now is V for any potential difference. This applies either to the voltage generated by a source or to the voltage drop across a passive component, such as a resistor.

In a practical circuit, the voltage determines how much current can be produced.

Practice Problems 1-5
(*answers on page 38*)
(a) How much potential difference is there between two identical charges?
(b) Which supplies a greater P.D., a 1.5-V battery or a 12-V battery?

1-6
CHARGE IN MOTION IS CURRENT

When the potential difference between two charges forces a third charge to move, the charge in motion is an *electrical current*. To produce current, therefore, charge must be moved by a potential difference.

In solid materials, such as copper wire, the free electrons are charges that can be forced to move with relative ease by a potential difference, since they require relatively little work to be moved. As illustrated in Fig. 1-8, if a potential difference is connected across two ends of a copper wire, the applied voltage forces the free electrons to move. This current is a drift of electrons, from the point of negative charge at one end, moving through the wire, and returning to the positive charge at the other end.

To illustrate the drift of free electrons through the wire shown in Fig. 1-9, each electron in the middle row is numbered, corresponding to a copper atom to which the free electron belongs. The electron at the left is labeled s to indicate that it comes from the negative charge of the source of potential dif-

[1] See Appendix D, Physics Units.

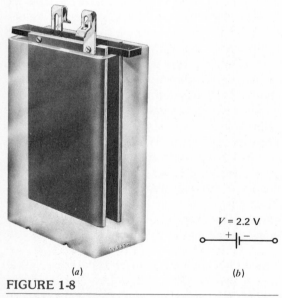

(a)   (b)

FIGURE 1-8

Chemical cell as voltage source. (*a*) Voltage output is potential difference between the two terminals. (*b*) Schematic symbol with longer line for positive side.

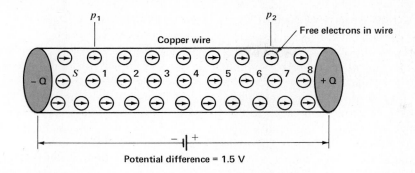

FIGURE 1-9

Potential difference across two ends of wire conductor causes drift of free electrons through the wire, producing electrical current.

ference. This one electron s is repelled from the negative charge $-Q$ at the left and is attracted by the positive charge $+Q$ at the right. Therefore, the potential difference of the voltage source can make electron s move toward atom 1. Now atom 1 has an extra electron. As a result, the free electron of atom 1 can then move to atom 2. In this way, there is a drift of free electrons from atom to atom. The final result is that the one free electron labeled 8 at the extreme right in Fig. 1-9 moves out from the wire to return to the positive charge of the voltage source.

Considering this case of just one electron moving, note that the electron returning to the positive side of the voltage source is not the electron labeled s that left the negative side. All electrons are the same, however, and have the same charge. Therefore, the drift of free electrons resulted in the charge of one electron moving through the wire. This charge in motion is the current. With more electrons drifting through the wire, the charge of many electrons moves, resulting in more current.

The current is a continuous flow of electrons. Only the electrons move, not the potential difference. For ordinary applications, where the wires are not long lines, the potential difference produces current instantaneously through the entire length of wire.

Furthermore, the current must be the same at all points of the wire at any time. Although a point nearer to the negative terminal of the voltage source has a greater repelling force on the free electrons, at this point the free electrons are farther from the positive terminal and have less attracting force. At the middle of the wire, the free electrons have equal forces of attraction and repulsion. Near the positive terminal, there is a greater attracting force on the free electrons but less repelling force from the negative terminal of the voltage source. In all cases, the total force causing motion of the free electrons is the same at any point of the wire, therefore resulting in the same current through all parts of the wire.

Potential Difference Is Necessary to Produce Current. The number of free electrons that can be forced to drift through the wire to produce the moving charge depends upon the amount of potential difference across the wire. With more applied voltage, the forces of attraction and repulsion can make more free electrons drift, producing more charge in motion. A larger amount of charge moving with the same speed means a higher value of current. Less applied voltage across the same wire results in a smaller amount of charge in motion, which is a smaller value of current. With zero potential difference across the wire, there is no current.

Two cases of zero potential difference and no current can be considered in order to emphasize the important fact that potential difference is needed to produce current. Assume the copper wire to be by itself, not connected to any voltage source, so that there is no potential difference across the wire. The free electrons in the wire can move from atom to atom, but this motion is random, without any organized drift through the wire. If the wire is considered as a whole, from one end to the other, the current is zero.

As another example, suppose that the two ends of the wire have the same potential. Then free electrons cannot move to either end, because both ends have the same force, and there is no current through the wire. A practical example of this case of zero potential difference would be to connect both ends of the wire to just one terminal of a battery. Each end of the wire would have the same potential and there would be no current. The conclusion, therefore, is that two connections are needed to two points at different potentials in order to produce the current.

The Ampere of Current. Since current is the movement of charge, the unit for stating the amount of current is defined in rate of flow of charge. When the charge moves at the rate of 6.25×10^{18} electrons flowing past a given point per second, the value of the current is one *ampere* (A). This is the same as one coulomb of charge per second. The ampere unit of current is named after André M. Ampère (1775–1836).

Referring back to Fig. 1-9, note that if 6.25×10^{18} free electrons move past p_1 in 1 s, the current is 1 A. Similarly, the current is 1 A at p_2 because the electron drift is the same throughout the wire. If twice as many electrons moved past either point in 1 s, the current would be 2 A.

The symbol for current is *I* or *i* for intensity, since the current is a measure of how intense or concentrated the electron flow is. Two amperes of current in a copper wire is a higher intensity than 1 A; a greater concentration of moving electrons results because of more electrons in motion, although all the electrons move with the same speed. Sometimes current is called *amperage*.

How Current Differs from Charge. Charge is a quantity of electricity accumulated in a dielectric. The charge is static electricity, at rest, without any motion. When the charge moves, usually in a conductor, the current *I* indicates the intensity of the electricity in motion. This characteristic is a fundamental definition of current:

$$I = \frac{Q}{T} \qquad (1\text{-}1)$$

where *I* is the current in amperes, *Q* is in coulombs, and the time *T* is in seconds (s). It does not matter whether the moving charge is positive or negative. The only question is how much charge moves and what its rate of motion is.

Example 5. The charge of 12 C moves past a given point every second. How much is the intensity of charge flow?

Answer. $\quad I = \dfrac{Q}{T} = \dfrac{12 \text{ C}}{1 \text{ s}}$

$\qquad\qquad I = 12 \text{ A}$

Example 6. The charge of 5 C moves past a given point in 0.1 s. How much is the current?

Answer. $I = \dfrac{Q}{T} = \dfrac{5\text{ C}}{0.1\text{ s}}$

$I = 50\text{ A}$

This fundamental definition of current can also be used to consider the charge as equal to the product of the current multiplied by the time. Or

$$Q = I \times T \qquad (1\text{-}2)$$

For instance, we can have a dielectric connected to conductors with a current of 0.4 A. If the current can deposit electrons for the time of 0.2 s, the accumulated charge in the dielectric will be

$Q = I \times T$
$ = 0.4 \times 0.2$
$Q = 0.08\text{ C}$

The formulas $Q = IT$ for charge and $I = Q/T$ for current illustrate the fundamental nature of Q as an accumulation of static charge in an insulator, while I measures the intensity of moving charges in a conductor.

The General Nature of Current. The moving charges that provide current in metal conductors like a copper wire are the free electrons of the copper atoms. In this case, the moving charges have negative polarity. The direction of motion between two terminals for this *electron current*, therefore, is toward the more positive end. It is important to note, however, that there are examples of positive charges in motion. Common applications include current in liquids, gases, and semiconductors. For the case of current resulting from the motion of positive charges, its direction is opposite from the direction of electron flow. Whether negative or positive charges move, though, the current is still defined fundamentally as Q/T.

Types of Electric Charges for Current. See Table 1-4. The most common charge is the electron. In metal conductors and solid materials in general, free electrons in the atoms can be forced to move by setting up a potential difference. Therefore current is produced. The direction of electron flow is from the negative terminal of the voltage source, through the external circuit, and returning to the positive source terminal (Fig. 1-10a). Also, electrons are released by thermionic emission from the heated cathode in a vacuum tube. Finally, N-type semiconductors such as silicon and germanium have unbound electrons as a result of doping with impurity elements that can provide valence electrons from the added atoms.

For P-type semiconductors, the silicon and germanium are doped with impurity elements

TABLE 1-4. Types of Electric Charges for Current

TYPE OF CHARGE	AMOUNT OF CHARGE	POLARITY	TYPE OF CURRENT	APPLICATIONS
Electron	$Q_e = 0.16 \times 10^{-18}$ C	Negative	Electron flow	In wire conductors, vacuum tubes, and N-type semiconductors
Ion	Q_e or multiples of Q_e	Positive or negative	Ion current	In liquids and gases
Hole	$Q_e = 0.16 \times 10^{-18}$ C	Positive	Hole current	In P-type semiconductors

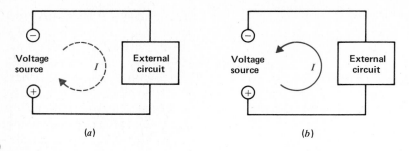

FIGURE 1-10

Direction of current *I*. (*a*) Dotted lines to show electron flow. (*b*) Flow of positive charges for hole current or ion current.

that cause a deficiency of electrons in the bonds between atoms. Each vacant space where an electron is missing is called a *hole charge*. The polarity is positive, opposite from the electron, but the amount of charge is exactly the same. In short, a hole charge is a deficiency of one valence electron in semiconductors. When hole charges move in a P-type semiconductor, they provide hole current. The direction of flow for the positive charges is from the positive terminal of the voltage source, through the external circuit, and returning to the negative source terminal (Fig. 1-10*b*).

An ion is an atom that has either lost or gained one or more valence electrons to become electrically charged. Therefore, the ion charge may be either positive or negative. The amount may be the charge of 1 electron Q_e, $2Q_e$, or $3Q_e$, etc. Ions can be produced by applying voltage to liquids and gases to produce ionization of the atoms. The ions are much less mobile than electrons or hole charges because an ion includes a complex atom with its nucleus.

Note that protons are not included as charge carriers for current in Table 1-4. The reason is that the protons are bound in the nucleus. They cannot be released except by nuclear forces. Therefore, a current of positive charges is a flow of either hole charges or positive ions. The hole charge has the same amount of charge as the proton, which is the same amount as an electron. However, the positive hole charge is in the valence structure of the atoms, not in the nucleus.

Magnetic Field Around an Electrical Current. When any current flows, it has an associated magnetic field. Figure 1-11 shows how iron filings line up in a circular field pattern corresponding to the magnetic lines of force. The magnetic field is in a plane perpendicular to the current. It should be noted that the iron filings are just a method of making the imaginary lines of force visible. The filings become magnetized

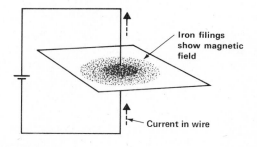

FIGURE 1-11

Magnetic field around any electrical current.

by the magnetic field. Both magnetic and electric fields can do the physical work of attraction or repulsion.

The magnetic field of any current is the basis for many electromagnetic applications, including magnets, relays, loudspeakers, transformers, and coils in general. Winding the conductor in the form of a coil concentrates the magnetic field. More details of electromagnetism are explained in Chap. 15, Electromagnetic Induction.

Practice Problems 1-6
(*answers on page 38*)
(a) The flow of 2 C/s of electron charges is how many amperes of current?
(b) The flow of 2 C/s of hole charges is how many amperes of current?
(c) How much is the current with zero potential difference?

1-7
RESISTANCE IS OPPOSITION TO CURRENT

The fact that a wire conducting current can become hot is evidence of the fact that the work done by the applied voltage in producing current must be accomplished against some form of opposition. This opposition, which limits the amount of current that can be produced by the applied voltage, is called *resistance*. Conductors have very little resistance; insulators have a large amount of resistance.

The atoms of a copper wire have a large number of free electrons, which can be moved easily by a potential difference. Therefore, the copper wire has little opposition to the flow of free electrons when voltage is applied, corresponding to a low value of resistance.

Carbon, however, has fewer free electrons than copper. When the same amount of voltage is applied to the carbon as to the copper, fewer electrons will flow. It should be noted that just as much current can be produced in the carbon by applying more voltage. For the same current, though, the higher applied voltage means that more work is necessary, causing more heat. Carbon opposes the current more than copper, therefore, and has a higher value of resistance.

The Ohm. The practical unit of resistance is the *ohm* (Ω). A resistance that develops 0.24 calorie[1] of heat when one ampere of current flows through it for one second has one ohm of opposition. As an example of a low resistance, a good conductor like copper wire can have a resistance of 0.01 Ω for a 1-ft length. The resistance-wire heating element in a 600-W toaster has a resistance of 24 Ω, and the tungsten filament in a 100-W light bulb has a resistance of 144 Ω, with a 120-V source.

Figure 1-12 shows a carbon-composition resistor. This type of resistance can be manufactured with a value from a few ohms to millions of ohms. The abbreviation for resistance is R or r. The symbol used for the ohm is the Greek letter *omega*, written as Ω. In diagrams, resistance is indicated by a zigzag line as shown by R in Fig. 1-12.

Conductance. The opposite of resistance is conductance. The less the resistance, the higher the conductance. Its symbol is G and the unit is the *siemens* (S), named after Ernst von Siemens, a European inventor. It is also called the *mho,* which is *ohm* spelled backward. Specifically, G is the reciprocal of R, or $G = 1/R$. For example, 5 Ω of resistance is equal to $\frac{1}{5}$ S of conductance.

Whether to use R or G for components is usually a matter of convenience. In general, R is easier to use in series circuits, because the se-

---

[1] One calorie is the quantity of heat that will raise the temperature of one gram of water by one degree Celsius. See Appendix D, Physics Units.

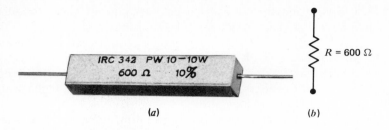

FIGURE 1-12

(a) Wirewound resistor with cement coating. (b) Schematic symbol for any type of resistance. (*International Resistance Co.*)

ries voltages are proportional to the resistances; G can be more convenient in parallel circuits, because the parallel currents are proportional to the conductances. (Series and parallel circuits are explained in Chaps. 3 and 4.)

Practice Problems 1-7
(*answers on page 38*)
(a) Which has more resistance, carbon or copper?
(b) With the same voltage applied, which resistance will allow more current, 4.7 Ω or 5000 Ω?
(c) What is the conductance in siemens or mho units for a 10-Ω R?

1-8
THE CLOSED CIRCUIT

In electrical applications requiring the use of current, the components are arranged in the form of a circuit, as shown in Fig. 1-13. A circuit can be defined as a path for current flow. The purpose of this circuit is to light the incandescent bulb. The bulb lights when the tungsten-filament wire inside is white hot, producing an incandescent glow.

By itself the tungsten filament cannot produce current. A source of potential difference is necessary. Since the battery produces a potential difference of 1.5 V across its two output terminals, this voltage is connected across the filament of the bulb by means of the two wires so that the applied voltage can produce current through the filament.

In Fig. 1-13c the schematic diagram of the circuit is shown. Here the components are represented by shorthand symbols. Note the symbols for the battery and resistance. The connecting wires are shown simply as straight lines because their resistance is small enough to be neglected. A resistance of less than 0.01 Ω for the wire is practically zero compared with the 300-Ω resistance of the bulb. If the resistance of the wire must be considered, the schematic diagram includes it as additional resistance in the same current path.

It should be noted that the schematic diagram does not look like the physical layout of the circuit. The schematic shows only the symbols for the components and their electrical connections.

Any electric circuit has three important characteristics:

1. There must be a source of potential difference. Without the applied voltage, current cannot flow.
2. There must be a complete path for current flow, from one side of the applied voltage source, through the external circuit, and

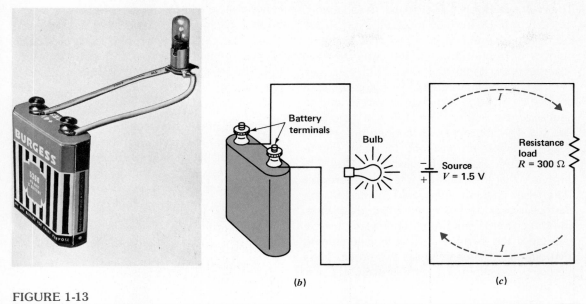

FIGURE 1-13

An electrical circuit with voltage source connected to a resistance load. (*a*) Photo of circuit. (*b*) Wiring diagram. (*c*) Schematic symbol.

returning to the other side of the voltage source.

3. The current path normally has resistance. The resistance is in the circuit for the purpose of either generating heat or limiting the amount of current.

How the Voltage Is Different from the Current. It is the current that moves through the circuit. The potential difference does not move. The voltage across the filament resistance makes electrons flow from one side to the other. While the current is flowing around the circuit, however, the potential difference remains across the filament to do the work of moving electrons through the resistance of the filament. As illustrated in Fig. 1-14, the voltage is the potential difference across the two ends of the resistance, while the current is the intensity of the electron flow past any one point in the circuit.

To illustrate the difference between V and I another way, suppose the circuit in Fig. 1-13 is opened by disconnecting the bulb. Now no current can flow because there is no closed path. Still, the battery has its potential difference. If you measure across the two terminals, the voltmeter will read 1.5 V even though the current is zero.

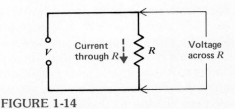

FIGURE 1-14

Comparison of voltage and current.

The Voltage Source Maintains the Current. As current flows in the circuit, electrons leave the negative terminal of the cell, and the same number of free electrons in the conductor are returned to the positive terminal. With electrons lost from the negative charge and gained by the positive charge, the two charges would tend to neutralize each other. The chemical action inside the dry cell, however, continuously separates electrons and protons to maintain the negative and positive charges on the outside terminals that provide the potential difference. Otherwise, the current would neutralize the charges, resulting in no potential difference, and the current would stop. Therefore, the dry cell keeps the current flowing by maintaining the potential difference across the circuit. Thus the cell is the generator, or voltage source, for the circuit.

The Circuit Is a Load on the Voltage Source. We can consider the circuit as a means whereby the energy of the voltage source is carried by means of the current through the filament of the bulb, where the electrical energy is used in producing heat energy. On this basis, the battery is the *source* in the circuit, since its voltage output represents the potential energy to be used. The part of the circuit connected to the voltage source is the *load resistance,* since it determines how much work the source will supply. In this case, the bulb's filament is the load resistance for the battery.

The resistance of the filament determines how much current the 1.5-V source will produce. Specifically, the current here is 0.005 A, equal to 1.5 V divided by 300 Ω. With more opposition, the same voltage will produce less current; less opposition allows more current.

The current that flows through the load resistance is the *load current.* Note that a lower value of ohms for the load resistance corresponds to a higher load current. Unless noted otherwise, the term *load* by itself can be assumed generally to mean the load current. Therefore, a heavy or big load electrically means a high value of load current, corresponding to a large amount of work supplied by the source.

Direction of the Electron Flow. As shown in Fig. 1-13c, the direction of the electron drift is from the negative side of the battery, through the load resistance R, and back to the positive terminal of the voltage source. Note that this is the direction in the external circuit connected across the output terminals of the voltage source.

Inside the battery, the electrons move to the negative terminal because this is how the voltage source produces its potential difference. The battery is doing the work of separating charges, accumulating electrons at the negative terminal and protons at the positive terminal. Then the potential difference across the two output terminals can do the work of moving electrons around the external circuit. In the circuit outside the voltage source, however, the direction of the electron flow is from a point of negative potential to a point of positive potential.

Conventional Current. The direction of moving positive charges, opposite from electron flow, is considered the conventional direction of current. In electrical engineering, circuits are usually analyzed with conventional current. The reason is based on the fact that, by the positive definitions of force and work, a positive potential is considered above a negative potential. So conventional current is a motion of positive charges "falling downhill" from a positive to a negative potential. The direction of conventional current, therefore, is the direction of positive charges in motion.

An example of positive charges moving in the direction of conventional current is hole current in P-type semiconductors. Also, a current of positive ions in liquids or gases moves in the opposite direction from electron flow. For instance, the current through the electrolyte inside a battery is ionization current.

Actually, either a positive or negative potential of the same value can do the same amount of work in moving charge. Any circuit can be analyzed either with electron flow or by conventional current in the opposite direction. In this book the current is considered as electron flow in the applications where electrons are the moving charges.

In summary, we can say that the closed circuit, normal circuit, or just a circuit is a closed path that has V to produce I with R to limit the amount of current. The circuit provides a means of using the energy of the battery as a voltage source. The battery has its potential difference V with or without the circuit. However, the battery alone is not doing any work in producing load current. The bulb alone has its resistance, but without current the bulb does not light. With the circuit, the voltage source is used for the purpose of producing current to light the bulb.

Open Circuit. When any part of the path is open or broken, the circuit is open because there is no continuity in the conducting path. The open can be in the connecting wires or in the bulb's filament as the load resistance. The resistance of an open circuit is infinitely high. The result is no current in an open circuit.

Short Circuit. In this case, the voltage source has a closed path across its terminals, but the resistance is practically zero. The result is too much current in a short circuit. Usually, the short circuit is a bypass across the load resistance. For instance, a short across the conducting wires for a bulb produces too much current in the wires but no current through the bulb. Then the bulb is shorted out. The bulb is not damaged, but the wires can become hot enough to burn unless the line has a fuse as a safety precaution against too much current.

Practice Problems 1-8
(*answers on page 38*)
Answer true or false for the circuit in Fig. 1-13.
(a) The bulb has a P.D. of 1.5 V across its filament only when connected to the voltage source.
(b) The battery has a P.D. of 1.5 V across its terminals only when connected to the bulb.

1-9
DIRECT CURRENT (DC) AND ALTERNATING CURRENT (AC)

The electron flow illustrated in the circuit of Fig. 1-13c is direct current because it has just one direction. The reason for the unidirectional current is that the battery maintains the same polarity of output voltage.

It is the flow of charges in just one direction and the fixed polarity of applied voltage that are the characteristics of a dc circuit. Actually, the current can be a motion of positive charges, rather than electrons, but the conventional direction of current does not change the fact that direct current has just one direction. Furthermore, the dc voltage source can change the amount of its output voltage, but if the same polarity is maintained, direct current will flow in just one direction, meeting the requirements of a dc circuit. A battery is a steady dc voltage source because it has fixed polarity and its output voltage is a steady value.

An alternating voltage source periodically reverses or alternates in polarity. The resulting alternating current, therefore, periodically reverses in direction. In terms of electron flow, the current always flows from the negative terminal of the voltage source, through the circuit, and back to the positive terminal, but when the generator alternates in polarity, the current must reverse its direction. The 60-cycle ac power line used in most homes is a common example. This frequency means that the voltage polarity and current direction go through 60 cycles of reversal per second.

The unit for 1 cycle per second is 1 hertz (Hz). Therefore 60 cycles per second is a frequency of 60 Hz.

The details of ac circuits are explained in Chap. 16 and the remainder of the book. Direct-current circuits are analyzed first because they usually are simpler. However, the principles of dc circuits also apply to ac circuits. Both types are important, as most electronic circuits include ac voltages and dc voltages. The waveforms for these two types of voltages are illustrated in Fig. 1-15. Their uses are compared in Table 1-5. Note that transistors and tubes require dc electrode voltages in order to amplify an ac signal voltage.

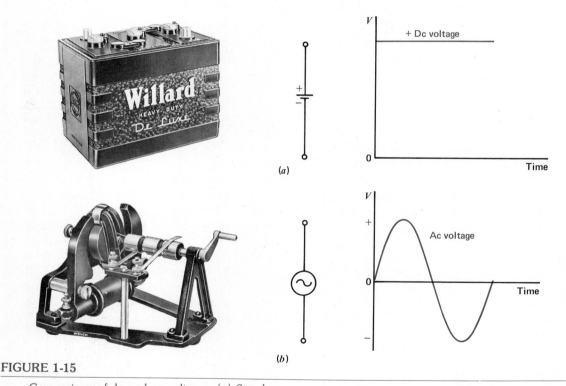

FIGURE 1-15

Comparison of dc and ac voltages. (*a*) Steady dc voltage of one polarity from a battery. (*b*) Sine-wave ac voltage with alternating polarity from small laboratory-type rotary ac generator. One complete cycle is shown. (*Sargent Welch Scientific Co.*)

TABLE 1-5. Comparison of DC Voltage and AC Voltage

DC VOLTAGE	AC VOLTAGE
Fixed polarity	Reverses in polarity
Can be steady or vary in magnitude	Varies between reversals in polarity
Steady value cannot be stepped up or down by a transformer	Can be stepped up or down for electrical power distribution
Electrode voltages for amplifiers	Signal input and output for amplifiers
Easier to measure	Easier to amplify
Heating effect the same for direct or alternating current	

Practice Problems 1-9
(*answers on page 38*)

Answer true or false.
(a) When the polarity of the applied voltage reverses, the direction of current flow also reverses.
(b) A battery is a dc voltage source because it cannot reverse the polarity across its output terminals.

1-10
SOURCES OF ELECTRICITY

There are electrons and protons in the atoms of all materials, but to do useful work the charges must be separated to produce a potential difference that can make current flow. Some of the more common methods of providing electrical effects are listed here.

Static Electricity by Friction. In this method, electrons in an insulator can be separated by the work of rubbing to produce opposite charges that remain in the dielectric.

Conversion of Chemical Energy. Wet or dry cells and batteries are the applications. Here a chemical reaction produces opposite charges on two dissimilar metals, which serve as the negative and positive terminals.

Ions. Atoms can be made to gain or lose orbital electrons, by either a chemical reaction or the electric field of an applied voltage. Such atoms with a net charge are ions. The ion charges can be either negative or positive. Ions are generally the charge carriers that provide ionization current in liquids and gases.

Electromagnetism. Electricity and magnetism are closely related. Any moving charge has an associated magnetic field; also, any changing magnetic field can produce current. A motor is an example of how current can react with a magnetic field to produce motion; a generator produces voltage by means of a conductor rotating in a magnetic field.

Photoelectricity. Some materials are photoelectric, meaning they can emit electrons when light strikes the surface. The element cesium is often used as a source of *photoelectrons*. Also, photovoltaic cells or solar cells use silicon to generate output voltage from the light input. In another effect, the resistance of the element selenium changes with light. Combined with a fixed voltage source, wide variations between *dark current* and *light current* can be produced. Such characteristics are the basis of many photoelectric devices, including television camera tubes, photoelectric cells, and phototransistors.

Thermal Emission. Some materials when heated can "boil off" electrons from the surface. Then these emitted electrons can be controlled to provide useful applications of electrical current. The emitting electrode is called a *cathode*, while an *anode* is used to collect the emitted electrons. A common material for thermionic cathodes is barium oxide, heated to a dull-red temperature. The vacuum tubes in radio and television receivers operate on this principle.

Doping of Semiconductors. Because of a valence of ±4, combined with a crystalline molecular structure, semiconductors such as germanium and silicon can be changed from the inert condition by adding small amounts of an impurity element. This technique of adding charges to a semiconductor is called *doping*. When electrons are added, the result is negative or N-type material; a deficiency of electrons makes the material positive or P-type. A deficiency of 1 electron in the semiconductor is called a *hole charge*. With one type of semiconductor between two opposite types, the result is a PNP or NPN transistor. The details of transistors are explained in Chap. 30.

Although there are so many different applications, remember that all electrons are the same, with identical charge and mass. Whether the electron flow results from a battery, rotary generator, or photoelectric device, and is controlled by a vacuum tube or transistor, the analysis of voltage, current, and resistance in the different types of circuits must follow the basic principles described here.

Practice Problems 1-10
(answers on page 38)
(a) The excess charges at the negative terminal of a battery are _____.
(b) The charges emitted from a heated cathode are _____.
(c) The charges in a P-type semiconductor are _____.
(d) In a liquid that is conducting current, the moving charges are _____.

Summary

1. Electricity is present in all matter in the form of electrons and protons.
2. The electron is the basic quantity of negative electricity, the proton of positive electricity. Both have the same amount of charge but opposite polarities. The charge of 6.25×10^{18} electrons or protons equals the practical unit of one coulomb.
3. Charges of the same polarity tend to repel each other; charges of opposite polarities attract. There must be a difference of charges for any force of attraction or repulsion.
4. Electrons tend to move toward protons because an electron has $1/1840$ the weight of a proton. Electrons in motion provide an electron current.
5. The atomic number of an element gives the number of protons in the nucleus of its atom, balanced by an equal number of orbital electrons.
6. The number of electrons in the outermost orbit is the valence of the element.
7. Table 1-6 summarizes the main features of electric circuits. In the

TABLE 1-6. Electrical Characteristics

CHARACTERISTIC	SYMBOL	UNIT	DESCRIPTION
Charge	Q or q^1	Coulomb (C)	Quantity of electrons or protons; $Q = I \times T$
Current	I or i^1	Ampere (A)	Charge in motion; $I = Q/T$
Voltage	V or $v^{1,2}$	Volt (V)	Potential difference between two unlike charges; makes charge move to produce I
Resistance	R or r^3	Ohm (Ω)	Opposition that reduces amount of current
Conductance	G or g^3	Siemens (S) or mho	Reciprocal of R, or $G = 1/R$

[1] Small letter q, i, or v is used for an instantaneous value of a varying charge, current, or voltage.
[2] E or e is sometimes used for a generated emf but the standard symbol is V or v for any potential difference in the international system of units (SI).
[3] Small letter r or g is used for internal resistance or conductance of transistors and tubes.

symbols, the small letters q, v, and i are used when the characteristic varies with respect to time. Also, the small letters r and g indicate internal characteristics of a source.

8. Types of negative charges include electrons and negative ions. Types of positive charges include protons, positive ions, and hole charges.
9. An electric circuit is a closed path for electron flow. Potential difference must be connected across the circuit to produce current. In the external circuit outside the voltage source, electrons flow from the negative terminal toward the positive terminal.
10. Direct current has just one direction as the dc voltage source has a fixed polarity. Alternating current periodically reverses in direction as the ac voltage source reverses its polarity.

Self-Examination (Answers at back of book.)

Answer true or false.

1. All matter has electricity in the form of electrons and protons in the atom.
2. The electron is the basic unit of negative charge.
3. A proton has the same amount of charge as the electron but opposite polarity.
4. Electrons are repelled from other electrons but are attracted to protons.
5. The force of attraction or repulsion between charges is in their electric field.

6. The nucleus is the massive stable part of an atom, with positive charge.
7. Neutrons add to the weight of the atom's nucleus but not to its electric charge.
8. An element with atomic number 12 has 12 orbital electrons.
9. This element has an electron valence of +2.
10. To produce current in a circuit, potential difference is connected across a closed path.
11. A dc voltage has fixed polarity while ac voltage periodically reverses its polarity.
12. The coulomb is a measure of the quantity of stored charge.
13. If a dielectric has 2 C of excess electrons, removing 3 C of electrons will leave the dielectric with the positive charge of 1 C.
14. A charge of 5 C flowing past a point each second is a current of 5 A.
15. A current of 7 A charging a dielectric will accumulate a charge of 14 C after 2 s.
16. A voltage source has two terminals with different charges.
17. An ion is a charged atom.
18. The resistance of a few feet of copper wire is practically zero.
19. The resistance of the rubber or plastic insulation on the wire is also practically zero.
20. A resistance of 600 Ω has a conductance of 6 S.

Essay Questions

1. Briefly define each of the following, giving its unit and symbol: charge, potential difference, current, resistance, and conductance.
2. Name two good conductors, two good insulators, and two semiconductors.
3. Explain briefly why there is no current in a light bulb unless it is connected across a source of applied voltage.
4. Give three differences between voltage and current.
5. In any circuit: (a) state two requirements for producing current; (b) give the direction of electron flow.
6. Show the atomic structure of the element sodium (Na) with atomic number 11. What is its electron valence?
7. Make up your own name for direct current and dc voltage to indicate how it differs from alternating voltage and current.
8. State the formulas for each of the following two statements: (a) Current is the time rate of change of charge. (b) Charge is current accumulated over a period of time.

9. Refer to Table 1-4. (*a*) Name two types of moving charges that provide current in the direction of electron flow. (*b*) Name two types that provide conventional current.
10. Why is it that protons are not considered a source of moving charges for current flow?
11. Give one difference and one similarity in comparing electric and magnetic fields.
12. Give three methods of providing electric charges, and give their practical applications.

Problems (Answers to odd-numbered problems at back of book.)

1. The charge of 8 C flows past a given point every 2 s. How much is the current in amperes?
2. The current of 4 A charges an insulator for 2 s. How much charge is accumulated?
3. Convert the following to siemens or mhos of conductance: (*a*) 1000 Ω; (*b*) 500 Ω; (*c*) 10 Ω; (*d*) 0.1 Ω.
4. Convert the following to ohms of resistance: (*a*) 0.001 S; (*b*) 0.002 S; (*c*) 0.1 S; (*d*) 10 S.
5. A battery can supply 11 J of energy to move 5 C of charge. How much is the voltage of the battery? (Hint: One volt equals one joule per coulomb.)
6. A material with a deficiency of 25×10^{18} electrons gains 31.25×10^{18} electrons. The excess electrons are then made to flow past a given point in 1 s. How much current is produced by the resultant electron flow?
7. Convert 5 S of conductance to ohms of resistance.
8. Connect the components in Fig. 1-16 to form an electrical circuit. Label the source voltage, with polarity, and the load resistance. Show the direction of electron flow. Determine the amount of current *I* by Ohm's law.

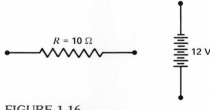

FIGURE 1-16
For Prob. 8.

Answers to Practice Problems

1-1   (a) negative
      (b) positive
      (c) true

1-2   (a) metals
      (b) silver
      (c) silicon

1-3   (a) 14
      (b) 1

1-4   (a) 6.25×10^{18}
      (b) 4 C

1-5   (a) zero
      (b) 12 V

1-6   (a) 2 A
      (b) 2 A
      (c) zero

1-7   (a) carbon
      (b) 4.7 Ω
      (c) $\frac{1}{10}$ S or mho

1-8   (a) T
      (b) F

1-9   (a) T
      (b) T

1-10 (a) electrons
      (b) electrons
      (c) holes
      (d) ions

Ohm's Law

Chapter 2

This unit explains how the amount of current *I* in a circuit depends on its resistance *R* and the applied voltage. Specifically, $I = V/R$, determined in 1828 by the experiments of George Simon Ohm. If you know any two of the factors *V*, *I*, and *R*, you can calculate the third. Ohm's law also determines the amount of electrical power in the circuit. More details are in the following topics:

2-1 The Current $I = V/R$
2-2 The Voltage $V = IR$
2-3 The Resistance $R = V/I$
2-4 Practical Units
2-5 Multiple and Submultiple Units
2-6 The Linear Proportion Between *V* and *I*
2-7 Power
2-8 Power Dissipation in Resistance
2-9 Electric Shock

2-1
THE CURRENT $I = V/R$

If we keep the same resistance in a circuit but vary the voltage, the current will vary. The circuit in Fig. 2-1 demonstrates this idea. The applied voltage *V* can be varied from 0 to 12 V, as an example. The bulb has a 12-V filament, which requires this much voltage for its normal current to light with normal intensity. The meter *I* indicates the amount of current in the circuit for the bulb.

With 12 V applied, the bulb lights, indicating normal current. When *V* is reduced to 10 V, there is less light because of less *I*. As *V* decreases, the bulb becomes dimmer. For zero volts applied there is no current and the bulb cannot light. In summary, the changing brilliance of the bulb shows that the current is varying with the changes in applied voltage.

For the general case of any *V* and *R*, Ohm's law is

$$I = \frac{V}{R} \qquad (2\text{-}1)$$

where *I* is the amount of current through the resistance *R* connected across the source of potential difference *V*. With volts as the practical unit for *V* and ohms for *R*, the amount of current *I* is in amperes. Therefore,

$$\text{Amperes} = \frac{\text{volts}}{\text{ohms}}$$

This formula says to simply divide the voltage across *R* by the ohms of resistance between the two points of potential difference to calculate the amperes of current through *R*. In Fig. 2-2, for instance, with 6 V applied across a 3-Ω resistance, by Ohm's law the amount of current *I* equals ⁶⁄₃ or 2 A.

Chapter 2
40 Ohm's Law

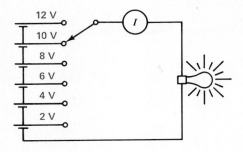

FIGURE 2-1

Increasing the applied voltage produces more current I to light the bulb with greater intensity.

High Voltage but Low Current. It is important to realize that with high voltage, the current can have a low value when there is a very large amount of resistance in the circuit. For example, 1000 V applied across 1,000,000 Ω results in a current of only $\frac{1}{1000}$ A. By Ohm's law,

$$I = \frac{V}{R}$$
$$= \frac{1000 \text{ V}}{1,000,000 \text{ Ω}}$$
$$= \frac{1}{1000}$$
$$I = 0.001 \text{ A}$$

Low Voltage but High Current. At the opposite extreme, a low value of voltage in a very low resistance circuit can produce a very large amount of current. A 6-V battery connected across a resistance of 0.01 Ω produces 600 A of current:

$$I = \frac{V}{R}$$
$$= \frac{6 \text{ V}}{0.01 \text{ Ω}}$$
$$I = 600 \text{ A}$$

(a)

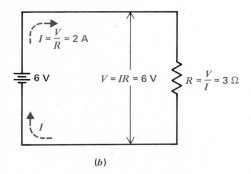

(b)

FIGURE 2-2

Using Ohm's law. (a) Voltage source applied across R. (b) Schematic diagram with values calculated by Ohm's law.

***I* Is Less with More *R*.** Note the values of *I* in the following two examples also:

Example 1. A heater with a resistance of 8 Ω is connected across the 120-V power line. How much is the current *I*?

Answer. $I = \dfrac{V}{R} = \dfrac{120 \text{ V}}{8 \text{ Ω}}$

$I = 15 \text{ A}$

Example 2. A small light bulb with a resistance of 2400 Ω is connected across the same 120-V power line. How much is the current *I*?

Answer. $I = \dfrac{V}{R} = \dfrac{120 \text{ V}}{2400 \text{ Ω}}$

$I = 0.05 \text{ A}$

Although both cases have the same 120 V applied, the current is much less in Example 2 because of the higher resistance.

*Practice Problems 2-1
(answers on page 53)*
(a) Calculate *I* for 100 V applied across 25 Ω.
(b) Calculate *I* for 50 V applied across 25 Ω.
(c) Calculate *I* for 100 V applied across 50 Ω.
(d) Calculate *I* for 100 V applied across 5 Ω.

2-2
THE VOLTAGE *V* = *IR*

Referring to Fig. 2-2, the amount of voltage across *R* must be the same as *V* because the resistance is connected directly across the battery. The numerical value of this *V* is equal to the product *I* × *R*.* For instance, the *IR* voltage in Fig. 2-2 is 2 A × 3 Ω, which equals the 6 V of

* For an explanation of how to invert factors from one side to the other side of an equation, see B. Grob, "Mathematics Outline and Review Problems for Basic Electronics," McGraw-Hill Book Company, New York.

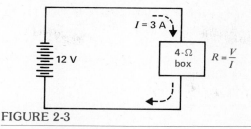

FIGURE 2-3

The resistance of any component is its *V/I* ratio.

the applied voltage. The formula is

$$V = IR \qquad (2\text{-}2)$$

With *I* in ampere units and *R* in ohms, their product *V* is in volts. Actually, this must be so because the *I* value equal to *V/R* is the amount that allows the *IR* product to be the same as the voltage across *R*.

Besides the numerical calculations possible with the *IR* formula, it is useful to consider that the *IR* product means voltage. Whenever there is current through a resistance, it must have a potential difference across its two ends equal to the *IR* product. If there were no potential difference, no electrons could flow to produce the current.

*Practice Problems 2-2
(answers on page 53)*
(a) Calculate *V* for 0.002 A through 1000 Ω.
(b) Calculate *V* for 0.004 A through 1000 Ω.
(c) Calculate *V* for 0.002 A through 2000 Ω.

2-3
THE RESISTANCE *R* = *V/I*

As the third and final version of Ohm's law, the three factors *V*, *I*, and *R* are related by the formula

$$R = \dfrac{V}{I} \qquad (2\text{-}3)$$

Chapter 2
Ohm's Law

In Fig. 2-2, R is 3 Ω because 6 V applied across the resistance produces 2 A through it. Whenever V and I are known, the resistance can be calculated as the voltage across R divided by the current through it.

Physically, a resistance can be considered as some material with elements having an atomic structure that allows free electrons to drift through it with more or less force applied. Electrically, though, a more practical way of considering resistance is simply as a V/I ratio. Anything that allows 1 A of current with 10 V applied has a resistance of 10 Ω. This V/I ratio of 10 Ω is its characteristic. If the voltage is doubled to 20 V, the current will also double to 2 A, providing the same V/I ratio of a 10-Ω resistance.

Furthermore, we do not need to know the physical construction of a resistance to analyze its effect in a circuit, so long as we know its V/I ratio. This idea is illustrated in Fig. 2-3. Here, a box with some unknown material in it is connected into a circuit where we can measure the 12 V applied across the box and the 3 A of current through it. The resistance is 12 V/3 A, or 4 Ω. There may be liquid, gas, metal, powder, or any other material in the box, but electrically it is just a 4-Ω resistance because its V/I ratio is 4.

Practice Problems 2-3
(answers on page 53)
(a) Calculate R for 12 V with 0.003 A.
(b) Calculate R for 12 V with 0.006 A.
(c) Calculate R for 12 V with 0.001 A.

2-4
PRACTICAL UNITS

The three forms of Ohm's law can be used to define the practical units of current, potential difference, and resistance as follows:

$$1 \text{ ampere} = \frac{1 \text{ volt}}{1 \text{ ohm}}$$

$$1 \text{ volt} = 1 \text{ ampere} \times 1 \text{ ohm}$$

$$1 \text{ ohm} = \frac{1 \text{ volt}}{1 \text{ ampere}}$$

One ampere is the amount of current through a one-ohm resistance that has one volt of potential difference applied across it.

One volt is the potential difference across a one-ohm resistance that has one ampere of current through it.

One ohm is the amount of opposition in a resistance that has a V/I ratio of 1, allowing one ampere of current with one volt applied.

In summary, the circle diagram in Fig. 2-4 for $V = IR$ can be helpful in using Ohm's law. Note that V is always at the top for $V = IR$, $V/R = I$, or $V/I = R$.

Practice Problems 2-4
(answers on page 53)
(a) Calculate V for 0.007 A through 5000 Ω.
(b) Calculate the amount of I for 12,000 V across 6,000,000 Ω.
(c) Calculate R for 8 V with 0.004 A.

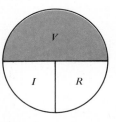

FIGURE 2-4

Circle to memorize $V = IR$, $V/I = R$, or $V/R = I$.

2-5
MULTIPLE AND SUBMULTIPLE UNITS

The basic units—ampere, volt, and ohm—are practical values in most electric power circuits, but in many electronics applications these units are either too small or too big. As examples, resistances can be a few million ohms, the output of a high-voltage supply in a television receiver is about 20,000 V, and current through tubes and transistors is generally thousandths or millionths of an ampere.

In such cases, it is helpful to use multiples and submultiples of the basic units. As shown in Table 2-1, these units are based on the decimal system of tens, hundreds, thousands, etc. The common conversions for V, I, and R are given here, but a complete listing of all the prefixes is in Appendix F. Note that capital M is used for 10^6 to distinguish from small m for 10^{-3}.

Example 3. The I of 8 mA flows through a 5-kΩ R. How much is the IR voltage?

Answer.
$$V = IR = 8 \times 10^{-3} \times 5 \times 10^3$$
$$= 8 \times 5$$
$$V = 40 \text{ V}$$

In general, milliamperes multiplied by kilohms results in volts for the answer, as 10^{-3} and 10^3 cancel.

Example 4. How much current is produced by 60 V across 12 kΩ?

Answer.
$$I = \frac{V}{R} = \frac{60}{12 \times 10^3} = 5 \times 10^{-3}$$
$$I = 5 \text{ mA}$$

Note that volts across kilohms produces milliamperes of current. Similarly, volts across megohms produces microamperes. These common combinations can be summarized as follows:

$$\text{volts} \div \text{kilohms} = \text{milliamperes}$$
$$\text{volts} \div \text{megohms} = \text{microamperes}$$
$$\text{kilohms} \times \text{milliamperes} = \text{volts}$$
$$\text{megohms} \times \text{microamperes} = \text{volts}$$

Practice Problems 2-5
(answers on page 53)
(a) Change the following to basic units with powers of 10: 6 mA, 5 kΩ, and 3 μA.
(b) Change the following to units with metric prefixes: 6×10^{-3} A, 5×10^3 Ω, and 3×10^{-6} A.

2-6
THE LINEAR PROPORTION BETWEEN V AND I

The Ohm's law formula $I = V/R$ states that V and I are directly proportional for any one value of R. This relation between V and I can be

TABLE 2-1. Conversion Factors

PREFIX	SYMBOL	RELATION TO BASIC UNIT	EXAMPLES
mega	M	1,000,000 or 1×10^6	5 MΩ (megohms) = 5,000,000 ohms = 5×10^6 ohms
kilo	k	1000 or 1×10^3	18 kV (kilovolts) = 18,000 volts = 18×10^3 volts
milli	m	0.001 or 1×10^{-3}	48 mA (milliamperes) = 48×10^{-3} ampere = 0.048 ampere
micro	μ	0.000 001 or 1×10^{-6}	15 μV (microvolts) = 15×10^{-6} volt = 0.000 015 volt

analyzed by using a fixed resistance of 2 Ω for R_L, as in Fig. 2-5. Then when V is varied, the meter shows I values directly proportional to V. For instance, with 12 V, I equals 6 A; for 10 V, the current is 5 A; an 8-V potential difference produces 4 A.

All the values of V and I are listed in the table in Fig. 2-5b and plotted in the graph in Fig. 2-5c. The I values are one-half the V values because R is 2 Ω. However, I is zero with zero volts applied.

Plotting the Graph. The voltage values for V are marked on the horizontal axis, called the x *axis* or *abscissa*. The current values I are on the vertical axis, called the y *axis* or *ordinate*.

Because the values for V and I depend on each other, they are variable factors. V is the independent variable here because we assign values of voltage and note the resulting current. Generally, the independent variable is plotted on the x axis, which is why the V values are shown here horizontally while the I values are on the ordinate.

The two scales need not be the same. The only requirement is that equal distances on either scale represent equal changes in magnitude. On the x axis here 2-V steps are chosen, while the y axis has 1-A scale divisions. The zero point at the origin is the reference.

The plotted points in the graph show the values in the table. For instance, the lowest point is 2 V horizontally from the origin, and 1 A up. Similarly, the next point is at the intersection of the 4-V mark and the 2-A mark. A line joining these two plotted points includes all values of I, for any value of V, with R constant at 2 Ω. This also applies to values not listed in the table. For instance, if we take the value of 7 V,

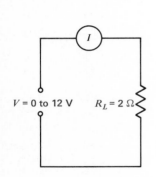

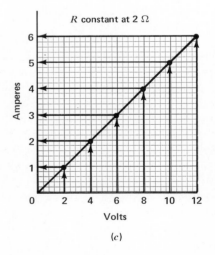

Volts	Ohms	Amperes
0	2	0
2	2	1
4	2	2
6	2	3
8	2	4
10	2	5
12	2	6

(a) (b) (c)

FIGURE 2-5
Experiment to show that I increases in direct proportion to V. (a) Circuit with variable V but constant R. (b) Table of increasing I values for higher V values. (c) Graph of V and I values. The linear voltampere characteristic shows a direct proportion between V and I.

up to the straight line and over to the *I* axis, the graph shows 3.5 A for *I*.

Voltampere Characteristic. The graph in Fig. 2-5c is called the voltampere characteristic of *R*. It shows how much current the resistor allows for different voltages. Multiple and submultiple units of *V* and *I* can be used, though. For transistors and tubes the units of *I* are often milliamperes or microamperes.

Linear Resistance. The straight-line graph in Fig. 2-5 shows *R* is a linear resistor. A linear resistance has a constant value of ohms. Its *R* does not change with the applied voltage. Then *V* and *I* are directly proportional. Doubling the value of *V* from 4 to 8 V results in twice the current, from 2 to 4 A. Similarly, three or four times the value of *V* will produce three or four times *I*, for a proportional increase in current.

Nonlinear Resistance. This type has a nonlinear voltampere characteristic. As an example, the resistance of the tungsten filament in a light bulb is nonlinear. The reason is that *R* increases with more current as the filament becomes hotter. Increasing the applied voltage does produce more current, but *I* does not increase in the same proportion as the increase in *V*.

Inverse Relation Between *I* and *V*. Whether *R* is linear or not, the current *I* is less for more *R*, with the applied voltage constant. This is an inverse relation, meaning that *I* goes down as *R* goes up. Remember that in the formula $I = V/R$, the resistance is in the denominator. A higher value of *R* actually lowers the value of the complete fraction.

As an example, let *V* be constant at 1 V. Then *I* is equal to the fraction $1/R$. As *R* increases, the values of *I* decrease. For *R* of 2 Ω, *I* is $\frac{1}{2}$ or 0.5 A. For a higher *R* of 10 Ω, *I* will be smaller at $\frac{1}{10}$ or 0.1 A.

Practice Problems 2-6
(answers on page 53)
Refer to the graph in Fig. 2-5c.
(a) Are the values of *I* on the *y* or *x* axis?
(b) Is this *R* linear or nonlinear?

2-7
POWER

The unit of electrical power is the *watt* (W), named after James Watt (1736–1819). One watt of power equals the work done in one second by one volt of potential difference in moving one coulomb of charge.

Remember that one coulomb per second is an ampere. Therefore power in watts equals the product of amperes times volts.

Power in watts = volts × amperes
$$P = V \times I \qquad (2\text{-}4)$$

When a 6-V battery produces 2 A in a circuit, for example, the battery is generating 12 W of power. Note the following additional examples:

Example 5. A toaster takes 10 A from the 120-V power line. How much power is used?

Answer. $P = V \times I = 120 \text{ V} \times 10 \text{ A}$
$P = 1200 \text{ W}$

Example 6. How much current flows in the filament of a 300-W bulb connected to the 120-V power line?

Answer. $P = V \times I$ or $I = P/V$. Then

$$I = \frac{300 \text{ W}}{120 \text{ V}}$$

$I = 2.5 \text{ A}$

Example 7. How much current flows in a 60-W bulb connected to the 120-V power line?

Answer. $P = V \times I$ or $I = P/V$. Then

$$I = \frac{60 \text{ W}}{120 \text{ V}}$$

$$I = 0.5 \text{ A}$$

Note that the lower-wattage bulb uses less current.

Work and Power. Work and energy are essentially the same with identical units. Power is different, however, because it is the time rate of doing work.

As an example of work, if you move 100 lb a distance of 10 ft, the work is 100 lb × 10 ft or 1000 ft · lb, regardless of how fast or how slowly the work is done. Note that the unit of work is foot-pounds, without any reference to time.

However, power equals the work divided by the time it takes to do the work. If it takes 1 s, the power in this example is 1000 ft · lb/s; if the work takes 2 s, the power is 1000 ft · lb in 2 s, or 500 ft · lb/s.

Similarly, electrical power is the time rate at which charge is forced to move by voltage. This is why the power in watts is the product of volts and amperes. The voltage states the amount of work per unit of charge; the current value includes the time rate at which the charge is moved.

Watts and Horsepower Units. A further example of how electrical power corresponds to mechanical power is the fact that

$$746 \text{ W} = 1 \text{ hp} = 550 \text{ ft} \cdot \text{lb/s}$$

This relation can be remembered more easily as 1 hp equals approximately ¾ kilowatt (kW). One kilowatt = 1000 W.

Practical Units of Power and Work. Starting with the watt, we can develop several other important units. The fundamental principle to remember is that power is the time rate of doing work, while work is power used during a period of time. The formulas are

$$\text{Power} = \frac{\text{work}}{\text{time}} \qquad (2\text{-}5)$$

and

$$\text{Work} = \text{power} \times \text{time} \qquad (2\text{-}6)$$

With the watt unit for power, one watt used during one second equals the work of one joule. Or one watt is one joule per second. Therefore, 1 W = 1 J/s. The joule is a basic practical unit of work or energy.[1]

A unit of work that can be used with individual electrons is the *electron volt* (eV). Note that the electron is charge while the volt is potential difference. Then 1 eV is the amount of work required to move an electron between two points having a potential difference of one volt. Since 6.25×10^{18} electrons equal 1 C and a joule is a volt-coulomb, there must be 6.25×10^{18} eV in 1 J.

Note that the electron volt or the joule unit of work is the product of charge times voltage, but the watt unit of power is the product of voltage times current. The division by time to convert work to power corresponds to the division by time that converts charge to current.

Kilowatthours. This is a unit commonly used for large amounts of electrical work or energy. The amount is calculated simply as the product of the power in kilowatts multiplied by the time in hours during which the power is used. As an example, if a light bulb uses 300 W or 0.3 kW for 4 hours (h), the amount of energy is 0.3 × 4, which equals 1.2 kWh.

We pay for electricity in kilowatthours of energy. The power-line voltage is constant at

---

[1] See Appendix D, Physics Units.

120 V. However, more appliances and light bulbs require more current because they all add in the main line to increase the power. Suppose the total load current on the main line equals 20 A. Then the power is 20 A × 120 V = 2400 W, or 2.4 kW. If this load current is used for 5 h, then the energy supplied equals $2.4 \times 5 = 12$ kWh. At 6 cents per kWh, the cost is $12 \times 0.06 = 0.72$, or 72 cents, just for 5 hours with a 20-A load.

Practice Problems 2-7
(answers on page 53)
(a) An electrical heater takes 15 A from the 120-V power line. Calculate the power.
(b) How much is the load current for a 100-W bulb connected to the 120-V power line?

2-8
POWER DISSIPATION IN RESISTANCE

When current flows in a resistance, heat is produced because friction between the moving free electrons and the atoms obstructs the path of electron flow. The heat is evidence that power is used in producing current. This is how a fuse opens, as heat resulting from excessive current melts the metal link in the fuse.

The power is generated by the source of applied voltage and consumed in the resistance in the form of heat. As much power as the resistance dissipates in heat must be supplied by the voltage source; otherwise, it cannot maintain the potential difference required to produce the current.

The correspondence between electrical power and heat is indicated by the fact that 1 W used during the time of 1 s is equivalent to 0.24 calorie of heat energy. The electrical energy converted to heat is considered to be dissipated or used up because the calories of heat cannot be returned to the circuit as electrical energy.

Since power is dissipated in the resistance of a circuit, it is convenient to express the power in terms of the resistance R. The $V \times I$ formula can be rearranged as follows:

Substituting IR for V,

$$P = V \times I = IR \times I$$
$$P = I^2 R \tag{2-7}$$

This is a common form for the power formula. For another form, substitute V/R for I. Then

$$P = V \times I = V \times \frac{V}{R}$$
$$P = \frac{V^2}{R} \tag{2-8}$$

In all the formulas, V is the voltage across R in ohms, producing the current I in amperes, for power in watts.

Any one of the three formulas can be used to calculate the power dissipated in a resistance. The one to be used is just a matter of convenience, depending on which factors are known.

In Fig. 2-6, for example, the power dissipated with 2 A through the resistance and 6 V across it is $2 \times 6 = 12$ W.

Or, calculating in terms of just the current and resistance, the power is the product of 2 squared, or 4, times 3, which equals 12 W.

Using the voltage and resistance, the power can be calculated as 6 squared, or 36, divided by 3, which also equals 12 W.

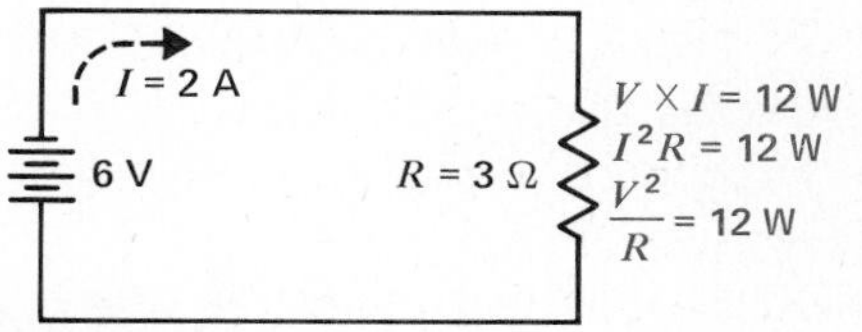

$I = 2$ A
6 V
$R = 3 \,\Omega$
$V \times I = 12$ W
$I^2 R = 12$ W
$\dfrac{V^2}{R} = 12$ W

FIGURE 2-6

Calculating the electrical power in a circuit as VI, $I^2 R$, or V^2/R.

No matter which formula is used, 12 W of power is dissipated, in the form of heat. This amount of power must be generated continuously by the battery in order to maintain the potential difference of 6 V that produces the 2-A current against the opposition of 3 Ω.

In some applications, the electrical power dissipation is desirable because the component must produce heat in order to do its job. For instance, a 600-W toaster must dissipate this amount of power to produce the necessary amount of heat. Similarly, a 300-W light bulb must dissipate this power to make the filament white-hot so that it will have the incandescent glow that furnishes the light. In other applications, however, the heat may be just an undesirable byproduct of the need to provide current through the resistance in a circuit. In any case, though, whenever there is current in a resistance, it dissipates power equal to I^2R.

Example 8. Calculate the power in a circuit where the source of 100 V produces 2 A in a 50-Ω R.

Answer. $P = I^2R = 4 \times 50$
$P = 200$ W

This means the source generates 200 W of power while the resistance dissipates 200 W in the form of heat.

Example 9. Calculate the power in a circuit where the same source of 100 V produces 4 A in a 25-Ω R.

Answer. $P = I^2R = 16 \times 25$
$P = 400$ W

Note the higher power in Example 9 because of more I, even though R is less than Example 8.

Components that utilize the power dissipated in their resistance, such as light bulbs and toasters, are generally rated in terms of power. The power rating is at normal applied voltage, which is usually the 120 V of the power line. For instance, a 600-W 120-V toaster has this rating because it dissipates 600 W in the resistance of the heating element when connected across 120 V.

In order to calculate I or R for components rated in terms of power at a specified voltage, it may be convenient to use the power formulas in different forms. There are three basic power formulas but nine combinations, as follows:

$$P = VI \qquad P = I^2R \qquad P = \frac{V^2}{R}$$

or $I = \dfrac{P}{V}$ or $R = \dfrac{P}{I^2}$ or $R = \dfrac{V^2}{P}$

or $V = \dfrac{P}{I}$ or $I = \sqrt{\dfrac{P}{R}}$ or $V = \sqrt{PR}$

Example 10. How much current is needed for a 600-W 120-V toaster?

Answer. $I = \dfrac{P}{V} = \dfrac{600}{120}$
$I = 5$ A

Example 11. How much is the resistance of a 600-W 120-V toaster?

Answer. $R = \dfrac{V^2}{P} = \dfrac{14{,}400}{600}$
$R = 24$ Ω

Example 12. How much current is needed for a 24-Ω R that dissipates 600 W?

Answer. $I = \sqrt{\dfrac{P}{R}} = \sqrt{\dfrac{600}{24}} = \sqrt{25}$
$I = 5$ A

Note that all these formulas are based on Ohm's law $V = IR$ and the power formula $P = V \times I$. The following example with a 300-W bulb illustrates this idea. Connected across the 120-V line, the 300-W filament requires 2.5 A, equal to P/V or $^{300}/_{120}$. The proof is that the VI product then is 120×2.5, which is 300 W.

Therefore, the resistance of the filament, equal to V/I, is $^{120}/_{2.5}$, or 48 Ω. If we use the formula $R = V^2/P$, or $^{14,400}/_{300}$, the answer is the same, 48 Ω.

In any case, when this bulb is connected across the 120-V line so that it can dissipate its rated power, it draws 2.5 A from the power line, and the resistance of its white-hot filament is 48 Ω.

Practice Problems 2-8
(answers on page 53)
(a) I is 2 A in a 5-Ω R. Calculate P.
(b) V is 10 V across a 5-Ω R. Calculate P.

2-9
ELECTRIC SHOCK

While you are working on electric circuits, there is often the possibility of receiving an electric shock by touching the "live" conductors when the power is on. The shock is a sudden involuntary contraction of the muscles, with a feeling of pain, caused by current through the body. If severe enough, the shock can be fatal. Safety first, therefore, should always be the rule.

The greatest shock hazard is from high-voltage circuits that can supply appreciable amounts of power. The resistance of the human body is also an important factor. If you hold a conducting wire in each hand, the resistance of the body across the conductors is about 10,000 to 50,000 Ω. Holding the conductors tighter lowers the resistance. If you hold only one conductor, your resistance is much higher. It follows that the higher the body resistance, the smaller the current that can flow through you.

A safety rule, therefore, is to work with only one hand if the power is on. Also, keep yourself insulated from earth ground when working on power-line circuits, since one side of the line is usually connected to earth. In addition, the metal chassis of radio and television receivers is often connected to the power-line ground. The final and best safety rule is to work on the circuits with the power disconnected if at all possible and make resistance tests.

Note that it is current through the body, not through the circuit, which causes the electric shock. This is why high-voltage circuits are most important, since sufficient potential difference can produce a dangerous amount of current through the relatively high resistance of the body. For instance, 500 V across a body resistance of 25,000 Ω produces 0.02 A, or 20 mA, which can be fatal. As little as 10 μA through the body can cause an electric shock. In an experiment[1] on electric shock to determine the current at which a person could release the live conductor, this value of "let-go" current was about 9 mA for men and 6 mA for women.

In addition to high voltage, the other important consideration in how dangerous the shock can be is the amount of power the source can supply. The current of 0.02 A through 25,000 Ω means the body resistance dissipates 10 W. If the source cannot supply 10 W, its output voltage drops with the excessive current load. Then the current is reduced to the amount corresponding to how much power the source can produce.

In summary, then, the greatest danger is from a source having an output of more than about 30 V with enough power to maintain the load current through the body when it is connected across the applied voltage. In general, components that can supply high power are physically big because of the need for dissipating heat.

[1] C. F. Dalziel and W. R. Lee, Lethal Electric Currents, *IEEE Spectrum,* February 1969.

Practice Problems 2-9
(*answers on page 53*)
Answer true or false.
(a) 120 V is more dangerous than 12 V for electric shock.
(b) Resistance tests with an ohmmeter can be made with power off in the circuit.

Summary

1. The three forms of Ohm's law are $I = V/R$, $V = IR$, and $R = V/I$. The basic units are volts for V, amperes for I, and ohms for R.
2. One ampere is the amount of current produced by one volt of potential difference across one ohm of resistance. This current of 1 A is the same as 1 C/s.
3. With constant R, the amount of I increases in direct proportion as V increases. This linear relation between V and I is shown by the graph in Fig. 2-5.
4. With constant V, the current I decreases as R increases. This is an inverse relation.
5. Power is the time rate of doing work or using energy. The unit is the watt. One watt equals $1\text{ V} \times 1\text{ A}$. Also, watts = joules per second.
6. The unit of work or energy is the joule. One joule equals $1\text{ W} \times 1\text{ s}$.
7. The most common multiples and submultiples of the practical units are listed in Table 2-1.
8. Voltage applied across your body can produce a dangerous electric shock. Whenever possible, shut off the power and make resistance tests. If the power must be on, use only one hand. Do not let the other hand rest on a conductor.
9. Table 2-2 summarizes the practical units used with Ohm's law.

TABLE 2-2. Practical Units of Electricity

UNIT	COULOMB	AMPERE	VOLT	WATT	OHM	MHO
Definition	6.25×10^{18} electrons	$\dfrac{\text{Coulomb}}{\text{second}}$	$\dfrac{\text{Joule}}{\text{coulomb}}$	$\dfrac{\text{Joule}}{\text{second}}$	$\dfrac{\text{Volt}}{\text{ampere}}$	$\dfrac{\text{Ampere}}{\text{volt}}$

Self-Examination (Answers at back of book.)

Fill in the missing answers.

1. With 10 V across 5 Ω R, the current I is _____ A.
2. When 10 V produces 2.5 A, R is _____ Ω.
3. With 8 A through a 2-Ω R, the IR voltage is _____ V.
4. The resistance of 500,000 Ω is _____ MΩ.
5. With 10 V across 5000 Ω R, the current I is _____ mA.
6. The power of 50 W = 2 A × _____ V.
7. The energy of 50 J = 2 C × _____ V.
8. The current drawn from the 120-V power line by a 1200-W toaster = _____ A.
9. The current of 400 μA = _____ mA.
10. With 12 V across a 2-Ω R, its power dissipation = _____ W.
11. A circuit has a 4-A I. If V is doubled and R is the same, I = _____ A.
12. A circuit has a 4-A I. If R is doubled and V is the same, I = _____ A.
13. A television receiver using 240 W from the 120-V power line draws current I = _____ A.
14. The rated current for a 500-W 120-V bulb = _____ A.
15. The resistance of the bulb in question 14 is _____ Ω.
16. The energy of 12.5×10^{18} eV = _____ J.
17. The current of 1200 mA = _____ A.
18. In a vacuum-tube amplifier circuit, the plate load resistor R_L of 50 kΩ has 150 V across it. Through R_L, then, the current = _____ mA.
19. In a transistor circuit, a 1-kΩ resistor R_1 has 200 μA through it. Across R_1, then, its voltage = _____ V.
20. In a transistor circuit, a 50-kΩ resistor R_2 has 6 V across it. Through R_2, then, its current = _____ mA.

Essay Questions

1. State the three forms of Ohm's law relating V, I, and R.
2. (a) Why does higher applied voltage with the same resistance result in more current? (b) Why does more resistance with the same applied voltage result in less current?
3. Calculate the resistance of a 300-W bulb connected across the 120-V power line, using two different methods to arrive at the same answer.
4. State which unit in each of the following pairs is larger: (a) volt or kilovolt; (b) ampere or milliampere; (c) ohm or megohm; (d) volt or

microvolt; (e) siemens or microsiemens; (f) electron volt or joule; (g) watt or kilowatt; (h) kilowatthour or joule; (i) volt or millivolt; (j) megohm or kilohm.
5. State two safety precautions to follow when working on electric circuits.
6. Referring back to the resistor shown in Fig. 1-12, suppose that it is not marked. How could you determine its resistance by Ohm's law? Show your calculations that result in the V/I ratio of 600 Ω. However, do not exceed the power rating of 10 W.
7. What is the difference between work and power? Give two units for each.
8. Refer to the two resistors in series with each other in Fig. 3-1 on page 55. If the current through R_1 is 2 A, how much would you guess is the current through R_2?
9. Prove that 1 kWh is equal to 3.6×10^6 J.
10. A circuit has a constant R of 5000 Ω, while V is varied from 0 to 50 V in 10-V steps. Make a table listing the values of I for each value of V. Then draw a graph plotting these values of milliamperes vs. volts. (This graph should be similar to Fig. 2-5c.)

Problems (Answers to odd-numbered problems at back of book.)

1. A 90-V source is connected across a 30-kΩ resistance. (a) Draw the schematic diagram. (b) How much current flows through the resistance? (c) How much current flows through the voltage source? (d) If the resistance is tripled, how much is the current in the circuit?
2. A 6-V battery is connected across a 2-Ω resistance. (a) Draw the schematic diagram. (b) Calculate the power dissipated in the resistance. (c) How much power is supplied by the battery? (d) If the resistance is doubled, how much is the power?
3. A vacuum-tube heater has 0.3 A of current with 6.3 V applied. (a) Draw the schematic diagram, showing the heater as a resistance. (b) How much is the resistance of the heater?
4. Convert the following units using powers of 10 where necessary: (a) 12 mA to amperes; (b) 5000 V to kilovolts; (c) ½ MΩ to ohms; (d) 100,000 Ω to megohms; (e) ½ A to milliamperes; (f) 9000 μS to siemens; (g) 1000 μA to milliamperes; (h) 5 kΩ to ohms; (i) 8 nanoseconds (ns) to seconds.
5. A current of 2 A flows through a 6-Ω resistance connected across a battery. (a) How much is the applied voltage of the battery? (b) How much power is dissipated in the resistance? (c) How much power is supplied by the battery?

6. (a) How much resistance allows 30-A current with 6 volts applied? (b) How much resistance allows 1-mA current with 10 kV applied? Why is it possible to have less current in (b) with the higher applied voltage?
7. A source of applied voltage produces 1 mA through a 10-MΩ resistance. How much is the applied voltage?
8. Calculate the current I, in ampere units, for the following examples: (a) 45 V applied across 68 kΩ; (b) 250 V across 12 MΩ; (c) 1200 W dissipated in 600 Ω.
9. Calculate the IR voltage for the following examples: (a) 68 μA through 22 MΩ; (b) 2.3 mA through 47 kΩ; (c) 237 A through 0.012 Ω.
10. Calculate the resistance R, in ohms, for the following examples: (a) 134 mA produced by 220 V; (b) 800 W dissipated with 120 V applied; (c) a conductance of 9000 μS.
11. Find the value of V in Fig. 2-7.

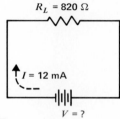

FIGURE 2-7

For Prob. 11.

Answers to Practice Problems

2-1	(a) 4 A	2-4	(c) 2000 Ω
	(b) 2 A	2-5	(a) See Prob. b
	(c) 2 A		(b) See Prob. a
	(d) 20 A	2-6	(a) y axis
2-2	(a) 2 V		(b) linear
	(b) 4 V	2-7	(a) 1.8 kW
	(c) 4 V		(b) 0.83 A
2-3	(a) 4000 Ω	2-8	(a) 20 W
	(b) 2000 Ω		(b) 20 W
	(c) 12,000 Ω	2-9	(a) T
2-4	(a) 35 V		(b) T
	(b) 0.002 A		

Series Circuits

Chapter 3

When the components in a circuit are connected in successive order with an end of each joined to an end of the next, as shown in Fig. 3-1, they form a series circuit. The resistors R_1 and R_2 are in series with each other and the battery. The result is only one path for electron flow. Therefore, the current I is the same in all the series components. This and other important characteristics are analyzed in the following topics:

3-1   Why I Is the Same in All Parts of a Series Circuit
3-2   Total R Equals the Sum of All Series Resistances
3-3   Series IR Voltage Drops
3-4   The Sum of Series IR Drops Equals the Applied V_T
3-5   Polarity of IR Voltage Drops
3-6   Polarities to Chassis Ground
3-7   Total Power in a Series Circuit
3-8   Series-Aiding and Series-Opposing Voltages
3-9   Analyzing Series Circuits
3-10 Effect of an Open Circuit in a Series Path

3-1
WHY I IS THE SAME IN ALL PARTS OF A SERIES CIRCUIT

An electrical current is a movement of charges between two points, produced by the applied voltage. In Fig. 3-2, the battery supplies the potential difference that forces electrons to drift from the negative terminal at A, toward B, through the connecting wires and resistances R_1, R_2, and R_3, back to the positive battery terminal at J.

At the negative battery terminal its negative charge repels electrons. Therefore, free electrons in the atoms of the wire at this terminal are repelled from A toward B. Similarly, free electrons at point B can then repel adjacent electrons, producing an electron drift toward C in the direction away from the negative battery terminal.

At the same time, the positive charge of the positive battery terminal attracts free electrons, causing electrons to drift toward I and J. As a result, the free electrons in the resistances R_1, R_2, and R_3 are forced to drift toward the positive terminal.

The positive terminal of the battery attracts electrons just as much as the negative side of the battery repels electrons. Therefore, the motion of free electrons in the circuit starts at the same time at the same speed in all parts of the circuit.

The electrons returning to the positive battery terminal are not the same electrons as those leaving the negative terminal. Free elec-

Chapter 3
Series Circuits 55

(a)

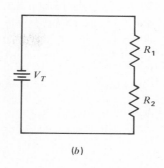
(b)

FIGURE 3-1

A series circuit. (a) Photo of wiring. (b) Schematic diagram.

trons in the wire are forced to move to the positive terminal because of the potential difference of the battery.

The free electrons moving away from one point are continuously replaced by free electrons flowing from an adjacent point in the series circuit. All electrons have the same speed as those leaving the battery. In all parts of the circuit, therefore, the electron drift is the same, with an equal number of electrons moving at

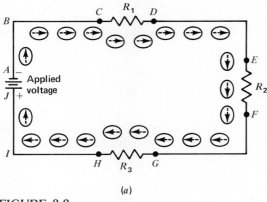
(a)

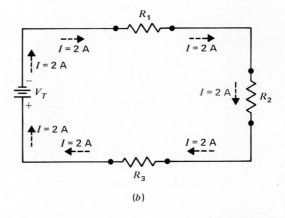
(b)

FIGURE 3-2

(a) Electron drift is the same at all points in a series circuit. (b) The current I is the same at all points in a series circuit.

Chapter 3
Series Circuits

one time with the same speed. That is why the current is the same in all parts of the series circuit.

In Fig. 3-2b, when the current is 2 A, for example, this is the value of the current through R_1, R_2, R_3, and the battery. Not only is the amount of current the same throughout, but in all parts of a series circuit the current cannot differ in any way because there is just one current path for the entire circuit.

The order in which components are connected in series does not affect the current. In Fig. 3-3b, resistances R_1 and R_2 are connected in reverse order compared with Fig. 3-3a, but in both cases they are in series. The current through each is the same because there is only one path for the electron flow. Similarly, R_3, R_4, and R_5 are in series and have the same current for the connections shown in Fig. 3-3c, d, and e. Furthermore, the resistances need not be equal.

The question of whether a component is first, second, or last in a series circuit has no meaning in terms of current. The reason is that I is the same amount at the same time in all the series components.

In fact, series components can be defined as those in the same current path. The path is from one side of the voltage source, through the series components, and back to the other side of the applied voltage. However, the series path must not have any point where the current can branch off to another path in parallel. This feature of series circuits applies not only to direct current, but also to alternating current of any frequency and for any waveshape.

Practice Problems 3-1
(answers on page 72)
(a) In Fig. 3-2, name five parts that have the I of 2 A.
(b) In Fig. 3-3e, when I in R_5 is 5 A, then I in R_3 is _____ A.

3-2
TOTAL R EQUALS THE SUM OF ALL SERIES RESISTANCES

When a series circuit is connected across a voltage source, as shown in Fig. 3-3, the free electrons forming the current must drift through all the series resistances. This path is the only way the electrons can return to the battery. With two or more resistances in the same current path, therefore, the total resistance across the voltage source is the opposition of all the resistances.

Specifically, the total resistance (R_T) of a series string is equal to the sum of the individual

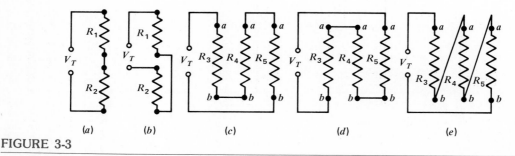

FIGURE 3-3
Series connections. R_1 and R_2 are in series in both (a) and (b). Also, R_3, R_4, and R_5 are in series in (c), (d), and (e).

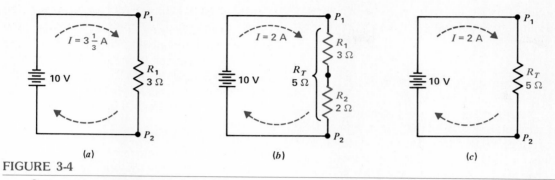

FIGURE 3-4
Series resistances are added. (a) R_1 alone is 3 Ω. (b) R_1 and R_2 in series total 5 Ω. (c) Total R_T is the same as one resistance of 5 Ω between P_1 and P_2.

resistances. This rule is illustrated in Fig. 3-4. In Fig. 3-4b, 2 Ω is added in series with the 3 Ω of Fig. 3-4a, producing the total resistance of 5 Ω. The total opposition of R_1 and R_2 limiting the amount of current is the same as though a 5-Ω resistance were used, as shown in the equivalent circuit in Fig. 3-4c.

Series String. A combination of series resistances is often called a *string*. The string resistance equals the sum of the individual resistances. For instance, R_1 and R_2 in Fig. 3-4 form a series string having the R_T of 5 Ω.

By Ohm's law, the amount of current between two points in a circuit equals the potential difference divided by the resistance between these points. As the entire string is connected across the voltage source, the current equals the voltage applied across the entire string divided by the total series resistance of the string. Between points P_1 and P_2 in Fig. 3-4, for example, 10 V is applied across 5 Ω in (b) and (c) to produce 2 A. This current flows through R_1 and R_2 in one series path.

Series Resistance Formula. In summary, the total resistance of a series string equals the sum of the individual resistances. The formula is

$$R_T = R_1 + R_2 + R_3 + \cdots + \text{etc.} \tag{3-1}$$

where R_T is the total resistance and R_1, R_2, and R_3 are individual series resistances. This formula applies to any number of resistances, whether equal or not, as long as they are in the same series string.

Note that R_T is the resistance to use in calculating the current in a series string. Then Ohm's law is

$$I = \frac{V_T}{R_T} \tag{3-2}$$

where R_T is the sum of all the resistances, V_T is the voltage applied across the total resistance, and I is the current in all parts of the string.

Example 1. Two resistances R_1 and R_2 of 5 Ω each and R_3 of 10 Ω are in series. How much is R_T?

Answer. $R_T = R_1 + R_2 + R_3 = 5 + 5 + 10$
$R_T = 20\ \Omega$

Example 2. With 80 V applied across the series string of Example 1, how much is the current in R_3?

Chapter 3
Series Circuits

Answer. $I = \dfrac{V_T}{R_T} = \dfrac{80\text{ V}}{20\text{ }\Omega}$

$I = 4\text{ A}$

This 4-A current is the same in R_3, R_2, R_1, or any part of the series circuit.

Note that adding series resistance reduces the current. In Fig. 3-4a the 3-Ω R_1 allows 10 V to produce $3\tfrac{1}{3}$ A. However, I is reduced to 2 A when the 2-Ω R_2 is added for a total series resistance of 5 Ω opposing the 10-V source.

Practice Problems 3-2
(answers on page 72)
(a) V is 10 V and R_1 is 5 kΩ. Calculate I.
(b) A 2-kΩ R_2 and 3-kΩ R_3 are added in series with R_1. Calculate R_T.
(c) Calculate I in R_1, R_2, and R_3.

3-3
SERIES *IR* VOLTAGE DROPS

With current through a resistance, by Ohm's law there is a voltage across the resistance equal to $I \times R$. This rule is illustrated in Fig. 3-5. Here the current equals 1 A through the 4-Ω R_1 and 6-Ω R_2 in series. The total R_T of 10 Ω is across the applied V_T of 10 V. The result is an *IR* voltage of 4 V equal to 1 A $\times$ 4 Ω across R_1 and 6 V equal to 1 A $\times$ 6 Ω across R_2.

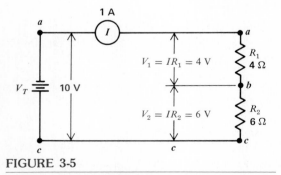

FIGURE 3-5

IR voltage drops in a series circuit.

The *IR* voltage across each resistance is called an *IR drop*, or a *voltage drop*, because it reduces the potential difference available for the remaining resistance in the series circuit. Note that the symbols V_1 and V_2 are used for the voltage drops across each resistor to distinguish them from the source V_T applied across both resistors.

In Fig. 3-5, the V_T of 10 V is applied across the total series resistance of R_1 and R_2. However, because of the *IR* voltage drop of 4 V across R_1, the potential difference across R_2 is only 6 V. The negative potential drops from 10 V at point *a*, with respect to the common reference point at *c*, down to 6 V at point *b*. The potential difference of 6 V between *b* and the reference at *c* is the voltage across R_2.

Similarly, there is an *IR* voltage drop of 6 V across R_2. The negative potential drops from 6 V at point *b* with respect to point *c*, down to 0 V at point *c* with respect to itself. The potential difference between any two points on the return line to the battery must be zero because the wire has practically zero resistance and therefore no *IR* drop.

It should be noted that voltage must be applied by a source of potential difference such as the battery in order to produce current and have an *IR* voltage drop across the resistance. With no current through a resistor, it has resistance only, but there is no potential difference across the two ends.

The *IR* drop of 4 V across R_1 in Fig. 3-5 represents that part of the applied voltage used to produce the current of 1 A through the 4-Ω resistance. Also, across R_2 the *IR* drop is 6 V because this much voltage allows 1 A in the 6-Ω resistance. The *IR* drop is more in R_2 because more potential difference is necessary to produce the same amount of current in the higher resistance. For series circuits, in general, the highest R has the largest *IR* voltage drop across it.

Practice Problems 3-3
(*answers on page 73*)
Refer to Fig. 3-5.
(a) How much is the sum of V_1 and V_2?
(b) Calculate I as V_T/R_T.
(c) How much is I through R_1?
(d) How much is I through R_2?

3-4
THE SUM OF SERIES *IR* DROPS EQUALS THE APPLIED V_T

The whole applied voltage is equal to the sum of its parts. For example, in Fig. 3-5, the individual voltage drops of 4 V and 6 V total the same 10 V produced by the battery. This relation for series circuits can be stated

$$V_T = V_1 + V_2 + V_3 + \cdots + \text{etc.} \qquad (3\text{-}3)$$

Where V_T is the applied voltage equal to the total of the individual *IR* drops.

Example 3. A voltage source produces an *IR* drop of 40 V across a 20-Ω R_1, 60 V across a 30-Ω R_2, and 180 V across a 90-Ω R_3, all in series. How much is the applied voltage?

Answer. $V_T = 40 + 60 + 180$
  $V_T = 280$ V

Note that the *IR* drop across each *R* results from the same current of 2 A, produced by 280 V across the total R_T of 140 Ω.

Example 4. An applied V_T of 120 V produces *IR* drops across two series resistors R_1 and R_2. If the drop across R_1 is 40 V, how much is the voltage across R_2?

Answer. Since V_1 and V_2 must total 120 V, and one is 40 V, the other must be the difference between 120 and 40 V. Or $V_2 = V_T - V_1$, which equals 120 − 40. Then $V_2 = 80$ V.

It really is logical that V_T is the sum of the series *IR* drops. *I* is the same in all the series components. The total of all the series voltages is needed to produce the same *I* in the total of all the series resistances as the *I* that each resistor voltage produces in its *R*.

Series Heater String for a Television Receiver. A common practical application of series circuits is shown in Fig. 3-6. This is a typical heater circuit for black-and-white television receivers that use tubes. All the heaters, including the picture tube, are in a series string. The heater current is alternating current from the ac power line, but the rules of series components apply the same way to dc circuits or ac circuits.

The first digits in the tube number give the required heater voltage for the rated amount of heater current. As an example, the 17BF11 amplifier tube needs 17 V across the two heater pins for 450 mA.

For picture tubes, however, the first digits give the screen size. For instance, the 16CWP4 picture tube has a 16-in screen, measured diagonally between opposite corners. The P4 is the phosphor number for a white screen; P22 indicates red, green, and blue phosphors for color television receivers. The heater voltage for most picture tubes is 6.3 V.

Each heater needs less than the 120 V from the power line, but they all use the same heater current for normal operation. Therefore, the heaters are in a series string connected across the voltage source. The idea is to make the sum of the series *IR* drops equal the power-line voltage, approximately. In Fig. 3-6, the series voltages are 17 + 33 + 23 + 15 + 11 + 8 + 6.3 + 3 + 6.3, which add up to 122.6 V.

Any one heater has its proportional part of the applied voltage. As determined by the heater resistance, its part of the applied voltage

Chapter 3
Series Circuits

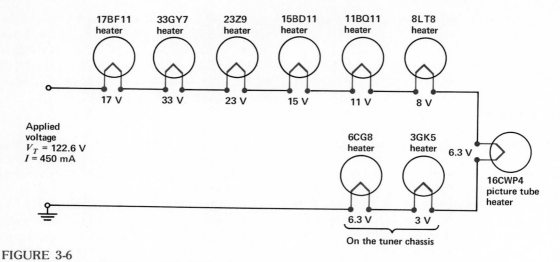

FIGURE 3-6
Typical string for tubes in a television receiver using series heaters.

is the required amount of heater voltage. The series circuit provides the same current for all the heaters, however. Tubes made for the series string have the same current rating but a higher heater voltage for tubes that need more power. The reason why there can be different voltages in the series string with the same current is that the heaters with higher resistance have more IR voltage drop.

How Series Circuits Are Used. The circuit in Fig. 3-6 shows how a series circuit is the way to connect components that all need the same current but with a voltage rating less than the source voltage. Then the applied voltage V_T is high enough to produce the required current I for the total resistance R_T.

Remember that V_T is across R_T, not each individual R. If you use only one R to calculate I, then use the individual voltage drop for that component. Then V/R for each component is the same value of I as V_T/R_T for the entire series circuit.

The Case of Zero IR Drop. Remember that the product of $I \times R$ can be zero when either I or R is zero. For a wire conductor used for connections, R is practically zero. Therefore, the IR voltage drop across the interconnecting wires is practically zero, even though the normal current is flowing. The other possibility is to have resistance but no current. Either the applied voltage is disconnected or there is an open circuit. A resistor by itself can have its normal resistance, but it cannot have an IR voltage drop unless I is flowing through R.

Practice Problems 3-4
(answers on page 73)
(a) A series circuit has 10-, 20-, and 30-V IR drops. How much is the applied voltage V_T of the source?

(b) 100 V is applied to R_1 and R_2 in series. If V_1 is 25 V, how much is V_2?

3-5
POLARITY OF IR VOLTAGE DROPS

When an IR voltage drop exists across a resistance, one end must be either more positive or more negative than the other end. Otherwise, without a potential difference no current could flow through the resistance to produce the IR drop. The polarity of this IR voltage can be associated with the direction of I through R. In brief, electrons flow into the negative side of the IR voltage and out the positive side (Fig. 3-7a).

If we want to consider conventional current with positive charges moving in the opposite direction from electron flow, the rule is reversed for the positive charges. See Fig. 3-7b. Here the positive charges for I are moving into the positive side of the IR voltage.

However, for either electron flow or conventional current the actual polarity of the IR drop is the same. In both a and b of Fig. 3-7, the top end of R in the diagrams is negative since this is the negative terminal of the source producing the current. After all, the resistor does not know which direction of current we are thinking of.

A series circuit with two IR voltage drops is

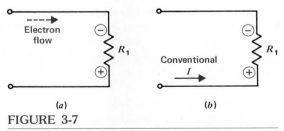

FIGURE 3-7

Polarity of IR voltage drops. (a) Electron flows into negative end. (b) Same polarity with positive charges into positive end.

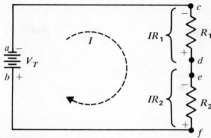

FIGURE 3-8

Two IR voltage drops in series. Electron flow shown for direction of I.

shown in Fig. 3-8. We can analyze these polarities in terms of electron flow. The electrons move from the negative terminal of the source V_T through R_1 from point c to d. Electrons move into c and out from d. Therefore c is the negative side of the voltage drop across R_1. Similarly, for the IR voltage drop across R_2, point e is the negative side, compared with point f.

A more fundamental way to consider the polarity of IR voltage drops in a circuit is the fact that between any two points the one nearer to the positive terminal of the voltage source is more positive; also, the point nearer to the negative terminal of the applied voltage is more negative. A point nearer to the terminal means there is less resistance in its path.

In Fig. 3-8 point c is nearer to the negative battery terminal than point d. The reason is that c has no resistance to a, while the path from d to a includes the resistance of R_1. Similarly, point f is nearer to the positive battery terminal than point e, which makes f more positive than e.

Notice that points d and e in Fig. 3-8 are marked with both plus and minus polarities. The plus polarity at d indicates it is more positive than c. This polarity, however, is shown just for the voltage across R_1. Point d cannot be more positive than points f and b. The positive terminal of the applied voltage must be the most positive point because the battery is generating the positive potential for the entire circuit.

Similarly, points a and c must have the most negative potential in the entire string, since this point is the negative terminal of the applied voltage. Actually, the plus polarity marked at d only means this end of R_1 is less negative than c, by the amount of voltage drop across R_1.

Consider the potential difference between e and d in Fig. 3-8, which is only a piece of wire. This voltage is zero because there is no resistance between these two points. Without any resistance here, the current cannot produce the IR drop necessary for a difference in potential. Points e and d are, therefore, the same electrically since they have the same potential.

When we go around the external circuit from the negative terminal of V_T, with electron flow, the voltage drops are drops in negative potential. For the opposite direction, starting from the positive terminal of V_T, the voltage drops are drops in positive potential. Either way, the voltage drop of each series R is its proportional part of V_T needed for the one value of current in all the resistances.

Practice Problems 3-5
(answers on page 73)
Refer to Fig. 3-8.
(a) Which point in the circuit is the most negative?
(b) Which point in the circuit is the most positive?
(c) Which is more negative, point d or f?

3-6
POLARITIES TO CHASSIS GROUND

In practical circuits, one side of the voltage source V_T is usually connected to chassis ground. The purpose is to simplify the wiring. On a plastic board with printed wiring, a rim of solder around the edge serves as the chassis ground return, as illustrated in Fig. 3-9a. Then only one terminal of the source voltage V_T is used for the *high side* of the wiring. The circuit components have return connections to the opposite side of V_T through the chassis ground conductor.

Either the negative or the positive terminal of V_T can be connected to the chassis ground return line. With the negative side grounded, V_T supplies positive voltage for the high side of the circuit (Fig. 3-9b). For the opposite case in Fig. 3-9c, the high side of the circuit has negative voltage with respect to chassis ground.

In Fig. 3-9, the two equal resistances divide the applied voltage equally. Then R_1 and R_2 each have a voltage drop of 10 V, equal to one-half the 20 V of V_T. The sum of the IR drops is $10 + 10 = 20$ V, equal to the total applied voltage.

Positive Voltages to Ground. In Fig. 3-9b point S is at $+20$ V. However, point J at the junction of R_1 and R_2 is at $+10$ V. The potential of $+10$ V is 10 V less than at S because of the 10-V drop across R_1. All these voltages are positive to chassis ground because the negative side of V_T and the ground return are really the same.

Negative Voltages to Ground. In Fig. 3-9c everything is the same as in (b) but with negative instead of positive voltages. Point S is at -20 V, since the positive side of the source voltage is grounded.

Practice Problems 3-6
(answers on page 73)
(a) In Fig. 3-9b, give each voltage to ground at points S, J, and G.
(b) In Fig. 3-9c, give each voltage to ground at points S, J, and G.

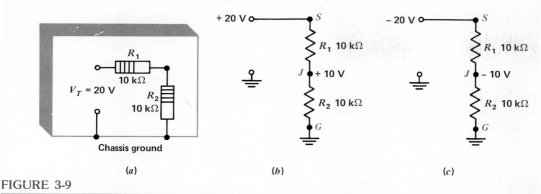

FIGURE 3-9

Polarity of *IR* voltage drops to chassis ground. (*a*) Wiring diagram with ground wiring around printed-circuit board. (*b*) Schematic diagram with $+20$ V for V_T with respect to chassis ground. (*c*) V_T is -20 V as positive side is grounded.

3-7 TOTAL POWER IN A SERIES CIRCUIT

The power needed to produce current in each series resistor is used up in the form of heat. Therefore, the total power used is the sum of the individual values of power dissipated in each part of the circuit. As a formula,

$$P_T = P_1 + P_2 + P_3 + \cdots + \text{etc.} \qquad (3\text{-}4)$$

As an example, in Fig. 3-10, R_1 dissipates 40 W for P_1, equal to 20 V $\times$ 2 A for the *VI* product. Or, the P_1 calculated as I^2R is $4 \times 10 = 40$ W. Also, the P_1 is V^2/R, or $400/10 = 40$ W.

Similarly, P_2 for R_2 is 80 W. This value is 40×2 for *VI*, 4×20 for I^2R, or $^{1600}/_{20}$ for V^2/R.

The total power dissipated by R_1 and R_2, then, is $40 + 80 = 120$ W. This power is generated by the source of applied voltage.

The total power can also be calculated as $V_T \times I$. The reason is that V_T is the sum of all the series voltages and *I* is the same in all the series components. In this case, then, P_T is $60 \times 2 = 120$ W, calculated as $V_T \times I$.

The total power here is 120 W, calculated either from the total voltage or from the sum of P_1 and P_2. This is the amount of power produced by the battery. The voltage source produces this power, equal to the amount used by the load.

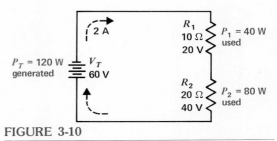

FIGURE 3-10

The sum of the individual powers P_1 and P_2 used in each resistance equals the total power P_T produced by the source.

Practice Problems 3-7
(answers on page 73)
(a) Each of three equal resistances dissipates 2 W. How much is P_T supplied by the source?
(b) A 1-kΩ R_1 and 40-kΩ R_2 are in series with a 50-V source. Which R dissipates more power?

3-8 SERIES-AIDING AND SERIES-OPPOSING VOLTAGES

Series-aiding voltages are connected with polarities that allow current in the same direction. In Fig. 3-11a, the 6-V of V_1 alone could produce 3-A electron flow from the negative terminal, with the 2-Ω R. Also, the 8-V of V_2 could produce 4 A in the same direction. The total I then is 7 A.

Instead of adding the currents, however, the voltages V_1 and V_2 can be added for a V_T of $6 + 8 = 14$ V. This 14 V produces 7 A in all parts of the series circuit with a resistance of 2 Ω.

Voltages are connected series-aiding with the plus terminal of one to the negative terminal of the next. They can be added for a total equivalent voltage. This idea applies in the same way to voltage sources, such as batteries, and to voltage drops across resistances. Any number of voltages can be added, as long as they are connected with series-aiding polarities.

Series-opposing voltages are subtracted, as shown in Fig. 3-11b. Notice here that the positive terminals of V_1 and V_2 are connected. Subtract the smaller from the larger value, and give the net V the polarity of the larger voltage. In this example, V_T is $8 - 6 = 2$ V. The polarity of V_T is the same as V_2 because it is larger than V_1.

If two series-opposing voltages are equal, the net voltage will be zero. In effect, one voltage balances out the other. The current I also is zero, without any net potential difference.

Practice Problems 3-8
(answers on page 73)
(a) V_1 of 40 V is series-aiding with V_2 of 60 V. How much is V_T?
(b) The same V_1 and V_2 are connected series-opposing. How much is V_T?

3-9 ANALYZING SERIES CIRCUITS

Refer to Fig. 3-12. Suppose that the source V_T of 50 V is known, with the 14-Ω R_1 and 6-Ω R_2. The problem is to find R_T, I, the individual voltage drops V_1 and V_2 across each resistor, and the power dissipated.

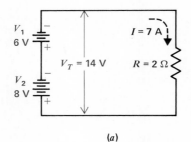
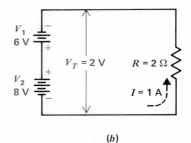

(a)   (b)

FIGURE 3-11

Voltages V_1 and V_2 in series. (a) Aiding. (b) Opposing.

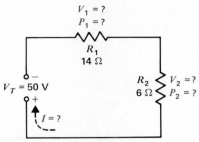

FIGURE 3-12
Analyzing a series circuit to find I, V_1, V_2, P_1, and P_2. See text for solution.

We must know the total resistance R_T to calculate I because the total applied voltage V_T is given. This V_T is applied across the total resistance R_T. In this example, R_T is $14 + 6 = 20\ \Omega$.

Now I can be calculated as V/R_T, or $^{50}/_{20}$, which equals 2.5 A. This 2.5-A I flows through R_1 and R_2.

The individual voltage drops are

$V_1 = IR_1 = 2.5 \times 14 = 35\ \text{V}$
$V_2 = IR_2 = 2.5 \times 6 = 15\ \text{V}$

Note that V_1 and V_2 total 50 V, equal to the applied V_T.

To find the power dissipated in each resistor,

$P_1 = V_1 \times I = 35 \times 2.5 = 87.5\ \text{W}$
$P_2 = V_2 \times I = 15 \times 2.5 = 37.5\ \text{W}$

These two values of dissipated power total 125 W. The power generated by the source equals $V_T \times I$ or 50×2.5, which is also 125 W.

General Methods for Series Circuits. For other types of problems with series circuits it is useful to remember the following:

1. When you know the I for one component, use this for I in all the components, as the current is the same in all parts of a series circuit.
2. To calculate I, the total V_T can be divided by the total R_T, or an individual IR drop can be divided by its R. For instance, the current in Fig. 3-12 could be calculated as V_2/R_2 or $^{15}/_6$, which equals the same 2.5 A for I. However, do not mix a total value for the entire circuit with an individual value for only part of the circuit.
3. When you know the individual voltage drops around the circuit, these can be added to equal the applied V_T. This also means a known voltage drop can be subtracted from the total V_T to find the remaining voltage drop.

These principles are illustrated by the problem in Fig. 3-13. In this circuit R_1 and R_2 are known but not R_3. However, the current through R_3 is given as 3 mA.

With just this information, all values in this circuit can be calculated. The I of 3 mA is the same in all three series resistances. Therefore,

$V_1 = 3\ \text{mA} \times 10\ \text{k}\Omega = 30\ \text{V}$
$V_2 = 3\ \text{mA} \times 30\ \text{k}\Omega = 90\ \text{V}$

The sum of V_1 and V_2 is $30 + 90 = 120$ V.

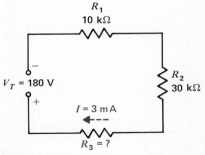

FIGURE 3-13
Find the resistance of R_3. See text for analysis of this series circuit.

This 120 V plus V_3 must total 180 V. Therefore, V_3 is $180 - 120 = 60$ V.

With 60 V for V_3, equal to IR_3, then R_3 must be $^{60}/_{0.003}$, equal to 20,000 Ω or 20 kΩ. The total circuit resistance is 60 kΩ, which results in the current of 3 mA with 180 V applied, as specified in the chapter.

Another way of doing this problem is to find R_T first. The equation $I = V_T/R_T$ can be inverted to calculate the R_T as V_T/I. With a 3-mA I and 180 V for V_T, the value of R_T must be 180 V/3 mA = 60 kΩ. Then R_3 is 60 kΩ − 40 kΩ = 20 kΩ.

The power dissipated in each resistance is 90 mW in R_1, 270 mW in R_2, and 180 mW in R_3. The total power is 540 mW.

Series Voltage-Dropping Resistors. A common application of series circuits is to use a resistance to drop the voltage from the source V_T to a lower value, as in Fig. 3-14. The load R_L here represents a transistor radio that operates normally with a 9-V battery. When the radio is on, the dc load current with 9 V applied is 18 mA. Therefore, the requirements are 9 V, at 18 mA as the load.

To operate this radio from 12.6 V, the voltage-dropping resistor R_S is inserted in series to provide a voltage drop V_S that will make V_L equal to 9 V. The required voltage drop across V_S is the difference between V_L and the higher V_T. As a formula,

$$V_S = V_T - V_L$$

$$V_S = 12.6 - 9 = 3.6 \text{ V}$$

Furthermore, this voltage drop of 3.6 V must be provided with a current of 18 mA, as the current is the same through R_S and R_L. To calculate R_S, then, it is 3.6 V/18 mA, which equals 0.2 kΩ or 200 Ω.

The power dissipated in R_L is calculated as $V_L \times I_L$. This value is 3.6 V × 18 mA, which equals 64.8 mW.

Circuit with Voltage Sources in Series. See Fig. 3-15. Note that V_1 and V_2 are series-opposing, with + to + through R_1. Their net effect then is 0 V. Therefore, V_T consists only of V_3, equal to 4.5 V. The total R is $2 + 1 + 2 = 5$ kΩ for R_T. Finally, I is V_T/R_T or 4.5 V/5 kΩ, which is equal to 0.9 mA.

Practice Problems 3-9
(answers on page 73)
Refer to Fig. 3-13.
(a) Calculate V_1 across R_1.
(b) Calculate V_2 across R_2.
(c) How much is V_3?

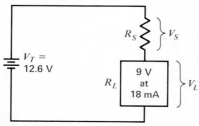

FIGURE 3-14

Series voltage-dropping resistor to drop V_T of 12.6 V to 9 V for R_L. See text for calculations.

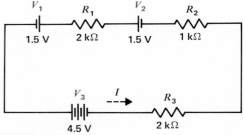

FIGURE 3-15

Find I for this series circuit with three voltage sources. See text for solution.

3-10
EFFECT OF AN OPEN CIRCUIT IN A SERIES PATH

An open is a break in the current path. The resistance of the open is very high because an insulator like air takes the place of a conducting part of the circuit. Remember that the current is the same in all parts of a series circuit. Therefore, an open in any part results in no current for the entire circuit. As illustrated in Fig. 3-16, the circuit is normal in (a), but in (b) there is no current in R_1, R_2, or R_3 because of the open in the series path.

The open between P_1 and P_2, or at any other point in the circuit, has practically infinite resistance because its opposition to electron flow is so great compared with the resistance of R_1, R_2, and R_3. Therefore, the value of current is practically zero, even though the battery produces its normal applied voltage of 40 V.

To take an example, suppose that the open between P_1 and P_2 has a resistance of 40 billion Ω. The resistance of the entire circuit is essentially 40 billion Ω, since the resistance of R_1, R_2, and R_3 can then be neglected compared with the resistance of the open. Such a high resistance is practically infinite ohms.

By Ohm's law, the current that results from 40 V applied across 40 billion Ω is one-billionth of an ampere, which is practically zero. This is the value of current in all parts of the series circuit. With practically no current, the IR voltage drop is practically zero across the 25 Ω of R_1, the 10 Ω of R_2, and the 5 Ω of R_3.

In summary, with an open in any part of a series circuit the current is zero in the entire circuit. There is no IR voltage drop across any of the series resistances, although the generator still maintains its output voltage.

The Case of Zero IR Drop. In Fig. 3-16b, each of the resistors in the open circuit has an IR drop of zero. The reason is that current of practically zero is the value in all the series components. Each R still has its resistance. However, with zero current the IR voltage is zero.

The Source Voltage V_T Is Still Present with Zero I. The open circuit in Fig. 3-16b illus-

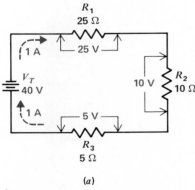
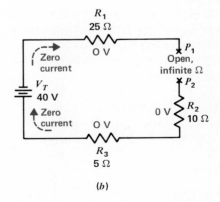

FIGURE 3-16

Effect of an open in a series circuit. (a) Normal closed circuit with current of 1 A. (b) Open in any part of the circuit results in no current in the entire circuit.

trates another example of how V and I are different forms of electricity. There is no current with the open circuit because there is no complete path outside the battery between its two terminals. However, the battery is generating a potential difference across the positive and negative terminals. This source voltage is present with or without current in the external circuit. If you measure V_T, the meter will read 40 V with the circuit closed or open.

The same idea applies to the 120-V ac voltage from the power line in the home. The 120-V potential difference is across the two terminals of the wall outlet. If you connect a lamp or appliance, current will flow in the circuit. When nothing is connected, though, the 120-V potential difference is still there at the outlet. If you should touch it, you will get an electric shock. The generator at the power station is maintaining the 120-V at the outlets as a source to produce current in any circuits that will be plugged in.

The Applied Voltage Is Across the Open Terminals. It is useful to note that the entire applied voltage is present across the open circuit. Between P_1 and P_2 in Fig. 3-16b, there is 40 V. The reason is that essentially all the resistance of the series circuit is between P_1 and P_2. Therefore, the resistance of the open circuit develops all the IR voltage drop.

---

[1] The voltage across an open circuit equals the applied voltage, even without any current, after the capacitance between the open terminals becomes charged by V, as described in Chap. 21, Capacitance.

The extremely small current of one-billionth of an ampere is not enough to develop any appreciable IR drop across R_1, R_2, and R_3. However, across the open[1] the resistance is 40 billion Ω. Therefore, the IR voltage across the open here is one-billionth of an ampere multiplied by 40 billion Ω, which equals 40 V.

We could also consider the open circuit as a proportional voltage divider. Since practically all the series resistance is between P_1 and P_2, all the applied voltage is across the open terminals.

The fact that the open terminals have the entire applied voltage indicates a good way to find an open component in a series string. If you measure the voltage across each good component, zero voltage will be normal. However, the component that has the full source voltage is the one that is open.

Open Heater String. As another example, refer back to the series heaters in Fig. 3-6 for a television receiver with tubes. If one heater opens, the entire string will be open. Then none of the tubes can operate, including the picture tube. The result is no picture and no sound, from the simple trouble of an open heater in the series string.

Practice Problems 3-10
(answers on page 73)
Refer to Fig. 3-6.
(a) How much is the normal voltage across the 15BD11 heater?
(b) How much is the voltage across the 15BD11 heater if the 3GK5 heater is open?

Summary

1. There is only one current I in a series circuit. $I = V_T/R_T$, where V_T is the voltage applied across the total series resistance R_T. This I is the same in all the series components.

2. The total resistance R_T of a series string is the sum of the individual resistances.
3. The applied voltage V_T equals the sum of the series IR voltage drops.
4. The negative side of an IR voltage drop is where electrons flow in, attracted to the positive side at the opposite end.
5. The sum of the individual values of power used in the individual resistances equals the total power supplied by the source.
6. Series-aiding voltages are added; series-opposing voltages are subtracted.
7. An open results in no current in all parts of the series circuit.
8. In an open circuit, the voltage across the two open terminals is equal to the applied voltage.

Self-Examination (Answers at back of book.)

Choose (a), (b), (c), or (d).

1. When two resistances are connected in series, (a) they must both have the same resistance value; (b) the voltage across each must be the same; (c) they must have different resistance values; (d) there is only one path for current through both resistances.
2. In Fig. 3-3c, if the current through R_5 is 1 A, then the current through R_3 must be (a) ⅓ A; (b) ½ A; (c) 1 A; (d) 3 A.
3. With a 10-kΩ resistance in series with a 2-kΩ resistance, the total R_T equals (a) 2 kΩ; (b) 8 kΩ; (c) 10 kΩ; (d) 12 kΩ.
4. With two 45-kΩ resistances in series across a 90-V battery, the voltage across each resistance equals (a) 30 V; (b) 45 V; (c) 90 V; (d) 180 V.
5. The sum of series IR voltage drops (a) is less than the smallest voltage drop; (b) equals the average value of all the voltage drops; (c) equals the applied voltage; (d) is usually more than the applied voltage.
6. R_1 and R_2 are in series with 90 V applied. If V_1 is 30 V, then V_2 must be (a) 30 V; (b) 90 V; (c) 45 V; (d) 60 V.
7. With a 4-Ω resistance and a 2-Ω resistance in series across a 6-V battery, the current (a) in the larger resistance is 1⅓ A; (b) in the smaller resistance is 3 A; (c) in both resistances is 1 A; (d) in both resistances is 2 A.
8. When one resistance in a series string is open, (a) the current is maximum in the normal resistances; (b) the current is zero in all the resistances; (c) the voltage is zero across the open resistance; (d) the current increases in the voltage source.

9. The resistance of an open series string is (a) zero; (b) infinite; (c) equal to the normal resistance of the string; (d) about double the normal resistance of the string.
10. A source of 100 V is applied across a 20-Ω R_1 and 30-Ω R_2 in series. V_1 is 40 V. The current in R_2 is (a) 5 A; (b) $3\frac{1}{3}$ A; (c) $1\frac{1}{3}$ A; (d) 2 A.

Essay Questions

1. Show how to connect two resistances in series with each other across a voltage source.
2. State three rules for the current, voltage, and resistance in a series circuit.
3. For a given amount of current, why does more resistance have a bigger voltage drop across it?
4. Two 300-W 120-V light bulbs are connected in series across a 240-V line. If the filament of one bulb burns open, will the other bulb light? Why? With the open, how much is the voltage across the source and across each bulb?
5. Prove that if $V_T = V_1 + V_2 + V_3$, then $R_T = R_1 + R_2 + R_3$.
6. State briefly a rule for determining polarity of the voltage drop across each resistor in a series circuit.
7. Redraw the circuit in Fig. 3-13, marking the polarity of V_1, V_2, and V_3.
8. State briefly a rule to determine when voltages are series-aiding.
9. Derive the formula $P_T = P_1 + P_2 + P_3$ from the fact that $V_T = V_1 + V_2 + V_3$.
10. In a series string, why does the largest R dissipate the most power?

Problems (Answers to odd-numbered problems at back of book.)

1. A circuit has 10 V applied across a 10-Ω resistance R_1. How much is the current in the circuit? How much resistance R_2 must be added in series with R_1 to reduce the current one-half? Show the schematic diagram of the circuit with R_1 and R_2.
2. Draw the schematic diagram of 20-, 30-, and 40-Ω resistances in series. (a) How much is the total resistance of the entire series string? (b) How much current flows in each resistance, with a voltage of 180 V applied across the series string? (c) Find the voltage drop

across each resistance. (d) Find the power dissipated in each resistance.
3. R_1 of 90 kΩ and an R_2 of 10 kΩ are in series across a 3-V source. (a) Draw the schematic diagram. (b) How much is V_2?
4. Draw a schematic diagram showing two resistances R_1 and R_2 in series across a 100-V source. (a) If the *IR* voltage drop across R_1 is 60 V, how much is the *IR* voltage drop across R_2? (b) Label the polarity of the voltage drops across R_1 and R_2. (c) If the current is 1 A through R_1, how much is the current through R_2? (d) How much is the resistance of R_1 and R_2? How much is the total resistance across the voltage source? (e) If the voltage source is disconnected, how much is the voltage across R_1 and across R_2?
5. Three 10-Ω resistances are in series across a voltage source. Show the schematic diagram. If the voltage across each resistor is 10 V, how much is the applied voltage? How much is the current in each resistance?
6. How much resistance R_1 must be added in series with a 100-Ω R_2 to limit the current to 0.3 A with 120 V applied? Show the schematic diagram. How much power is dissipated in each resistance?
7. Find the total R_T of the following resistances in series: 2 MΩ, 0.5 MΩ, 47 kΩ, 5 kΩ, and 470 Ω.
8. Referring to Fig. 3-6, calculate the resistance of each heater with its normal load current. How much is the total normal resistance of the heater string? How much is the resistance with one heater open?
9. Draw the circuit with values for three equal series resistances across a 90-V source, where each *R* has one-third the applied voltage and the current in the circuit is 2 mA.
10. A 100-W bulb normally takes 0.833 A, and a 200-W bulb takes 1.666 A from the 120-V power line. If these two bulbs were connected in series across a 240-V power line, prove that the current would be 1.111 A in both bulbs, assuming the resistances remain constant.
11. Referring to Fig. 3-9, calculate *I* in R_1 and R_2 for the diagrams in (a), (b), and (c).
12. In Fig. 3-17, calculate I, V_1, V_2, P_1, P_2, and P_T. (Note: R_1 and R_2 are in series with V_T even though the source is shown at the right instead of at the left.)
13. If R_1 is increased to 8 kΩ in Fig. 3-17, what will be the new *I*?
14. In Fig. 3-18, find R_1. Why is *I* in the direction shown?
15. In Fig. 3-19, find R_2.
16. Figure 3-20 shows the circuit for keeping a 12.6-V car battery charged from a 15-V dc generator. Calculate *I* and show the direction of electron flow.

Chapter 3
Series Circuits

17. In Fig. 3-21, find V_2. Show polarity for V_1, V_2, and V_3.
18. In Fig. 3-22, find V_T. Show polarity for V_T, V_1, V_2, and V_3.

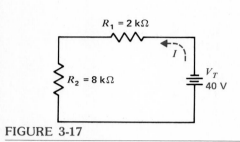

FIGURE 3-17
For Probs. 12 and 13.

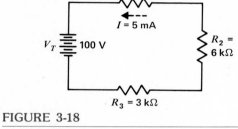

FIGURE 3-18
For Prob. 14.

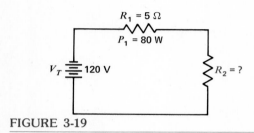

FIGURE 3-19
For Prob. 15.

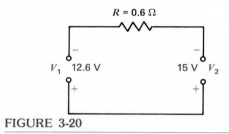

FIGURE 3-20
For Prob. 16.

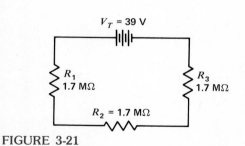

FIGURE 3-21
For Prob. 17.

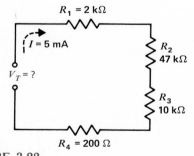

FIGURE 3-22
For Prob. 18.

Answers to Practice Problems

3-1 (a) R_1, R_2, R_3, V_T and the wires
    (b) 5 A

3-2 (a) 2 mA
    (b) 10 kΩ

3-2 (c) 1 mA
3-3 (a) 10 V
    (b) 1 A
    (c) 1 A
    (d) 1 A
3-4 (a) 60 V
    (b) 75 V
3-5 (a) point *a* or *c*
    (b) point *b* or *f*
    (c) point *d*
3-6 (a) *S* is +20 V
        J is +10 V
        G is 0 V
3-6 (b) *S* is −20 V
        J is −10 V
        G is 0 V
3-7 (a) 6 W
    (b) 40-kΩ R_2
3-8 (a) 100 V
    (b) 20 V
3-9 (a) $V_1 = 30$ V
    (b) $V_2 = 90$ V
    (c) $V_3 = 60$ V
3-10 (a) 15 V
     (b) 0 V

Parallel Circuits

Chapter 4

When two or more components are connected across one voltage source, as shown in Fig. 4-1, they form a parallel circuit. The resistors R_1 and R_2 are in parallel with each other and with the battery. Each parallel path is then a branch, with its own individual current. Parallel circuits, therefore, have one common voltage across all the branches but individual branch currents that can be different. These characteristics are opposite from series circuits that have one common current but individual voltage drops that can be different. The important features of parallel circuits are explained in the following topics:

4-1 The Applied Voltage V_A Is the Same Across Parallel Branches
4-2 Each Branch I Equals V_A/R
4-3 The Main-Line I_T Equals the Sum of the Branch Currents
4-4 Resistances in Parallel
4-5 Conductances in Parallel
4-6 Total Power in Parallel Circuits
4-7 Analyzing Parallel Circuits
4-8 Effect of an Open Branch in Parallel Circuits
4-9 Effect of a Short Circuit across Parallel Branches

4-1
THE APPLIED VOLTAGE V_A IS THE SAME ACROSS PARALLEL BRANCHES

In Fig. 4-1b, the points a, b, c, and e are really equivalent to a direct connection at the negative terminal of the battery because the connecting wires have practically no resistance. Similarly, points h, g, d, and f are the same as a direct connection at the positive battery terminal. Since R_1 and R_2 are directly connected across the two terminals of the battery, both resistances must have the same potential difference as the battery. It follows that the voltage is the same across components connected in parallel. The parallel circuit arrangement is used, therefore, to connect components that require the same voltage.

A common application of parallel circuits is typical house wiring to the power line, with many lights and appliances connected across the 120-V source (Fig. 4-2). The wall receptacle has the potential difference of 120 V across each pair of terminals. Therefore, any resistance connected to an outlet has the applied voltage of 120 V. The light bulb is connected to one outlet and the toaster to another outlet, but both have the same applied voltage of 120 V. Therefore, each operates independently of any other appliance, with all the individual branch circuits connected across the 120-V line.

Practice Problems 4-1
(answers on page 90)
(a) In Fig. 4-1, how much is the common voltage across R_1 and R_2?

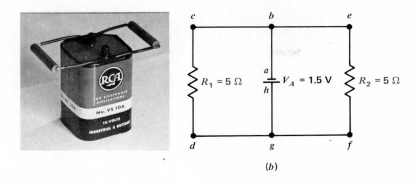

FIGURE 4-1
A parallel circuit. (a) Photograph of wiring. (b) Schematic diagram.

(b) In Fig. 4-2, how much is the common voltage across the bulb and the toaster?

4-2
EACH BRANCH I EQUALS V_A/R

In applying Ohm's law, it is important to note that the current equals the voltage applied across the circuit divided by the resistance between the two points where that voltage is applied. In Fig. 4-3, 10 V is applied across the 5 Ω of R_2, resulting in the current of 2 A between points e and f through R_2. The battery voltage is also applied across the parallel resistance of R_1, applying 10 V across 10 Ω. Through R_1, therefore, the current is 1 A between points c and d. The current has a different value through R_1, with the same applied voltage, because the resistance is different. These values are calculated as follows:

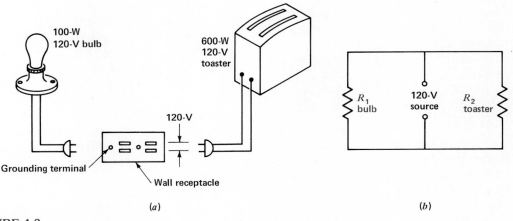

FIGURE 4-2
Light bulb and toaster connected in parallel to the 120-V line. (a) Wiring diagram. (b) Schematic diagram.

Chapter 4
Parallel Circuits

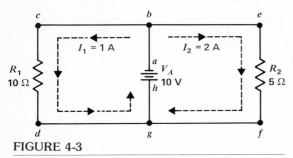

FIGURE 4-3
The current in each parallel branch equals the applied voltage divided by each branch R.

$$I_1 = \frac{V_A}{R_1} = \frac{10}{10} = 1 \text{ A}$$

$$I_2 = \frac{V_A}{R_2} = \frac{10}{5} = 2 \text{ A}$$

Just as in a circuit with just one resistance, any branch that has less R allows more I. If R_1 and R_2 were equal, however, the two branch currents would have the same value. For instance, in Fig. 4-1b each branch has its own current equal to $1.5 \text{ V}/5 \text{ }\Omega = 0.3$ A.

The I can be different in parallel circuits having different R because V is the same across all the branches. Any voltage source generates a potential difference across its two terminals. This voltage does not move. Only I flows around the circuit. The source voltage is available to make electrons move around any closed path connected to the generator terminals. How much I is in the separate paths depends on the amount of R in each branch.

Practice Problems 4-2
(*answers on page 90*)
Refer to Fig. 4-3.
(a) How much is the voltage across R_1?
(b) How much is I_1 through R_1?
(c) How much is the voltage across R_2?
(d) How much is I_2 through R_2?

4-3
THE MAIN-LINE I_T EQUALS THE SUM OF THE BRANCH CURRENTS

Components to be connected in parallel are usually wired directly across each other, with the entire parallel combination connected to the voltage source, as illustrated in Fig. 4-4. This circuit is equivalent to wiring each parallel branch directly to the voltage source, as shown in Fig. 4-1, when the connecting wires have essentially zero resistance.

The advantage of having only one pair of connecting leads to the source for all the parallel branches is that usually less wire is necessary. The pair of leads connecting all the branches to the terminals of the voltage source is the *main line*. In Fig. 4-4, the wires from g to a on the negative side and from b to f in the return path form the main line.

In Fig. 4-4b, with 20 Ω of resistance for R_1 connected across the 20-V battery, the current through R_1 must be 20 V/20 Ω = 1 A. This current is electron flow from the negative terminal of the source, through R_1, and back to the positive battery terminal. Similarly, the R_2 branch of 10 Ω across the battery has its own branch current of 20 V/10 Ω = 2 A. This current flows from the negative terminal of the source, through R_2, and back to the positive terminal, since it is a separate path for electron flow.

All the current in the circuit, however, must come from one side of the voltage source and return to the opposite side for a complete path. In the main line, therefore, the amount of current is equal to the total of the branch currents.

For example, in Fig. 4-4b, the total current in the line from point g to point a is 3 A. The total current at branch point a subdivides into its component branch currents for each of the branch resistances. Through the path R_1 from a

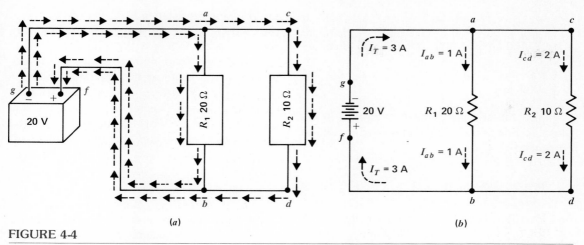

FIGURE 4-4
The main-line current equals the sum of the branch currents. From g to a is the negative side and from b to f is the positive side of the main line. (*a*) Wiring diagram. Arrows inside lines indicate current for R_1; arrows outside lines indicate current for R_2. (*b*) Schematic diagram. I_T is total line current.

to b the current is 1 A. The other branch path $acdb$ through R_2 has a current of 2 A. At the branch point b, the electron flow from both parallel branches combines, so that the current in the main-line return path from b to f has the same value of 3 A as in the other side of the main line.

The formula for the total current I_T in the main line is

$$I_T = I_1 + I_2 + I_3 + \cdots + \text{etc.} \qquad (4\text{-}1)$$

This rule applies for any number of parallel branches, whether the resistances are equal or unequal.

Example 1. An R_1 of 20 Ω, an R_2 of 40 Ω, and an R_3 of 60 Ω are connected in parallel across the 120-V power line. How much is the total line current I_T?

Answer. I_1 for the R_1 branch is $^{120}\!/_{20}$ or 6 A. Similarly I_2 is $^{120}\!/_{40}$ or 3 A, and I_3 is $^{120}\!/_{60}$ or 2 A. The total current in the main line is

$$I_T = I_1 + I_2 + I_3 = 6 + 3 + 2$$
$$I_T = 11 \text{ A}$$

Example 2. Two branches R_1 and R_2 across the 120-V power line draw a total line current I_T of 15 A. The R_1 branch takes 10 A. How much is the current I_2 in the R_2 branch?

Answer. $I_2 = I_T - I_1 = 15 - 10$
$I_2 = 5 \text{ A}$

With two branch currents, one must equal the difference between I_T and the other branch current.

Example 3. Three parallel branch currents are 0.1 A, 500 mA, and 800 μA. Calculate I_T.

Chapter 4
Parallel Circuits

Answer. All values must be in the same units to be added. Converted to milliamperes, therefore 0.1 A = 100 mA and 800 µA = 0.8 mA. Then

$I_T = 100 + 500 + 0.8$
$I_T = 600.8$ mA

You can add the currents in A, mA, or µA units, as long as the same unit is used for all the currents.

Practice Problems 4-3
(answers on page 90)
(a) Parallel branch currents are 1 A for I_1, 2 A for I_2, and 3 A for I_3. Calculate I_T.
(b) $I_T = 6$ A for three branch currents. I_1 is 1 A and I_2 is 2 A. Calculate I_3.

4-4
RESISTANCES IN PARALLEL

The total resistance across the main line in a parallel circuit can be found by Ohm's law: *Divide the common voltage across the parallel resistances by the total current of all the branches.* Referring to Fig. 4-5a, note that the parallel resistance of R_1 with R_2, indicated by the combined resistance R_T, is the opposition to the total current in the main line. In this example V_A/I_T is 60 V/3 A = 20 Ω for R_T.

The total load connected to the source voltage is the same as though one equivalent resistance of 20 Ω were connected across the main line. This is illustrated by the equivalent circuit in Fig. 4-5b. For any number of parallel resistances of any value, therefore,

$$R_T = \frac{V_A}{I_T} \tag{4-2}$$

where I_T is the sum of all the branch currents and R_T is the equivalent resistance of all the parallel branches across the voltage source V_A.

Example 4. Two branches, each with a 5-A current, are connected across a 90-V source. How much is the equivalent total resistance R_T?

Answer. The total line current I_T is 5 + 5 = 10 A. Then,

$R_T = \dfrac{V_A}{I_T} = \dfrac{90}{10}$
$R_T = 9$ Ω

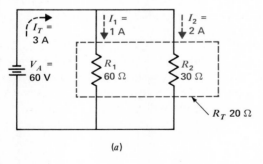
(a)

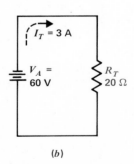
(b)

FIGURE 4-5

Resistances in parallel. (a) Combined parallel resistance of R_1 and R_2 is the total resistance R_T in the main line. (b) Equivalent circuit showing combined R_T drawing the same 3-A I_T as the parallel combination of R_1 and R_2.

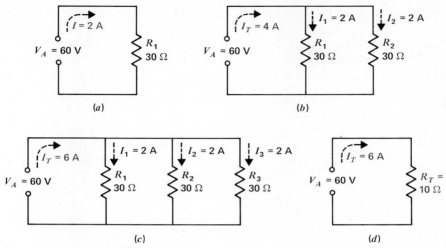

FIGURE 4-6

How adding parallel branches increases I_T but reduces R_T. (a) One branch. (b) Two branches. (c) Three branches. (d) Equivalent of (c).

Parallel Bank. A combination of parallel branches is often called a *bank*. In Fig. 4-5, the bank consists of the 60-Ω R_1 and 30-Ω R_2 in parallel. Their combined parallel resistance R_T is the bank resistance, equal to 20 Ω in this example.

When a circuit has more current with the same applied voltage, this greater value of I corresponds to less R because of their inverse relation. Therefore, the combination of parallel resistances R_T for the bank is always less than the smallest individual branch resistance. The reason is that I_T must be more than any one branch current.

Why R_T Is Less Than Any Branch R. It may seem unusual at first that putting more resistance into a circuit lowers the equivalent resistance. This feature of parallel circuits is illustrated in Fig. 4-6. Note that equal resistances of 30 Ω each are added across the source voltage, one branch at a time. The circuit in (a) has just R_1, which allows 2 A with 60 V applied. In (b) the R_2 branch is added across the same V_A. This branch also has 2 A. Now the parallel circuit has a 4-A total line current because of $I_1 + I_2$. Then the third branch is added in (c), which also takes 2 A for I_3. The combined circuit with three branches therefore requires a total load current of 6 A, which is supplied by the voltage source.

The combined resistance across the source then is V_A/I_T, which is $^{60}/_6$ or 10 Ω. This equivalent resistance R_T, representing the entire load on the voltage source, is shown in (d). More resistance branches reduce the combined resistance of the parallel circuit because more current is required from the same voltage source.

Reciprocal Resistance Formula. We can derive this formula from the fact that I_T is the sum of all the branch currents, or,

$$I_T = I_1 + I_2 + I_3 + \cdots + \text{etc.}$$

However, I_T is V/R_T. Also, each I is V/R. Substituting V/R_T for I_T on the left side of the equation and V/R for each branch I on the right side, the result is

$$\frac{V}{R_T} = \frac{V}{R_1} + \frac{V}{R_2} + \frac{V}{R_3} + \cdots + \text{etc.}$$

Dividing by V because it is the same across all the resistances

$$\frac{1}{R_T} = \frac{1}{R_1} + \frac{1}{R_2} + \frac{1}{R_3} + \cdots + \text{etc.} \qquad (4\text{-}3)$$

This reciprocal formula applies to any number of parallel resistances of any value. Using the values in Fig. 4-7a as an example,

$$\frac{1}{R_T} = \frac{1}{20} + \frac{1}{10} + \frac{1}{10} = \frac{1}{20} + \frac{2}{20} + \frac{2}{20} = \frac{5}{20}$$

$$R_T = \frac{20}{5}$$

$$R_T = 4\ \Omega$$

Notice that the value for $1/R_T$ must be inverted to obtain R_T when using Formula (4-3) because it gives the reciprocal of R_T.

Total-Current Method. Figure 4-7b shows how this same problem can be calculated in terms of total current instead of by the reciprocal formula, if it is easier to work without fractions. Although the applied voltage is not known always, any convenient value can be assumed because it cancels in the calculations. It is usually simplest to assume an applied voltage of the same numerical value as the highest resistance. Then one assumed branch current will automatically be 1 A and the other branch currents will be more, eliminating fractions less than 1 in the calculations.

For the example in Fig. 4-7b, the highest branch R is 20 Ω. Therefore, assume 20 V for the applied voltage. Then the branch currents are 1 A in R_1, 2 A in R_2, and 2 A in R_3. Their sum is $1 + 2 + 2 = 5$ A for I_T. The combined resistance R_T across the main line is V_A/I_T or 20 V/5 A = 4 Ω. This is the same value calculated with the reciprocal resistance formula.

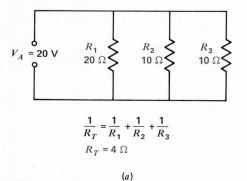
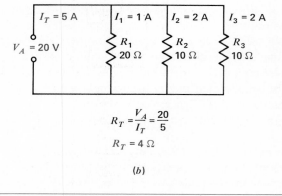

FIGURE 4-7

Combining parallel resistances using (a) reciprocal resistance formula and (b) total line current method with an assumed line voltage of 20 V.

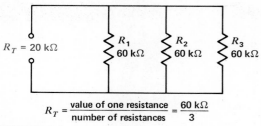

$$R_T = \frac{\text{value of one resistance}}{\text{number of resistances}} = \frac{60 \text{ k}\Omega}{3}$$

FIGURE 4-8

For equal branch resistances, R divided by the number of branches is the R_T. Here R_T is $60 \text{ k}\Omega/3 = 20 \text{ k}\Omega$.

Special Case of Equal R in All Branches. If R is equal in all branches, the combined R_T equals the value of one branch resistance divided by the number of branches. This rule is illustrated in Fig. 4-8 where three 60-kΩ resistances in parallel equal 20 kΩ.

The rule applies to any number of parallel resistances, but they must all be equal. As another example, five 60-Ω resistances in parallel have the combined resistance of $^{60}/_5$, or 12 Ω. A common application is two equal resistors wired in a parallel bank for R_T one-half each R.

Special Case of Only Two Branches. When there are two parallel resistances and they are not equal, it is usually quicker to calculate the combined resistance by the method shown in Fig. 4-9. This rule says that the combination of two parallel resistances is their product divided by their sum.

$$R_T = \frac{R_1 \times R_2}{R_1 + R_2} \qquad (4\text{-}4)$$

where R_T is in the same units as all the individual resistances. For the example in Fig. 4-9,

$$R_T = \frac{R_1 \times R_2}{R_1 + R_2} = \frac{40 \times 60}{40 + 60} = \frac{2400}{100}$$
$$R_T = 24 \ \Omega$$

Each R can have any value but there must be only two resistances. Note that this method gives R_T directly, not its reciprocal. If you use the reciprocal formula for this example, the answer will be $1/R_T = \frac{1}{24}$, which is the same value as R_T equals 24 Ω.

Short-Cut Calculations. Figure 4-10 shows how these special rules can help in reducing parallel branches to a simpler equivalent circuit. In (a), the 60-Ω R_1 and R_4 are equal and in parallel. Therefore, they are equivalent to the 30-Ω R_{14} in (b). Similarly, the 20-Ω R_2 and R_3 are equivalent to the 10 Ω of R_{23}. The circuit in (a) is equivalent to the simpler circuit in (b) with just the two parallel resistances of 30 and 10 Ω.

Finally, the combined resistance for these two equals their product divided by the sum, which is 300/40 or 7.5 Ω, as shown in (c). This value of R_T in (c) is equivalent to the combination of the four branches in (a). If you connect a voltage source across either circuit, the generator current in the main line will be the same for both cases.

The order of connections for parallel resistances does not matter in determining R_T. There is no question as to which is first or last because they are all across the same voltage source.

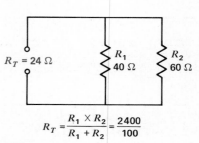

$$R_T = \frac{R_1 \times R_2}{R_1 + R_2} = \frac{2400}{100}$$

FIGURE 4-9

For only two parallel resistances, R_T is the product divided by the sum. Here R_T is $2400/100 = 24 \ \Omega$.

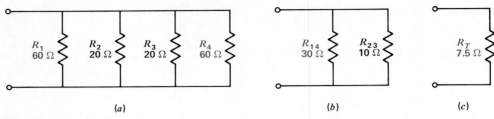

(a)   (b)   (c)

FIGURE 4-10

An example of parallel-resistance calculations. (a) Circuit with four branches. (b) Circuit combined into two branches. (c) Equivalent circuit reduced to one R_T.

Finding an Unknown Branch Resistance. In some cases with two parallel resistors, it is useful to be able to determine what size R_x to connect in parallel with a known R in order to obtain a required value of R_T. Then the factors can be transposed as follows:

$$R_x = \frac{R \times R_T}{R - R_T} \quad (4\text{-}5)$$

Example 5. What R_x in parallel with 40 Ω will provide an R_T of 24 Ω?

Answer. $R_x = \dfrac{R \times R_T}{R - R_T} = \dfrac{40 \times 24}{40 - 24} = \dfrac{960}{16}$

$R_x = 60$ Ω

This problem corresponds to the circuit in Fig. 4-9.

Note that Formula (4-5) for R_x has a product over a difference. R_T is subtracted because it is the smallest R.

Example 6. What R in parallel with 50 kΩ will provide an R_T of 25 kΩ?

Answer. $R = 50$ kΩ

Two equal resistances in parallel have R_T equal to one-half the R of either one.

Practice Problems 4-4
(answers on page 90)
(a) Find R_T for three 4.7-MΩ resistances in parallel.
(b) Find R_T for 3 MΩ in parallel with 2 MΩ.

4-5
CONDUCTANCES IN PARALLEL
Since conductance G is equal to $1/R$, the reciprocal resistance Formula (4-3) can be stated for conductance as

$$G_T = G_1 + G_2 + G_3 + \cdots + \text{etc.} \quad (4\text{-}6)$$

With R in ohms, G is in siemens units or mhos. For the example in Fig. 4-11,

$$G_T = 0.05 + 0.2 + 0.5 = 0.75 \text{ S or } 0.75 \text{ mho}$$

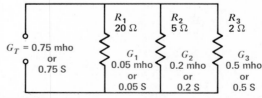

FIGURE 4-11

Conductances in parallel are added for the total G_T.

Notice that adding the conductances does not require reciprocals. Actually, each value of G is the reciprocal of R.

Working with G may be more convenient than working with R in parallel circuits since it will avoid using the reciprocal formula for R_T. Each branch current is directly proportional to its conductance. This idea corresponds to the fact that in series circuits each voltage drop is directly proportional to each series resistance.

The reason why parallel conductances are added directly can be illustrated by assuming a 1-V source across all the branches. Then calculating the values of $1/R$ for the conductances is the same as calculating the branch currents. These values are added for the total I_T or G_T.

Practice Problems 4-5
(answers on page 90)
(a) G_1 is 2 S and G_2 in parallel is 5 S. Calculate G_T.
(b) G_1 is 0.05 μS, G_2 is 0.2 μS, and G_3 is 0.5 μS, all in parallel. Find G_T and R_T.

4-6
TOTAL POWER IN PARALLEL CIRCUITS

Since the power dissipated in the branch resistances must come from the voltage source, the total power equals the sum of the individual values of power in each branch. This rule is illustrated in Fig. 4-12. We can also use this

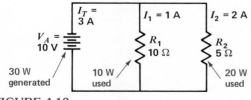

FIGURE 4-12
The sum of the individual powers P_1 and P_2 used in each resistance equals the total power P_T produced by the source.

circuit as an example of how to apply the rules of current, voltage, and resistance for a parallel circuit.

The applied 10 V is across the 10-Ω R_1 and the 5-Ω R_2 in Fig. 4-12. The branch current I_1 then is V_A/R_1 or $^{10}/_{10}$, which equals 1 A. Similarly I_2 is $^{10}/_5$, or 2 A. The total I_T is $1 + 2 = 3$ A. If we want to find R_T, it equals V_A/I_T or $^{10}/_3$, which is $3\frac{1}{3}$ Ω.

The power dissipated in each branch R is $V_A \times I$. In the R_1 branch, I_1 is $^{10}/_{10} = 1$ A. Then P_1 is $V_A \times I_1$ or $10 \times 1 = 10$ W.

For the R_2 branch, I_2 is $^{10}/_5 = 2$ A. Then P_2 is $V_A \times I_2$ or $10 \times 2 = 20$ W.

Adding P_1 and P_2, the answer is $10 + 20 = 30$ W. This P_T is the total power dissipated in both branches.

This value of 30 W for P_T is also the total power supplied by the voltage source by means of its total line current I_T. With this method, the total power is $V_A \times I_T$ or $10 \times 3 = 30$ W for P_T. The 30 W of power supplied by the voltage source is dissipated or used up in the branch resistances.

Note that in both parallel and series circuits the sum of the individual values of power dissipated in the circuit equals the total power generated by the source. This can be stated as a formula

$$P_T = P_1 + P_2 + P_3 + \cdots + \text{etc.} \qquad (4\text{-}7)$$

The series or parallel connections can alter the distribution of voltage or current, but power is the rate at which energy is supplied. The circuit arrangement cannot change the fact that all the energy in the circuit comes from the source.

Practice Problems 4-6
(answers on page 90)
(a) Two parallel branches each have 2 A at 120 V. Calculate P_T.

(b) Three parallel branches of 10, 20, and 30 Ω have 60 V applied. Calculate P_T.

4-7
ANALYZING PARALLEL CIRCUITS

For many types of problems with parallel circuits it is useful to remember the following points:

1. When you know the voltage across one branch, this voltage is across all the branches. There can be only one voltage across branch points with the same potential difference.
2. If you know I_T and one of the branch currents I_1, you can find I_2 by subtracting from I_T. Since $I_T = I_1 + I_2$, it is also true that $I_2 = I_T - I_1$ or $I_1 = I_T - I_2$.

The circuit in Fig. 4-13 illustrates these points. The problem is to find the applied voltage V_A and the value of R_3. Of the three branch resistances, only R_1 and R_2 are known. However, since I_2 is given as 2 A, the I_2R_2 voltage must be $2 \times 60 = 120$ V.

Although the applied voltage is not given, this must also be 120 V. The voltage across all the parallel branches is the same 120 V that is across the R_2 branch.

Now I_1 can be calculated as V_A/R_1. This is $120/30 = 4$ A for I_1.

I_T is given as 7 A. The two branches take $2 + 4 = 6$ A. The third branch current through R_3 must be $7 - 6 = 1$ A for I_3.

Now R_3 can be calculated as V_A/I_3. This is $120/1 = 120$ Ω for R_3.

Practice Problems 4-7
(answers on page 90)
Refer to Fig. 4-13.
(a) How much is V_2 across R_2?
(b) Calculate I_1 through R_1.

4-8
EFFECT OF AN OPEN BRANCH IN PARALLEL CIRCUITS

An open in any circuit is an infinite resistance that results in no current. However, in parallel circuits there is a difference between an open in the main line and an open in a parallel branch. These two cases are illustrated in Fig. 4-14. In (a) the open in the main line prevents any electron flow in the line to all the branches. The current is zero in every branch, therefore, and none of the bulbs can light.

However, in Fig. 4-14b the open is in the branch circuit for bulb 1. The open branch circuit has no current, then, and this bulb cannot light. The current in all the other parallel branches is normal, though, because each is connected to the voltage source. Therefore, the other bulbs light.

The circuit in Fig. 4-14b applies to television receivers with tubes where the heaters are wired in parallel. If one heater opens, the other tubes will still operate normally. Usually, you can see which heater is not red-hot.

These circuits show the advantage of wiring components in parallel. An open in one component only opens one branch, while the other parallel branches have their normal voltage and current.

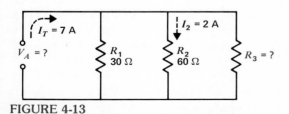

FIGURE 4-13

Analyzing a parallel circuit. What are the values for V_A and R_3? See solution in text.

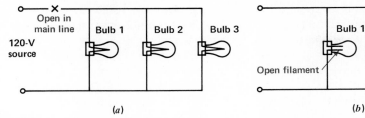

FIGURE 4-14
Open in parallel circuits. (a) Open in main line—no current and no light in all bulbs. (b) Open in one branch—bulb 1 is dark, but the two other bulbs operate normally.

Practice Problems 4-8
(answers on page 90)
(a) How much is the R of an open filament or heater?
(b) In Fig. 4-14b, if only bulb 3 is open, which bulbs will light?

4-9
EFFECT OF A SHORT CIRCUIT ACROSS PARALLEL BRANCHES

A short circuit has practically zero resistance. Its effect, therefore, is to allow excessive current. Consider the example in Fig. 4-15. Suppose the conducting wire at point a should accidentally contact the wire at point b. Since the wire is an excellent conductor, the short circuit results in practically zero resistance from points a to b. These two points are connected directly across the voltage source. With no opposition, the applied voltage could produce an infinitely high value of current through this current path.

The Short-Circuit Current. Practically, the amount of current is limited by the small resistance of the wire. Also, the source usually cannot maintain its output voltage while supplying much more than its rated load current. Still the amount of current can be dangerously high. For instance, the short-circuit current might be more than 100 A instead of the normal line current of 1 A illustrated in Fig. 4-15. This excessive current flows in the voltage source V, in the line to the short circuit at point a, through the short circuit, and in the line returning to the source from b. Because of the large amount of current, the wires can become hot enough to ignite and burn. There should be a fuse that would open if there is too much current in the main line because of a short circuit across any of the branches.

The Short-Circuited Components Have No Current. For the short circuit in Fig. 4-15, the I is 0 A in the parallel resistors R_1 and R_2. The reason is that the short circuit is a parallel path with practically zero resistance. Then all the current flows in this path, bypassing the resis-

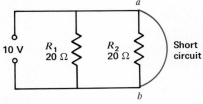

FIGURE 4-15
A short circuit across one parallel branch must short all the branches.

tors R_1 and R_2. Therefore R_1 and R_2 are short-circuited or *shorted-out* of the circuit. They cannot function without their normal current. If they were filament resistances of light bulbs or heaters for tubes, they would not light without any current.

The short-circuited components are not damaged, however. They do not even have any current passing through them. Assuming the short circuit has not damaged the voltage source and the wiring for the circuit, the components can operate again when the circuit is restored to normal by removing the short circuit.

All the Parallel Branches Are Short-Circuited. If there were only one R in Fig. 4-15, two, three, or more resistances, or any number of parallel components, they would all be shorted out by the short circuit across points *a* and *b*. Therefore, a short circuit across one branch in a parallel circuit shorts out all the parallel branches.

This idea also applies to a short circuit across the voltage source in any type of circuit. Then the entire circuit is shorted out.

Practice Problems 4-9
(answers on page 90)
Refer to Fig. 4-15.
(a) How much is the R of the short circuit between *a* and *b*?
(b) How much is I_1 in R_1 with the short circuit across R_2?

Summary

1. There is only one voltage V_A across all components in parallel.
2. The current in each branch I_b equals the voltage V_A across the branch divided by the branch resistance R_b. Or $I_b = V_A/R_b$.
3. The total line current equals the sum of all the branch currents. Or $I_T = I_1 + I_2 + I_3 + \cdots +$ etc.
4. The equivalent resistance R_T of parallel branches is less than the smallest branch resistance, since all the branches must take more current from the source than any one branch.
5. For only *two* parallel resistances of any value, $R_T = R_1 R_2/(R_1 + R_2)$.
6. For any number of *equal* parallel resistances, R_T is the value of one resistance divided by the number of resistances.
7. For the general case of any number of branches, calculate R_T as V_A/I_T or use the reciprocal resistance formula.
8. For any number of conductances in parallel, their values are added for G_T, in the same way as adding parallel branch currents.
9. The sum of the individual values of power dissipated in parallel resistances equals the total power produced by the source. Or $P_T = P_1 + P_2 + P_3 + \cdots +$ etc.
10. An open in one branch results in no current through that branch, but

the other branches can have their normal current. However, an open in the main line results in no current for any of the branches.
11. A short circuit has zero resistance, resulting in excessive current. When one branch is short-circuited, all the parallel paths are also short-circuited. The entire current is in the short circuit and bypasses the short-circuited branches.

Self-Examination (Answers at back of book.)

Choose (a), (b), (c), or (d).

1. With two resistances connected in parallel: (a) the current through each must be the same; (b) the voltage across each must be the same; (c) their combined resistance equals the sum of the individual values; (d) each must have the same resistance value.
2. With 100 V applied across ten 50-Ω resistances in parallel, the current through each resistance equals (a) 2 A; (b) 10 A; (c) 50 A; (d) 100 A.
3. With three 1-kΩ resistances connected in parallel, their combined equivalent resistance equals (a) $\frac{1}{3}$ kΩ; (b) 1 kΩ; (c) 2 kΩ; (d) 3 kΩ.
4. A 1-Ω resistance in parallel with a 2-Ω resistance provides a combined equivalent resistance of (a) 3 Ω; (b) 1 Ω; (c) 2 Ω; (d) $\frac{2}{3}$ Ω.
5. With resistances of 100, 200, 300, 400, and 500 Ω in parallel, R_T is (a) less than 100 Ω; (b) more than 1 MΩ; (c) about 500 Ω; (d) about 1 kΩ.
6. With two resistances connected in parallel, if each dissipates 10 W, the total power supplied by the voltage source equals (a) 5 W; (b) 10 W; (c) 20 W; (d) 100 W.
7. With eight 10-MΩ resistances connected in parallel across a 10-V source, the main-line current equals (a) 0.1 μA; (b) $\frac{1}{8}$ μA; (c) 8 μA; (d) 10 μA.
8. A parallel circuit with 20 V applied across two branches has a total line current of 5 A. One branch resistance equals 5 Ω. The other branch resistance equals (a) 5 Ω; (b) 20 Ω; (c) 25 Ω; (d) 100 Ω.
9. Three 100-W light bulbs are connected in parallel across the 120-V power line. If one bulb opens, how many bulbs can light? (a) None; (b) one; (c) two; (d) all.
10. If a parallel circuit is open in the main line, the current (a) increases in each branch; (b) is zero in all the branches; (c) is zero only in the branch that has highest resistance; (d) increases in the branch that has lowest resistance.

Essay Questions

1. Draw a wiring diagram showing three resistances connected in parallel across a battery. Indicate each branch and the main line.
2. State two rules for the voltage and current values in a parallel circuit.
3. Explain briefly why the current is the same in both sides of the main line that connects the voltage source to the parallel branches.
4. (a) Show how to connect three equal resistances for a combined equivalent resistance one-third of the value of one resistance. (b) Show how to connect three equal resistances for a combined equivalent resistance three times the value of one resistance.
5. Why can the current in parallel branches be different when they all have the same applied voltage?
6. Why does the current increase in the voltage source as more parallel branches are added to the circuit?
7. Show the algebra for deriving the formula $R_T = R_1 R_2/(R_1 + R_2)$ from the reciprocal formula for two resistances.
8. Draw the circuit of five heaters for vacuum tubes wired in parallel across a 6.3-V ac source.
9. State briefly why the total power equals the sum of the individual values of power, whether a series circuit or parallel circuit is used.
10. Explain why an open in the main line disables all the branches, but an open in one branch affects only that branch current.
11. Give two differences between an open circuit and a short circuit.
12. List as many differences as you can in comparing series circuits with parallel circuits.

Problems (Answers to odd-numbered problems at back of book.)

1. A 15-Ω R_1 and a 45-Ω R_2 are connected in parallel across a 45-V battery. (a) Draw the schematic diagram. (b) How much is the voltage across R_1 and R_2? (c) How much is the current in R_1 and R_2? (d) How much is the main-line current? (e) Calculate R_T.
2. For the circuit in question 1, how much is the total power supplied by the battery?
3. A parallel circuit has three branch resistances of 20, 10, and 5 Ω for R_1, R_2, and R_3. The current through the 20-Ω branch is 1 A. (a) Draw the schematic diagram. (b) How much is the voltage applied across all the branches? (c) Find the current through the 10-Ω branch and the 5-Ω branch.
4. (a) Draw the schematic diagram of a parallel circuit with three branch

resistances, each having 10 V applied and a 2-A branch current. (b) How much is I_T? (c) How much is R_T?

5. Referring to Fig. 4-12, assume that R_2 opens. (a) How much is the current in the R_2 branch? (b) How much is the current in the R_1 branch? (c) How much is the line current? (d) How much is the total resistance of the circuit? (e) How much power is generated by the battery?

6. Two resistances R_1 and R_2 are in parallel across a 100-V source. The total line current is 10 A. The current I_1 through R_1 is 4 A. Draw a schematic diagram of the circuit, giving the values of currents I_1 and I_2 and resistances R_1 and R_2 in both branches. How much is the combined equivalent resistance of both branches across the voltage source?

7. Find the R_T for the following groups of branch resistances: (a) 10 Ω and 25 Ω; (b) five 10-kΩ resistances; (c) two 500-Ω resistances; (d) 100 Ω, 200 Ω, and 300 Ω; (e) two 5-kΩ and two 2-kΩ resistances; (f) four 40-kΩ and two 20-kΩ resistances.

8. How much parallel R_x must be connected across a 100-kΩ resistance to reduce R_T to (a) 50 kΩ; (b) 25 kΩ; (c) 10 kΩ?

9. Find the total conductance in siemens for the following branches: $G_1 = 9000$ μS; $G_2 = 7000$ μS; $G_3 = 22{,}000$ μS.

10. Referring to Fig. 4-11, calculate R_T by combining resistances. Show that this R_T equals $1/G_T$, where G_T is 0.75 S.

11. Find R_3 in Fig. 4-16.

12. In Fig. 4-17: (a) find each branch current and show the direction of electron flow; (b) calculate I_T; (c) calculate R_T; (d) calculate P_1, P_2, P_3, and P_T.

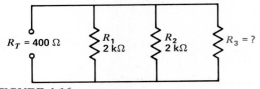

FIGURE 4-16
    For Prob. 11.

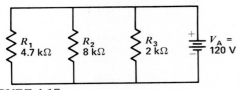

FIGURE 4-17
    For Prob. 12.

Answers to Practice Problems

4-1 (a) 1.5 V
    (b) 120 V
4-2 (a) 10 V
    (b) 1 A
    (c) 10 V
    (d) 2 A
4-3 (a) $I_T = 6$ A
    (b) $I_3 = 3$ A
4-4 (a) $R_T = 1.57$ MΩ
    (b) $R_T = 1.2$ MΩ
4-5 (a) $G_T = 7$ S or 7 mhos
4-5 (b) $G_T = 0.75$ μS or 0.75 μmho and $R_T = 1.33$ MΩ
4-6 (a) 480 W
    (b) 660 W
4-7 (a) 120 V
    (b) $I_1 = 4$ A
4-8 (a) Infinite ohms
    (b) Bulbs 1 and 2
4-9 (a) 0 Ω
    (b) $I_1 = 0$ A

Series-Parallel Circuits

Chapter 5

In many circuits, some components are connected in series to have the same current, while others are in parallel for the same voltage. Figure 5-1 shows an example. Such a circuit is used where it is necessary to provide different amounts of current and voltage with one source of applied voltage. The main features of series-parallel circuits are explained in the following topics:

5-1 Finding R_T for Series-Parallel Resistances
5-2 Resistance Strings in Parallel
5-3 Resistance Banks in Series
5-4 Resistance Banks and Strings in Series-Parallel
5-5 Analyzing Series-Parallel Circuits
5-6 Wheatstone Bridge
5-7 Chassis-Ground Connections
5-8 Voltages Measured to Chassis Ground
5-9 Opens and Shorts in Series-Parallel Circuits

5-1 FINDING R_T FOR SERIES-PARALLEL RESISTANCES

In Fig. 5-1, R_1 is in series with R_2. Also, R_3 is in parallel with R_4. However, R_2 is *not* in series with R_3 or R_4. The reason is the branch point A where the current through R_2 divides for R_3 and R_4. As a result, the current through R_3 must be less than the current through R_2. Therefore, R_2 and R_3 cannot be in series because they do not have the same current. For the same reason, R_4 also cannot be in series with R_2.

To find R_T, we add the series resistances and combine the parallel resistances.

In Fig. 5-1c, the 0.5-kΩ R_1 and 0.5-kΩ R_2 in series total 1 kΩ for R_{1-2}. Also, the 1-kΩ R_3 in parallel with the 1-kΩ R_4 can be combined for an equivalent resistance of 0.5 kΩ for R_{3-4}, as shown in Fig. 5-1d.

This parallel R_{3-4} combination of 0.5 kΩ is then added to the series R_{1-2} combination of 1 kΩ, for the final R_T value of 1.5 kΩ.

With R_T known, we can find I_T in the main line produced by the 1.5-V source V_T. Then I_T is V_T/R_T or 1.5 V/1.5 kΩ = 1 mA. This 1-mA I_T is the current through R_1 and R_2.

At branch point A, the 1-mA I_T divides into the two branch currents of 0.5 mA each for R_3 and R_4. At branch point B, the two 0.5-mA

Chapter 5
Series-Parallel Circuits

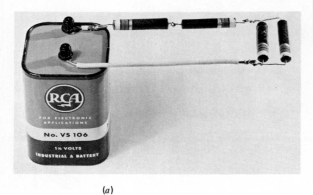
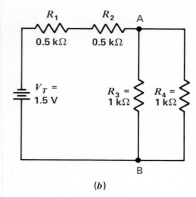
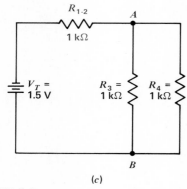
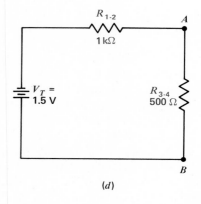

FIGURE 5-1

Series-parallel circuit. (*a*) Photograph of wiring. (*b*) Schematic diagram. (*c*) R_1 and R_2 in series added for R_{1-2}. (*d*) R_3 and R_4 in parallel combined as R_{3-4}.

branch currents combine to equal the 1-mA I_T in the main line, returning to the source V_T.

Practice Problems 5-1
(answers on page 112)
Refer to Fig. 5-1.
(*a*) Calculate the series R of R_1 and R_2.
(*b*) Calculate the parallel R of R_3 and R_4.
(*c*) Calculate R_T across the source V_T.

5-2 RESISTANCE STRINGS IN PARALLEL

More details about the voltages and currents in a series-parallel circuit are illustrated by the example in Fig. 5-2. Suppose there are four 120-V 100-W light bulbs to be wired, with a voltage source that produces 240 V. Each bulb needs 120 V for normal brilliance. If the bulbs

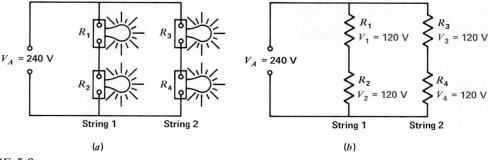

FIGURE 5-2

Two identical series strings in parallel. All bulbs have a 120-V 100-W rating. (*a*) Wiring diagram. (*b*) Schematic diagram.

were connected across the source, each would have the applied voltage of 240 V, causing excessive current in all the bulbs that could result in burned-out filaments.

If the four bulbs were connected in series, each would have a potential difference of 60 V, equal to one-fourth the applied voltage. With too low a voltage, there would be insufficient current for normal operation and the bulbs would not operate at normal brilliance.

However, two bulbs in series across the 240-V line provide 120 V for each filament, which is the normal operating voltage. Therefore, the four bulbs are wired in strings of two in series, with the two strings in parallel across the 240-V source. Both strings have 240 V applied. In each string two series bulbs divide the applied voltage equally to provide the required 120 V for the filaments.

Another example is illustrated in Fig. 5-3. This circuit has just two parallel branches where one branch includes R_1 in series with R_2. The other branch has just the one resistance R_3. Ohm's law can be applied to each branch.

Branch Currents I_1 and I_2. Each branch current equals the voltage applied across the branch divided by the total resistance in the branch. In branch 1, R_1 and R_2 total 12 Ω. With 12 V applied, this branch current I_1 is $^{12}/_{12} = 1$ A. Branch 2 has only the 6-Ω R_3. Then I_2 in this branch is $^{12}/_6 = 2$ A.

Series Voltage Drops in a Branch. For any one resistance in a string, the current in the string multiplied by the resistance equals the *IR* voltage drop across that particular resistance. Also, the sum of the series *IR* drops in the string equals the voltage across the entire string.

In string 1, the I_1R_1 drop equals 8 V, while the I_1R_2 drop is 4 V. These drops of 8 and 4 V add to equal the 12 V applied. The voltage across the R_3 branch is also 12 V.

Calculating I_T. The total line current equals the sum of the branch currents for all the parallel strings. Here I_T is 3 A, equal to the sum of 1 A in branch 1 and 2 A in branch 2.

Chapter 5
Series-Parallel Circuits

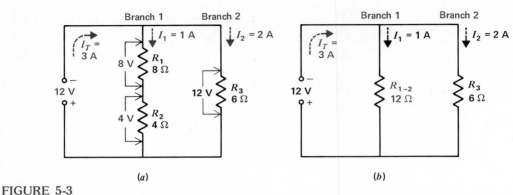

FIGURE 5-3 Series string in parallel with another branch. (*a*) Schematic diagram. (*b*) Equivalent circuit.

Calculating R_T. The resistance of the total series-parallel circuit across the voltage source equals the applied voltage divided by the total line current. In Fig. 5-3, R_T equals 12 V/3 A, or 4 Ω. This resistance can also be calculated as 12 Ω in parallel with 6 Ω, equivalent to one combined resistance of $^{72}/_{18} = 4$ Ω, for the product over the sum.

Applying Ohm's Law. There can be any number of parallel strings and more than two series resistances in a string. Still, Ohm's law can be used in the same way for the series and parallel parts of the circuit. The series parts have the same current. The parallel parts have the same voltage. Remember that for V/R the R must include all the resistance across the two terminals of V.

Practice Problems 5-2
(answers on page 112)
Refer to Fig. 5-3*a*.
(*a*) If I in R_2 were 6 A, what would I in R_1 be?
(*b*) If the source voltage were 72 V, what would V_3 be across R_3?

5-3 RESISTANCE BANKS IN SERIES

In Fig. 5-4*a*, the group of parallel resistances R_2 and R_3 is a bank. This is in series with R_1 because the total current of the bank must go through R_1.

The circuit here has R_2 and R_3 in parallel in one bank so that these two resistances will have the same potential difference of 20 V across them. The source applies 24 V, but there is a 4-V drop across R_1.

The two series voltage drops of 4 V across R_1 and 20 V across the bank add to equal the applied voltage of 24 V. The purpose of a circuit like this is to provide the same voltage for two or more resistances in a bank, where the bank voltage must be less than the applied voltage by the amount of IR drop across any series resistance.

To find the resistance of the entire circuit, combine the parallel resistances in each bank and add the series resistance. As shown in Fig. 5-4*b*, the two 10-Ω resistances R_2 and R_3 in parallel are equivalent to 5 Ω. Since the bank resistance of 5 Ω is in series with 1 Ω for R_1, the total resistance is 6 Ω across the 24-V source. Therefore, the main-line current is 24 V/6 Ω, which equals 4 A.

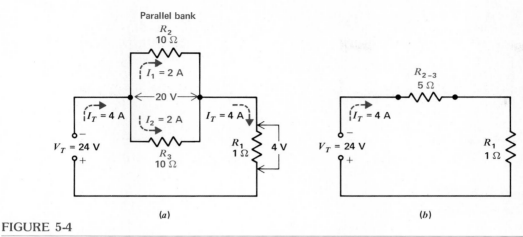

FIGURE 5-4
Parallel bank of R_2 and R_3 in series with R_1. (a) Schematic diagram. (b) Equivalent circuit.

The total line current of 4 A divides into two parts of 2 A each in the parallel resistances R_2 and R_3. Note that each branch current equals the bank voltage divided by the branch resistance. For this bank, $^{20}/_{10} = 2$ A for each branch.

The branch currents are combined in the line to provide the total 4 A in R_1. This is the same total current flowing in the main line, in the source, into the bank, and out of the bank.

There can be more than two parallel resistances in a bank and any number of banks in series. Still, Ohm's law can be applied the same way to the series and parallel parts of the circuit. The general procedure for circuits of this type is to find the equivalent resistance of each bank and then add all the series resistances.

Practice Problems 5-3
(answers on page 112)
Refer to Fig. 5-4a.
(a) If V_2 across R_2 were 40 V, what would V_3 across R_3 be?
(b) If I in R_2 were 4 A, with 4 A in R_3, what would I in R_1 be?

5-4
RESISTANCE BANKS AND STRINGS IN SERIES-PARALLEL

In the solution of such circuits, the most important fact to know is which components are in series with each other and what parts of the circuit are parallel branches. The series components must be in one current path without any branch points. A branch point such as point A or B in Fig. 5-5 is common to two or more current paths. For instance, R_1 and R_6 are *not* in series with each other. They do not have the same current, because the current in R_1 divides at point A into its two component branch currents. Similarly, R_5 is not in series with R_2, because of the branch point B.

To find the currents and voltages in Fig. 5-5, first find R_T in order to calculate the main-line current I_T as V_T/R_T. In calculating R_T, start reducing the branch farthest from the source and work toward the applied voltage. The reason for following this order is that you cannot tell how much resistance is in series with R_1 and R_2 until the parallel branches are reduced to their equivalent resistance. If no

Chapter 5
Series-Parallel Circuits

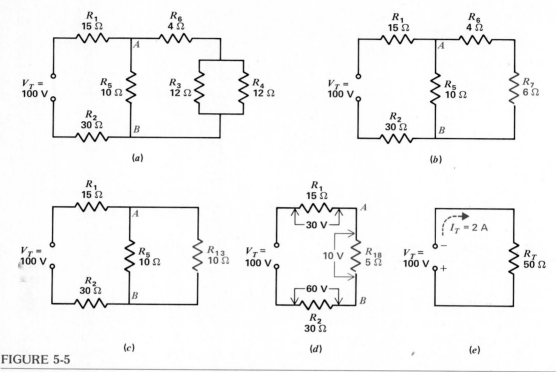

FIGURE 5-5

Reducing a series-parallel circuit to an equivalent series circuit to find the R_T. (a) Actual circuit. (b) R_3 and R_4 in parallel equal R_7. (c) R_7 and R_6 in series equal R_{13}. (d) R_{13} and R_5 in parallel equal R_{18}. (e) R_{18}, R_1, and R_2 in series are added for the total circuit resistance of 50 Ω.

source voltage is shown, R_T can still be calculated from the outside in toward the open terminals where a source would be connected.

To calculate R_T in Fig. 5-5, the steps are as follows:

1. The bank of the 12-Ω R_3 and 12-Ω R_4 in parallel in (a) is equal to the 6-Ω R_7 in (b).
2. The 6-Ω R_7 and 4-Ω R_6 in series in the same current path total 10 Ω for R_{13} in (c).
3. The 10-Ω R_{13} is in parallel with the 10-Ω R_5, across the branch points A and B. Their equivalent resistance then is the 5-Ω R_{18} in (d).
4. Now the circuit in (d) has just the 15-Ω R_1, 5-Ω R_{18}, and 30-Ω R_2 in series. These resistances total 50 Ω for R_T, as shown in (e).
5. With a 50-Ω R_T across the 100-V source, the line current I_T is $^{100}\!/_{50} = 2$ A.

To see the individual currents and voltages, we can use the I_T of 2 A for the equivalent circuit in (*d*). Now we work from the source *V* out toward the branches. The reason is that I_T can be used to find the voltage drops in the main line. The *IR* voltage drops here are:

$$V_1 = I_T R_1 = 2 \times 15 = 30 \text{ V}$$
$$V_{18} = I_T R_{18} = 2 \times 5 = 10 \text{ V}$$
$$V_2 = I_T R_2 = 2 \times 30 = 60 \text{ V}$$

The 10-V drop across R_{18} is actually the potential difference between branch points A and B. This means 10 V across R_5 and R_{13} in (*c*). The 10 V produces 1 A in the 10-Ω R_5 branch. The same 10 V is also across the R_{13} branch.

Remember that the R_{13} branch is actually the string of R_6 in series with the $R_3 R_4$ bank. Since this branch resistance is 10 Ω, with 10 V across it, the branch current here is 1 A. The 1 A through the 4 Ω of R_6 produces a voltage drop of 4 V. The remaining 6-V *IR* drop is across the $R_3 R_4$ bank. With 6 V across the 12-Ω R_3, its current is ½ A; the current is also ½ A in R_4.

Tracing all the current paths from the source, the main-line current through R_1 is 2 A. At the branch point A, this current divides into 1 A for R_5 and 1 A for the string with R_6. There is a 1-A branch current in R_6, but it subdivides in the bank with ½ A in R_3 and ½ A in R_4. At the branch point B, the total bank current of 1 A combines with the 1 A through the R_5 branch, resulting in a 2-A total line current through R_2, the same as through R_1 in the opposite side of the line.

Practice Problems 5-4
(*answers on page 112*)
Refer to Fig. 5-5a.
(a) Which *R* is in series with R_2?
(b) Which *R* is in parallel with R_3?
(c) Which *R* is in series with the $R_3 R_4$ bank?

5-5
ANALYZING SERIES-PARALLEL CIRCUITS

The circuits in Figs. 5-6 to 5-9 will be solved now. The following principles are illustrated:

1. With parallel strings across the main line, the branch currents and I_T can be found without R_T (see Figs. 5-6 and 5-7).
2. When parallel strings have series resistance in the main line, R_T must be calculated to find I_T, assuming no branch currents are known (see Fig. 5-9).
3. The source voltage is applied across the R_T of the entire circuit, producing an I_T that flows only in the main line.
4. Any individual series *R* has its own *IR* drop that must be less that the total V_T. In addition, any individual branch current must be less than I_T.

Solution for Fig. 5-6. The problem here is to calculate the branch currents I_1 and I_{2-3}, total line current I_T, and the voltage drops V_1, V_2, V_3. This order will be used for the calculations, because we can find the branch currents from the 90 V across the known branch resistances.

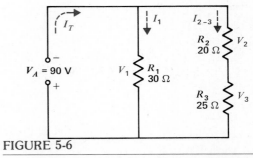

FIGURE 5-6

Find all the currents and voltages. See text for solution by calculating the branch currents first.

In the 30-Ω branch of R_1, the branch current is $90/30 = 3$ A for I_1. The other branch resistance, with a 20-Ω R_2 and a 25-Ω R_3, totals 45 Ω. This branch current then is $90/45 = 2$ A for $I_{2\text{-}3}$. In the main line, I_T is $3 + 2$, which equals 5 A.

For the branch voltages, V_1 must be the same as V_A, equal to 90 V. Or $V_1 = I_1 R_1$, which is $3 \times 30 = 90$ V.

In the other branch, the 2-A $I_{2\text{-}3}$ flows through the 20-Ω R_2 and the 25-Ω R_3. Therefore, V_2 is $2 \times 20 = 40$ V. Also, V_3 is $2 \times 25 = 50$ V. Note that these 40-V and 50-V series IR drops in one branch add to equal the 90-V source.

If we want to know R_T, it can be calculated as V_A/I_T. Then 90 V/5 A equals 18 Ω. Or R_T can be calculated by combining the branch resistances of 30 Ω in parallel with 45 Ω. Then R_T is $(30 \times 45)/(30 + 45)$. This answer is $1350/75$, which equals the same value of 18 Ω for R_T.

Solution for Fig. 5-7. To find the applied voltage first, the I_1 branch current is given. This 3-A current through the 10-Ω R_1 produces a 30-V drop V_1 across R_1. The same 3-A current through the 20-Ω R_2 produces 60 V for V_2 across R_2. The 30-V and 60-V drops are in series with each other across the applied voltage. Therefore, V_A equals the sum of $30 + 60$, or 90 V. This 90 V is also across the other branch combining R_3 and R_4 in series.

The other branch current I_2 in Fig. 5-7 must be 4 A, equal to the 7-A I_T minus the 3-A I_1. With 4 A for I_2, the voltage drop across the 12-Ω R_3 equals 48 V for V_3. Then the voltage across R_4 is $90 - 48$, or 42 V for V_4, as the sum of V_3 and V_4 must equal the applied 90 V.

Finally, with 42 V across R_4 and 4 A through it, this resistance equals $42/4$, or 10.5 Ω. Note that 10.5 Ω for R_4 added to the 12 Ω of R_3 equals 22.5 Ω, which allows $90/22.5$ or a 4-A branch current for I_2.

Solution for Fig. 5-8. The division of branch currents also applies to Fig. 5-8, but the main principle here is that the voltage must be the same across R_1 and R_2 in parallel. For the branch currents, I_2 is 2 A, equal to the 6-A I_T minus the 4-A I_1. The voltage across the 10-Ω R_1 is 4×10, or 40 V. This same voltage is also across R_2. With 40 V across R_2 and 2 A through it, R_2 equals $40/2$ or 20 Ω.

If we want to find V_T in Fig. 5-8, it can be calculated as 100 V. The 6-A I_T through R_3 produces a voltage drop of 60 V for V_3. Also, the voltage across the parallel bank with R_1 and

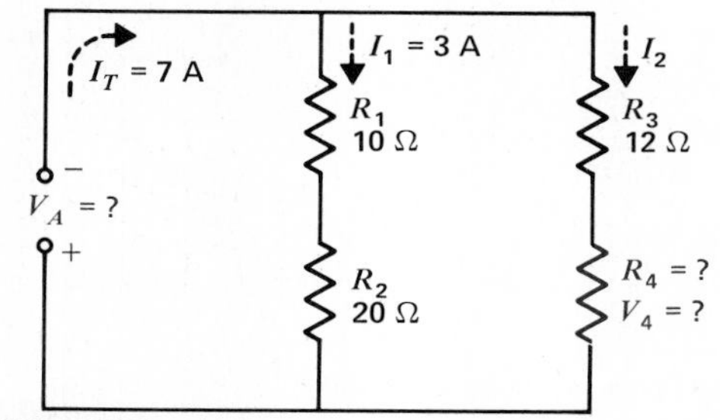

FIGURE 5-7

Find the applied voltage V_A, V_4, and R_4. See text for solution by calculating I_2 and the branch voltage.

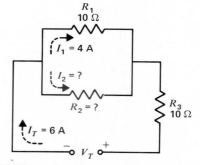

FIGURE 5-8

Find R_2 and I_2. See text for solution.

R_2 has been calculated as 40 V. This 40 V across the bank in series with 60 V across R_3 totals 100 V for the applied voltage.

Solution for Fig. 5-9. In order to find all the current and voltage drops, we need R_T to calculate I_T through R_6 in the main line. Combining resistances for R_T, we start with R_1 and R_2 and work in toward the source. Add the 8-Ω R_1 and 8-Ω R_2 in series with each other for 16 Ω. This 16 Ω combined with the 16-Ω R_3 in parallel equals 8 Ω between points c and d. Add this 8 Ω to the series 12-Ω R_4 for 20 Ω. This 20 Ω combined with the parallel 20-Ω R_5 equals 10 Ω between points a and b. Add this 10 Ω in series with the 10-Ω R_6, to make R_T of 20 Ω for the entire series-parallel circuit.

I_T in the main line is V_T/R_T, or $^{80}/_{20}$, which equals 4 A. This 4-A I_T flows through the 10-Ω R_6, producing a 40-V IR drop for V_6.

Now that we know I_T and V_6 in the main line, we use these values to calculate all the other voltages and currents. Start from the main line, where we know the current, and work outward from the source. To find V_5, the IR drop of 40 V for V_6 in the main line is subtracted from the source voltage. The reason is that V_5 and V_6 must add to equal the 80 V of V_T. Then V_5 is 80 − 40 = 40 V.

V_5 and V_6 happen to be equal at 40 V each. They split the 80 V in half because the 10-Ω R_6 equals the combined resistance of 10 Ω between points a and b.

With V_5 known to be 40 V, then I_5 through the 20-Ω R_5 is $^{40}/_{20}$ = 2 A. Since I_5 is 2 A and I_T is 4 A, I_4 must be 2 A also, equal to the difference between I_T and I_5. At the branch point a, the 4-A I_T divides into 2 A through R_5 and 2 A through R_4.

The 2-A I_4 through the 12-Ω R_4 produces an IR drop equal to 2 × 12 = 24 V for V_4. It should be noted now that V_4 and V_3 must add to equal V_5. The reason is that both V_5 and the path with V_4 and V_3 are across the same two points ab or ad. Since the potential difference across any two points is the same regardless of the paths, $V_5 = V_4 + V_3$. To find V_3 now, we can subtract the 24 V of V_4 from the 40 V of V_5. Then 40 − 24 = 16 V for V_3.

With 16 V for V_3 across the 16-Ω R_3, its current I_3 is 1 A. Also I_{1-2} in the branch with R_1 and R_2 is equal to 1 A. The 2-A I_4 into branch point c divides into the two equal branch currents of 1 A each because of the equal branch resistances.

Finally, with 1 A through the 8-Ω R_2 and 8-Ω R_1, their voltage drops are $V_2 = 8$ V and $V_1 = 8$ V. Note that the 8 V of V_1 in series with

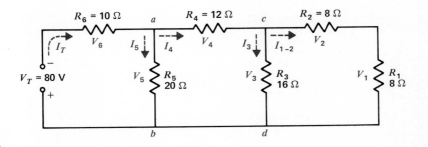

FIGURE 5-9

Find all currents and voltages. See text for solution in which R_T and then I_T are calculated to find V_6 first.

the 8 V of V_2 add to equal the 16-V potential difference V_3 between points c and d.

All the answers for the solution of Fig. 5-9 are summarized below:

$R_T = 20\ \Omega$ $I_T = 4$ A $V_6 = 40$ V
$V_5 = 40$ V $I_5 = 2$ A $I_4 = 2$ A
$V_4 = 24$ V $V_3 = 16$ V $I_3 = 1$ A
$I_{1\text{-}2} = 1$ A $V_2 = 8$ V $V_1 = 8$ V

Practice Problems 5-5
(answers on page 112)

(a) In Fig. 5-6, which R is in series with R_2?
(b) In Fig. 5-6, which R is across V_A?
(c) In Fig. 5-7, how much is I_2?
(d) In Fig. 5-8, how much is V_3?

5-6
WHEATSTONE[1] BRIDGE

A bridge circuit has four terminals, two for input voltage and two for output. The purpose is to have a circuit where the voltage drops can be balanced to provide zero voltage across the output terminals, with voltage applied across the input. In Fig. 5-10 the input terminals are C and D, while the output terminals are A and B.

The bridge circuit has many uses for comparison measurements. In the Wheatstone bridge, an unknown resistance R_X is balanced against a standard accurate resistor R_S for precise measurement of resistance.

In Fig. 5-10, S_1 applies battery voltage to the four resistors in the bridge. To balance the bridge, the value of R_S is varied. Balance is indicated by zero current in the galvanometer G. S_2 is a spring switch that is closed just to check the meter reading.

The reason for zero current in the meter can be seen by analysis of the voltage drops across the resistors. R_S in series with R_X forms a voltage divider across V_T; the parallel string of R_1 in series with R_2 is also a voltage divider across the same source. When the voltage division is in the same ratio for both strings, the voltage drop across R_S equals the voltage across R_2. Also, the voltage across R_X then equals the voltage across R_1. In this case, points A and B must be at the same potential. The difference of potential across the meter then must be zero, and there is no deflection.

At balance, the equal voltage ratios in the two branches of the Wheatstone bridge can be stated as

[1] Sir Charles Wheatstone (1802–1875), English physicist and inventor.

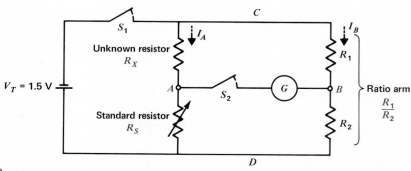

FIGURE 5-10
Wheatstone-bridge circuit.

$$\frac{I_A R_X}{I_A R_S} = \frac{I_B R_1}{I_B R_2} \quad \text{or} \quad \frac{R_X}{R_S} = \frac{R_1}{R_2}$$

Note that I_A and I_B cancel. Now, inverting R_S to the right side of the equation,

$$R_X = R_S \times \frac{R_1}{R_2} \qquad (5\text{-}1)$$

Usually, the total resistance of R_1 and R_2 is fixed, but any desired ratio can be chosen by moving point B on the ratio arm. The bridge is balanced by varying R_S for zero current in the meter. At balance, then, the value of R_X can be determined by multiplying R_S by the ratio of R_1/R_2. As an example, if the ratio is $1/100$ and R_S is 248 Ω, the value of R_X equals 248 × 0.01, or 2.48 Ω.

The balanced bridge circuit can be analyzed as simply two series resistance strings in parallel when the current is zero through the meter. Without any current between A and B this path is effectively open. When current flows through the meter path, however, the bridge circuit must be analyzed by Kirchhoff's laws or network theorems, as described in Chaps. 8 and 9.

Practice Problems 5-6
(answers on page 112)
(a) A bridge circuit has how many *pairs* of terminals?
(b) In Fig. 5-10, how much is V_{AB} at balance?

5-7
CHASSIS-GROUND CONNECTIONS

In the wiring of practical circuits, one side of the voltage source is usually grounded. For the 120-V ac power line in residential wiring, the ground is actually earth ground, usually by connection to a metal cold-water pipe. For electronic equipment, the ground just indicates a metal chassis, which is used as a common return for connections to the source. With printed wiring on a plastic board instead of a metal chassis, a conducting path around the entire board is used as a common return for chassis ground. The chassis ground may or may not be connected to earth ground. In either case the grounded side is called the "cold side" or "low side" of the applied voltage, while the ungrounded side is the "hot side" or "high side."

Grounding One Side of the Source Voltage. Three examples are shown in Fig. 5-11. In (a) one side of the 120-V ac power line is grounded. Note the symbol $\equiv$ for earth ground. This symbol also indicates a chassis ground that is connected to one side of the voltage source. In electronic equipment, black wire is generally used for chassis ground returns and red wire for the high side of the voltage source. See Table F-1 in Appendix F.

In Fig. 5-11b and c the 12-V battery is used as an example of a voltage source connected to chassis ground but not to earth. For instance, in an automobile one side of the battery is connected to the metal frame of the car. In (b), the negative side is grounded, while in (c) the positive side is grounded. Some people have the idea that ground must always be negative, but this is not necessarily so.

The reason for connecting one side of the 120-V ac power line to earth ground is to reduce the possibility of electric shock. However, chassis ground in electronic equipment is mainly a common-return connection. Where the equipment operates from the power line, the metal chassis should be at ground potential, not connected to the hot side of the ac outlet. This connection reduces the possibility of electric shock from the chassis. Also, hum from the power line is reduced in audio, radio, and television equipment.

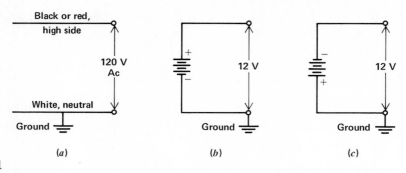

FIGURE 5-11
Grounding one side of a voltage source. (a) The ac power line. (b) Negative side of battery connected to chassis ground. (c) Positive side of battery connected to chassis ground.

Practice Problems 5-7
(answers on page 112)
(a) In Fig. 5-11b, give the voltage to ground with polarity.
(b) Do the same for Fig. 5-11c.

5-8
VOLTAGES MEASURED TO CHASSIS GROUND

When a circuit has the chassis as a common return, we generally measure the voltages with respect to chassis. Let us consider the voltage divider in Fig. 5-12 without any ground in (a), and then analyze the effect of grounding different points on the divider. It is important to realize that this circuit operates the same way with or without the ground. The only factor that changes is the reference point for measuring the voltages.

In Fig. 5-12a, the three 10-Ω resistances R_1, R_2, and R_3 divide the 30-V source equally. Then each voltage drop is $30/3 = 10$ V for V_1, V_2, and V_3. The polarity is positive at the top and negative at the bottom, the same as V_T.

If we want to consider the current, I is $30/30 = 1$ A. Each IR drop is $1 \times 10 = 10$ V for V_1, V_2, and V_3.

Positive Voltages to Negative Ground. In Fig. 5-12b, the negative side of V_T is grounded and the bottom end of R_1 is also grounded to complete the circuit. The ground is at point A. Note that the individual voltages V_1, V_2, and V_3 are still 10 V each. Also the current is still 1 A. The direction is also the same, from the negative side of V_T, through the metal chassis, to the bottom end of R_1. The only effect of the chassis ground here is to provide a conducting path from one side of the source to one side of the load.

With the ground in Fig. 5-12b, though, it is useful to consider the voltages with respect to chassis ground. In other words, the ground at point A will now be the reference for all voltages. When a voltage is indicated for only one point in a circuit, generally the other point is assumed to be chassis ground. We must have two points for a potential difference.

Let us consider the voltages at points B, C, and D. The voltage at B to ground is V_{BA}. This double subscript notation shows that we measure at B with respect to A. In general, the first letter indicates the point of measurement and the second letter is the reference point.

Then V_{BA} is $+10$ V. The positive sign is

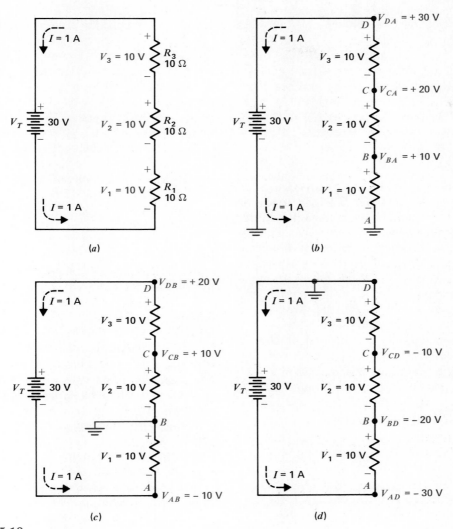

FIGURE 5-12

Voltages to chassis ground. (*a*) Voltage divider without ground. (*b*) With negative side of source V_T grounded, all voltages are positive to the chassis ground. (*c*) Positive and negative voltages with respect to the ground at point *B*. (*d*) With positive side of source grounded, all voltages are negative to chassis ground.

used here to emphasize the polarity. The value of 10 V for V_{BA} is the same as V_1 across R_1 because points B and A are across R_1. However, V_1 as the voltage across R_1 really cannot be given any polarity without a reference point.

When we consider the voltage at C, then, V_{CA} is $+20$ V. This voltage equals $V_1 + V_2$, connected with series-aiding polarities. Also, for point D at the top, V_{DA} is $+30$ V for $V_1 + V_2 + V_3$.

Positive and Negative Voltages to a Grounded Tap. In Fig. 5-12c point B in the divider is grounded. The purpose is to have the divider supply negative and positive voltages with respect to chassis ground. The negative voltage here is V_{AB}, which equals -10 V. This value is the same 10 V of V_1, but V_{AB} is the voltage at the negative end A with respect to the positive end B. The other voltages in the divider are $V_{CB} = +10$ V and $V_{DB} = +20$ V.

We can consider the ground at B as a dividing point for positive and negative voltages. For all points toward the positive side of V_T, any voltage is positive to ground. Going the other way, at all points toward the negative side of V_T, any voltage is negative to ground.

Negative Voltages to Positive Ground. In Fig. 5-12d, point D at the top of the divider is grounded, which is the same as grounding the positive side of the source V_T. The voltage source here is inverted, compared with (a), as the opposite side is grounded. In (d), all the voltages on the divider are negative to ground. Here, $V_{CD} = -10$ V, while $V_{BD} = -20$ V and $V_{AD} = -30$ V. Any point in the circuit must be more negative than the positive terminal of the source, even when this terminal is grounded.

Practice Problems 5-8
(answers on page 112)
Refer to Fig. 5-12c and give the voltage with polarity for
(a) A to ground.
(b) B to ground.
(c) D to ground.
(d) V_{DA} across V_T.

5-9
OPENS AND SHORTS IN SERIES-PARALLEL CIRCUITS

A short circuit has practically zero resistance. Its effect, therefore, is to allow excessive current. An open circuit has the opposite effect because an open circuit has infinitely high resistance with practically zero current. Furthermore, in series-parallel circuits an open or short circuit in one path changes the circuit for the other re-

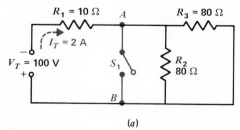
(a)

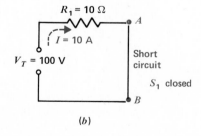
(b)

FIGURE 5-13

Effect of a short circuit with series-parallel connections. (a) Normal circuit with S_1 open. (b) Circuit with short between A and B when S_1 is closed. R_2 and R_3 are short-circuited.

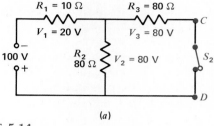
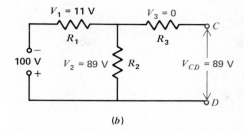

FIGURE 5-14

Effect of an open in a series-parallel circuit. (a) Normal circuit with S_2 closed. (b) Series circuit with R_1 and R_2 when S_2 is open. R_3 in the open path has no current and zero IR voltage drop.

sistances. For example, in Fig. 5-13, the series-parallel circuit in (a) becomes a series circuit with only R_1 when there is a short circuit between terminals A and B. As an example of an open circuit, the series-parallel circuit in Fig. 5-14a becomes a series circuit with just R_1 and R_2 when there is an open circuit between terminals C and D.

Effect of a Short Circuit. We can solve the series-parallel circuit in Fig. 5-13a in order to see the effect of the short circuit. For the normal circuit, with S_1 open, R_2 and R_3 are in parallel. Although R_3 is drawn horizontally, both ends are across R_2. The switch S_1 has no effect as a parallel branch here because it is open.

The combined resistance of the 80-Ω R_2 in parallel with the 80-Ω R_3 is equivalent to 40 Ω. This 40 Ω for the bank resistance is in series with the 10-Ω R_1. Then R_T is $40 + 10 = 50$ Ω.

In the main line I_T is $^{100}/_{50} = 2$ A. Then V_1 across the 10-Ω R_1 in the main line is $2 \times 10 = 20$ V. The remaining 80 V is across R_2 and R_3 as a parallel bank. As a result, $V_2 = 80$ V and $V_3 = 80$ V.

Now consider the effect of closing switch S_1. A closed switch has zero resistance. Not only is R_2 short-circuited, but R_3 in the bank with R_2 is also short-circuited. The closed switch short-circuits everything connected between terminals A and B. The result is the series circuit shown in Fig. 5-13b.

Now the 10-Ω R_1 is the only opposition to current. I equals V/R_1, which is $^{100}/_{10} = 10$ A. This 10 A flows through R_1, the closed switch, and the source. With 10 A through R_1, instead of its normal 2 A, the excessive current can cause excessive heat in R_1. There is no current through R_2 and R_3, as they are short-circuited out of the path for current.

Effect of an Open Circuit. Figure 5-14a shows the same series-parallel circuit as Fig. 5-13a, except that switch S_2 is used now to connect R_3 in parallel with R_2. With S_2 closed for normal operation, all currents and voltages have the values calculated for the series-parallel circuit. However, let us consider the effect of opening S_2, as shown in Fig. 5-14b. An open switch has infinitely high resistance. Now there is an open between terminals C and D. Furthermore, because R_3 is in the open path its 80 Ω cannot be considered in parallel with R_2.

The circuit with S_2 open in Fig. 5-14b is really the same as having just R_1 and R_2 in series with the 100-V source. The open path with R_3 has no effect as a parallel branch. The reason is that no current flows through R_3.

We can consider R_1 and R_2 in series as a voltage divider, where each IR drop is proportional to its resistance. The total series R is $80 + 10 = 90\ \Omega$. The 10-Ω R_1 is $^{10}\!/_{90}$ or $^1\!/_9$ of the total R and the applied V_T. Then V_1 is $^1\!/_9 \times 100\text{ V} = 11\text{ V}$ and V_2 is $^8\!/_9 \times 100\text{ V} = 89\text{ V}$, approximately. The 11-V drop for V_1 and 89-V drop for V_2 add to equal the 100 V of the applied voltage.

Note that V_3 is zero. Without any current through R_3, it cannot have any voltage drop.

Furthermore, the voltage across the open terminals C and D is the same 89 V as the potential difference V_2 across R_2. Since there is no voltage drop across R_3, terminal C has the same potential as the top terminal of R_2. Terminal D is directly connected to the bottom end of R_2. Therefore, the potential difference from C to D is the 89 V across R_2.

Practice Problems 5-9
(answers on page 112)
(a) In Fig. 5-13, the short circuit increases I_T from 2 A to what value?
(b) In Fig. 5-14, the open branch reduces I_T from 2 A to what value?

Summary

1. Table 5-1 summarizes the main characteristics of series and parallel circuits. In circuits combining series and parallel connections, the components in one current path without any branch points are in series; the parts of the circuit connected across the same two branch points are in parallel.
2. To calculate R_T in a series-parallel circuit with R in the main line, combine resistances from the outside back toward the source.
3. Chassis ground is commonly used as a return connection to one side of the source voltage. Voltages measured to chassis ground can have either negative or positive polarity.
4. When the potential is the same at the two ends of a resistance, its voltage is zero. Or if no current flows through a resistance, it cannot have any IR voltage drop.

Self-Examination (Answers at back of book.)

Choose (a), (b), (c), or (d).

1. In the series-parallel circuit in Fig. 5-1b: (a) R_1 is in series with R_3; (b) R_2 is in series with R_3; (c) R_4 is in parallel with R_3; (d) R_1 is in parallel with R_3.

TABLE 5-1. Comparison of Series and Parallel Circuits

SERIES CIRCUIT	PARALLEL CIRCUIT
Current the same in all components	Voltage the same across all branches
V across each series R is $I \times R$	I in each branch R is V/R
$V_T = V_1 + V_2 + V_3 + \cdots +$ etc.	$I_T = I_1 + I_2 + I_3 + \cdots +$ etc.
$R_T = R_1 + R_2 + R_3 + \cdots +$ etc.	$G_T = G_1 + G_2 + G_3 + \cdots +$ etc.
R_T must be more than the largest individual R	R_T must be less than the smallest branch R
$P_T = P_1 + P_2 + P_3 + \cdots +$ etc.	$P_T = P_1 + P_2 + P_3 + \cdots +$ etc.
Applied voltage is divided into IR voltage drops	Main-line current is divided into branch currents
The largest IR drop is across the largest series R	The largest branch I is in the smallest parallel R
Open in one component causes entire circuit to be open	Open in one branch does not prevent I in other branches

2. In the series-parallel circuit in Fig. 5-2b: (a) R_1 is in parallel with R_3; (b) R_2 is in parallel with R_4; (c) R_1 is in series with R_2; (d) R_2 is in series with R_4.
3. In the series-parallel circuit in Fig. 5-5, the total of all the branch currents into branch point A and out of branch point B equals (a) $\frac{1}{2}$ A; (b) 1 A; (c) 2 A; (d) 4 A.
4. In the circuit in Fig. 5-2 with four 120-V 100-W light bulbs, the resistance of one bulb equals (a) 72 Ω; (b) 100 Ω; (c) 144 Ω; (d) 120 Ω.
5. In the series-parallel circuit in Fig. 5-4a: (a) R_2 is in series with R_3; (b) R_1 is in series with R_3; (c) the equivalent resistance of the R_2R_3 bank is in parallel with R_1; (d) the equivalent resistance of the R_2R_3 bank is in series with R_1.
6. In a series circuit with unequal resistances: (a) the lowest R has the highest V; (b) the highest R has the highest V; (c) the lowest R has the most I; (d) the highest R has the most I.

7. In a parallel bank with unequal branch resistances: (a) the current is highest in the highest R; (b) the current is equal in all the branches; (c) the voltage is highest across the lowest R; (d) the current is highest in the lowest R.
8. In Fig. 5-14, with S_2 open, R_T equals (a) 90 Ω; (b) 100 Ω; (c) 50 Ω; (d) 10 Ω.
9. In Fig. 5-12c, V_{DA} equals (a) +10 V; (b) −20 V; (c) −30 V; (d) +30 V.
10. In the Wheatstone bridge of Fig. 5-10, at balance: (a) $I_A = 0$; (b) $I_B = 0$; (c) $V_2 = 0$; (d) $V_{AB} = 0$.

Essay Questions

1. In a series-parallel circuit, how can you tell which resistances are in series with each other and which are in parallel?
2. Draw a schematic diagram showing two resistances in a bank that is in series with one resistance.
3. Draw a diagram showing how to connect three resistances of equal value so that the combined resistance will be $1\frac{1}{2}$ times the resistance of one unit.
4. Draw a diagram showing two strings in parallel across a voltage source, where each string has three series resistances.
5. Explain why components are connected in series-parallel, showing a circuit as an example of your explanation.
6. Give two differences between a short circuit and an open circuit.
7. Explain the difference between voltage division and current division.
8. Show an example where a voltage is negative with respect to chassis ground.
9. Draw a circuit with nine 40-V 100-W bulbs connected to a 120-V source.
10. (a) Two 10-Ω resistors are in series with a 100-V source. If a third 10-Ω R is added in series, explain why I will decrease. (b) The same two 10-Ω resistors are in parallel with the 100-V source. If a third 10-Ω R is added in parallel, explain why I_T will increase.

Problems (Answers to odd-numbered problems at back of book.)

1. Refer to Fig. 5-1. (a) Calculate the total resistance of the circuit if all resistances are 10 Ω. (b) How much is the main-line current if V_T equals 100 V?

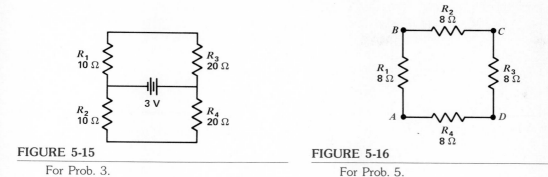

FIGURE 5-15
For Prob. 3.

FIGURE 5-16
For Prob. 5.

2. In Fig. 5-2, calculate the total power supplied by the source for the four 100-W bulbs.
3. Refer to the diagram in Fig. 5-15. (a) Why is R_1 in series with R_3 but not with R_2? (b) Find the total circuit resistance across the battery.
4. Two 60-Ω resistances R_1 and R_2 in parallel require 60 V across the bank with 1 A through each branch. Show how to connect a series resistance R_3 in the main line to drop an applied voltage of 100 V to 60 V across the bank. (a) How much is the required voltage across R_3? (b) How much is the required current through R_3? (c) How much is the required resistance of R_3? (d) If R_3 opens, how much is the voltage across R_1 and R_2? (e) If R_1 opens, what are the voltages across R_2 and R_3?
5. Refer to the diagram in Fig. 5-16. (a) Calculate R across points AD. (b) How much is R across points AD with R_4 open?
6. Show how to connect four 100-Ω resistances in a series-parallel circuit with a combined resistance equal to 100 Ω. (a) If the combination is connected across a 100-V source, how much power is supplied by the source? (b) How much power is dissipated in each resistance?
7. The following four resistors are in series with a 32-V source: R_1 is 24 Ω, R_2 is 8 Ω, R_3 is 72 Ω, and R_4 is 240 Ω. (a) Find the voltage drop across each resistor. (b) Calculate the power dissipated in each resistor. (c) Which resistor has the most voltage drop? (d) Which resistor dissipates the most power?
8. The same four resistors are in parallel with the 32-V source. (a) Find the branch current in each resistor. (b) Calculate the power dissipated in each resistor. (c) Which resistor has the most branch current? (d) Which resistor dissipates the most power?
9. Find R_1 and R_2 for a voltage divider that takes 10 mA from a 200-V source, with 50 V across R_2.

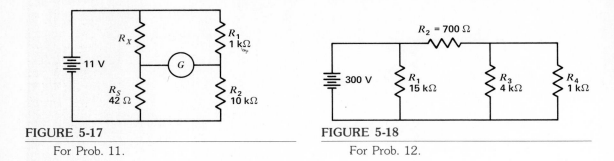

FIGURE 5-17
For Prob. 11.

FIGURE 5-18
For Prob. 12.

10. Refer to Fig. 8-2 in Chap. 8. Show the calculations for R_T, I_T, and each of the individual voltages and currents.
11. In the Wheatstone-bridge circuit of Fig. 5-17, find each voltage, label polarity, and calculate R_X. The bridge is balanced.
12. In Fig. 5-18, find each V and I for the four resistors.
13. In Fig. 5-19, calculate R_T.
14. In Fig. 5-20, find V_6.
15. Refer to Fig. 5-21. (a) Calculate V_2. (b) Find V_2 when R_3 is open.
16. In Fig. 5-22, find I and V for the five resistors and calculate V_T.
17. Refer to Fig. 5-23. (a) Find V_1, V_2, V_3, I_1, I_2, I_3, and I_T in the circuit as shown. (b) Now connect point G to ground. Give the voltages, with polarity, at terminals A, B, and C with respect to ground. In addition, give the values of I_1, I_2, I_3, and I_T with point G grounded.
18. In Fig. 5-24, give the voltages at points A, B, and C with polarity to ground when (a) point A is grounded; (b) point B is grounded; (c) point C is grounded.

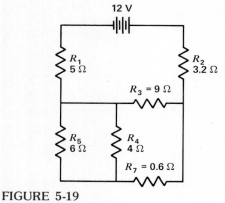

FIGURE 5-19
For Prob. 13.

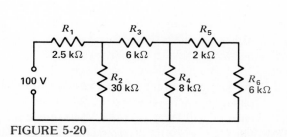

FIGURE 5-20
For Prob. 14.

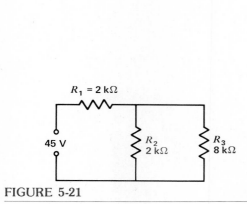

FIGURE 5-21

For Prob. 15.

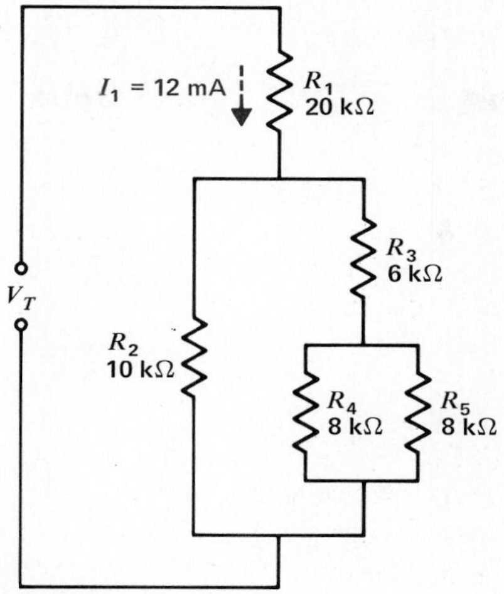

FIGURE 5-22

For Prob. 16.

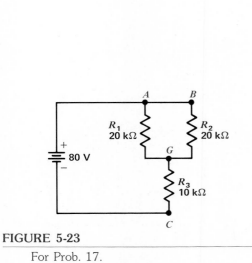

FIGURE 5-23

For Prob. 17.

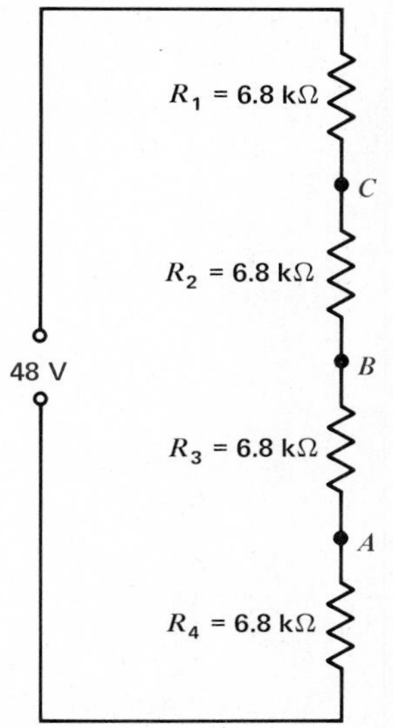

FIGURE 5-24

For Prob. 18.

Answers to Practice Problems

5-1 (a) $R = 1 \text{ k}\Omega$
    (b) $R = 0.5 \text{ k}\Omega$
    (c) $R_T = 1.5 \text{ k}\Omega$
5-2 (a) $I = 6 \text{ A}$
    (b) $V_A = 72 \text{ V}$
5-3 (a) $V_3 = 40 \text{ V}$
    (b) $I = 8 \text{ A}$
5-4 (a) R_1
    (b) R_4
    (c) R_6
5-5 (a) R_3
    (b) R_1
5-5 (c) $I_2 = 4 \text{ A}$
    (d) $V_3 = 60 \text{ V}$
5-6 (a) Two
    (b) 0 V
5-7 (a) $+12 \text{ V}$
    (b) -12 V
5-8 (a) -10 V
    (b) 0 V
    (c) $+20 \text{ V}$
    (d) $+30 \text{ V}$
5-9 (a) $I = 10 \text{ A}$
    (b) $I = 1.1 \text{ A}$

Review of Chapters 1 to 5

Summary

1. The electron is the basic quantity of negative electricity; the proton is the basic quantity of positive electricity. Both have the same charge but opposite polarities.
2. A quantity of electrons is a negative charge; a deficiency of electrons is a positive charge. Like charges repel each other; unlike charges attract.
3. Charge is measured in coulombs; 6.25×10^{18} electrons equals one coulomb. Charge in motion is current. One coulomb per second equals one ampere of current.
4. Potential difference is measured in volts. One volt produces one ampere of current against the opposition of one ohm of resistance.
5. The three forms of Ohm's law are $I = V/R$, $V = IR$, and $R = V/I$.
6. Power in watts equals VI, I^2R, or V^2/R, with V, I, and R in volts, amperes, and ohms, respectively.
7. The most common multiples and submultiples of the practical units are *mega* or M for 10^6, *micro* or μ for 10^{-6}, *kilo* or k for 10^3, and *milli* or m for 10^{-3}.
8. For series resistances: (*a*) the current is the same in all resistances; (*b*) the *IR* drops can be different with unequal resistances; (*c*) the applied voltage equals the sum of the series *IR* drops; (*d*) the total resistance equals the sum of the individual resistances; (*e*) an open circuit in one resistance results in no current through the entire series circuit.
9. For parallel resistances: (*a*) the voltage is the same across all resistances; (*b*) the branch currents can be different with unequal resistances; (*c*) the total line current equals the sum of the parallel branch currents; (*d*) the combined resistance of parallel branches is less than the smallest resistance, as determined by the reciprocal Formula (4-3); (*e*) an open circuit in one branch does not open the other

branches; (f) a short circuit across one branch short-circuits all the branches.
10. In series-parallel circuits, the resistances in one current path without any branch points are in series; all the rules of series resistances apply. The resistances across the same two branch points are in parallel; all the rules of parallel resistances apply.

Review Self-Examination (Answers at back of book.)

Choose (a), (b), (c), or (d).

1. In which of the following circuits will the voltage source produce the most current? (a) 10 V across a 10-Ω resistance; (b) 10 V across two 10-Ω resistances in series; (c) 10 V across two 10-Ω resistances in parallel; (d) 1000 V across a 1-MΩ resistance.
2. Three 120-V 100-W bulbs are in parallel across the 120-V power line. If one bulb burns open: (a) the other two bulbs cannot light; (b) all three bulbs light; (c) the other two bulbs can light; (d) there is excessive current in the main line.
3. A circuit allows 1 mA of current to flow with 1 V applied. The conductance of the circuit equals (a) 0.002 Ω; (b) 0.005 µS; (c) 1000 µS; (d) 1 S.
4. If 2 A of current is allowed to accumulate charge for 5 s, the resultant charge equals (a) 2 C; (b) 10 C; (c) 5 A; (d) 10 A.
5. A potential difference applied across a 1-MΩ resistor produces 1 mA of current. The applied voltage equals (a) 1 µV; (b) 1 mV; (c) 1 kV; (d) 1,000,000 V.
6. A string of two 1000-Ω resistances is in series with a parallel bank of two 1000-Ω resistances. The total resistance of the series-parallel circuit equals (a) 250 Ω; (b) 2500 Ω; (c) 3000 Ω; (d) 4000 Ω.
7. In the circuit of question 6, one of the resistances in the series string opens. Then the current in the parallel bank (a) increases slightly in both branches; (b) equals zero in one branch but is maximum in the other branch; (c) is maximum in both branches; (d) equals zero in both branches.
8. With 100 V applied across a 10,000-Ω resistance, the power dissipation equals (a) 1 mW; (b) 1 W; (c) 100 W; (d) 1 kW.
9. Ten volts is applied across R_1, R_2, and R_3 in series, producing 1 A in the series circuit. R_1 equals 6 Ω and R_2 equals 2 Ω. Therefore, R_3 equals (a) 2 Ω; (b) 4 Ω; (c) 10 Ω; (d) 12 Ω.

10. A 5-V source and 3-V source are connected with series-opposing polarities. The combined voltage across both sources equals (a) 5 V; (b) 3 V; (c) 2 V; (d) 8 V.
11. In a circuit with three parallel branches, if one branch opens, the main-line current will be (a) more; (b) less; (c) the same; (d) infinite.
12. A 10-Ω R_1 and a 20-Ω R_2 are in series with a 30-V source. If R_1 opens, the voltage drop across R_2 will be (a) zero; (b) 20 V; (c) 30 V; (d) infinite.
13. V_1 of 40 V is connected series-opposing with V_2 of 50 V. The total voltage across both components is: (a) 10 V; (b) 40 V; (c) 50 V; (d) 90 V.
14. Two series voltage drops V_1 and V_2 total 100 V for V_T. When V_1 is 60 V, then V_2 must equal: (a) 40 V; (b) 60 V; (c) 100 V; (d) 160 V.
15. Two parallel branch currents I_1 and I_2 total 100 mA for I_T. When I_1 is 60 mA, then I_2 must equal: (a) 40 mA; (b) 60 mA; (c) 100 mA; (d) 160 mA.

References (Additional references at back of book.)

Cooke, N. M., and H. F. R. Adams: "Basic Mathematics for Electronics," 4th ed., McGraw-Hill Book Company, New York.

De France, J. J.: "Electrical Fundamentals," Prentice-Hall, Inc., Englewood Cliffs, N.J.

Grob, B.: "Mathematics Outline and Review Problems for Basic Electronics," McGraw-Hill Book Company, New York.

Mittelstadt, W. S.: "Basic Slide Rule Operation," McGraw-Hill Book Company, New York.

Oppenheimer, S. L., and J. P. Borchers: "Direct and Alternating Currents," McGraw-Hill Book Company, New York.

"Periodic Chart of the Atoms," Sargent Welch Scientific Co., Skokie, Ill. 60076.

Slater, A. L.: "The Slide Rule," Holt, Rinehart and Winston, Inc., New York.

Timbie, W. H.: "Basic Electricity for Communications," John Wiley & Sons, Inc., New York.

Laboratory Manuals

Tinnell, R. W.: "Experiments in Electricity," McGraw-Hill Book Company, New York.

Zbar, P. B.: "Basic Electricity," 4th ed., McGraw-Hill Book Company, New York.

Voltage Dividers and Current Dividers

Chapter 6

Any series circuit is a voltage divider. The *IR* voltage drops are proportional parts of the applied voltage. Also, any parallel circuit is a current divider. Each branch current is part of the total line current, but in inverse proportion to the branch resistance. Special formulas can be used for the voltage and current division as short cuts in the calculations. The voltage division formula gives the series voltages even when the current is not known. Also, the current division formula gives the branch currents even when the branch voltage is not known. Finally, we consider a series voltage divider with parallel branches that have load currents. The design of such a loaded voltage divider can be applied to the important case of tapped voltages from the power supply in electronic equipment. The topics here are

6-1 Series Voltage Dividers
6-2 Current Divider with Two Parallel Resistances
6-3 Current Division by Parallel Conductances
6-4 Series Voltage Divider with Parallel Load Current
6-5 Design of a Loaded Voltage Divider

6-1 SERIES VOLTAGE DIVIDERS

The current is the same in all the resistances in a series circuit. Also, the voltage drops equal the product of *I* times *R*. Therefore, the *IR* voltages are proportional to the series resistances. A higher resistance has a greater *IR* voltage than a smaller resistance in the same series circuit; equal resistances have the same amount of *IR* drop. If R_1 is double R_2, then V_1 will be double V_2.

The series string can be considered as a *voltage divider*. Each resistance provides an *IR* drop *V* equal to its proportional part of the applied voltage. Stated as a formula,

$$V = \frac{R}{R_T} \times V_T \qquad (6\text{-}1)$$

Example 1. Three 50-kΩ resistors R_1, R_2, and R_3 are in series across an applied voltage of 180 V. How much is the *IR* voltage drop across each resistor?

Answer. 60 V. Since R_1, R_2, and R_3 are equal, each has one-third the total resistance of the circuit and one-third the total applied voltage. Using the formula,

$$V = \frac{R}{R_T} \times V_T = \frac{50 \text{ k}\Omega}{150 \text{ k}\Omega} \times 180 \text{ V}$$
$$= \frac{1}{3} \times 180 \text{ V}$$
$$V = 60 \text{ V}$$

Note that R and R_T must be in the same units for the proportion. Then V is in the same units as V_T.

The circuit in Fig. 6-1 illustrates another example of a proportional voltage divider. Let the problem be to find the voltage across R_3. We can either calculate this voltage as IR_3 or determine its proportional part of V_T. We arrive at the same answer both ways.

In the proportional method, V_3 is $^{20}/_{100}$ of the applied voltage because R_3 is 20 kΩ and R_T is 100 kΩ. Therefore, V_3 is $^{20}/_{100} \times 200$, or $^{1}/_{5} \times 200$, which is equal to 40 V.

In the same way, V_2 can be calculated as $^{30}/_{100} \times 200$, or $^{3}/_{10} \times 200 = 60$ V. Also, V_1 is $^{50}/_{100}$, or $^{1}/_{2} \times 200 = 100$ V. The sum of V_1, V_2, and V_3 is 40 + 60 + 100 = 200 V to equal V_T.

If we want to solve for the current in Fig. 6-1, I is V_T/R_T, or 200 V/100 kΩ = 2 mA. The *IR* drop V_3 then is 2 mA × 20 kΩ = 40 V. Also, V_2 is 2 mA × 30 kΩ = 60 V and V_1 is 2 mA × 50 kΩ = 100 V. These are the same values calculated by Formula (6-1) for proportional voltage dividers.

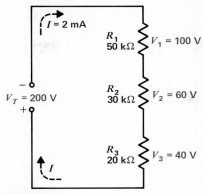

FIGURE 6-1

Series string as proportional voltage divider. Each V_R is R/R_T of V_T.

With two series voltages it is not necessary to calculate both. After you find one, subtract from V_T to find the other.

The fact that series voltage drops are proportional to the resistances means that a very small R has a negligible *IR* drop in series with a much larger R. An example is shown in Fig. 6-2. Here the 1 kΩ of R_1 is in series with the 999 kΩ of R_2, with a V_T of 1000 V.

By the voltage divider formula, V_1 is $^{1}/_{1000} \times 1000$ V = 1 V. Also, V_2 is $^{99}/_{1000} \times 1000$ V = 999 V. This value is practically the entire applied voltage.

Furthermore, the current 1 mA through R_1 and R_2 is determined almost entirely by the 999 kΩ of R_2. This *I* for R_T is 1000 V/1000 kΩ = 1 mA. The 999-kΩ R_2 alone would allow 1.001 mA for the current.

The advantage of using the divider formula is that we can find the proportional voltage drops from V_T and the series resistances without knowing *I*. For odd values of *R*, calculating *I* sometimes takes more time than finding the voltages. In many cases we can see the voltage division approximately without any written calculations.

Chapter 6
Voltage Dividers and Current Dividers

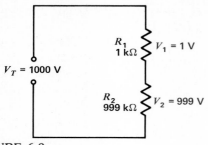

FIGURE 6-2

Very small R_1 in series with large R_2. V_1 is very small compared with V_2.

Practice Problems 6-1
(*answers on page 125*)
Refer to Fig. 6-1.
(a) How much is R_T?
(b) What fraction of the applied voltage is V_3?

6-2
CURRENT DIVIDER WITH TWO PARALLEL RESISTANCES

It is often necessary to find the individual branch currents in a bank from the resistances and I_T, but without knowing the voltage across the bank. This problem can be solved by using the fact that currents divide inversely as the branch resistances. An example is shown in Fig. 6-3. The formulas for the two branch currents are

$$I_1 = \frac{R_2}{R_1 + R_2} \times I_T \qquad (6\text{-}2)$$

or

$$I_2 = \frac{R_1}{R_1 + R_2} \times I_T$$

Notice that the formula for each branch I has the opposite R in the numerator. The reason is that each branch current is inversely proportional to the branch resistance. The denominator is the same in both formulas, equal to the sum of the two branch resistances.

To calculate the currents in Fig. 6-3, with a 30-A I_T, a 2-Ω R_1, and a 4-Ω R_2,

$$I_1 = \frac{4}{2+4} \times 30$$
$$= \frac{4}{6} \times 30 = \frac{2}{3} \times 30$$
$$I_1 = 20 \text{ A}$$

For the other branch,

$$I_2 = \frac{2}{2+4} \times 30$$
$$= \frac{2}{6} \times 30 = \frac{1}{3} \times 30$$
$$I_2 = 10 \text{ A}$$

With all the resistances in the same units, the branch currents are in the units of I_T. For instance, kilohms of R and milliamperes of I can be used.

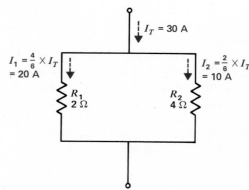

FIGURE 6-3

Current division with two branch resistances. Each I is inversely proportional to R. The smaller R has more I.

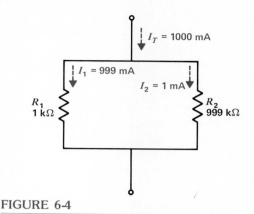

FIGURE 6-4
Very large R_2 in parallel with R_1. I_2 is very small compared with I_1.

Actually, it is not necessary to calculate both currents. After one I is calculated, the other can be found by subtracting from I_T.

Notice that the division of branch currents in a parallel bank is opposite from the voltage division of resistance in a series string. With series resistances, a higher resistance develops a larger IR voltage proportional to its R; with parallel branches, a lower resistance takes more branch current equal to V/R.

In Fig. 6-3, the 20-A I_1 is double the 10-A I_2 because the 2-Ω R_1 is one-half the 4-Ω R_2. This is an inverse proportion of I to R.

The inverse relation between I and R in a parallel bank means that a very large R has little effect with a much smaller R in parallel. As an example, Fig. 6-4 shows a 999-kΩ R_2 in parallel with a 1-kΩ R_1 dividing the I_T of 1000 mA. The branch currents are calculated as follows:

$$I_1 = \frac{999}{1000} \times 1000 \text{ mA} = 999 \text{ mA}$$

$$I_2 = \frac{1}{1000} \times 1000 \text{ mA} = 1 \text{ mA}$$

The 999 mA for I_1 is almost the entire line current of 1000 mA because R_1 is so small compared with R_2.

The current divider Formula (6-2) can be used for only two branch resistances. The reason is the inverse relation between each branch I and its R. In comparison, the voltage divider Formula (6-1) can be used for any number of series resistances because of the direct proportion between each voltage drop V and its R.

For more branches, it is possible to combine the branches in order to work with only two divided currents at a time. However, a better method is to use parallel conductances, because I and G are directly proportional, as explained in the next section.

Practice Problems 6-2
(answers on page 125)
Refer to Fig. 6-3.
(a) What is the ratio of R_2 to R_1?
(b) What is the ratio of I_2 to I_1?

6-3
CURRENT DIVISION BY PARALLEL CONDUCTANCES

Remember that the conductance G is $1/R$. Therefore, conductance and current are directly proportional. More conductance allows more current, for the same V. With any number of parallel branches, each branch current is

$$I = \left(\frac{G}{G_T}\right) I_T \qquad (6\text{-}3)$$

where G is the conductance of one branch and G_T is the sum of all the parallel conductances. The unit for G is the siemens (S) or mho.

Note that Formula (6-3), for dividing branch currents in proportion to G, has the same form as Formula (6-1), for dividing series voltages in proportion to R. The reason is that both formulas specify a direct proportion.

As an example of using Formula (6-3), we can go back to Fig. 6-3 and find the branch currents with G instead of R. For the 2-Ω R_1, the G is $\frac{1}{2} = 0.5$ S, or 0.5 mho. The 4-Ω R_2 has G of $\frac{1}{4} = 0.25$ S. Then G_T is $0.5 + 0.25 = 0.75$ S. The I_R is 30 mA. For the branch currents

$$I_1 = \frac{G_1}{G_T} \times I_T$$

$$= \frac{0.50}{0.75} \times 30 \text{ mA} = \frac{2}{3} \times 30 \text{ mA}$$

$$I_1 = 20 \text{ mA}$$

This 20 mA is the same I_1 calculated before. Also, I_2 is $30 - 20 = 10$ mA.

A circuit with three branches is shown in Fig. 6-5. Here R_1 is 10 Ω and $G_1 = \frac{1}{10}$ or 0.1 S. Also, G_2 is $\frac{1}{2} = 0.5$ S and G_3 is $\frac{1}{5} = 0.2$ S. Then G_T is $0.1 + 0.5 + 0.2 = 0.8$ S. The I_T is 40 mA. To find the I_1 branch current

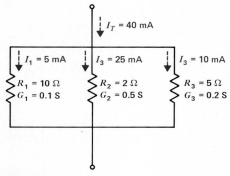

FIGURE 6-5

Current divider with three branch conductances G_1, G_2, and G_3. The S unit is the siemens (formerly called the *mho*). Each branch I is directly proportional to the branch G.

$$I_1 = \frac{G_1}{G_T} \times I_T$$

$$= \frac{0.1}{0.8} \times 40 \text{ mA}$$

$$I_1 = 5 \text{ mA}$$

In the same way, I_2 is calculated as $\frac{5}{8} \times 40 = 25$ mA and I_3 is $\frac{2}{8} \times 40 = 10$ mA. The sum is $5 + 25 + 10 = 40$ mA for I_T.

Although three branches are shown here, Formula (6-3) can be used for any number of parallel conductances because of the direct proportion between I and G.

Practice Problems 6-3
(answers on page 125)
Refer to Fig. 6-3.
(a) What is the ratio of G_3 to G_1?
(b) What is the ratio of I_3 to I_1?

6-4
SERIES VOLTAGE DIVIDER WITH PARALLEL LOAD CURRENT

The voltage dividers shown so far illustrate just a series string without any branch currents. Actually, though, a voltage divider is often used to tap off part of the applied voltage V_T for a load that needs less voltage than V_T. Then the added load is a parallel branch across part of the divider. Figure 6-6 shows how the loaded voltage at the tap is reduced from its potential without the branch current.

Why the Loaded Voltage Decreases. We can start with (a) in Fig. 6-6, which shows an $R_1 R_2$ voltage divider alone. R_1 and R_2 simply form a proportional series divider across the 60-V source V. R_1 is 40 kΩ and R_2 is 20 kΩ, making R_T 60 kΩ. I is 60 V/60 kΩ = 1 mA.

Then V_1 across R_1 is $\frac{40}{60} \times 60 \text{ V} = 40 \text{ V}$. Also, V_2 is $\frac{20}{60} \times 60 \text{ V} = 20 \text{ V}$. Note that

Chapter 6
Voltage Dividers and Current Dividers

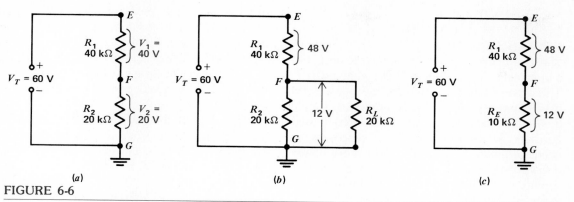

FIGURE 6-6
Effect of a parallel load on part of a series voltage divider. (a) R_1 and R_2 in series without any branch current. (b) Reduced voltage across R_2 with the parallel R_L. (c) Equivalent circuit of the loaded voltage divider.

$V_1 + V_2$ is $40 + 20 = 60$ V, which is the total applied voltage.

However, in (b) the 20-kΩ branch of R_L changes the equivalent resistance at tap F to ground. This change in the proportions of R changes the voltage division. Now the resistance from F to G is 10 kΩ, equal to the 20-kΩ R_2 and R_L in parallel. This equivalent bank resistance is shown as the 10-kΩ R_E in (c).

R_1 is still the same 40 kΩ because it has no parallel branch. The new R_T for the divider in (c) is 40 kΩ + 10 kΩ = 50 kΩ. As a result, V_E from F to G is now $^{10}/_{50} \times 60$ V = 12 V. Therefore, the voltage across R_2 and R_L in parallel is reduced to 12 V. This is the voltage at the tap F for R_L.

Note that V_1 across R_1 increases to 48 V in (c). Now V_1 is $^{40}/_{50} \times 60$ V = 48 V. The sum of $V_1 + V_2$ is $12 + 48 = 60$ V, still equal to the applied voltage.

Path of Current for R_L. All the current in the circuit must come from the source V_T. Trace the electron flow for R_L. It starts from the negative side of V_T, through R_L, to the tap at F, and returns through R_1 in the divider to the positive side of V_T. This current I_L goes through R_1 but not R_2.

Bleeder Current. In addition, both R_1 and R_2 have their own current from the source. This current through all the resistances in the divider is bleeder current I_B. The electron flow for I_B is from the negative side of V_T, through R_2 and R_1, and back to the positive side of V_T.

The bleeder current is a steady drain on the source. However, I_B has the advantage of reducing variations in the total current in the voltage source for different values of load current.

In summary, then, for the three resistances in Fig. 6-6, (a) R_L has just its load current I_L; (b) R_2 has only the bleeder current I_B; (c) R_1 has both I_L and I_B.

Practice Problems 6-4
(answers on page 125)
Refer to Fig. 6-6.
(a) What is the proportion of R_2/R_T in (a)?
(b) What is the proportion of R_e/R_T in (c)?

Chapter 6
Voltage Dividers and Current Dividers

6-5 DESIGN OF A LOADED VOLTAGE DIVIDER

These principles can be applied to the design of a practical voltage divider, as shown in Fig. 6-7. This type of circuit is used for the output of a power supply in electronic equipment to supply different voltages at the taps, with different load currents. For instance, load D can represent the collector-emitter circuit for one or more power transistors that need +100 V for the collector supply. Also, the tap at E can be the 40-V collector supply for medium power transistors. Finally, the 20-V tap at F can be for base-emitter bias current in the power transistors and collector voltage for smaller transistors.

Note the load specifications in Fig. 6-7. Load F needs 18 V from point F to chassis ground. When the 18 V is supplied by this part of the divider, a 36-mA branch current will flow through the load. Similarly, 40 V is needed at tap E for 54 mA of I_E in load E. Also, 100 V is available at D with a load current I_D of 180 mA. The total load current here is $36 + 54 + 180 = 270$ mA.

In addition, the bleeder current I_B through the entire divider is generally specified at about 10 percent of the load current. For the example here, I_B is taken as 30 mA to make a total line current I_T of $270 + 30 = 300$ mA from the source. Remember that the 30-mA I_B flows through R_1, R_2, and R_3.

The design problem in Fig. 6-7 is to find the values of R_1, R_2, and R_3 needed to provide the specified voltages. Each R is calculated as its ratio of V/I. However, the question is what are

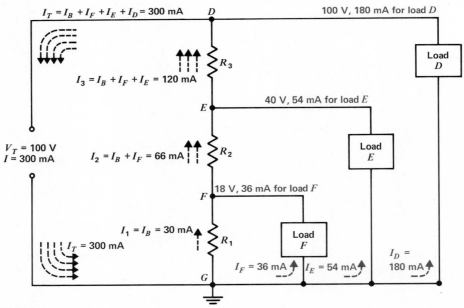

FIGURE 6-7

Voltage divider for different voltages and load currents from the source V_T. See text for design calculations to find R_1, R_2, and R_3.

the correct values of V and I to use for each part of the divider.

Find the Current in Each R. We start with R_1 because its current is only the 30-mA bleeder current I_B. No load current flows through R_1. Therefore I_1 through R_1 equals 30 mA.

The 36-mA current I_F for load F returns to the source through R_2 and R_3. Considering just R_2 now, its current is the I_F load current and the 30-mA bleeder current I_B. Therefore, I_2 through R_2 is $36 + 30 = 66$ mA.

The 54-mA current I_E for load E returns to the source through R_3 alone. However, R_3 also has the 36-mA I_F and the 30-mA I_B. Therefore I_3 through R_3 is $54 + 36 + 30 = 120$ mA. The values for I_1, I_2, and I_3 are summarized in Table 6-1.

Note that the load current I_D for load D at the top of the diagram does not flow through R_3 or any of the resistors in the divider. However, the I_D of 180 mA is the main load current through the source of applied voltage. The 120 mA of bleeder and load currents plus the 180-mA I_D load add to equal 300 mA for I_T in the main line of the power supply.

TABLE 6-1. Design Values for Voltage Divider in Fig. 6-7

	CURRENT, mA	VOLTAGE, V	RESISTANCE, Ω
R_1	30	18	600
R_2	66	22	333
R_3	120	60	500

Calculate the Voltage Across Each R. The voltages at the taps in Fig. 6-7 give the potential to chassis ground. However, we need the voltage across the two ends of each R. For R_1, the voltage V_1 is the indicated 18 V to ground because one end of R_1 is grounded. However, across R_2 the voltage is the difference between the 40-V potential at point E and the 18 V at F. Therefore V_2 is $40 - 18 = 22$ V. Similarly, V_3 is calculated as 100 V at point D minus the 40 V at E, or, V_3 is $100 - 40 = 60$ V. These values for V_1, V_2, and V_3 are summarized in Table 6-1.

Calculating Each R. Now we can calculate the resistance of R_1, R_2, and R_3 as each V/I ratio. For the values listed in Table 6-1,

$$R_1 = \frac{V_1}{I_1} = \frac{18 \text{ V}}{30 \text{ mA}} = 0.6 \text{ k}\Omega = 600 \text{ }\Omega$$

$$R_2 = \frac{V_2}{I_2} = \frac{22 \text{ V}}{66 \text{ mA}} = 0.333 \text{ k}\Omega = 333 \text{ }\Omega$$

$$R_3 = \frac{V_3}{I_3} = \frac{60 \text{ V}}{120 \text{ mA}} = 0.5 \text{ k}\Omega = 500 \text{ }\Omega$$

When these values are used for R_1, R_2, and R_3 and connected in a voltage divider across the source of 100 V, as in Fig. 6-7, each load will have the specified voltage at its rated current.

Practice Problems 6-5
(answers on page 125)
Refer to Fig. 6-7.
(a) How much is the bleeder current I_B through R_1, R_2, and R_3?
(b) How much is the voltage for load E at tap E to ground?
(c) How much is V_2 across R_2?

Summary

1. In a series circuit V_T is divided into IR voltage drops proportional to the resistances. Each $V_R = (R/R_T) \times V_T$, for any number of series resistances. The largest series R has the largest voltage drop.

2. In a parallel circuit, I_T is divided into branch currents. Each I is inversely proportional to the branch R. The inverse division of branch currents is given by Formula (6-2), for two resistances only. The smaller branch R has the larger branch current.
3. For any number of parallel branches, I_T is divided into branch currents directly proportional to each conductance G. Each $I = (G/G_T) \times I_T$.
4. A series voltage divider is often tapped for a parallel load, as in Fig. 6-6. Then the voltage at the tap is reduced because of the load current.
5. The design of a loaded voltage divider, as in Fig. 6-7, involves calculating each R. Find the I and potential difference V for each R. Then $R = V/I$.

Self-Examination (Answers at back of book.)

Answer true or false.

1. In a series voltage divider, each IR voltage is proportional to its R.
2. With parallel branches, each branch I is inversely proportional to its R.
3. With parallel branches, each branch I is directly proportional to its G.
4. Formula (6-2) for parallel current dividers can be used for three or more resistances.
5. Formula (6-3) for parallel current dividers can be used for five or more branch conductances.
6. In the series voltage divider of Fig. 6-1, V_1 is 2.5 times V_3 because R_1 is 2.5 times R_3.
7. In the parallel current divider of Fig. 6-3, I_1 is double I_2 because R_1 is one-half R_2.
8. In the parallel current divider of Fig. 6-5, I_3 is five times I_1 because G_3 is five times G_1.
9. In Fig. 6-6b, the branch current I_L flows through R_L, R_2, and R_1.
10. In Fig. 6-7, the bleeder current I_3 flows through R_1, R_2, and R_3.

Essay Questions

1. Define a series voltage divider.
2. Define a parallel current divider.
3. Give two differences between a series voltage divider and a parallel current divider.

4. Give three differences between Formula (6-2) for branch resistances and Formula (6-3) for branch conductances.
5. Define bleeder current.
6. What is the main difference between the circuits in (a) and (b) of Fig. 6-6?
7. Referring to Fig. 6-1, why is V_1 series-aiding with V_2 and V_3 but in series opposition to V_T? Show polarity of each IR drop.
8. Show the algebra for deriving Formula (6-2) for each branch current in a parallel bank of two resistances. [Hint: The voltage across the bank is $I_T \times R_T$ and R_T is $R_1 R_2/(R_1 + R_2)$.]

Problems (Answers to odd-numbered problems at back of book.)

1. A 200-Ω R_1 is in series with a 400-Ω R_2 and a 2-kΩ R_3. The applied voltage is 52 V. Calculate V_1, V_2, and V_3.
2. Find R_1 and R_2 for a voltage divider that takes 10 mA from a 200-V source, with 50 V across R_2. There are no load-current branches.
3. How much is the bleeder current through R_1 and R_2 in Fig. 6-6b?
4. I_T is 7 mA for two branches. R_1 is 20 kΩ and R_2 is 56 kΩ. Find I_1 and I_2 in this current-divider circuit.
5. Three parallel branches have $G_1 = 1000$ μS, $G_2 = 2000$ μS, $G_3 = 10,000$ μS. I_T is 39 mA. Find I_1, I_2, and I_3.
6. Referring to Fig. 6-3, find R_T for the two branch resistances and calculate the voltage across the bank as $I_T R_T$.
7. For the voltage divider in Fig. 6-7, how much is the equivalent resistance for load D, load E, and load F?
8. Referring to the voltage divider in Fig. 6-7, calculate the power dissipated in R_1, R_2, and R_3.
9. Design a voltage divider similar to that in Fig. 6-7 with R_1, R_2, and R_3 across a 48-V source and the following loads: 48 V at 800 mA, 28 V at 300 mA, and 9 V at 100 mA. Use the bleeder current I_B of 120 mA.

Answers to Practice Problems

6-1 (a) $R_T = 100$ kΩ
     (b) $V_3 = (^2\!/_{10}) \times V_T$
6-2 (a) 2 to 1
     (b) 1 to 2
6-3 (a) 2 to 1
     (b) 2 to 1
6-4 (a) $^1\!/_3$
     (b) $^1\!/_5$
6-5 (a) $I_B = 30$ mA
     (b) $V_{EG} = 40$ V
     (c) $V_2 = 22$ V

Direct-Current Meters

Chapter 7

Voltage, current, and resistance measurements are generally made with a combination volt-ohm-milliammeter (VOM) like the one in Fig. 7-1. To measure voltage, connect the voltmeter test leads across the two points of potential difference, as in (a). Similarly, when using the ohmmeter, connect the two leads across the resistance to be measured, as in (b), but turn the power off. No power is needed in the circuit being tested because the ohmmeter has its own internal battery. To measure current, the meter is connected as a series component in the circuit. A combination meter with all three functions is generally used as a multitester to check V, I, and R when troubleshooting electronic circuits. The details of these meter measurements are explained in the following topics:

7-1 Moving-Coil Meter
7-2 Measurement of Current
7-3 Meter Shunts
7-4 The Ayrton or Universal Shunt
7-5 Voltmeters
7-6 Loading Effect of a Voltmeter
7-7 Ohmmeters
7-8 Multimeters
7-9 Digital Meters
7-10 Meter Applications
7-11 Checking Continuity with the Ohmmeter

7-1 MOVING-COIL METER

This type of meter movement, shown in Fig. 7-2, is generally used in a VOM. The construction consists essentially of a coil of fine wire on a drum mounted between the poles of a permanent magnet. When direct current flows in the coil, the magnetic field of the current reacts with the field of the magnet. The resultant force turns the drum with its pointer. The amount of deflection indicates the amount of current in the coil. Correct polarity allows the pointer to read up-scale, to the right; the opposite polarity forces the pointer off-scale, to the left.

The pointer deflection is directly proportional to the amount of current in the coil. If 100 μA is the current needed for full-scale de-

Chapter 7
Direct-Current Meters

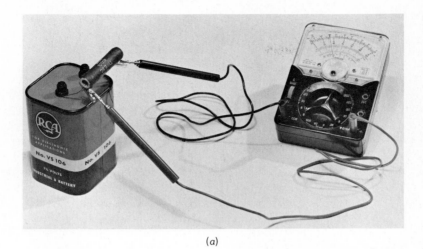

(a)

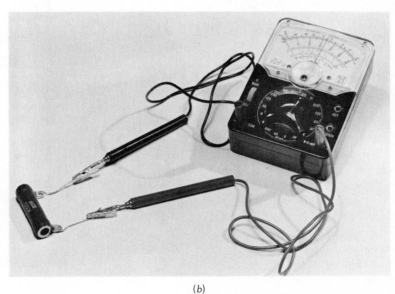

(b)

FIGURE 7-1

Using a VOM for voltage and resistance measurements. (a) To read voltage, connect the voltmeter test leads across the potential difference being measured. Observe polarity for voltage. (b) To read resistance, connect the ohmmeter test leads across R, but with the power off. Polarity of meter leads does not matter for resistance.

Chapter 7
Direct-Current Meters

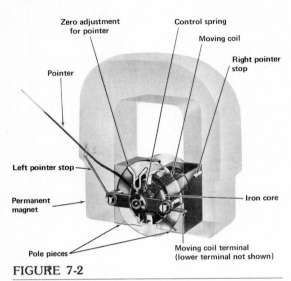

FIGURE 7-2

Construction of moving-coil meter. (*Weston Electrical Instrument Corp.*)

flection, 50 μA in the coil will produce a half-scale deflection. The accuracy of the moving-coil meter mechanism is 0.1 to 2 percent.

The moving-coil principle is applied in several meter types which have different names. A *galvanometer* is an extremely sensitive instrument for measuring very small values of current. Laboratory-type galvanometers, which include a suspended moving coil with an optical system to magnify small deflection, can measure a small fraction of one microampere. A *ballistic galvanometer* is used for reading the value of a small momentary current, to measure electric charge. The suspended moving-coil arrangement of a galvanometer is often called a *D'Arsonval movement,* after its inventor, who patented this meter movement in 1881. The practical, commercial moving-coil meter in Fig. 7-2 is a *Weston movement.*

Values of I_M. The full-scale deflection current I_M is the amount needed to deflect the pointer all the way to the right to the last mark on the printed scale. Typical values of I_M are from about 10 μA to 30 mA for Weston movements. Smaller currents require more wire in the moving coil, so that the magnetic field of the current can be strong enough to react with the permanent magnet to move the pointer. Fine wire must be used to reduce the weight of the moving coil. For the opposite case, higher currents require thicker wire, which can make the coil too heavy. However, the current range of the movement can be increased to almost any value by using meter shunts, as explained in Sec. 7-3.

As an example, I_M is 50 μA for the microammeter shown in Fig. 7-3. Notice the mirror along the scale to eliminate parallax. You read the meter where the pointer and its mirror re-

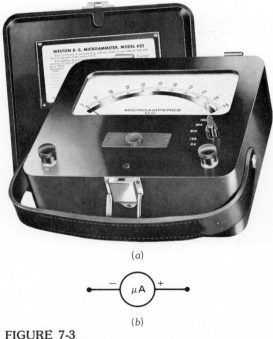

FIGURE 7-3

Microammeter with mirror on scale to minimize parallax error. (*a*) Photo of meter. (*b*) Schematic symbol. (*Weston Electrical Instrument Corp.*)

flection are one. This eliminates the optical error of parallax when you look at the meter from the side. The schematic symbol for a current meter is a circle, as in (b).

Values of r_M. This is the internal resistance of the wire of the moving coil. Typical values range from 1.2 Ω for a 30-mA movement to 2000 Ω for a 50-μA movement. A movement with a smaller I_M has a higher r_M because many turns of fine wire are needed. An average value of r_M for a 1-mA movement is about 120 Ω.

Taut-band Meters. The meter movement can be constructed with the moving coil and pointer suspended by a metal band, instead of the pivot and jewel design with a restoring spring. Both types of movements have similar operating characteristics. However, taut-band meters generally have lower values of r_M because a smaller coil can be used to force the pointer up-scale.

Practice Problems 7-1
(answers on page 158)
(a) Is a voltmeter connected in parallel or series?
(b) Is a milliammeter connected in parallel or series?

7-2
MEASUREMENT OF CURRENT

Whether we are measuring amperes, milliamperes, or microamperes, two important facts to remember are:

1. The current meter must be in series in the circuit where the current is to be measured. The amount of deflection depends on the current through the meter. In a series circuit, the current is the same through all series components. Therefore, the current to be measured must be made to flow through the meter as a series component in the circuit.
2. A dc meter must be connected in the correct polarity for the meter to read up-scale. Reversed polarity makes the meter read down-scale, forcing the pointer against the stop at the left, which can bend the pointer.

How to Connect a Current Meter in Series. As illustrated in Fig. 7-4, the circuit must be

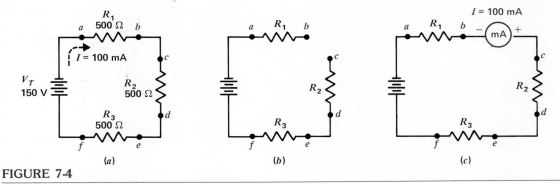

FIGURE 7-4

Inserting a current meter in series. (a) Circuit without meter. (b) Circuit opened between points b and c for meter. (c) Meter connected between R_1 and R_2 in series with the circuit.

opened at one point in order to insert the current meter in series in the circuit. Since R_1, R_2, R_3, and the meter are all in series, the current is the same in each and the meter reads the current in any part of the series circuit. If V_T is 150 V with a total series resistance of 1500 Ω, the current is 0.1 A, or 100 mA. This value is the current in R_1, R_2, R_3, and the battery, as shown in (a). Note that in (b), the circuit is opened at the junction of R_1 and R_2 for insertion of the meter. In (c), the meter completes the series circuit to read the current of 100 mA. The meter inserted in series at any point in the circuit would read the same current.

How to Connect a DC Meter in the Correct Polarity. A dc meter has its terminals marked for polarity, either with + and − signs or red for plus and black for minus. Electrons must flow into the negative side through the movement and out from the positive side for the meter to read up-scale.

To have the meter polarity correct, always connect its negative terminal to the point in the circuit that has a path back to the negative side of the voltage source, *without going through the meter*. Similarly, the positive terminal of the meter returns to the positive terminal of the voltage source (Fig. 7-5). Here the negative terminal of the meter is joined to R_2 because this path with R_1 connects to the negative terminal of the battery. The positive meter terminal is connected to R_3. Electrons in the circuit will flow through R_1 and R_2 into the negative side of the meter, through the movement, and out from the meter and return through R_3 to the positive battery terminal.

A Current Meter Should Have Very Low Resistance. Referring back to Fig. 7-4, the milliammeter in (c) reads 100 mA because its resistance is negligible compared with the total series R of 1500 Ω. Then I is the same with or without the meter.

In general, a current meter should have very low R compared with the circuit where the current is being measured. We take an arbitrary figure of $\frac{1}{100}$. For the circuit in Fig. 7-4, then, the meter resistance should be less than $\frac{1500}{100} = 15$ Ω. Actually, a meter for 100 mA would have an internal R of about 1 Ω or less because of its internal shunt resistor. The higher the current range of the meter, the smaller its resistance.

An extreme case of a current meter with too much R is shown in Fig. 7-6. Here the series R_T is doubled when the meter is inserted in the circuit. The result is one-half the actual I in the circuit without the meter.

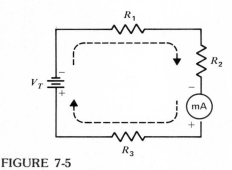

FIGURE 7-5
Correct polarity for a dc meter.

Practice Problems 7-2
(answers on page 158)
(a) In Fig. 7-4, how much will the milliammeter read when inserted at point a?
(b) In Fig. 7-5, which R is connected to the positive side of the meter to make it read up-scale?
(c) Should a current meter have very low or very high resistance?
(d) Should a voltmeter have very high or very low resistance?

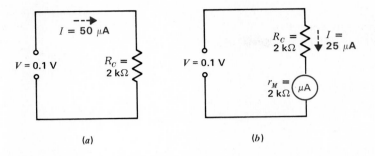

FIGURE 7-6

Example of a current meter having too high a resistance. (*a*) Circuit without the meter has an I of 50 µA. (*b*) Meter resistance reduces I to 25 µA.

7-3 METER SHUNTS

A meter shunt is a precision resistor connected across the meter movement for the purpose of shunting, or bypassing, a specific fraction of the circuit's current around the meter movement. The combination then provides a current meter with an extended range. The shunts are usually inside the meter case. In addition, the schematic symbol for the current meter usually does not show the shunt.

In current measurements, the parallel bank of the movement with its shunt is still connected as a current meter in series in the circuit (Fig. 7-7). It should be noted that a meter with an internal shunt has the scale calibrated to take into account the current through the shunt and the meter movement. Therefore, the scale reads total circuit current.

Resistance of the Meter Shunt. In Fig. 7-7*b*, the 25-mA movement has a resistance of 1.2 Ω, which is the resistance of the moving coil r_M. To double the range, the shunt resistance R_S is made equal to the 1.2 Ω of the movement. When the meter is connected in series in a circuit where the current is 50 mA, this total current into one terminal of the meter divides equally between the shunt and the meter movement. At the opposite meter terminal, these two branch currents combine to provide the 50 mA equal to the circuit current.

Inside the meter, the current is 25 mA through the shunt and 25 mA through the moving coil. Since it is a 25-mA movement, this current produces full-scale deflection. The scale is doubled, however, reading 50 mA, to account for the additional 25 mA through the shunt. Therefore, the scale reading indicates total current at the meter terminals, not just coil current. The movement with its shunt, then, is a 50-mA meter. Its internal resistance is $1.2 \times \frac{1}{2} = 0.6$ Ω.

Another example is shown in Fig. 7-8. In general, the shunt resistance for any range can be calculated with Ohm's law from the formula

$$R_S = \frac{V_M}{I_S} \qquad (7\text{-}1)$$

R_S is the resistance of the shunt and I_S is the current through it.

V_M is equal to $I_M \times r_M$. This is the voltage across both the shunt and the meter movement, which are in parallel.

Chapter 7
Direct-Current Meters

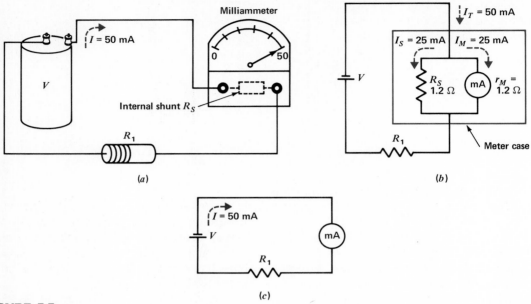

FIGURE 7-7

Effect of a shunt in bypassing current around the meter movement to extend its range from 25 to 50 mA. (*a*) Wiring diagram. (*b*) Schematic showing effect of shunt R_S. With $R_S = r_M$ the current range is doubled. (*c*) Circuit with 50-mA meter.

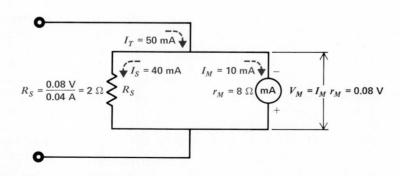

FIGURE 7-8

Calculating the resistance of a meter shunt $R_S = V_M/I_S$.

Calculating I_S. This current through the shunt alone is the difference between the total current I_T through the meter and the divided current I_M through the movement. Or

$$I_S = I_T - I_M \tag{7-2}$$

Use the values of current for full-scale deflection, as these are known. In Fig. 7-8, $I_S = 50 - 10 = 40$ mA, or 0.04 A.

Calculating R_S. The complete procedure for using the formula $R_S = V_M/I_S$ can be as follows:

1. Find V_M. Calculate this for full-scale deflection as $I_M \times r_M$. In Fig. 7-8, with a 10-mA full-scale current through the 8-Ω movement, V_M is $0.01 \times 8 = 0.08$ V.
2. Find I_S. For the values in Fig. 7-8, $I_S = 50 - 10 = 40$ mA = 0.04 A.
3. Divide V_M by I_S to find R_S. Here, $R_S = 0.08/0.04 = 2$ Ω.

This shunt enables the 10-mA movement to be used for the extended range of 0 to 50 mA.

Note that R_S and r_M are inversely proportional to their full-scale currents. The 2 Ω for R_S equals one-fourth the 8 Ω of r_M because the shunt current of 40 mA is four times the 10 mA through the movement for full-scale deflection.

Example 1. A shunt extends the range of a 50-μA meter movement to 1 mA. How much is the current through the shunt at full-scale deflection?

Answer. All the currents must be in the same units for Formula (7-2). To avoid fractions, use 1000 μA for the 1-mA I_T. Then

$I_S = I_T - I_M$
$\quad = 1000 \ \mu\text{A} - 50 \ \mu\text{A}$
$I_S = 950 \ \mu\text{A}$

Example 2. A 50-μA meter movement has r_M of 1000 Ω. What R_S is needed to extend the range to 500 μA?

Answer. The shunt current I_S is $500 - 50$, or 450 μA. Then

$$R_S = \frac{V_M}{I_S}$$
$$= \frac{50 \times 10^{-6} \ \text{A} \times 10^3 \ \Omega}{450 \times 10^{-6} \ \text{A}} = \frac{50{,}000}{450} = \frac{1000}{9}$$
$$R_S = 111.1 \ \Omega$$

The shunts usually are precision wire-wound resistors. For very low values, a short wire of precise size can be used.

Practice Problems 7-3
(answers on page 158)
A 50-μA movement with a 900-Ω r_M has a shunt R_S for a range of 500 μA.
(a) How much is I_S?
(b) How much is V_M?

7-4
THE AYRTON OR UNIVERSAL SHUNT

In Fig. 7-9, R_1, R_2, and R_3 are used in series-parallel combinations with the meter movements for different current ranges. The circuit is called an *Ayrton shunt* or *universal shunt*. This method is generally used for multiple current ranges in a VOM because the series-parallel circuit provides a safe method of switching between current ranges without danger of excessive current through the meter movement.

The wide contact on the switch arm in (a) indicates it makes the next connection before breaking the old contact. This short-circuiting type switch protects the meter movement by providing a shunt at all times during the switching to change ranges.

Chapter 7
134 Direct-Current Meters

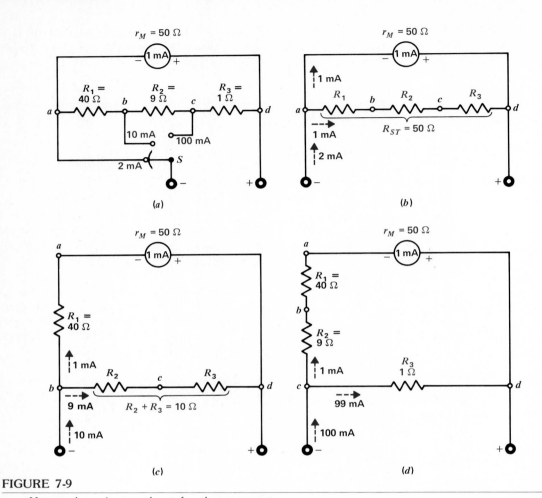

FIGURE 7-9
Universal or Ayrton shunt for three current ranges. *I* values shown for full-scale deflection. (a) Actual circuit with switch *S* to choose different ranges. (b) Circuit for 2-mA range; (c) 10-mA range; (d) 100-mA range.

The universal shunt consists of R_1, R_2, and R_3 in Fig. 7-9. How they are connected as a shunt is determined by the switch *S* for the different current ranges. Their total resistance (R_{ST}) is $40 + 9 + 1 = 50\ \Omega$. This resistance is used as a shunt in parallel with r_M for the 2-mA range in (b). For the higher current ranges in (c) and (d), part of R_{ST} is connected in series with r_M while the remainder of R_{ST} is in parallel as a shunt path.

The values in Fig. 7-9 are calculated as follows: Since the 2-mA range in (b) is double the 1-mA current rating of the meter movement, the shunt resistance must equal the r_M of

50 Ω so that 1 mA can flow in each of the two parallel paths. Therefore, R_{ST} is equal to the 50 Ω of r_M.

For the 10-mA range in (c), 9 mA must flow through the shunt path and 1 mA through the meter path. Now r_M has R_1 in series with it, in the path *bad*. The shunt now includes R_2 in series with R_3, in the path *bcd*. Remember that the voltage is the same across the two parallel paths *bad* and *bcd*. The current is 1 mA in one path and 9 mA in the other path. To calculate R_1 we can equate the voltage across the two paths:

$$1 \text{ mA} \times (R_1 + r_M) = 9 \text{ mA} \times (R_2 + R_3)$$

We know r_M is 50 Ω. We also know R_{ST} is 50 Ω. We do not know R_1, R_2, or R_3, but $(R_2 + R_3)$ must be 50 Ω minus R_1. Therefore,

$$1 \text{ mA} \times (R_1 + 50) = 9 \text{ mA} \times (50 - R_1)$$

Solving for R_1,

$$R_1 + 50 = 450 - 9R_1$$
$$10R_1 = 400$$
$$R_1 = 40 \text{ Ω}$$

Not only do we know now that R_1 is 40 Ω, but we also know that $(R_2 + R_3)$ must be 10 Ω, as they all must add up to 50 Ω. This value of 10 Ω for $(R_2 + R_3)$ is used for the next step in the calculations.

For the 100-mA range in (d), 1 mA flows through R_1, R_2, and r_M in the path *cbad*, and 99 mA through R_3 in the path *cd*. The voltage is the same across both paths. To calculate R_2,

$$1 \text{ mA} \times (R_1 + R_2 + r_M) = 99 \text{ mA} \times (R_3)$$

We know R_1 is 40 Ω. Then

$$40 + R_2 + 50 = 99 R_3$$

If $(R_2 + R_3)$ is 10 Ω, then R_3 must be $(10 - R_2)$. Substituting $(10 - R_2)$ for R_3, the equation then becomes:

$$40 + R_2 + 50 = 99 \times (10 - R_2)$$
$$R_2 + 90 = 990 - 99R_2$$
$$100R_2 = 900$$
$$R_2 = 9 \text{ Ω}$$

Finally, R_3 must be 1 Ω.

The total of $R_1 + R_2 + R_3$ equals $40 + 9 + 1$, which equals the 50 Ω of R_{ST}.

As a proof of the resistance values, note that in (b), 1 mA in each 50-Ω branch produces 50 mV across both parallel branches. In (c), 1 mA in the 90-Ω branch with the meter produces 90 mV from b to d, while 9 mA through the 10 Ω of $R_1 + R_2$ produces the same 90 mV. In (d), 99 mA through the 1-Ω R_3 produces 99 mV, while 1 mA through the 99 Ω in path *cbad* produces the same 99 mV.

Practice Problems 7-4
(answers on page 158)
Refer to Fig. 7-9 and give the full-scale I through the meter movement on
(a) 2-mA range.
(b) 100-mA range.

7-5
VOLTMETERS

Although a meter movement responds only to current in the moving coil, it is commonly used for measuring voltage by the addition of a high resistance in series with the movement (Fig. 7-10). The series resistance must be much higher than the coil resistance in order to limit the current through the coil. The combination of the meter movement with its added series resistance then forms a voltmeter. The series resistor, called a *multiplier,* is usually connected inside the voltmeter case.

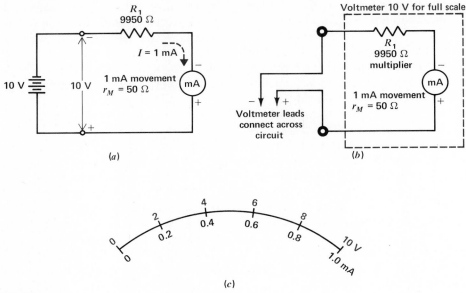

FIGURE 7-10

A multiplier resistor added in series with the meter movement forms a voltmeter. (*a*) The multiplier R_1 allows full-scale meter deflection with 10 V applied. (*b*) The voltmeter leads can be connected across a circuit to measure 0 to 10 V. (*c*) 10-V scale and the corresponding 1-mA scale.

Since a voltmeter has high resistance, it must be connected in parallel to measure the potential difference across two points in a circuit. Otherwise, the high-resistance multiplier would add so much series resistance that the current in the circuit would be reduced to a very low value. Connected in parallel, though, the high resistance of the voltmeter is an advantage. The higher the voltmeter resistance, the smaller the effect of its parallel connection on the circuit being tested.

The circuit is not opened to connect the voltmeter in parallel. Because of this convenience, it is common practice to make voltmeter tests in troubleshooting. The voltage measurements apply the same way to either an IR drop or a generated emf.

The correct polarity must be observed in using a dc voltmeter. Connect the negative voltmeter lead to the negative side of the potential difference being measured and the positive lead to the positive side.

Multiplier Resistance. Figure 7-10 illustrates how the meter movement and its multiplier R_1 form a voltmeter. With 10 V applied by the battery in (*a*), there must be 10,000 Ω of resistance to limit the current to 1 mA for full-scale deflection of the meter movement. Since the movement has a 50-Ω resistance, 9950 Ω is added in series, resulting in a 10,000-Ω total resistance. Then I is 10 V/10 kΩ = 1 mA.

With 1 mA in the movement, the full-scale deflection can be calibrated as 10 V on the meter scale, as long as the 9950-Ω multiplier is included in series with the movement. The multiplier can be connected on either side of the movement.

If the battery is taken away, as in (b), the movement with its multiplier forms a voltmeter that can indicate a potential difference of 0 to 10 V applied across its terminals. When the voltmeter leads are connected across a potential difference of 10 V in a dc circuit, the resulting 1-mA current through the meter movement produces full-scale deflection and the reading is 10 V. In (c) the 10-V scale is shown corresponding to the 1-mA range of the movement.

If the voltmeter is connected across a 5-V potential difference, the current in the movement is ½ mA, the deflection is one-half of full scale, and the reading is 5 V. Zero voltage across the terminals means no current in the movement, and the voltmeter reads zero. In summary, then, any potential difference up to 10 V, whether an IR voltage drop or generated emf, can be applied across the meter terminals. The meter will indicate less than 10 V in the same ratio that the meter current is less than 1 mA.

The resistance of a multiplier can be calculated from the formula

$$R_{mult} = \frac{\text{full-scale } V}{\text{full-scale } I} - r_M \qquad (7\text{-}3)$$

Applying this formula to the example of R_1 in Fig. 7-10 gives

$$R_{mult} = \frac{10 \text{ V}}{0.001 \text{ A}} - 50 \text{ Ω}$$
$$= 10,000 - 50$$
$$R_{mult} = 9950 \text{ Ω}$$

We can take another example for the same 10-V scale but with a 50-µA meter movement, which is commonly used. Now the multiplier resistance is much higher, though, because less I is needed for full-scale deflection. Let the resistance of the 50-µA movement be 2000 Ω. Then

$$R_{mult} = \frac{10 \text{ V}}{0.000\ 050 \text{ A}} - 2000 \text{ Ω}$$
$$= 200,000 - 2000$$
$$R_{mult} = 198,000 \text{ Ω}$$

Multiple Voltmeter Ranges. Voltmeters often have several multipliers which are used with one meter movement. A range switch selects one multiplier for the required scale. The higher the voltage range is, the higher the multiplier resistance, in essentially the same proportion as the ranges.

Figure 7-11 illustrates two ranges. When the switch is on the 10-V range, multiplier R_1 is connected in series with the 1-mA movement. Then you read the 10-V scale on the meter face. With the range switch on 25 V, R_2 is then the multiplier, and the measured voltage is read on the 25-V scale.

Several examples of using these two scales are listed in Table 7-1. Note that voltages less than 10 V can be read on either scale. It is preferable, however, to have the pointer read on the middle third of the scale. That is why the scales are usually multiples of 10 and 2.5 or 3.

Range Switch. With multiple ranges, the setting of the selector switch is the voltage that produces full-scale deflection (Fig. 7-12). One scale is generally used for ranges that are multiples of 10. If the range switch is set for 250 V in Fig. 7-12, read the top scale as is. With the range switch at 25 V, however, the readings on the 250-V scale are divided by 10.

Chapter 7
Direct-Current Meters

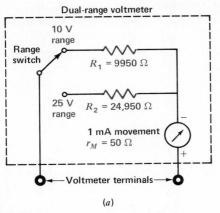
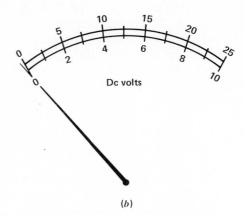

FIGURE 7-11
Voltmeter with a range of either 10 or 25 V. (a) Range switch selects scale by connecting either R_1 or R_2 as the series multiplier. (b) Both voltage ranges on the face of the meter.

Similarly, the 100-V scale is used for the 100-V range and the 10-V range. In Fig. 7-12 the pointer indicates 30 V when the switch is on the 100-V range; this reading on the 10-V range is 3 V.

Typical Multiple Voltmeter Circuit. Another example of multiple voltage ranges is shown in Fig. 7-13, with a typical switching arrangement. R_1 is the series multiplier for the lowest voltage range of 2.5 V. When higher resistance is needed for the higher ranges, the switch adds the required series resistors.

The meter in Fig. 7-13 requires 50 µA for full-scale deflection. For the 2.5-V range, a series resistance of $2.5/(50 \times 10^{-6})$, or 50,000 Ω, is needed. Since r_M is 2000 Ω, the value of R_1 is $50,000 - 2000$, which equals 48,000 Ω or 48 kΩ.

For the 10-V range, a series resistance of $10/(50 \times 10^{-6})$, or 200,000 Ω, is needed. Since $R_1 + r_M$ provide 50,000 Ω, R_2 is made

TABLE 7-1. Multiple Voltage-Scale Readings for Fig. 7-11

10-V SCALE, $R_V{}^* = 10,000$ Ω			25-V SCALE, $R_V{}^* = 25,000$ Ω		
METER, mA	DEFLECTION	SCALE READING, V	METER, mA	DEFLECTION	SCALE READING, V
0	0	0	0	0	0
0.5	½	5	0.2	2/10	5
1.0	Full scale	10	0.4	4/10	10
			0.5	½	12.5
			1.0	Full scale	25

*R_V is total voltmeter resistance of multiplier and meter movement.

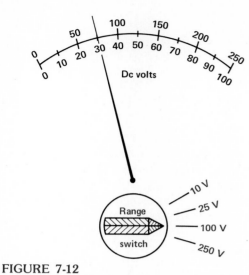

FIGURE 7-12
The range switch selects the voltage that can produce full-scale deflection. The reading shown here is 30 V on the 100-V range.

on the highest range for 5000 V. This method of adding series multipliers for higher voltage ranges is the circuit generally used in commercial multimeters.

Voltmeter Resistance. The high resistance of a voltmeter with a multiplier is essentially the value of the multiplier resistance. Since the multiplier is changed for each range, the voltmeter resistance changes.

Table 7-2 shows how the voltmeter resistance increases for the higher ranges. The middle column lists the total internal resistance R_V, including R_{mult} and r_M, for the voltmeter circuit in Fig. 7-13. With a 50-μA movement, R_V increases from 50 kΩ on the 2.5-V range to 20 MΩ on the 1000-V range. It should be noted that R_V has these values on each range whether you read full-scale or not.

Ohms-per-Volt Rating. To indicate the voltmeter's resistance independently of the range, voltmeters are generally rated in ohms of resistance needed for 1 V of deflection. This value is the ohms-per-volt rating of the voltmeter. As an example, see the last column in Table 7-2. The

150,000 Ω for a total of 200,000 Ω series resistance on the 10-V range. Similarly, additional resistors are switched in to increase the multiplier resistance for the higher voltage ranges. Note the separate jack and extra multiplier R_6

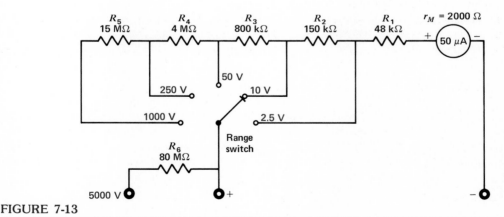

FIGURE 7-13
Typical voltmeter circuit for multiple ranges. (*Simpson VOM Model 260*)

Chapter 7
Direct-Current Meters

TABLE 7-2. Characteristics of a Voltmeter Using a 50-μA Movement

FULL-SCALE VOLTAGE (V_F)	$R_V = R_{\text{mult}} + r_M$	OHMS PER VOLT = R_V/V_F
2.5	50 kΩ	20,000 Ω/V
10	200 kΩ	20,000 Ω/V
50	1 MΩ	20,000 Ω/V
250	5 MΩ	20,000 Ω/V
1000	20 MΩ	20,000 Ω/V

values in the top row show that this meter needs 50,000 Ω R_V for 2.5 V of full-scale deflection. The resistance per 1 V of deflection then is 50,000/2.5, which equals 20,000 Ω/V.

The ohms-per-volt value is the same for all ranges. The reason is that this characteristic is determined by the full-scale current I_M of the meter movement. To calculate the ohms-per-volt rating, take the reciprocal of I_M in ampere units. For example, a 1-mA movement results in 1/0.001 or 1000 Ω/V; a 50-μA movement allows 20,000 Ω/V, and 20 μA allows 50,000 Ω/V. The ohms-per-volt rating is also called the *sensitivity* of the voltmeter.

A high value of ohms per volt means a high voltmeter resistance R_V. In fact R_V can be calculated as the product of the ohms-per-volt rating and the full-scale voltage of each range. For instance across the second row in Table 7-2, on the 10-V range with a 20,000 Ω/V rating,

$$R_V = 10 \text{ V} \times \frac{20,000 \text{ }\Omega}{\text{volts}}$$

$$R_V = 200,000 \text{ }\Omega$$

These values are for dc volts only. The sensitivity for ac voltage is made lower, generally, to prevent erratic meter deflection produced by stray magnetic fields before the meter leads are connected into the circuit. Usually the ohms-per-volt rating of a voltmeter is printed on the meter face.

The sensitivity of 1000 Ω/V with a 1-mA movement used to be common for dc voltmeters, but 20,000 Ω/V with a 50-μA movement is generally used now. Higher sensitivity is an advantage, not only for less voltmeter loading, but lower voltage ranges and higher ohmmeter ranges can be obtained.

Practice Problems 7-5
(answers on page 158)
Refer to Fig. 7-13 to calculate the voltmeter resistance R_V on
(a) 2.5-V range.
(b) 50-V range.

7-6
LOADING EFFECT OF A VOLTMETER

When the voltmeter resistance is not high enough, connecting it across a circuit can reduce the measured voltage, compared with the voltage present without the voltmeter. This effect is called *loading down* the circuit, since the measured voltage decreases because of the additional load current for the meter.

Loading Effect. This loading effect can be appreciable in high-resistance circuits, as shown in Fig. 7-14. In (a), without the voltmeter, R_1 and R_2 form a voltage divider across the applied voltage of 120 V. The two equal resistances of 100 kΩ each divide the applied voltage equally, with 60 V across each.

When the voltmeter in (b) is connected across R_2 to measure its potential difference, however, the voltage division changes. The

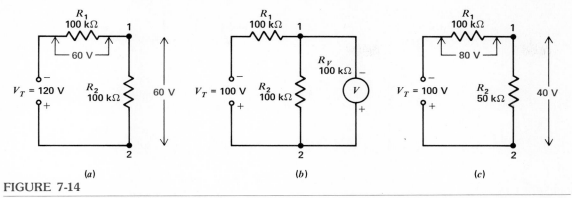

FIGURE 7-14
Loading effect of a voltmeter. (a) High-resistance series circuit. (b) Voltmeter across one of the series resistances. (c) Reduced resistance and voltage between points 1 and 2 caused by the voltmeter resistance as a parallel branch.

voltmeter resistance R_V of 100 kΩ is the value for a 1000 ohms-per-volt meter on the 100-V range. Now the voltmeter in parallel with R_2 draws additional current and the equivalent resistance between the measured points 1 and 2 is reduced from 100,000 to 50,000 Ω. This resistance is one-third the total circuit resistance, and the measured voltage across points 1 and 2 drops to 40 V, as shown in (c).

As additional current drawn by the voltmeter flows through the other series resistance R_1, this voltage goes up to 80 V.

Similarly, if the voltmeter were connected across R_1, this voltage would go down to 40 V, with the voltage across R_2 rising to 80 V. When the voltmeter is disconnected, the circuit returns to the condition in (a) with 60 V across both R_1 and R_2.

The loading effect is minimized by using a voltmeter having a resistance much greater than the resistance across which the voltage is measured. As shown in Fig. 7-15, with a voltmeter resistance of 11 MΩ, its loading effect is negligible. Because R_V is so high, it does not change the voltage division in the circuit. The 11 MΩ of the meter in parallel with the 100,000 Ω for R_2 results in an equivalent resistance practically equal to 100,000 Ω.

With multiple ranges on a VOM, the voltmeter resistance changes with the range selected. Higher ranges require more multiplier resistance, increasing the voltmeter resistance for less loading. As examples, a 20,000 ohms-per-volt meter on the 250-V range has an internal resistance R_V of 20,000 × 250, or 5 MΩ. However, on the 2.5-V range the same meter has an R_V of 20,000 × 2.5, which is only 50,000 Ω.

On any one range, though, the voltmeter resistance is constant whether you read full-scale or less than full-scale deflection. The reason is that the multiplier resistance set by the range switch is the same for any reading on that range.

Correction for Loading Effect. The following formula can be used:

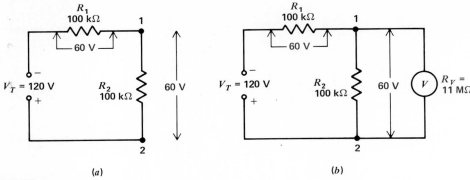

FIGURE 7-15
Negligible loading with a high-resistance voltmeter. (a) High-resistance series circuit. (b) Same voltages in circuit with voltmeter connected.

Actual reading + Correction
$$V = V_M + \frac{R_1 R_2}{R_V(R_1 + R_2)} V_M \quad (7\text{-}4)$$

V is the corrected reading the voltmeter would show if it had infinitely high resistance. V_M is the actual voltage reading. R_1 and R_2 are the voltage-dividing resistances in the circuit, without the voltmeter resistance R_V. As an example, in Fig. 7-14,

$$V = 40\text{ V} + \frac{100\text{ k}\Omega \times 100\text{ k}\Omega}{100\text{ k}\Omega \times (200\text{ k}\Omega)} \times 40\text{ V}$$
$$= 40 + \left(\frac{1}{2}\right) \times 40$$
$$= 40 + 20$$
$$V = 60\text{ V}$$

The loading effect of a voltmeter reading too low because R_V is too low as a parallel resistance corresponds to the case of a current meter reading too low because r_M is too high as a series resistance. Both of these effects illustrate the general problem of trying to make any measurement without changing the circuit being measured.

Practice Problems 7-6
(answers on page 158)
With the voltmeter across R_2 in Fig. 7-14, what are the values for
(a) V_1
(b) V_2

7-7
OHMMETERS
Basically, an ohmmeter consists of an internal battery, the meter movement, and a current-limiting resistance, as illustrated in Fig. 7-16. For measuring resistance, the ohmmeter leads are connected across an external resistance to be measured, with power off in the circuit being tested. Then only the ohmmeter battery produces current for deflecting the meter movement. Since the amount of current through the meter depends on the external resistance, the scale can be calibrated in ohms.

Chapter 7
Direct-Current Meters 143

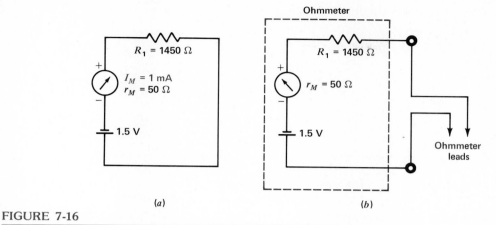

FIGURE 7-16
Ohmmeter circuit. (*a*) Equivalent closed circuit when ohmmeter leads are short-circuited for zero ohms of external resistance. (*b*) Circuit with ohmmeter leads open.

The amount of deflection on the ohms scale indicates directly the measured resistance. The ohmmeter reads up-scale regardless of the polarity of the leads because the polarity of the internal battery determines the direction of current through the meter movement.

Series-Ohmmeter Circuit. In Fig. 7-16*a*, the circuit has 1500 Ω for ($R_1 + r_M$). Then the 1.5-V cell produces 1 mA, deflecting the moving coil full scale. When these components are enclosed in a case, as in (*b*), the series circuit forms an ohmmeter.

If the leads are short-circuited together or connected across a short circuit, 1 mA flows. The meter movement is deflected full scale to the right. This ohmmeter reading is 0 Ω.

With the ohmmeter leads open, not touching each other, the current is zero. The ohmmeter indicates infinitely high resistance or an open circuit across its terminals.

Therefore, the meter face can be marked zero ohms at the right for full-scale deflection and infinite ohms at the left for no deflection. In-between values of resistance result in less than 1 mA through the meter movement. The corresponding deflection on the ohms scale indicates how much resistance is across the ohmmeter terminals.

Back-off Ohmmeter Scale. Table 7-3 and Fig. 7-17 illustrate the calibration of an ohmmeter scale in terms of meter current. The current equals V/R_T. V is the fixed applied voltage of 1.5 V supplied by the internal battery. R_T is the total resistance of R_x and the ohmmeter's internal resistance. Note that R_x is the external resistance to be measured.

The ohmmeter's internal resistance R_i is constant at 50 + 1450, or 1500 Ω here. If R_x also equals 1500 Ω, for example, R_T equals 3000 Ω. The current then is 1.5 V/3000 Ω, or 0.5 mA, resulting in half-scale deflection for the 1-mA movement. Therefore, the center of the ohms scale is marked for 1500 Ω. Similarly, the amount of current and meter deflection can be

TABLE 7-3. Calibration of Ohmmeter in Fig. 7-17

EXTERNAL R_x, Ω	INTERNAL $R_i = R_1 + r_M$, Ω	$R_T = R_x + R_i$, Ω	$I = V/R_T$, mA	DEFLECTION	SCALE READING, Ω
0	1500	1500	1	Full scale	0
750	1500	2250	⅔ = 0.67	⅔ scale	750
1500	1500	3000	½ = 0.5	½ scale	1500
3000	1500	4500	⅓ = 0.33	⅓ scale	3000
150,000	1500	151,500	0.01	1/100 scale	150,000
500,000	1500	501,500	0	None	∞

calculated for any value of the external resistance R_x.

Note that the ohms scale increases from right to left. This arrangement is called a *back-off scale*, with ohms values increasing to the left as the current backs off from full-scale deflection. The back-off scale is a characteristic of any ohmmeter where the internal battery is in series with the meter movement. Then more external R_x decreases the meter current.

A back-off ohmmeter scale is expanded at the right near zero ohms and crowded at the left near infinite ohms. This nonlinear scale results from the relation of $I = V/R$ with V constant at 1.5 V. Specifically, the back-off ohms scale represents the graph of a hyperbolic curve for the reciprocal function $y = 1/x$.

The highest resistance that can be indicated by the ohmmeter is about 100 times its total internal resistance. Therefore, the infinity mark on the ohms scale, or the "lazy eight" symbol ∞ for infinity, is only relative. It just means that the measured resistance is infinitely greater than the ohmmeter resistance.

For instance, if a 500,000-Ω resistor in good condition were measured with the ohmmeter in Fig. 7-17, it would indicate infinite resistance because this ohmmeter cannot measure as high as 500,000 Ω. To read higher values of resistance, the battery voltage can be

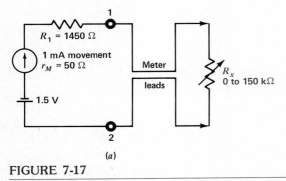

(a)

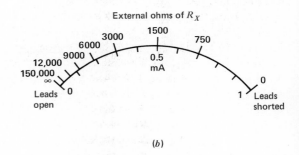

(b)

FIGURE 7-17

Back-off ohmmeter. (*a*) Series ohmmeter circuit with an external resistor R_x. (*b*) Ohms scale reads higher resistances from right to left, as more R_x decreases I_M (see Table 7-3).

increased to provide more current, or a more sensitive meter movement is necessary to provide deflection with less current.

Multiple Ohmmeter Ranges. Commercial multimeters provide for resistance measurements from less than 1 Ω up to many megohms, in several ranges. The range switch in Fig. 7-18 shows the multiplying factors for the ohms scale. On the $R \times 1$ range, for low-resistance measurements, read the ohms scale directly. In the example here, the pointer indicates 12 Ω. When the range switch is on $R \times 100$, multiply the scale reading by 100; this reading would then be 12×100 or 1200 Ω. On the $R \times 10,000$ range, the pointer would indicate 120,000 Ω.

A multiplying factor is given, instead of full-scale resistance, for each ohms range because the highest resistance is infinite on all the ohms ranges. This method for ohms should not be confused with the full-scale values for voltage ranges. For the ohmmeter ranges, always multiply the scale reading by the $R \times$ factor. On voltage ranges, you may have to multiply or divide the scale reading to match the full-scale voltage with the value on the range switch.

Typical Ohmmeter Circuit. For high-ohms ranges a sensitive meter is necessary to read the low values of I with the high values of R_x. For the case of low ohms, however, less sensitivity is needed for the higher currents. These opposite requirements are solved by using a meter shunt across the meter movement and changing the shunt resistance for the multiple ohmmeter ranges. In Fig. 7-19, R_S is the meter shunt.

To analyze the ohmmeter circuit in Fig. 7-19, three conditions are shown. All are for the $R \times 1$ range with 12-Ω R_S. Figure 7-19a shows the internal circuit, before the ohmmeter is adjusted for zero ohms. In (b) the test leads are short-circuited. Then there are two paths for branch current produced by the battery V. One branch is R_S. The other branch includes R_b, R_A, and the meter movement. The 1.5 V is across both branches.

To allow 50 μA through the meter, R_A is adjusted to 5000 Ω. Then the total resistance in this branch is 23 kΩ + 5 kΩ + 2 kΩ, which equals 30 kΩ. With 30 kΩ across 1.5 V, I_M equals 50 μA. Therefore, R_A is adjusted for full-scale deflection to read zero ohms with the test leads short-circuited.

In (c), assume a resistance R_x being measured is 12 Ω, equal to R_S. Then the meter current is practically 25 μA for half-scale deflection. The center ohms reading on the $R \times 1$ scale, therefore, is 12 Ω. For higher values of R_x, the meter current decreases to indicate higher resistances on the back-off ohms scale.

For higher ohms ranges, the resistance of the R_S branch is increased. The half-scale ohms reading on each range is equal to the resistance of the R_S branch. A higher battery voltage can also be used for the highest ohms range.

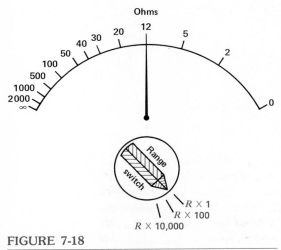

FIGURE 7-18

Multiple ohmmeter ranges. Multiply reading by the factor set on range switch.

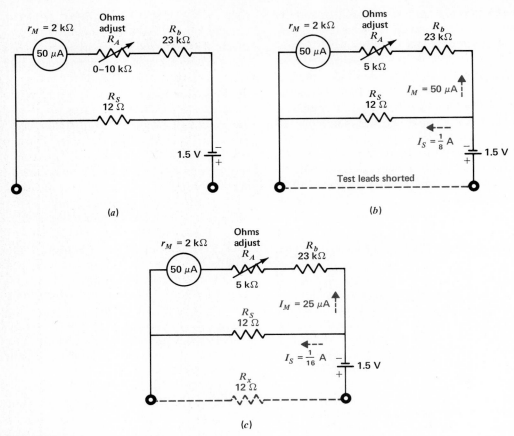

FIGURE 7-19

Typical ohmmeter circuit for $R \times 1$ range. (a) Circuit before zero-ohms adjustment. (b) Test leads short-circuited to adjust for zero ohms. (c) Measuring external resistance R_x. The value of 12 Ω for R_x results in half-scale deflection, as shown on the ohms scale in Fig. 7-18.

On any range, R_A is adjusted for full-scale deflection to read zero ohms with the test leads short-circuited. This variable resistor is the *ohms adjust* or *zero-ohms* adjustment.

Zero-Ohms Adjustment. To compensate for lower voltage output as the internal battery ages, an ohmmeter includes a variable resistor such as R_A in Fig. 7-19, to calibrate the ohms scale. A back-off ohmmeter is always adjusted for zero ohms. With the test leads short-circuited, vary the ZERO OHMS control on the front panel of the meter until the pointer is exactly on zero at the right edge of the ohms scale. Then the ohms readings are correct for the entire scale.

This type of ohmmeter must be zeroed again every time you change the range. The reason is that the internal circuit is changed.

When the adjustment cannot deflect the pointer all the way to zero at the right edge, it usually means the battery voltage is too low and the internal dry cells must be replaced. Usually, this trouble shows up first on the $R \times 1$ range, which takes the most current from the battery. The ohmmeter battery in a typical VOM can be seen in Fig. 7-20.

Shunt-Ohmmeter Circuit. In this circuit, the internal battery, meter movement, and external R_x are in three parallel paths. The main advantage is a low-ohms scale that reads from left to right. However, the shunt ohmmeter circuit is seldom used because of constant current drain on the internal battery.

Characteristics of the Series Ohmmeter. The main features are:

1. The circuit includes an internal battery and a variable ohms-adjust control R_A in series with the meter movement.
2. The back-off scale has zero ohms at the right edge, corresponding to a full-scale I_M, and infinite ohms at the left, corresponding to a zero I_M.
3. Short-circuit the ohmmeter leads and adjust R_A for zero ohms. This zero adjustment must be changed for each ohmmeter range, as the value of R_A required for the full-scale I_M is different.
4. Lower ohms ranges are obtained by decreasing the meter sensitivity with a shunt. Higher ohms ranges are obtained by increasing the meter sensitivity and using a higher battery voltage.

FIGURE 7-20

Typical VOM with back cover off to show shunts, multipliers, and batteries for ohmmeter. Separate 7.5-V cell is for high-ohms ranges. (*Triplett Model 630*)

Practice Problems 7-7
(answers on page 158)
(a) The ohmmeter reads 40 Ω on the $R \times 10$ range. How much is R_x?
(b) A voltmeter reads 40 V on the 300-V scale, but with the range switch on 30 V. How much is the measured voltage?

7-8
MULTIMETERS

The main types are the volt-ohm-milliammeter (VOM) in Fig. 7-21, vacuum-tube voltmeter (VTVM) in Fig. 7-22, and the digital multimeter

FIGURE 7-21

Front view of typical VOM. Height is 7 in. Note switch positions at left side for clamp-on ammeter and at top for low-power ohms. (*Simpson Model 260*)

(DMM) in Fig. 7-23. Table 7-4 compares the features of the VOM and VTVM, while more details of the DMM are described in the next section. Actually, the use of a digital meter is similar to a VTVM, but the digital display is much easier to read without mistakes.

The VOM is probably the most common multimeter. It is simple, compact, and portable because the only power it uses is the battery for the ohmmeter.

The VTVM needs an internal power supply because the meter movement is in a bridge circuit with an amplifier. Transistors can be used instead of tubes, but the unit must be plugged into the ac power line for operation or a battery must be used for the power supply.

The main advantage of the VTVM is its high input resistance, usually 11 MΩ on all dc

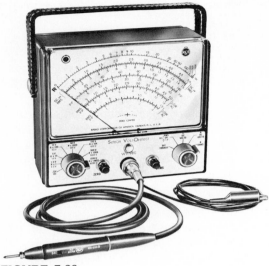

FIGURE 7-22

Typical VTVM. Height is 7 in. Range switch is at left and selector switch at right. Zero-adjust control at left is for voltmeter; ohms-adjust control is at right. (*RCA Model WV-98 C*)

voltage ranges. In transistor circuits, though, the resistances to be measured are usually in the kΩ range or less. The VTVM also has very high ohms ranges, since the voltmeter circuit is used with an internal battery for the ohmmeter function. The VTVM cannot be used as a current meter because of its high input resistance.

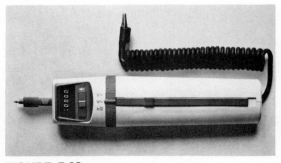

FIGURE 7-23

Digital volt-ohmmeter. (*Hewlett Packard*)

TABLE 7-4. Comparison of VOM and VTVM

VOM	VTVM
Dc voltmeter R_V changes with range. $R_V = \Omega/V \times$ full-scale voltage	Very high dc voltmeter R_V of 11 MΩ or more, the same on all ranges
Measures ac volts	Measures ac volts
Measures direct current	Does not measure direct current
Does not measure alternating current	Does not measure alternating current
Ohms ranges are usually up to $R \times 10,000\ \Omega$	Ohms ranges are usually up to $R \times 1$ MΩ
Zero-ohms adjustment must be changed for each range	Ohms adjustment is the same for all ranges
Portable; black lead need not be connected to chassis ground	Usually plugged into the power line; black lead must be at chassis ground

The VOM measures direct current but not alternating current. In general, measuring alternating current requires an ac meter movement or a clamp-on accessory probe.

All the multimeters measure ac voltage easily. The reason is that the ac voltage to be measured is rectified to provide dc voltage for the meter.

For either a VOM or VTVM it is important to have a low-voltage scale of 1 V or less, full-scale deflection. This range is needed for measuring dc bias voltages of 0.2 to 0.6 V in transistor circuits.

Low-Power Ohms. Another good feature for transistor circuits is an ohmmeter that does not have enough battery voltage to bias a semiconductor junction into the ON or conducting condition. The limit is 0.2 V or less. For the VOM in Fig. 7-21, the open-circuit voltage is 0.1 V on the low-power ohms ranges. These are also low-ohms ranges of $R \times 1$ and $R \times 10$. The high-ohms ranges need more battery voltage.

Decibel Scale. Most multimeters have an ac voltage scale calibrated in decibels (dB), for measuring ac signals. The decibel is a unit for comparisons of power levels or voltage levels. The mark of 0 dB on the scale indicates the reference level, which is usually 0.775 V for 1 mW across 600 Ω. Positive dB values above the zero mark indicate ac voltages more than 0.775 V; negative dB values are less than the reference level.

Amp-Clamp Probe. The problem of opening a circuit to measure I can be eliminated by using a probe with a clamp that fits around the current-carrying wire. Its magnetic field is used to indicate the current. The clamp in Fig. 7-24 is an accessory probe for the VOM in Fig. 7-21.

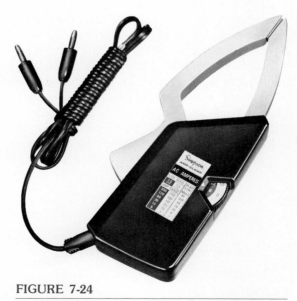

FIGURE 7-24

Amp-clamp accessory for VOM in Fig. 7-21. (*Simpson Electric Company*)

This probe measures just ac amperes, generally for the 60-Hz power line.

High-Voltage Probe. The accessory probe in Fig. 7-25 can be used with a multimeter to measure dc voltages up to 30 kV. One application is measuring the anode voltage of 20 to 30 kV for the picture tube in a television receiver. The probe basically is just an external multiplier resistance for the voltmeter. The required R for a 30-kV probe is 580 MΩ with a 20-kΩ/V meter on the 1000-V range.

Practice Problems 7-8
(answers on page 158)
(a) Which type of multimeter is simple, portable, and has the lowest cost?
(b) How much is the typical input R of a VTVM on all dc voltage ranges?

7-9
DIGITAL METERS

These have become popular because the digital readout is displayed automatically with decimal point, polarity, and units (Fig. 7-23). The conventional meter with a pointer is an *analog* meter; that is, the deflection corresponds directly to the measured quantity. Digital meters are much easier to use, mainly because they reduce the human error that often occurs in reading the different scales on an analog meter. The digital VOM in Fig. 7-23 is a small, portable

FIGURE 7-25

High-voltage probe accessory for VOM. (*Triplett Corp.*)

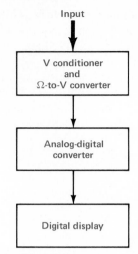

FIGURE 7-26

Basic functions in a digital multimeter for volts and ohms.

unit, but most digital meters are larger than analog meters.

The functional blocks in a digital VOM are shown in Fig. 7-26. First the input must be put into a form that fits the needs of the analog-digital (A-D) converter. As an example, the dc voltage range of 0 to 1 V may be required for the digital conversion circuit. When the input is ac voltage, it is converted to dc voltage. For measuring resistance, a direct current is supplied that converts R to an IR voltage.

When the dc voltage is too high, it is divided down for 0 to 1 V. When the dc voltage is too low, it is increased by a dc amplifier. This control voltage can then be compared with a fixed reference voltage in the meter. The input resistance for the dc voltage range is 10 MΩ.

Examples of the digital readout are -4.92 or $+4.92$ V for dc voltage, 250.0-V ac for ac voltage, and 397 Ω for the ohmmeter. How many "9s" can be displayed specifies the number of digits for the DVM. The display generally uses light-emitting diodes (LED).

The A-D converter has a clock generator which produces timing pulses and a circuit which counts the pulses. A special digital logic IC chip is the heart of the instrument. The IC input includes timing pulses from the clock and voltage from a circuit comparing the meter input with the reference voltage. The output of the IC chip drives the digital display unit.

Practice Problems 7-9
(*answers on page 158*)
Answer true or false.
(a) In the digital VOM and analog VTVM, R is converted to IR voltage for ohms measurements.
(b) An A-D converter has a clock circuit to generate timing pulses.

7-10
METER APPLICATIONS

Table 7-5 summarizes the main points to remember in using a voltmeter, ohmmeter, or milliammeter. These rules apply whether the meter is a single unit or one function on a multimeter. Also, the voltage and current tests apply to either dc or ac circuits.

To avoid excessive current through the meter, it is good practice to start on a high range when measuring an unknown value of voltage or current. It is very important not to make the mistake of connecting a current meter in parallel, because usually this mistake ruins

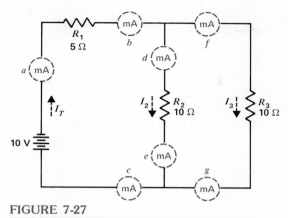

FIGURE 7-27

Inserting a current meter in a series-parallel circuit. At a, b, or c, meter reads I_T; at d or e, meter reads I_2; at f or g, meter reads I_3.

the meter. The mistake of connecting a voltmeter in series does not damage the meter, but the reading is wrong.

If the ohmmeter is connected to a circuit where power is on, the meter can be damaged, besides giving the wrong reading. An ohmmeter has its own internal battery, and the power must be off in the circuit being tested.

Connecting a Current Meter in the Circuit. In a series-parallel circuit, the current meter must be inserted in a branch to read branch current. In the main line, the meter reads the total current. These different connections are illustrated in Fig. 7-27. The meters are shown

TABLE 7-5. Direct-Current Meters

VOLTMETER	MILLIAMMETER OR AMMETER	OHMMETER
Power on in circuit	Power on in circuit	Power off in circuit
Connect in parallel	Connect in series	Connect in parallel
High internal R	Low internal R	Has internal battery
Has internal series multipliers; higher R for higher ranges	Has internal shunts; lower resistance for higher current ranges	Higher battery voltage and more sensitive meter for higher ohms ranges

by dashed lines to illustrate the different connections where one could be connected to read the respective currents.

If the circuit is opened at point *a* to insert the meter in series in the main line here, it will read total line current I_T through R_1. A meter at *b* or *c* will read the same line current.

In order to read the branch current through R_2, this R must be disconnected from its junction with the main line at either end. A meter inserted at *d* or *e*, therefore, will read the R_2 branch current I_2. Similarly, a meter at *f* or *g* will read the R_3 branch current I_3.

Calculating *I* from Measured Voltage. The inconvenience of opening the circuit to measure current can often be eliminated by the use of Ohm's law. The voltage and resistance can be measured without opening the circuit and the current calculated as V/R. In the example in Fig. 7-28, when the voltage across R_2 is 15 V and its resistance is 15 Ω, the current through R_2 must be 1 A. When values are checked during troubleshooting, if the voltage and resistance are normal, so is the current.

This technique can also be convenient for determining *I* in low-resistance circuits where the resistance of a microammeter may be too high. Instead of measuring *I*, measure *V* and *R* to calculate *I* as V/R.

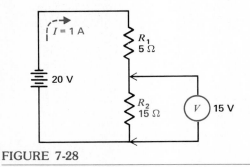

FIGURE 7-28

With 15 V measured across 15 Ω, the current *I* is $V/R = 1$ A.

Furthermore, if necessary, we can insert a known resistance R_S in series in the circuit, temporarily, just to measure V_S. Then *I* is calculated as V_S/R_S. The resistance of R_S, however, must be small enough to have little effect on R_T and *I* in the series circuit.

This technique is often used with oscilloscopes to produce a voltage waveform of *IR* which has the same waveform as the current in a resistor. The oscilloscope must be connected as a voltmeter because of its high input resistance.

Checking Fuses. Turn the power off or remove the fuse from the circuit to check with an ohmmeter. A good fuse reads 0 Ω. A blown fuse is open, which reads infinity on the ohmmeter.

A fuse can also be checked with the power on in the circuit by using a voltmeter. Connect the voltmeter across the two terminals of the fuse. A good fuse reads 0 V because there is practically no *IR* drop. With an open fuse, though, the voltmeter reading is equal to the full value of the applied voltage. Having the full applied voltage seems to be a good idea, but it should not be across the fuse.

Voltage Tests for an Open Circuit. Figure 7-29 shows four equal resistors in series with a 100-V source. A ground return is shown here because voltage measurements are usually made to chassis ground. Normally, each resistor would have an *IR* drop of 25 V. Then, at point *B* the voltmeter to ground should read $100 - 25 = 75$ V. Also, the voltage at *C* should be 50 V, with 25 V at *D*.

However, the circuit is shown with an open in R_3, toward the end of the series string of voltages to ground. Now, when you measure at *B*, the reading is 100 V, equal to the applied voltage. This full voltage at *B* shows the series circuit is open without any *IR* drop across R_1. The question is, however, which *R* has the

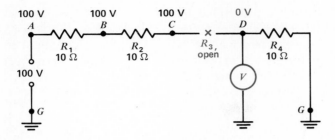

FIGURE 7-29

Voltage tests to localize an open circuit. R_3 is normally 10 Ω. Reading of 0 V at point D shows R_3 is open.

open? Continue the voltage measurements to ground until you find 0 V. In this example, the open is in R_3, between the 100 V at C and 0 V at D.

The points that read the full applied voltage have a path back to the source of voltage. The first point that reads 0 V has no path back to the high side of the source. Therefore, the open circuit must be between points C and D in Fig. 7-29.

Practice Problems 7-10
(answers on page 158)
(a) Which type of meter requires an internal battery?
(b) How much is the normal voltage across a good fuse?

7-11
CHECKING CONTINUITY WITH THE OHMMETER

A wire conductor that is continuous without a break has practically zero ohms of resistance. Therefore, the ohmmeter can be useful in testing for continuity. This test should be done on the lowest ohms range. There are many applications. A wire conductor can have an internal break, which is not visible because of the insulated cover, or the wire can have a bad connection at the terminal. Checking for zero ohms between any two points along the conductor tests continuity. A break in the conducting path is evident from a reading of infinite resistance, showing an open circuit.

As another application of checking continuity, suppose there is a cable of wires harnessed together as illustrated in Fig. 7-30, where the individual wires cannot be seen, but it is desired to find the conductor that connects to terminal A. This is done by checking continuity for each conductor to point A. The wire that has zero ohms to A is the one connected to this terminal. Often the individual wires are color-coded, but it may be necessary to check the continuity of each lead.

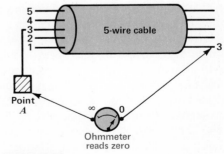

FIGURE 7-30

Continuity from A to 3 shows this wire is connected to terminal A.

An additional technique that can be helpful is illustrated in Fig. 7-31. Here it is desired to check the continuity of the two-wire line, but its ends are too far apart for the ohmmeter leads to reach. The two conductors are temporarily short-circuited at one end, however, so that the continuity of both wires can be checked at the other end.

In summary, then, the ohmmeter is helpful in checking the continuity of any wire conductor. This check includes resistance-wire heating elements, like the wires in a toaster or the filament of an incandescent bulb. Their cold resistance is normally just a few ohms. Infinite resistance means that the wire element is open. Similarly, a good fuse has practically zero resistance; a burned-out fuse has infinite resistance, meaning it is open. Also, any coil for transformers, solenoids, and motors will have infinite resistance if the winding is open.

Practice Problems 7-11
(answers on page 158)
(a) On a back-off ohmmeter, is zero ohms at the left or right edge?
(b) What is the ohmmeter reading for an open circuit?

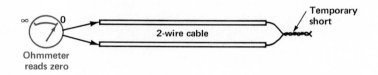

FIGURE 7-31

Temporary short circuit at one end of a long two-wire line to check continuity from opposite end.

Summary

1. Direct current in a moving-coil meter deflects the coil in proportion to the amount of current.
2. A current meter is a low-resistance meter connected in series to read the amount of current in the circuit.
3. A meter shunt R_S in parallel with the meter movement extends the range of a current meter [see Formula (7-1)].
4. A voltmeter consists of the meter movement in series with a high-resistance multiplier. The voltmeter with its multiplier is connected across two points to measure their potential difference in volts. The multiplier R can be calculated from Formula (7-3).
5. The ohms-per-volt rating of a voltmeter with series multipliers specifies the sensitivity on all voltage ranges. It equals the reciprocal of the full-scale deflection current of the meter. A typical value is 20,000 Ω/V for a voltmeter using a 50-μA movement. The higher the ohms-per-volt rating, the better.

6. Voltmeter resistance R_V is higher for higher ranges because of higher-resistance multipliers. Multiply the ohms-per-volt rating by the voltage range to calculate R_V.
7. An ohmmeter consists of an internal battery in series with the meter movement. Power must be off in the circuit being checked with an ohmmeter. The series ohmmeter has a back-off scale with zero ohms at the right edge and infinity at the left. Adjust for zero ohms with the leads short-circuited each time the ohms range is changed.
8. The VOM (Fig. 7-21) is a portable multimeter to measure volts, ohms, and milliamperes.
9. The VTVM (Fig. 7-22) is a multimeter with a high R of 11 MΩ for all dc voltage ranges, resulting in practically no loading effect. Also, its R ranges allow readings up to 1000 MΩ.
10. The digital VOM (Fig. 7-25) generally has an input resistance of 10 MΩ on all dc voltage ranges.
11. In checking wire conductors, the ohmmeter reads 0 Ω or very low R for normal continuity and infinite ohms for an open.

Self-Examination (Answers at back of book.)

Choose (a), (b), (c), or (d).

1. To connect a current meter in series: (a) open the circuit at one point and use the meter to complete the circuit; (b) open the circuit at the positive and negative terminals of the voltage source; (c) short-circuit the resistance to be checked and connect the meter across it; (d) open the circuit at one point and connect the meter to one end.
2. To connect a voltmeter in parallel to read an IR drop: (a) open the circuit at one end and use the meter to complete the circuit; (b) open the circuit at two points and connect the meter across both points; (c) allow the circuit to remain as is and connect the meter across the resistance; (d) allow the circuit to remain closed but disconnect the voltage source.
3. A shunt for a milliammeter (a) extends the range and reduces the meter resistance; (b) extends the range and increases the meter resistance; (c) decreases the range and the meter resistance; (d) decreases the range but increases the meter resistance.
4. For a 50-μA movement with 2000-Ω r_M, its voltage V_M at full-scale deflection is (a) 0.1 V; (b) 0.2 V; (c) 0.5 V; (d) 250 μV.
5. A voltmeter using a 20-μA meter movement has a sensitivity of (a) 1000 Ω/V; (b) 20,000 Ω/V; (c) 50,000 Ω/V; (d) 11 MΩ/V.

6. When using an ohmmeter, disconnect the applied voltage from the circuit being checked because: (a) the voltage source will increase the resistance; (b) the current will decrease the resistance; (c) the ohmmeter has its own internal battery; (d) no current is needed for the meter movement.
7. A multiplier for a voltmeter is (a) a high resistance in series with the meter movement; (b) a high resistance in parallel with the meter movement; (c) usually less than 1 Ω in series with the meter movement; (d) usually less than 1 Ω in parallel with the meter movement.
8. To double the current range of a 50-µA 2000-Ω meter movement, the shunt resistance is (a) 40 Ω; (b) 50 Ω; (c) 2000 Ω; (d) 18,000 Ω.
9. With a 50-µA movement, a VOM has an input resistance of 6 MΩ on the dc voltage range of (a) 3; (b) 12; (c) 60; (d) 300.
10. For a 1-V range, a 50-µA movement with an internal R of 2000 Ω needs a multiplier resistance of (a) 1 kΩ; (b) 3 kΩ; (c) 18 kΩ; (d) 50 kΩ.

Essay Questions

1. (a) Why is a milliammeter connected in series in a circuit? (b) Why should the milliammeter have low resistance?
2. (a) Why is a voltmeter connected in parallel in a circuit? (b) Why should the voltmeter have high resistance?
3. A circuit has a battery across two resistances in series. (a) Draw a diagram showing how to connect a milliammeter in the correct polarity to read current through the junction of the two resistances. (b) Draw a diagram showing how to connect a voltmeter in the correct polarity to read the voltage across one resistance.
4. Explain briefly why a meter shunt equal to the resistance of the moving coil doubles the current range.
5. Describe how to adjust the ZERO OHMS control on a back-off ohmmeter.
6. Describe how to set the ZERO VOLTS and OHMS ADJUST controls on a VTVM.
7. Give two advantages of the digital VOM in Fig. 7-23 compared with the conventional VOM in Fig. 7-21.
8. What is the function of the ZERO OHMS control in the circuit of a back-off ohmmeter?
9. State two precautions to be observed when you use a milliammeter.
10. State two precautions to be observed when you use an ohmmeter.
11. The resistance of a voltmeter R_V is 300 kΩ on the 300-V range when

measuring 300 V. Why is R_V still 300 kΩ when measuring 250 V on the same range?

12. Redraw the schematic diagram in Fig. 5-1b, in Chap. 5, showing a milliammeter to read line current through R_1 and R_2, a meter for R_3 branch current, and a meter for R_4 branch current. Label polarities on each meter.

Problems (Answers to odd-numbered problems at back of book.)

1. Calculate the shunt resistance needed to extend the range of a 50-Ω 1-mA movement to (a) 2 mA; (b) 10 mA; (c) 100 mA; (d) In each case, how much current is indicated by half-scale deflection?
2. With a 50-Ω 1-mA movement, calculate the multiplier resistances needed for ranges of (a) 10 V; (b) 30 V; (c) 100 V; (d) 300 V. How much voltage is indicated by half-scale deflection for each range?
3. A voltmeter reads 30 V across a 100-Ω resistance. (a) How much is the current in the resistor? (b) If the current through the same resistance were doubled, how much would its IR voltage be?
4. A voltmeter has a sensitivity of 10,000 Ω/V on all ranges. (a) How much is the total voltmeter resistance on the 5-V range? (b) On the 50-V range? (c) On the 500-V range? (d) How much is the voltmeter resistance for a reading of 225 V on the 500-V range?
5. A 50-μA meter movement has an internal resistance of 1000 Ω. (a) Calculate the multiplier resistance needed for voltmeter ranges of 10, 30, and 500 V. (b) How much is the ohms-per-volt sensitivity rating on all ranges? (c) How much is the voltmeter resistance on the 500-V range?
6. For the same meter movement as in Prob. 5, calculate the shunt resistances needed for current ranges of 10, 30, and 500 mA. How much is the resistance of the meter with its shunt on each range? (Note: 1 mA = 1000 μA.)
7. Referring to the universal shunt in Fig. 7-9, calculate the required values of R_1, R_2, and R_3 for a 50-μA, 2000-Ω movement to provide current ranges of 1.2, 12, and 120 mA.
8. Referring to the voltmeter loading problem in Fig. 7-14, exactly how much voltage would be indicated by a 20,000 ohms-per-volt meter on its 100-V range?
9. Refer to the ohmmeter in Fig. 7-17. Assume that the movement is shunted to become a 10-mA meter. (a) Calculate the value of R_1 that would be required for full-scale deflection with the ohmmeter leads

short-circuited. (b) How much would the half-scale reading be on the ohms scale?
10. Refer to Fig. 7-13. (a) How much is the total voltmeter resistance using the 5000-V jack with the range switch on the 1000-V position? (b) How much is the ohms-per-volt sensitivity? (c) Why must the range switch be on the 1000-V position?
11. In Fig. 7-14, if the voltmeter is connected across R_1 instead of R_2, what will the values be for V_1 and V_2?

Answers to Practice Problems

7-1   (a) Parallel
      (b) Series
7-2   (a) $I = 100$ mA
      (b) R_3
      (c) Low
      (d) High
7-3   (a) $I_S = 450$ μA
      (b) $V_M = 0.045$ V
7-4   (a) $I_M = 1$ mA
      (b) $I_M = 1$ mA
7-5   (a) $R_V = 50$ kΩ
      (b) $R_V = 1$ MΩ

7-6   (a) $V_1 = 80$ V
      (b) $V_2 = 40$ V
7-7   (a) $R_x = 400$ Ω
      (b) $V = 4$ V
7-8   (a) VOM
      (b) 11 MΩ
7-9   (a) True
      (b) True
7-10   (a) Ohmmeter
      (b) 0 V
7-11   (a) Right edge
      (b) ∞ ohms

Review of Chapters 6 and 7

Summary

1. In a series voltage divider the IR drop across each resistance is proportional to its R. A larger R has a larger voltage drop. Each $V = (R/R_T) \times V_T$. In this way, the series voltage drops can be calculated from V_T without I.
2. In a parallel current divider, each branch current is inversely proportional to its R. A smaller R has more branch current. For only two resistances, we can use the inverse relation $I_1 = [R_2/(R_1 + R_2)] \times I_T$. In this way, the parallel branch currents can be calculated from I_T without V.
3. In a parallel current divider, each branch current is directly proportional to its conductance G. A larger G has more branch current. For any number of parallel resistances, each branch $I = (G/G_T) \times I_T$.
4. A milliammeter or ammeter is a low-resistance meter connected in series in a circuit to measure current.
5. Different current ranges are obtained by meter shunts in parallel with the meter.
6. A voltmeter is a high-resistance meter connected across the voltage to be measured.
7. Different voltage ranges are obtained by multipliers in series with the meter.
8. An ohmmeter has an internal battery to indicate the resistance of a component across its two terminals, with external power off.
9. In making resistance tests, remember $R = 0\ \Omega$ for continuity or a short circuit, but the resistance of an open circuit is infinite.

Review Self-Examination (Answers at back of book.)

Answer true or false.

1. The internal R of a milliammeter must be low to have minimum effect on I in the circuit.
2. The internal R of a voltmeter must be high to have minimum current through the meter.
3. Power must be off when checking resistance in a circuit because the ohmmeter has its own internal battery.
4. In the series voltage divider in Fig. 7-29, the normal voltage from point B to ground is 75 V.
5. In Fig. 7-29, the normal voltage across R_1, between A and B, is 75 V.
6. The highest ohms range is best for checking continuity with an ohmmeter.
7. With four equal resistors in a series voltage divider with V_T of 44.4 V, each IR drop is 11.1 V.
8. With four equal resistors in parallel with I_T of 44.4 mA, each branch current is 11.1 mA.
9. Series voltage drops divide V_T in direct proportion to each series R.
10. Parallel currents divide I_T in direct proportion to each branch R.

References (Additional references at back of book.)

Herrick, C. N.: "Instruments and Measurements for Electronics," McGraw-Hill Book Company, New York.

Prensky, S. D.: "Electronic Instrumentation," Prentice-Hall, Inc., Englewood Cliffs, N.J.

Zbar, P. B.: "Basic Electricity," McGraw-Hill Book Company, New York.

Kirchhoff's Laws

Chapter 8

Many types of circuits have components that are not in series, in parallel, or in series-parallel. One example is a circuit with two voltages applied in different branches, as in Fig. 8-3. Another example is an unbalanced bridge circuit. Where the rules of series and parallel circuits cannot be applied, more general methods of analysis become necessary. These methods include the application of Kirchhoff's laws, as described here, and the network theorems explained in Chap. 9. Any circuit can be solved by Kirchhoff's laws because they do not depend on series or parallel connections. Stated in 1847 by the German physicist Gustav R. Kirchhoff, the two basic rules for voltage and current are:

1. *The algebraic sum of the voltage sources and IR voltage drops must total zero around any closed path.*
2. *At any point in a circuit the algebraic sum of the currents directed in and out must total zero.*

Specific methods for applying these basic rules are explained in the following topics:

8-1   Kirchhoff's Current Law
8-2   Kirchhoff's Voltage Law
8-3   Method of Branch Currents
8-4   Node-Voltage Analysis
8-5   Method of Mesh Currents

8-1 KIRCHHOFF'S CURRENT LAW

The algebraic sum of the currents entering and leaving any point in a circuit must equal zero. Or stated another way: *The algebraic sum of the currents into any point of the circuit must equal the algebraic sum of the currents out of that point.* Otherwise, charge would accumulate at the point, instead of having a conducting path.

Algebraic Signs. In using Kirchhoff's laws to solve circuits it is necessary to adopt conventions that determine the algebraic signs for current and voltage terms. A convenient system for currents is: *Consider all currents into a branch point as positive and all currents directed away from that point as negative.*

As an example, in Fig. 8-1 we can write the currents as:

$$I_A + I_B - I_C = 0$$

or

$$5\,\text{A} + 3\,\text{A} - 8\,\text{A} = 0$$

Chapter 8
Kirchhoff's Laws

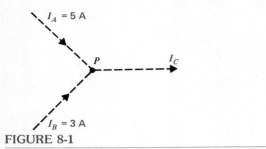

FIGURE 8-1

Current I_C out from P equals 5 A + 3 A into P.

I_A and I_B are positive terms because these currents flow into P, but I_C directed out is negative.

Current Equations. For a circuit application, refer to point c at the top of the diagram in Fig. 8-2. The 6-A I_T into point c divides into the 2-A I_3 and 4-A $I_{4\text{-}5}$, both directed out. Note that $I_{4\text{-}5}$ is the current through R_4 and R_5. The algebraic equation is

$$I_T - I_3 - I_{4\text{-}5} = 0$$

Substituting the values for these currents,

$$6\text{ A} - 2\text{ A} - 4\text{ A} = 0$$

For the opposite directions, refer to point d at the bottom of Fig. 8-2. Here the branch currents into d combine to equal the main-line current I_T returning to the voltage source. Now I_T is directed out from d, with I_3 and $I_{4\text{-}5}$ directed in. The algebraic equation is

$$-I_T + I_3 + I_{4\text{-}5} = 0$$
$$-6\text{ A} + 2\text{ A} + 4\text{ A} = 0$$

The $I_{\text{in}} = I_{\text{out}}$. Note that at either point c or point d in Fig. 8-2, the sum of the 2-A and 4-A branch currents must equal the 6-A total line current. Therefore, Kirchhoff's current law can be stated more briefly as: $I_{\text{in}} = I_{\text{out}}$. In Fig. 8-2, the equations of current can be written:

At point c: $6\text{ A} = 2\text{ A} + 4\text{ A}$
At point d: $2\text{ A} + 4\text{ A} = 6\text{ A}$

Kirchhoff's current law is really the basis for the practical rule in parallel circuits that the total line current must equal the sum of the branch currents.

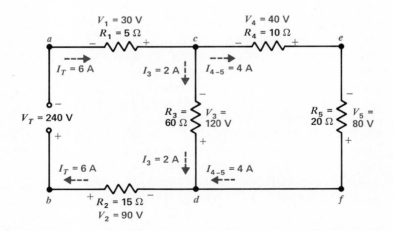

FIGURE 8-2

Series-parallel circuit to illustrate Kirchhoff's laws. See text for current and voltage equations.

Practice Problems 8-1
(answers on page 174)
(a) With a 1-A I_1, 2-A I_2, and 3-A I_3 into a point, how much is I out?
(b) I_1 into a point is 3 A. I out is 7 A. How much is I_2 in?

8-2
KIRCHHOFF'S VOLTAGE LAW

The algebraic sum of the voltages around any closed path is zero. If you start from any point at one potential and come back to the same point and the same potential, the difference of potential must be zero.

Algebraic Signs. In determining the algebraic signs for voltage terms, first mark the polarity of each voltage, as shown in Fig. 8-2. A convenient system then is: *Go around any closed path and consider any voltage whose plus terminal is reached first as positive, and vice versa.* This method applies to voltage drops and voltage sources. Also, the direction can be clockwise or counterclockwise. In any case, if you come back to the starting point, the algebraic sum of all the voltage terms must be zero.

If you do not come back to the start, then the algebraic sum is the voltage between the start and finish points.

You can follow any closed path. The reason is that the net voltage between any two points in a circuit is the same regardless of the path used in determining the potential difference.

Loop Equations. Any closed path is called a *loop*. A loop equation specifies the voltages around the loop.

Figure 8-2 has three loops. The outside loop, starting from point *a* at the top, through *cefdb*, and back to *a*, includes the voltage drops V_1, V_4, V_5, and V_2, and the source V_T.

The inside loop *acdba* includes V_1, V_3, V_2, and V_T. The other inside loop, *cefdc* with V_4, V_5, and V_3, does not include the voltage source.

Consider the voltage equation for the inside loop with V_T. In the clockwise direction, starting from point *a*, the algebraic sum of the voltages is

$$-V_1 - V_3 - V_2 + V_T = 0$$

or

$$-30 \text{ V} - 120 \text{ V} - 90 \text{ V} + 240 \text{ V} = 0$$

V_1, V_3, and V_2 have the negative sign, because for each of these voltages the negative terminal is reached first. However, the source V_T is a positive term because its plus terminal is reached first, going in the same direction.

For the opposite direction, going counterclockwise, in the same loop from point *b* at the bottom, V_2, V_3, and V_1 have positive values and V_T is negative. Then

$$V_2 + V_3 + V_1 - V_T = 0$$

or

$$90 \text{ V} + 120 \text{ V} + 30 \text{ V} - 240 \text{ V} = 0$$

When we transpose the negative term of -240 V, the equation becomes

$$90 \text{ V} + 120 \text{ V} + 30 \text{ V} = 240 \text{ V}$$

This equation states that the sum of the voltage drops equals the applied voltage.

$\Sigma V = V_T$. The Greek letter Σ means "sum of." In either direction, for any loop, the sum of the IR voltage drops V must equal the applied voltage V_T. In Fig. 8-2, for the inside loop with the source V_T,

Chapter 8
Kirchhoff's Laws

Counterclockwise from b:
$$90\text{ V} + 120\text{ V} + 30\text{ V} = 240\text{ V}$$

This system does not contradict the rule for algebraic signs. If 240 V were in the left side of the equation, this term would have a negative sign.

Stating a loop equation as $\Sigma V = V_T$ eliminates the step of transposing the negative terms from one side to the other to make them positive. In this form, the loop equations show that Kirchhoff's voltage law is really the basis for the practical rule in series circuits that the sum of the voltage drops must equal the applied voltage.

When a loop does not have any voltage source, the algebraic sum of the IR voltage drops alone must total zero. For instance, in Fig. 8-2, for the loop $cefdc$ without the source V_T, going clockwise from point c, the loop equation of voltages is

$$-V_4 - V_5 + V_3 = 0$$
$$-40\text{ V} - 80\text{ V} + 120\text{ V} = 0$$
$$0 = 0$$

Notice that V_3 is positive now, because its plus terminal is reached first by going clockwise from d to c in this loop.

Practice Problems 8-2
(answers on page 174)
Refer to Fig. 8-2.
(a) For loop $cefd$, what is the total with -40 V for V_4 and -80 V for V_5?
(b) For loop $cefdc$, what is the total with -40 V for V_4, -80 V for V_5, and including 120 V for V_3?

8-3
METHOD OF BRANCH CURRENTS

Now we can use Kirchhoff's laws to analyze the circuit in Fig. 8-3. The problem is to find the currents and voltages for the three resistors.

First, indicate current directions and mark the voltage polarity across each resistor consistent with the assumed current. Remember that electron flow in a resistor produces negative polarity where the current enters. In Fig. 8-3, we assume that the source V_1 produces electron flow from left to right through R_1, while V_2 produces electron flow from right to left through R_2.

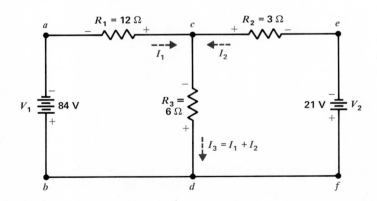

FIGURE 8-3
Application of Kirchhoff's laws to a circuit with two sources in different branches. See text for solution by finding the branch currents.

The three different currents in R_1, R_2, and R_3 are indicated as I_1, I_2, and I_3. However, three unknowns would require three equations for the solution. From Kirchhoff's current law, $I_3 = I_1 + I_2$, as the current out of point c must equal the current in. The current through R_3, therefore, can be specified as $I_1 + I_2$.

With two unknowns, two independent equations are needed to solve for I_1 and I_2. These equations are obtained by writing two Kirchhoff's voltage law equations around two loops. There are three loops in Fig. 8-3, the outside loop and two inside loops, but we need only two. The inside loops are used for the solution here.

Writing the Loop Equations. For the loop with V_1, start at point b, at the bottom left, and go clockwise through V_1, V_{R_1}, and V_{R_3}. This equation for loop 1 is

$$84 - V_{R_1} - V_{R_3} = 0$$

For the loop with V_2, start at point f, at the lower right, and go counterclockwise through V_2, V_{R_2}, and V_{R_3}. This equation for loop 2 is

$$21 - V_{R_2} - V_{R_3} = 0$$

Using the known values of R_1, R_2, and R_3 to specify the IR voltage drops,

$$V_{R_1} = I_1 R_1 = 12 I_1$$
$$V_{R_2} = I_2 R_2 = 3 I_2$$
$$V_{R_3} = (I_1 + I_2) R_3 = 6(I_1 + I_2)$$

Substituting these values in the voltage equation for loop 1,

$$84 - 12 I_1 - 6(I_1 + I_2) = 0$$

Also, in loop 2,

$$21 - 3 I_2 - 6(I_1 + I_2) = 0$$

Multiplying $(I_1 + I_2)$ by 6 and combining terms and transposing, the two equations are

$$-18 I_1 - 6 I_2 = -84$$
$$-6 I_1 - 9 I_2 = -21$$

Divide the top equation by -6 and the bottom equation by -3 to make coefficients smaller and to have all positive terms. The two equations in their simplest form then become

$$3 I_1 + I_2 = 14$$
$$2 I_1 + 3 I_2 = 7$$

Solving for the Currents. These two equations in the two unknowns I_1 and I_2 contain the solution of the network. It should be noted that the equations include *every* resistance in the circuit. I_1 and I_2 can be calculated by any of the methods for the solution of simultaneous equations. Using the method of elimination, multiply the top equation by 3 to make the I_2 terms the same in both equations. Then

$$9 I_1 + 3 I_2 = 42$$
$$2 I_1 + 3 I_2 = 7$$

Subtract the bottom equation from the top equation, term by term, to eliminate I_2. Then, since the I_2 term becomes zero,

$$7 I_1 = 35$$
$$I_1 = 5 \text{ A}$$

The 5-A I_1 is the current through R_1. Its direction is from a to c, as assumed, because the answer for I_1 is positive.

To calculate I_2 substitute 5 for I_1 in either of the two loop equations. Using the bottom equation for the substitution,

$$2(5) + 3I_2 = 7$$
$$3I_2 = 7 - 10$$
$$3I_2 = -3$$
$$I_2 = -1 \text{ A}$$

The negative sign for I_2 means this current is opposite to the assumed direction. Therefore, I_2 flows through R_2 from c to e instead of the direction from e to c.

Why the Solution for I_2 is Negative. In Fig. 8-3, I_2 was assumed from points e to c through R_2 because V_2 produces electron flow in this direction. However, the other voltage source V_1 produces electron flow through R_2 in the opposite direction from points c to e. This solution of -1 A for I_2 shows that the current through R_2 produced by V_1 is more than the current produced by V_2. The net result is 1 A through R_2 from c to e.

The actual direction of I_2 is shown in Fig. 8-4 with all the values for the solution of this circuit. Notice that the polarity of V_{R_2} is reversed from the assumed polarity in Fig. 8-3. Since the net electron flow through R_2 is actually from c to e, the end of R_2 at c is the negative end. However, the polarity of V_2 is the same in both diagrams, because it is a voltage source, which generates its own polarity.

To calculate I_3 through R_3,

$$I_3 = I_1 + I_2$$
$$= 5 + (-1)$$
$$I_3 = 4 \text{ A}$$

The 4 A for I_3 is in the assumed direction from c to d. Although the negative sign for I_2 only means a reversed direction, its algebraic value of -1 must be used for substitution in the algebraic equations written for the assumed direction.

Calculating the Voltages. With all the currents known, the voltage across each resistor can be calculated as follows:

$$V_{R_1} = I_1 R_1 = 5 \times 12 = 60 \text{ V}$$
$$V_{R_2} = I_2 R_2 = 1 \times 3 = 3 \text{ V}$$
$$V_{R_3} = I_3 R_3 = 4 \times 6 = 24 \text{ V}$$

All the currents are taken as positive, in the correct direction, to calculate the voltages. Then

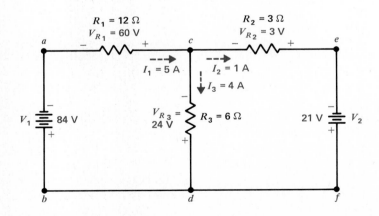

FIGURE 8-4
Solution of Fig. 8-3 with all currents and voltages.

the polarity of each IR drop is determined from the actual direction of current, with electron flow into the negative end (see Fig. 8-4). Notice that V_{R_3} and V_{R_2} have opposing polarities in loop 2. Then the sum of $+3$ V and -24 V equals the -21 V of V_2.

Checking the Solution. As a summary of all the answers for this problem, Fig. 8-4 shows the network with all the currents and voltages. The polarity of each V is marked from the known directions. In checking the answers, we can see whether Kirchhoff's current and voltage laws are satisfied:

At point c: $\quad 5a = 4a + 1a$
At point d: $\quad 4a + 1a = 5a$
Around the loop with V_1:
$\quad 84\text{ V} - 60\text{ V} - 24\text{ V} = 0 \quad$ clockwise from b
Around the loop with V_2:
$\quad 21\text{ V} + 3\text{ V} - 24\text{ V} = 0 \quad$ counterclockwise from f

It should be noted that the circuit has been solved using only the two Kirchhoff laws, without any of the special rules for series and parallel circuits. Any circuit can be solved just by applying Kirchhoff's laws for the voltages around a loop and the currents at a branch point.

Practice Problems 8-3
(answers on page 174)
Refer to Fig. 8-4.
(a) How much is the voltage around loop *cefd*?
(b) How much is the voltage around loop *cefdc*?

8-4
NODE-VOLTAGE ANALYSIS

In the method of branch currents, these currents are used for specifying the voltage drops around the loops. Then loop equations are written to satisfy Kirchhoff's voltage law. Solving the loop equations, we can calculate the unknown branch currents.

Another method uses the voltage drops to specify the currents at a branch point, also called a *node*. Then node equations of currents are written to satisfy Kirchhoff's current law. Solving the node equations, we can calculate the unknown node voltages. This method of node-voltage analysis often is shorter than the method of branch currents.

A node is simply a common connection for two or more components. A *principal node* has three or more connections. In effect, a principal node is just a junction or branch point, where currents can divide or combine. Therefore, we can always write an equation of currents at a principal node. In Fig. 8-5, points N and G are principal nodes.

However, one node must be the reference for specifying the voltage at any other node. In Fig. 8-5, point G connected to chassis ground is the reference node. Therefore, we need only write one current equation for the other node N. In general, the number of current equations required to solve a circuit is one less than its principal nodes.

Writing the Node Equations. The circuit of Fig. 8-3, earlier solved by the method of branch currents, is redrawn in Fig. 8-5 to be solved now by node-voltage analysis. The problem here is to find the node voltage V_N from N to G. Once this voltage is known, all the other voltages and currents can be determined.

The currents in and out of node N are specified as follows: I_1 is the only current through the 12-Ω R_1. Therefore, I_1 is V_{R_1}/R_1 or $V_{R_1}/12\ \Omega$. Similarly I_2 is $V_{R_2}/3\ \Omega$. Finally, I_3 is $V_{R_3}/6\ \Omega$.

Note that V_{R_3} is the node voltage V_N that we are to calculate. Therefore, I_3 can also be stated

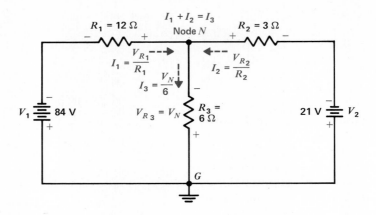

FIGURE 8-5

Node-voltage analysis for the same circuit as in Fig. 8-3. See text for solution by finding V_N across R_3 from the principal node N to ground.

as $V_N/6\,\Omega$. The equation of currents at node N is

$$I_1 + I_2 = I_3$$

or

$$\frac{V_{R_1}}{12} + \frac{V_{R_2}}{3} = \frac{V_N}{6}$$

There are three unknowns here, but V_{R_1} and V_{R_2} can be specified in terms of V_N and the known values of V_1 and V_2. We can use Kirchhoff's voltage law, because the applied voltage V must equal the algebraic sum of the voltage drops. For the loop with V_1 of 84 V,

$$V_{R_1} + V_N = 84 \quad \text{or} \quad V_{R_1} = 84 - V_N$$

For the loop with V_2 of 21 V,

$$V_{R_2} + V_N = 21 \quad \text{or} \quad V_{R_2} = 21 - V_N$$

Now substitute these values of V_{R_1} and V_{R_2} in the equation of currents:

$$I_1 = I_2 + I_3$$
$$\frac{V_{R_1}}{R_1} = \frac{V_{R_2}}{R_2} + \frac{V_{R_3}}{R_3}$$

Using the value of each V in terms of V_N.

$$\frac{84 - V_N}{12} + \frac{21 - V_N}{3} = \frac{V_N}{6}$$

This equation has only the one unknown, V_N. Clearing fractions by multiplying each term by 12, the equation is

$$(84 - V_N) + 4(21 - V_N) = 2V_N$$
$$84 - V_N + 84 - 4V_N = 2V_N$$
$$-7V_N = -168$$
$$V_N = 24 \text{ V}$$

This answer of 24 V for V_N is the same as that calculated for V_{R_3} by the method of branch currents. The positive value means the direction of I_3 is correct, making V_N negative at the top of R_3 in Fig. 8-5.

Calculating All Voltages and Currents. The reason for finding the voltage at a node, rather than some other voltage, is the fact that a node voltage must be common to two loops. As a result, the node voltage can be used for calculating all the voltages in the loops. In Fig. 8-5, with a V_N of 24 V, then V_{R_1} must be $84 - 24 = 60$ V. Also, I_1 is 60 V/12 Ω, which equals 5 A.

To find V_{R_2}, it must be $21 - 24$, which equals -3 V. The negative answer means that I_2 is opposite to the assumed direction and the polarity of V_{R_2} is the reverse of the signs shown across R_2 in Fig. 8-5. The correct directions are shown in the solution for the circuit in Fig. 8-4. The magnitude of I_2 is 3 V/3 Ω, which equals 1 A.

The following comparisons can be helpful in using node equations and loop equations. A node equation applies Kirchhoff's current law to the currents in and out of a node point. However, the currents are specified as V/R so that the equation of currents can be solved to find a node voltage.

A loop equation applies Kirchhoff's voltage law to the voltages around a closed path. However, the voltages are specified as IR so that the equation of voltages can be solved to find a loop current. This procedure with voltage equations is used for the method of branch currents explained before with Fig. 8-3 and to the method of mesh currents to be described next with Fig. 8-6.

Practice Problems 8-4
(answers on page 174)
(a) Figure 8-5 has how many principal nodes?
(b) How many node equations are necessary to solve a circuit with three principal nodes?

8-5
METHOD OF MESH CURRENTS

A mesh is the simplest possible closed path. The circuit in Fig. 8-6 has the two meshes *acdba* and *cefdc*. The outside path *acefdba* is a loop but not a mesh. Each mesh is like a single window frame. There is only one path without any branches.

A mesh current is assumed to flow around a mesh without dividing. In Fig. 8-6, the mesh current I_A flows through V_1, R_1, and R_3; mesh

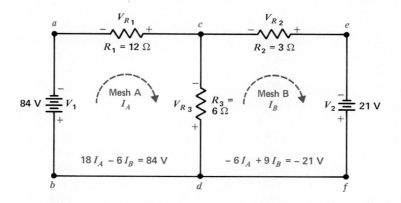

FIGURE 8-6

The same circuit as Fig. 8-3 analyzed as two meshes. See text for solution by calculating the assumed mesh currents I_A and I_B.

current I_B flows through V_2, R_2, and R_3. A resistance common to two meshes, such as R_3, has two mesh currents, which are I_A and I_B here.

The fact that a mesh current does not divide at a branch point is the difference between mesh currents and branch currents. A mesh current is an assumed current, while a branch current is the actual current. However, when the mesh currents are known, all the individual currents and voltages can be determined.

As an example, Fig. 8-6, which has the same circuit as Fig. 8-3, will now be solved by using the assumed mesh currents I_A and I_B. The mesh equations are

$$18I_A - 6I_B = 84 \text{ V} \quad \text{in mesh } A$$
$$-6I_A + 9I_B = -21 \text{ V} \quad \text{in mesh } B$$

Writing the Mesh Equations. The number of meshes equals the number of mesh currents, which is the number of equations required. Here two equations are used for I_A and I_B in the two meshes.

The assumed current is usually taken in the same direction around each mesh, in order to be consistent. Generally, the clockwise direction is used, as shown for I_A and I_B in Fig. 8-6.

In each mesh equation, the algebraic sum of the voltage drops equals the applied voltage.

The voltage drops are added going around a mesh in the same direction as its mesh current. Any voltage drop in a mesh produced by its own mesh current is considered positive because it is added in the direction of the mesh current.

Since all the voltage drops of a mesh current in its own mesh must have the same positive sign, they can be written collectively as one voltage drop, by adding all the resistances in the mesh. For instance, in the first equation, for mesh A, the total resistance equals $12 + 6$, or 18 Ω. Therefore, the voltage drop for I_A is $18I_A$ in mesh A.

In the second equation, for mesh B, the total resistance is $3 + 6$, or 9 Ω, making the total voltage drop $9I_B$ for I_B in mesh B. You can add all the resistances in a mesh for one R_T, because they can be considered in series for the assumed mesh current.

Any resistance common to two meshes has two opposite mesh currents. In Fig. 8-6, I_A flows down while I_B is up through the common R_3, with both currents clockwise. As a result, a common resistance has two opposing voltage drops. One voltage is positive for the current of the mesh whose equation is being written. The opposing voltage is negative for the current of the adjacent mesh.

In mesh A, the common 6-Ω R_3 has the opposing voltages $6I_A$ and $-6I_B$. The $6I_A$ of R_3 adds to the $12I_A$ of R_1 for the total positive voltage drop of $18I_A$ in mesh A. With the opposing voltage of $-6I_B$, then the equation for mesh A is $18I_A - 6I_B = 84$ V.

The same idea applies to mesh B. However, now the voltage $6I_B$ is positive because the equation is for mesh B. The $-6I_A$ voltage is negative here because I_A is for the adjacent mesh. The $6I_B$ adds to the $3I_B$ of R_2 for the total positive voltage drop of $9I_B$ in mesh B. With the opposing voltage of $-6I_A$, the equation for mesh B then is $-6I_A + 9I_B = -21$ V.

The algebraic sign of the source voltage in a mesh depends on its polarity. When the assumed mesh current flows into the positive terminal, as for V_1 in Fig. 8-6, it is considered positive for the right-hand side of the mesh equation. This direction of electron flow produces voltage drops that must add to equal the applied voltage.

With the mesh current into the negative terminal, as for V_2 in Fig. 8-5, it is considered negative. This is why V_2 is -21 V in the equation for mesh B. Then V_2 is actually a load for the larger applied voltage of V_1, instead of V_2 being the source. When a mesh has no source voltage, the algebraic sum of the voltage drops must equal zero.

Solving the Mesh Equations to Find the Mesh Currents. The two equations for the two meshes in Fig. 8-6 are

$$18I_A - 6I_B = 84$$
$$-6I_A + 9I_B = -21$$

These equations have the same coefficients as in the voltage equations written for the branch currents, but the signs are different. The reason is that the directions of the assumed mesh currents are not the same as for the branch currents.

The solution will give the same answers for either method, but you must be consistent in algebraic signs. Either use the rules for meshes with mesh currents or the rules of loops with branch currents, but do not mix the two methods.

For smaller coefficients, divide the first equation by 2 and the second equation by 3. Then

$$9I_A - 3I_B = 42$$
$$-2I_A + 3I_B = -7$$

Add the equations, term by term, to eliminate I_B. Then

$$7I_A = 35$$
$$I_A = 5 \text{ A}$$

To calculate I_B, substitute 5 for I_A in the second equation:

$$-2(5) + 3I_B = -7$$
$$3I_B = -7 + 10 = 3$$
$$I_B = 1 \text{ A}$$

The positive solutions mean that the electron flow for both I_A and I_B is actually clockwise, as assumed. When the solution for a mesh current is negative, its direction is opposite to the assumed direction.

Finding the Branch Currents and Voltage Drops. Referring to Fig. 8-6, the 5-A I_A is the only current through R_1. Therefore, I_A and I_1 are the same. Then V_{R_1} across the 12-Ω R_1 is 5×12 or 60 V. The polarity of V_{R_1} is marked negative at the left, with the electron flow into this side.

Similarly, the 1-A I_B is the only current through R_2. The direction of this electron flow through R_2 is from left to right. Note that this value of 1 A for I_B clockwise is the same as -1 A for I_2, assumed in the opposite direction in Fig. 8-3. Then V_{R_2} across the 3-Ω R_2 is 1×3 or 3 V, with the left side negative.

The current I_3 through R_3, common to both meshes, consists of I_A and I_B. Then I_3 is $5 - 1$ or 4 A. The currents are subtracted because I_A and I_B are in opposing directions through R_3. When all the mesh currents are taken one way, they will always be in opposite directions through any resistance common to two meshes.

The direction of the net 4-A I_3 through R_3 is downward, the same as I_A, because it is larger than I_B. Then, V_{R_3} across the 6-Ω R_3 is $4 \times 6 = 24$ V, with the top negative.

The Set of Mesh Equations. The system for algebraic signs of the voltages in the mesh equations is different from the method used with branch currents, but the end result is the same. The advantage of mesh currents is the pattern of algebraic signs for the voltages, without the need for tracing any branch currents. This feature is especially helpful in a more elaborate circuit, such as the one in Fig. 8-7 that has three meshes. We can use Fig. 8-7 for more

Chapter 8
Kirchhoff's Laws

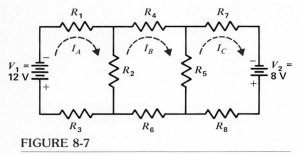

FIGURE 8-7
A circuit with three meshes. Each R is $2\ \Omega$. See text for mesh equations.

practice in writing mesh equations, without doing the numerical work of solving a set[1] of three equations. Each R is $2\ \Omega$.

In mesh A: $\qquad 6I_A - 2I_B + 0 = 12$

[1] A set with any number of simultaneous linear equations, for any number of meshes, can be solved by determinants. This procedure is shown in mathematics textbooks (see the bibliography at the end of this book).

In mesh B: $\qquad -2I_A + 8I_B - 2I_C = 0$
In mesh C: $\qquad 0 - 2I_B + 6I_C = -8$

The zero term in equations A and C represents a missing mesh current. Only mesh B has all three mesh currents.

In summary, the only positive IR voltage in a mesh is for the R_T of each mesh current in its own mesh. All other voltage drops for any adjacent mesh current across a common resistance are always negative. This system of algebraic signs for the voltage drops is the same whether the source voltage in the mesh is zero, positive, or negative.

Practice Problems 8-5
(answers on page 174)
Answer true or false.
(a) A network with four mesh currents needs four mesh equations for a solution.
(b) An R common to two meshes has opposing mesh currents.

Summary

1. Kirchhoff's voltage law states that the algebraic sum of all voltages around any closed path must equal zero. Or the sum of the voltage drops equals the applied voltage.
2. Kirchhoff's current law states that the algebraic sum of all currents directed in and out at any point in a closed path must equal zero. Or the current in equals the current out.
3. A closed path is a loop. The method of using algebraic equations for the voltages around the loops to calculate the branch currents is illustrated in Fig. 8-3.
4. A principal node is a branch point where currents divide or combine. The method of using algebraic equations for the currents at a node to calculate each node voltage is illustrated in Fig. 8-5.
5. A mesh is the simplest possible loop. A mesh current is assumed to flow around the mesh without branching. The method of using algebraic equations for the voltages around the meshes to calculate the mesh currents is illustrated in Fig. 8-6.

Self-Examination (Answers at back of book.)

Answer true or false.

1. The algebraic sum of all voltages around any mesh or any loop must equal zero.
2. A mesh with two resistors has two mesh currents.
3. With $I_1 = 3$ A and $I_2 = 2$ A directed into a node, the current I_3 directed out must equal 5 A.
4. In a loop without any voltage source, the algebraic sum of the voltage drops must equal zero.
5. The algebraic sum of $+40$ V and -10 V equals $+30$ V.
6. A principal node is a junction where branch currents can divide or combine.
7. In the node-voltage method, the number of equations of current equals the number of principal nodes.
8. In the mesh-current method, the number of equations of voltage equals the number of meshes.
9. When all mesh currents are clockwise, or all counterclockwise, any resistor common to two meshes has two currents in opposite directions.
10. The rules of series voltages and parallel currents are based on Kirchhoff's laws.

Essay Questions

1. State Kirchhoff's current law in two ways.
2. State Kirchhoff's voltage law in two ways.
3. What is the difference between a loop and a mesh?
4. What is the difference between a branch current and a mesh current?
5. Define a principal node.
6. Define a node voltage.
7. Use the values in Fig. 8-4 to show that the algebraic sum is zero for all voltages around the outside loop *acefdba*.
8. Use the values in Fig. 8-4 to show that the algebraic sum is zero for all the currents in and out at node *c* and at node *d*.

Problems (Answers to odd-numbered problems at back of book.)

1. Find the current I_1 through R_1 in Fig. 8-8 by the method of mesh currents.

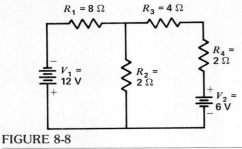

FIGURE 8-8
For Probs. 1, 2, 3, 4, and 5.

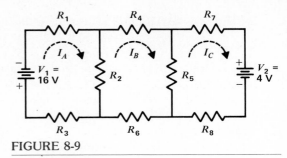

FIGURE 8-9
For Prob. 6.

2. Find the voltage V_2 across R_2 in Fig. 8-8 by the method of node-voltage analysis.
3. Find all the currents and voltages in Fig. 8-8 by the method of branch currents.
4. Check your answers for Prob. 3 by showing that the algebraic sum is zero for the voltages in the three paths.
5. Reverse the polarity of V_2 in Fig. 8-8 and calculate the new I_1, compared with Prob. 1. (Hint: Use mesh currents to eliminate the need for tracing the branch currents.)
6. Write the mesh equations for the circuit in Fig. 8-9. Each R is 1 Ω. No solution is necessary.

Answers to Practice Problems

8-1 (a) 6 A
    (b) 4 A
8-2 (a) 120 V
    (b) 0 V
8-3 (a) 24 V
8-3 (b) 0 V
8-4 (a) Two
    (b) Two
8-5 (a) T
    (b) T

Network Theorems

Chapter 9

A network is just a combination of components, such as resistances, interconnected in any way. However, networks generally need more than the rules of series and parallel circuits for analysis. Kirchhoff's laws can always be applied for any circuit connections. The network theorems, though, usually provide shorter methods of solving the circuit. The reason is that the theorems enable us to convert the network into a simpler circuit, equivalent to the original. Then the equivalent circuit can be solved by the rules of series and parallel circuits.

Only the applications are given here, although all the network theorems can be derived from Kirchhoff's laws. It should also be noted that resistance networks with batteries are shown as examples but the theorems can be applied to both dc and ac networks. The network theorems explained here are

9-1 Superposition
9-2 Thevenin's Theorem
9-3 Thevenizing a Circuit with Two Voltage Sources
9-4 Thevenizing a Bridge Circuit
9-5 Norton's Theorem
9-6 Thevenin-Norton Conversions
9-7 Conversion of Voltage and Current Sources
9-8 Millman's Theorem
9-9 T and π Networks

9-1
SUPERPOSITION

This theorem is very useful because it extends the use of Ohm's law to circuits that have more than one source. In brief, we can calculate the effect of one source at a time and then superimpose the results of all the sources. As a definition, the superposition theorem states that: *In a network with two or more sources the current or voltage for any component is the algebraic sum of the effects produced by each source acting separately.*

In order to use one source at a time, all other sources are "killed" temporarily. This means disabling the source so that it cannot generate voltage or current, without changing the resistance of the circuit. A voltage source such as a battery is killed by assuming a short circuit across its two terminals.

Voltage Divider with Two Sources. The problem in Fig. 9-1 is to find the voltage at P to chassis ground for the circuit in (a). The method is to calculate the voltage at P contributed by

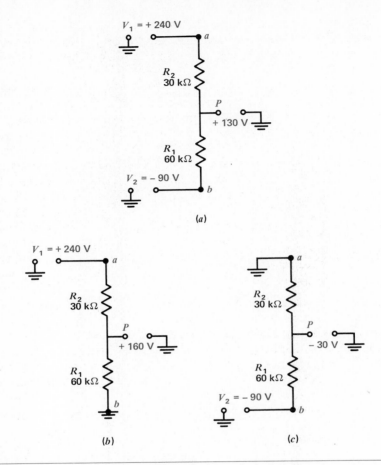

FIGURE 9-1

Superposition applied to a voltage divider with two sources. (a) Actual circuit with $+130\,\text{V}$ between P and chassis ground. (b) V_1 producing $+160\,\text{V}$ at P. (c) V_2 producing $-30\,\text{V}$ at P.

each source separately, as in (b) and (c), and then superimpose these voltages.

To find the effect of V_1 first, short-circuit V_2 as shown in Fig. 9-1b. Note that the bottom of R_1 then becomes connected to chassis ground because of the short circuit across V_2. As a result, R_2 and R_1 form a series voltage divider for the V_1 source.

Furthermore, the voltage across R_1 becomes the same as the voltage from P to ground. To find this V_{R_1} across R_1 as the contribution of the V_1 source, we use the voltage divider formula:

$$V_{R_1} = \frac{R_1}{R_1 + R_2} \times V_1$$

$$= \frac{60 \text{ k}\Omega}{30 \text{ k}\Omega + 60 \text{ k}\Omega} \times 240 \text{ V} = \frac{2}{3} \times 240 \text{ V}$$
$$V_{R_1} = 160 \text{ V}$$

This voltage is positive at P because V_1 is positive.

Next find the effect of V_2 alone, with V_1 short-circuited as shown in Fig. 9-1c. Then point a at the top of R_2 becomes grounded. R_1 and R_2 form a series voltage divider again, but here the R_2 voltage is the voltage at P to ground.

With one side of R_2 grounded and the other side to point P, V_{R_2} is the voltage to calculate. Again we have a series divider but this time for the negative voltage V_2. Using the voltage divider formula for V_{R_2} as the contribution of V_2 to the voltage at P,

$$V_{R_2} = \frac{R_2}{R_1 + R_2} \times V_2$$
$$= \frac{30 \text{ k}\Omega}{30 \text{ k}\Omega + 60 \text{ k}\Omega} \times -90 \text{ V} = \frac{1}{3} \times -90 \text{ V}$$
$$V_{R_2} = -30 \text{ V}$$

This voltage is negative at P because V_2 is negative.

Finally, the total voltage at P is:

$$V_P = V_1 + V_2 = 160 - 30$$
$$V_P = 130 \text{ V}$$

This algebraic sum is positive for the net V_P because the positive V_1 is larger than the negative V_2.

By means of superpositions, therefore, this problem was reduced to two series voltage dividers. The same procedure can be used with more than two sources. Also, each voltage divider can have any number of series resistances.

Requirements for Superposition. All the components must be linear and bilateral in order to superimpose currents and voltages. *Linear* means that the current is proportional to the applied voltage. Then the currents calculated for different source voltages can be superimposed.

Bilateral means that the current is the same amount for opposite polarities of the source voltage. Then the values for opposite directions of current can be combined algebraically. Networks with resistors, capacitors, and air-core inductors are generally linear and bilateral. These are also *passive components*, meaning they do not amplify or rectify. *Active components*, such as transistors, semiconductor diodes, and electron tubes are never bilateral and often are not linear.

Practice Problems 9-1
(answers on page 200)
(a) In Fig. 9-1b, which R is shown grounded at one end?
(b) In Fig. 9-1c, which R is shown grounded at one end?

9-2
THEVENIN'S THEOREM

Named after M. L. Thevenin, a French engineer, this theorem is very useful in simplifying the voltages in a network. By Thevenin's theorem many sources and components, no matter how they are interconnected, can be represented by an equivalent series circuit with respect to any pair of terminals in the network (see Fig. 9-2). Imagine that the block at the left contains a network connected to terminals a and b. Thevenin's theorem states that *the entire network connected to a and b can be replaced by a single voltage source V_{Th} in series with a single resistance R_{Th}, connected to the same two terminals.*

Chapter 9
178 Network Theorems

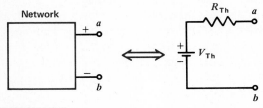

FIGURE 9-2
Any network in the block at the left can be reduced to the Thevenin equivalent circuit at the right.

V_{Th} is the open-circuit voltage across the ab terminals. This means, find the voltage that the network produces across the two terminals with an open circuit between a and b. The polarity of V_{Th} is such that it will produce current from a to b in the same direction as in the original network.

R_{Th} is the open-circuit resistance across terminals a and b, but with all the sources killed. This means, find the resistance looking back into the network from the ab terminals. Although the terminals are open, an ohmmeter across ab would read the value of R_{Th} as the resistance of the remaining paths in the network, without any sources operating.

Thevenizing a Circuit. As an example, refer to Fig. 9-3a, where we want to find the voltage V_L across the 2-Ω R_L and its current I_L. To use Thevenin's theorem, mentally disconnect R_L. The two open ends then become the a and b terminals. Now we find the Thevenin equivalent of the remainder of the circuit that is still connected to a and b. In general, open the part of the circuit to be analyzed and then "thevenize"

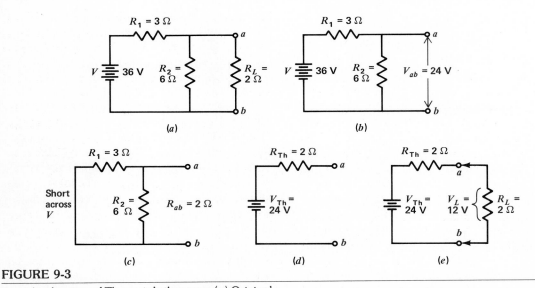

FIGURE 9-3
Application of Thevenin's theorem. (a) Original circuit with terminals a and b across R_L. (b) Disconnect R_L to find V_{ab} is 24 V. (c) Short-circuit V to find R_{ab} is 2 Ω. (d) Thevenin equivalent. (e) R_L reconnected at terminals a and b to find V_L is 12 V.

the remainder of the circuit connected to the two open terminals.

Our only problem now is to find the value of the open-circuit voltage V_{Th} across ab and the equivalent resistance R_{Th}. The Thevenin equivalent always consists of a single voltage source in series with a single resistance as in Fig. 9-3d.

The effect of opening R_L is shown in Fig. 9-3b. As a result, the 3-Ω R_1 and 6-Ω R_2 form a series voltage divider, without R_L.

Furthermore, the voltage across R_2 now is the same as the open-circuit voltage across terminals a and b. Therefore, V_{R_2} with R_L open is V_{ab}. This is the V_{Th} we need for the Thevenin equivalent circuit. Using the voltage divider formula,

$$V_{R_2} = 6/9 \times 36 \text{ V} = 24 \text{ V}$$

or

$$V_{R_2} = 24 \text{ V} = V_{ab} = V_{Th}$$

The polarity of this voltage is positive at the a terminal.

To find R_{Th}, the 2-Ω R_L is still disconnected. However, now the source V is short-circuited. So the circuit looks like (c) in Fig. 9-3. The 3-Ω R_1 is now in parallel with the 6-Ω R_2, as they are both connected across the same two points. This combined resistance is $18/9$ or 2 Ω for the value of R_{Th}.

As shown in Fig. 9-3d, the Thevenin circuit to the left of terminals a and b then consists of the equivalent voltage V_{Th} equal to 24 V, in series with the equivalent series resistance R_{Th} equal to 2 Ω. This Thevenin equivalent applies for any value of R_L because R_L was disconnected. We are actually thevenizing the circuit that feeds the open ab terminals.

To find V_L and I_L we can finally reconnect R_L to the ab terminals of the Thevenin equivalent circuit, as shown in Fig. 9-3e. Then R_L is in series with R_{Th} and V_{Th}. Using the voltage divider formula for the 2-Ω R_{Th} and 2-Ω R_L, $V_L = \frac{1}{2} \times 24$ V $= 12$ V. To find I_L as V_L/R_L, the value is 12 V/2 Ω, which equals 6 A.

These answers of 6 A for I_L and 12 V for V_L apply to R_L in the original circuit in Fig. 9-3a and the equivalent circuit in Fig. 9-3e. Note that the 6-A I_L also flows through R_{Th}.

The same answers could be obtained by solving the series-parallel circuit in Fig. 9-3a, using Ohm's law. However, the advantage of thevenizing the circuit is that the effect of R_L can be calculated easily for different values. Suppose that R_L were changed to 4 Ω. In the Thevenin circuit, the new value of V_L would be $4/6 \times 24$ V $= 16$ V. The new I_L would be 16 V/4 Ω, which equals 4 A. In the original circuit, a complete new solution would be required each time R_L was changed.

Looking Back from the ab Terminals. Which way to look at the resistance of a series-parallel circuit depends on where the source is connected. In general, we calculate the total resistance from the outside terminals of the circuit in toward the source, as the reference.

When the source is short-circuited for thevenizing a circuit, the a and b terminals become the reference. Looking back from a and b to calculate R_{Th}, the viewpoint becomes reversed from the way V_{Th} is determined.

For R_{Th}, imagine that a source could be connected across a and b and calculate the total resistance working from the outside in toward the ab terminals. Actually an ohmmeter across ab would read this resistance.

This idea of reversing the reference is illustrated in Fig. 9-4. The circuit in (a) has the ab terminals open, ready to be thevenized. This circuit is similar to Fig. 9-3 but with the 4-Ω R_3 inserted between R_2 and terminal a. The interesting point is that R_3 does not change the value of V_{ab} produced by the source V, but R_3 does

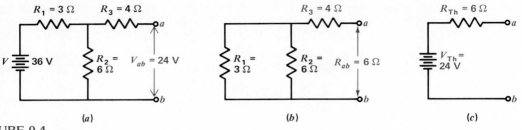

FIGURE 9-4 Thevenizing the circuit of Fig. 9-3b, but with a 4-Ω R_3 in series with the a terminal. (a) V_{ab} is still 24 V. (b) R_{ab} is $2 + 4 = 6$ Ω. (c) Thevenin equivalent.

increase the value of R_{Th}. When we look back from terminals a and b, the 4 Ω of R_3 is in series with 2-Ω to make R_{Th} 6 Ω, as shown in (b).

Let us consider why V_{ab} is the same 24 V with or without R_3. Since R_3 is connected to the open terminal a, the source V cannot produce current in R_3. Therefore, R_3 has no IR drop. A voltmeter would read the same 24 V across R_2 and from a to b. Since V_{ab} equals 24 V, this is the value of V_{Th}.

Now consider why R_3 does change the value of R_{Th}. Remember that we must work from the outside in toward ab to calculate the total resistance. Then the 3-Ω R_1 and 6-Ω R_2 are in parallel for a combined resistance of 2 Ω. Furthermore, this 2 Ω is in series with the 4-Ω R_3 because R_3 is in the main line from the ab terminals. Then R_{Th} is $2 + 4 = 6$ Ω. As shown in Fig. 9-4c, the Thevenin equivalent circuit consists of $V_{Th} = 24$ V and $R_{Th} = 6$ Ω.

Practice Problems 9-2
(answers on page 200)
Answer true or false. For a Thevenin equivalent circuit,
(a) The a and b terminals are open to find both V_{Th} and R_{Th}.
(b) The source voltage is killed only to find R_{Th}.

9-3
THEVENIZING A CIRCUIT WITH TWO VOLTAGE SOURCES

The circuit in Fig. 9-5 has already been solved by Kirchhoff's laws, but we can use Thevenin's theorem to find the current I_3 through the middle resistance R_3. As shown in Fig. 9-5a, first mark the terminals a and b across R_3. In Fig. 9-5b, R_3 is disconnected. To calculate V_{Th}, find V_{ab} across the open terminals.

Superposition Method. With two sources we can use superposition to calculate V_{ab}. First short-circuit V_2. Then the 84 V of V_1 is divided between R_1 and R_2. The voltage across R_2 is between terminals a and b. To calculate this divided voltage across R_2,

$$V_{R_2} = 3/15 \times V_1 = 1/5 \times (-84)$$
$$V_{R_2} = -16.8 \text{ V}$$

This is only the contribution of V_1 to V_{ab}. The polarity is negative at terminal a.

To find the voltage that V_2 produces between a and b, short-circuit V_1. Then the voltage across R_1 is connected from a to b. To calculate this divided voltage across R_1,

$$V_{R_1} = 12/15 \times V_2 = 4/5 \times (-21)$$
$$V_{R_1} = -16.8 \text{ V}$$

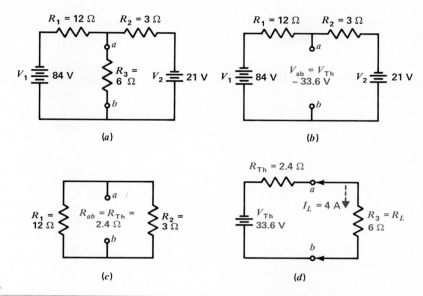

FIGURE 9-5
Thevenizing a circuit with two voltage sources V_1 and V_2. (a) Original circuit with a and b terminals across the middle resistor R_3. (b) Disconnect R_3 to find V_{ab} is -33.6 V. (c) Short-circuit V_1 and V_2 to find $R_{ab} = 2.4$ Ω. (d) Thevenin equivalent with R_L reconnected to ab terminals.

Both V_1 and V_2 produce -16.8 V across the ab terminals with the same polarity. Therefore, they are added.

The resultant value of $V_{ab} = -33.6$ V, shown in Fig. 9-5b, is the value of V_{Th}. The negative polarity means the a terminal is negative with respect to b.

To calculate R_{Th}, short-circuit the sources V_1 and V_2, as shown in Fig. 9-5c. Then the 12-Ω R_1 and 3-Ω R_2 are in parallel across the ab terminals. Their combined resistance is $^{36}/_{15}$, or 2.4 Ω, which is the value of R_{Th}.

The final result is the Thevenin equivalent in Fig. 9-5d with an R_{Th} of 2.4 Ω and a V_{Th} of 33.6 V, negative toward the a terminal.

In order to find the current through R_3, it is reconnected as a load resistance across terminals a and b. Then V_{Th} produces current through the total resistance of 2.4 Ω for R_{Th} and 6 Ω for R_3:

$$I_3 = \frac{V_{Th}}{R_{Th} + R_3} = \frac{33.6}{2.4 + 6} = \frac{33.6}{8.4} = 4 \text{ A}$$

This answer of 4 A for I_3 is the same value calculated before, using Kirchhoff's laws, in Fig. 8-4.

It should be noted that this circuit can be solved by superposition alone, without using Thevenin's theorem, if R_3 is not disconnected. However, opening the ab terminals for the Thevenin equivalent simplifies the superposition, as the circuit then has only series voltage dividers without any parallel current paths. In general, a circuit can often be simplified by disconnecting a component to open the *ab* terminals for Thevenin's theorem.

Short-cut Method. The circuit in Fig. 9-5*b* with two voltage sources feeding the *a* and *b* terminals can be thevenized quicker by using the following formulas for V_{Th} and R_{Th}:

$$V_{Th} = \frac{V_1 R_2 + V_2 R_1}{R_1 + R_2}$$
$$= \frac{(-84)(3) + (-21)(12)}{12 + 3} = \frac{-252 - 252}{15}$$
$$V_{Th} = \frac{-504}{15} = -33.6 \text{ V}$$

V_1 and V_2 are considered negative because the top is negative, compared with our reference at the bottom of the diagram.

To find R_{Th} the two resistances in series with the sources are combined in parallel:

$$R_{Th} = \frac{12 \times 3}{12 + 3} = \frac{36}{15} = 2.4 \text{ }\Omega$$

Practice Problems 9-3
(answers on page 200)
In the Thevenin equivalent circuit in Fig. 9-5d,
(a) How much is R_T?
(b) How much is V_{R_L}?

9-4
THEVENIZING A BRIDGE CIRCUIT

As another example of Thevenin's theorem, we can find the current through the 2-Ω R_L at the center of the bridge circuit in Fig. 9-6*a*. When R_L is disconnected to open the *ab* terminals, the result is shown in (*b*). Notice how the circuit has become simpler because of the open. Instead of the unbalanced bridge in (*a*) which would require Kirchhoff's laws for a solution, the Thevenin equivalent in (*b*) consists of just two voltage dividers. Both the $R_3 R_4$ divider and $R_1 R_2$ divider are across the same 30-V source.

Since the open terminal *a* is at the junction of R_3 and R_4, this divider can be used to find the potential at point *a*. Similarly the potential at terminal *b* can be found from the $R_1 R_2$ divider. Then V_{ab} is the difference between the potentials at terminals *a* and *b*.

Note the voltages for the two dividers. In the divider with the 3-Ω R_3 and 6-Ω R_4, the bottom voltage V_{R_4} is $6/9 \times 30 = 20$ V. Then V_{R_3} at the top is 10 V because both must add up to equal the 30-V source. The polarities are marked negative at the top, the same as *V*.

Similarly, in the divider with the 6-Ω R_1 and 4-Ω R_2, the bottom voltage V_{R_2} is $4/10 \times 30 = 12$ V. Then V_{R_1} at the top is 18 V, as the two must add up to equal the 30-V source. The polarities are also negative at the top, the same as *V*.

Now we can determine the potentials at terminals *a* and *b*, with respect to a common reference, in order to find V_{ab}. Imagine that the positive side of the source *V* is connected to a chassis ground. Then we would use the bottom line in the diagram as our reference for voltages. Note that V_{R_4} at the bottom of the $R_3 R_4$ divider is the same as the potential of terminal *a*, with respect to ground. This value is -20 V, with terminal *a* negative.

Similarly, V_{R_2} in the $R_1 R_2$ divider is the potential at *b* with respect to ground. This value is -12 V, with terminal *b* negative. As a result, V_{ab} is the difference between the -20 V at *a* and the -12 V at *b*, both with respect to the common ground reference.

The potential difference V_{ab} then equals $-20 - (-12)$, or $-20 + 12$, equaling -8 V. Terminal *a* is 8 V more negative than *b*. Therefore, V_{Th} is 8 V, with the negative side toward the *a* terminal, as shown in the Thevenin equivalent in Fig. 9-6*d*.

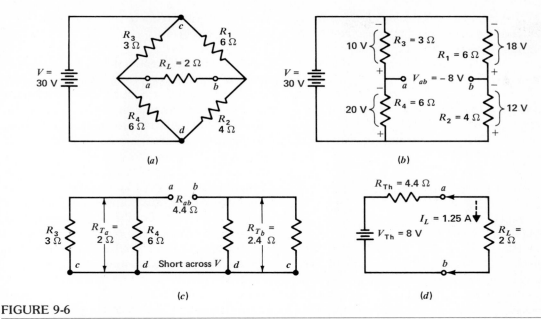

FIGURE 9-6

Thevenizing a bridge circuit. (a) Original circuit with a and b terminals across the middle resistor R_L. (b) Disconnect R_L to find V_{ab} of -8 V. (c) With V short-circuited, R_{ab} is $2 + 2.4 = 4.4 \ \Omega$. (d) Thevenin equivalent with R_L reconnected to ab terminals.

V_{ab} can also be found as the difference between V_{R_3} and V_{R_1} in Fig. 9-6b. In this case V_{R_3} is 10 V and V_{R_1} is 18 V, both being positive with respect to the top line connected to the negative side of the source V. The potential difference between terminals a and b then is $10 - 18$, which also equals -8 V. Note that V_{ab} must have the same value no matter which path is used to determine the voltage.

To find R_{Th}, the 30-V source is short-circuited while the ab terminals are still open. Then the circuit looks like Fig. 9-6c. Looking back from the ab terminals, the 3-Ω R_3 and 6-Ω R_4 are in parallel for a combined resistance R_{T_a} of $^{18}\!/_9$ or 2 Ω. The reason is that R_3 and R_4 are joined at terminal a, while their opposite ends are connected by the short circuit across the source V. Similarly, the 6-Ω R_1 and 4-Ω R_2 are in parallel for a combined resistance R_{T_b} of $^{24}\!/_{10} = 2.4 \ \Omega$. Furthermore, the short circuit across the source now provides a path that connects R_{T_a} and R_{T_b} in series. The entire resistance is $2 + 2.4 = 4.4 \ \Omega$ for R_{ab} or R_{Th}.

The Thevenin equivalent in Fig. 9-6d represents the bridge circuit feeding the open ab terminals, with 8 V for V_{Th} and 4.4 Ω for R_{Th}. Now connect the 2-Ω R_L to the ab terminals in order to calculate I_L. This current is

$$I_L = \frac{V_{Th}}{R_{Th} + R_L} = \frac{8}{4.4 + 2} = \frac{8}{6.4} = 1.25 \text{ A}$$

This 1.25 A is the current through the 2-Ω R_L at the center of the unbalanced bridge in Fig. 9-6a. Furthermore, the amount of I_L for any value of R_L in (a) can be calculated from the equivalent circuit in (d).

> *Practice Problems 9-4*
> *(answers on page 200)*
> In the Thevenin equivalent circuit in Fig. 9-6d,
> (a) How much is R_T?
> (b) How much is V_{R_L}?

9-5
NORTON'S THEOREM

Named after E. L. Norton, a scientist with Bell Telephone Laboratories, this theorem is used for simplifying a network in terms of currents instead of voltages. In many cases, analyzing the division of currents may be easier than voltage analysis. For current analysis, therefore, Norton's theorem can be used to reduce a network to a simple parallel circuit, with a current source. The idea of a current source is that it supplies a total line current to be divided among parallel branches, corresponding to a voltage source applying a total voltage to be divided among series components. This comparison is illustrated in Fig. 9-7.

Example of a Current Source. A source of electrical energy supplying voltage is often shown with a series resistance, which represents the internal resistance of the source as in Fig. 9-7a. This method corresponds to showing an actual voltage source, such as a battery for dc circuits. However, the source may be represented also as a current source with a parallel resistance, as in Fig. 9-7b. Just as a voltage source is rated at, say, 10 V, a current source may be rated at 2 A. For the purpose of analyzing parallel branches, the concept of a current source may be more convenient than a voltage source.

If the current I in Fig. 9-7b is a 2-A source, it supplies 2 A no matter what is connected across the output terminals a and b. Without anything connected across a and b, all the 2 A flows through the shunt R. When a load resistance R_L is connected across a and b, then the 2-A I divides according to the current-division rules for parallel branches.

Remember that parallel currents divide inversely to branch resistances but directly with

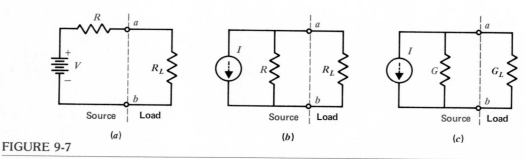

FIGURE 9-7

General forms for a voltage or current source connected to a load R_L across terminals a and b. (a) V source with series R. (b) I source with parallel R. (c) I source with parallel conductance G.

conductances. For this reason it may be preferable to consider the current source shunted by the conductance G, as shown in (c). We can always convert between resistance and conductance, because $1/R$ in ohms is equal to G in siemens.

The symbol for a current source is a circle with an arrow inside, as shown in Fig. 9-7b and c, to show the direction of current. This direction must be the same as the current produced by the polarity of the corresponding voltage source. Remember that a source produces electron flow out from the negative terminal.

An important difference between voltage and current sources is that a current source is killed by making it open, compared with short-circuiting a voltage source. Opening a current source kills its ability to supply current without affecting any parallel branches. A voltage source is short-circuited to kill its ability to supply voltage without affecting any series components.

The Norton Equivalent Circuit. As illustrated in Fig. 9-8, Norton's theorem states that the entire network connected to terminals a and b can be replaced by a single current source I_N in parallel with a single resistance R_N. The value of I_N is equal to the short-circuit current through the ab terminals. This means, find the current that the network would produce through a and b with a short circuit across these two terminals.

The value of R_N is the resistance looking back from the open ab terminals. These terminals are not short-circuited for R_N but are open, as in calculating R_{Th} for Thevenin's theorem. Actually, the single resistor is the same for both the Norton and Thevenin equivalent circuits. In the Norton case, this value of R_{ab} is R_N in parallel with the current source; in the Thevenin case, it is R_{Th} in series with the voltage source.

Nortonizing a Circuit. As an example, let us recalculate the current I_L in Fig. 9-9a, which was solved before by Thevenin's theorem. The first step in applying Norton's theorem is to imagine a short circuit across the ab terminals, as shown in (b). How much current is flowing in the short circuit? Note that a short circuit across ab short-circuits R_L and the parallel R_2. Then the only resistance in the circuit is the 3-Ω R_1 in series with the 36-V source as shown in (c). The short-circuit current, therefore, is

$$I_N = \frac{36 \text{ V}}{3 \text{ Ω}} = 12 \text{ A}$$

This 12-A I_N is the total current available from the current source in the Norton equivalent shown in Fig. 9-9e.

To find R_N, remove the short circuit across a and b and consider the terminals open, without R_L. Now the source V is considered to be short-circuited. As shown in Fig. 9-9d, the re-

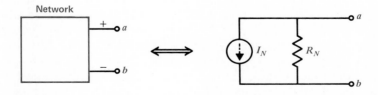

FIGURE 9-8

Any network in the block at the left can be reduced to the Norton equivalent circuit at the right.

Chapter 9
Network Theorems

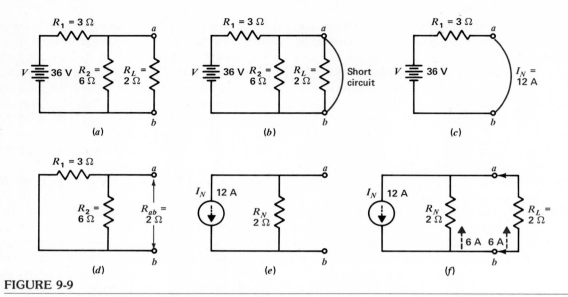

FIGURE 9-9

Same circuit as shown in Fig. 9-3 but solved by Norton's theorem. (a) Original circuit. (b) Short circuit across a and b terminals. (c) Short-circuit I_N is $36/3 = 12$ A. (d) Open ab terminals, but short-circuit V to find $R_{ab} = 2\ \Omega$, the same as R_{Th}. (e) Norton equivalent circuit. (f) R_L reconnected to ab terminals to find I_L is 6 A.

sistance seen looking back from the ab terminals is 6 Ω in parallel with 3 Ω, which equals 2 Ω for the value of R_N.

The resultant Norton equivalent is shown in Fig. 9-9e. It consists of a 12-A current source I_N shunted by the 2-Ω R_N. The arrow on the current source shows the direction of electron flow from terminal b to terminal a, as in the original circuit.

Finally, to calculate I_L, replace the 2-Ω R_L between the ab terminals, as shown in Fig. 9-9f. The current source still delivers 12 A, but now that current divides between the two branches of R_N and R_L. Since these two resistances are equal, the 12-A I_N divides into 6 A for each branch, and I_L is equal to 6 A. This value is the same current we calculated in Fig. 9-3, by Thevenin's theorem. Also, V_L can be calculated as $I_L R_L$, or 6 A × 2 Ω, which equals 12 V.

Looking at the Short-circuit Current. In some cases, there may be a question of which current is I_N when the ab terminals are short-circuited. Imagine that a wire jumper is connected between a and b to short-circuit these terminals. Then I_N must be the current that flows in this wire between terminals a and b.

Remember that any components directly across these two terminals are also short-circuited by the wire jumper. Then these parallel paths have no effect. However, any components in series with terminal a or terminal b are in series with the wire jumper. Therefore, the short-circuit current I_N also flows through the series components.

An example of a resistor in series with the short circuit across the ab terminals is shown in Fig. 9-10. The idea here is that the short-circuit I_N is a branch current, not the main-line current. In (a) the short circuit connects R_3 across R_2. Also, the short-circuit current I_N is now the same as the current I_3 through R_3. Note that I_3 is only a branch current.

To calculate I_3, the circuit is solved by Ohm's law. The parallel combination of R_2 with R_3 equals $^{72}/_{18}$ or 4 Ω. The R_T is $4 + 4 = 8$ Ω. Then I_T is 48 V/8 Ω = 6 A.

This I_T of 6 A in the main line divides into 4 A for R_2 and 2 A for R_3. The 2-A I_3 for R_3 flows through the short-circuited ab terminals. Therefore, this current of 2 A is the value of I_N.

To find R_N in Fig. 9-10b, the short circuit is removed from the ab terminals. Now the source V is short-circuited. Looking back from the open ab terminals, the 4-Ω R_1 is in parallel with the 6-Ω R_2. This combination is $^{24}/_{10} = 2.4$ Ω. The 2.4 Ω is in series with the 12-Ω R_3 to make $R_{ab} = 2.4 + 12 = 14.4$ Ω.

The final Norton equivalent is shown in Fig. 9-10c. I_N is 2 A because this branch current in the original circuit is the current that flows through R_3 and the short-circuited ab terminals. R_N is 14.4 Ω looking back from the open ab terminals with the source V short-circuited the same way as for R_{Th}.

Practice Problems 9-5
(answers on page 200)
Answer true or false. For a Norton equivalent circuit,
(a) The a and b terminals are short-circuited to find I_N.
(b) The a and b terminals are open to find R_N.

9-6
THEVENIN-NORTON CONVERSIONS

Thevenin's theorem says that any network can be represented by a voltage source and series resistance, while Norton's theorem says that the same network can be represented by a current source and shunt resistance. It must be possible, therefore, to convert directly from a Thevenin form to a Norton form and vice versa. Such conversions are often useful.

Norton from Thevenin. Consider the Thevenin equivalent circuit in Fig. 9-11a. What

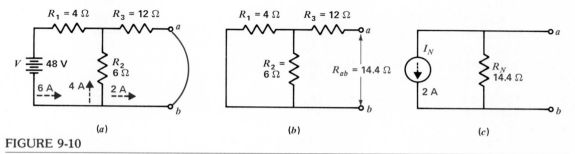

(a)   (b)   (c)

FIGURE 9-10

Nortonizing a circuit where the short-circuit I_N is a branch current. (a) I_N is 2 A through short-circuited ab terminals and R_3. (b) $R_N = R_{ab} = 14.4$ Ω. (c) Norton equivalent.

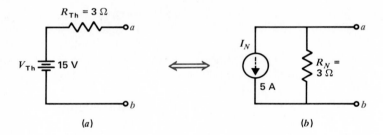

FIGURE 9-11
Thevenin equivalent circuit in (a) corresponds to the Norton equivalent in (b).

is its Norton equivalent? Just apply Norton's theorem the same as for any other circuit. The short-circuit current through the *ab* terminals is

$$I_N = \frac{V_{Th}}{R_{Th}} = \frac{15 \text{ V}}{3 \text{ }\Omega} = 5 \text{ A}$$

The resistance, looking back from the open *ab* terminals with the source V_{Th} short-circuited, is equal to the 3 Ω of R_{Th}. Therefore, the Norton equivalent consists of a current source that supplies the short-circuit current of 5 A, shunted by the same 3-Ω resistance that is in series in the Thevenin circuit. The results are shown in Fig. 9-11b.

Thevenin from Norton. For the opposite conversion, we can start with the Norton circuit of Fig. 9-11b and get back to the original Thevenin circuit. To do this, apply Thevenin's theorem, the same as for any other circuit. First, we find the Thevenin resistance by looking back from the open *ab* terminals. An important principle here, though, is that while a voltage source is short-circuited to find R_{Th}, a current source is an open circuit. Therefore, we have just the 3-Ω R_N, in parallel with the infinite resistance of the open current source. The combined resistance then is 3 Ω.

In general, the resistance R_N always has the same value as R_{Th}. The only difference is that R_N is connected in parallel with I_N, but R_{Th} is in series with V_{Th}.

Now all that is required is to calculate the open-circuit voltage in Fig. 9-11b to find the equivalent V_{Th}. Note that with terminals *a* and *b* open, all the current of the current source flows through the 3-Ω R_N. Then the open-circuit voltage across the *ab* terminals is

$$I_N R_N = 5 \text{ A} \times 3 \text{ }\Omega = 15 \text{ V} = V_{Th}$$

As a result, we have the original Thevenin circuit, consisting of the 15-V source V_{Th} in series with the 3-Ω R_{Th}.

Conversion Formulas. In summary, the following formulas can be used for these conversions:

Thevenin from Norton

$$R_{Th} = R_N$$
$$V_{Th} = I_N \times R_N$$

Norton from Thevenin

$$R_N = R_{Th}$$
$$I_N = V_{Th} \div R_{Th}$$

Another example of these conversions is shown in Fig. 9-12b and c.

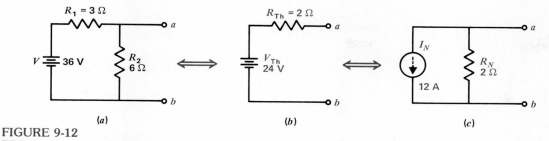

FIGURE 9-12

Example of Thevenin-Norton conversions. (a) Original circuit, the same as Figs. 9-3a and 9-9a. (b) Thevenin equivalent. (c) Norton equivalent.

*Practice Problems 9-6
(answers on page 200)
Answer true or false. In Thevenin-Norton conversions,*
(a) R_N and R_{Th} have the same value.
(b) I_N is V_{Th}/R_{Th}.
(c) V_{Th} is $I_N \times R_N$.

9-7 CONVERSION OF VOLTAGE AND CURRENT SOURCES

Norton conversion is a specific example of the general principle that any voltage source with its series resistance can be converted to an equivalent current source with the same resistance in parallel. In Fig. 9-13, the voltage source in (a) is equivalent to the current source in (b). Just divide the source V by its series R to calculate the value of I for the equivalent current source shunted by the same R. Either source will supply the same current and voltage for any components connected across the ab terminals.

Conversion of voltage and current sources can often simplify circuits, especially with two or more sources. Current sources are easier for parallel connections where we can add or divide currents. Voltage sources are easier for series connections, where we can add or divide voltages.

Two Sources in Parallel Branches. Referring to Fig. 9-14a, assume that the problem is to find I_3 through the middle resistor R_3. Note that V_1 with R_1 and V_2 with R_2 are branches in parallel with R_3. All three branches are connected across the ab terminals.

When we convert V_1 and V_2 to current sources in (b), the circuit has all parallel branches. I_1 is $^{84}/_{12}$ or 7 A, while I_2 is $^{21}/_{3}$, which also happens to be 7 A. I_1 has its parallel R of 12 Ω while I_2 has its parallel R of 3 Ω.

Furthermore, I_1 and I_2 can be combined for the one equivalent current source I_T shown in (c). Since both sources produce current in the

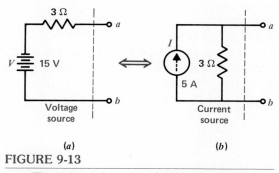

FIGURE 9-13

The voltage source in (a) corresponds to the current source in (b).

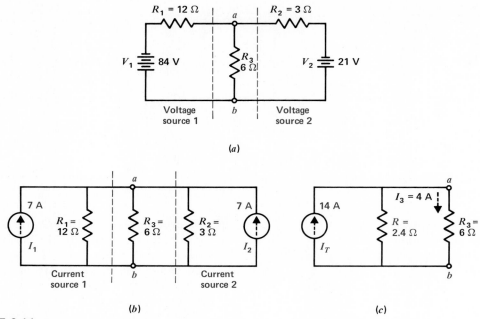

FIGURE 9-14
Converting voltage sources in parallel branches to current sources that can be combined. (a) Original circuit. (b) V_1 and V_2 converted to parallel current sources I_1 and I_2. (c) Circuit with one combined current source I_T.

same direction through R_L, they both are added for $I_T = 7 + 7 = 14$ A.

The shunt R for the 14-A combined source is the combined resistance of the 12-Ω R_1 and the 3-Ω R_2 in parallel. This R equals $^{36}/_{15}$ or 2.4 Ω, as shown in (c).

To find I_L, we can use the current-divider formula for the 6- and 2.4-Ω branches, dividing the 14-A I_T from the current source. Then

$$I_L = \frac{2.4}{2.4 + 6} \times 14 = \frac{33.6}{8.4} = 4 \text{ A}$$

The voltage V_{R_3} across the ab terminals is $I_L R_L$, which equals $4 \times 6 = 24$ V. These are the same values calculated for V_{R_3} and I_3 by Kirchhoff's laws in Fig. 8-4 and by Thevenin's theorem in Fig. 9-5.

Two Sources in Series. Referring to Fig. 9-15, assume that the problem is to find the current I_L through the load resistance R_L between terminals a and b. This circuit has the two current sources I_1 and I_2 in series with each other.

The problem here can be simplified by converting I_1 and I_2 to the series voltage sources V_1 and V_2 shown in (b). The 2-A I_1 with its shunt 4-Ω R_1 is equivalent to 4×2 or 8 V for V_1 with a 4-Ω series resistance. Similarly, the 5-A I_2 with its shunt 2-Ω R_2 is equivalent to 5×2, or 10 V, for V_2 with a 2-Ω series resistance. The polarities of V_1 and V_2 produce electron flow in the same direction as I_1 and I_2.

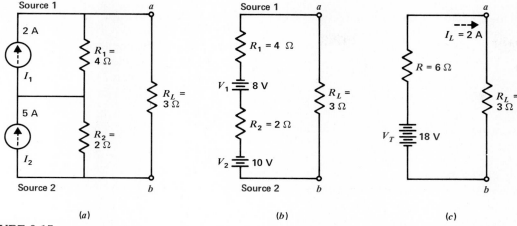

FIGURE 9-15

Converting current sources in series to voltage sources that can be combined. (a) Original circuit. (b) I_1 and I_2 converted to series V_1 and V_2. (c) Circuit with one combined voltage source V_T.

The series voltages can now be combined as in Fig. 9-15c. The 8 V of V_1 and 10 V of V_2 are added because they are series-aiding, resulting in the total V_T of 18 V. And, the resistances of 4 Ω for R_1 and 2 Ω for R_2 are added for a combined R of 6 Ω. This is the series resistance of the 18-V source V_T connected across terminals a and b.

The total resistance of the circuit in (c) is R plus R_L, or $6 + 3 = 9$ Ω. With 18 V applied, $I_L = {}^{18}\!/_9 = 2$ A through R_L between the ab terminals.

Practice Problems 9-7
(answers on page 200)
A voltage source has 21 V in series with 3 Ω. For the equivalent current source,
(a) How much is I?
(b) How much is the shunt R?

9-8
MILLMAN'S THEOREM

This theorem provides a shortcut for finding the common voltage across any number of parallel branches with different voltage sources. A typical example is shown in Fig. 9-16. For all the branches, the ends at point y are connected to chassis ground. Furthermore, the opposite ends of all the branches are also connected to the common point x. The voltage Vxy, therefore, is the common voltage across all the branches.

Finding the value of Vxy gives the net effect

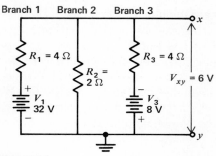

FIGURE 9-16

Example of Millman's theorem to find Vxy, the common voltage across branches with separate voltage sources.

of all the sources in determining the voltage at x with respect to chassis ground. To calculate this voltage

$$V_{xy} = \frac{V_1/R_1 + V_2/R_2 + V_3/R_3}{1/R_1 + 1/R_2 + 1/R_3} \cdots \text{etc.} \quad (9\text{-}1)$$

This formula is derived from converting the voltage sources to current sources and combining the results. The numerator with V/R terms is the sum of the parallel current sources. The denominator with $1/R$ terms is the sum of the parallel conductances. The net V_{xy} then is in the form of I/G or $I \times R$, which is in units of voltage.

Calculating V_{xy}. For the values in Fig. 9-16,

$$V_{xy} = \frac{32/4 + 0/2 - 8/4}{1/4 + 1/2 + 1/4} = \frac{8 + 0 - 2}{1}$$

$$V_{xy} = 6 \text{ V}$$

Note that in branch 3, V_3 is considered negative because it would make point x negative. However, all the resistances are positive. The positive answer for V_{xy} means that point x is positive with respect to y.

In branch 2, V_2 is zero because this branch has no voltage source. However, R_2 is still used in the denominator.

This method can be used for any number of branches but they must all be in parallel, without any series resistances between the branches. In a branch with several resistances, they can be combined as one R_T. When a branch has more than one voltage source, they can be combined algebraically for one V_T.

Applications of Millman's Theorem. In many cases, a circuit can be redrawn to show the parallel branches and their common voltage V_{xy}. Then with V_{xy} known the entire circuit can be analyzed quickly. For instance, Fig. 9-17 has been solved before by other methods. For Millman's theorem the common voltage V_{xy} across all the branches is the same as V_3 across R_3. This voltage is calculated with Formula (9-1), as follows:

$$V_{xy} = \frac{-84/12 + 0/6 - 21/3}{1/12 + 1/6 + 1/3}$$

$$= \frac{-7 + 0 - 7}{7/12} = \frac{-14}{7/12} = -14 \times \frac{12}{7}$$

$$V_{xy} = -24 \text{ V} = V_3$$

The negative sign means that point x is the negative side of V_{xy}. With V_3 known to be 24 V across the 6-Ω R_3, I_3 must be $^{24}\!/_6 = 4$ A. Similarly, all the voltages and currents in this circuit can then be calculated. (See Fig. 8-4 in Chap. 8.)

As another application, the example of superposition in Fig. 9-1 has been redrawn in Fig. 9-18 to show the parallel branches with a common voltage V_{xy} to be calculated by Millman's theorem. Then

$$V_{xy} = \frac{240 \text{ V}/30 \text{ k}\Omega - 90 \text{ V}/60 \text{ k}\Omega}{1/(30 \text{ k}\Omega) + 1/(60 \text{ k}\Omega)}$$

$$= \frac{8 \text{ mA} - 1.5 \text{ mA}}{3/(60 \text{ k}\Omega)} = 6.5 \times \frac{60}{3} = \frac{390}{3}$$

$$V_{xy} = 130 \text{ V} = V_P$$

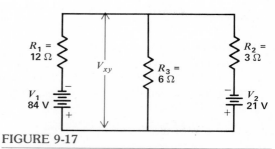

FIGURE 9-17

The same circuit as in Fig. 8-4 for Kirchhoff's laws, but shown with parallel branches to calculate V_{xy} by Millman's theorem.

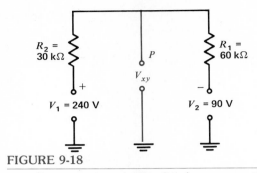

FIGURE 9-18

Same circuit as in Fig. 9-1 for superposition, but shown with parallel branches to calculate V_{xy} by Millman's theorem.

This answer of 130 V from point P to ground, using Millman's theorem, is the same value calculated before by superposition.

*Practice Problems 9-8
(answers on page 200)*
For the example of Millman's theorem in Fig. 9-16,
(a) How much is V_{R_2}?
(b) How much is V_{R_3}?

9-9
T AND π NETWORKS

The network in Fig. 9-19 is called a T (tee) or Y (wye) network, as suggested by its shape. T and Y are different names for the same network, the only difference being that the R_2 and R_3 arms are at an angle in the Y.

The network in Fig. 9-20 is called a π (pi) or Δ (delta) network, as the shape is similar to these Greek letters. Actually, the network can have R_A either at the top or bottom, between R_C and R_B. Note that when the single point c of the delta is separated into two points c' and c'' for the π network, the connections are not really changed. The π and Δ are different names for the same network.

Conversion Formulas. In the analysis of networks, it is often helpful to convert a Δ to Y or vice versa. Either it may be impossible to solve the circuit without the conversion, or the conversion makes the solution simpler. The formulas for these transformations are given here. All are derived from Kirchhoff's laws. Note that letters are used as subscripts for R_A, R_B, and R_C in the Δ while the resistances are numbered R_1, R_2, and R_3 in the Y.

Conversions of Y to Δ, or T to π (9-2)

$$R_A = \frac{R_1 R_2 + R_2 R_3 + R_3 R_1}{R_1}$$

$$R_B = \frac{R_1 R_2 + R_2 R_3 + R_3 R_1}{R_2}$$

$$R_C = \frac{R_1 R_2 + R_2 R_3 + R_3 R_1}{R_3}$$

or

$$R_\Delta = \frac{\Sigma \text{ all cross products in Y}}{\text{opposite } R \text{ in Y}}$$

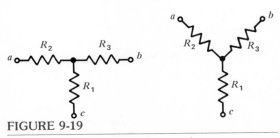

FIGURE 9-19

The form of a T or Y network.

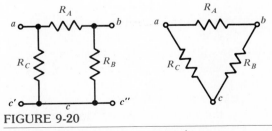

FIGURE 9-20

The form of a π or Δ network.

These formulas can be used to convert a Y network to an equivalent Δ, or a T network to π. Both networks will have the same resistance across any pair of terminals.

The three formulas have the same general form, indicated at the bottom as one basic rule. The symbol Σ is the Greek capital letter sigma, meaning "sum of."

For the opposite conversion:

Conversion of Δ to Y or π to T   (9-3)

$$R_1 = \frac{R_B R_C}{R_A + R_B + R_C}$$

$$R_2 = \frac{R_C R_A}{R_A + R_B + R_C}$$

$$R_3 = \frac{R_A R_B}{R_A + R_B + R_C}$$

or

$$R_Y = \frac{\text{product of two adjacent } R \text{ in } \Delta}{\Sigma \text{ all } R \text{ in } \Delta}$$

As an aid in using these formulas, the following scheme is useful. Place the Y inside the Δ, as shown in Fig. 9-21. Notice that the Δ has three closed sides, while the Y has three open arms. Also note how resistors can be considered opposite each other in the two networks. For instance, the open arm R_1 is opposite the closed side R_A; R_2 is opposite R_B; and R_3 is opposite R_C.

Furthermore, each resistor in an open arm has two adjacent resistors in the closed sides. For R_1, its adjacent resistors are R_B and R_C; also, R_C and R_A are adjacent to R_2, while R_A and R_B are adjacent to R_3.

In the formulas for the Y-to-Δ conversion, each side of the delta is found by first taking all

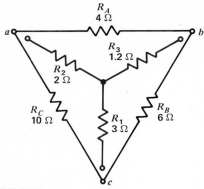

FIGURE 9-21
Conversion between Y and Δ networks. See text for conversion formulas.

possible cross products of the arms of the wye, using two arms at a time. There are three such cross products. The sum of the three cross products is then divided by the opposite arm to find the value of each side in the delta. Note that the numerator remains the same for the sum of the three cross products. However, each side of the delta is calculated by dividing this sum by the opposite arm.

For the case of the Δ-to-Y conversion, each arm of the wye is found by taking the product of the two adjacent sides in the delta and dividing by the sum of the three sides of the delta. The product of two adjacent resistors excludes the opposite resistor. The denominator for the sum of the three sides remains the same in the three formulas. However, each arm is calculated by dividing this sum into each cross product.

An Example of Conversion. The values shown for the equivalent Y and Δ in Fig. 9-21 are calculated as follows: Starting with 4, 6, and 10 Ω for sides R_A, R_B, and R_C, respectively, in the delta, the corresponding arms in the wye are:

$$R_1 = \frac{R_B R_C}{R_A + R_B + R_C}$$
$$= \frac{6 \times 10}{4 + 6 + 10} = \frac{60}{20} = 3\,\Omega$$
$$R_2 = \frac{R_C R_A}{20}$$
$$= \frac{10 \times 4}{20} = \frac{40}{20} = 2\,\Omega$$
$$R_3 = \frac{R_A R_B}{20}$$
$$= \frac{4 \times 6}{20} = \frac{24}{20} = 1.2\,\Omega$$

As a check on these values, we can calculate the equivalent delta for this wye. Starting with values of 3, 2, and 1.2 Ω for R_1, R_2, and R_3, respectively, in the wye, the corresponding values in the delta are:

$$R_A = \frac{R_1 R_2 + R_2 R_3 + R_3 R_1}{R_1} = \frac{6 + 2.4 + 3.6}{3}$$
$$= \frac{12}{3} = 4\,\Omega$$
$$R_B = \frac{12}{R_2} = \frac{12}{2} = 6\,\Omega$$
$$R_C = \frac{12}{R_3} = \frac{12}{1.2} = 10\,\Omega$$

These results show that the Y and Δ networks in Fig. 9-21 are equivalent to each other when they have the values obtained with the conversion formulas.

Simplifying a Bridge Circuit. As an example of the use of such transformations, consider the bridge circuit of Fig. 9-22. The total current I_T from the battery is desired. Therefore, we must find the total resistance R_T.

One approach is to note that the bridge in *a* consists of two deltas connected between terminals P_1 and P_2. One of them can be replaced by an equivalent wye. We use the bottom delta with R_A across the top, in the same form as Fig. 9-21. We then replace this delta $R_A R_B R_C$ by an equivalent wye $R_1 R_2 R_3$ as shown in *b*. Using the conversion formulas,

$$R_1 = \frac{R_B R_C}{R_A + R_B + R_C} = \frac{24}{12} = 2\,\Omega$$
$$R_2 = \frac{R_C R_A}{12} = \frac{12}{12} = 1\,\Omega$$
$$R_3 = \frac{R_A R_B}{12} = \frac{8}{12} = \frac{2}{3}\,\Omega$$

We next use these values for R_1, R_2, and R_3 in an equivalent wye to replace the original delta. Then the resistances form the series-parallel circuit shown in (*c*). The combined resistance of the two parallel branches here is $4 \times 6\frac{2}{3}$ divided by $10\frac{2}{3}$, which equals $\frac{80}{32}$, or 2.5 Ω. Adding this 2.5 Ω to the series R_1 of 2 Ω, the total resistance is 4.5 Ω in (*d*).

This 4.5 Ω is the R_T for the entire bridge circuit between terminals P_3 and P_4 connected to the source V. Then I_T is 30 V/4.5 Ω, which equals $6\frac{2}{3}$ A supplied by the source.

Another approach to finding R_T for the bridge circuit in Fig. 9-22*a* is to recognize that the bridge also consists of two T or Y networks between terminals P_3 and P_4. One of them can be transformed into an equivalent delta. The result is another series-parallel circuit but with the same R_T of 4.5 Ω.

Practice Problems 9-9
(answers on page 200)
In the standard form for conversion,
(a) Which R in the Y is opposite R_A in the Δ?
(b) Which two resistors in the Δ are adjacent to R_1 in the Y?

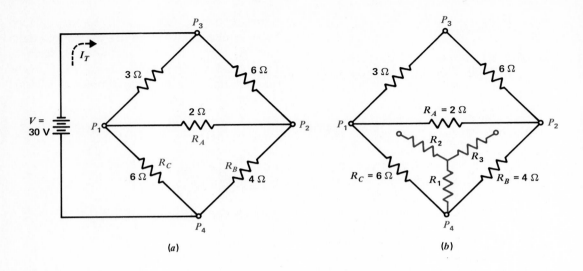
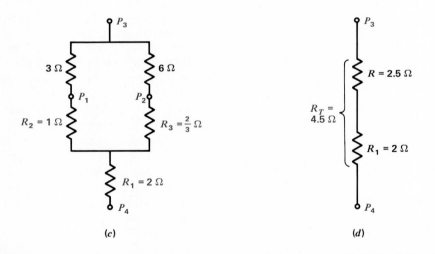

FIGURE 9-22

Solving a bridge circuit by Δ-to-Y conversion. (a) Original circuit. (b) How Y of $R_1R_2R_3$ corresponds to Δ of $R_AR_BR_C$. (c) The Y substituted for the Δ network. Result is a series-parallel circuit with same R_T as the original bridge circuit. (d) R_T is 4.5 Ω between points P_3 and P_4.

Summary

1. *Superposition theorem.* In a linear, bilateral network having more than one source, the current and voltage in any part of the network can be found by adding algebraically the effect of each source separately. All other sources are temporarily killed by short-circuiting voltage sources and opening current sources.
2. *Thevenin's theorem.* Any network with two open terminals a and b can be replaced by a single voltage source V_{Th} in series with a single resistance R_{Th} driving the ab terminals. V_{Th} is the voltage produced by the network across terminals a and b. R_{Th} is the resistance across the open ab terminals, with all sources killed.
3. *Norton's theorem.* Any two-terminal network can be replaced by a single current source I_N in parallel with a single resistance R_N. The value of I_N is the current produced by the network through the short-circuited terminals. R_N is the resistance across the open terminals with all sources killed.
4. *Millman's theorem.* The common voltage across parallel branches with different V sources can be determined with Formula (9-1).
5. A voltage source V with its series R can be converted to an equivalent current source I with parallel R, or vice versa. The value of I is V/R, or V is $I \times R$. The value of R is the same for both sources. However, R is in series with V but in parallel with I.
6. The comparison between delta and wye networks is illustrated in Fig. 9-21. To convert from one network to the other, Formula (9-2) or (9-3) is used.

Self-Examination (Answers at back of book.)

Answer true or false.

1. V_{Th} is an open-circuit voltage.
2. I_N is a short-circuit current.
3. R_{Th} and R_N have the same value.
4. A voltage source has series resistance.
5. A current source has parallel resistance.
6. A voltage source is killed by short-circuiting the terminals.
7. A current source is killed by opening the source.
8. A π network is the same as a T network.

Chapter 9
Network Theorems

9. Millman's theorem is useful for parallel branches with different voltage sources.
10. A 10-V source has a 2-Ω series R. Its equivalent current source is 2 A in parallel with 10 Ω.

Essay Questions

1. State the superposition theorem.
2. In applying the superposition theorem, how do we kill or disable voltage sources and current sources?
3. State the method of calculating V_{Th} and R_{Th} for a Thevenin equivalent circuit.
4. State the method of calculating I_N and R_N for a Norton equivalent circuit.
4. How is a voltage source converted to a current source, and vice versa?
6. For what type of circuit is Millman's theorem used?
7. Draw a delta network and a wye network and give the six formulas needed to convert from one to the other.

Problems (Answers to odd-numbered problems at back of book.)

1. Refer to Fig. 9-23. Show the Thevenin equivalent and calculate V_L.
2. Show the Norton equivalent of Fig. 9-23 and calculate I_L.
3. In Fig. 9-23, convert V and R_1 to a current source and calculate I_L.
4. Use Ohm's law to solve Fig. 9-23 as a series-parallel circuit in order to calculate V_L and I_L. (Note: R_L is not opened for Ohm's law.)

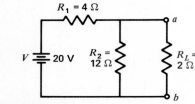

FIGURE 9-23

For Probs. 1, 2, 3, 4, and 5.

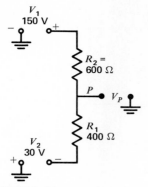

FIGURE 9-24

For Probs. 6 and 7.

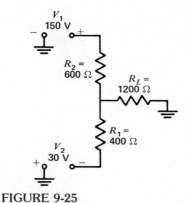

FIGURE 9-25

For Prob. 8.

5. Why is the value of V_L across the *ab* terminals in Prob. 4 not the same as V_{ab} for the Thevenin equivalent circuit in Prob. 1?
6. Refer to Fig. 9-24. Determine V_P by superposition.
7. Redraw Fig. 9-24 as two parallel branches to calculate V_P by Millman's theorem.
8. Refer to Fig. 9-25. Calculate V_L across R_L by Millman's theorem and also by superposition.
9. Show the Thevenin equivalent of Fig. 9-26, where the *ab* terminals are across the middle resistor R_2. Then calculate V_{R_2}.
10. In Fig. 9-26, solve for V_{R_2} by superposition.
11. In Fig. 9-26, solve for V_{R_2} by Millman's theorem.
12. In Fig. 9-27, solve for all the currents by Kirchhoff's laws.
13. In Fig. 9-27, find V_3 by Millman's theorem.
14. Convert the T network in Fig. 9-28 to an equivalent π network.
15. Convert the π network in Fig. 9-29 to an equivalent T network.
16. Show the Thevenin and Norton equivalent circuits for the diagram in Fig. 9-30.

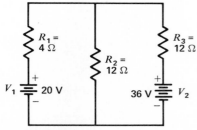

FIGURE 9-26

For Probs. 9, 10, and 11.

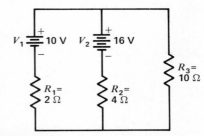

FIGURE 9-27

For Probs. 12 and 13.

Chapter 9
200 Network Theorems

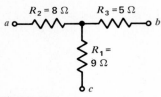

FIGURE 9-28
For Prob. 14.

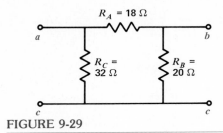

FIGURE 9-29
For Prob. 15.

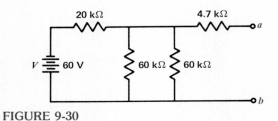

FIGURE 9-30
For Prob. 16.

Answers to Practice Problems

9-1 (a) R_1
    (b) R_2
9-2 (a) T
    (b) T
9-3 (a) 8.4 Ω
    (b) 24 V
9-4 (a) 6.4 Ω
    (b) 2.5 V
9-5 (a) T
    (b) T
9-6 (a) T
    (b) T
    (c) T
9-7 (a) 7 A
    (b) 3 Ω
9-8 (a) 6 V
    (b) 14 V
9-9 (a) R_1
    (b) R_B and R_C

Review of Chapters 8 and 9

Summary

1. Methods of applying Kirchhoff's laws include (a) equations of voltages using the branch currents in the loops to specify the voltages; (b) equations of currents at a node using the node voltages to specify the node currents; (c) equations of voltages using assumed mesh currents to specify the voltages.
2. Methods of reducing a network to a simple equivalent circuit include (a) superposition theorem using one source at a time; (b) Thevenin's theorem to convert the network to a series circuit with one source; (c) Norton's theorem to convert the network to a parallel circuit with one source; (d) Millman's theorem to find the common voltage across parallel branches with different sources; (e) delta-wye conversions to transform a network into a series-parallel circuit.

Review Self-Examination (Answers at back of book.)

Answer true or false.

1. In Fig. 8-3, V_3 can be found by using Kirchhoff's laws with either branch currents or mesh currents.
2. In Fig. 8-3, V_3 can be found by superposition, thevenizing, or using Millman's theorem.
3. In Fig. 9-6, I_L cannot be found by delta-wye conversion because R_L disappears in the transformation.
4. In Fig. 9-6, I_L can be calculated with Kirchhoff's laws, using mesh currents for three meshes.
5. With superposition, we can use Ohm's law for circuits that have more than one source.

6. A Thevenin equivalent is a parallel circuit.
7. A Norton equivalent is a series circuit.
8. Either a Thevenin or a Norton equivalent of a network will produce the same current in any load across the *ab* terminals.
9. A Thevenin-Norton conversion means converting a voltage source to a current source.
10. The units are volts for (volts/ohms) ÷ siemens.
11. A node voltage is a voltage between current nodes.
12. A π network can be converted to an equivalent T network.
13. A 10-V source with 10-Ω series R will supply 5 V to a 10-Ω load R_L.
14. A 10-A source with 10-Ω parallel R will supply 5 A to a 10-Ω load R_L.
15. Current sources in parallel can be added when they supply current in the same direction through R_L.

References (Additional references at back of book.)

Cutler, P.: "Outline for DC Circuit Analysis," McGraw-Hill Book Company, New York.

Hayt and Kemmerling: "Engineering Circuit Analysis," McGraw-Hill Book Company, New York.

Lippin, G.: "Circuit Problems and Solutions," Hayden Book Company, New York.

Mueller, G. V.: "Introduction to Electrical Engineering," McGraw-Hill Book Company, New York.

Romanowitz, A. H.: "Electric Fundamentals and Circuit Analysis," John Wiley & Sons, New York.

Conductors and Insulators

Chapter 10

Conductors have very low resistance. Less than 1 Ω for 10 ft of copper wire is a typical value. The function of the wire conductor is to connect a source of applied voltage to a load resistance with minimum IR voltage drop in the conductor. Then all the applied voltage can produce current in the load resistance.

At the opposite extreme, materials having a very high resistance of many megohms are insulators. Some common examples are air, paper, mica, glass, plastics, rubber, cotton, and shellac or varnish.

Between the extremes of conductors and insulators are semiconductor materials such as carbon, silicon, and germanium. Carbon is used in the manufacture of resistors. Silicon and germanium are used for transistors. The following topics are covered:

10-1   Function of the Conductor
10-2   Standard Wire Gage Sizes
10-3   Types of Wire Conductors
10-4   Printed Wiring
10-5   Switches
10-6   Fuses
10-7   Pilot Lamps
10-8   Wire Resistance
10-9   Temperature Coefficient of Resistance
10-10 Ion Current in Liquids and Gases
10-11 Electrons and Hole Charges in Semiconductors
10-12 Insulators

10-1 FUNCTION OF THE CONDUCTOR

In Fig. 10-1, the resistance of the two 10-ft lengths of copper-wire conductors is approximately 0.6 Ω. This is negligibly small compared with the 144-Ω resistance for the tungsten filament in the bulb. When the current of approximately 0.9 A flows in the bulb and the series conductors, the IR voltage drop across the conductors is 0.54 V with 119.5 V across the bulb. Practically all the applied voltage is across the filament of the bulb. Since the bulb then has its

Chapter 10
Conductors and Insulators

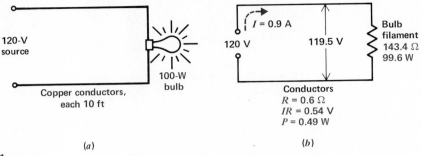

FIGURE 10-1
The conductors should have minimum resistance to light the bulb with full brilliance. (a) Wiring diagram. (b) Schematic diagram.

rated voltage of 120 V, approximately, it will dissipate its rated power of 100 W and light with full brilliance.

The current in the wire conductors and the bulb is the same, since they are in series. However, the IR voltage drop in the conductor is practically zero because its R is almost zero.

Also, the I^2R power dissipated in the conductor is negligibly small, allowing the conductor to operate without becoming hot. Therefore, the conductor delivers energy from the source to the load with minimum loss, by means of electron flow in the copper wires.

Although the resistance of wire conductors is very small, for some cases of excessive current the resultant IR drop can be appreciable. The complaint that the size of a television picture shrinks at night is one example. With many lights and possibly other appliances on, the high value of current can produce too much voltage drop in the power line. A 30-V IR drop results in only 90 V at the load, which is low enough to reduce the picture size. As additional examples, excessive IR drop in the line and low voltage at the load can be the cause of a toaster that does not heat quickly or an electric motor that does not start properly.

Practice Problems 10-1
(*answers on page 223*)
Refer to Fig. 10-1.
(a) How much is R for the 20 ft of copper wire?
(b) How much is the IR voltage drop for the wire conductors?

10-2
STANDARD WIRE GAGE SIZES

Table 10-1 lists the standard wire sizes in the system known as the American Wire Gage (AWG), or Brown and Sharpe (B&S) gage. The gage numbers specify the size of round wire in terms of its diameter and cross-sectional circular area. Note the following:

1. As the gage numbers increase from 1 to 40, the diameter and circular area decrease. Higher gage numbers indicate thinner wire sizes.
2. The circular area doubles for every three gage sizes. For example, No. 10 wire has approximately twice the area of No. 13 wire.
3. The higher the gage number and the thinner the wire, the greater the resistance of the wire for any given length.

Chapter 10
Conductors and Insulators

TABLE 10-1. Copper Wire Table

GAGE NO.	DIAMETER, MIL	CIRCULAR-MIL AREA	OHMS PER 1000 FT OF COPPER WIRE AT 25°C*	GAGE NO.	DIAMETER, MIL	CIRCULAR-MIL AREA	OHMS PER 1000 FT OF COPPER WIRE AT 25°C*
1	289.3	83,690	0.1264	21	28.46	810.1	13.05
2	257.6	66,370	0.1593	22	25.35	642.4	16.46
3	229.4	52,640	0.2009	23	22.57	509.5	20.76
4	204.3	41,740	0.2533	24	20.10	404.0	26.17
5	181.9	33,100	0.3195	25	17.90	320.4	33.00
6	162.0	26,250	0.4028	26	15.94	254.1	41.62
7	144.3	20,820	0.5080	27	14.20	201.5	52.48
8	128.5	16,510	0.6405	28	12.64	159.8	66.17
9	114.4	13,090	0.8077	29	11.26	126.7	83.44
10	101.9	10,380	1.018	30	10.03	100.5	105.2
11	90.74	8234	1.284	31	8.928	79.70	132.7
12	80.81	6530	1.619	32	7.950	63.21	167.3
13	71.96	5178	2.042	33	7.080	50.13	211.0
14	64.08	4107	2.575	34	6.305	39.75	266.0
15	57.07	3257	3.247	35	5.615	31.52	335.0
16	50.82	2583	4.094	36	5.000	25.00	423.0
17	45.26	2048	5.163	37	4.453	19.83	533.4
18	40.30	1624	6.510	38	3.965	15.72	672.6
19	35.89	1288	8.210	39	3.531	12.47	848.1
20	31.96	1022	10.35	40	3.145	9.88	1069

*20 to 25°C or 68 to 77°F is considered average room temperature.

In typical applications, hookup wire for radio receiver circuits with current in the order of milliamperes is generally about No. 22 gage. For this size, 0.5 to 1 A is the maximum current the wire can carry without heating.

House wiring for circuits where the current is 5 to 15 A is about No. 14 gage. Minimum sizes for house wiring are set by Fire Underwriters requirements in most localities. A gauge for measuring wire size is shown in Fig. 10-2.

Circular Mils. The cross-sectional area of round wire is measured in circular mils, abbreviated cmil. A mil is one-thousandth of an inch, or 0.001 in. One circular mil is the cross-sectional area of a wire with a diameter of 1 mil.

The number of circular mils in any circular area is equal to the square of the diameter in mils.

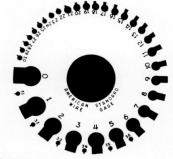

FIGURE 10-2

Wire gage, actual size. (*Hammel, Riglander & Co. Inc.*)

Example 1. What is the area in circular mils of a wire with a diameter of 0.005 in?

Answer. We must convert the diameter to mils. Since 0.005 in equals 5 mil,

Circular mil area = (5 mil)^2
Area = 25 cmil

Note that the circular mil is a unit of area, obtained by squaring the diameter, while the mil is a linear unit of length equal to thousandths of an inch. Therefore, the circular-mil area increases as the square of the diameter. As illustrated in Fig. 10-3, doubling the diameter quadruples the area. Circular mils are convenient for round wire because the cross section is specified without using the formula πr^2 or $\pi d^2/4$ for the area of a circle.

Practice Problems 10-2
(answers on page 223)
(a) How much is R for 1 ft of No. 22 wire?
(b) What is the cross-sectional area in cmil for wire with a diameter of 0.025 in?

10-3
TYPES OF WIRE CONDUCTORS

Most wire conductors are copper, although aluminum and silver are also used. Generally the copper is tinned with a thin coating of solder,[1] which gives it a silvery appearance. The wire can be solid or stranded, as shown in Fig. 10-4a and b.

Stranded wire is flexible and less likely to break open. Sizes for stranded wire are equivalent to the sum of the areas for the individual strands. For instance, two strands of No. 30 wire are equivalent to solid No. 27 wire.

Two or more conductors in a common covering form a wire *cable*. The two-conductor line in Fig. 10-4d is called *coaxial cable*. The metallic braid is one conductor, which is con-

[1] See Appendix H for more information about solder.

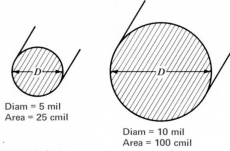

Diam = 5 mil
Area = 25 cmil

Diam = 10 mil
Area = 100 cmil

FIGURE 10-3

Cross-sectional areas for round wire. Double the diameter equals four times the circular area.

nected to ground to shield the inner conductor against external magnetic interference.

Constant spacing between two conductors provides a *transmission line,* such as a coaxial cable. The twin-lead transmission line in Fig. 10-4e is commonly used in television for connecting the antenna to the receiver. Gage No. 20 wire is generally used for the copper conductors.

Very thin wire such as No. 30 often has an insulating coating of enamel or shellac. It may look like copper, but the coating must be scraped off at the ends to make a good connection to the wire.

Heavier wires generally are in an insulating sleeve, which may be rubber, cotton, or one of many plastics. General-purpose wire for connecting electronic components is generally plastic-coated hookup wire of No. 20 gage. Hookup wire that is bare should be enclosed in a hollow insulating sleeve called *spaghetti*.

Practice Problems 10-3
(answers on page 223)
Answer true or false.
(a) The plastic coating on wire conductors has very high resistance.
(b) Coaxial cable is a shielded transmission line.

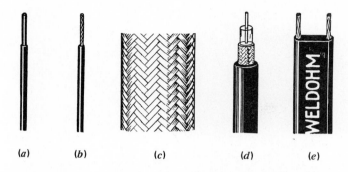

FIGURE 10-4
Types of wire conductors: (a) solid; (b) stranded; (c) braid; (d) coaxial cable; (e) twin-lead cable.

10-4
PRINTED WIRING

Most electronic circuits are mounted on a plastic insulating board with printed wiring, as shown in Fig. 10-5. This is a printed-circuit (PC) or printed-wiring (PW) board. One side has the components, such as resistors, capacitors, coils, tubes, transistors, and diodes. The other side has the conducting paths printed with silver or copper on the board, instead of using wires. Sockets, small metal eyelets, or just holes in the board are used to connect the components to the wiring. With a bright light on one side, you can see through to the opposite side to trace the connections. However, the circuit is usually drawn on the PC board.

It is important not to use too much heat in soldering[1] or desoldering. Otherwise the printed wiring can be lifted off the board. Use a small iron of about 25 to 35 W rating. When soldering semiconductor diodes and transistors, hold the lead with pliers or connect an alligator clip as a heat sink to conduct heat away from the semiconductor junction.

In some cases, defective R, L, and C components can be replaced without disturbing the printed wiring. Just break the old component in the middle with diagonal cutting pliers. Then solder the new component to the old leads. However, the best way is to desolder the leads and solder the new component into the printed wiring.

For desoldering, use a solder-sucker tool, with a soldering iron, to clean each terminal. Another method is to use wire braid. Put the braid on the joint and heat until the solder runs up into the braid. The terminal must be clean enough to lift out the component easily without damaging the PC board.

A small crack in the printed wiring can be repaired by soldering a short length of bare wire over the open. If a larger section of printed wiring is open, or if the board is cracked, you can bridge the open with a length of hookup wire soldered at two convenient end terminals of the printed wiring.

Practice Problems 10-4
(answers on page 223)

(a) Which is the best size iron to use on a PC board, 25, 100, or 150 W?
(b) How much is the resistance of a printed-wire conductor with a break in the middle?

---

[1] More details of solder, soldering, and desoldering are described in Appendix H.

Chapter 10
Conductors and Insulators

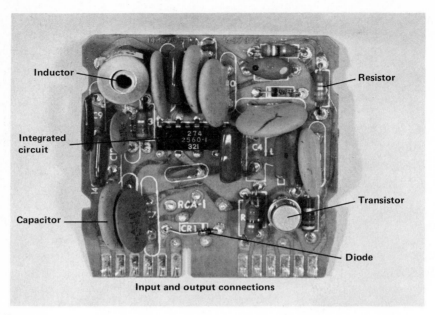

FIGURE 10-5

Printed-wiring board from television receiver. Components are mounted on front, with printed wiring on opposite side. (*RCA*)

10-5 SWITCHES

As shown in Fig. 10-6, switches are commonly used to open or close a circuit. Closed is the ON, or *make,* position; open is the OFF, or *break,* position.

The switch is in series with the voltage source and its load. In the ON position, the

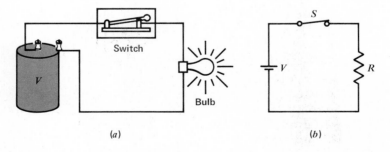

FIGURE 10-6

Single-pole single-throw switch to open or close one circuit. (*a*) Wiring diagram with knife switch. (*b*) Schematic with general symbol for a switch S.

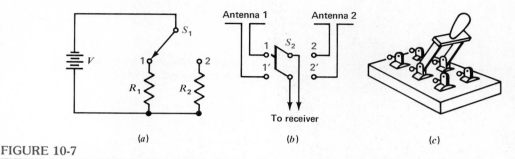

FIGURE 10-7

Switch applications. (*a*) SPDT switch to make one connection to either of two circuits. (*b*) DPDT switch to make two connections to either of two circuits. (*c*) Construction of DPDT knife switch.

closed switch has very little resistance. Then maximum current can flow in the load, with practically zero voltage drop across the switch. Open, the switch has infinite resistance, and no current flows in the circuit.

Note that the switch is in just one side of the line, but the entire series circuit is open when the switch is turned off. In the open position, the applied voltage is across the switch contacts. Therefore, the insulation must be good enough to withstand this amount of voltage without arcing.

The switch in Fig. 10-6 is a single-pole single-throw (SPST) switch. It provides an ON or OFF position for one circuit. Two connections are necessary.

Figure 10-7 shows double-throw switches for two circuits. S_1 in (*a*) is single-pole double-throw (SPDT) to switch one side of the circuit. This switching can be done because R_1 and R_2 both have a common line. Three connections are necessary, one for the common line and one for each of the circuits to be switched.

In Fig. 10-7*b* S_2 is double-pole double-throw (DPDT) to switch both sides of two circuits. This switching is done because there is no common return line for the two separate antennas. Six connections are necessary here, two for each of the circuits to be switched and two for the center contacts.

Figure 10-8 illustrates a toggle switch, and Fig. 10-9 illustrates a rotary switch. Additional types include the knife switch and the pushbutton switch. A spring switch that is normally closed is indicated as *NC*; normally open is *NO*.

FIGURE 10-8

DPDT toggle switch. Length is 1 in. Note six soldering lugs. (*J-B-T Instruments Inc.*)

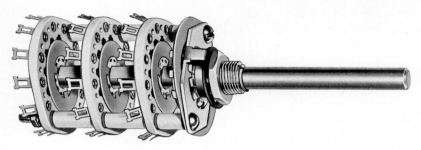

FIGURE 10-9

Rotary switch with three wafers or decks on a common shaft. (*Centralab*)

In general, larger switches are necessary for high-current circuits, where the contacts must be heavy for minimum resistance. In high-voltage circuits, wider spacing is needed for maximum insulation between contacts in the open position.

Practice Problems 10-5
(*answers on page 223*)
(a) How much is the *IR* voltage drop across a closed switch?
(b) How many connections are needed for a SPDT switch?

10-6 FUSES

Many circuits have a fuse in series as a protection against an overload resulting from a short circuit. Excessive current melts the fuse element, blowing the fuse and opening the series circuit. The purpose is to let the fuse blow before the components are damaged. The blown fuse can easily be replaced by a new one, after the overload has been eliminated. A glass-cartridge fuse with holders is shown in Fig. 10-10. This is a type 3AG fuse, with a diameter of $\frac{1}{4}$ in and length of $1\frac{1}{4}$ in. AG is an abbreviation of "automobile glass" since that was one of the first applications of fuses in a glass holder to make the wire link visible.

The metal fuse element may be made of aluminum, tin-coated copper, or nickel. Fuses are available in a current rating from $\frac{1}{500}$ A to hundreds of amperes. The thinner the wire element in the fuse, the smaller is its current rating. A 2-in length of No. 28 wire can serve as a 2-A fuse. As typical applications, the rating for plug fuses in each branch of house wiring is

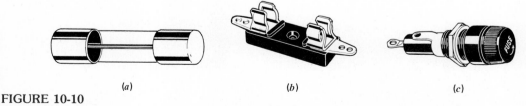

FIGURE 10-10

Fuses. (*a*) Glass-cartridge type. (*b*) and (*c*) Fuse holders.

often 15 A; the high-voltage circuit in a television receiver is usually protected by a glass-cartridge ¼-A fuse.

Slow-blow Fuses. These have a coiled construction. They are designed to open only on a continued overload, such as a short circuit. The purpose of coiled construction is to prevent the fuse from blowing on just a temporary current surge. As an example, a 1-A slow-blow fuse will hold a 400 percent overload in current up to 2 s.

Circuit Breakers. These have a thermal element in the form of a spring. The spring expands with heat and trips open the circuit. The circuit breaker can be reset for normal operation, however, after the short has been eliminated and the thermal element cools down.

Wire Links. A short length of bare wire is often used as a fuse in television receivers. For instance, a 2-in length of No. 24 gage wire can hold a current of 500 mA but burn open with an overload. The wire link can be mounted between two terminal strips on the chassis. Or, the wire link may be wrapped over a small insulator to make a separate component.

Testing Fuses. With glass fuses, you can usually see if the wire element inside is burned open. When measured with an ohmmeter, a good fuse has practically zero resistance. An open fuse reads infinite ohms. Power must be off or the fuse must be out of the circuit to test a fuse with an ohmmeter.

When you test with a voltmeter, a good fuse has zero volts across its two terminals (Fig. 10-11a). If you read appreciable voltage across the fuse, this means it is open. In fact, the full applied voltage is across the open fuse in a series circuit, as shown in (b). This is why fuses also have a voltage rating, which gives the maximum voltage without arcing in the open fuse.

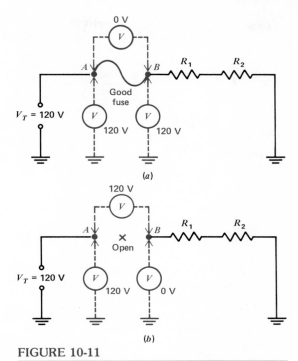

FIGURE 10-11
When a fuse opens, the applied voltage is across the fuse terminals. (a) Circuit closed with good fuse. Note schematic symbol. (b) Fuse open.

Referring to Fig. 10-11, notice the results when measuring the voltages to ground at the two fuse terminals. In (a), the voltage is the same 120 V at both ends because there is no voltage drop across the good fuse. In (b), however, terminal B reads 0 V, as this end is disconnected from V_T because of the open fuse. These tests apply to either dc or ac voltages.

Practice Problems 10-6
(answers on page 223)
(a) How much is the resistance of a good fuse?
(b) How much is the IR voltage drop across a good fuse?

Chapter 10
Conductors and Insulators

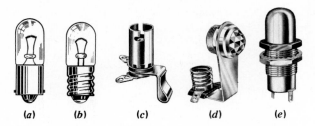

FIGURE 10-12

Panel lamps and sockets. Diameter of base is ½ in. (a) Bayonet base; (b) screw-type base; (c) bayonet socket; (d) screw-type socket with glass bezel; (e) miniature lamp and socket. (*Dialight Corp.*)

10-7
PILOT LAMPS

A small incandescent bulb is often used as a panel lamp to light a dial or as a pilot lamp that lights to show that the equipment is on. Typical lamps and sockets are shown in Fig. 10-12. The side wall of the socket is one connection. In the bayonet socket in (c), the center connection has a spring to hold the bulb tight. To insert or remove the bulb, press down and twist slightly. The mounting for a pilot lamp usually has a colored glass cover called a *bezel,* as shown in (d).

Power is applied to the pilot lamp when the ON-OFF switch is turned on. The lamp is usually connected as a parallel branch so that the equipment can operate if the bulb burns out. Most bulbs are rated at 6.3 V. However, some are for 120-V operation, directly from the power line. A neon bulb can also be used on 120 V as a panel lamp.

*Practice Problems 10-7
(answers on page 223)
Answer true or false.*
(a) A bezel is a glass cover for a pilot lamp.
(b) The pilot lamp is usually connected as a parallel branch.

10-8
WIRE RESISTANCE

The longer a wire, the higher is its resistance. More work must be done to make electrons drift from one end to the other. However, the thicker the wire, the less the resistance, since there are more free electrons in the cross-sectional area. As a formula,

$$R = \rho \frac{l}{A} \qquad (10\text{-}1)$$

where R is the total resistance, l the length, A the cross-sectional area, and ρ the specific resistance or resistivity. The factor ρ then enables different materials to be compared for resistance according to their nature without regard to different lengths or areas. Higher values of ρ mean more resistance. Note that ρ is Greek letter *rho,* corresponding to *r.*

Specific Resistance. Table 10-2 lists resistance values for different metals having the standard wire size of a 1-ft length with a cross-sectional area of 1 cmil. This rating is the *specific resistance* of the metal, in circular-mil ohms per foot. Since silver, copper, gold, and aluminum are the best conductors, they have

TABLE 10-2. Properties of Conducting Materials*

MATERIAL	DESCRIPTION AND SYMBOL	ρ = SPECIFIC RESISTANCE, AT 20°C, CMIL·Ω/FT	TEMPERATURE COEFFICIENT, PER °C, α	MELTING POINT, °C
Aluminum	Element (Al)	17	0.004	660
Carbon	Element (C)	†	−0.0003	3000
Constantan	55% Cu, 45% Ni, alloy	295	0 (average)	1210
Copper	Element (Cu)	10.4	0.004	1083
Gold	Element (Au)	14	0.004	1063
Iron	Element (Fe)	58	0.006	1535
Manganin	84% Cu, 12% Mn, 4% Ni, alloy	270	0 (average)	910
Nichrome	65% Ni, 23% Fe, 12% Cr, alloy	676	0.0002	1350
Nickel	Element (Ni)	52	0.005	1452
Silver	Element (Ag)	9.8	0.004	961
Steel	99.5% Fe, 0.5% C, alloy	100	0.003	1480
Tungsten	Element (W)	33.8	0.005	3370

*Listings approximate only, since precise values depend on exact composition of material.
†Carbon has about 2500 to 7500 times the resistance of copper. Graphite is a form of carbon.

the lowest values of specific resistance. Tungsten and iron have a much higher resistance.

Example 2. How much is the resistance of 100 ft of No. 20 copper wire? Note that from Table 10-1, the cross-sectional area for No. 20 wire is 1022 cmil; from Table 10-2, the ρ for copper is 10.4. Using Formula (10-1) gives

Answer.
$$R = \rho \frac{l}{A} = 10.4 \frac{\text{cmil} \cdot \Omega}{\text{ft}} \times \frac{100 \text{ ft}}{1022 \text{ cmil}}$$
$$R = 1 \, \Omega$$

All the units cancel except the ohms for R. Note that 1 Ω for 100 ft is approximately $\frac{1}{10}$ the resistance of 10.35 Ω for 1000 ft of No. 20 copper wire listed in Table 10-1, showing that the resistance is proportional to length.

Example 3. How much is the resistance of a 100-ft length of No. 23 copper wire?

Answer.
$$R = \rho \frac{l}{A} = 10.4 \frac{\text{cmil} \cdot \Omega}{\text{ft}} \times \frac{100 \text{ ft}}{509.5 \text{ cmil}}$$
$$R = 2 \, \Omega$$

Note that the increase of 3 in gage size provides one-half the circular area and double the resistance, approximately, for the same wire length.

Units of Ohm-Centimeters for ρ. Except for wire conductors, specific resistances are usually compared for the standard size of a 1-cm cube. Then ρ is specified in $\Omega \cdot$ cm for the unit cross-sectional area of 1 cm^2.

As an example, pure germanium has $\rho = 55\ \Omega \cdot \text{cm}$, as listed in Table 10-3. This value means that R is 55 Ω for a cube with a cross-sectional area of 1 cm² and length of 1 cm.

For other sizes, use Formula (10-1) with l in cm and A in cm². Then all the units of size cancel to give R in ohms.

Example 4. How much is the resistance for a slab of germanium 0.2 cm long with a cross-sectional area of 1 cm²?

Answer.
$$R = \rho \frac{l}{A} = 55\ \Omega \cdot \text{cm} \times \frac{0.2\ \text{cm}}{1\ \text{cm}^2}$$
$$R = 11\ \Omega$$

Types of Resistance Wire. For applications in heating elements, as in a toaster, an incandescent light bulb, or a heater, it is necessary to use wire that has more resistance than good conductors like silver, copper, or aluminum. Higher resistance is preferable so that the required amount of I^2R power dissipated as heat in the wire can be obtained without excessive current. Typical materials for resistance wire are the elements tungsten, nickel, or iron and alloys[1] such as manganin, Nichrome, and constantan. These types are generally called resist-

[1] An *alloy* is a fusion of elements, without chemical action between them. Metals are commonly alloyed to alter their physical characteristics.

TABLE 10-3. Comparison of Specific Resistances

MATERIAL	ρ, $\Omega \cdot \text{CM}$, AT 25°C	DESCRIPTION
Silver	1.6×10^{-6}	Conductor
Germanium	55	Semiconductor
Silicon	55,000	Semiconductor
Mica	2×10^{12}	Insulator

ance wire because R is greater than for copper wire, for the same length.

Practice Problems 10-8
(answers on page 223)
(a) Does Nichrome wire have less or more resistance than copper wire?
(b) For 100 ft of No. 14 copper wire, R is 0.26 Ω. How much is R for 1000 ft?

10-9 TEMPERATURE COEFFICIENT OF RESISTANCE

This factor with the symbol alpha (α) states how much the resistance changes for a change in temperature. A positive value for α means R increases with temperature; with a negative α, R decreases; zero for α means R is constant. Some typical values of α, for metals and for carbon, are listed in Table 10-2 in the fourth column.

Positive α. All metals in their pure form, such as copper and tungsten, have a positive temperature coefficient. The α for tungsten, for example, is 0.005. Although α is not exactly constant, an increase in wire resistance caused by a rise in temperature can be calculated approximately from the formula

$$R_t = R_o + R_o\ (\alpha\ \Delta t) \qquad (10\text{-}2)$$

where R_o is the resistance at 20°C, R_t is the higher resistance at the higher temperature, and Δt is the temperature rise above 20°C.

Example 5. A tungsten wire has a 14-Ω R at 20°C. Calculate its resistance at 120°C.

Answer. The temperature rise Δt here is 100°C; α is 0.005. Substituting in Formula (10-2),

$$R_t = 14 + 14(0.005 \times 100)$$
$$= 14 + 7$$
$$R_t = 21\ \Omega$$

The added resistance of 7 Ω increases the wire resistance by 50 percent because of the 100°C rise in temperature.

In practical terms, a positive α means that heat increases R in wire conductors. Then I is reduced, with a specified applied voltage.

Negative α. Note that carbon has a negative temperature coefficient. In general α is negative for all semiconductors, including germanium and silicon. Also, all electrolyte solutions, such as sulfuric acid and water, have a negative α.

A negative value of α means less resistance at higher temperatures. The resistance of semiconductor diodes and transistors, therefore, can be reduced appreciably when they become hot with normal load current.

The negative α has a practical application in the use of carbon *thermistors*. A thermistor can be connected as a series component to decrease its resistance to compensate for the increased hot resistance of wire conductors.

Zero α. This means R is constant with changes in temperature. The metal alloys constantan and manganin, for example, have the value of zero for α. They can be used for precision wirewound resistors, which do not change resistance when the temperature increases.

Hot Resistance. With resistance wire made of tungsten, Nichrome, iron, or nickel, there is usually a big difference in the amount of resistance the wire has when hot in normal operation and when cold without its normal load current. The reason is that the resistance increases for higher temperatures, with a positive temperature coefficient for these materials, as shown in Table 10-2.

As an example, the tungsten filament of a 100-W 120-V incandescent bulb has a current of 0.833 A when the bulb lights with normal brilliance at its rated power, since $I = P/V$. By Ohm's law, the hot resistance is V/I, or 120 V/0.833 A, which equals 144 Ω. If, however, the filament resistance is measured with an ohmmeter when the bulb is not lit, the cold resistance is only about 10 Ω.

The Nichrome heater elements in appliances and the tungsten heaters in vacuum tubes also become several hundred degrees hotter in normal operation. In these cases, only the cold resistance can be measured with an ohmmeter. The hot resistance must be calculated from voltage and current measurements with the normal value of load current.

Example 6. The heater of a vacuum tube has 6.3 V with its normal load current of 0.3 A. How much is the hot resistance?

Answer.
$$R = \frac{V}{I} = \frac{6.3}{0.3}$$
$$R = 21 \text{ Ω}$$

It should be noted that the cold resistance of this heater measured with an ohmmeter equals 2 Ω, which is about one-tenth of the hot resistance.

Superconductivity. The opposite effect of hot resistance is to cool a metal down to very low temperatures to reduce its resistance. Near absolute zero at 0 K or −273°C, some metals abruptly lose practically all their resistance. As an example, the metal tin when cooled by liquid helium becomes superconductive at 3.7 K. Tremendous currents can be produced, resulting in very strong electromagnetic fields. Such work at very low temperatures,[1] near absolute zero, is called *cryogenics*.

[1] See Appendix D, Physics Units, for a description of different temperature scales.

Practice Problems 10-9
(answers on page 223)
Answer true or false.
(a) Metal conductors have more R at higher temperatures.
(b) A thermistor has a negative temperature coefficient.

10-10
ION CURRENT IN LIQUIDS AND GASES

We usually think of metal wire for a conductor, but there are other possibilities. Liquids such as salt water or dilute sulfuric acid can also allow the movement of electric charges. For gases, consider the neon glow lamp, where neon serves as a conductor.

The mechanism may be different for conduction in metal wire, liquids, or gases, but in any case the current is a motion of charges. Furthermore, either positive or negative charges can be the carriers that provide electrical current. The amount of current is Q/T. For one coulomb of charge per second, the current is one ampere.

In solid materials like the metals, the atoms are not free to move among each other. Therefore, conduction of electricity must take place by the drift of free electrons. Each atom remains neutral, neither gaining nor losing charge, but the metals are good conductors because they have plenty of free electrons that can be forced to drift through the solid substance.

In liquids and gases, however, each atom is able to move freely among all the other atoms because the substance is not solid. As a result, the atoms can easily take on electrons or lose electrons, particularly the valence electrons in the outside shell. The result is an atom that is no longer electrically neutral. Adding one or more electrons produces a negative charge; the loss of one or more electrons results in a positive charge. The charged atoms are called *ions*. Such charged particles are commonly formed in liquids and gases.

The Ion. An ion is an atom that has a net electric charge, either positive or negative, resulting from a loss or gain of electrons. See Fig. 10-13. In (a), the sodium atom is neutral, with 11 positive charges in the nucleus balanced by 11 electrons in the outside shells. This atom has only 1 electron in the shell farthest from the nucleus. When the sodium is in a liquid solution, this 1 electron can easily leave the atom. The reason may be another atom close by that needs 1 electron for a stable ring of 8 electrons in its outside shell. Notice that if the sodium atom loses 1 valence electron, the atom will still have an outside ring of 8 electrons, as shown in (b). This sodium atom now is a positive ion, with a charge equal to 1 proton. An ion still has the characteristics of the element because the nucleus is not changed.

Current of Ions. Just as in electron flow, opposite ion charges are attracted to each other, while like charges repel. The resultant motion of ions provides electrical current. In liquids and gases, therefore, conduction of electricity results mainly from the movement of ions. This motion of ion charges is called *ionization cur-*

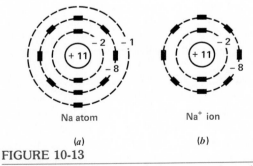

FIGURE 10-13

Formation of ions. (a) Normal sodium (Na) atom. (b) Positively charged sodium (Na$^+$) ion.

rent. Since an ion includes the nucleus of the atom, the ion charge is much heavier than an electron charge and moves with less velocity. We can say that ion charges are less mobile than electron charges.

The direction of ionization current can be the same as electron flow or the opposite. When negative ions move, they are attracted to the positive terminal of an applied voltage, in the same direction as electron flow. However, when positive ions move, this ionization current is in the opposite direction, toward the negative terminal of an applied voltage.

For either direction, though, the amount of ionization current is determined by the rate at which the charge moves. If 3 C of positive ion charges move past a given point per second, the current is 3 A, the same as 3 C of negative ions or 3 C of electron charges.

Ionization in Liquids. Ions are usually formed in liquids when salts or acids are dissolved in water. Salt water is a good conductor because of ionization, but pure distilled water is an insulator. In addition, metals immersed in acids or alkaline solutions produce ionization. Liquids that are good conductors because of ionization are called *electrolytes*. In general, electrolytes have a negative value of α, as more ionization at higher temperatures lowers the resistance.

Ionization in Gases. Gases have a minimum striking or ionization potential, which is the lowest applied voltage that will ionize the gas. Before ionization the gas is an insulator, but the ionization current makes the ionized gas a low resistance. The ionized gas usually glows. Argon, for instance, emits blue light when the gas is ionized. Ionized neon gas glows red. The amount of voltage needed to reach the striking potential varies with different gases and depends on the gas pressure. For example, a neon glow lamp for use as a night light ionizes at approximately 70 V.

Ionic Bonds. The sodium ion in Fig. 10-14 has a charge of +1 because it is missing 1 electron. If such positive ions are placed near negative ions with a charge of −1, there will be an electrical attraction to form an ionic bond.

A common example is the combination of sodium (Na) ions and chlorine (Cl) ions to form table salt (NaCl), as shown in Fig. 10-14. Notice that the 1 outer electron of the Na atom can fit into the 7-electron shell of the Cl atom. When these two elements are combined, the Na atom

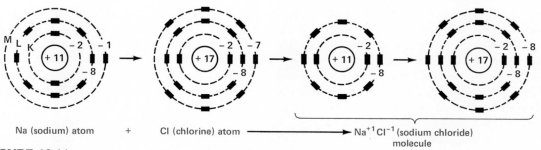

FIGURE 10-14

Ionic bond between atoms of sodium (Na) and chlorine (Cl) to form a molecule of sodium chloride (NaCl).

gives up 1 electron to form a positive ion, with a stable L shell having 8 electrons; also, the Cl atom adds this 1 electron to form a negative ion, with a stable M shell having 8 electrons. The two opposite types of ions are bound in NaCl because of the strong attractive force between opposite charges close together.

The ions in NaCl can separate in water to make salt water a conductor of electricity, while pure water is not. When current flows in salt water, then, the moving charges must be ions, as another example of ionization current.

Practice Problems 10-10
(answers on page 223)
(a) How much is I for 2 C/s of positive ion charges?
(b) Which have the greatest mobility, positive ions, negative ions, or electrons?

10-11
ELECTRONS AND HOLE CHARGES IN SEMICONDUCTORS

The semiconductor materials like germanium and silicon are in a class by themselves as conductors, because the charge carriers for current flow are neither ions nor free valence electrons. With a valence of ±4 for these elements, the tendency to gain or lose electrons to form a stable 8 shell is the same either way. As a result, these elements tend to share their outer electrons in pairs of atoms.

An example is illustrated in Fig. 10-15, for two silicon (Si) atoms, each sharing its 4 valence electrons with the other atom to form one Si_2 molecule. This type of combination of atoms sharing their outer electrons to form a stable molecule is called a *covalent bond*.

The covalent-bond structure in germanium and silicon is the basis for their use in transistors. The reason why is that, although the covalent-bond structure is electrically neutral, it permits charges to be added by *doping* the

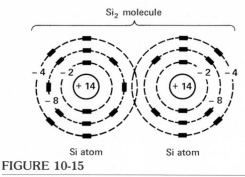

FIGURE 10-15
Covalent bond between silicon (Si) atoms.

semiconductor with a small amount of impurity atoms.

As a specific example, silicon, with a valence of 4, is combined with phosphorus, with a valence of 5. Then the doped germanium has covalent bonds with an excess of 1 electron for each impurity atom of phosphorus. The result is a negative, or N-type, semiconductor.

For the opposite case, silicon can be doped with aluminum, which has a valence of 3. Then covalent bonds formed with the impurity atoms have 7 outer electrons, instead of the 8 with a pair of silicon atoms.

The 1 missing electron for each covalent bond with an impurity atom corresponds to a positive charge called a *hole*. The amount of charge for each hole is 0.16×10^{-18} C, the same as for an electron, but of opposite polarity. This type of doping results in a P-type semiconductor with positive hole charges.

For either N- or P-type semiconductors the charges can be made to move by an applied voltage that produces current. When electrons move, the current direction is the same as for electron flow. When the positive hole charges move, the direction is opposite from electron current. For either electrons or hole charges, when 1 C moves past a given point in 1 s, the amount of current is 1 A. However, electrons have greater mobility than hole charges.

Chapter 10
Conductors and Insulators

For semiconductor diodes, the P and N types are combined. In a PNP transistor, the N type is between two P types. The opposite case, a P type between two N types, results in the NPN transistor. More details of semiconductors in general, as important solid-state components, and transistors in particular, as commonly used for amplifier circuits, are given in Chap. 30, which explains the operation of semiconductor diodes and transistors. Solid-state integrated circuits are in Chap. 31.

Practice Problems 10-11
(*answers on page 223*)
(a) What is the polarity of the hole charges in P-type doped semiconductors?
(b) What is the electron valence of silicon and germanium?
(c) What are the charge carriers in N-type semiconductors?

10-12
INSULATORS

Substances that have very high resistance, of the order of many megohms, are classed as insulators. With such high resistance, an insulator cannot conduct appreciable current when voltage is applied. As a result, insulators can have either of two functions. One is to isolate conductors to eliminate conduction between them. The other is to store an electric charge when voltage is applied.

An insulator maintains its charge because electrons cannot flow to neutralize the charge. The insulators are commonly called *dielectric materials,* therefore, meaning that they can store a charge.

Among the best insulators, or dielectrics, are air, vacuum, rubber, wax, shellac, glass, mica, porcelain, oil, dry paper, textile fibers, and plastics such as Bakelite, formica, and polystyrene. Pure water is a good insulator, but salt water is not. Moist earth is a fairly good conductor, while dry, sandy earth is an insulator.

For any insulator, a high enough voltage can be applied to break down the internal structure of the material, forcing the dielectric to conduct. This dielectric breakdown is usually the result of an arc, which ruptures the physical structure of the material, making it useless as an insulator. Table 10-4 compares several insulators in terms of dielectric strength, which is the voltage breakdown rating. The higher the dielectric strength, the better the insulator, since it is less likely to break down with a high value of applied voltage. The breakdown voltages in Table 10-4 are approximate values for the standard thickness of 1 mil, or 0.001 in. More thickness allows a higher breakdown-voltage rating. Note that the value of 20 V/mil for air or vacuum is the same as 20 kV/in.

TABLE 10-4. Voltage Breakdown of Insulators

MATERIAL	DIELECTRIC STRENGTH, V/MIL	MATERIAL	DIELECTRIC STRENGTH, V/MIL
Air or vacuum	20	Paraffin wax	200–300
Bakelite	300–550	Phenol, molded	300–700
Fiber	150–180	Polystyrene	500–760
Glass	335–2000	Porcelain	40–150
Mica	600–1500	Rubber, hard	450
Paper	1250	Shellac	900
Paraffin oil	380		

Insulator Discharge Current. An insulator in contact with a voltage source stores charge, producing a potential on the insulator. The charge tends to remain on the insulator, but it can be discharged by one of the following methods:

1. Conduction through a conducting path. For instance, a wire across the charged insulator provides a discharge path. Then the discharged dielectric has no potential.
2. Brush discharge. As an example, high voltage on a sharp pointed wire can discharge through the surrounding atmosphere by ionization of the air molecules. This may be visible in the dark as a bluish or reddish glow, called the *corona effect*.
3. Spark discharge. This is a result of breakdown in the insulator because of a high potential difference, rupturing the dielectric. The current that flows across the insulator at the instant of breakdown causes the spark.

Corona is undesirable as it reduces the potential by brush discharge into the surrounding air. In addition, the corona often indicates the beginning of a spark discharge. A potential of the order of kilovolts is usually necessary for corona, as the breakdown voltage for air is approximately 20 kV/in. To reduce the corona effect, conductors that have high voltage should be smooth, rounded, and thick. This equalizes the potential difference from all points on the conductor to the surrounding air. Any sharp point can have a more intense field, making it more susceptible to corona and eventual spark discharge.

Practice Problems 10-12
(answers on page 223)
(a) Which has a higher voltage breakdown rating, air or mica?
(b) Can 30 kV arc across an air gap of 1 in?

Summary

1. A conductor has very low resistance. All the metals are good conductors, the best being silver, copper, and aluminum. Copper is generally used for wire conductors.
2. The sizes for copper wire are specified by the American Wire Gage. Higher gage numbers mean thinner wire. Typical sizes are No. 22 gage hookup wire for electronic circuits and No. 14 for house wiring.
3. The cross-sectional area of round wire is measured in circular mils. One mil is 0.001 in. The area in circular mils equals the diameter in mils squared.
4. $R = \rho(l/A)$. The factor ρ is specific resistance. Wire resistance increases directly with length l, but decreases inversely with the cross-sectional area A, or the square of the diameter.
5. A switch inserted in one side of a circuit opens the entire series circuit. When open, the switch has the applied voltage across it.
6. A fuse protects the circuit components against overload, as excessive current melts the fuse element to open the entire series circuit. A

good fuse has very low resistance and practically zero voltage across it.
7. Ionization in liquids and gases produces atoms that are not electrically neutral. These are ions. Negative ions have an excess of electrons; positive ions have a deficiency of electrons. In liquids and gases, electrical current is a result of movement of the ions.
8. In the semiconductors, such as germanium and silicon, the charge carriers are electrons in N type and positive hole charges in P type. One hole charge is 0.16×10^{-18} C, the same as one electron.
9. The resistance of pure metals increases with temperature. For semiconductors and liquid electrolytes, the resistance decreases at higher temperatures.
10. An insulator has very high resistance. Common insulating materials are air, vacuum, rubber, paper, glass, porcelain, shellac, and plastics. Insulators are also called dielectrics.

Self-Examination (Answers at back of book.)

Choose (a), (b), (c), or (d).

1. A 10-ft length of copper-wire conductor gage No. 20 has a total resistance of (a) less than 1 Ω; (b) 5 Ω; (c) 10.4 Ω; (d) approximately 1 MΩ.
2. A copper-wire conductor with 0.2-in diameter has an area of (a) 200 cmil; (b) 400 cmil; (c) 20,000 cmil; (d) 40,000 cmil.
3. If a wire conductor of 0.1-Ω resistance is doubled in length, its resistance becomes (a) 0.01 Ω; (b) 0.02 Ω; (c) 0.05 Ω; (d) 0.2 Ω.
4. If two wire conductors are tied in parallel, their total resistance is (a) double the resistance of one wire; (b) one-half the resistance of one wire; (c) the same as one wire; (d) two-thirds the resistance of one wire.
5. The hot resistance of the tungsten filament in a bulb is higher than its cold resistance because the filament's temperature coefficient is (a) negative; (b) positive; (c) zero; (d) about 10 Ω per degree.
6. A closed switch has a resistance of (a) zero; (b) infinity; (c) about 100 Ω at room temperature; (d) at least 1000 Ω.
7. An open fuse has a resistance of (a) zero; (b) infinity; (c) about 100 Ω at room temperature; (d) at least 1000 Ω.
8. Insulating materials have the function of (a) conducting very large currents; (b) preventing an open circuit between the voltage source and the load; (c) preventing a short circuit between conducting wires; (d) storing very high currents.

9. An ion is (*a*) a free electron; (*b*) a proton; (*c*) an atom with unbalanced charges; (*d*) a nucleus without protons.
10. Ionization current in liquids and gases results from a flow of (*a*) free electrons; (*b*) protons; (*c*) positive or negative ions; (*d*) ions that are lighter in weight than electrons.

Essay Questions

1. Name three good metal conductors in their order of resistance. Give one application.
2. Name four insulators. Give one application.
3. Name two semiconductors. Give one application.
4. Name two types of resistance wire. Give one application.
5. What is meant by the dielectric strength of an insulator?
6. Why does ionization occur more readily in liquids and gases, compared with the solid metals? Give an example of ionization current.
7. Define the following: ion, ionic bond, covalent bond, molecule.
8. Draw a circuit with two bulbs, a battery, and an SPDT switch that determines which bulb lights.
9. Why is it not possible to measure the hot resistance of a filament with an ohmmeter?
10. Give one way in which negative ion charges are similar to electron charges and one way in which they are different.

Problems (Answers to odd-numbered problems at back of book.)

1. A copper wire has a diameter of 0.032 in. (*a*) How much is its circular-mil area? (*b*) What is its AWG size? (*c*) How much is the resistance of a 100-ft length?
2. Draw the schematic diagram of a resistance in series with an open SPST switch and a 100-V source. (*a*) With the switch open, how much is the voltage across the resistance? How much across the open switch? (*b*) With the switch closed, how much is the voltage across the switch and across the resistance? (*c*) Do the voltage drops around the series circuit add to equal the applied voltage in both cases?
3. Draw the schematic diagram of a fuse in series with the resistance of a 100-W 120-V bulb connected to a 120-V source. (*a*) What size fuse

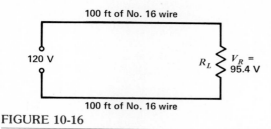

FIGURE 10-16
For Probs. 13 and 14.

can be used? (*b*) How much is the voltage across the good fuse? (*c*) How much is the voltage across the fuse if it is open?

4. Compare the resistance of two conductors: 100 ft of No. 10 gage copper wire and 200 ft of No. 7 gage copper wire.
5. How much is the hot resistance of a 300-W 120-V bulb operating with normal load current?
6. How much is the resistance of a slab of silicon 0.1 cm long with a cross-sectional area of 1 cm²?
7. A cable with two lengths of No. 10 copper wire is short-circuited at one end. The resistance reading at the open end is 10 Ω. What is the cable length in feet? (Temperature is 25°C.)
8. (*a*) How many hole charges are needed to equal 1 C? (*b*) How many electrons? (*c*) How many ions with a negative charge of 1 electron?
9. (*a*) If a copper wire has a resistance of 4 Ω at 25°C, how much is its resistance at 75°C? (*b*) If the wire is No. 10 gage, what is its length in feet?
10. A coil is wound with 3000 turns of No. 20 copper wire. If the average amount of wire in a turn is 4 in, how much is the total resistance of the coil? What will be its resistance if No. 30 wire is used instead? (Temperature is 25°C.)
11. Calculate the voltage drop across 1000 ft of No. 10 gage wire connected to a 2-A load.
12. What is the smallest size of copper wire that will limit the line drop to 5 V, with 120 V applied and a 6-A load? The total line length is 200 ft.
13. Refer to Fig. 10-16. Calculate the load current I for the IR drop of 24.6 V that reduces V_R to 95.4 V with the 120-V supply.
14. From Fig. 10-16, calculate the value of R_L.

Answers to Practice Problems

10-1   (*a*) $R = 0.6 \, \Omega$ 10-2   (*a*) $0.016 \, \Omega$
         (*b*) $IR = 0.54 \, V$ (*b*) 625 cmil

Chapter 10
Conductors and Insulators

10-3	(a) True		10-8	(b) 2.6 Ω
	(b) True		10-9	(a) True
10-4	(a) 25 W			(b) True
	(b) Infinite ohms		10-10	(a) $I = 2$ A
10-5	(a) Zero			(b) Electrons
	(b) Three		10-11	(a) Positive
10-6	(a) Zero			(b) Four
	(b) Zero			(c) Electrons
10-7	(a) True		10-12	(a) Mica
	(b) True			(b) Yes
10-8	(a) More			

Resistors

Chapter 11

In many applications, resistance must be inserted into a circuit. The purpose is either to reduce the current or to produce a desired IR voltage drop. The components for these uses, manufactured with a specific R, are *resistors*. Actually, resistors are probably the most common component in all kinds of electronic equipment, from a small AM radio to a color television receiver. The main types are the carbon resistors shown in Fig. 11-2 and the wirewound resistors in Fig. 11-1. More details are explained in the following topics:

11-1 Resistor Types
11-2 Variable Resistors
11-3 Potentiometers and Rheostats
11-4 Resistor Color Coding
11-5 Power Rating of Resistors
11-6 Choosing the Resistor for a Circuit
11-7 Series and Parallel Combinations of Resistors
11-8 Resistor Troubles

11-1 RESISTOR TYPES

The two main characteristics of a resistor are its R in ohms and the wattage rating. Resistors are available in a very wide range of R values, from a fraction of an ohm to many megohms. The power rating may be as high as several hundred watts or as low as $\frac{1}{10}$ W.

The power rating is important because it specifies the maximum wattage the resistor can dissipate without excessive heat. Dissipation means that the power is wasted as I^2R loss, since the resultant heat is not used. Too much heat can make the resistor burn open.

Wirewound resistors are used where the power dissipation is about 5 W or more. For 2 W or less, carbon resistors are preferable because they are smaller and cost less. Most common in electronic equipment are small carbon resistors with a power rating of 1 W or less. Usually, higher R values have smaller wattage ratings because they have less current.

Both carbon and wirewound resistors can be either fixed or variable. A fixed resistor has a specific R that cannot be adjusted. A variable resistor can be adjusted for any value between zero ohms and its maximum R. Carbon-composition variable resistors are commonly used for controls, such as the volume control in a radio or the contrast control in a television receiver. An application for a variable wirewound resistor is to divide the voltage from a power supply.

Wirewound Resistors. In this construction, resistance wire such as Advance or manganin is wrapped around an insulating core. Insulating materials commonly used are porcelain, cement, phenolic materials like Bakelite, or just plain pressed paper. The wire is bare, but usually the entire unit is encased in an insulator. Examples are shown in Fig. 11-1. The length of wire used and its specific resistivity determine the resistance of the unit.

Chapter 11
Resistors

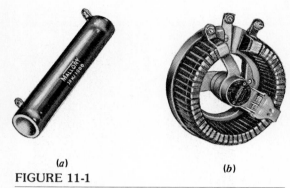

FIGURE 11-1

Wirewound resistors. (a) Fixed R, 5-in long, 50 W. (b) Variable R, 2-in diameter, 50 W. (P. R. Mallory)

Since they are generally for low-resistance high-power applications, wirewound resistors are available in power ratings from 5 W to several hundred watts, with a resistance range of less than 1 Ω to several thousand ohms. In addition, wirewound precision resistors are used where accurate, stable resistance values are required, as for meter shunts and multipliers.

Carbon-Composition Resistors. This type is made of finely divided carbon or graphite mixed with a powdered insulating material in the proportions needed for the desired resistance value. As shown in Fig. 11-2, the resistor element is usually enclosed in a plastic case for insulation and mechanical strength. Joined to the two ends of the carbon resistance element are metal caps with leads of tinned copper wire for soldering the resistor connections into a circuit. These are called *axial leads* because they come straight out from the ends. Carbon resistors are commonly available in resistance values of 1 Ω to 20 MΩ. Their power rating is generally $\frac{1}{10}$, $\frac{1}{8}$, $\frac{1}{4}$, $\frac{1}{2}$, 1, or 2 W.

Carbon-Film Resistors. These have a carbon-film coating around an insulator, like alu-

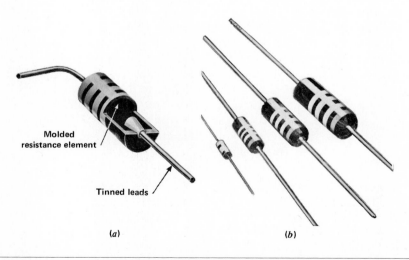

FIGURE 11-2

Carbon-composition resistors. (a) Internal construction. Tinned leads have coating of solder. (b) Sizes of $\frac{1}{4}$, $\frac{1}{2}$, 1, and 2 W with length of 0.7 in. (Allen-Bradley)

mina. The cost is less than for the hot-molded carbon-composition type.

Metal-Film Resistors. These have a conductive coating on a glass substrate. The advantage is more exact resistance values.

Cermet-Film Resistors. These have a carbon coating fired onto a solid ceramic substrate. The purpose is to have more precise R values and greater stability with heat. They are often made in a small square, with leads to fit a PC board. A flat package with multiple leads can be used for several resistors in one unit.

Practice Problems 11-1
(*answers on page 239*)
State whether the following would be a carbon or wirewound resistor.
(a) R is 100 kΩ, with a 1-W power rating.
(b) R is 10 Ω, with a 10-W power rating.

11-2
VARIABLE RESISTORS

These can be wirewound, as in Fig. 11-1, or the carbon type shown in Fig. 11-3. Inside the metal case, the control has a circular disk that is the carbon-composition resistance element. Joined to the two ends are the outside soldering-lug terminals 1 and 3. The middle lug 2 is connected to the variable arm contacting the resis-

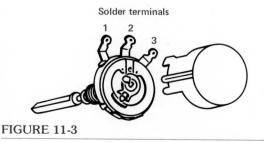

FIGURE 11-3
Variable carbon-resistance control. Diameter $\tfrac{3}{4}$ in.

tor element by a metal spring wiper. As the shaft of the control is rotated, the variable arm moves the wiper to make contact at different points.

When the contact moves closer to one end, the resistance decreases between this end and the variable arm. The variable resistance is zero when the wiper contacts this end but is maximum with the variable arm at the opposite end. Between the two outside ends, the resistance is not variable but always has the maximum resistance of the control.

Carbon controls are available with a total resistance value from 1000 Ω to 5 MΩ, approximately. Their power rating is usually $\tfrac{1}{2}$ to 2 W. A carbon control is often combined with an ON-OFF switch, which is a separate unit operated by the control shaft. A common example is the power ON-OFF switch and volume control for a receiver.

Tapered Controls. The way that R varies with shaft rotation is called the *taper* of the control. With a linear taper, one-half rotation changes R by one-half the maximum value. For a nonlinear taper, though, R can change more gradually at one end, with bigger changes at the opposite end of the control. An audio taper, as an example, allows smaller changes in R at low-volume settings.

Decade Resistance Box. As shown in Fig. 11-4, the *decade box* is a convenient arrangement for obtaining any one resistance within a wide range of values. Inside the box are five series strings of resistors, with one string for each dial switch.

The first dial connects in resistances of 1 to 9 Ω and is the *units* dial. The second dial has units of ten from 10 to 90 Ω and is the *tens* dial. The *hundreds* dial has resistances from 100 to 900 Ω, the fourth dial provides a resistance of 1000 to 9000 Ω, and the top dial is 10,000 to 90,000 Ω.

FIGURE 11-4

Decade resistance box. (*Heath Company*)

The five dial sections are connected in series. Then any value from 1 to 99,999 Ω can be obtained.

Practice Problems 11-2
(answers on page 239)
(a) In Fig. 11-3, which terminal provides a variable R?
(b) Is an audio taper linear or nonlinear?

11-3 POTENTIOMETERS AND RHEOSTATS

These are variable resistances, either carbon or wirewound, used to control the voltage and current in a circuit. The controls can be used for either dc or ac applications.

A rheostat, a variable R with two terminals connected in series with the load, is used to adjust the current in the circuit. Decreasing the series R increases I. The power rating must be enough for the highest current in the rheostat.

A potentiometer, generally called *pot* for short, has three terminals. The fixed maximum R across the two ends is connected across a voltage source. Then the variable arm is used to vary the voltage division between the center terminal and the ends.

Potentiometer Circuit. The purpose of the circuit in Fig. 11-5 is to tap off a variable part of the 100 V from the source. Consider this as input to the potentiometer and the variable voltage as output. Two pairs of connections to the three terminals are necessary, with one terminal common to the input and output. One pair connects the source voltage to the outside terminals 1 and 3. The other pair of connections is between the variable arm at the center terminal and an end terminal.

When the variable arm is at the middle value of R in Fig. 11-5, the 50 V is tapped off between 2 and 1 as one-half the 100-V input. The other 50 V is between 2 and 3.

As the control is turned up to move the variable arm closer to 3, more of the input voltage is available between 2 and 1. With the control at maximum R, the voltage between 2 and 1 is the entire 100 V.

When the variable arm is at minimum R, rotated to terminal 1, the output between 2 and 1 is zero. Now all the applied voltage is between 2 and 3. Note that the source is not shorted, since the maximum R of the potentiometer is always across the input, regardless of where the variable arm is set.

Examples of small potentiometers are shown in Fig. 11-6. A precision potentiometer with a screwdriver adjustment is shown in

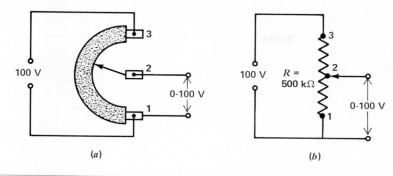

FIGURE 11-5

Potentiometer connected across source to function as a voltage divider. (*a*) Wiring diagram. (*b*) Schematic diagram.

Fig. 11-7. In this type, a wirewound resistance element may be used to provide a precise number of turns for a specific R.

Potentiometer Used As a Rheostat. Commercial rheostats are generally wirewound high-wattage resistors. However, in many applications for electronic circuits a small, low-wattage rheostat is needed. One example is a continuous tone control in a receiver, which needs the variable series resistance of a rheostat but dissipates very little power. A method of wiring a potentiometer as a rheostat is to connect just one end of the control and the variable arm, using only two terminals. The third terminal is open, not connected to anything. Another method is to wire the unused terminal to the center terminal, which converts the potentiometer to a rheostat.

Either end of the potentiometer can be used for the rheostat. The direction of increasing R with shaft rotation reverses, however, for

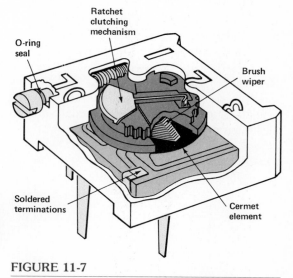

FIGURE 11-7

Precision potentiometer with screw adjustment. Height is $3/8$ in. (*Spectral Electronics Corporation*)

FIGURE 11-6

Small potentiometers. Diameter $1/2$ to $3/4$ in. (*Centralab*)

connections at opposite ends. Also, the taper is reversed on a nonlinear control.

Practice Problems 11-3
(answers on page 239)
(a) How many terminal lugs are used on a rheostat?
(b) How many terminal lugs are used on a potentiometer?
(c) In Fig. 11-5, with a 500-kΩ linear potentiometer, how much is the output voltage with 400 kΩ between lugs 1 and 2?

11-4
RESISTOR COLOR CODING

Because carbon resistors are small physically, they are color-coded to indicate their resistance value in ohms. The basis of this system is the use of colors for numerical values, as listed in Table 11-1. In memorizing these colors, remember that the dark colors, black and brown, correspond to the lowest numbers, zero and one, through lighter colors, to white for nine. The color coding is standardized by the Electronics Industries Association (EIA). These codes are summarized in Appendix G.

Resistance Color Bands. This code is the most common system used for color-coding insulated carbon resistors having axial leads, as illustrated in Fig. 11-8. Color bands are printed at one end of the insulating body, which is usually tan. Reading from left to right, the first color band close to the edge indicates the first digit in the numerical value of resistance. The second band gives the second digit. The third band is the decimal multiplier, giving the number of zeros after the two digits. The resulting number is the resistance in ohms.

As examples, in Fig. 11-9a the first stripe is yellow for 4; the second stripe is violet for 7; the orange multiplier means "add three zeros to 47," or "this multiplier is 10^3." Therefore, this resistance value is 47×10^3, or 47,000 Ω.

The example in Fig. 11-9b illustrates that black for the third stripe just means "do not add any zeros to the first two figures." Since this resistor has yellow, violet, and black stripes, the resistance value is 47 Ω.

Resistors Under 10 Ω. These resistors have a third stripe of gold or silver, which are fractional decimal multipliers. When the third stripe is gold, multiply the first two digits by 0.1. Silver is a multiplier of 0.01. For example, in Fig. 11-9c the color stripes are yellow, violet, and gold. The resistance is 47×0.1, therefore, which equals 4.7 Ω. If the multiplier were silver, the resistance would be 47×0.01, or 0.47 Ω.

Gold and silver are fractional multipliers only in the third stripe. However, gold and silver are most often used as a fourth stripe to indicate how accurate the resistance value is.

TABLE 11-1. Color Code

COLOR	VALUE	COLOR	VALUE
Black	0	Green	5
Brown	1	Blue	6
Red	2	Violet	7
Orange	3	Gray	8
Yellow	4	White	9

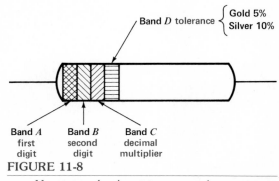

FIGURE 11-8
How to read color stripes on carbon resistors.

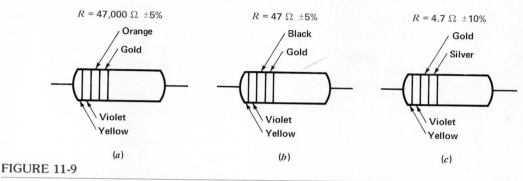

FIGURE 11-9

Examples of color-coded *R* values.

Resistor Tolerance. The amount by which the actual resistance can be different from the color-coded value is the *tolerance,* usually given in percent. For instance, a 100,000-Ω resistor with ±10 percent tolerance can have a resistance 10 percent below or above the coded value. This resistance, therefore, is between 90,000 and 110,000 Ω. The inexact value of carbon resistors is a disadvantage resulting from their economical construction, but in most circuits 5 to 10 percent variation in resistance can be tolerated.

As illustrated in Fig. 11-8, silver in the fourth band indicates a tolerance of ±10 percent; gold is ±5 percent. When there is no fourth stripe, the tolerance is ±20 percent. The smaller the tolerance, the more accurate the resistance value and the higher the cost. Some resistors may have a fifth color band to designate a military failure rate, indicating reliability.

Wirewound Resistor Marking. Usually, wirewound resistors are physically big enough to have their resistance value and tolerance printed on the insulating case. The tolerance of most is ±5 percent, except for precision resistors, which have a tolerance of ±1 percent or less. Some small wirewound resistors may be coded with stripes, however, like carbon resistors. In this case, the first stripe is double the width of the others to indicate a wirewound resistor.

Preferred Resistance Values. In order to minimize the problem of manufacturing different resistance values for an almost unlimited variety of circuits, certain values of fixed carbon-composition resistors are manufactured in large quantities so that they are cheaper and more easily available than unusual sizes. For resistors of ±10 percent tolerance, the preferred values are 10, 12, 15, 18, 22, 27, 33, 39, 47, 56, 68, and 82 and their decimal multiples, such as 820, 8200, 82,000, or 820,000. In this way, a preferred value is available for any resistance within 10 percent. For more accurate resistors of low tolerance, there are additional preferred values. (See Appendix G, Table G-4.)

*Practice Problems 11-4
(answers on page 239)*
(a) What is the color for 5?
(b) What is the tolerance for a silver stripe?
(c) What is the multiplier for yellow in the third stripe?

11-5
POWER RATING OF RESISTORS

In addition to having the required ohms value, a resistor should have a wattage rating high enough to dissipate the I^2R power produced by the current flowing through the resistance, without becoming too hot. Carbon resistors in normal operation are often quite warm, up to a maximum temperature of about 85°C, which is close to the 100°C boiling point of water. Carbon resistors should not be so hot, however, that they "sweat" beads of liquid on the insulating case. Wirewound resistors operate at very high temperatures, a typical value being 300°C for the maximum temperature. If a resistor becomes too hot because of excessive power dissipation, it can change appreciably in resistance value or burn open.

The power rating is a physical property depending on the resistor construction, especially physical size. Note the following:

1. A larger physical size indicates a higher power rating.
2. Higher-wattage resistors can operate at higher temperatures.
3. Wirewound resistors are physically larger with higher wattage ratings than carbon resistors.

For a common example, a 2-W carbon resistor is about 1 in long with $\frac{1}{4}$ in diameter.

For both types, a higher power rating allows a higher voltage rating. This rating gives the highest voltage that may be applied across the resistor without internal arcing. In wirewound resistors, excessive voltage can produce an arc between turns; in carbon resistors, the arc is between carbon granules.

Shelf Life. Resistors keep their characteristics almost indefinitely, when not used. Without any current in a circuit to heat the resistor, the resistor has practically no change with age. The shelf life of resistors is usually no problem, therefore. Actually, the only components that should be used fresh from manufacture are batteries and electrolytic capacitors.

Practice Problems 11-5
(*answers on page 239*)
Answer true or false.
(a) A 5-Ω 50-W resistor is physically larger than a 5-MΩ 1-W resistor.
(b) Resistors should not operate above a temperature of 0°C.

11-6
CHOOSING THE RESISTOR FOR A CIRCUIT

In determining what size resistor to use, the first requirement is to have the amount of resistance needed. In Fig. 11-10a, for example, suppose that a resistor is to be inserted in series with R_1 for the purpose of limiting the current through the 900-Ω resistance to 0.1 A with a 100-V source. Since the total resistance required is 1000 Ω, the 100-Ω resistor R_2 is added in series with R_1. The current through both R_1 and R_2 is then 0.1 A.

The I^2R power dissipated in R_2 is 1 W, equal to 0.01 × 100. However, a 2-W resistor would normally be used. This safety factor of 2 in the power rating is common practice with carbon resistors so that they will not become too hot in normal operation.

A resistor with a higher wattage rating but the same resistance would allow the circuit to operate normally and last longer without breaking down. However, it can be inconvenient when the next larger size is wirewound and physically bigger. Wirewound resistors can operate closer to their power rating, assuming adequate ventilation, because of their higher maximum operating temperature, compared with carbon resistors.

Another example is shown in Fig. 11-10b. The 10-MΩ resistor R_4 is used with R_3 to pro-

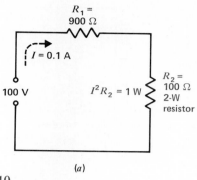 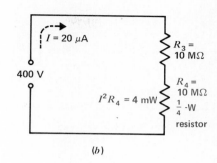

FIGURE 11-10 Examples of power rating for resistors. (a) R_2 dissipates 1 W, but a 2-W resistor is used for a safety factor of 2. (b) R_4 dissipates 0.004 W, but a $\frac{1}{4}$-W resistor is used.

vide an IR voltage drop equal to one-half the applied voltage of 400 V. Since R_3 and R_4 are equal, each has an IR voltage drop of 200 V. Their total series resistance of 20 MΩ limits the current in the circuit to 20 µA.

The I^2R power dissipated in R_4 is 4 mW. The wattage rating used, however, is $\frac{1}{4}$ W. In this case, the wattage rating is much higher than the actual amount of power dissipated in the resistor.

Notice the small amount of power dissipated in this circuit, with an applied voltage of 400 V. The reason is that the very high resistance limits the current to a low value.

In general, using a resistor with a high enough wattage rating automatically provides the required voltage rating. The exception, however, is a low-current high-voltage circuit where the applied voltage is of the order of a kilovolt or more.

Practice Problems 11-6
(answers on page 239)

(a) In Fig. 11-10a, how much is the product of $V_2 \times I$?
(b) In Fig. 11-10b, how much is the product of $V_4 \times I$?

11-7
SERIES AND PARALLEL COMBINATIONS OF RESISTORS

In some cases two or more resistors are combined in series or parallel to obtain a desired resistance value with a higher wattage rating. Several examples are shown in Fig. 11-11.

The total resistance depends on the series and parallel connections. However, the combination has a power rating equal to the sum of the individual wattage ratings, whether the resistors are in series or in parallel. The reason is that the total physical size increases with each added resistor. Such combinations can be used to obtain a higher power rating.

In Fig. 11-11a, the two equal resistors in series double the resistance. Also, the power rating of the combination is twice the value for one resistor.

In Fig. 11-11b, the two equal resistors in parallel have one-half the resistance. However, the combined power rating is still twice the value for one resistor.

In Fig. 11-11c, the series-parallel combination of four resistors makes R_T the same as each R. However, the total power rating is four times the value for one resistor.

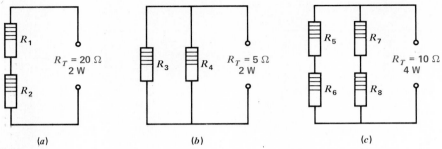

FIGURE 11-11

Total resistance and wattage rating for series and parallel combinations. All resistors are 10 Ω with 1-W rating. (a) In series, add wattage ratings. (b) In parallel, add wattage ratings. (c) In series-parallel, add wattage ratings.

Practice Problems 11-7
(*answers on page 239*)

(a) Two 5-kΩ 5-W resistors in series have R_T of 10 kΩ and what power rating?
(b) Two 5-kΩ 5-W resistors in parallel have R_T of 2.5 kΩ and what power rating?

11-8
RESISTOR TROUBLES

The most common trouble in resistors is an open. When the open resistor is a series component, there is no current in the entire series path and it cannot operate. Carbon-composition resistors are often partially open, with much higher resistance than the color-coded value.

Noisy Controls. In applications such as volume and tone controls, carbon controls are preferred because the smoother change in resistance results in less noise when the variable arm is rotated. With use, however, the resistance element becomes worn by the wiper contact, making the control noisy. When a volume or tone control makes a scratchy noise as the shaft is rotated, it indicates a worn-out resistance element.

Checking Resistors with an Ohmmeter. Since the ohmmeter has its own voltage source, it is always used without any external power applied to the resistance being measured. Just connect the ohmmeter leads across the resistance to be measured.

An open resistor reads infinitely high ohms. For some reason, infinite ohms is often confused with zero ohms. Remember, though, that infinite ohms means an open circuit. The current is zero but the resistance is infinitely high. Furthermore, it is practically impossible for a resistor to become short-circuited in itself. The resistor may be short-circuited by some other part of the circuit. However, the construction of resistors is such that the trouble they develop is an open, with infinitely high ohms.

The ohmmeter must have an ohms scale capable of reading the resistance value, or the resistor cannot be checked. In checking a 10-MΩ resistor, for instance, if the highest reading is 1 MΩ, the ohmmeter will indicate infinite resistance, even if the resistor has its normal value of 10 MΩ. An ohms scale of 100 MΩ or more should be used for checking such high resistances.

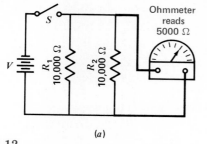 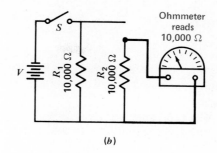

(a) (b)

FIGURE 11-12
Parallel resistance of R_1 can lower ohmmeter reading for testing R_2. (a) R_1 and R_2 in parallel. (b) R_2 isolated from R_1.

Similarly, in checking resistance values of less than 10 Ω, a low-ohms scale of about 100 Ω or less is necessary. Otherwise, the ohmmeter will read a normally low resistance value as zero ohms.

When checking resistance in a circuit, it is important to be sure there are no parallel paths across the resistor being measured. Otherwise, the measured resistance can be much lower than the actual resistor value, as illustrated in Fig. 11-12a. Here, the ohmmeter reads the resistance of R_2 in parallel with R_1. To check across R_2 alone, one end is disconnected, as in b.

For very high resistances, it is important not to touch the ohmmeter leads. There is no danger of shock, but the body resistance of about 50,000 Ω as a parallel path will lower the ohmmeter reading.

Practice Problems 11-8
(answers on page 239)
(a) What is the ohmmeter reading for a short circuit?
(b) What is the ohmmeter reading for an open resistor?

TABLE 11-2. Comparison of Resistor Types

CARBON-COMPOSITION RESISTORS	WIREWOUND RESISTORS
Carbon granules in binder	Turns of resistance wire
R up to 20 MΩ	R down to a fraction of 1 Ω
Color-coded for resistance value	Resistance printed on unit
For low-current circuits; power ratings of $\frac{1}{10}$ to 2 W	For high-current circuits; ratings of 5 to over 100 W
Variable potentiometers and rheostats up to 5 MΩ, for controls such as volume and tone in receivers	Low-resistance rheostats for varying current; potentiometers up to 50 kΩ for voltage divider in power supply

Summary

1. The two main types of resistors are carbon-composition and wirewound. Their characteristics are compared in Table 11-2. The schematic symbols for fixed and variable resistances are summarized in Fig. 11-13.
2. A rheostat is a variable series resistance with two terminals to adjust the amount of current in a circuit.
3. A potentiometer is a variable voltage divider with three terminals.
4. Carbon resistors are practically always color-coded, as in Fig. 11-8 and 11-9, to indicate the resistance value.
5. The wattage rating of carbon resistors depends mainly on their physical size, larger resistors being able to dissipate more power. The power rating is not part of the color coding but may be printed on the resistor or judged from its size.
6. With carbon resistors the wattage rating should be about double the actual I^2R power dissipation for a safety factor of 2 or more.
7. Carbon resistors can be combined for a higher wattage rating. The total power rating is the sum of the individual wattage values, whether in series or in parallel. In series, though, the total resistance increases; in parallel, the combined resistance decreases.
8. The most common trouble in resistors is an open. An ohmmeter reads infinite ohms across the open resistor, assuming there is no parallel path.

Self-Examination (Answers at back of book.)

Choose (a), (b), (c), or (d).

1. Which of the following are typical resistance and power-dissipation values for a wirewound resistor? (a) 1 MΩ, ⅓ W; (b) 500 Ω, 1 W; (c) 50,000 Ω, 1 W; (d) 10 Ω, 50 W.
2. Which of the following are typical resistance and power-dissipation values for a carbon-composition resistor? (a) 100,000 Ω, 1 W; (b) 5 Ω, 5 W; (c) 10,000 Ω, 10 W; (d) 1000 Ω, 100 W.
3. For a carbon-composition resistor color-coded with yellow, violet, orange, and silver stripes from left to right, the resistance and tolerance are (a) 740 Ω ±5 percent; (b) 4700 Ω ±10 percent; (c) 7400 Ω ±1 percent; (d) 47,000 Ω ±10 percent.
4. For a carbon-composition resistor color-coded with green, black, gold, and silver stripes from left to right, the resistance and tolerance

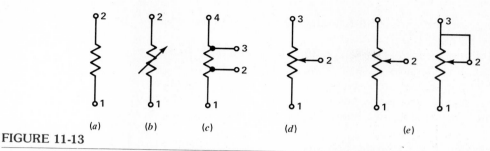

FIGURE 11-13
Schematic symbols for types of resistance. (a) Fixed R. (b) Any type of variable R. (c) Tapped R. (d) Potentiometer. (e) Potentiometer used as rheostat.

are (a) 0.5 Ω ±5 percent; (b) 0.5 Ω ±10 percent; (c) 5 Ω ±10 percent; (d) 50 Ω ±10 percent.
5. A resistor with the color-coded value of 100 Ω and ±20 percent tolerance can have an actual resistance between (a) 80 and 120 Ω; (b) 90 and 110 Ω; (c) 98 and 102 Ω; (d) 100 and 120 Ω.
6. Two 1000-Ω 1-W resistors are connected in parallel. Their combined resistance value and wattage rating is (a) 500 Ω, 1 W; (b) 500 Ω, 2 W; (c) 1000 Ω, 2 W; (d) 2000 Ω, 2 W.
7. A resistor is to be connected across a 45-V battery to provide 1 mA of current. The required resistance with a suitable wattage rating is (a) 4.5 Ω, 1 W; (b) 45 Ω, 10 W; (c) 450 Ω, 2 W; (d) 45,000 Ω, ¼ W.
8. Which of the following is a preferred resistor value? (a) 47; (b) 520; (c) 43,000; (d) 54,321.
9. When checked with an ohmmeter, an open resistor reads (a) zero; (b) infinite; (c) high but within the tolerance; (d) low but not zero.
10. One precaution in checking resistors with an ohmmeter is: (a) Check high resistances on the lowest ohms range. (b) Check low resistances on the highest ohms range. (c) Disconnect all parallel resistance paths. (d) Check high resistances with your fingers touching the test leads.

Essay Questions

1. Show how to connect two 1000-Ω 1-W resistors to obtain 2000 Ω with a power rating of 2 W.
2. State the colors corresponding to digits 1 to 9, inclusive.

3. Give the color coding for the following resistance values: 1 MΩ, 33,000 Ω, 8200 Ω, 150 Ω, and 68 Ω.
4. A 50-Ω rheostat R_1 is in series with a 25-Ω R_2, with a 50-V source. Draw a graph of I against R_1 as it is varied in 10-Ω steps.
5. Why do high-resistance carbon-composition resistors, of the order of megohms, usually have a low power rating of 1 W or less?
6. Name three factors that determine the resistance of a wirewound resistor.
7. Describe briefly how you would check a 5-MΩ resistor to see if it is open. State two precautions to make sure the check is not misleading.
8. Show how to connect resistors for the following examples: (a) two 20-kΩ 1-W resistors for a total of 10,000 Ω with a power rating of 2 W; (b) two 20-kΩ 1-W resistors for a total of 40,000 Ω with a power rating of 2 W; (c) four 10-kΩ 1-W resistors for a total of 10,000 Ω with a power rating of 4 W; (d) three 10-kΩ resistors for a total of 15,000 Ω.
9. Show the schematic diagram of a 50-kΩ potentiometer to tap off a variable voltage from a 45-V battery.

Problems (Answers to odd-numbered problems at back of book.)

1. A current of 1 mA flows through a 1-MΩ 2-W carbon resistor. (a) How much power is dissipated as heat in the resistor? (b) How much is the maximum power that can be dissipated without excessive heat?
2. A resistor is to be connected across a 10-V battery for the desired current of 1 mA. (a) What size resistance is required? (b) How much is its power dissipation in this circuit? (c) State the wattage rating of the resistor to be used. (d) Can this be a carbon resistor?
3. Give the resistance and tolerance for the following examples of resistors with color stripes: (a) yellow, violet, yellow, and silver; (b) red, red, green, and silver; (c) orange, orange, black, and gold; (d) white, brown, brown, and gold; (e) red, red, gold, and gold; (f) brown, black, orange, no tolerance band.
4. For resistors color-coded in the body-end-dot system shown in Appendix G, what is the resistance for the following: (a) body brown, end black, dot orange; (b) resistor completely red?

5. Determine the resistance and power rating of a carbon-composition resistor to fit the following requirements: 5-V *IR* drop, with 100-mA current and a safety factor of 2 for power dissipation.
6. Fill in the resistance values of the total *R* for the following dial settings on the decade box in Fig. 11-4.

Dial Settings

$R \times 10^4$	$R \times 10^3$	$R \times 10^2$	$R \times 10$	R	TOTAL R
9	6	7	4	2	
0	5	6	8	3	
6	7	0	5	4	
1	2	3	4	5	
5	4	3	2	1	

Answers to Practice Problems

11-1 (a) Carbon
    (b) Wirewound
11-2 (a) Terminal 2
    (b) Nonlinear
11-3 (a) Two
    (b) Three
    (c) 80 V
11-4 (a) Green
    (b) ±10 percent

11-4 (c) 10^4
11-5 (a) T
    (b) F
11-6 (a) 1 W
    (b) 4 mW
11-7 (a) 10 W
    (b) 10 W
11-8 (a) 0 Ω
    (b) Infinite ohms

Batteries

Chapter 12

A battery is a group of cells which produce electrical energy from their internal chemical reaction. Specifically, the battery is a source of steady dc voltage. More generally, a battery is a good example of a generator with an internal resistance that affects the output voltage.

This chapter describes the main types, such as dry batteries for portable equipment and the lead-acid wet cell used in automobile batteries and many industrial applications (Fig. 12-1). Most common is the carbon-zinc dry cell with an output of 1.5 V. The output of the lead-acid cell is 2.2 V. Additional types often used are the 1.35-V mercury cell and the 1.25-V nickel-cadmium cell.

The chemical battery has always been important as a dc voltage source for radio and electronic equipment. The reason is that to amplify a signal an amplifier needs dc operating potentials in order to conduct. Originally, all radio receivers used batteries. Then rectifier power supplies were developed to convert the ac power line to dc voltage, eliminating the need for batteries. However, now batteries are used more than ever for transistorized portable equipment which can operate without being connected to the ac power line. The topics here explaining batteries are:

12-1 Functions of Batteries
12-2 The Voltaic Cell
12-3 Carbon-Zinc Dry Cell
12-4 Series and Parallel Cells
12-5 Lead-Acid Wet Cell
12-6 Types of Electromotive Cells
12-7 Internal Resistance of a Generator
12-8 Matching a Load Resistance to the Generator

12-1
FUNCTIONS OF BATTERIES

From the old days of radio, dry batteries are still called A, B, or C batteries, according to their original functions in vacuum-tube operation. The A battery supplies current to heat the filament so that electrons can be emitted from the cathode. A typical rating for an A battery is 4.5 V with a load current of 150 mA. The C battery was used for a small negative grid bias of -1.5 to -7.5 V, with practically no current drain.

The A and C batteries are seldom used, but the function of the B battery is the same now as it always was. For a vacuum-tube amplifier, the B battery supplies positive dc plate voltage to attract the electrons emitted from the cathode. This acceleration of electrons to the anode or plate is necessary for the flow of plate current. In fact, the plate supply for amplifier tubes is still called B+ voltage, whether it comes from a battery or a rectifier power supply.

Furthermore, transistors also need a steady dc voltage for the collector electrode,

Chapter 12
Batteries

FIGURE 12-1
Typical batteries. (*a*) Miniature 9-V battery for transistor radio. (*b*) Lead-acid 12-V automobile battery, with six cells in series.

which corresponds to the plate in a tube. The collector voltage may be positive or negative, depending on whether the transistor is NPN or PNP. Positive collector voltage, also called $B+$ voltage, is needed for NPN transistors, which are most common. Common values of $B+$ voltage for transistor batteries are 4.5 and 9 V. Note that these values are multiples of 1.5 V, which is the output voltage of the basic carbon-zinc dry cell.

Practice Problems 12-1
(*answers on page 262*)
(*a*) How much is the output voltage of the lead-acid wet cell?
(*b*) How much is the output voltage of the carbon-zinc dry cell?

12-2
THE VOLTAIC CELL

When two different conducting materials are immersed in an electrolyte, as illustrated in Fig. 12-2*a*, the chemical reaction of forming a new solution results in the separation of charges. This arrangement for converting chemical energy into electrical energy is a *voltaic cell*.

The charged conductors are the electrodes, serving as the terminals of the cell for connection to an external circuit. Then the potential difference resulting from the separated charges enables the cell to function as a source of applied voltage. As shown in Fig. 12-2*b*, the voltage across the cell's terminals forces current to flow in the external circuit to light the bulb. Electrons from the negative terminal of the cell flow through the external circuit and return to the positive terminal. The chemical action in the

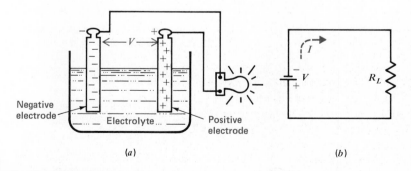

FIGURE 12-2
Voltaic cell converting chemical energy into electrical energy. (*a*) Electrodes in electrolyte. (*b*) Schematic diagram.

cell continuously separates charges to maintain the terminal voltage that produces current in the circuit.

Separation of Charges in the Cell. When metals dissolve in water or an electrolyte, the chemical action of forming the solution causes separation or dissociation of molecules, which results in electrically charged ions. Figure 12-3 illustrates the action of zinc dissolving in an ammonium chloride solution as the electrolyte. Where the electrode contacts the solution, molecules of zinc dissolve to form a different compound, which is zinc ammonium chloride. This chemical reaction requires zinc ions. Each zinc ion in solution is positive, with a deficiency of 2 electrons. In the solution process, then, each molecule of zinc dissolving in the electrolyte leaves 2 surplus electrons on the zinc electrode.

In Fig. 12-3, zinc ions are shown in solution with two positive charges, corresponding to the 2 electrons left behind. As a result, the entire solid zinc conductor has excess electrons that make the electrode negative with respect to the solution. With a second electrode, other than zinc, immersed in the electrolyte, the result is a voltaic cell with a difference of potential between the two electrodes.

When the potential difference of a cell is used to produce current in an external circuit, electrons leave the negative electrode, go through the external load resistance, and are added to the positive electrode. This current tends to neutralize the charges generated by the cell. The chemical reaction can increase, however, to maintain the separation of charges as the cell works harder to produce its output voltage. With the negative zinc electrode in Fig. 12-3, as more electrons are lost through the external circuit, more ions can be formed by the zinc dissolving in solution. Eventually, all the zinc will dissolve, and the cell then cannot separate charges to produce output voltage.

Primary Cells. In a primary cell, the chemical process of forming the solution is not reversible. For instance, zinc can dissolve in ammonium chloride, but the process cannot be reversed to form the zinc electrode from the solution. The many forms of dry cells, such as the flashlight cell, are examples of a primary cell.

Secondary Cells. Here the chemical action is reversible. The electrodes can dissolve in solution with current in one direction, or the current can be reversed to make the solution build up the electrodes. When the electrodes are going into solution, the cell is *discharging* as the current tends to neutralize the separated charges. Then the cell is used as a voltage source to produce current in a load resistance.

For the opposite case, reversing the current to re-form the plates is *charging* the cell. The charging current must be supplied by an external voltage source, with the cell then serving just as a load resistance. Since a secondary cell can be recharged, it is also a *storage cell*. The most common form of storage cell is the lead–sulfuric acid cell generally used in automobile batteries.

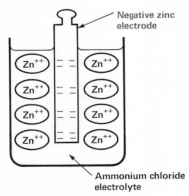

FIGURE 12-3

Zinc electrode being charged negative by ionization in electrolyte.

Cell Requirements. The two conductors used for electrodes in the electrolyte must be different to form a voltaic cell. With the same electrodes, each could become charged, but there would be no difference of potential across the terminals. The voltage output of one voltaic cell is approximately 1 to 2 V, depending only on the materials used for the electrodes and the electrolyte. The physical size of the cell determines its current capacity, since larger electrodes can supply more current.

*Practice Problems 12-2
(answers on page 262)
Answer true or false.*
(a) The lead-acid cell can be recharged.
(b) The carbon-zinc cell is a storage cell.

12-3
CARBON-ZINC DRY CELL

Figure 12-4 shows an internal view. The can is zinc, which functions both as a container to hold the electrolyte and as the negative electrode. The positive electrode is a carbon rod down the center but not low enough to touch the zinc.

The electrolyte is a solution of ammonium chloride, called *sal ammoniac,* and water. The electrolyte is not in liquid form, however, but is a paste produced by saturating granulated carbon and powdered manganese dioxide with the ammonium chloride solution. Although not completely dry, the cell can be mounted in any position, while a cell with liquid electrolyte must be upright.

Polarization. When zinc dissolves in ammonium chloride, the ammonia molecules, which contain nitrogen and hydrogen, release hydrogen. The resulting collection of hydrogen around the carbon electrode is called *polarization* of the cell. Polarization reduces the output voltage of the dry cell.

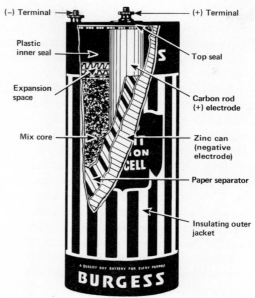

FIGURE 12-4
Construction of carbon-zinc dry cell. This is No. 6 size, 6 in high.

To minimize polarization, hydrogen gas must be removed from the carbon electrode. This is the function of the manganese dioxide, which is the *depolarizer* or depolarizing agent, in the dry cell. Rich in oxygen, the manganese dioxide releases oxygen that combines with the hydrogen to form water. As a result, the dry cell can supply more output voltage, without the collection of hydrogen gas around the carbon.

Local Action. If the zinc electrode contains impurities, such as iron and carbon, small voltaic cells are formed that do not add to the output voltage of the cell. This local action at the zinc electrode does use up the zinc, however. To minimize local action, the zinc electrode is generally coated with mercury, a process called *amalgamation*.

Shelf Life. Because of local action and drying of the paste electrolyte, a dry cell slowly loses its ability to produce output voltage, even when out of use and stored on a shelf. For this reason, dry cells are usually dated and should be used when fresh from manufacture. The shelf life for dry cells of medium size is about a year. For very small sizes, such as penlight cells, the shelf life may be only a few months.

Operating Characteristics. The output voltage of a carbon–zinc–sal ammoniac dry cell is 1.4 to 1.6 V, regardless of size. However, larger sizes with more zinc, electrolyte, and depolarizer have a higher current rating. Maximum ratings for continuous operation range from a few milliamperes for the very small cells to $\frac{1}{4}$ A for the relatively large No. 6 dry cell, which is 6 in high, with a diameter of $2\frac{1}{2}$ in. The size D flashlight cell has a current rating of 50 mA for approximately 60 h of service. This size is $2\frac{1}{4}$ in high, with a diameter of $1\frac{1}{4}$ in.

Current Drain Depends on the Load Resistance. It is important to note that the current rating is only a guide to typical values permissible for normal service life. The amount of current produced by a cell connected to a load resistance equals V/R, by Ohm's law.

Figure 12-5 illustrates three different cases of using the applied voltage of 1.5 V from a No. 6 dry cell. In (a), the load resistance R_1 of 6 Ω across the 1.5-V source voltage allows a value of current equal to $^{1.5}/_6$, which is 0.25 A, or 250 mA. This amount happens to be the same as the maximum current rating.

In (b), though, the higher value of 150 Ω for R_2, as the load resistance, limits the current to $^{1.5}/_{150}$, equal to 0.010 A, or 10 mA. This value is less than the current rating, which means only that the cell can be in service a longer time. It need not work so hard to produce the 1.5-V output with less current drain.

In (c), the 5-Ω load resistance R_3 is low enough to result in more current than the rated value of the cell. This much current is not desirable for long cell life, but assuming that the 1.5-V output is maintained, the current must be 1.5 V/5 Ω, which equals 0.3 A. In summary, then, the current drain in a cell connected to a load resistance is equal to the output voltage of the cell divided by the resistance in the circuit.

Practice Problems 12-3
(answers on page 262)
Answer true or false.
(a) The voltage output of a size D cell is less than that of a No. 6 cell.
(b) A cell rated at 250 mA will produce this I for any value of R_L.

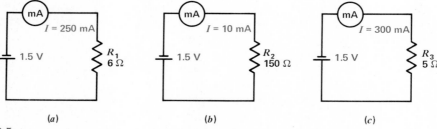

FIGURE 12-5

The current drain from the cell depends on the load resistance. (a) V/R_1 equals normal current rating of $\frac{1}{4}$ A or 250 mA. (b) V/R_2 equals I of 10 mA. (c) V/R_3 equals I of 300 mA.

12-4
SERIES AND PARALLEL CELLS

An applied voltage higher than the emf of one cell can be obtained by connecting cells in series. The total voltage available across the battery of cells is equal to the sum of the individual values for each cell. Parallel cells have the same voltage as one cell but have more current capacity.

Figure 12-6 shows the series-aiding connections for three dry cells. Here the three 1.5-V cells in series provide a total battery voltage of 4.5 V. Notice that the two end terminals A and B are left open to serve as the plus and minus terminals of the battery. These terminals are used for connecting the battery to the load circuit, as shown in (c).

In the lead-acid battery in Fig. 12-1b short heavy metal straps connect the cells in series. The current capacity of a battery with series cells is the same as for one cell because the same current flows through all the series cells.

Parallel Connections. For more current capacity, the battery has cells in parallel as shown in Fig. 12-7. All the positive terminals are strapped together, as are all the negative terminals. Any point on the positive side can be the plus terminal of the battery, and any point on the negative side is the negative terminal.

The parallel connection is equivalent to increasing the size of the electrodes and electrolyte, which increases the current capacity. The voltage output of the battery, however, is the same as for one cell.

Identical cells in parallel all supply equal parts of the load current. For example, with three identical parallel cells producing a load current of 300 mA, each cell has a drain of 100 mA. Bad cells should not be connected in parallel with good cells, however, since the cells in good condition will supply more current, which may overload the good cells. In addition, a cell with lower output voltage will act as a load resistance, draining excessive current from the cells that have higher output voltage.

Series-Parallel Connections. In order to provide higher output and more current capacity, cells can be connected in series-parallel combinations. Figure 12-8 shows four No. 6 cells in series-parallel to form a battery that has a 3-V output with a current capacity of $\frac{1}{2}$ A. Two of the 1.5-V cells in series provide 3 V total output voltage. This series string has a current capacity of $\frac{1}{4}$ A, however, as for one cell.

To double the current capacity, another string is connected in parallel. The two strings in parallel have the same 3-V output as one string, but with a current capacity of $\frac{1}{2}$ A instead of the $\frac{1}{4}$ A for one string.

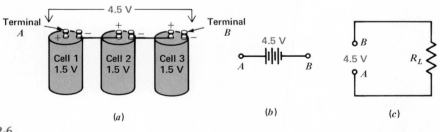

FIGURE 12-6
Cells in series to add voltages. (a) Wiring. (b) Schematic symbol. (c) Battery connected to load resistance.

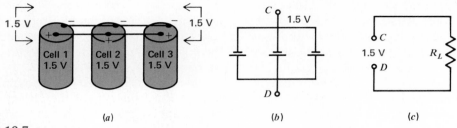

FIGURE 12-7

Cells in parallel to increase rating for maximum current. (*a*) Wiring. (*b*) Schematic symbol. (*c*) Battery connected to load resistance.

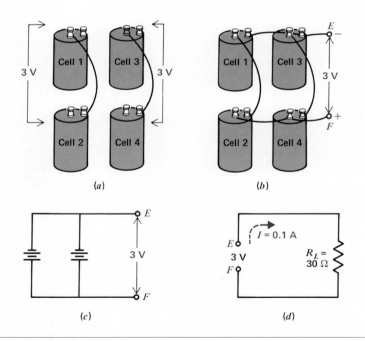

FIGURE 12-8

Cells in series-parallel combinations. (*a*) Wiring two 3-V strings, each with two 1.5-V cells in series. (*b*) Wiring the two 3-V strings in parallel. (*c*) Schematic symbol. (*d*) Battery connected to load resistance.

Referring to the circuit in Fig. 12-8d, note that the 3-V source produces 100 mA, or 0.1 A, equal to 3 V/30 Ω. The rated current capacity is 500 mA, but the actual load current is only 0.1 A.

Practice Problems 12-4
(answers on page 262)
(a) How many series cells are needed for 9-V output?
(b) How many parallel cells are needed to double the current rating?

12-5
LEAD-ACID WET CELL

Where high values of load current are necessary, the lead-acid cell is the type most commonly used. In the application of battery power to start an automobile, for example, the load current at starting time is 200 to 300 A. One lead-acid cell has an output of 2 to 2.2 V, but they are generally used in a series combination of three for a 6-V battery and six in series for a 12-V battery.

Since it is a storage cell, charging can be done repeatedly to restore the output voltage, as long as the cell is in good physical condition. Heat with excessive discharging and charging, however, shortens the useful life to about 3 years for an automobile battery.

Construction. The cutaway view of a lead-acid cell (Fig. 12-9) shows the component parts and how they are assembled. Inside the cell, the positive and negative electrodes consist of a group of plates welded to a connecting strap. The plates are immersed in a dilute solution of sulfuric acid for the electrolyte, which contains 8 parts of water to 3 parts of concentrated sulfuric acid.

Each plate is a grid or framework, made of a lead-antimony alloy. This construction enables the active material, which is lead oxide, to

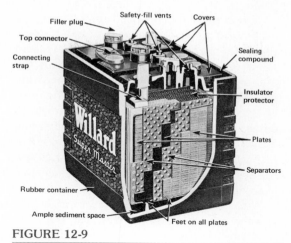

FIGURE 12-9
Construction of lead-acid cell.

be pasted into the grid. In manufacture of the cell, after the plates are dried, a forming charge produces the positive and negative electrodes. In the forming process, the active material in the positive plate is changed to lead peroxide to produce the positive electrode. The negative electrode is spongy lead.

Chemical Action. Sulfuric acid is a combination of hydrogen and sulfate ions. When the cell discharges, lead peroxide of the positive electrode unites with hydrogen ions to form water and with sulfate ions to form lead sulfate. The lead sulfate is also produced by combining lead on the negative plate with sulfate ions. Therefore, the net result of discharge is to produce more water, which dilutes the electrolyte, and to form lead sulfate on the plates.

As discharge continues, the sulfate fills the pores of the grids, retarding circulation of acid in the active material. Lead sulfate is the white powder often seen on the outside of old batteries. When the combination of weak electrolyte and sulfation on the plates lowers the output voltage, charging is necessary.

On charge, the reversed direction of the ions flowing in the electrolyte results in a reversal of the chemical reactions. Now the lead sulfate on the positive plate reacts with the water and sulfate ions to produce lead peroxide and sulfuric acid. This action re-forms the positive electrode and strengthens the electrolyte by adding sulfuric acid. At the same time, charging enables the lead sulfate on the negative plate to react with hydrogen ions, which also forms sulfuric acid while re-forming lead on the negative electrode.

As a result, the charging current can restore the cell to full output with lead peroxide on the positive plates, spongy lead on the negative plates, and the required concentration of sulfuric acid in the electrolyte. The chemical formula for the lead-acid cell is

$$Pb + PbO_2 + 2H_2SO_4 \underset{\text{Discharge}}{\overset{\text{Charge}}{\rightleftarrows}} 2PbSO_4 + 2H_2O$$

On discharge, the lead (Pb) and lead peroxide (PbO_2) electrodes supply Pb ions that combine with the sulfate ions (SO_4) to form lead sulfate ($PbSO_4$) and water (H_2O). On charge, with reverse current through the electrolyte, the chemical action is reversed. Then the Pb ions from the lead sulfate re-form the lead peroxide electrode. Also, the SO_4 ions combine with the H_2 ions to produce more sulfuric acid.

Current Ratings. Lead-acid batteries are rated in terms of how much discharge current can be supplied continuously for a specified period of time. The output voltage must be maintained above a minimum level, which is 1.5 to 1.8 V per cell. A common rating is ampere-hours (Ah) based on an 8-h discharge. Typical values are 100 to 300 Ah. For example, a 120-Ah battery can supply load current of $120/8$, or 15 A, based on an 8-h discharge. The battery can supply less current for a longer time or more current for a shorter time.

Note that the ampere-hour unit specifies coulombs of charge. As an example, 200 Ah corresponds to 200 A × 3600 s, which equals 720,000 A·s, or 7.2×10^5 C.

The ratings of a lead-acid battery are for a temperature of 80°F, approximately. Higher temperatures increase the chemical reaction, but operation above 110°F shortens the battery life. Note that 80°F equals 26.7°C. See Appendix D for temperature scales.

Lower temperatures reduce the voltage and current output. The ampere-hour capacity is reduced approximately 0.75 percent for each decrease of 1°F. At 0°F the available output is only 60 percent of the rating. In cold weather, therefore, it is important to have an automobile battery up to full charge. In addition, the electrolyte freezes more easily when diluted by water in the discharged condition.

Specific Gravity. The state of discharge for a lead-acid cell is generally checked by measuring the specific gravity of the electrolyte. This is a ratio comparing the weight of a substance with the weight of water. For instance, concentrated sulfuric acid is 1.835 times as heavy as an equal volume of water. Therefore, its specific gravity is 1.835. The specific gravity of water is 1, since it is the reference.

In a fully charged cell, the mixture of water and sulfuric acid results in a specific gravity of 1.280, approximately, at room temperatures of 70 to 80°F. As the cell discharges, the water formed dilutes the acid and lowers the specific gravity. When the specific gravity is down to about 1.150, the cell is completely discharged.

Specific-gravity readings are taken with a battery hydrometer, which has a calibrated float that will rest higher in an electrolyte of higher specific gravity. The decimal point is usually omitted for convenience. For example, a specific-gravity value of 1.250 is simply "twelve-fifty." A hydrometer reading of 1260 to 1280 indicates full charge, approximately 1250 is half-charge, and 1150 to 1200 is complete discharge.

Charging the Lead-Acid Battery. The requirements are illustrated in Fig. 12-10. An external dc voltage source is necessary to produce current in one direction. Also, the charging voltage must be more than the battery emf. Approximately 2.5 V per cell is enough to overcome the cell emf so that the charging voltage can produce current opposite to the direction of discharge current.

The reversed directions of charge current and discharge current are shown in Fig. 12-10a and b. In (a), the battery is the voltage source V_B producing current in the load resistance R_L. The direction of electrons in the external circuit is from the negative battery terminal, through R_L, and back to the positive terminal. Inside the battery, the ion current is from the positive electrode to the negative electrode.

In (b), however, the battery is effectively a load resistance across the external charging voltage source V_C. The net voltage available to produce charging current is the difference between the source and battery voltages, which equals 3 V here. The resultant electron current flows from the negative terminal of the charging voltage V_C to the less negative electrode of the battery and returns from the positive electrode of the battery to the more positive terminal of the charging voltage.

In the circuit outside the battery, the current in (b) is in the reverse direction from the current in (a). Inside the battery, the ion current produced in the electrolyte by the charging voltage is also opposite to the direction of ion current on discharge.

Note that the reversal of current is obtained just by connecting the battery and charging voltage with + to + and − to − as shown in (b). The charging current is reversed because the battery effectively becomes a load resistance for the charging voltage source when V_C is higher than V_B.

Battery Chargers. There are many commercial types that can be plugged into the ac power line to convert a 120-V ac input to the dc voltage output required for charging. Figure 12-11 shows one type that supplies a 6- to 12-V dc output. Since automobile radios operate on this battery voltage, the 6- to 12-V output of the eliminator can serve as a voltage source for testing these receivers on the service bench without a battery. In addition, the dc voltage output can be used to charge a battery.

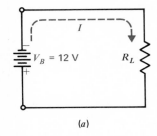
(a)

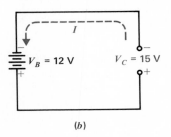
(b)

FIGURE 12-10

Reversed directions for charge and discharge current of battery. (a) V_B discharges to supply current for R_L. (b) Battery is load resistance for the source of charging voltage V_C.

Chapter 12
Batteries

FIGURE 12-11
Battery eliminator or charger. (*Heath Company*)

Practice Problems 12-5
(*answers on page 262*)
(a) How many lead-acid cells in series are needed for a 12-V battery?
(b) How many hours are needed with a 5-A charging current to supply 100 Ah of charge?

12-6
TYPES OF ELECTROMOTIVE CELLS

Most common are the zinc-carbon dry cell for primary batteries and the lead-acid wet cell for secondary or storage batteries. A primary battery cannot be recharged since its chemical reaction is not reversible. Sometimes the output can be increased temporarily by resting a primary battery or improving the depolarizing action at the electrodes. However, a secondary battery can be recharged repeatedly to bring the output up to the normal value when new.

Additional types are listed in Table 12-1 for primary cells and 12-2 for secondary cells, wet and dry. The specified voltages are nominal values without a load. Output voltage with load current normally drops 10 to 20 percent. More details on construction of recently developed types are given below.

Manganese-Alkaline Cell. The cathode is manganese dioxide and the anode is zinc in a leakproof steel can (Fig. 12-12). The electrolyte is potassium hydroxide, or sodium hydroxide, which is alkaline with negative hydroxyl (OH)

TABLE 12-1. Primary cells

TYPE	VOLTAGE	NOTES
Carbon-zinc	1.5	Most common type; A, B, and C batteries; for flashlights and toys; lowest cost but low current capacity.
Zinc-chloride	1.5	Similar to carbon-zinc type but higher current capacity.
Manganese-alkaline	1.5	Hydroxide electrolyte; high current capacity.
Mercury	1.35	Cathode is mercuric oxide; stable voltage reference.
Mercury	1.4	Cathode is mercuric oxide with manganese dioxide; miniature button sizes for cameras, electronic watches, and hearing aids; high current capacity; long shelf life.
Silver oxide	1.5	Hydroxide electrolyte; miniature button sizes.
Lithium	2.95	Long life; high cost; high current capacity and energy density.

TABLE 12-2. Secondary cells

TYPE	VOLTAGE	NOTES
Lead-acid	2.2	Wet electrolyte; lowest cost; very low r_i and very high current ratings.
Gelled-electrolyte, lead-acid	2.1	Moist electrolyte; sealed unit; needs no water added; calcium-lead in small sizes.
Nickel-iron	1.36	Edison cell; wet hydroxide electrolyte; industrial uses.
Nickel-cadmium	1.25	Most common rechargeable dry battery.
Silver-cadmium	1.1	Rechargeable dry cell; high current rating and energy density.
Silver-zinc	1.86	Rechargeable dry cell; high efficiency.

ions, while an acid electrolyte has positive hydrogen (H) ions. High conductivity of the electrolyte results in higher current ratings than the carbon-zinc cell. This is a primary cell, but some alkaline batteries can be re-energized a few times. Output is 1.5 V.

Mercury Cell. The cathode is mercuric oxide, with a zinc anode in an alkaline electrolyte. The anode and cathode are compacted powders in a liquid electrolyte, but they are in a double-sealed can assembly. With pure mercuric oxide for the cathode, voltage output is extremely stable at 1.35 V. This type of cell can be used as a voltage-reference source. The general-purpose type with an output of 1.4 V uses manganese dioxide with the mercuric oxide. Both types are primary cells. The flat pellet construction is illustrated in Fig. 12-13.

Silver Oxide Cell. As shown in Fig. 12-14, the cathode is silver oxide, with a powdered zinc anode in an alkaline electrolyte. This type is generally made in the button size for hearing aids, cameras, and electronic watches. Output of this primary cell is 1.5 V.

Zinc-Air Cell. This is an experimental fuel cell that extracts oxygen from the air to serve as the cathode. Output is 1.45 V.

Nickel-Cadmium Cell. This is a dry cell but rechargeable. It is the most common dry storage battery.

In the charged condition, the anode is nickel hydroxide with a cathode of metallic cadmium in an alkaline electrolyte of potassium hydroxide in a sealed can (Fig. 12-15). Output is 1.25 V.

The formula for the chemical reaction is

$$Cd + 2NiOH + 2KOH \rightleftharpoons Cd(OH)_2 + NiO + 2KOH$$

FIGURE 12-12

Alkaline-type battery. Output is 1.5 V.

Chapter 12
Batteries

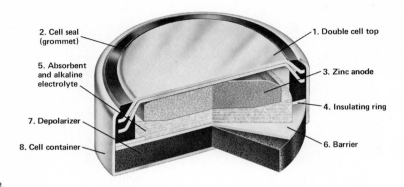

FIGURE 12-13

Construction of pellet type of mercury cell. (*Mallory Battery Company*)

This dry cell is a true storage cell with a reversible chemical reaction which allows recharging up to 1000 times. The standard type may take as long as 15 h to recharge, but fast-charging units are available. New batteries may need recharging before use.

Solar Battery. This contains photosensitive silicon cells. The semiconductor units generate voltage output when exposed to light. Typical output is 0.26 V per cell.

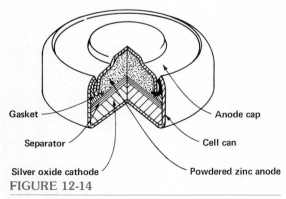

FIGURE 12-14

Button cell using silver oxide. Diameter is 0.3 to 0.5 in. Output is 1.5 V. (*Union Carbide Corporation*)

Practice Problems 12-6
(*answers on page 262*)

(a) How much is the voltage output of the nickel-cadmium cell?
(b) Is the mercury 1.35-V reference cell rechargeable?

12-7
INTERNAL RESISTANCE OF A GENERATOR

Any source that produces voltage output continuously is a generator. It may be a cell separating charges by chemical action or a rotary generator converting motion and magnetism into voltage output, for common examples. In any case, all generators have internal resistance, which is labeled r_i in Fig. 12-16.

The internal resistance r_i is important when a generator supplies load current because its internal Ir_i drop subtracts from the generated emf, resulting in lower voltage across the output terminals. Physically, r_i may be the resistance of the wire in a rotary generator, or in a chemical cell r_i may be the resistance of the electrolyte between electrodes. More generally, the internal resistance r_i is the opposition to load current inside the generator.

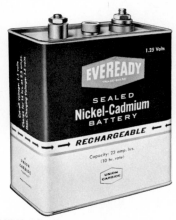

FIGURE 12-15

Nickel-cadmium battery. Output is 1.25 V.

Since any current in the generator must flow through the internal resistance, r_i is in series with the generated voltage, as shown in Fig. 12-16c. It may be of interest to note that, with just one load resistance connected across a generator, they are in series with each other because R_L is in series with r_i.

If there is a short circuit across the generator, its r_i prevents the current from becoming

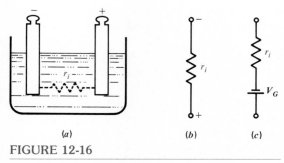

FIGURE 12-16

Internal resistance r_i is in series with the generated voltage. (a) Physical arrangement for a voltaic cell. (b) Schematic. (c) Equivalent circuit.

infinitely high. As an example, if a 1.5-V battery is temporarily short-circuited, the short-circuit current I_{sc} is about 10 A. Then r_i is V/I_{sc}, which equals $1.5/10$, or 0.15 Ω for the internal resistance of the battery. These are typical values for a No. 6 dry cell.

Why the Terminal Voltage Drops with More Load Current. Figure 12-17 illustrates how the output of a 100-V dry battery can drop to 90 V because of the internal 10-V Ir_i drop. In (a), the voltage across the output terminals is equal to the 100 V of V_G because there is no load current on an open circuit. With no current, the voltage drop across r_i is zero. Then the full generated voltage is available across the output terminals. This is the generated emf, open-circuit voltage, or no-load voltage.

We cannot connect the test leads inside the source to measure V_G. However, measuring this no-load voltage without any load current provides a method of determining the internally generated emf.

In Fig. 12-17b with a load, however, the current of 0.1 A flows to produce a drop of 10 V across the 100 Ω of r_i. Note that R_T is $900 + 100 = 1000$ Ω. Then I_L equals $100/1000$, which is 0.1 A.

As a result, the voltage output V_L equals $100 - 10 = 90$ V. This terminal voltage or load voltage is available across the output terminals when the generator is in a closed circuit with load current. The 10-V Ir_i drop is subtracted from V_G because they are series-opposing voltages.

The graph in Fig. 12-18 shows how the terminal voltage V_L drops with increasing load current I_L. The reason is the greater internal voltage drop across r_i, as shown by the calculated values listed in Table 12-3.

Across the top row, infinite ohms for R_L means an open circuit. Then I_L is zero, there is no internal drop V_i, and V_L is the same as V_G.

Chapter 12
254 Batteries

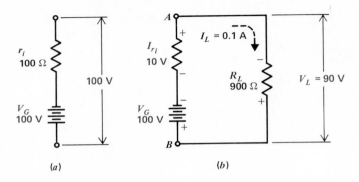

FIGURE 12-17
Internal voltage drop decreases terminal voltage of generator. (*a*) Open-circuit voltage is the V_G of 100 V because there is no load current. (*b*) Terminal voltage V_L between A and B is reduced to 90 V because of the 10-V drop across 100-Ω r_i with 0.1 A of I_L.

Across the bottom row, zero ohms for R_L means a short circuit. Then the short-circuit current of 1 A results in zero output voltage because the entire generator voltage is dropped across the internal resistance. Or we can say that with a short circuit of zero ohms across the load, the current is limited to V_G/r_i.

The lower the internal resistance of a generator, the better it is in terms of being able to produce full output voltage when supplying current for a load. For example, the very low r_i, about 0.01 Ω, for a 12-V lead-acid battery is the reason it can supply high values of load current and maintain its output voltage.

For the opposite case, a higher r_i means that the terminal voltage of a generator is much less with load current. As an example, an old dry battery with a 500-Ω r_i would appear normal when measured by a voltmeter but be useless because of low voltage when normal load current flows in an actual circuit.

How to Measure r_i. The internal resistance of any generator can be measured indirectly by determining how much the output voltage drops for a specified amount of load current. The difference between the no-load voltage and the load voltage is the amount of internal voltage drop $I_L r_i$. Dividing by I_L gives the value of r_i. As a formula,

$$r_i = \frac{V_{NL} - V_L}{I_L} \qquad (12\text{-}1)$$

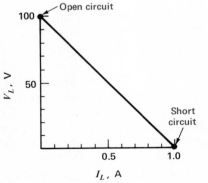

FIGURE 12-18
How terminal voltage V_L drops with more load current I_L. Graph is plotted for values in Table 12-3.

Chapter 12
Batteries   255

TABLE 12-3.   For Fig. 12-17. How V_L drops with more I_L

V_G, V	r_i, Ω	R_L, Ω	$R_T = R_L + r_i$, Ω	$I_L = V_G/R_T$, A	$V_i = I_L r_i$, V	$V_L = V_G - V_i$, V
100	100	∞	∞	0	0	100
100	100	900	1000	0.1	10	90
100	100	600	700	0.143	14.3	85.7
100	100	300	400	0.25	25	75
100	100	100	200	0.5	50	50
100	100	0	100	1.0	100	0

Example 1. Calculate r_i if the output of a generator drops from 100 V with zero load current to 80 V with a 2-A I_L.

Answer. $r_i = \dfrac{100-80}{2} = \dfrac{20}{2}$

$r_i = 10\ \Omega$

A convenient technique for measuring r_i is to use a variable load resistance R_L. Vary R_L until the load voltage is one-half the no-load voltage. This value of R_L is also the value of r_i, since they must be equal to divide the generator voltage equally. For the same 100-V generator with the 10-Ω r_i used in Example 1, if a 10-Ω R_L were used, the load voltage would be 50 V, equal to one-half the no-load voltage.

You can solve this circuit by Ohm's law to see that I_L is 5 A with a 20-Ω R_T. Then the two voltage drops of 50 V each add to equal the 100 V of the generator.

Constant-Voltage Generator. A generator with very low internal resistance is considered a constant-voltage source. Then the output voltage remains essentially the same when the load current changes. This idea is illustrated in Fig. 12-19a for a 6-V lead-acid battery with an r_i of 0.005 Ω. If the load current varies over the wide range of 1 to 100 A, for any of these values, the internal Ir_i drop across 0.005 Ω is less than 0.5 V.

The voltage output of 5.5 to 6.0 A can be considered constant, compared with the wide variations in load current. The relatively constant V_L is shown by the graph in (b).

Constant-Current Generator. The opposite case is a generator that has very high resistance compared with the external load resistance, resulting in constant current although the output voltage varies. Examples are found in vacuum-tube circuits, where a tube can be a generator having internal resistance as high as 1 MΩ.

The constant-current generator shown in Fig. 12-20 has such high resistance, with an r_i of 0.9 MΩ, that it is the main factor determining how much current can be produced by V_G. Here R_L varies in a 3:1 range from 50 to 150 kΩ. Since the current is determined by the total resistance of R_L and r_i in series, however, I is essentially constant at 1.05 to 0.95 mA, or approximately 1 mA. This relatively constant I_L is shown by the graph in (b).

Note that the terminal voltage V_L varies approximately in the same 3:1 range as R_L. Also, the output voltage is much less than the generator voltage because of its high internal resistance compared with R_L. This is a necessary condition, however, in a circuit with a constant-current generator.

A more common example is to insert a series resistance to keep the current constant,

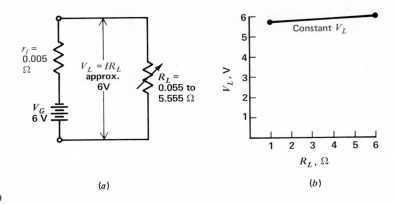

FIGURE 12-19

Constant-voltage generator with low r_i. The V_L stays approximately the same 6 V, as I varies with R_L. (a) Circuit. (b). Graph.

as shown in Fig. 12-21a. R_1 must be very high compared with R_L. In this example, I_L is 50 μA with 50 V applied, and R_T is practically equal to the 1 MΩ of R_1. The value of R_L can vary over a range as great as 10:1 without changing R_T or I_L appreciably.

The circuit with an equivalent constant-current source is shown in (b). Note the arrow symbol for a current source. As far as R_L is concerned, its terminals A and B can be considered as receiving 50 V in series with 1 MΩ or 50 μA in shunt with 1 MΩ.

Practice Problems 12-7
(answers on page 262)
(a) Does a constant-voltage source have high or low internal resistance?
(b) Does the internal voltage drop of a voltage source aid or oppose the generated voltage?

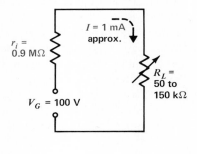
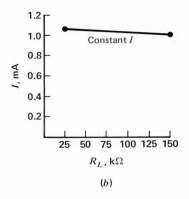

FIGURE 12-20

Constant-current generator with high r_i. The I_L stays approximately the same 1 mA, as V_L varies with R_L. (a) Circuit. (b) Graph.

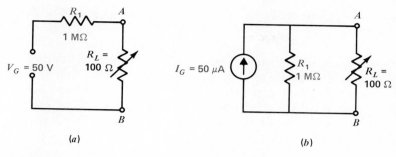

FIGURE 12-21

Voltage source in (a) equivalent to current source in (b) for R_L across terminals A and B.

12-8
MATCHING A LOAD RESISTANCE TO THE GENERATOR

In the diagram in Fig. 12-22, when R_L equals r_i, the load and generator are matched. The matching is significant because the generator then produces maximum power in R_L, as verified by the values listed in Table 12-4.

Maximum Power in R_L. When R_L is 100 Ω, matching the 100 Ω of r_i, maximum power is transferred from the generator to the load. With higher resistance for R_L, the output voltage V_L is higher, but the current is reduced.

Lower resistance for R_L allows more current, but V_L is less. When r_i and R_L both equal 100 Ω, the combinations of current and voltage produce the maximum power of 100 W across R_L.

With generators having very low resistance, however, matching is often impractical. For example, if a 6-V lead-acid battery with a 0.003-Ω internal resistance were connected to a 0.003-Ω load resistance, the battery could be

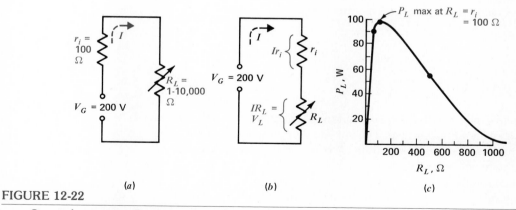

FIGURE 12-22

Circuit for varying R_L to match r_i. (a) Schematic diagram. (b) Equivalent voltage divider for voltage output across R_L. (c) Graph of power output P_L vs. R_L.

damaged by excessive current as high as 1000 A.

Maximum Voltage across R_L. If maximum voltage is desired, rather than power, the load should have as high a resistance as possible. Note that R_L and r_i form a voltage divider for the generator voltage, as illustrated in Fig. 12-22. The values tabulated for IR_L in Table 12-4 show how the output voltage V_L increases with higher values of R_L.

Maximum Efficiency. Note also that the efficiency increases as R_L increases because there is less current, resulting in less power lost in r_i. When R_L equals r_i, the efficiency is only 50 percent, since one-half the total generated power is dissipated in r_i, the internal resistance of the generator. In conclusion, then, matching the load and generator resistances is desirable when the load requires maximum power rather than maximum voltage or efficiency, assuming that the match does not result in excessive current.

Practice Problems 12-8
(answers on page 262)
Answer true or false.
(a) When $R_L = r_i$, the P_L is maximum.
(b) V_L is maximum when R_L is maximum.

TABLE 12-4. Effect of Load Resistance on Generator Output*

	R_L, Ω	$I = V_G/R_T$, A	Ir_i, V	IR_L, V	P_L, W	P_i, W	P_T, W	EFFICIENCY = P_L/P_T, %
	1	1.98	198	2	4	396	400	1
	50	1.33	133	67	89	178	267	33
$R_L = r_i \rightarrow$	100	1	100	100	100	100	200	50
	500	0.33	33	167	55	11	66	83
	1000	0.18	18	180	32	3.24	35.24	91
	10,000	0.02	2	198	4	0.04	4.04	99

*Values calculated approximately for circuit in Fig. 12-22, with $V_G = 200$ V and $r_i = 100$ Ω.

Summary

1. A voltaic cell consists of two different conductors as electrodes immersed in an electrolyte. The voltage output depends only on the chemicals in the cell. The current capacity increases with larger sizes. A primary cell cannot be recharged. A secondary or storage cell can be recharged.
2. A battery is a group of cells in series or in parallel. With cells in series, their voltages add, but the current capacity is the same as one cell. With cells in parallel, the voltage output is the same as that of one cell, but the total current capacity is the sum of the individual values.
3. The zinc–carbon–sal ammoniac dry cell is the most common type of

primary cell. Zinc is the negative electrode; carbon is the positive electrode. Its output voltage is approximately 1.5 V.
4. The lead-acid cell is the most common form of storage battery. The positive electrode is lead peroxide; spongy lead is the negative electrode. Both are in a dilute solution of sulfuric acid for the electrolyte. The voltage output is approximately 2.2 V per cell.
5. To charge a lead-acid battery, connect it to a dc voltage equal to approximately 2.5 V per cell. Connecting the positive terminal of the battery to the positive side of the charging source and the negative terminal to the negative side results in charging current through the battery.
6. The mercury cell is a primary cell with an output of 1.35 or 1.4 V.
7. The nickel-cadmium cell is a dry cell that is rechargeable, with an output of 1.25 V.
8. A constant-voltage generator has very low internal resistance. Output voltage is relatively constant with changing values of load because of the small internal voltage drop.
9. A constant-current generator has a very high internal resistance. This determines the constant value of current in the generator circuit relatively independent of the load resistance.
10. Any generator has an internal resistance r_i. With load current I_L, the internal $I_L r_i$ drop reduces the voltage across the output terminals. When I_L makes the terminal voltage drop to one-half the zero load voltage, the external R_L equals the internal r_i.
11. Matching a load to a generator means making the R_L equal to the generator's r_i. The result is maximum power delivered to the load from the generator.

Self-Examination (Answers at back of book.)

Choose (a), (b), (c), or (d).

1. Which of the following is false? (a) A lead-acid cell can be recharged. (b) A primary cell has an irreversible chemical reaction. (c) A storage cell has a reversible chemical reaction. (d) A carbon-zinc cell has unlimited shelf life.
2. The output of a lead-acid cell is (a) 1.25 V; (b) 1.35 V; (c) 2.2 V; (d) 6 V.
3. The current in a chemical cell is a movement of (a) positive hole charges; (b) positive and negative ions; (c) positive ions only; (d) negative ions only.

4. Cells are connected in series to (a) increase the voltage output; (b) decrease the voltage output; (c) decrease the internal resistance; (d) increase the current capacity.
5. Cells are connected in parallel to (a) increase the voltage output; (b) increase the internal resistance; (c) decrease the current capacity; (d) increase the current capacity.
6. Which of the following is a dry storage cell? (a) Edison cell; (b) carbon-zinc cell; (c) mercury cell; (d) nickel-cadmium cell.
7. When R_L equals the generator r_i, which of the following is maximum? (a) Power in R_L; (b) current; (c) voltage across R_L; (d) efficiency of the circuit.
8. Five carbon-zinc cells in series have an output of (a) 1.5 V; (b) 5.0 V; (c) 7.5 V; (d) 11.0 V.
9. A constant-voltage generator has (a) low internal resistance; (b) high internal resistance; (c) minimum efficiency; (d) minimum current capacity.
10. A generator has a 100-V output on open circuit, which drops to 50 V with a load current of 50 mA and an R_L of 1000 Ω. The internal resistance r_i equals (a) 25 Ω; (b) 50 Ω; (c) 100 Ω; (d) 1000 Ω.

Essay Questions

1. Draw a sketch illustrating the construction of a carbon-zinc dry cell. Indicate the negative and positive electrodes and the electrolyte.
2. Draw a sketch illustrating construction of the lead-acid cell. Indicate the negative and positive electrodes and the electrolyte.
3. Show the wiring for the following batteries: (a) six lead-acid cells for a voltage output of approximately 12 V; (b) six standard No. 6 dry cells for a voltage output of 4.5 V with a current capacity of ½ A.
4. (a) What is the advantage of connecting cells in series? (b) What is connected to the end terminals of the series cells?
5. (a) What is the advantage of connecting cells in parallel? (b) Why can the load be connected across any one of the parallel cells?
6. How many cells are necessary in a battery to double the voltage and current rating of a single cell? Show the wiring diagram.
7. Draw a diagram showing two 12-V lead-acid batteries being charged by a 15-V source.
8. Why is a generator with very low internal resistance called a constant-voltage source?
9. Why does discharge current lower the specific gravity in a lead-acid cell?

10. Would you consider the lead-acid battery a constant-current source or constant-voltage source? Why?
11. List five types of chemical cells, giving two features of each.
12. Referring to Fig. 12-19b, draw the corresponding graph that shows how I varies with R_L.
13. Referring to Fig. 12-20b, draw the corresponding graph that shows how V_L varies with R_L.
14. Referring to Fig. 12-22c, draw the corresponding graph that shows how V_L varies with R_L.

Problems (Answers to odd-numbered problems at back of book.)

1. A 1.5-V No. 6 carbon-zinc dry cell is connected across an R_L of 1000 Ω. How much current flows in the circuit?
2. Draw the wiring diagram for six No. 6 cells providing a 3-V output with a current capacity of ¾ A. Draw the schematic diagram of this battery connected across a 10-Ω resistance. (a) How much current flows in the circuit? (b) How much power is dissipated in the resistance? (c) How much power is supplied by the battery?
3. A 6-V lead-acid battery has an internal resistance of 0.01 Ω. How much current will flow if the battery has a short circuit?
4. How much is the specific gravity of a solution with equal parts of sulfuric acid and water?
5. A lead-acid battery discharges at the rate of 8 A for 10 h. (a) How many coulombs of charge must be put back into the battery to restore the original charge, assuming 100 percent efficiency? (b) How long will this recharging take, with a charging current of 2 A?
6. The output voltage of a battery drops from 90 V at zero load to 60 V with a load current of 50 mA. (a) How much is the internal r_i of the battery? (b) How much is R_L for this load current? (c) How much R_L reduces the load voltage to one-half the no-load voltage?
7. A 100-V source with an internal resistance of 10 kΩ is connected to a variable load resistance R_L. Tabulate I, V_L, and power in R_L for values of 1 kΩ, 5 kΩ, 10 kΩ, 15 kΩ, and 20 kΩ.
8. The output voltage of a source reads 60 V with a VTVM. When a meter with 1000 Ω/V sensitivity is used, the reading is 50 V on the 100-V range. How much is the internal resistance of the source?
9. A generator has an open-circuit emf of 180 V. Its terminal voltage drops to 150 V with an R_L of 5 kΩ. Calculate r_i.
10. Referring to Fig. 12-22, calculate P_L when R_L is 200 Ω. Compare this value with the maximum P_L at $R_L = r_i = 100$ Ω.

Answers to Practice Problems

12-1	(a) 2.2 V		12-5	(a) Six
	(b) 1.5 V			(b) 20 h
12-2	(a) T		12-6	(a) 1.25 V
	(b) F			(b) No
12-3	(a) F		12-7	(a) High
	(b) F			(b) Opposes
12-4	(a) Six		12-8	(a) T
	(b) Two			(b) T

Review of Chapters 10 to 12

Summary

1. A conductor has very low resistance. Silver and copper are the best conductors, with copper generally used for wire.
2. The gage numbers for copper wire are listed in Table 10-1.
3. Resistance wire for use in heating elements and filaments has a much higher R when hot, compared with its cold resistance. The hot resistance, equal to V/I, cannot be measured with an ohmmeter.
4. An applied voltage can produce charged ions in liquids and gases. The ions may be either positive or negative.
5. Insulators or dielectrics have very high resistance. Examples are air, vacuum, paper, glass, rubber, shellac, wood, and plastics.
6. The two main types of resistors are wirewound and carbon-composition. The wirewound type has relatively low R for power applications up to several hundred watts; the carbon type has high R with power ratings up to 2 W.
7. The color coding of carbon resistors is illustrated in Figs. 11-5 and 11-6, for the R in ohms. The power rating depends on physical size.
8. An open resistor is a common trouble. The reading is infinite ohms with an ohmmeter.
9. The main types of cells used for batteries are the carbon-zinc dry cell, lead-acid wet cell, and nickel-cadmium dry cell.
10. The carbon-zinc type is a primary cell, which cannot be recharged. The lead-acid and nickel-cadmium types are storage cells or secondary cells, which can be recharged. Additional types of cells are listed in Table 12-1 for primary cells and Table 12-2 for storage cells.
11. With cells or batteries connected in series, the total voltage equals the sum of the individual values. The connections are series-aiding with the + terminal of one to the − terminal of the next. The current rating for the series combination is the same as each cell.
12. For cells or batteries in parallel, the voltage across all is the same as one. However, the current rating of the combination equals the sum of the individual values.

Review Self-Examination (Answers at back of book.)

Choose (a), (b), (c), or (d).

1. Which of the following is the best conductor? (a) Carbon; (b) silicon; (c) rubber; (d) copper.
2. A 1-W 1-kΩ carbon resistor with current of 2 mA dissipates a power of (a) 4 W; (b) 4 mW; (c) 2 W; (d) 2 mW.
3. Which of the following cells has a reversible chemical reaction? (a) Carbon-zinc; (b) lead-acid; (c) silver oxide; (d) mercury oxide.
4. A tungsten filament measures 10 Ω with an ohmmeter. In a circuit with 100 V applied, 2 A flows. The hot resistance of the filament equals (a) 2 Ω; (b) 10 Ω; (c) 50 Ω; (d) 100 Ω.
5. Three resistors R_1, R_2, and R_3 are in series across a 100-V source. If R_2 opens: (a) the voltage across R_2 is zero; (b) the total resistance of R_1, R_2, and R_3 decreases; (c) the voltage across R_1 is 100 V; (d) the voltage across R_2 is 100 V.
6. The current flowing between electrodes inside a lead-acid battery is (a) electron current; (b) proton current; (c) ionization current; (d) polarization current.
7. With 300 V applied across two equal resistors in series, 1 mA of current flows. Typical values for each resistor to be used here are (a) 100 Ω, 1 W; (b) 150,000 Ω, ½ W; (c) 150,000 Ω, 50 W; (d) 300,000 Ω, 2 W.
8. A carbon resistor is coded with brown, green, yellow, and gold stripes from left to right. Its value is (a) 15,000 Ω $\pm$10 percent; (b) 68,000 Ω $\pm$10 percent; (c) 150,000 Ω $\pm$5 percent; (d) 500,000 Ω $\pm$5 percent.
9. Thirty zinc-carbon dry cells are connected in series. The total voltage output is (a) 1.5 V; (b) 30 V; (c) 45 V; (d) 60 V.
10. A 45-V source with an internal resistance of 2 Ω is connected across a wirewound resistor. Maximum power will be dissipated in the resistor when its R is (a) zero; (b) 2 Ω; (c) 45 Ω; (d) infinity.

References (Additional references at back of book.)

Crouse, W. H.: "Electrical Appliance Servicing," McGraw-Hill Book Company, New York.

Richter, H. P.: "Practical Electrical Wiring," McGraw-Hill Book Company, New York.

Slurzberg, M., and W. Osterheld: "Essentials of Electricity and Electronics," McGraw-Hill Book Company, New York.

Marcus, A., and C. Thompson: "Electricity for Technicians," Prentice-Hall, Englewood Cliffs, N.J.

Magnetism

Chapter 13

Electrical effects exist in two forms, voltage and current. In terms of voltage, separated electric charges have the potential to do mechanical work in attracting or repelling charges. Similarly, any electrical current has an associated magnetic field that can do the work of attraction or repulsion. Materials made of iron, nickel, and cobalt, particularly, concentrate their magnetic effects at opposite ends, where the magnetic material meets a nonmagnetic medium such as air.

As shown in Fig. 13-1, these points of concentrated magnetic strength are north and south poles. The opposite magnetic poles correspond to the idea of opposite polarities of electric charges. The name *magnetism* is derived from the iron oxide mineral *magnetite*. Ferromagnetism refers specifically to the magnetic properties of iron. The topics explained here are:

13-1   The Magnetic Field
13-2   Magnetic Flux ϕ
13-3   Flux Density B
13-4   Induction by the Magnetic Field
13-5   Air Gap of a Magnet
13-6   Types of Magnets
13-7   Ferrites
13-8   Magnetic Shielding
13-9   The Hall Effect

13-1
THE MAGNETIC FIELD

The practical effects of ferromagnetism result from the magnetic field of force between the two poles at opposite ends of the magnet. Although the magnetic field is invisible, evidence of its force can be seen when small iron filings are sprinkled on a glass or paper sheet placed over a bar magnet (Fig. 13-2a). Each iron filing becomes a small bar magnet. If the sheet is tapped gently to overcome friction so that the filings can move, they become aligned by the magnetic field.

Many filings cling to the ends of the magnet, showing that the magnetic field is strongest at the poles. The field exists in all directions but decreases in strength inversely as the square of the distance from the poles of the magnet.

Field Lines. In order to visualize the magnetic field without iron filings, we show the field as lines of force, as in Fig. 13-2b. The direction of the lines outside the magnet shows the path a north pole would follow in the field, repelled away from the north pole of the magnet and attracted to its south pole. Although we cannot actually have a unit north pole by itself, the field can be explored by noting how the north pole moves on a small compass needle.

The magnet can be considered as the generator for an external magnetic field, provided by the two opposite magnetic poles at the ends.

Chapter 13
Magnetism

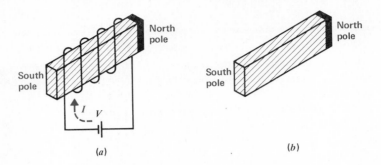

FIGURE 13-1

Poles of a magnet. (a) Electromagnet (EM) produced by current from battery. (b) Permanent magnet (PM) without any external current source.

This idea corresponds to the two opposite terminals on a battery as the source for an external electric field provided by opposite charges.

Magnetic field lines are unaffected by nonmagnetic materials such as air, vacuum, paper, glass, wood, or plastics. When these materials are placed in the magnetic field of a magnet, the field lines are the same as though the material were not there.

However, the magnetic field lines become concentrated when a magnetic substance like iron is placed in the field. Inside the iron, the field lines are more dense, compared with the field in air.

North and South Magnetic Poles. The earth itself is a huge natural magnet, with its greatest strength at the north and south poles. Because of the earth's magnetic poles, if a small bar magnet is suspended so that it can turn easily, one end will always point north. This end of the bar magnet is defined as the *north-seeking pole*, as shown in Fig. 13-3. The opposite end is the *south-seeking pole*. When polarity is indicated on a magnet, the north-seeking end is the north pole (N) and the opposite end is the south pole (S). It should be noted that the magnetic north pole differs by about 15° from true geographic north on the axis of the earth's rotation.

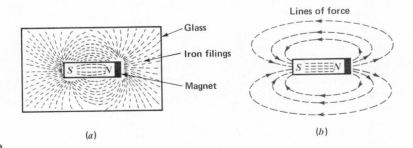

FIGURE 13-2

Magnetic field of force around a bar magnet. (a) Field outlined by iron filings. (b) Field indicated by lines of force.

Chapter 13
Magnetism

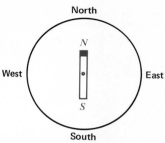

FIGURE 13-3

Definition of north and south poles of bar magnet.

Similar to the force between electric charges is a force between magnetic poles causing attraction of opposite poles and repulsion between similar poles:

1. A north pole (N) and a south pole (S) tend to attract each other.

2. A north pole (N) tends to repel another north pole (N), while a south pole (S) tends to repel another south pole (S).

These forces are illustrated by the field of iron filings between opposite poles in Fig. 13.4a and between similar poles in (b).

*Practice Problems 13-1
(answers on page 279)
Answer true or false.*
(a) On a magnet, the north-seeking pole is labeled N.
(b) Like poles have a force of repulsion.

13-2
MAGNETIC FLUX ϕ

The entire group of magnetic field lines, which can be considered to flow outward from the north pole of a magnet, is called *magnetic flux*.

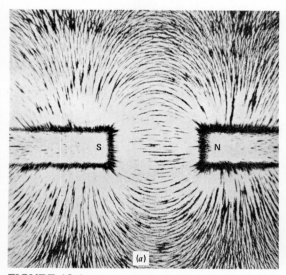
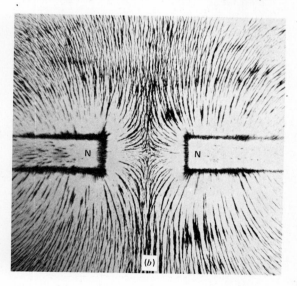

FIGURE 13-4

Photos of magnetic field patterns produced by iron filings. (a) Field between opposite poles. The north and south poles could be reversed. (b) Field between similar poles. The two north poles could be south poles.

Its symbol is the Greek letter ϕ (phi). A strong magnetic field has more lines of force and more flux than a weak magnetic field.

The Maxwell. One maxwell (Mx) unit equals one magnetic field line. In Fig. 13-5, as an example, the flux illustrated is 6 Mx because there are 6 field lines flowing in or out for each pole. A 1-lb magnet can provide a magnetic flux ϕ of about 5000 Mx. This unit is named for James Clerk Maxwell (1831–1879), an important Scottish mathematical physicist who contributed much to electrical and field theory.

The Weber. This is a larger unit of magnetic flux. One weber (Wb) equals 1×10^8 lines or maxwells. Since the weber is a large unit for typical fields, the microweber unit can be used. Then $1 \mu\text{Wb} = 10^{-6}$ Wb. This unit is named for Wilhelm Weber (1804–1890), a German physicist.

To convert microwebers to lines, multiply by the conversion factor 10^8 lines per weber, as follows:

$$1 \mu\text{Wb} = 1 \times 10^{-6} \text{ Wb} \times 10^8 \frac{\text{lines}}{\text{Wb}}$$

$$= 1 \times 10^2 \text{ lines}$$

$1 \mu\text{Wb} = 100$ lines or Mx

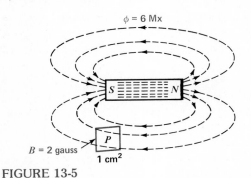

FIGURE 13-5

Total flux ϕ is 6 lines or Mx. Flux density B at point P is 2 lines per cm^2 or 2 G.

Note that the conversion is arranged to make the weber units cancel, since we want maxwell units in the answer. For the same 1-lb magnet producing the magnetic flux of 5000 Mx, this flux corresponds to 50 μWb.

Systems of Magnetic Units. As explained in Appendix D, Physics Units, units can be defined basically in two ways. The centimeter-gram-second (cgs) system defines small units. The meter-kilogram-second (mks) system is for larger units of a more practical size. Furthermore, the Système International (SI) units are in mks dimensions for an international standard of practical units, based on the ampere of current. More details of the SI units for electricity and magnetism are described on page 290. For the units of magnetic flux ϕ, the maxwell is a cgs unit, while the weber is an mks or SI unit.

Practice Problems 13-2
(answers on page 279)
The magnetic flux ϕ of 2000 lines is
(a) How many Mx?
(b) How many μWb?

13-3
FLUX DENSITY B

As shown in Fig. 13-5, the flux density is the number of magnetic field lines per unit area of a section perpendicular to the direction of flux. As a formula:

$$B = \frac{\phi}{A} \qquad (13\text{-}1)$$

where ϕ is the flux through an area A, and the flux density is B.

The Gauss. In the cgs system, this unit is one line per square centimeter, or 1 Mx/cm^2. As an example, in Fig. 13-5, the total flux ϕ is 6 lines, or 6 Mx. At point P in this field, however, the flux density B is 2 G because there are 2 lines per cm^2. The flux density has a higher value close to the poles where the flux lines are more crowded.

As an example of flux density, B for a 1-lb magnet would be 1000 G at the poles. This unit is named for Karl F. Gauss (1777–1855), a German mathematician.

Example 1. With a flux of 10,000 Mx through a perpendicular area of 5 cm^2, what is the flux density in gauss?

Answer. $B = \dfrac{\phi}{A} = \dfrac{10{,}000 \text{ Mx}}{5 \text{ cm}^2} = 2000 \dfrac{\text{Mx}}{\text{cm}^2}$

$B = 2000$ G

As typical values, B for the earth's magnetic field can be about 0.2 G; a large laboratory magnet produces B of 50,000 G. Since the gauss is so small, it is often used in kilogauss units, where $1 \text{ kG} = 10^3$ G.

The Tesla. In SI, the unit of flux density B is webers per square meter (Wb/m^2). One weber per square meter is called a *tesla,* abbreviated T. This unit is named for Nikola Tesla (1857–1943), a Yugoslav-born American inventor in electricity and magnetism.

When converting between cgs and mks units note that

$1 \text{ m} = 100$ cm
$1 \text{ m}^2 = 10{,}000$ or 10^4 cm^2
$1 \text{ cm}^2 = 0.0001$ or 10^{-4} m^2

As an example, 5 cm^2 is 0.0005 m^2.

Example 2. With a flux of 400 μWb through an area of 0.0005 m^2, what is the flux density in tesla units?

Answer. $B = \dfrac{\phi}{A} = \dfrac{400 \times 10^{-6} \text{ Wb}}{5 \times 10^{-4} \text{ m}^2}$

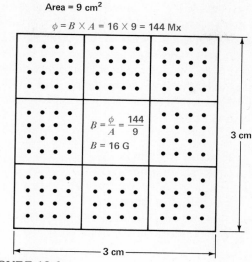

FIGURE 13-6

Comparison of total flux ϕ and flux density B. Total area of 9 cm^2 has 144 lines or Mx. For 1 cm^2 the flux density is 144 Mx/9 cm^2 = 16 G.

$= \dfrac{400}{5} \times 10^{-2} = 80 \times 10^{-2} \dfrac{\text{Wb}}{\text{m}^2}$

$B = 0.80$ T

It should be noted that all these units for flux or flux density apply to the magnetic field produced either by a permanent magnet (PM) or by an electromagnet (EM). Also, the weber and tesla in the mks system are practical units of magnetism, closely related to the practical volt and ampere units of electricity.

Comparison of Flux and Flux Density. Remember that the flux ϕ includes total area, while the flux density B is for a specified unit area. The difference between ϕ and B is illustrated in Fig. 13-6 with cgs units. The total area A here is 9 cm^2, equal to 3 cm $\times$ 3 cm. For one unit box of 1 cm^2, 16 lines are shown. Therefore, the flux density B is 16 lines or maxwells per cm^2, which equals 16 G. The total area includes nine of these boxes. Therefore, the total flux ϕ is 144 lines or maxwells, equal to 9×16 for $B \times A$.

For the opposite case, if the total flux ϕ is given as 144 lines or maxwells, the flux density is found by dividing 144 by 9 cm². This division of $^{144}/_9$ equals 16 lines or maxwells per cm², which is 16 G.

Practice Problems 13-3
(answers on page 279)
(a) The ϕ is 9000 Mx through 3 cm². How much is B in gauss units?
(b) How much is B in tesla units for ϕ of 90 μWb through 0.0003 m²?

13-4 INDUCTION BY THE MAGNETIC FIELD

The electrical effect of one body on another without any physical contact between them is called induction. For instance, a permanent magnet can induce an unmagnetized iron bar to become a magnet, without the two touching. The iron bar then becomes a magnet, as shown in Fig. 13-7. What happens is that the magnetic lines of force generated by the permanent magnet make the internal molecular magnets in the iron bar line up in the same direction, instead of the random directions in unmagnetized iron. The magnetized iron bar then has magnetic poles at the ends, as a result of the magnetic induction.

Note that the induced poles in the iron have opposite polarity from the poles of the magnet. Since opposite poles attract, the iron bar will be attracted. Any magnet attracts to itself all magnetic materials by induction.

Although the two bars in Fig. 13-7 are not touching, the iron bar is in the magnetic flux of the permanent magnet. It is the invisible magnetic field that links the two magnets, enabling one to affect the other. Actually, this idea of magnetic flux extending outward from the magnetic poles is the basis for many inductive effects in ac circuits. More generally, the magnetic field between magnetic poles and the electric field between electric charges form the basis for wireless radio transmission and reception.

Polarity of Induced Poles. Note that the north pole of the permanent magnet in Fig. 13-7 induces an opposite south pole at this end of the iron bar. If the permanent magnet were reversed, its south pole would induce a north pole. The closest induced pole will always be of opposite polarity. This is the reason why either end of a magnet can attract another magnetic material to itself. No matter which pole is used, it will induce an opposite pole, and the opposite poles are attracted.

Relative Permeability. Soft iron, as an example, is very effective in concentrating magnetic field lines, by induction in the iron. This ability to concentrate magnetic flux is called permeability. Any material that is easily magnetized has high permeability, therefore, as the field lines are concentrated because of induction.

Numerical values of permeability for different materials can be assigned in comparison with air or vacuum. For example, if the flux density in air is 1 G but an iron coil in the same position in the same field has a flux density of

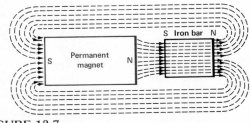

FIGURE 13-7
Magnetizing an iron bar by induction.

200 G, the relative permeability of the iron coil equals $^{200}/_1$, or 200.

The symbol for relative permeability is μ_r (mu), where the subscript r indicates relative permeability. Typical values for μ_r are 100 to 9000 for iron and steel. There are no units, because μ_r is a comparison of two flux densities and the units cancel. The symbol K_m may also be used for relative permeability, to indicate this characteristic of a material for a magnetic field, corresponding to K_ϵ for an electric field.

Practice Problems 13-4
(answers on page 279)
Answer true or false.
(a) Induced poles always have opposite polarity from the inducing poles.
(b) The relative permeability of air or vacuum is approximately 300.

13-5
AIR GAP OF A MAGNET

As shown in Fig. 13-8, the air space between poles of a magnet is its air gap. The shorter the air gap, the stronger the field in the gap for a given pole strength. Since air is not magnetic and cannot concentrate magnetic lines, a larger air gap only provides additional space for the magnetic lines to spread out.

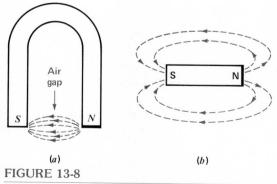

FIGURE 13-8

The horseshoe magnet in (a) has a smaller air gap than the bar magnet in (b).

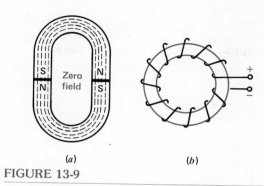

FIGURE 13-9

Examples of a closed magnetic ring without any air gap. (a) Two PM horseshoe magnets with opposite poles touching. (b) Toroid electromagnet.

Referring to Fig. 13-8, note that the horseshoe magnet has more crowded magnetic lines in the air gap, compared with the widely separated lines around the bar magnet in (b). Actually, the horseshoe magnet can be considered as a bar magnet bent around to place the opposite poles closer. Then the magnetic lines of the poles reinforce each other in the air gap. The purpose of a short air gap is to concentrate the magnetic field outside the magnet, for maximum induction in a magnetic material placed in the gap.

Ring Magnet without Air Gap. When it is desired to concentrate magnetic lines within a magnet, however, it can be formed as a closed magnetic loop. This method is illustrated in Fig. 13-9a by the two permanent horseshoe magnets placed in a closed loop with opposite poles touching. Since the loop has no open ends, there can be no air gap and no poles. The north and south poles of each magnet cancel as opposite poles touch.

Each magnet has its magnetic lines inside, plus the magnetic lines of the other magnet, but outside the magnets the lines cancel because they are in opposite directions. The effect of the closed magnetic loop, therefore, is maximum concentration of magnetic lines in the magnet with minimum lines outside.

The same effect of a closed magnetic loop is obtained with the *toroid* or ring magnet in Fig. 13-9b, made in the form of a doughnut. Iron is often used for the core. This type of electromagnet has maximum strength in the iron ring, with little flux outside. As a result, the toroidal magnet is less sensitive to induction from external magnetic fields and, conversely, has little magnetic effect outside the coil.

It should be noted that, even if the winding is over only a small part of the ring, practically all the flux is in the iron core because its permeability is so much greater than air. The small part of the field in the air is called *leakage flux*.

Keeper for a Magnet. The principle of the closed magnetic ring is used to protect permanent magnets in storage. In Fig. 13-10a, four PM bar magnets are in a closed loop while (b) shows a stacked pair. Additional even pairs can be stacked this way with opposite poles touching. The closed loop in (c) shows one permanent horseshoe magnet with a soft iron *keeper* across the air gap. The keeper maintains the strength of the permanent magnet as it becomes magnetized by induction to form a closed loop. Then any external magnetic field is just concentrated in the closed loop without inducing opposite poles in the permanent magnet. If permanent magnets are not stored this way, the polarity can be reversed with induced poles produced by a strong external field from a dc source; an alternating field can demagnetize the magnet.

Practice Problems 13-5
(answers on page 279)
Answer true or false.
(a) A short air gap has a stronger field than a large air gap, for the same magnetizing force.
(b) A toroid magnet has no air gap.

13-6
TYPES OF MAGNETS

The two broad classes are permanent magnets and electromagnets. An electromagnet needs current from an external source to maintain its magnetic field. With a permanent magnet, not only is its magnetic field present without any external current, but the magnet can maintain its strength indefinitely.

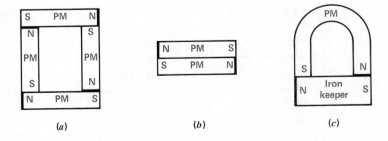

FIGURE 13-10

Storing permanent magnets in a closed loop, with opposite poles touching. (a) Four bar magnets. (b) Two bar magnets. (c) Horseshoe with iron keeper across air gap.

Electromagnets. Current in a wire conductor has an associated magnetic field. If the wire is wrapped in the form of a coil, as in Fig. 13-11, the current and its magnetic field become concentrated in a smaller space, resulting in a stronger field. With the length much greater than its width, the coil is called a *solenoid*. It acts like a bar magnet, with opposite poles at the ends.

More current and more turns make a stronger magnetic field. Also, the iron core concentrates magnetic lines inside the coil. Soft iron is generally used for the core because it is easily magnetized and demagnetized.

The coil in Fig. 13-11, with the switch closed and current in the coil, is an electromagnet that can pick up the steel nail shown. If the switch is opened, the magnetic field is reduced to zero and the nail will drop off. This ability of an electromagnet to provide a strong magnetic force of attraction that can be turned on or off easily has many applications in lifting magnets, buzzers, bells or chimes, and relays. A relay is a switch with contacts that are opened or closed by an electromagnet.

Another common application is magnetic tape recording. The tape is coated with fine particles of iron oxide. The recording head is a coil that produces a magnetic field in proportion

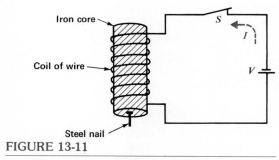

FIGURE 13-11

Electromagnet holding nail when switch S is closed for current in coil.

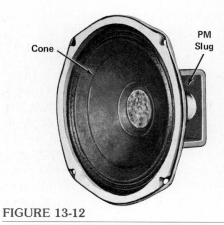

FIGURE 13-12

Example of a PM loudspeaker.

to the current. As the tape passes through the air gap of the head, small areas of the coating become magnetized by induction. On playback, the moving magnetic tape produces variations in electrical current.

Permanent Magnets. These are made of hard magnetic materials, such as cobalt steel, magnetized by induction in the manufacturing process. A very strong field is needed for induction in these materials. When the magnetizing field is removed, however, a residual induction makes the material a permanent magnet. A common PM material is *alnico*, a commercial alloy of aluminum, nickel, and iron, with cobalt, copper, and titanium added to produce about 12 grades. The Alnico V grade is often used for PM loudspeakers (Fig. 13-12). In this application, a typical size of PM slug for a steady magnetic field is a few ounces to about 5 lb, with a flux B of 500 to 25,000 lines or maxwells. One advantage of a PM loudspeaker is that only two connecting leads are needed for the voice coil, as the steady magnetic field of the PM slug is obtained without any field-coil winding.

Commercial permanent magnets will last indefinitely if not subjected to high temperatures, to physical shock, or to a strong demagnetizing field. If the magnet becomes hot, however, the molecular structure can be rearranged, resulting in loss of magnetism that is not recovered after cooling. The point at which a magnetic material loses its ferromagnetic properties is the *Curie temperature.* For iron, this temperature is about 800°C, when the relative permeability drops to unity. A permanent magnet does not become exhausted with use, as its magnetic properties are determined by the structure of the internal atoms and molecules.

Classification of Magnetic Materials. When we consider materials simply as either magnetic or nonmagnetic, this division is really based on the strong magnetic properties of iron. However, weak magnetic materials can be important in some applications. For this reason, a more exact classification includes the following three groups:

1. *Ferromagnetic materials.* These include iron, steel, nickel, cobalt, and commercial alloys such as alnico and Permalloy. They become strongly magnetized, in the same direction as the magnetizing field, with high values of permeability from 50 to 5000. Permalloy has μ_r of 100,000 but is easily saturated at relatively low values of flux density.
2. *Paramagnetic materials.* These include aluminum, platinum, manganese, and chromium. The permeability is slightly more than 1. They become weakly magnetized in the same direction as the magnetizing field.
3. *Diamagnetic materials.* These include bismuth, antimony, copper, zinc, mercury, gold, and silver. The permeability is less than 1. They become weakly magnetized but in the opposite direction from the magnetizing field.

The basis of all magnetic effects is the magnetic field associated with electric charges in motion. Within the atom, the motion of its orbital electrons generates a magnetic field. There are two kinds of electron motion in the atom. First is the electron revolving in its orbit. This motion provides a diamagnetic effect. However, this magnetic effect is weak because thermal agitation at normal room temperature results in random directions that neutralize each other.

More effective is the magnetic effect from the motion of each electron spinning on its own axis. The spinning electron serves as a tiny permanent magnet. Opposite spins provide opposite polarities. Two electrons spinning in opposite directions form a pair, neutralizing the magnetic fields. In the atoms of ferromagnetic materials, however, there are many unpaired electrons with spins in the same direction, resulting in a strong magnetic effect.

In terms of molecular structure, iron atoms are grouped in microscopically small arrangements called *domains.* Each domain is an elementary *dipole magnet,* with two opposite poles. In crystal form, the iron atoms have domains that are parallel to the axes of the crystal. Still, the domains can point in different directions, because of the different axes. When the material becomes magnetized by an external magnetic field, though, the domains become aligned in the same direction. With PM materials, the alignment remains after the external field is removed.

*Practice Problems 13-6
(answers on page 279)
Answer true or false.*
(a) An electromagnet needs current to maintain its magnetic field.

(b) A relay coil is an electromagnet.
(c) Steel is a diamagnetic material.

13-7
FERRITES

This is the name for recently developed nonmetallic materials that have the ferromagnetic properties of iron. The ferrites have very high permeability, like iron. However, a ferrite is a ceramic material, while iron is a conductor. The permeability of ferrites is in the range of 50 to 3000. The specific resistance is $10^5 \, \Omega \cdot \text{cm}$, which makes the ferrite an insulator.

A common application is a ferrite core, usually adjustable, in the coils for rf transformers. The ferrite core is much more efficient than iron when the current alternates at a high frequency. The reason is that less I^2R power is lost by eddy currents in the core because of its very high resistance.

A ferrite core is used in small coils and transformers for signal frequencies up to 20 MHz, approximately. The high permeability means the transformer can be very small. However, the ferrites are easily saturated at low values of magnetizing current. This disadvantage means the ferrites are not used for power transformers.

Another application is in ferrite beads (Fig. 13-13). A bare wire is used as a string for one or more beads. The bead concentrates the magnetic field of the current in the wire. This construction serves as a simple, economical rf choke, instead of a coil. The purpose of the choke is to reduce the current just for an undesired radio frequency.

Practice Problems 13-7
(answers on page 279)
(a) Which has more R, the ferrites or soft iron?
(b) Which has more I^2R losses, an insulator or a conductor?

13-8
MAGNETIC SHIELDING

The idea of preventing one component from affecting another through their common electric or magnetic field is called *shielding*. Examples are the metal cover on a vacuum tube, the braided copper-wire shield enclosing the inner conductor of a coaxial cable, or a shield of magnetic material enclosing a cathode-ray tube.

The problem in shielding is to prevent one component from inducing an effect in the shielded component. The shielding materials are always metals, but there is a difference in using good conductors with low resistance like copper and aluminum or using good magnetic materials like soft iron or Mumetal. A good conductor is best for two shielding functions. One is to prevent induction of static electric charges. The other is to shield against the induction of a varying magnetic field. For static charges, the shield provides opposite induced charges, which prevent induction inside the shield. For a varying magnetic field, the shield has induced currents that oppose the inducing field. Then there is little net field strength to produce induction inside the shield.

The best shield for a steady magnetic field is a good magnetic material of high permeability. A steady field is produced by a permanent magnet, a coil with steady direct current, or the earth's magnetic field. With a magnetic shield of high permeability, it concentrates the magnetic flux. Then there is little flux to induce poles in a

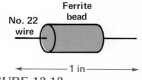

FIGURE 13-13
Ferrite bead equivalent to coil with 20 μH of inductance at 10 MHz.

component inside the shield. The shield can be considered as a short circuit for the lines of magnetic flux.

Practice Problems 13-8
(answers on page 279)
Answer true or false.
(a) Magnetic material with high permeability is a good shield for a steady magnetic field.
(b) A conductor is a good shield against a varying magnetic field.

13-9
THE HALL EFFECT

In 1879, E. H. Hall observed that a small voltage is generated across a conductor carrying current in an external magnetic field. The Hall voltage was very small with typical conductors, and little use was made of this effect. However, with the development of semiconductors, larger values of Hall voltage can be generated. The semiconductor material indium arsenide (InAs) is generally used. As illustrated in Fig. 13-13, the InAs element inserted in the magnetic field can generate 60 mV with B equal to 10 kG and an I of 100 mA. The applied flux must be perpendicular to the direction of current. With current in the direction of the length of conductor, the generated voltage is developed across the width.

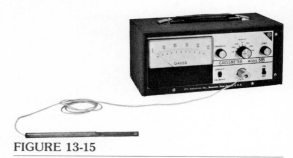

FIGURE 13-15

Gaussmeter to measure flux density, with probe containing indium-arsenide element. (*RFL Industries Inc.*)

The amount of Hall voltage v_H is directly proportional to the value of flux density B. This means that values of B can be measured by means of v_H. As an example, the gaussmeter in Fig. 13-14 uses an InAs probe in the magnetic field to generate a proportional Hall voltage v_H. This value of v_H is then read by the meter, which is calibrated in gauss. The original calibration is made in terms of a reference magnet having a specified flux density.

Practice Problems 13-9
(answers on page 279)
(a) In Fig. 13-14, how much is the generated Hall voltage?
(b) Does the gaussmeter in Fig. 13-15 measure flux or flux density?

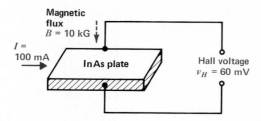

FIGURE 13-14

The Hall effect. The voltage v_H generated across the element is proportional to the perpendicular flux density B.

Summary

1. Iron, nickel, and cobalt are common examples of magnetic materials. Air, paper, wood, and plastics are nonmagnetic, meaning there is no effect with a magnetic field.
2. The pole of a magnet that seeks the magnetic north pole of the earth is called a north pole; the opposite pole is a south pole.
3. Opposite magnetic poles have a force of attraction; similar poles repel.
4. An electromagnet needs current from an external source to provide a magnetic field. Permanent magnets retain their magnetism indefinitely.
5. Any magnet has an invisible field of force outside the magnet, indicated by magnetic field lines. Their direction is from the north to south pole outside the magnet.
6. The open ends of a magnet where it meets a nonmagnetic material provide magnetic poles. At opposite open ends, the poles have opposite polarity.
7. A magnet with an air gap has opposite poles with magnetic lines of force across the gap. A closed magnetic ring has no poles. Practically all the magnetic lines are in the ring.
8. Magnetic induction enables the field of a magnet to induce magnetic poles in a magnetic material without touching.
9. Permeability is the ability to concentrate magnetic flux. A good magnetic material has high permeability, similar to the idea of high conductance for a good conductor of electricity.
10. Magnetic shielding means isolating a component from a magnetic field. The best shield against a steady magnetic field is a material with high permeability.
11. The Hall voltage is a small emf generated across the width of a conductor carrying current through its length, when magnetic flux is applied perpendicular to the current. This effect is generally used in the gaussmeter to measure flux density.
12. Table 13-1 summarizes the units of magnetic flux ϕ and flux density B.

TABLE 13-1. Magnetic Flux ϕ and Flux Density B

NAME	SYMBOL	CGS UNITS	MKS or SI UNITS
Flux, or total lines	$\phi = B \times \text{area}$	1 maxwell (Mx) = 1 line	1 weber (Wb) = 10^8 Mx
Flux density, or lines per unit area	$B = \dfrac{\phi}{\text{area}}$	1 gauss (G) = $\dfrac{1 \text{ Mx}}{\text{cm}^2}$	1 tesla (T) = $\dfrac{1 \text{ Wb}}{\text{m}^2}$

Self-Examination (Answers at back of book.)

Answer true or false.

1. Iron and steel are ferromagnetic materials with high permeability.
2. Ferrites are magnetic but have high resistance.
3. Air, vacuum, wood, paper, and plastics have practically no effect on magnetic flux.
4. Aluminum is ferromagnetic.
5. Magnetic poles exist on opposite sides of an air gap.
6. A closed magnetic ring has no poles and no air gap.
7. A magnet can pick up a steel nail by magnetic induction.
8. Induced poles are always opposite from the original field poles.
9. Soft iron concentrates magnetic flux by means of induction.
10. Without current, an electromagnet has practically no magnetic field.
11. The total flux ϕ of 5000 lines equals 5 Mx.
12. A flux ϕ of 5000 Mx through a cross-sectional area of 5 cm² has a flux density B of 1000 G or 1 kG.
13. The flux density B of 1000 G equals 1000 lines per cm².
14. A magnetic pole is a terminal where a magnetic material meets a nonmagnetic material.
15. High permeability for magnetic flux corresponds to high resistance for a conductor of current.

Essay Questions

1. Name two magnetic materials and three nonmagnetic materials.
2. Explain briefly the difference between a permanent magnet and an electromagnet.
3. Draw a horseshoe magnet, with its magnetic field. Label the magnetic poles, indicate the air gap, and show the direction of flux.
4. Define the following: relative permeability, shielding, induction, Hall voltage.
5. Give the symbol, cgs unit, mks unit, and SI unit for magnetic flux and for flux density.
6. How would you determine the north and south poles of a bar magnet, using a magnetic compass?
7. Referring to Fig. 13-11, why can either end of the magnet pick up the nail?
8. What is the difference between flux ϕ and flux density B?

Problems (Answers to odd-numbered problems at back of book.)

1. A magnetic pole produces 5000 field lines. How much is the flux ϕ in maxwells and webers?
2. If the area of this pole in Prob. 1 is 5 cm², calculate the flux density B in gauss units.
3. Calculate B in tesla units for a 50-μWb flux through an area of 5×10^{-4} m².
4. Convert 1000 G to tesla units.
5. For a flux density B of 3 kG at a pole with a cross-sectional area of 8 cm², how much is the total flux ϕ in maxwell units?
6. Convert 24,000 Mx to weber units.
7. The flux density is 0.002 T in the air core of an electromagnet. When an iron core is inserted, the flux density in the core is 0.6 T. How much is the relative permeability μ_r of the iron core?
8. Draw the diagram of an electromagnet operated from a 12-V battery, in series with a switch. (a) If the coil resistance is 60 Ω, how much is the current in the coil with the switch closed? (b) Why is the magnetic field reduced to zero when the switch is opened?
9. Derive the conversion of 1 μWb = 100 Mx from the fact that 1 μWb = 10^{-6} Wb and 1 Wb = 10^8 Mx.
10. Derive the relation 1 T = 10^4 G. (Note: 1 m² = 10,000 cm².)

Answers to Practice Problems

13-1	(a) T	13-6	(a) T
	(b) T		(b) T
13-2	(a) 2000 Mx		(c) F
	(b) 20 μWb	13-7	(a) Ferrites
13-3	(a) 3000 G		(b) Conductor
	(b) 0.3 T	13-8	(a) T
13-4	(a) T		(b) T
	(b) F	13-9	(a) 60 mV
13-5	(a) T		(b) Flux density
	(b) T		

Magnetic Units

Chapter 14

A magnetic field is always associated with charges in motion. Therefore, the magnetic units can be derived from the current that produces the field.

The current in a conductor and its magnetic flux through the medium outside the conductor are related as follows: (1) The current I supplies a magnetizing force, or magnetomotive force (mmf) that increases with the amount of I. (2) The mmf results in a magnetic field intensity H that decreases with the length of conductor, as the field is less concentrated with more length. (3) The field intensity H produces a flux density B that increases with the permeability of the medium.

More details of magnetic units based on the ampere of current are explained in the following topics:

14-1 Ampere-turns (NI)
14-2 Field Intensity (H)
14-3 Permeability (μ)
14-4 B-H Magnetization Curve
14-5 Magnetic Hysteresis
14-6 Ohm's Law for Magnetic Circuits
14-7 Relations between Magnetic Units
14-8 Comparison of Magnetic and Electric Fields

14-1 AMPERE-TURNS (NI)

With a coil magnet, the strength of the magnetic field depends on how much current flows in the turns of the coil. The more current, the stronger is the magnetic field. Also, more turns in a specific length concentrate the field. The coil serves as a bar magnet, with opposite poles at the ends, providing a magnetic field proportional to the ampere-turns. As a formula,

$$\text{Ampere-turns} = NI \qquad (14\text{-}1)$$

where N is the number of turns, multiplied by the current I in amperes. The quantity NI is the magnetizing force, magnetic potential, or magnetomotive force (mmf).

As shown in Fig. 14-1, a solenoid with 5 turns and 2 amperes has the same magnetizing force as 10 turns with 1 ampere, as the product of the amperes and turns is 10 for both cases. With thinner wire, more turns can be used in a given space. The amount of current is determined by the resistance of the wire and the source voltage. How many ampere-turns are necessary depends on the required magnetic field strength.

Example 1. Calculate the ampere-turns for a coil with 2000 turns and a 5-mA current.

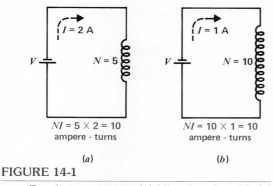

FIGURE 14-1
Equal ampere-turns. (a) $NI = 5 \times 2 = 10$. (b) $NI = 10 \times 1 = 10$.

Answer. $NI = 2000 \times 5 \times 10^{-3}$
$NI = 10$ ampere-turns

Example 2. A coil with 4 amperes is to provide the magnetizing force of 600 ampere-turns. How many turns are necessary?

Answer. $N = \dfrac{NI}{I} = \dfrac{600}{4}$
$N = 150$ turns

Example 3. A coil with 400 turns must provide 800 ampere-turns of magnetizing force. How much current is necessary?

Answer. $I = \dfrac{NI}{N} = \dfrac{800}{400}$
$I = 2$ amperes

Example 4. The wire in a solenoid of 250 turns has a resistance of 3 Ω (a) How much is the current with the coil connected to a 6-V battery? (b) Calculate the ampere-turns.

Answer.

(a) $I = \dfrac{V}{R} = \dfrac{6}{3}$
$I = 2$ amperes

(b) $NI = 250 \times 2$
$NI = 500$ ampere-turns

The practical unit for NI is the ampere-turn. The standard abbreviation for ampere-turn is A, the same as for the ampere, since the number of turns in a coil usually is constant but the current can be varied.

The ampere-turn is an SI unit. The cgs unit of mmf is the *gilbert*,[1] abbreviated Gb. One ampere-turn equals 1.26 Gb. The number 1.26 is approximately $4\pi/10$, derived from the surface area of a sphere, which is $4\pi r^2$.

To convert NI to gilberts, multiply the ampere-turns by the constant conversion factor 1.26 Gb/1 A. As an example, 1000 A is the same mmf as 1260 Gb. The calculations are

$$1000 \text{ A} \times 1.26 \dfrac{\text{Gb}}{1 \text{ A}} = 1260 \text{ Gb}$$

Note that the units of A for ampere-turns cancel in the conversion.

Practice Problems 14-1
(answers on page 293)

(a) NI is 243 ampere-turns. I is doubled from 2 to 4 A with the same turns. How much is NI?
(b) Convert 500 ampere-turns to gilberts.

14-2
FIELD INTENSITY (H)

The ampere-turns of mmf specify the magnetizing force, but the intensity of the magnetic field depends on how long the coil is. At any point in space, a specific value of ampere-turns for a long coil must produce less field intensity than a short coil that concentrates the same NI. Specifically, the field intensity H in mks units is:

$$H = \dfrac{NI \text{ ampere-turns}}{l \text{ meters}} \qquad (14\text{-}2)$$

[1] William Gilbert (1540–1603) was an English scientist who investigated the magnetism of the earth.

This formula is for a solenoid. H is the intensity at the center of an air core. With an iron core, H is the intensity through the entire core. By means of units for H, the magnetic field intensity can be specified for either electromagnets or permanent magnets, since both provide the same kind of magnetic field.

The length in formula (14-2) is between poles. In Fig. 14-2a, the length is 1 m between the poles at the ends of the coil. In (b), also, l is 1 m between the ends of the iron core. In (c), though, l is 2 m between the poles at the ends of the iron core, although the winding is only 1 m long.

The examples in Fig. 14-2 illustrate the following comparisons:

1. In all three cases, the mmf equal to NI is 1000 ampere-turns or 1000 A.
2. In both (a) and (b), H equals 1000 A/m. In (a), this H is the intensity at the center of the air core; in (b) this H is the intensity through the entire iron core.
3. In (c), because l is 2 m, H is 1000/2, or 500 A/m. This H is the intensity in the entire iron core.

Units for H are basically mmf per unit of length. In practical units, H is ampere-turns per meter. The cgs unit for H is the *oersted*,[1] abbreviated Oe, which equals one gilbert of mmf per centimeter.

To convert SI units of A/m to cgs units of Oe, multiply by the conversion factor 0.0126 Oe per 1 A/m. As an example, 1000 A/m is the same H as 12.6 Oe. The calculations are

$$1000\, \frac{A}{m} \times 0.0126\, \frac{Oe}{1\, A/m} = 12.6\, Oe$$

[1] H. C. Oersted (1777–1851), a Danish physicist, discovered electromagnetism.

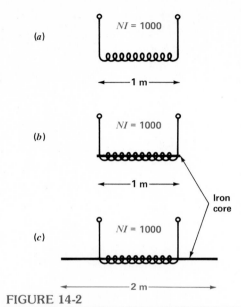

FIGURE 14-2

Relation between mmf and field intensity: H = mmf/length. (a) H is 1000 A/m in air. (b) H is 1000 A/m in the iron core. (c) H is $1000/2 = 500$ A/m in the longer iron core.

Note that the units of A and m cancel. The m in the conversion factor becomes inverted to the numerator.

Practice Problems 14-2
(answers on page 293)
(a) H is 250 ampere-turns. The length is doubled from 0.2 to 0.4 m for the same NI. How much is H?
(b) Convert 500 A/m to oersted units.

14-3
PERMEABILITY (μ)

Whether we say H is 1000 A/m or 12.6 Oe, these units specify how much field intensity is available to produce magnetic flux. However, the amount of flux actually produced by H depends on the material in the field. A good magnetic material with high relative permeability can concentrate flux and produce a large value of flux density B, for a specified H. These factors are related by the formula:

$$B = \mu \times H \quad (14\text{-}3)$$

or

$$\mu = \frac{B}{H} \quad (14\text{-}4)$$

Using SI units, B is the flux density in webers per square meter, or teslas; H is the field intensity in ampere-turns per meter. In the cgs system the units are gauss for B and oersted for H. The factor μ is the absolute permeability, not referred to any other material, in units of B/H.

In the cgs system the units of gauss for B and oersteds for H have been defined to give μ the value of 1 G/Oe, for vacuum, air, or space. This simplification means that B and H have the same numerical values in air or vacuum. For instance, the field intensity H of 12.6 Oe produces the flux density of 12.6 G, in air.

Furthermore, the values of relative permeability μ_r are the same as absolute permeability in B/H units in the cgs system. The reason is that μ is 1 for air or vacuum, used as the reference for comparison. As an example, if μ_r for an iron sample is 600, the absolute μ is also 600 G/Oe.

In SI, however, the permeability of air or vacuum is not 1. Specifically, this value is $4\pi \times 10^{-7}$, or 1.26×10^{-6}, with the symbol μ_0. Therefore, values of relative permeability μ_r must be multiplied by 1.26×10^{-6} for μ_0 to calculate μ as B/H in SI units. For $\mu_r = 100$, the SI value of μ is:

$$\mu = \mu_r \times \mu_0$$
$$= 100 \times 1.26 \times 10^{-6} \frac{T}{A/m}$$

$$\mu = 126 \times 10^{-6} \frac{T}{A/m}$$

Example 5. A magnetic material has a μ_r of 500. Calculate the absolute μ as B/H (a) in cgs units, and (b) in SI units.

Answer

(a) $\mu = \mu_r \times \mu_0$ in cgs units. Then

$$\mu = 500 \times 1 \frac{G}{Oe}$$

$$\mu = 500 \frac{G}{Oe}$$

(b) $\mu = \mu_r \times \mu_0$ in SI units. Then

$$\mu = 500 \times 1.26 \times 10^{-6} \frac{T}{A/m}$$

$$\mu = 630 \times 10^{-6} \frac{T}{A/m}$$

Example 6. For this example of $\mu = 630 \times 10^{-6}$ in SI units, calculate the flux density B that will be produced by the field intensity H equal to 1000 A/m.

Answer. $B = \mu H = \left(630 \times 10^{-6} \frac{T}{A/m}\right)\left(1000 \frac{A}{m}\right)$
$= 630 \times 10^{-3}$ T
$B = 0.63$ T

Note that the ampere-turns and meter units cancel, leaving only the tesla unit for the flux density B.

Practice Problems 14-3
(answers on page 293)
(a) What is the value of μ_r for air, vacuum, or space?
(b) An iron core has 200 times more flux density than air for the same field intensity H. How much is μ_r?
(c) An iron core produces 200 G of flux density for 1 Oe of field intensity H. How much is μ?

14-4
B-H MAGNETIZATION CURVE

The *B-H* curve in Fig. 14-3 is often used to show how much flux density *B* results from increasing the amount of field intensity *H*. This curve is for soft iron, plotted for the values in Table 14-1, but similar curves can be obtained for all magnetic materials.

Calculating H and B. The values in Table 14-1 are calculated as follows:

1. The current *I* in the coil equals *V/R*. For a 10-Ω coil resistance with 20 V applied, *I* is 2 A as listed in the top row of Table 14-1. Increasing values of *V* produce more current in the coil.
2. The ampere-turns *NI* of magnetizing force increase with more current. Since the turns are constant at 100, the values of *NI* increase from 200 for 2 A in the top row to 1000 for 10 A in the bottom row.
3. The field intensity *H* increases with higher *NI*. The values of *H* are in mks units of ampere-turns per meter. These values equal *NI*/0.2, as the length is 0.2 m. Therefore, each *NI* is just divided by 0.2, or multiplied by 5, for the corresponding values of *H*. Since *H* increases in the same proportion as *I*, sometimes the horizontal axis on a *B-H* curve is calibrated only in amperes, instead of in *H* units.
4. The flux density *B* depends on the field

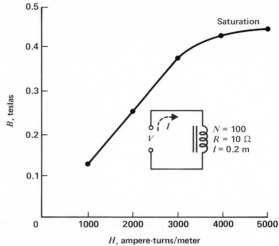

FIGURE 14-3

B-H magnetization curve for soft iron. No values shown near zero, where μ may vary with previous magnetization.

intensity *H* and permeability of the iron. The values of *B* in the last column are obtained by multiplying μ × *H*. However, with SI units the values of μ_r listed must be multiplied by 1.26×10^{-6} to obtain μ × *H* in teslas.

Saturation. Note that the permeability decreases for the highest values of *H*. With less μ, the iron core cannot provide proportional increases in *B* for increasing values of *H*. In the graph, for values of *H* above 4000 A/m, ap-

TABLE 14-1. *B-H* Values for Fig. 14-3

V, VOLTS	R, OHMS	I = V/R, AMPERES	NI	N, TURNS	l, m	H, NI/m	μ_r	B = μ × H, T
20	10	2	200	100	0.2	1000	100	0.126
40	10	4	400	100	0.2	2000	100	0.252
60	10	6	600	100	0.2	3000	100	0.378
80	10	8	800	100	0.2	4000	85	0.428
100	10	10	1000	100	0.2	5000	70	0.441

proximately, the values of B increase at a much slower rate, making the curve relatively flat at the top. The effect of little change in flux density when the field intensity increases is called *saturation*.

The reason is that the iron becomes saturated with magnetic lines of induction. After most of the molecular dipoles and the magnetic domains are aligned by the magnetizing force, very little additional induction can be produced. When the value of μ is specified for a magnetic material, it is usually the highest value before saturation.

*Practice Problems 14-4
(answers on page 293)
Refer to Fig. 14-3.*
(a) How much is B, in tesla units, for 1500 ampere-turns per meter?
(b) What value of H starts to produce saturation?

14-5
MAGNETIC HYSTERESIS

Hysteresis means "a lagging behind." With respect to the magnetic flux in an iron core of an electromagnet, the flux lags the increases or decreases of magnetizing force. The hysteresis results from the fact that the magnetic dipoles are not perfectly elastic. Once aligned by an external magnetizing force, the dipoles do not return exactly to their original positions when the force is removed. The effect is the same as if the dipoles were forced to move against an internal friction between molecules. Furthermore, if the magnetizing force is reversed in direction by reversal of the current in an electromagnet, the flux produced in the opposite direction lags behind the reversed magnetizing force.

Hysteresis Loss. When the magnetizing force reverses thousands or millions of times per second, as with rapidly reversing alternating current, the hysteresis can cause a considerable loss of energy. A large part of the magnetizing force is then used just for overcoming the internal friction of the molecular dipoles. The work done by the magnetizing force against this internal friction produces heat. This energy wasted in heat as the molecular dipoles lag the magnetizing force is called hysteresis loss. For steel and other hard magnetic materials, the hysteresis losses are much higher than in soft magnetic materials like iron.

When the magnetizing force varies at a slow rate, the hysteresis losses can be considered negligible. An example is an electromagnet with direct current that is simply turned on and off, or the magnetizing force of an alternating current that reverses sixty times per second or less. The faster the magnetizing force changes, however, the greater is the hysteresis effect.

Hysteresis Loop To show the hysteresis characteristics of a magnetic material, its values of flux density B are plotted for a periodically reversing magnetizing force. See Fig. 14-4. This curve is the hysteresis loop of the material. The larger the area enclosed by the curve, the greater the hysteresis loss. The hysteresis loop is actually a B-H curve with an ac magnetizing force.

On the vertical axis, values of flux density B are indicated. The units can be gauss, or teslas.

The horizontal axis indicates values of field intensity H. On this axis the units can be oersteds, ampere-turns per meter, ampere-turns, or just magnetizing current, as all factors are constant except I.

Opposite directions of current result in the opposite directions of $+H$ and $-H$ for the field lines. Similarly, opposite polarities are indicated for flux density as $+B$ or $-B$.

The current starts from zero at the center, when the material is unmagnetized. Then posi-

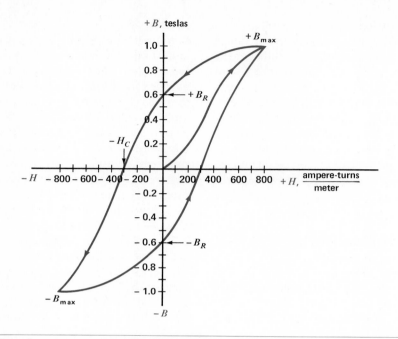

FIGURE 14-4

Hysteresis loop for magnetic materials. This is a B-H curve like in Fig. 14-3, but H alternates in polarity with alternating current.

tive H values increase B to saturation at $+B_{max}$. Next H decreases to zero, but B drops to the value of B_R, instead of zero, because of hysteresis. When H becomes negative, B drops to zero and continues to $-B_{max}$, which is saturation in the opposite direction from $+B_{max}$ because of the reversed magnetizing current.

Then as the $-H$ values decrease, the flux density is reduced to $-B_R$. Finally, the loop is completed with positive values of H producing saturation at B_{max} again. The curve does not return to the zero origin at the center because of hysteresis. As the magnetizing force periodically reverses, the values of flux density are repeated to trace out the hysteresis loop.

The value of either $+B_R$ or $-B_R$, which is the flux density remaining after the magnetizing force has been reduced to zero, is the *residual induction* of a magnetic material, also called its *retentivity*. In Fig. 14-4, the residual induction is 0.6 T, in either the positive or negative direction.

The value of $-H_C$, which equals the magnetizing force that must be applied in the reverse direction to reduce the flux density to zero, is the *coercive force* of the material. In Fig. 14-4, the coercive force $-H_C$ is 300 A/m.

Demagnetization. In order to demagnetize a magnetic material completely, the residual induction B_R must be reduced to zero. This usually cannot be accomplished by a reversed dc magnetizing force, because the material then would just become magnetized with opposite polarity. The practical way is to magnetize and demagnetize the material with a continuously decreasing hysteresis loop. This can be done with a magnetic field produced by alternating current. Then as the magnetic field and the material are moved away from each other, or the current amplitude is reduced, the hysteresis loop becomes smaller and smaller. Finally, with the weakest field, the loop collapses practically to zero, resulting in zero residual induction.

This method of demagnetization is also called *degaussing.* One application is degaussing the metal electrodes in a color picture tube, with a degaussing coil providing alternating current from the power line. Another example is erasing the recorded signal on magnetic tape by demagnetizing with an ac bias current.

Practice Problems 14-5
(answers on page 293)
Answer true or false.
(a) Hysteresis loss increases with higher frequencies.
(b) Degaussing is done with alternating current.

14-6
OHM'S LAW FOR MAGNETIC CIRCUITS

In comparison with electric circuits, the magnetic flux ϕ corresponds to current. The flux ϕ is produced by ampere-turns NI of magnetomotive force. Therefore, the mmf corresponds to voltage.

Opposition to the production of flux in a material is called its *reluctance,* comparable with resistance. The symbol for reluctance is $\mathcal{R}$. Reluctance is inversely proportional to permeability. Iron has high permeability and low reluctance. Air or vacuum has low permeability and high reluctance.

In Fig. 14-5, the ampere-turns of the coil produce magnetic flux throughout the magnetic path. The reluctance is the total opposition to the flux ϕ. In (a), there is little reluctance in the closed iron path, and few ampere-turns are necessary. In (b), however, the air gap has high reluctance, which requires many more ampere-turns for the same flux as in (a).

The three factors—flux, ampere-turns, and reluctance—are related as follows:

$$\phi = \frac{\text{mmf}}{\mathcal{R}} \qquad (14\text{-}5)$$

which is known as Ohm's law for magnetic circuits, corresponding to $I = V/R$. The mmf is considered to produce flux ϕ in a magnetic

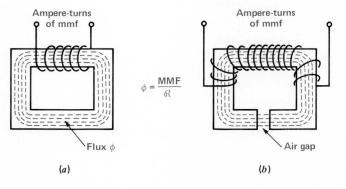

FIGURE 14-5
Two examples of a magnetic circuit. (a) Closed iron path having low reluctance requiring little mmf. (b) Higher-reluctance path with air gap requiring more mmf.

material against the opposition of its reluctance $\mathcal{R}$. This relationship corresponds to emf or voltage producing current in a conducting material against the opposition of its resistance.

Remember that the units for the flux ϕ are maxwells and webers. These units measure total lines, as distinguished from flux density B, which equals lines per unit area.

There are no specific units for reluctance, but it can be considered as an mmf/ϕ ratio, just as resistance is a V/I ratio. Then $\mathcal{R}$ is ampere-turns per weber in SI units, or gilberts per maxwell in the cgs system.

The units for mmf are either gilberts in the cgs system or ampere-turns in SI. Note that the symbol A is used for ampere-turns in the following examples.

Example 7. A coil has an mmf of 600 A and reluctance of 2×10^6 A/Wb. Calculate the total flux ϕ in microwebers.

Answer. $\phi = \dfrac{\text{mmf}}{\mathcal{R}} = \dfrac{600 \text{ A}}{2 \times 10^6 \text{ A/Wb}}$

$\phi = 300 \times 10^{-6}$ Wb $= 300$ μWb

Example 8. A magnetic material has a total flux ϕ of 80 μWb with an mmf of 160 A. Calculate the reluctance in ampere-turns per weber.

Answer. $\mathcal{R} = \dfrac{\text{mmf}}{\phi} = \dfrac{160 \text{ A}}{80 \times 10^{-6} \text{ Wb}}$

$\mathcal{R} = 2 \times 10^6 \dfrac{\text{A}}{\text{Wb}}$

Practice Problems 14-6
(answers on page 293)
Answer true or false.
(a) Air has higher reluctance than soft iron.
(b) More reluctance means more flux for a specified mmf.

14-7
RELATIONS BETWEEN MAGNETIC UNITS

The following examples show how the values of NI, H, ϕ, B, and $\mathcal{R}$ depend on each other. These calculations are in SI units, which are generally used for magnetic circuits.

Example 9. For a coil having 50 turns and 2 amperes, how much is the mmf?

Answer. mmf $= NI = 50 \times 2$
mmf $= 100$ A

The value of 100 ampere-turns for NI is the mmf producing the magnetic field, with either an air core or an iron core.

Example 10. If this coil is on an iron core with a length of 0.2 m, how much is the field intensity H throughout the iron?

Answer. $H = \dfrac{\text{mmf}}{l} = \dfrac{100 \text{ A}}{0.2 \text{ m}}$

$H = 500$ A/m

This is an example of calculating the field intensity of the external magnetic field from the mmf of the current in the coil.

Example 11. If this iron core with an H of 500 A/m has a relative permeability μ_r of 200, calculate the flux density B in teslas.

Answer. $B = \mu H = \mu_r \times 1.26 \times 10^{-6} \times H$

$= 200 \times 1.26 \times 10^{-6} \dfrac{\text{T}}{\text{A/m}} \times \dfrac{500 \text{ A}}{\text{m}}$

$B = 0.126$ T

Example 12. For this iron core with a flux density B of 0.126 T, if its cross-sectional area is 2×10^{-4} m^2, calculate the amount of flux ϕ in the core.

Answer. Use the relations between flux and density: $\phi = B \times$ Area. Since $B = 0.126$ T, or 0.126 Wb/m^2, then

$$\phi = B \times \text{Area}$$
$$= 0.126 \frac{\text{Wb}}{\text{m}^2} \times 2 \times 10^{-4} \text{ m}^2$$
$$= 0.252 \times 10^{-4} \text{ Wb} = 25.2 \times 10^{-6} \text{ Wb}$$
$$\phi = 25.2 \text{ }\mu\text{Wb}$$

Example 13. With the mmf 100 A for the coil in Fig. 14-5a and a value for ϕ of approximately 25×10^{-6} Wb in the iron core, calculate its reluctance $\mathcal{R}$.

Answer. Using Ohm's law for magnetic circuits,

$$\mathcal{R} = \frac{\text{mmf}}{\phi} = \frac{100 \text{ A}}{25 \times 10^{-6} \text{ Wb}}$$

$$\mathcal{R} = 4 \times 10^6 \text{ A/Wb}$$

Example 14. If the reluctance of the path with an air gap in Fig. 14-5b were 400×10^6 A/Wb, how much mmf would be required for the same flux of 25 μWb?

Answer. $\text{mmf} = \phi \times \mathcal{R}$

$$= 25 \times 10^{-6} \text{ Wb} \times 400 \times 10^6 \frac{\text{A}}{\text{Wb}}$$

$$\text{mmf} = 10{,}000 \text{ A}$$

Notice that the 10,000 ampere-turns of mmf here is 100 times more than the 100 ampere-turns in Example 13, because of the higher reluctance with an air gap. This idea corresponds to the higher voltage needed to produce the same current in a higher resistance.

Practice Problems 14-7
(answers on page 293)
Answer true or false.
(a) More I in a coil produces more mmf for a specified number of turns.
(b) More length for the coil produces more field intensity H for a specified mmf.
(c) Higher permeability in the core produces more flux density B for a specified H.

14-8
COMPARISON OF MAGNETIC AND ELECTRIC FIELDS

As shown in Fig. 14-6a, there is an external field of lines of force between two electric charges, similar to the magnetic field between the magnetic poles in (b). We cannot see the force of attraction and repulsion, just as the force of gravity is invisible, but the force is evident in the work it can do. For both fields, the force tends to make opposite polarities attract and similar polarities repel.

The electric lines show the path an electron would follow in the field; the magnetic lines show how a north pole would move. The entire

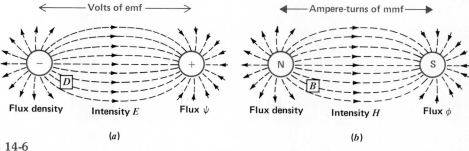

FIGURE 14-6
Comparison of electric and magnetic fields.
(a) Attraction between opposite static charges.
(b) Attraction between opposite magnetic poles.

group of electric lines of force of the static charges is called *electrostatic flux*. Its symbol is the Greek letter ψ (psi), corresponding to ϕ for magnetic flux.

In general, magnetic flux is associated with moving charges, or current, while electrostatic flux is associated with the voltage between static charges. For electric circuits, the application of magnetic flux is often a coil of wire, which is the construction of an inductor. With current, the wire has a magnetic field. As a coil, the wire's magnetic flux is concentrated in the coil. Furthermore, when the magnetic field varies, the change in magnetic flux produces an induced voltage, as explained in Chap. 15, Electromagnetic Induction, and Chap. 18, Inductance.

For the case of an electric field, the application is often an insulator between two conducting plates, which is the construction of a capacitor. With voltage across the insulator, it has an electric field. As a capacitor, the insulator's electric field is concentrated between the plates. Furthermore, when the electric field varies, the result is induced current through any conducting path connected to the capacitor. More details are explained in Chap. 21, Capacitance.

Coulomb's Law. The electric lines of force in Fig. 14-6a illustrate the force on an electron in the field. The amount of force between two charges is given by Coulomb's law:

$$F = 9 \times 10^9 \times \frac{q_1\,q_2}{r^2} \qquad (14\text{-}6)$$

where q_1 and q_2 are in coulomb units, F is in newtons, and r is the distance in meters between the charges. The constant factor 9×10^9 converts the values to SI units of newtons for the force in air or vacuum.

Coulomb's law states that the force increases with the amount of charge, but decreases as the square of the distance between charges. Typical values of q are in microcoulombs, since the coulomb is a very large unit of charge.

International System of Units. In order to provide a closer relation between practical units for both electricity and magnetism, these mks units were standardized in 1960 by international agreement. The abbreviation is *SI*, for *système international*. Table 14-2 lists the magnetic SI units. The corresponding electrical SI units include the coulomb, which is used for both electric flux and charge, the ampere for current, the volt for potential, and the ohm for resistance. The henry unit for inductance and the farad unit for capacitance are also SI units in the mks system.

In Table 14-2, note that the reciprocal of reluctance is *permeance*, corresponding to conductance as the reciprocal of resistance. The SI unit for conductance is the siemens (S), replacing the mho; both equal $1/(1\ \Omega)$. This unit is named after Ernst von Siemens, a European inventor.

As another comparison, the permeability μ of a magnetic material with magnetic flux corresponds to the electric *permittivity* ϵ of an insulator with electric flux. Just as permeability is the ability of a magnetic material to concentrate magnetic flux, permittivity is the ability of an insulator to concentrate electric flux. The symbol K_ϵ is used for relative permittivity, corresponding to K_m for relative permeability.

*Practice Problems 14-8
(answers on page 293)
Give the SI units for the following:*
(a) Voltage potential.
(b) Magnetic potential.
(c) Electrical current.
(d) Magnetic flux.

TABLE 14-2. International System of MKS Units (SI) for Magnetism

QUANTITY	SYMBOL	UNIT
Flux	ϕ	Weber (Wb)
Flux density	B	Wb/m^2 = tesla (T)
Potential	Mmf	Ampere-turn (A)
Field intensity	H	Ampere-turn per meter (A/m)
Reluctance	$\mathcal{R}$	Ampere-turn per weber (A/Wb)
Permeance	$\rho = \dfrac{1}{\mathcal{R}}$	Weber per ampere-turn (Wb/A)
Relative μ	μ_r or K_m	None, pure number
Permeability	$\mu = \mu_r \times 1.26 \times 10^{-6}$	$\dfrac{B}{H} = \dfrac{\text{tesla (T)}}{\text{ampere-turn per meter (A/m)}}$

Summary

Table 14-2 summarizes the magnetic units and their definitions.

Self-Examination (Answers at back of book.)

Answer true or false

1. A current of 4 amperes through 200 turns provides an mmf of $NI = 200$ ampere-turns.
2. For the mmf of 200 NI with 100 turns, a current of 2 amperes is necessary.
3. An mmf of 200 A across a flux path of 0.1 m provides the field intensity H of 2000 A/m.
4. A magnetic material with relative permeability μ_r of 100 has an absolute permeability μ of 126 G/Oe in cgs units.
5. There are no units for relative permeability μ_r.
6. Hysteresis losses are greater in soft iron than in air.
7. Magnetic saturation means that flux density B does not increase in proportion to increases in field intensity H.
8. The units for a B-H curve can be teslas plotted against ampere-turns.
9. In Ohm's law for magnetic circuits, reluctance $\mathcal{R}$ is the opposition to flux ϕ.
10. Ampere-turns of mmf between magnetic poles do not depend on the length of the coil.

Chapter 14
Magnetic Units

Essay Questions

1. In Ohm's law for magnetic circuits, what magnetic quantities correspond to V, I, and R?
2. Why can reluctance and permeability be considered opposite characteristics?
3. Give the SI magnetic unit and symbol for each of the following: (a) flux; (b) flux density; (c) field intensity; (d) absolute permeability.
4. What cgs units correspond to the following mks units? (a) weber; (b) tesla; (c) ampere-turn; (d) ampere-turn per meter.
5. Define the following: (a) saturation; (b) relative permeability; (c) relative permittivity.
6. Explain briefly how to demagnetize a metal object that has become temporarily magnetized.
7. Draw a B-H curve with μ, N, L, and V the same as in Fig. 14-3, but with a coil resistance of 5 Ω.
8. Give the formula for Coulomb's law of the force between electrostatic charges, with SI units.

Problems (Answers to odd-numbered problems at back of book.)

1. A coil of 2000 turns with a 100-mA current has a length of 0.2 m. (a) Calculate the mmf in ampere-turns. (b) Calculate the field intensity H in ampere-turns per meter.
2. If the current is increased to 400 mA in the coil of Prob. 1, calculate the increased values of mmf and H.
3. An iron core has a flux density B of 3600 G with an H of 12 Oe. Calculate (a) the permeability μ in cgs units; (b) the permeability μ in SI units; (c) the relative permeability μ_r of the iron core.
4. A coil of 250 turns with a 400-mA current is 0.2 m long with an iron coil of the same length. Calculate the following in mks units: (a) mmf; (b) H; (c) B in the iron core with a μ_r of 200; (d) B with an air core instead of the iron core.
5. Referring to the B-H curve in Fig. 14-3, calculate the μ in SI units for the iron core at: (a) H of 3000 A/m; (b) 5000 A/m.
6. Referring to the hysteresis loop in Fig. 14-4, give the values of (a) residual induction B_R, and (b) coercive force $-H_C$.
7. A battery is connected across a coil of 100 turns and a 20-Ω R, with an iron core 0.2 m long. (a) Draw the circuit diagram. (b) How much battery voltage is needed for 200 ampere-turns? (c) Calculate H in

the iron core in ampere-turns per meter. (d) Calculate B in teslas in the iron core if its μ_r is 300. (e) Calculate ϕ in webers at each pole with an area of 8×10^{-4} m². (f) How much is the reluctance $\mathcal{R}$ of the iron core, in ampere-turns per weber?

8. In cgs units, how much is the flux density B in gauss, for a field intensity H of 24 Oe, with μ of 500?
9. Calculate the force, in newtons, between two 4-μC charges separated by 0.1 m in air or vacuum.

Answers to Practice Problems

14-1	(a) 486	14-5	(b) T
	(b) 530 Gb	14-6	(a) Reluctance
14-2	(a) 125		(b) mmf
	(b) 6.3 Oe	14-7	(a) T
14-3	(a) 1		(b) F
	(b) 200		(c) T
	(c) 200 G/Oe	14-8	(a) Volt
14-4	(a) 0.2 T		(b) Ampere-turn
	(b) 4000 A/m		(c) Ampere
14-5	(a) T		(d) Weber

Electro-magnetic Induction

Chapter 15

The link between electricity and magnetism was discovered in 1824 by Oersted, who found that current in a wire could move a magnetic compass needle. A few years later the opposite effect was discovered: A magnetic field in motion forces electrons to move, producing current. This important effect was studied by Faraday, Henry, and Lenz.[1] Electromagnetism, therefore, includes the magnetic effects of electrical current.

Electrons in motion have an associated magnetic field; a moving magnetic field forces electrons to move, producing current. These electromagnetic effects have many practical applications that are the basis for motors and generators. Perhaps most important, the fundamental reaction of an inductance to a change in current is an electromagnetic effect. The details of electromagnetism are analyzed in the following topics:

15-1 Magnetic Field Around an Electrical Current
15-2 Magnetic Polarity of a Coil
15-3 Motor Action between Two Magnetic Fields
15-4 Induced Current
15-5 Lenz' Law
15-6 Generating an Induced Voltage
15-7 Faraday's Law of Induced Voltage

15-1
MAGNETIC FIELD AROUND AN ELECTRICAL CURRENT

In Fig. 15-1, the iron filings aligned in concentric rings around the conductor show the magnetic field of the current in the wire. The iron filings are dense next to the conductor, showing that the field is strongest at this point. Furthermore, the field strength decreases inversely as the square of the distance from the conductor. It is important to note the following two factors about the magnetic lines of force:

1. The magnetic lines are circular, as the field is symmetrical with respect to the wire in the center.
2. The magnetic field with circular lines of force is in a plane perpendicular to the current in the wire.

---

[1] Michael Faraday (1791–1867), eminent British physicist and pioneer in electromagnetism; Joseph Henry (1797–1878), American physicist; H. F. E. Lenz (1804–1865), Russian physicist.

Chapter 15
Electromagnetic Induction

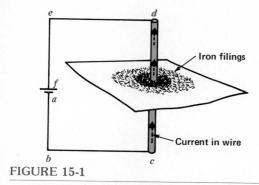

FIGURE 15-1

Iron filings in magnetic field around current in a conductor.

From points *c* to *d* in the wire, its circular magnetic field is in the horizontal plane because the wire is vertical. Also, the vertical conductor between points *ef* and *ab* has the associated magnetic field in the horizontal plane. Where the conductor is horizontal, as from *b* to *c* and *d* to *e*, the magnetic field is in a vertical plane.

These two requirements of a circular magnetic field in a perpendicular plane apply to any charge in motion. Whether electron flow or a motion of positive charges is considered, the associated magnetic field must be at right angles to the direction of current.

In addition, the current need not be in a wire conductor. As an example, the beam of moving electrons in the vacuum of a cathode-ray tube has an associated magnetic field. In all cases, the magnetic field has circular lines of force in a plane perpendicular to the direction of motion of the electric charge.

Clockwise and Counterclockwise Fields. With circular lines of force, the magnetic field would tend to move a magnetic pole in a circular path. Therefore, the direction of the lines must be considered as either clockwise or counterclockwise. This idea is illustrated in Fig. 15-2, showing how a north pole would move in the circular field.

The directions are tested with a magnetic compass needle. When the compass is in front of the wire, the north pole on the needle points up. On the opposite side, the compass points down. If the compass were placed at the top, its needle would point toward the back of the wire; below the wire, the compass would point forward.

Combining all these directions, the result is the circular magnetic field shown, with counterclockwise lines of force. This direction has the magnetic lines upward at the front of the conductor and downward at the back.

Instead of testing every conductor with a magnetic compass, however, we can use the

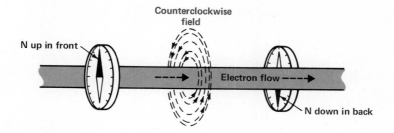

FIGURE 15-2

Rule for determining direction of counterclockwise field around straight conductor. The reverse direction of electron flow would provide clockwise field.

Chapter 15
Electromagnetic Induction

following rule to determine the circular direction of the magnetic field: *If you look along the wire in the direction of electron flow, the magnetic field is counterclockwise.* In Fig. 15-2, the line of electron flow is from left to right. Facing this way, you can assume the circular magnetic flux in a perpendicular plane has lines of force in the counterclockwise direction.

The opposite direction of current produces a reversed field. Then the magnetic lines of force have clockwise rotation. If the charges were moving from right to left in Fig. 15-2, the associated magnetic field would be in the opposite direction, with clockwise lines of force.

Fields Aiding or Canceling. When the magnetic lines of two fields are in the same direction, the lines of force aid each other, making the field stronger. With magnetic lines in opposite directions, the fields cancel.

In Fig. 15-3 the fields are shown for two conductors with opposite directions of current. The dot in the middle of the field at the left indicates the tip of an arrowhead to show current up from the paper. The cross symbolizes the back of an arrow to indicate current into the paper.

Notice that the magnetic lines *between the conductors* are in the same direction, although one field is clockwise and the other counterclockwise. Therefore, the fields aid here, making a stronger total field. On either side of the conductors, the two fields are opposite in direction and tend to cancel each other. The net result, then, is to strengthen the field in the space between the conductors.

*Practice Problems 15-1
(answers on page 309)*
Answer true or false.
(a) Magnetic field lines around a conductor are circular in a perpendicular plane.
(b) In Fig. 15-3, the field is strongest between the conductors.

15-2
MAGNETIC POLARITY OF A COIL

Bending a straight conductor around in the form of a loop, as shown in Fig. 15-4, has two effects. First, the magnetic field lines are more dense inside the loop. The total number of lines is the same as for the straight conductor, but in the loop the lines are concentrated in a smaller space. Furthermore, all the lines are aiding in

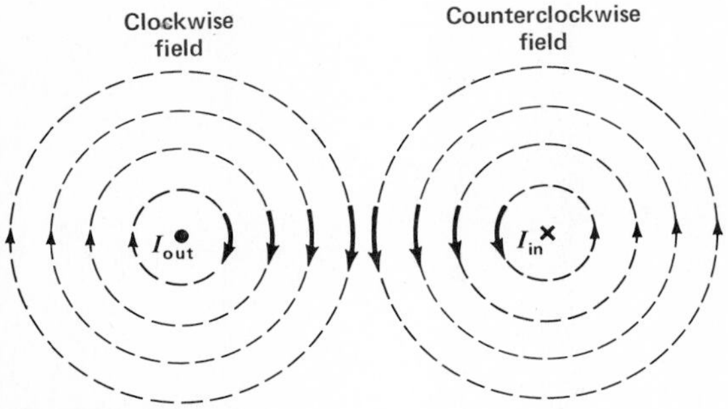

FIGURE 15-3
Magnetic fields aiding between parallel conductors with opposite directions of current.

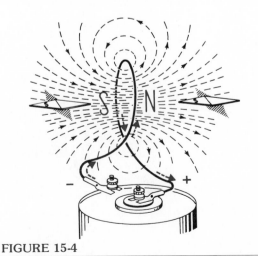

FIGURE 15-4
Magnetic poles of a current loop.

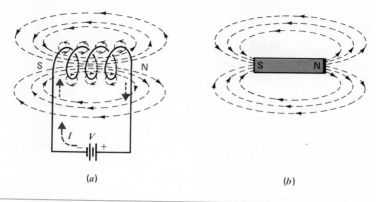

FIGURE 15-5

Magnetic poles of a solenoid. (a) Coil winding. (b) Equivalent bar magnet.

the same direction inside the loop. This makes the loop field effectively the same as a bar magnet with opposite poles at opposite faces of the loop.

Solenoid as a Bar Magnet. A coil of wire conductor with more than one turn is generally called a *solenoid*. An ideal solenoid, however, has a length much greater than its diameter. Like a single loop, the solenoid concentrates the magnetic field inside the coil and provides opposite magnetic poles at the ends. These effects are multiplied, however, by the number of turns as the magnetic field lines aid each other in the same direction inside the coil. Outside the coil, the field corresponds to a bar magnet with north and south poles at opposite ends, as illustrated in Fig. 15-5.

Magnetic Polarity. To determine the magnetic polarity, use the *left-hand rule* illustrated in Fig. 15-6: *If the coil is grasped with the fingers of the left hand curled in the direction of electron flow around the coil, the thumb points to the north pole of the coil.* The left hand is used here because the current is electron flow.

The solenoid acts like a bar magnet whether it has an iron core or not. Adding an iron core increases the flux density inside the coil. In addition, the field strength then is uniform for the entire length of the core. The polarity is the same, however, for air-core or iron-core coils.

The magnetic polarity depends on the direction of current flow and the direction of winding. The current is determined by the connections to the voltage source. Electron flow is from the negative side of the voltage source, through the coil, and back to the positive terminal.

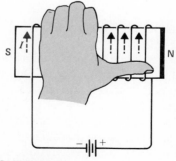

FIGURE 15-6

Left-hand rule for north pole of a coil with current *I*.

Chapter 15
Electromagnetic Induction

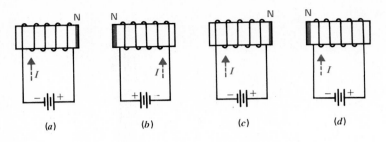

FIGURE 15-7

Four examples for determining the magnetic polarity of a coil. In (b) the battery polarity is opposite from (a) to reverse the direction of current. In (d) the direction of winding is reversed from (c).

The direction of winding can be over and under, starting from one end of the coil, or under and over with respect to the same starting point. Reversing either the direction of winding or the direction of current reverses the magnetic poles of the solenoid. See Fig. 15-7. With both reversed, though, the polarity is the same.

Practice Problems 15-2
(answers on page 309)
(a) In Fig. 15-5, if the battery is reversed, will the north pole be at the left or right?
(b) If one end of a solenoid is a north pole, is the opposite end a north or south pole?

15-3
MOTOR ACTION BETWEEN TWO MAGNETIC FIELDS

The physical motion resulting from the forces of magnetic fields is called *motor action*. One example is the simple attraction or repulsion between bar magnets.

We know that like poles repel and unlike poles attract. It can also be considered that fields in the same direction repel and opposite fields attract.

Consider the repulsion between two north poles illustrated in Fig. 15-8. Similar poles have fields in the same direction. Therefore, the similar fields of the two like poles repel each other.

A more fundamental reason for motor action, however, is the fact that the force in a magnetic field tends to produce motion from a stronger field toward a weaker field. In Fig. 15-8, note that the field intensity is greatest in the space between the two north poles. Here the

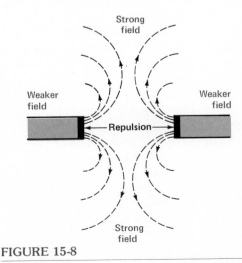

FIGURE 15-8

Repulsion between north poles of two bar magnets, showing motion from stronger field to weaker field.

field lines of similar poles in both magnets reinforce in the same direction. Farther away the field intensity is less, for essentially one magnet only. As a result there is a difference in field strength, providing a net force that tends to produce motion. The direction of motion is always toward the weaker field.

To remember the directions, we can consider that the stronger field moves to the weaker field, tending to equalize the field intensity. Otherwise, the motion would make the strong field stronger and the weak field weaker. This must be impossible, because then the magnetic field would multiply its own strength without any work being added.

Force on a Straight Conductor in a Magnetic Field. Current in a conductor has its associated magnetic field. When this conductor is placed in another magnetic field from a separate source, the two fields can react to produce motor action. The conductor must be perpendicular to the magnetic field, however, as illustrated in Fig. 15-9. This way, the perpendicular magnetic field of the current then is in the same plane as the external magnetic field.

Unless the two fields are in the same plane, they cannot affect each other. In the same plane, however, lines of force in the same direction reinforce to make a stronger field, while lines in the opposite direction cancel and result in a weaker field.

To summarize these directions:

1. With the conductor at 90°, or perpendicular to the external field, the reaction between the two magnetic fields is maximum.
2. With the conductor at 0°, or parallel to the external field, there is no effect between them.
3. When the conductor is at an angle between 0 and 90°, only the perpendicular component is effective.

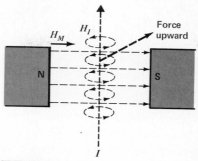

FIGURE 15-9

Motor action of current in a straight conductor in an external magnetic field. The net force of the resultant magnetic field here moves the conductor upward to the weaker field.

In Fig. 15-9, electrons flow in the wire conductor in the plane of the paper, from the bottom to top of the page. This flow provides the counterclockwise field H_I around the wire, in a perpendicular plane cutting through the paper. The external field H_M has lines of force from left to right in the plane of the paper. Then lines of force in the two fields are parallel above and below the wire.

Below the conductor, its field lines are left to right in the same direction as the external field. Therefore, these lines reinforce to produce a stronger field. Above the conductor the lines of the two fields are in opposite directions, causing a weaker field. As a result, the net force of the stronger field makes the conductor move upward out of the page, toward the weaker field.

If electrons flow in the reverse direction in the conductor, or if the external field is reversed, the motor action will be in the opposite direction. Reversing both the field and the current results in the same direction of motion.

Rotation of a Current Loop in a Magnetic Field. With a loop of wire in the magnetic field, opposite sides of the loop have current in opposite directions. Then the associated magnetic fields are opposite. The resulting forces are upward on one side and downward on the other side on the loop, making it rotate. This effect of a force in producing rotation is called *torque*.

The principle of motor action between magnetic fields producing rotational torque is the basis of all electric motors. Also, the moving-coil meter described in Sec. 7-1 is a similar application. Since the torque is proportional to current, the amount of rotation indicates how much current flows through the coil.

Practice Problems 15-3 *(answers on page 309)*
Answer true or false.
(a) In Fig. 15-8, the field is strongest between the two north poles.
(b) In Fig. 15-9, if both the magnetic field and current are reversed, the motion will still be upward.

15-4
INDUCED CURRENT

Just as electrons in motion provide an associated magnetic field, when magnetic flux moves, the motion of magnetic lines cutting across a conductor forces free electrons in the conductor to move, producing current. This action is called *induction* because there is no physical connection between the magnet and the conductor. The induced current is a result of generator action as the mechanical work put into moving the magnetic field is converted into electrical energy when current flows in the conductor.

Referring to Fig. 15-10, let the conductor AB be placed at right angles to the flux in the air gap of the horseshoe magnet. Then, when the magnet is moved up or down, its flux cuts across the conductor. The action of magnetic flux cutting across the conductor generates cur-

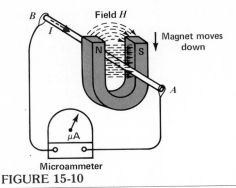

FIGURE 15-10
Induced current produced by magnetic flux cutting across a conductor.

rent. The fact that current flows is indicated by the microammeter.

When the magnet is moved downward, current flows in the direction shown. If the magnet is moved upward, current will flow in the opposite direction. Without motion, there is no current.

Direction of Motion. The motion is necessary in order to have the flux lines of the magnetic field cut across the conductor. This cutting can be accomplished by motion of either the field or the conductor. When the conductor is moved upward or downward, it cuts across the flux. The generator action is the same as moving the field, except that the relative motion is opposite. Moving the conductor upward, for instance, corresponds to moving the magnet downward.

Conductor Perpendicular to External Flux. In order to have electromagnetic induction, the conductor and the magnetic lines of flux must be perpendicular to each other. Then the motion makes the flux cut through the cross-sectional area of the conductor. As shown in Fig. 15-10, the conductor is at right angles to the lines of force in the field H.

The reason the conductor must be perpendicular is to make its induced current have an associated magnetic field in the same plane as the external flux. If the field of the induced current does not react with the external field, there can be no induced current.

How Induced Current Is Generated. The induced current can be considered the result of motor action between the external field H and the magnetic field of free electrons in every cross-sectional area of the wire. Without an external field, the free electrons move at random without any specific direction and they have no net magnetic field. When the conductor is in the magnetic field H, there still is no induction without relative motion, since the magnetic fields for the free electrons are not disturbed. When the field or conductor moves, however, there must be a reaction opposing the motion. The reaction is a flow of free electrons resulting from motor action on the electrons.

Referring to Fig. 15-10, for example, the induced current must flow in the direction shown because the field is moved downward, pulling the magnet away from the conductor. The induced current of electrons then has a clockwise field with lines of force aiding H above the conductor and canceling H below. With motor action between the two magnetic fields tending to move the conductor toward the weaker field, the conductor will be forced downward, staying with the magnet to oppose the work of pulling the magnet away from the conductor.

The effect of electromagnetic induction is increased where a coil is used for the conductor. Then the turns concentrate more conductor length in a smaller area. As illustrated in Fig. 15-11, moving the magnet into the coil enables the flux to cut across many turns of conductors.

*Practice Problems 15-4
(answers on page 309)*
Answer true or false. Refer to Fig. 15-10.
(a) If the conductor is moved up, instead of the magnet down, the induced current will flow in the same direction.
(b) The electron flow through the meter is from terminal A to B.

15-5
LENZ' LAW

This basic principle is used to determine the direction of an induced voltage or current. Based on the principle of conservation of energy, Lenz' law simply states that the direction of the induced current must be such that its own magnetic field will oppose the action that produced the induced current.

In Fig. 15-11, for example, the induced current has the direction that produces a north pole at the left to oppose the motion by repulsion of the north pole being moved in. This is why it takes some work to push the permanent magnet into the coil. The work expended in moving the permanent magnet is the source of energy for the current induced in the coil.

Using Lenz' law, we can start with the fact that the left end of the coil in Fig. 15-11 must be a north pole to oppose the motion. Then the direction of the induced current is determined by the left-hand rule for electron flow. If the fingers coil around the direction of electron flow

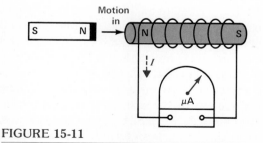

FIGURE 15-11

Induced current produced by magnetic flux cutting across turns of wire in a coil.

shown, under and over the winding, the thumb will point to the left for the north pole.

For the opposite case, suppose that the north pole of the permanent magnet in Fig. 15-11 is moved away from the coil. Then the induced pole at the left end of the coil must be a south pole, by Lenz' law. The induced south pole will attract the north pole to oppose the motion of the magnet being moved away. For a south pole at the left end of the coil, then, the electron flow will be reversed from the direction shown in Fig. 15-11. We could actually generate an alternating current in the coil by moving the magnet periodically in and out.

Practice Problems 15-5
(answers on page 309)
Refer to Fig. 15-11.
(a) If the north end of the magnet is moved away from the coil, will its left side be north or south?
(b) If the south end of the magnet is moved in, will the left end of the coil be north or south?

15-6
GENERATING AN INDUCED VOLTAGE

Consider the case of magnetic flux cutting a conductor that is not in a closed circuit, as shown in Fig. 15-12. The motion of flux across the conductor forces free electrons to move, but with an open circuit, the displaced electrons produce opposite electric charges at the two open ends.

For the directions shown, free electrons in the conductor are forced to move to point A. Since the end is open, electrons accumulate here. Point A then develops a negative potential.

At the same time, point B loses electrons and becomes charged positive. The result is a potential difference across the two ends, pro-

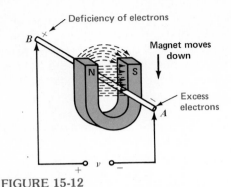

FIGURE 15-12

Voltage induced across open ends of conductor cut by magnetic flux.

vided by the separation of electric charges in the conductor.

The potential difference is an electromotive force, generated by the work of cutting across the flux. You can measure this potential difference with a voltmeter. However, a conductor cannot store electric charge. Therefore, the voltage is present only while the motion of flux cutting across the conductor is producing the induced voltage.

Induced Voltage Across a Coil. With a coil, as in Fig. 15-13a, the induced emf is increased by the number of turns. Each turn cut by flux adds to the induced voltage, since they all force free electrons to accumulate at the negative end of the coil, with a deficiency of electrons at the positive end.

The polarity of the induced voltage follows from the direction of induced current. The end of the conductor to which the electrons go and where they accumulate is the negative side of the induced voltage. The opposite end with a deficiency of electrons is the positive side. The total emf across the coil is the sum of the induced voltages, since all the turns are in series.

Furthermore, the total induced voltage acts in series with the coil, as illustrated by the equivalent circuit in Fig. 15-13b, showing the

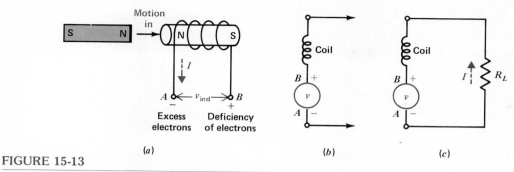

FIGURE 15-13

Voltage induced across coil cut by magnetic flux. (a) Motion of flux generating voltage across coil. (b) Induced voltage acts in series with coil. (c) Induced voltage is a source that can produce current in external load circuit connected across coil.

induced voltage as a separate generator. This generator represents a voltage source with a potential difference resulting from the separation of charges produced by electromagnetic induction. The source v then can produce current in an external load circuit connected across the negative and positive terminals, as shown in Fig. 15-13c.

The induced voltage is in series with the coil because current produced by the generated emf must flow through all the turns. An induced voltage of 10 V, for example, with R_L equal to 5 Ω, results in a current of 2 A, which flows through the coil, the equivalent generator v, and the load resistance R_L.

The direction of current in Fig. 15-13c shows electron flow around the circuit. Outside the source v, the electrons move from its negative terminal, through R_L, and back to the positive terminal of v because of its potential difference.

Inside the generator, however, the electron flow is from the + terminal to the − terminal. This direction of electron flow results from the fact that the left end of the coil in (a) must be a north pole by Lenz' law, to oppose the north pole being moved in.

Notice how motors and generators are similar in using the motion of a magnetic field, but with opposite applications. In a motor, current is supplied for an associated magnetic field to react with the external flux to produce motion of the conductor. In a generator, motion must be supplied so that the flux and conductor can cut across each other to induce voltage across the ends of the conductor.

Practice Problems 15-6
(answers on page 309)
Refer to Fig. 15-13.
(a) Is terminal A or B the negative side of the induced voltage?
(b) Is terminal A or B the negative side of V_{R_L}?

15-7
FARADAY'S LAW OF INDUCED VOLTAGE

The voltage induced by magnetic flux cutting the turns of a coil depends upon the number of turns and how fast the flux moves across the conductor. Either the flux or the conductor can move. Specifically, the amount of induced voltage is determined by the following three factors:

1. *Amount of flux.* The more magnetic lines of force that cut across the conductor, the higher is the amount of induced voltage.
2. *Number of turns.* The more turns in a coil, the higher the induced voltage. The v_{ind} is the sum of all the individual voltages generated in each turn in series.
3. *Time rate of cutting.* The faster the flux cuts a conductor, the higher the induced voltage. Then more lines of force cut the conductor within a specific period of time.

These factors are of fundamental importance in many applications. Any conductor with current will have voltage induced in it by a change in current and its associated magnetic flux.

The amount of induced voltage can be calculated by Faraday's law:

$$v_{ind} = N \frac{d\phi \text{ (webers)}}{dt \text{ (seconds)}} \quad (15\text{-}1)$$

where N is the number of turns and $d\phi/dt$ specifies how fast the flux ϕ cuts across the conductor.

With $d\phi/dt$ in webers per second, the induced voltage is in volt units. As an example, if the magnetic flux cuts across 300 turns at the rate of 2 Wb/s, $v_{ind} = 300 \times 2$, or 600 V. It is assumed that the flux links all the turns, which is true with an iron core.

Time Rate of Change. The symbol d in $d\phi$ and dt is an abbreviation for delta (Δ), which means a *change*.[1] The $d\phi$ means a change in the flux ϕ, while dt means a change in time.

[1] In calculus, dt represents only an infinitesimally small change, but we are using this symbol for rate of change in general.

As an example, if the flux ϕ is 4 W at one time but then changes to 6 Wb, the change in flux is 2 Wb for $d\phi$. The same idea applies to a decrease as well as an increase. If the flux changed from 6 to 4 Wb, $d\phi$ would still be 2 Wb. However, an increase is usually considered a change in the positive direction, with an upward slope, while a decrease has a negative slope downward.

Similarly, dt means a change in time. If we consider the flux at a time 2 s after the start, and at a later time 3 s after the start, the change in time is $3 - 2$, or 1 s for dt. Time always increases in the positive direction.

Combining the two factors of $d\phi$ and dt, we can say that for magnetic flux increasing by 2 Wb in 1 s, $d\phi/dt$ equals $\frac{2}{1}$, or 2 Wb/s, which states the time rate of change of the magnetic flux. For 300 turns cut by the changing flux of 2 Wb/s, then, the induced voltage is 300×2, or 600 V.

Analysis of Induced Voltage as $N\, d\phi/dt$. This fundamental concept of voltage induced by a change in flux is illustrated by the graphs in Fig. 15-14, for the values listed in Table 15-1. The linear rise in (a) shows values of flux ϕ increasing at a uniform rate. In this case, the curve goes up 2 Wb for every 1-s interval of time. The slope of this curve, then, equal to $d\phi/dt$, is 2 Wb/s. Note that, although ϕ increases, the rate of change is constant because the linear rise has a constant slope.

For induced voltage, only the $d\phi/dt$ factor is important, not the actual value of flux. To emphasize this basic concept, the graph in (b) shows the $d\phi/dt$ values alone. This graph is just a straight horizontal line for the constant value of 2 Wb/s.

The induced-voltage graph in (c) is also a straight horizontal line. Since $v_{ind} = N(d\phi/dt)$, the graph of induced voltage is just the $d\phi/dt$ values multiplied by the number of turns. The

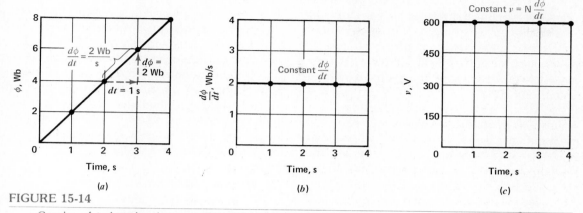

FIGURE 15-14
Graphs of induced voltage produced by flux changes $d\phi/dt$ in a coil with an N of 300 turns. (a) Linear increase of flux ϕ. (b) Constant rate of change $d\phi/dt$ at 2 Wb/s. (c) Constant induced voltage of 600 V.

result is a constant 600 V, with 300 turns cut by flux changing at the constant rate of 2 Wb/s.

The example illustrated here can be different in several ways without changing the basic fact that the induced voltage is equal to $N(d\phi/dt)$. First, the number of turns or the $d\phi/dt$ values can be greater than the values assumed here, or less. More turns will provide more induced voltage, while fewer turns mean less voltage. Similarly, a higher value for $d\phi/dt$ results in more induced voltage.

Note that two factors are included in $d\phi/dt$. Its value can be increased by a higher value of $d\phi$ or a smaller value of dt. As an example, the value of 2 Wb/s for $d\phi/dt$ can be doubled by either increasing $d\phi$ to 4 Wb or reducing dt to $\frac{1}{2}$ s. Then $d\phi/dt$ is $\frac{4}{1}$ or $\frac{2}{0.5}$, which equals 4 Wb/s in either case. The same flux changing within a shorter time means a faster rate of flux cutting for a higher value of $d\phi/dt$ and more induced voltage.

For the opposite case, a smaller value of $d\phi/dt$, with less flux or a slower rate of change, results in a smaller value of induced voltage. When $d\phi/dt$ decreases, the induced voltage has opposite polarity, compared with an increase.

Finally, it should be noted that the $d\phi/dt$ graph in Fig. 15-14b has the constant value of

TABLE 15-1. Induced-Voltage Calculations for Fig. 15-14

ϕ, Wb	$d\phi$, Wb	t, s	dt, s	$d\phi/dt$, Wb/s	N, TURNS	$N(d\phi/dt)$, V
2	2	1	1	2	300	600
4	2	2	1	2	300	600
6	2	3	1	2	300	600
8	2	4	1	2	300	600

2 Wb/s because the flux is increasing at a linear rate. However, the flux need not have a uniform rate of change. Then the $d\phi/dt$ values will not be constant. In any case, though, the values of $d\phi/dt$ at all instants of time will determine the instantaneous values of the induced voltage equal to $N(d\phi/dt)$.

Polarity of the Induced Voltage. The polarity is determined by Lenz' law. The induced voltage has the polarity that opposes the change causing the induction. Sometimes this fact is indicated by using a negative sign for v_{ind} in Formula (15-1). However, the absolute polarity depends on whether the flux is increasing or decreasing, the method of winding, and which end of the coil is the reference.

When all these factors are considered, v_{ind} has the polarity such that the current it produces and the associated magnetic field will oppose the change in flux producing the induced voltage. If the external flux increases, the magnetic field of the induced current will be in the opposite direction. If the external field decreases, the magnetic field of the induced current will be in the same direction as the external field to oppose the change by sustaining the flux. In short, the induced voltage has the polarity that opposes the change.

Practice Problems 15-7
(answers on page 309)
(a) The magnetic flux of 10 Wb changes to 8 Wb in 1 s. How much is $d\phi/dt$?
(b) The flux of 10 μWb changes to 8 μWb in 1 μs. How much is $d\phi/dt$?

Summary

1. Current in a straight conductor has an associated magnetic field with circular lines of force in a plane perpendicular to the conductor. The direction of the circular field is counterclockwise when you look along the conductor in the direction of electron flow.
2. With two fields in the same plane, produced by either current or a permanent magnet, lines of force in the same direction aid each other to provide a stronger field. Lines of force in opposite directions cancel and result in a weaker field.
3. A solenoid is a long, narrow coil of wire which concentrates the conductor and its associated magnetic field. Because the fields for all turns aid inside the coil and cancel outside, a solenoid has a resultant electromagnetic field like a bar magnet with north and south poles at opposite ends.
4. The left-hand rule for polarity of an electromagnet says that when your fingers curl around the turns in the direction of electron flow, the thumb points to the north pole.
5. Motor action is the motion that results from the net force of two fields that can aid or cancel each other. The direction of the resultant force is always from the stronger field to the weaker field.
6. Generator action refers to induced voltage. For N turns,

$v_{ind} = N(d\phi/dt)$, with $d\phi/dt$ in webers per second. There must be a change in the flux to produce induced voltage.
7. Lenz' law states that the polarity of the induced voltage will oppose the change in magnetic flux causing the induction.
8. The faster the flux changes, the higher is the induced voltage.
9. When the flux changes at a constant rate, the induced voltage has a constant value.

Self-Examination (Answers at back of book.)

Answer true or false.

1. A vertical wire with electron flow downward through this page has an associated magnetic field counterclockwise in the plane of the paper.
2. Lines of force of two magnetic fields in the same direction aid each other to produce a stronger resultant field.
3. Motor action always tends to produce motion toward the weaker field.
4. In Fig. 15-6, if the battery connections are reversed, the magnetic poles of the coil will be reversed.
5. A solenoid is a long, narrow coil that acts as a bar magnet only when current flows.
6. A torque is a force tending to cause rotation.
7. In Fig. 15-9, if the poles of the external field are reversed, the motor action will be downward.
8. In Fig. 15-10, if the conductor is moved down, instead of the magnet, the induced current flows in the opposite direction.
9. An induced voltage is present only while the flux is changing.
10. Faraday's law determines the amount of induced voltage.
11. Lenz' law determines the polarity of an induced voltage.
12. Induced voltage increases with a faster rate of flux cutting.
13. An induced voltage is effectively in series with the turns of the coil in which the voltage is produced.
14. A decrease in flux will induce a voltage of opposite polarity from an increase in flux, with the same direction of field lines in both cases.
15. The flux of 1000 lines increasing to 1001 lines in 1 s produces a flux change $d\phi/dt$ of 1 line per s.
16. The flux of 2 lines increasing to 3 lines in 1 μs corresponds to a flux change $d\phi/dt$ of 1,000,000 lines per s.
17. In question 16, $d\phi/dt$ equals 0.01 Wb/s.
18. The induced voltage will be much greater for the example in question 16, compared with question 15, with the same number of turns.

19. The more turns in a coil, the higher is its induced voltage.
20. In Fig. 15-14, the flux ϕ is increasing, but its rate of change $d\phi/dt$ is constant.

Essay Questions

1. Draw a diagram showing two conductors connecting a battery to a load resistance through a closed switch. (a) Show the magnetic field of the current in the negative side of the line and in the positive side. (b) Where do the two fields aid? Where do they oppose?
2. State the rule for determining the magnetic polarity of a solenoid. (a) How can the polarity be reversed? (b) Why are there no magnetic poles when the current through the coil is zero?
3. Why does the motor action between two magnetic fields result in motion toward the weaker field?
4. Why does current in a conductor perpendicular to this page have a magnetic field in the plane of the paper?
5. Why must the conductor and external field be perpendicular to each other in order to have motor action or to generate induced voltage?
6. Explain briefly how either motor action or generator action can be obtained with the same conductor in a magnetic field.
7. Assume that a conductor being cut by the flux of an expanding magnetic field has 10 V induced with the top end positive. Now analyze the effect of the following changes: (a) The magnetic flux continues to expand but at a slower rate. How does this affect the amount of induced voltage and its polarity? (b) The magnetic flux is constant, neither increasing nor decreasing. How much is the induced voltage? (c) The magnetic flux contracts, cutting across the conductor with the opposite direction of motion. How does this affect the polarity of the induced voltage?
8. Redraw the graph in Fig. 15-14c for the case of 500 turns, with all other factors the same.
9. Redraw the circuit with the coil and battery in Fig. 15-6, showing two different ways to reverse the magnetic polarity.
10. Referring to Fig. 15-14, suppose that the flux decreases from 8 Wb to zero at the same rate as the increase. Tabulate all the values as in Table 15-1 and draw the three graphs corresponding to those in Fig. 15-14.

Problems (Answers to odd-numbered problems at back of book.)

1. A magnetic flux of 900 Mx cuts across a coil of 1000 turns in 1 μs. How much is the voltage induced in the coil? [1 Mx = 10^{-8} Wb]
2. Refer to Fig. 15-13. (a) Show the induced voltage here connected to a load resistance R_L of 100 Ω. (b) If the induced voltage is 100 V, how much current flows in R_L? (c) Give one way to reverse the polarity of the induced voltage. (d) Why will this method reverse the direction of current through R_L?
3. Calculate the rate of flux change $d\phi/dt$ in webers per second for the following: (a) 6 Wb increasing to 8 Wb in 1 s; (b) 8 Wb decreasing to 6 Wb in 1 s; (c) 5000 Mx increasing to 6000 Mx in 5 μs.
4. Calculate the induced voltage produced in 400 turns by each of the flux changes in Prob. 3.
5. Draw a circuit with a 20-V battery connected to a 100-Ω coil of 400 turns with an iron core 0.2 m long. Using SI magnetic units, calculate (a) I; (b) NI; (c) the field intensity H; (d) the flux density B in a core with a μ_r of 500; (e) the total flux ϕ at each pole with an area of 6×10^{-4} m²; (f) show the direction of winding and magnetic polarity of the coil.
6. For the coil in Prob. 5: (a) If the iron core is removed, how much will the flux be in the air-core coil? (b) How much induced voltage would be produced by this change in flux while the core is being moved out in 1 s? (c) How much is the induced voltage after the core is removed?

Answers to Practice Problems

15-1	(a) T	15-4	(b) T
	(b) T	15-5	(a) South
15-2	(a) Left		(b) South
	(b) South pole	15-6	(a) A
15-3	(a) T		(b) A
	(b) T	15-7	(a) 2 Wb/s
15-4	(a) T		(b) 2 Wb/s

Alternating Voltage and Current

Chapter 16

This unit begins the analysis of ac voltage, as used in the ac power line, and the alternating current it produces in an ac circuit. Figure 16-1 shows the variations and polarity reversals of the power-line voltage. Audio and radio signals are also important examples of ac voltages.

We can utilize the rules for dc circuits as an introduction to the analysis of ac circuits. All the dc principles of series and parallel circuits with Ohm's-law calculations still apply. However, the new factor to consider with an ac source is that the voltage alternately reverses its polarity, producing current that reverses in direction. Most important, the voltage and current are always changing instead of remaining at a steady value.

This characteristic of varying values is the reason why ac circuits have so many useful applications. For instance, a transformer can operate only with alternating current. This is just one example of inductance L in ac circuits, where the changing magnetic flux of a varying current can produce induced voltage. The details of inductance follow in Chaps. 18, 19, and 20.

A similar but opposite effect in ac circuits is capacitance C. The C is important with the changing electric field of a varying voltage. Just as L has a big effect with an alternating current, the C has an effect which depends on alternating voltage. The details of capacitance are explained in Chaps. 21, 22, and 23.

The L and C are additional factors, besides R, in ac circuits. The main difference between them is that R is the same in either a dc or ac circuit, but the effects of L and C depend on having an ac source. How fast the ac variations occur determines the frequency and allows a greater or smaller reaction by L and C. Therefore, the effect is different for different frequencies. One important application is a resonant circuit with L and C which is tuned to a particular frequency. All applications of tuning in radio and television are examples of resonance in an LC circuit.

Chapter 16
Alternating Voltage and Current

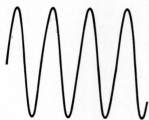

FIGURE 16-1

Oscilloscope photo of ac power-line voltage with frequency of 60 Hz. Four cycles shown.

In general, electronic circuits are combinations of R, L, and C with both direct current and alternating current. The audio and radio signals are ac voltages or currents. However, the amplifiers using transistors or tubes need dc voltages in order to conduct any current at all. The result is direct current with a superimposed ac signal. Furthermore, the combination can include variations with different frequencies in the one circuit. To begin the analysis of ac circuits, the special features of alternating voltage and current are explained in the following topics:

16-1 Alternating-Voltage Generator
16-2 The Sine Wave
16-3 Alternating Current
16-4 Voltage and Current Values for a Sine Wave
16-5 Frequency
16-6 Period
16-7 Wavelength
16-8 Phase Angle
16-9 The Time Factor in Frequency and Phase
16-10 AC Circuits with Resistance
16-11 Nonsinusoidal AC Waveforms
16-12 Harmonic Frequencies

16-1
ALTERNATING-VOLTAGE GENERATOR

We can define an ac voltage as one that continuously varies in magnitude and periodically reverses in polarity. In Fig. 16-1, the variations up and down on the waveform show the changes in magnitude. The zero axis is a horizontal line across the center. Then voltages above the center have positive polarity, while the values below center are negative.

Figure 16-2 illustrates how such a voltage waveform is produced by a rotary generator. The conductor loop rotates through the magnetic field to generate the induced ac voltage across its open terminals. The magnetic flux shown here is vertical, with lines of force down in the plane of the paper.

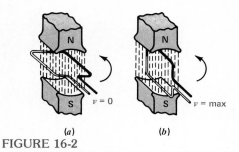

FIGURE 16-2

Loop rotating in magnetic field to produce alternating induced voltage v. (a) Loop conductors moving parallel to field results in zero voltage. (b) Loop conductors cutting across field produce maximum induced voltage.

In (a) the loop is in its horizontal starting position in a plane perpendicular to the paper. When the loop rotates counterclockwise, the two longer conductors move around a circle. Note that in the flat position shown, the two long conductors of the loop move vertically up or down through the paper but parallel to the vertical flux lines. In this position, motion of the loop does not induce a voltage because the conductors are not cutting across the flux.

When the loop rotates through the upright position in (b), however, the conductors cut across the flux, producing maximum induced voltage. The shorter connecting wires in the loop do not have any appreciable voltage induced in them.

Each of the longer conductors has opposite polarity of induced voltage because the one at the top is moving to the left while the bottom conductor is moving to the right. The amount of voltage varies from zero to maximum as the loop moves from a flat position to upright, where it can cut across the flux. Also, the polarity at the terminals of the loop reverses as the motion of each conductor reverses during each half-revolution.

With one revolution of the loop in a complete circle back to the starting position, therefore, the induced voltage provides a potential difference v across the loop, varying in the same way as the wave of voltage shown in Fig. 16-1. If the loop rotates at the speed of 60 revolutions per second, the ac voltage will have the frequency of 60 Hz.

The Cycle. One complete revolution of the loop around the circle is a *cycle*. In Fig. 16-3, the generator loop is shown in its position at each quarter-turn during one complete cycle. The corresponding wave of induced voltage also goes through one cycle. Although not shown, the magnetic field is from top to bottom of the page as in Fig. 16-2.

At position A in Fig. 16-3, the loop is flat and moves parallel to the magnetic field, so that the induced voltage is zero. Counterclockwise rotation of the loop moves the dark conductor to the top at position B, where it cuts across the field to produce maximum induced voltage. The polarity of the induced voltage here makes the open end of the dark conductor positive. This conductor at the top is cutting across the flux from right to left. At the same time, the opposite conductor below is moving from left to right, causing its induced voltage to have opposite polarity. Therefore, maximum induced voltage is produced at this time across the two open ends of the loop. Now the top conductor is positive with respect to the bottom conductor.

In the graph of induced voltage values below the loop in Fig. 16-3, the polarity of the dark conductor is shown with respect to the other conductor. Positive voltage is shown above the zero axis in the graph. As the dark conductor rotates from its starting position parallel to the flux toward the top position, where it cuts maximum flux, more and more induced voltage is produced, with positive polarity.

When the loop rotates through the next quarter-turn, it returns to the flat position shown in C, where it cannot cut across flux.

Chapter 16
Alternating Voltage and Current

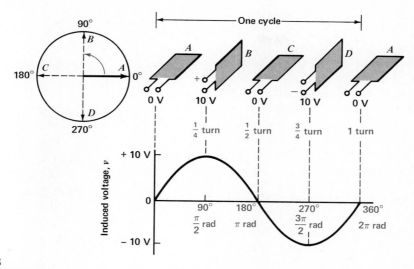

FIGURE 16-3

One cycle of alternating voltage generated by rotating loop. Magnetic field not shown here is from top to bottom of page, as in Fig. 16-2.

Therefore, the graph of induced voltage values decreases from its maximum value to zero at the half-turn, just as it is zero at the start. The half-cycle of revolution is called an *alternation*.

The next quarter-turn of the loop moves it to the position shown at D in Fig. 16-3, where the loop cuts across the flux again for maximum induced voltage. Note, however, that here the dark conductor is moving left to right at the bottom of the loop. This motion is reversed from the direction it had when it was at the top, moving right to left. Because of the reversed direction of motion during the second half-revolution, the induced voltage has opposite polarity, with the dark conductor negative. This polarity is shown in the graph as negative voltage below the zero axis. The maximum value of induced voltage at the three-quarter turn is the same as at the first quarter-turn but with opposite polarity.

When the loop completes the last quarter-turn in the cycle, the induced voltage returns to zero as the loop returns to its flat position at A, the same as at the start. This cycle of values of induced voltage is repeated as the loop continues to rotate, with one complete cycle of voltage values, as shown, for each circle of revolution.

Note that zero at the start and zero after the half-turn of an alternation are not the same. At the start, the voltage is zero because the loop is flat, but the dark conductor is moving upward in the direction that produces positive voltage. After one half-cycle, the voltage is zero with the loop flat, but the dark conductor is moving downward in the direction that produces negative voltage. After one complete cycle, the loop and its corresponding waveform of induced voltage are the same as at the start. *A cycle can be defined, therefore, as including the variations between two successive points having the same value and varying in the same direction.*

Angular Measure. Because the cycle of voltage in Fig. 16-3 corresponds to rotation of the loop around a circle, it is convenient to consider parts of the cycle in angles. The complete circle includes 360°. One half-cycle, or one alternation, is 180° of revolution. A quarter-turn is 90°. The circle next to the loop positions in Fig. 16-3 illustrates the angular rotation of the dark conductor as it rotates counterclockwise from 0 to 90 to 180° for one half-cycle and then to 270°, returning to 360° to complete the cycle. Therefore, one cycle corresponds to 360°.

Radian Measure. In angular measure it is convenient to use a specific unit angle called the *radian* (abbreviated rad), which is an angle equal to 57.3°. Its convenience is due to the fact that a radian is the angular part of the circle that includes an arc equal to the radius r of the circle, as shown in Fig. 16-4. The circumference around the circle equals $2\pi r$. A circle includes 2π rad, then, as each radian angle includes one length r of the circumference. Therefore, one cycle equals 2π rad.

As shown in the graph in Fig. 16-3, divisions of the cycle can be indicated by angles in either degrees or radians. Zero degrees is also zero radians, 360° is 2π rad, 180° is π rad, 90° is $\pi/2$ rad, and 270° is π rad plus $\pi/2$ rad, which equals $3\pi/2$ rad.

The constant 2π in circular measure is numerically equal to 6.2832. This is double the value of 3.1416 for π. The π is a symbol for the ratio of the circumference to the diameter for any circle, which always has the numerical value of 3.1416. The fact that 2π rad is 360° can be shown numerically as $2 \times 3.1416 \times 57.3° = 360°$.

Practice Problems 16-1
(answers on page 336)
Refer to Fig. 16-3.
(a) How much is the induced voltage at $\pi/2$ rad?

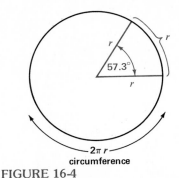

FIGURE 16-4
One radian is the angle equal to 57.3°. The complete circle has 2π rad.

(b) How many degrees are in a complete cycle?

16-2
THE SINE WAVE

The voltage waveform in Figs. 16-1 and 16-3 is called a *sine wave, sinusoidal wave,* or *sinusoid* because the amount of induced voltage is proportional to the sine of the angle of rotation in the circular motion producing the voltage. The sine is a trigonometric function[1] of an angle equal to the ratio of the opposite side to the hypotenuse. This numerical ratio increases from zero for 0° to a maximum value of 1 for 90° as the side opposite the angle becomes larger.

The voltage waveform produced by the circular motion of the loop is a sine wave, because the induced voltage increases to a maximum at 90°, when the loop is vertical, in the same way that the sine of the angle of rotation increases to a maximum at 90°. The induced voltage and sine of the angle correspond for the

---
[1] See Appendix E for an explanation of the sine, cosine, and tangent functions of an angle. More details are in B. Grob, "Mathematics Outline and Review Problems for Basic Electronics," McGraw-Hill Book Company, New York.

TABLE 16-1. Values in a Sine Wave

ANGLE θ		SIN θ	LOOP VOLTAGE
DEGREES	RADIANS		
0	0	0	Zero
30	$\dfrac{\pi}{6}$	0.500	50% of maximum
45	$\dfrac{\pi}{4}$	0.707	70.7% of maximum
60	$\dfrac{\pi}{3}$	0.866	86.6% of maximum
90	$\dfrac{\pi}{2}$	1.000	Positive maximum value
180	π	0	Zero
270	$\dfrac{3\pi}{2}$	−1.000	Negative maximum value
360	2π	0	Zero

full 360° of the cycle. Table 16-1 lists the numerical values of the sine for several important angles, to illustrate the specific characteristics of a sine wave.

Notice that the sine wave reaches ½ its maximum value in 30°, which is only ⅓ of 90°. This fact means that the sine wave has a sharper slope of changing values when the wave is near the zero axis, compared with the more gradual changes near the maximum value.

The instantaneous value of a sine-wave voltage for any angle of rotation is expressed by the formula

$$v = V_M \sin \theta \qquad (16\text{-}1)$$

where θ (Greek letter *theta*) is the angle, sin is the abbreviation for its sine, V_M is the maximum voltage value, and v is the instantaneous value for any angle.

Example 1. A sine wave of voltage varies from zero to a maximum of 100 V. How much is the voltage at the instant of 30° of the cycle? 45°? 90°? 270°?

Answer.
$v = V_M \sin \theta = 100 \sin \theta$
At 30°: $v = V_M \sin 30° = 100 \times 0.5$
$v = 50$ V
At 45°: $v = V_M \sin 45° = 100 \times 0.707$
$v = 70.7$ V
At 90°: $v = V_M \sin 90° = 100 \times 1$
$v = 100$ V
At 270°: $v = V_M \sin 270° = 100 \times -1$
$v = -100$ V

The value of −100 V at 270° is the same as that at 90° but with opposite polarity.

Between zero at 0° and maximum at 90° the amplitudes of a sine wave increase exactly as the sine value for the angle of rotation. These values are for the first quadrant in the circle. From 90 to 180°, in the second quadrant, the values decrease as a mirror image of the first 90°. The values in the third and fourth quadrants, from 180 to 360°, are exactly the same as 0 to 180° but with opposite sign. At 360° the waveform is back to 0° to repeat its values every 360°.

In summary, the characteristics of the sine-wave ac waveform are:

Chapter 16
Alternating Voltage and Current

1. The cycle includes 360° or 2π rad.
2. The polarity reverses each half-cycle.
3. The maximum values are at 90 and 270°.
4. The zero values are at 0 and 180°.
5. The waveform changes its values the fastest when it crosses the zero axis.
6. The waveform changes its values the slowest when it is at its maximum value. The values must stop increasing before they can decrease.

A perfect example of the sine-wave ac waveform is the 60-Hz power-line voltage in Fig. 16-1.

Practice Problems 16-2
(answers on page 336)
A sine-wave voltage has a peak value of 170. What is its value at
(a) 30°.
(b) 270°.
(c) 360°.

16-3
ALTERNATING CURRENT

When a sine wave of alternating voltage is connected across a load resistance, the current that flows in the circuit is also a sine wave. In Fig. 16-5 let the sine-wave voltage at the left in the diagram be applied across R of 100 Ω. The resulting sine wave of alternating current is shown at the right in the diagram. Note that the frequency is the same for v and i.

During the first half-cycle of v in Fig. 16-5, terminal 1 is positive with respect to terminal 2. Since the direction of electron flow is from the negative side of v, through R, and back to the positive side of v, current flows in the direction indicated by arrow a for the first half-cycle. This direction is taken as the positive direction of current in the graph for i, corresponding to positive values of v.

The amount of current is equal to v/R. If several instantaneous values are taken, when v is zero, i is zero; when v is 50 V, i equals 50 V/100, or 0.5 A; when v is 100 V, i equals 100 V/100, or 1 A. For all values of applied voltage with positive polarity, therefore, the current is in one direction, increasing to its maximum value and decreasing to zero, just like the voltage.

On the next half-cycle, the polarity of the alternating voltage reverses. Then terminal 1 is negative with respect to terminal 2. With reversed voltage polarity, current flows in the opposite direction. Electron flow is from terminal 1 of the voltage source, which is now the negative side, through R, and back to terminal 2. This direction of current, as indicated by arrow b in Fig. 16-5, is negative.

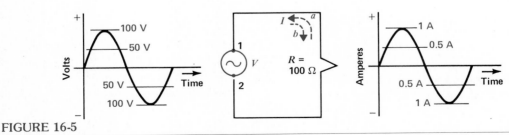

FIGURE 16-5
A sine wave of alternating voltage applied across R produces a sine wave of alternating current in the circuit.

The negative values of i in the graph have the same values as in the first half-cycle, corresponding to the reversed values of applied voltage. As a result, the alternating current in the circuit has sine-wave variations corresponding exactly to the sine-wave alternating voltage.

Only the waveforms for v and i can be compared. There is no comparison between relative values, because the current and voltage are different quantities.

It is important to note that the negative half-cycle of applied voltage is just as useful as the positive half-cycle in producing current. The only difference is that the reversed polarity of voltage produces the opposite direction of current.

Furthermore, the negative half-cycle of current is just as effective as the positive values when heating the filament to light a bulb. With positive values, electrons flow through the filament in one direction. Negative values produce electron flow in the opposite direction. In both cases, electrons flow from the negative side of the voltage source, through the filament, and return to the positive side of the source. For either direction, the current heats the filament. The direction does not matter, since it is just the motion of electrons against resistance that produces power dissipation. In short, resistance R has the same effect in reducing I for either direct current or alternating current.

*Practice Problems 16-3
(answers on page 336)
Refer to Fig. 16-5.*
(a) When v is 70.7 V, how much is i?
(b) How much is i at 30°?

16-4
VOLTAGE AND CURRENT VALUES FOR A SINE WAVE

Since an alternating sine wave of voltage or current has many instantaneous values through the cycle, it is convenient to define specific magnitudes for comparing one wave with another. The peak, average, or root-mean-square (rms) value can be specified as indicated in Fig. 16-6. These values can be used for either current or voltage.

Peak Value. This is the maximum value V_M or I_M. For example, specifying that a sine wave has a peak value of 170 V states how much it is, since all other values during the cycle follow a

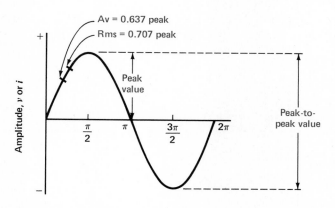

FIGURE 16-6

Amplitude values for a sine wave of voltage or current.

sine wave. The peak value applies to either the positive or the negative peak.

In order to include both peak amplitudes, the *peak-to-peak* (p-p) *value* may be specified. For the same example, the peak-to-peak value is 340 V, double the peak value of 170 V, since the positive and negative peaks are symmetrical. It should be noted, though, that the two opposite peak values cannot occur at the same time. Furthermore, in some waveforms the two peaks are not equal.

Average Value. This is an arithmetical average of all the values in a sine wave for one alternation, or half-cycle. The half-cycle is used for the average because over a full cycle the average value is zero, which is useless for comparison purposes. If the sine values for all angles up to 180°, for one alternation, are added and then divided by the number of values, this average equals 0.637. These calculations are shown in Table 16-2.

Since the peak value of the sine is 1 and the average equals 0.637, then

$$\text{Average value} = 0.637 \times \text{peak value} \quad (16\text{-}2)$$

With a peak of 170 V, for example, the average value is 0.637 × 170 V, which equals approximately 108 V.

Root-Mean-Square, or Effective, Value. The most common method of specifying the amount of a sine wave of voltage or current is by stating its value at 45°, which is 70.7 percent of the peak. This is its *root-mean-square* value, abbreviated rms. Therefore,

$$\text{rms value} = 0.707 \times \text{peak value} \quad (16\text{-}3)$$

or

$$V_{rms} = 0.707\, V_{max} \quad \text{and} \quad I_{rms} = 0.707\, I_{max}$$

TABLE 16-2. Derivation of Average and RMS Values for a Sine-wave Alternation

INTERVAL	ANGLE θ	SIN θ	(SIN θ)²
1	15°	0.26	0.07
2	30°	0.50	0.25
3	45°	0.71	0.50
4	60°	0.87	0.75
5	75°	0.97	0.93
6	90°	1.00	1.00
7*	105°	0.97	0.93
8	120°	0.87	0.75
9	135°	0.71	0.50
10	150°	0.50	0.25
11	165°	0.26	0.07
12	180°	0.00	0.00
	Total	7.62	6.00
	Average →	$\frac{7.62}{12} = 0.635$†	$\sqrt{\frac{6}{12}} = \sqrt{0.5} = 0.707$

*For angles between 90 and 180°, sin θ = sin (180° − θ).
†More intervals and precise values are needed for the exact average of 0.637.

With a peak of 170 V, for example, the rms value is 0.707 × 170, or 120 V, approximately. This is the voltage of the commercial ac power line, which is always given in rms value.

It is often necessary to convert from rms to peak value. This can be done by inverting Formula (16-3), as follows:

$$\text{Peak} = \frac{1}{0.707} \times \text{rms} = 1.414 \times \text{rms} \quad (16\text{-}4)$$

or

$$V_{max} = 1.414\, V_{rms} \quad \text{and} \quad I_{max} = 1.414\, I_{rms}$$

Dividing by 0.707 is the same as multiplying by 1.414.

For example, the commercial power line voltage with an rms value of 120 V has a peak value of 120 × 1.414, which equals 170 V, approximately. Its peak-to-peak value is 2 × 170, or 340 V, which is double the peak value. As a formula,

$$\text{Peak-to-peak value} = 2.828 \times \text{rms value} \quad (16\text{-}5)$$

The factor 0.707 for rms value is derived as the square root of the average (mean) of all the squares of the sine values. If we take the sine for each angle in the cycle, square each value, add all the squares, divide by the number of values added to obtain the average square, and then take the square root of this mean value, the answer is 0.707. These calculations are shown in Table 16-2 for one alternation from 0 to 180°. The results are the same for the opposite alternation.

The advantage of the rms value derived in terms of the squares of the voltage or current values is that it provides a measure based on the ability of the sine wave to produce power, which is I^2R or V^2/R. As a result, the rms value of an alternating sine wave corresponds to the same amount of direct current or voltage in heating power. An alternating voltage with an rms value of 120 V, for instance, is just as effective in heating the filament of a light bulb as 120 V from a steady dc voltage source. For this reason, the rms value is also the *effective* value.

Unless indicated otherwise, all sine-wave ac measurements are in rms values. The capital letters V and I are used, corresponding to the symbols for dc values. As an example, V = 120 V for the ac power-line voltage.

The ratio of the rms to average values is the *form factor*. For a sine wave, this ratio is 0.707/0.637 = 1.11.

Note that sine waves can have different amplitudes but still follow the sinusoidal waveform. Figure 16-7 compares a low-amplitude voltage with a high-amplitude voltage. Although different in amplitude, they are both sine waves. In each wave, the rms value is 0.707 of the peak value.

Practice Problems 16-4
(answers on page 336)
(a) Convert 170 V peak to rms value.
(b) Convert 10 V rms to peak value.

16-5
FREQUENCY

The number of cycles per second is the *frequency*, with the symbol f. In Fig. 16-3, if the

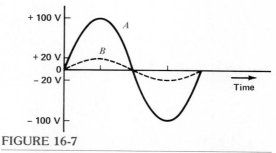

FIGURE 16-7

Waveforms A and B have different amplitudes, but both are sine waves.

loop rotates through 60 complete revolutions, or cycles, during 1 s, the frequency of the generated voltage is 60 cps, or 60 Hz. You see only one cycle of the sine waveform, instead of 60 cycles, because the time interval shown here is $\frac{1}{60}$ s. Note that the factor of time is involved. More cycles per second means a higher frequency and less time for one cycle, as illustrated in Fig. 16-8. Then the changes in values are faster for higher frequencies.

A complete cycle is measured between two successive points that have the same value and direction. In Fig. 16-8 the cycle is between successive points where the waveform is zero and ready to increase in the positive direction. Or the cycle can be measured between successive peaks.

On the time scale of 1 s, waveform a goes through one cycle, while waveform b has much faster variations, with four complete cycles during 1 s. Both waveforms are sine waves, even though each has a different frequency.

In comparing sine waves, the amplitude has no relation to frequency. Two waveforms can have the same frequency with different amplitudes (Fig. 16-7), the same amplitude but different frequencies (Fig. 16-8), or different amplitudes and frequencies. The amplitude indicates how much the voltage or current is, while the frequency indicates the time rate of change of the amplitude variations, in cycles per second.

Frequency Units. The unit called the *hertz* (Hz), named after H. Hertz, is used for cycles per second. Then 60 cps = 60 Hz. All the metric prefixes can be used. As examples:

1 kilocycle per second = 1×10^3 Hz = 1 kHz
1 megacycle per second = 1×10^6 Hz = 1 MHz
1 gigacycle per second = 1×10^9 Hz = 1 GHz

Audio and Radio Frequencies. The entire frequency range of alternating voltage or current from 1 Hz to many megahertz can be considered in two broad groups: audio frequencies (af) and radio frequencies (rf). *Audio* is a Latin word meaning "I hear." The audio range includes frequencies that can be heard in the form of sound waves by the human ear. This range of audible frequencies is approximately 16 to 16,000 Hz.

The higher the frequency, the higher the pitch or tone of the sound. High audio frequencies, about 3000 Hz and above, can be considered to provide *treble* tone. Low audio frequencies, about 300 Hz and below, provide *bass* tone.

Loudness is determined by amplitude. The greater the amplitude of the af variation, the louder is its corresponding sound.

Alternating current and voltage above the audio range provide rf variations, since electrical variations of high frequency can be transmitted by electromagnetic radio waves. The more common frequency bands for radio

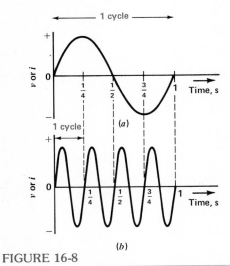

FIGURE 16-8

Number of cycles per second (Hz) is the frequency. (*a*) The $f = 1$ Hz. (*b*) The $f = 4$ Hz.

TABLE 16-3. Examples of Common Frequencies

FREQUENCY	USE
60 Hz	Ac power line
50–15,000 Hz	Audio equipment
535–1605 kHz	AM radio band
54–60 MHz	TV channel 2
88–108 MHz	FM radio band

broadcasting are listed in Appendix B. Some applications are listed here, though, in Table 16-3.

Sonic and Supersonic Frequencies. These terms refer to sound waves, which are variations in pressure generated by mechanical vibrations, rather than electrical variations. The velocity of transmission for sound waves equals 1130 ft/s, through dry air at 20°C. Sound waves above the audible range of frequencies are called *supersonic* waves. The range of frequencies for supersonic applications, therefore, is from 16,000 Hz up to several megahertz. Sound waves in the audible range of frequencies below 16,000 Hz can be considered *sonic* or sound frequencies, reserving *audio* for electrical variations that can be heard when converted to sound waves.

*Practice Problems 16-5
(answers on page 336)*

(a) What is the frequency of the bottom waveform in Fig. 16-8?
(b) Convert 1605 kHz to megahertz.

16-6
PERIOD

The amount of time for one cycle is the *period*. Its symbol is T for time. With a frequency of 60 Hz, as an example, the time for one cycle is $\frac{1}{60}$ s. Therefore, the period is $\frac{1}{60}$ s in this case. The frequency and period are reciprocals of each other:

$$T = \frac{1}{f} \quad \text{or} \quad f = \frac{1}{T} \quad (16\text{-}6)$$

The higher the frequency, the shorter the period. In Fig. 16-8, the period for wave *a* with a frequency of 1 Hz is 1 s, while the higher-frequency wave of 4 Hz in (*b*) has the period of $\frac{1}{4}$ s for a complete cycle.

Units of Time. The second is the basic unit, but for higher frequencies and shorter periods, smaller units of time are convenient. Those used most often are:

$T = 1$ millisecond $= 1$ ms $= 1 \times 10^{-3}$ s
$T = 1$ microsecond $= 1$ μs $= 1 \times 10^{-6}$ s
$T = 1$ nanosecond $= 1$ ns $= 1 \times 10^{-9}$ s

These units of time for period are reciprocals of the corresponding units for frequency. The reciprocal of frequency in kilohertz gives the period T in milliseconds; the reciprocal of megahertz is microseconds; the reciprocal of gigahertz is nanoseconds.

Example 2. An alternating current varies through one complete cycle in $\frac{1}{1000}$ s. Calculate the period and frequency.

Answer.
$$T = \frac{1}{1000} \text{ s}$$
$$f = \frac{1}{T} = \frac{1}{\frac{1}{1000}} = \frac{1000}{1} = 1000$$
$$f = 1000 \text{ Hz}$$

Example 3. Calculate the period for the two frequencies of 1 MHz and 2 MHz.

Answer.

(a) For 1 MHz,
$$T = \frac{1}{1 \times 10^6} = 1 \times 10^{-6}$$
$$T = 1 \text{ } \mu\text{s}$$

(b) For 2 MHz,

$$T = \frac{1}{2 \times 10^6} = 0.5 \times 10^{-6}$$

$$T = 0.5 \ \mu s$$

Practice Problems 16-6
(answers on page 336)
(a) $T = \frac{1}{400}$ s. Calculate f.
(b) $f = 400$ Hz. Calculate T.

16-7 WAVELENGTH

When a periodic variation is considered with respect to distance, one cycle includes the *wavelength,* which is the length of one complete wave or cycle (Fig. 16-9). For example, when a radio wave is transmitted, variations in the electromagnetic field travel through space. Also, with sound waves, the variations in air pressure corresponding to the sound wave move through air. In these applications, the distance traveled by the wave in one cycle is the wavelength. The wavelength depends upon the frequency of the variation and its velocity of transmission:

$$\lambda = \frac{\text{velocity}}{\text{frequency}} \qquad (16\text{-}7)$$

where λ (lambda) is the symbol for one complete wavelength.

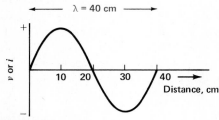

FIGURE 16-9
Wavelength λ is the distance traveled by the wave in one cycle.

Wavelength of Radio Waves. For electromagnetic radio waves, the velocity in air or vacuum is 186,000 mi/s, or 3×10^{10} cm/s, which is the speed of light. Therefore,

$$\lambda \text{ (cm)} = \frac{3 \times 10^{10} \text{ cm/s}}{f(\text{Hz})} \qquad (16\text{-}8)$$

Note that the higher the frequency is, the shorter the wavelength. For instance, the short-wave radio broadcast band of 5.95 to 26.1 MHz includes higher frequencies than the standard radio broadcast band of 540 to 1620 kHz.

Example 4. Calculate λ for a radio wave with f of 30 GHz.

Answer.
$$\lambda = \frac{3 \times 10^{10} \text{ cm/s}}{30 \times 10^9 \text{ Hz}} = \frac{3}{30} \times 10 \text{ cm}$$
$$= 0.1 \times 10$$
$$\lambda = 1 \text{ cm}$$

Such short wavelengths are called *microwaves.* This range includes λ of 1 m or less, for frequencies of 300 MHz or more.

Example 5. The length of a TV antenna is $\lambda/2$ for radio waves with f of 60 MHz. What is the antenna length in centimeters and feet?

Answer.

(a) $\lambda = \dfrac{3 \times 10^{10} \text{ cm/s}}{60 \times 10^6 \text{ Hz}} = \dfrac{1}{20} \times 10^4 \text{ cm}$

$\qquad = 0.05 \times 10^4$

$\qquad \lambda = 500 \text{ cm}$

Then, $\lambda/2 = {}^{500}\!/_{2} = 250$ cm.

(b) Since 2.54 cm = 1 in,

$\lambda/2 = \dfrac{250 \text{ cm}}{2.54 \text{ cm/in}} = 100$ in   approximately

$\lambda/2 = \dfrac{100 \text{ in}}{12 \text{ in/ft}} = 8\frac{1}{3}$ ft

This half-wave dipole antenna is made with two quarter-wave poles, each 50 in long.

Example 6. For the 6-m band used in amateur radio, what is the corresponding frequency?

Answer. The formula $\lambda = v/f$ can be inverted to $f = v/\lambda$. Then

$$f = \frac{3 \times 10 \text{ cm/s}}{6 \text{ m}} = \frac{3 \times 10^{10} \text{ cm/s}}{6 \times 10^2 \text{ cm}}$$

$$= \frac{3}{6} \times 10^8 \times \frac{1}{\text{s}} = 0.5 \times 10^8 \text{ Hz}$$

$$f = 50 \times 10^6 \text{ Hz} \quad \text{or} \quad 50 \text{ MHz}$$

Wavelength of Sound Waves. The velocity is much lower, compared with radio waves, because sound waves result from mechanical vibrations rather than electrical variations. For average conditions the velocity of sound waves in air equals 1130 ft/s. To calculate the wavelength, therefore,

$$\lambda = \frac{1130 \text{ ft/s}}{f \text{ Hz}} \tag{16-9}$$

This formula can also be used for supersonic waves. Although their frequencies are too high to be audible, supersonic waves are still sound waves rather than radio waves.

Example 7. What is the wavelength of the sound waves produced by a loudspeaker at a frequency of 100 Hz?

Answer. $\quad \lambda = \dfrac{1130 \text{ ft/s}}{100 \text{ Hz}}$

$\qquad \lambda = 11.3 \text{ ft}$

Example 8. For supersonic waves at a frequency of 34.44 kHz, calculate the wavelength in feet and in centimeters.

Answer. $\quad \lambda = \dfrac{1130}{34.44 \times 10^3} = 32.8 \times 10^{-3}$

$\qquad \lambda = 0.0328 \text{ ft}$

To convert to inches:

$0.0328 \text{ ft} \times 12 = 0.3936 \text{ in}$

To convert to centimeters:

$0.3936 \text{ in} \times 2.54 = 1 \text{ cm} \quad$ approximately

Note that for sound waves with a frequency of 34.44 kHz in this example, the wavelength is the same 1 cm as radio waves with the much higher frequency of 30 GHz, as calculated in Example 4. The reason is that radio waves have a much higher velocity.

*Practice Problems 16-7
(answers on page 336)*
Answer true or false.
(a) The higher the frequency, the shorter the wavelength λ.
(b) The higher the frequency, the longer the period T.
(c) The velocity of propagation for radio waves in free space is 3×10^{10} cm/s.

16-8
PHASE ANGLE

Referring back to Fig. 16-3, suppose that the generator started its cycle at point *B*, where maximum voltage output is produced, instead of starting at the point of zero output. If we compare the two cases, the two output voltage waves would be as in Fig. 16-10. Each is the same waveform of alternating voltage, but wave *B* starts at maximum, while wave *A* starts at zero. The complete cycle of wave *B* through 360° takes it to the maximum value from which it started. Wave *A* starts and finishes its cycle at zero. With respect to time, therefore, wave *B* is ahead of wave *A* in its values of generated voltage. The amount it leads in time equals one quarter-revolution, which is 90°. This angular difference is the phase angle between waves *B*

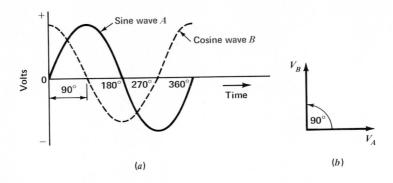

FIGURE 16-10
Two voltage waveforms 90° out of phase. (a) Wave B leads wave A by 90°. (b) Phasors V_B and V_A for the two voltages with phase angle $\theta = 90°$.

and A. Wave B leads wave A by the phase angle of 90°.

The 90° phase angle between waves B and A is maintained throughout the complete cycle and in all successive cycles, as long as they both have the same frequency. At any instant of time for wave B, it has the value that A will have 90° later. For instance, at 180° wave A is at zero, but B is already at its negative maximum value, where wave A will be later at 270°.

In order to compare the phase angle between two waves, they must have the same frequency. Otherwise, the relative phase keeps changing. Also, they must have sine-wave variations, as this is the only kind of waveform that is measured in angular units of time. The amplitudes can be different for the two waves, although they are shown the same here. We can compare the phase of two voltages, two currents, or a current with a voltage.

The 90° Phase Angle. The two waves in Fig. 16-10 represent a sine wave and a cosine wave 90° out of phase with each other. The 90° phase angle means that one has its maximum amplitude when the other is at zero value. Wave A starts at zero, corresponding to the sine of 0°, has its peak amplitude at 90 and 270°, and is back to zero after one cycle of 360°. Wave B starts at its peak value, corresponding to the cosine of 0°, has its zero value at 90 and 270°, and is back to the peak value after one cycle of 360°.

However, wave B can also be considered a sine wave that starts 90° before wave A in time. This phase angle of 90° for current and voltage waveforms has many applications in sine-wave ac circuits with inductance or capacitance.

The sine and cosine waveforms really have the same variations, but displaced by 90°. In fact, both waveforms are called *sinusoids*. The 90° angle is called *quadrature phase*.

Phase-Angle Diagrams. To compare phases of alternating currents and voltages, it is much more convenient to use phasor diagrams corresponding to the voltage and current waveforms, as shown in Fig. 16-10b. The arrows here represent the phasor quantities corresponding to the generator voltage.

A phasor is a quantity that has magnitude and direction. The length of the arrow indicates the magnitude of the alternating voltage, in rms, peak, or any ac value as long as the same measure is used for all the phasors. The angle of the arrow with respect to the horizontal axis indicates the phase angle.

The terms *phasor* and *vector* are used for a quantity that has direction, requiring an angle to specify the value completely. However, a vector quantity has direction in space while a phasor quantity varies in time. As an example of a vector, a mechanical force can be represented by a vector arrow at a specific angle, with respect to either the horizontal or vertical direction.

For phasor arrows, the angles shown represent differences in time. One sinusoid is chosen as the reference. Then the timing of the variations in another sinusoid can be compared to the reference by means of the angle between the phasor arrows.

The phasor corresponds to the entire cycle of voltage, but is shown only at one angle, such as the starting point, since the complete cycle is known to be a sine wave. Without the extra details of a whole cycle, phasors represent the alternating voltage or current in a compact form that is easier for comparing phase angles.

In Fig. 16-10b, for instance, the phasor V_A represents the voltage wave A, with a phase angle of 0°. This angle can be considered as the plane of the loop in the rotary generator where it starts with zero output voltage. The phasor V_B is vertical to show the phase angle of 90° for this voltage wave, corresponding to the vertical generator loop at the start of its cycle. The angle between the two phasors is the phase angle.

The symbol for a phase angle is θ (theta). In Fig. 16-10, $\theta = 90°$.

Phase-Angle Reference. The phase angle of one wave can be specified only with respect to another as reference. How the phasors are drawn to show the phase angle depends on which phase is chosen as the reference. Generally, the reference phasor is horizontal, corresponding to 0°. Two possibilities are shown in Fig. 16-11. In (a) the voltage wave A or its phasor V_A is the reference. Then the phasor V_B is 90° counterclockwise. This method is standard practice, using counterclockwise rotation as the positive direction for angles. Also, a leading angle is positive. In this case, then, V_B is 90° counterclockwise from the reference V_A to show that wave B leads wave A by 90°.

However, wave B is shown as the reference in (b). Now V_B is the horizontal phasor. In order to have the same phase angle, V_A must be 90° clockwise, or −90° from V_B. This arrangement shows that negative angles, clockwise from the 0° reference, are used to show a lagging phase angle. The reference determines whether the phase angle is considered leading or lagging in time.

The phase is not actually changed by the method of showing it. In Fig. 16-11, V_A and V_B are 90° out of phase. V_B leads V_A by 90° in time. There is no fundamental difference whether we say V_B is ahead of V_A by +90° or V_A is behind V_B by −90°.

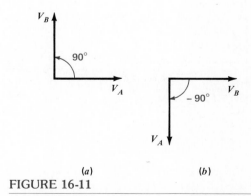

FIGURE 16-11
Leading and lagging phase angles of 90°.
(a) V_A is the reference, and V_B leads by +90°.
(b) V_B is the reference, and V_A lags by −90°.

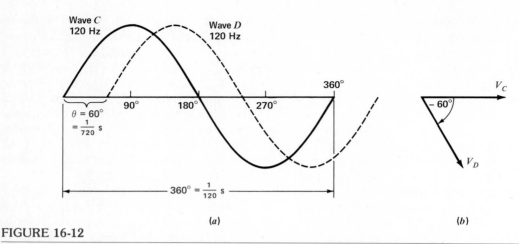

FIGURE 16-12
Phase angle of 60° is the time for $^{60}/_{360}$ or $\frac{1}{6}$ of the cycle. (*a*) Waveforms. (*b*) Phasor diagram.

Two waves and their corresponding phasors can be out of phase by any angle, either less or more than 90°. For instance, a phase angle of 60° is shown in Fig. 16-12. For the waveforms in (*a*), wave *D* is behind *C* by 60° in time. For the phasors in (*b*), this lag is shown by the phase angle of −60°.

In-phase Waveforms. A phase angle of 0° means the two waves are in phase (Fig. 16-13). Then the amplitudes add.

Out-of-phase Waveforms. An angle of 180° means opposite phase, or the two waveforms are exactly out of phase (Fig. 16-14). Then the amplitudes are opposing. Equal values of opposite phase cancel each other.

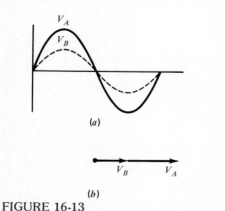

FIGURE 16-13
Two waves in phase with angle of 0°. (*a*) Waveforms. (*b*) Phasor diagram.

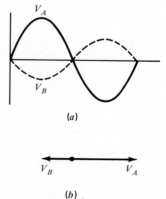

FIGURE 16-14
Two waves out of phase with angle of 180°. (*a*) Waveforms. (*b*) Phasor diagram.

*Practice Problems 16-8
(answers on page 336)
Give the phase angle in*
(a) Fig. 16-10.
(b) Fig. 16-12.
(c) Fig. 16-13.

16-9
THE TIME FACTOR IN FREQUENCY AND PHASE

It is important to remember that the waveforms we are showing are just graphs drawn on paper. The physical factors represented are variations in amplitude, usually on the vertical scale, with respect to equal intervals on the horizontal scale, which can represent either distance or time. To show wavelength, as in Fig. 16-9, the cycles of amplitude variations are plotted against distance or length units. To show frequency, the cycles of amplitude variations are shown with respect to time in angular measure. The angle of 360° represents the time for one cycle, or the period T.

As an example of how frequency involves time, a waveform with stable frequency is actually used in electronic equipment as a clock reference for very small units of time. Assume a voltage waveform with the frequency of 10 MHz. The period T is 0.1 μs. Every cycle is repeated at 0.1-μs intervals, therefore. When each cycle of voltage variations is used to indicate time, then, the result is effectively a clock that measures 0.1-μs units. Even smaller units of time can be measured with higher frequencies. In everyday applications, an electric clock connected to the power line keeps correct time because it is controlled by the exact frequency of 60 Hz.

Furthermore, the phase angle between two waves of the same frequency indicates a specific difference in time. As an example, Fig. 16-12 shows a phase angle of 60°, with wave C leading wave D. They both have the same frequency of 120 Hz. The period T for each wave then is $\frac{1}{120}$ s. Since 60° is one-sixth of the complete cycle of 360°, this phase angle represents one-sixth of the complete period of $\frac{1}{120}$ s. Multiplying $\frac{1}{6} \times \frac{1}{120}$, the answer is $\frac{1}{720}$ s for the time corresponding to the phase angle of 60°. If we consider wave D lagging wave C by 60°, this lag is a time delay of $\frac{1}{720}$ s.

More generally, the time for a phase angle θ can be calculated as

$$t = \frac{\theta}{360} \times \frac{1}{f} \tag{16-10}$$

With f in Hz and θ in degrees, then t is in seconds. The formula gives the time of the phase angle as its proportional part of the total period of one cycle. For the example of θ equal to 60° with f at 120 Hz,

$$t = \frac{\theta}{360} \times \frac{1}{f} = \frac{60}{360} \times \frac{1}{120} = \frac{1}{6} \times \frac{1}{120}$$

$$t = \frac{1}{720} \text{ s}$$

*Practice Problems 16-9
(answers on page 336)*
(a) In Fig. 16-12, how much time corresponds to 180°?
(b) For two waves with the frequency of 1 MHz, how much time is the phase angle of 36°?

16-10
AC CIRCUITS WITH RESISTANCE

An ac circuit has an ac voltage source. Note the symbol in Fig. 16-15 used for any source of sine-wave alternating voltage. This voltage connected across an external load resistance produces alternating current of the same waveform, frequency, and phase as the applied voltage.

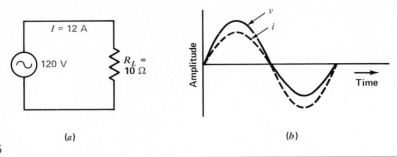

(a) (b)

FIGURE 16-15

Ac circuit with R alone. (a) Schematic diagram, with rms values of V and I. (b) Waveforms showing I in phase with V.

The amount of current equals V/R by Ohm's law. When V is an rms value, I is also an rms value. For any instantaneous value of V during the cycle, the value of I is for the corresponding instant of time.

In an ac circuit with only resistance, the current variations are in phase with the applied voltage, as shown in Fig. 16-16. This in-phase relationship between V and I means that such an ac circuit can be analyzed by the same methods used for dc circuits, since there is no phase angle to consider. Components that have R alone include resistors, the filaments for incandescent light bulbs, and vacuum-tube heaters.

The calculations in ac circuits are generally in rms values, unless otherwise noted. In Fig. 16-15a, for example, the 120 V applied across the 10-Ω R produces rms current of $120/10$, or $I = 12$ A. The rms power dissipation is I^2R, or $144 \times 10 = 1440$ W.

Series AC Circuit with R. In Fig. 16-16, R_T is 30 Ω, equal to the sum of 10 Ω for R_1 plus 20 Ω for R_2. The I is V/R_T, or $120/30$, for $I = 4$ A. The current is the same in all parts of a series circuit. This principle applies for either an ac or dc source and for R, L, or C components.

With 4 A through the 10-Ω R_1, its IR voltage drop is 4×10, or $V_1 = 40$ V. The same 4 A through the 20-Ω R_2 produces an IR voltage drop of 80 V for V_2. Note that the sum of V_1 and V_2 in series equals the 120 V applied.

Parallel AC Circuit with R. In Fig. 16-17, the 10-Ω R_1 and 20-Ω R_2 are in parallel across the 120-V ac source. Therefore the voltage across the parallel branches is the same as the applied voltage.

Each branch current, then, is V divided by each branch R. For R_1, its branch current is $120/10$, or $I_1 = 12$ A. In the R_2 branch, the current is $120/20$, or $I_2 = 6$ A. The total line current I_T is $12 + 6 = 18$ A.

Series-Parallel AC Circuit with R. See Fig. 16-18. The 20-Ω R_2 and 20-Ω R_3 are in parallel

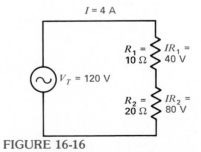

FIGURE 16-16

Series ac circuit with R.

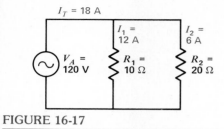

FIGURE 16-17
Parallel ac circuit with R.

for a bank resistance of 10 Ω. This 10-Ω bank is in series with the 20-Ω R_1 for a total of 30 Ω for R_T across the 120-V source. Therefore, the main-line current I_T is $120/30 = 4$ A.

The voltage drop across R_1 in the main line is $I_T \times R_1$, or $4 \times 20 = 80$ V for V_1. Subtracting this 80-V drop from the source V of 120 V, the remaining 40 V is across the bank of R_2 and R_1 in parallel. Since these resistances are equal, the 4-A I_T divides equally with 2 A in the R_2 branch and 2 A in the R_3 branch.

AC Circuits with R and Reactance for L or C. The opposition of inductance and capacitance to sine-wave alternating current is called *reactance,* indicated as X. The symbols are X_L for inductive reactance and X_C for capacitive reactance. The opposition is measured in ohms,

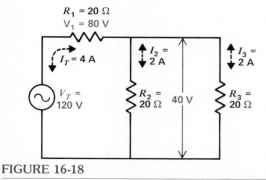

FIGURE 16-18
Series-parallel ac circuit with R.

like resistance, but reactance has a phase angle of ±90°. The X_L is at +90° while X_C is at −90°.

Resistance has a phase angle of 0°. Reactance has a phase angle of ±90°. Therefore, when R and X are combined, the phase angle of the ac circuit is between 0 and +90°, or between 0 and −90°.

Furthermore, the ohms of resistance and reactance must be combined by phasor addition because of the 90° phase angle between the two. The resultant sum is called *impedance,* with the symbol Z, which is the total opposition of resistance and reactance to a sine-wave alternating current. The methods of combining R, X_L, and X_C to find the total Z and phase angle θ are explained in detail in Chap. 25, Alternating-current Circuits.

*Practice Problems 16-10
(answers on page 336)
Calculate R_T in*
(a) Fig. 16-16.
(b) Fig. 16-17.
(c) Fig. 16-18.

16-11
NONSINUSOIDAL AC WAVEFORMS

The sine wave is the basic waveform for ac variations for several reasons. This waveform is produced by a rotary generator, as the output is proportional to the angle of rotation. In addition, electronic oscillator circuits with inductance and capacitance naturally produce sine-wave variations.

Because of its derivation from circular motion, any sine wave can be analyzed in angular measure, either in degrees from 0 to 360° or in radians from 0 to 2π rad.

Another feature of a sine wave is its basic simplicity, as the rate of change for the amplitude variations corresponds to a cosine wave

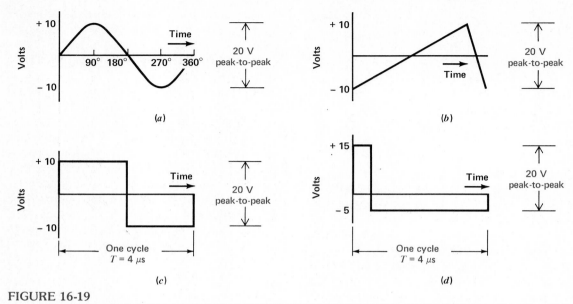

FIGURE 16-19

Comparison of sine wave with nonsinusoidal ac waveforms. (*a*) Sine wave. (*b*) Sawtooth wave. (*c*) Symmetrical square wave. (*d*) Unsymmetrical rectangular wave.

which is similar but 90° out of phase. The sine wave is the only waveform that has this characteristic of a rate of change with the same waveform as the original changes in amplitude.

In many electronic applications, however, other waveshapes are important. Any waveform that is not a sine or cosine wave is a *nonsinusoidal waveform*. Common examples are the square wave and sawtooth wave in Fig. 16-19.

With nonsinusoidal waveforms, for either voltage or current, there are important differences and similarities to consider. Note the following comparisons with sine waves.

1. In all cases, the cycle is measured between two points having the same amplitude and varying in the same direction. The period is the time for one cycle. In Fig. 16-19, T for any of the waveforms is 4 μs and the corresponding frequency is $\frac{1}{4}$ MHz = 0.25 MHz.

2. Peak amplitude is measured from the zero axis to the maximum positive or negative value. However, peak-to-peak is better for measuring nonsinusoidal waveshapes because they can have unsymmetrical peaks, as in (*d*). For all the waveforms shown here, though, the peak-to-peak (p-p) amplitude is 20 V.

3. The rms value 0.707 of maximum applies only to sine waves, as this factor is derived from the sine values in the angular measure used only for the sine waveform.

4. Phase angles apply only to sine waves, as angular measure is used only for sine waves. Note that the horizontal axis for time is divided into angles for the sine wave

in (a), but there are no angles shown for the nonsinusoidal waveshapes.

The sawtooth wave in (b) represents a voltage that slowly increases, with a uniform or linear rate of change, to its peak value, and then drops sharply to its starting value. This waveform is also called a *ramp voltage*. It is also often referred to as a *time base* because of its constant rate of change.

Note that one complete cycle includes the slow rise and the fast drop in voltage. In this example, the period T for a complete cycle is 4 μs. Therefore, these sawtooth cycles are repeated at the frequency of $\frac{1}{4}$ MHz, which equals 0.25 MHz. The sawtooth waveform of voltage or current is often used for horizontal deflection of the electron beam in the cathode-ray tube (CRT) for oscilloscopes and TV receivers.

The square wave in (c) represents a switching voltage. First, the 10-V peak is instantaneously applied in positive polarity. This voltage remains on for 2 μs, which is one half-cycle. Then the voltage is instantaneously reduced to zero and applied in reverse polarity for another 2 μs. The complete cycle then takes 4 μs and the frequency is $\frac{1}{4}$ MHz.

The rectangular waveshape in (d) is similar, but the positive and negative half-cycles are not symmetrical, either in amplitude or in time. However, the frequency is the same 0.25 MHz and the peak-to-peak amplitude is the same 20 V, as in all the waveshapes.

Practice Problems 16-11
(answers on page 336)
(a) In Fig. 16-19c, for how much time is the waveform at +10 V?
(b) In Fig. 16-19d, what voltage is the positive peak amplitude?

16-12
HARMONIC FREQUENCIES

Consider a repetitive nonsinusoidal waveform, such as a 100-Hz square wave. Its fundamental rate of repetition is 100 Hz. Exact multiples of the fundamental frequency are called harmonic frequencies. The second harmonic is 200 Hz, the third harmonic is 300 Hz, etc. Even multiples are even harmonics while odd multiples are odd harmonics.

Harmonics are useful in analyzing distorted sine waves or nonsinusoidal waveforms. Such waveforms consist of a pure sine wave at the fundamental frequency plus harmonic frequency components. For example, Fig. 16-20 illustrates how a square wave corresponds to a fundamental sine wave with odd harmonics. Typical audio waveforms include odd and even harmonics. It is the harmonic components that make one source of sound different from another with the same fundamental frequency.

Another unit for frequency multiples is the *octave*, which is a range of 2:1. Doubling the frequency range from 100 to 200 Hz, or 200 to 400 Hz and 400 to 800 Hz, are examples of raising the frequency by one octave. The reason for this name is that an octave in music includes eight consecutive tones.

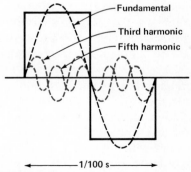

FIGURE 16-20

Fundamental and harmonic frequencies for a 100-Hz square wave.

*Practice Problems 16-12
(answers on page 336)*

(a) What frequency is the fourth harmonic of 12 MHz?

(b) What frequency is one octave above 220 Hz?

Summary

1. Alternating voltage continuously varies in magnitude and reverses in polarity. Alternating voltage applied across a load resistance produces alternating current in the circuit.
2. A complete set of values repeated periodically is one cycle of the ac waveform. The cycle can be measured from any one point on the wave to the next successive point having the same value and varying in the same direction. One cycle includes 360° in angular measure, or 2π rad.
3. The rms value of a sine wave is $0.707 \times$ peak value.
4. The peak amplitude, at 90 and 270° in the cycle, is $1.414 \times$ rms value.
5. Peak-to-peak value is double the peak amplitude, or $2.828 \times$ rms for a symmetrical ac waveform.
6. The average value is $0.637 \times$ peak value.
7. The frequency equals the number of cycles per second. One cps is 1 Hz; 1 MHz equals 1 million cps. The audio-frequency (af) range is 16 to 16,000 Hz. Higher frequencies up to 300,000 MHz are radio frequencies (rf).
8. The amount of time for one cycle is the period T. The period and frequency are reciprocals: $T = 1/f$, or $f = 1/T$. The higher the frequency, the shorter the period.
9. Wavelength λ is the distance a wave travels in one cycle. The higher the frequency, the shorter the wavelength. The wavelength also depends on the velocity at which the wave travels: $\lambda =$ velocity/frequency.
10. Phase angle is the angular difference in time between corresponding values in the cycles for two waveforms of the same frequency.
11. When one sine wave has its maximum value while the other is at zero, the two waves are 90° out of phase. Two waveforms with zero phase angle between them are in phase; a 180° phase angle means opposite phase.
12. The length of a phasor arrow indicates amplitude, while the angle corresponds to the phase. Leading phase is shown by counterclockwise angles.

13. Sine-wave alternating voltage V applied across a load resistance R produces alternating current I in the circuit. The current has the same waveform, frequency, and phase as the applied voltage because of the resistive load. The amount of $I = V/R$.
14. The sawtooth wave and square wave are two common examples of nonsinusoidal waveforms. These amplitudes are usually measured in peak-to-peak value.
15. Harmonic frequencies are exact multiples of the fundamental frequency.

Self-Examination (Answers at back of book.)

Answer true or false.

1. Ac voltage varies in magnitude and reverses in polarity.
2. Dc voltage always has one polarity.
3. Sine-wave alternating current flows in a load resistor with sine-wave voltage applied.
4. When two waves are 90° out of phase, one has its peak value when the other is at zero.
5. When two waves are in phase, they have their peak values at the same time.
6. The positive peak of a sine wave cannot occur at the same time as the negative peak.
7. The angle of 90° is the same as π rad.
8. A period of 2 μs corresponds to a higher frequency than T of 1 μs.
9. A wavelength of 2 ft corresponds to a lower frequency than a wavelength of 1 ft.
10. When we compare the phase between two waveforms, they must have the same frequency.

Fill in the missing answers.
11. For the rms voltage of 10 V, the p-p value is _____ V.
12. With 120 V rms across 100 Ω R_L, the rms current equals _____ A.
13. For a peak value of 100 V, the rms value is _____ V.
14. The wavelength of a 1000-kHz radio wave is _____ cm.
15. The period of a 1000-kHz voltage is _____ ms.
16. The period of $\frac{1}{60}$ s corresponds to a frequency of _____ Hz.
17. The frequency of 100 MHz corresponds to a period of _____ μs.
18. The square wave in Fig. 16-19c has the frequency of _____ MHz.
19. The rms voltage for the sine wave in Fig. 16-19c is _____ V.

20. The ac voltage across R_2 in Fig. 16-18 is _____ V.
21. For an audio signal with a T of 0.001 s, its frequency is _____ Hz.
22. For the 60-Hz ac power-line voltage, the third harmonic is _____ Hz.
23. For a 10-V average value, the rms value is _____ V.
24. For a 340-V p-p value, the rms value is _____ V.

Essay Questions

1. (a) Define an alternating voltage. (b) Define an alternating current. (c) Why does ac voltage applied across a load resistance produce alternating current in the circuit?
2. (a) State two characteristics of a sine wave of voltage. (b) Why does the rms value of 0.707 × peak value apply just to sine waves?
3. Draw two cycles of an ac sawtooth voltage waveform, with a peak-to-peak amplitude of 40 V. Do the same for a square wave.
4. Give the angle, in degrees and radians, for each of the following: one cycle, one half-cycle, one quarter-cycle, three quarter-cycles.
5. The peak value of a sine wave is 1 V. How much is its average value? Rms value? Effective value? Peak-to-peak value?
6. State the following ranges in Hz: (a) audio frequencies; (b) radio frequencies; (c) standard AM radio broadcast band; (d) FM broadcast band; (e) VHF band; (f) microwave band. (Hint: See Appendix B.)
7. Make a graph with two waves, one with a frequency of 500 kHz and the other with 1000 kHz. Mark the horizontal axis in time and label each wave.
8. Draw the sine waves and phasor diagrams to show (a) two waves 180° out of phase; (b) two waves 90° out of phase.
9. Give the voltage value for the 60-Hz ac line voltage with an rms value of 120 V at each of the following times in a cycle: 0°, 30°, 45°, 90°, 180°, 270°, 360°.
10. (a) The phase angle of 90° equals how many radians? (b) For two sine waves 90° out of phase with each other, compare their amplitudes at 0°, 90°, 180°, 270°, and 360°.
11. Tabulate the sine and cosine values every 30° from 0 to 360° and draw the corresponding sine wave and cosine wave.
12. Draw a graph of the values for $(\sin \theta)^2$ plotted against θ for every 30° from 0 to 360°.
13. Why is the wavelength of a supersonic wave at 34.44 kHz the same 1 cm as for the much higher-frequency radio wave at 30 GHz?
14. Draw the sine waves and phasors to show wave V_1 leading wave V_2 by 45°.

15. Why are amplitudes for nonsinusoidal waveforms generally measured in peak-to-peak values, rather than rms or average value?
16. Define harmonic frequencies, giving numerical values.
17. Define one octave, with an example of numerical values.
18. Which do you consider more important for applications of alternating current—the polarity reversals or the variations in value?

Problems (Answers to odd-numbered problems at back of book.)

1. The 60-Hz power-line voltage of 120 V is applied across a resistance of 10 Ω. (a) How much is the rms current in the circuit? (b) What is the frequency of the current? (c) What is the phase angle between the current and the voltage? (d) How much dc applied voltage would be necessary for the same heating effect in the resistance?
2. What is the frequency for the following ac variations? (a) 10 cycles in 1 s; (b) 1 cycle in $\frac{1}{10}$ s; (c) 50 cycles in 1 s; (d) 50 cycles in $\frac{1}{2}$ s; (e) 50 cycles in 5 s.
3. Calculate the time delay for a phase angle of 45° at the frequency of (a) 500 Hz; (b) 2 MHz.
4. Calculate the period T for the following frequencies: (a) 500 Hz; (b) 5 MHz; (c) 5 GHz.
5. Calculate the frequency for the following periods: (a) 0.05 s; (b) 5 ms; (c) 5 µs; (d) 5 ns.
6. Referring to Fig. 16-18, calculate the I^2R power dissipated in R_1, R_2, and R_3.
7. Referring to Fig. 16-19, give the positive and negative peak values for each waveform in (a), (b), (c), and (d).
8. An ac circuit has a 5-MΩ resistor R_1 in series with a 10-MΩ resistor R_2 across a 200-V source. Calculate I, V_1, V_2, P_1, and P_2.
9. The same two resistors as in Prob. 8 are in parallel. Calculate I_1, I_2, V_1, V_2, P_1, and P_2.
10. A series-parallel ac circuit has two branches across the 60-Hz 120-V power line. One branch has a 10-Ω R_1 in series with a 20-Ω R_2. The other branch has a 10-MΩ R_3 in series with a 20-MΩ R_4. Find V_1, V_2, V_3, and V_4.
11. How much rms I does a 300-W 120-V bulb take from the 120-V 60-Hz power line? What is the frequency and phase of I, compared with V?
12. In Fig. 16-21, calculate V_{rms}, period T, and frequency f.

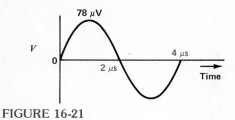

FIGURE 16-21

For Prob. 12.

Answers to Practice Problems

16-1	(a) 10 V		16-7	(b) F
	(b) 360°			(c) T
16-2	(a) 85 V		16-8	(a) 90°
	(b) −170 V			(b) 60°
	(c) 0 V			(c) 0°
16-3	(a) 0.707 A		16-9	(a) $\frac{1}{240}$ s
	(b) 0.5 A			(b) 0.1 μs
16-4	(a) 120 V rms		16-10	(a) 30 Ω
	(b) 14.14 V peak			(b) 6.66 Ω
16-5	(a) 4 Hz			(c) 30 Ω
	(b) 1.605 MHz		16-11	(a) 2 μs
16-6	(a) 400 Hz			(b) 15 V
	(b) $\frac{1}{400}$ s		16-12	(a) 48 MHz
16-7	(a) T			(b) 440 Hz

The 60-Hz AC Power Line

Chapter 17

Practically all homes in the United States are supplied alternating voltage at 115 to 125 V, rms, with the frequency of exactly 60 Hz. This is a sine-wave voltage produced by a rotary generator. The electrical power is distributed by power lines from the generating station to the main line in the home. Here the 120-V line is connected to all the wall outlets and electrical equipment in parallel. This 120-V source of commercial electrical power is the *60-Hz power line,* or *mains,* indicating that it is the main line for all the parallel branches. Some practical features of the ac power line are described in the following topics:

17-1 Advantages of 120-V 60-Hz AC Power
17-2 Motors and Generators
17-3 Fluorescent Lighting
17-4 Residential Wiring
17-5 Three-phase Power

17-1 ADVANTAGES OF 120-V 60-Hz AC POWER

With an rms value of 120 V, the ac power is equivalent to 120-V dc power in heating effect. With a higher value than 120 V there would be more danger of a fatal electric shock. Lower voltages would be less efficient in supplying power.

A higher voltage has the advantage of less I^2R loss, since the same power can be produced with less I. The I^2R power increases as the square of the current. For industrial applications, where large amounts of power are used, the main line is often 208 to 240 V, three-phase. The three-phase ac power is more efficient for the operation of large motors.

The advantage of ac over dc power is greater efficiency in distribution from the generating station. Ac voltages can easily be stepped up by means of a transformer, with very little loss, but a transformer cannot operate on direct current. The reason is that a transformer needs a varying current in the primary winding, with its varying magnetic field, to produce induced voltage in the secondary winding.

Therefore, the alternating voltage at the generating station can be stepped up to values as high as 80 kV for high-voltage distribution lines. These high-voltage lines supply large amounts of power with much less current and less I^2R loss, compared with a 120-V line. At the home, the lower voltage required is supplied from a step-down transformer.

Furthermore, ac voltage in the home at 120 V can easily be stepped up or down by a relatively small power transformer, in electronic equipment operating from the power line. The input voltage to the transformer is stepped up when the secondary winding has more turns than the primary winding. Or, less turns in the secondary results in voltage step-down.

The frequency of 60 Hz is convenient for commercial ac power. Much lower frequencies would require transformers that would be too big. Also, too low a frequency for alternating current in a lamp could cause the light to flicker. Too high a frequency results in excessive iron-core losses from eddy currents and hysteresis in the transformer. It should be noted that the frequency of the ac power mains in most European countries is 50 Hz.

All power companies in the continental United States, except in Texas, are tied to a power grid that maintains the nominal ac power-line frequency between 59.98 and 60.02 Hz. The grid frequency is compared with the time standard provided by the Bureau of Standards radio station WWV at Fort Collins, Colorado. As a result, the 60-Hz power-line frequency is maintained accurate to ±0.033 percent. This accuracy makes the power-line voltage a good secondary standard to check frequencies based on 60 Hz.

*Practice Problems 17-1
(answers on page 347)*
Answer true or false.
(a) The frequency of the ac power-line voltage is 60 Hz ± 0.033 percent.
(b) A power transformer can step up or down an ac voltage but not a dc voltage.
(c) The 120 V of the ac power line is an rms value.
(d) The audio frequency range includes the value of 60 Hz.

17-2
MOTORS AND GENERATORS

A generator converts mechanical energy into electrical energy; a motor does the opposite, converting electricity into rotary motion. The main parts in the assembly of motors and generators are essentially the same (Fig. 17-1).

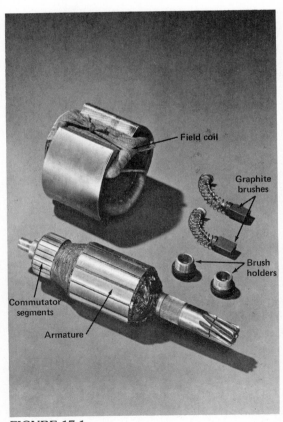

FIGURE 17-1

Main parts of a dc motor.

Armature. In a generator, the armature connects to the external circuit to provide the generator output voltage. Or, in a motor, the armature connects to the electrical source that drives the motor. The armature is often constructed in the form of a drum, using many conductor loops for increased output. In Fig. 17-1 the rotating armature is the *rotor* part of the assembly.

Field Winding. This electromagnet provides the flux cut by the rotor. In a motor, current for the field is produced by the same source that supplies the armature. In a generator, the field current may be obtained from a separate exciter source, or from its own armature output. Residual magnetism in the iron yoke of the field allows this *self-excited generator* to start.

The field coil may be connected in series with the armature, in parallel, or in a series-parallel *compound winding*. When the field winding is stationary, it is the *stator* part of the assembly.

Slip Rings. In an ac machine, two slip rings or *collector rings* enable the rotating loop to be connected to the stationary wire leads for the external circuit.

Brushes. These graphite connectors are spring-mounted to brush against the spinning rings on the rotor. The stationary external leads are connected to the brushes for connection to the rotating loop. Constant rubbing slowly wears down the brushes, and they must be replaced after they are worn.

Commutator. A dc machine has a commutator ring instead of the slip rings. As shown in Fig. 17-1, the commutator ring has segments, with one pair for each loop in the armature. Each of the commutator segments is insulated from the other by mica.

The commutator converts the ac machine to dc operation. In a generator, the commutator segments reverse the loop connections to the brushes every half-cycle to maintain constant polarity of output voltage. For a dc motor, the commutator segments allow the dc source to produce torque in one direction.

Brushes are necessary with a commutator ring. The two stationary brushes contact opposite segments on the rotating commutator. Graphite brushes are used for very low resistance.

AC Induction Motor. This type, for alternating current only, does not have any brushes. The stator is connected directly to the ac source. Then alternating current in the stator winding induces current in the rotor without any physical connection between them. The magnetic field of the current induced in the rotor reacts with the stator field to produce rotation. Ac induction motors are economical and rugged, without any troublesome brush arcing.

With a single-phase source, however, a starting torque must be provided for an ac induction motor. One method uses a starting capacitor in series with a separate starting coil. The capacitor supplies an out-of-phase current just for starting, and then is switched out. Another method of starting uses shaded poles. A solid copper ring on the main field pole makes the magnetic field unsymmetrical to allow starting.

The rotor of an ac induction motor may be wirewound or the squirrel-cage type. This rotor is constructed with a frame of metal bars.

Universal Motor. This type operates on either alternating or direct current because the field and armature are in series. Its construction is like a dc motor with the rotating armature connected to a commutator and brushes. The universal motor is commonly used for small machines such as portable drills and food mixers.

Chapter 17
The 60-Hz AC Power Line

Alternators. Ac generators are alternators. For large power requirements the alternator usually has a rotor field, while the armature is the stator. This method eliminates slip-ring connections with its arcing problems in the high-voltage output.

Practice Problems 17-2
(answers on page 347)
Answer true or false.
(a) In Fig. 17-1 the commutator segments are on the armature.
(b) Motor brushes are made of graphite for very low resistance.
(c) A starting capacitor is used with dc motors that have small brushes.

17-3
FLUORESCENT LIGHTING

Fluorescent lamps produce light with more efficiency and less heat than incandescent bulbs. The operation is illustrated by the basic construction of a fluorescent lamp in Fig. 17-2. Only one end is shown, but the other end is the same.

Two electrical effects result in the visible light output. First, electric discharge or ionization of mercury vapor produces ultraviolet light, which is not visible. However, this radiation energizes the phosphor coating on the inside wall, which reradiates light in the visible spectrum.

Droplets of mercury in the glass tube provide the mercury vapor. During lamp operation, the mercury vaporizes. Sufficient voltage applied across the two cathode terminals ionizes the mercury vapor.

The cathodes at both ends are preheated to emit electrons. This thermionic emission provides free electrons to excite the mercury atoms to produce ultraviolet radiation. The preheating of the cathodes is used for easier starting of the lamp discharge.

The phosphor coating on the inside wall of the tube is a powdered chemical, such as magnesium tungstate, which emits a bluish-white light. Different phosphors can be used for different colors and shades of white. The wavelength for ultraviolet light is 254×10^{-9} m, while bluish-white is at 360 to 720×10^{-9} m.

The phosphor atoms are energized by the ultraviolet radiation from the mercury vapor, resulting in visible light. The fluorescent characteristic means the phosphor can glow in a different wavelength of light without a rise in temperature for the phosphor coating.

Lamp Circuit. See Fig. 17-3. First the starter switch S_1 is closed to heat the cathodes for 1 to 2 s. With S_1 closed, the circuit, which is in series with the 120-V ac power line, includes the two

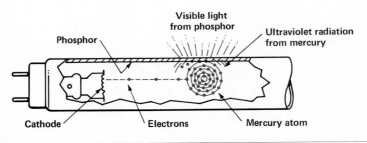

FIGURE 17-2
How light is produced from a fluorescent lamp with preheated cathodes. (*GTE Sylvania*)

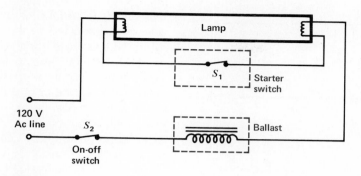

FIGURE 17-3
Basic circuit for lighting fluorescent lamp after switch S_1 has been closed to preheat the cathodes.

cathodes at opposite ends of the lamp and a ballast coil or resistor to limit the current. The preheating produces thermionic emission of electrons to help in ionizing the mercury vapor.

Then the starter switch is opened. The heating circuit is open, but now voltage is applied across the bulb to produce ionization. The ballast is still in the circuit to limit the ionization current. When the ballast is a coil, which is the usual case, it produces a higher induced voltage at the instant the starter switch is opened to help in producing ionization.

After the bulb is ignited, it stays on with the mercury vapor ionized and the ballast limiting the ionization current. The bulb is turned off with the main ON/OFF switch S_2. Approximate values for a 20-W fluorescent lamp connected to the 120-V ac power line are 0.35 A through the bulb and ballast, with about 65 V across the bulb and 55 V across the ballast.

The starter is generally a separate switch in a small metal can that is plugged into the light fixture. The switch has bimetallic contacts that open after 1 to 2 s of heating.

Types of Lamps. The circuit in Fig. 17-3 is for the older preheat lamps. This type needs a separate starter switch for the preheat circuit.

A later type is the *instant-start lamp* which does not require any cathode heating circuit. This type usually has a slimline base with only one pin at each end.

Most common now for home use, though, is the *rapid-start lamp*. This type uses cathode heating, but there is no starter. The ballast for a rapid-start system includes a separate winding to heat the cathodes continuously.

Practice Problems 17-3
(answers on page 347)
(a) In a fluorescent lamp, what chemical produces the ionization?
(b) Which part of the lamp produces thermionic emission of electrons?
(c) In Fig. 17-3 is S_1 open or closed when the lamp is producing light output?

17-4
RESIDENTIAL WIRING

Most homes have at the service entrance the three-wire single-phase power lines illustrated in Fig. 17-4. The three wires including the grounded neutral can be used for either 240 or 120 V.

Note the color coding. The grounded neutral is white, or bare wire is used. Each high side

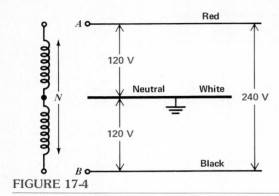

FIGURE 17-4

Three-wire single-phase power lines for 240 or 120 V. Alternator output illustrated at left.

can use any color except white and green but usually is black[1] or red. White and green are reserved for grounding.

From either the red or black high side to the neutral, 120 V is available for separate branch circuits to the lights and outlets. Across the red and black wires, 240 V is available for high-power appliances. This three-wire service with a grounded neutral is called the *Edison system*.

The electrical service is commonly rated for 100 A. At 240 V, then, the power available is $10 \times 240 = 24{,}000$ W, or 24 kW.

The main wires to the service entrance, where the power enters the house, are generally gage No. 4 to 8. Sizes 6 and heavier are always stranded wire. The branch circuits at 120 V, usually rated at 15 A, use gage No. 8 to 14 wire. Each branch has its own fuse or circuit breaker. A main switch should be included to cut off all power from the service entrance.

The neutral wire is grounded at the service entrance to a water pipe or a metal rod driven into the earth. All 120-V branches must have one side connected to the grounded neutral. White wire is used for these connections. In addition, all the metal boxes for outlets, switches, and lights must have a continuous ground to each other and to the neutral. The wire cable usually has a bare wire for this grounding of the boxes.

Cables commonly used are armored sheath with the trade name BX and flexible cable with the trade name Romex. Each has two wires for the neutral and high-side connections. In addition, the metal sheath is used for continuous grounding between boxes. The flexible cable usually has an extra bare wire for grounding.

The purpose of grounding is safety against electric shock. Switches and fuses are never in the ground side of the line, in order to maintain the ground connections.

Rules for grounding and wiring are specified by local electrical codes. The National Electrical Code standards are available from the National Board of Fire Underwriters. One general requirement is that all wire connections must be in a metal box with a cover.

Outlet Boxes. Figure 17-5 illustrates how out-

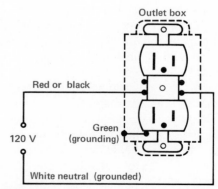

FIGURE 17-5

Wiring of duplex outlet. Extra terminals can be used for the next outlet.

---

[1] It should be noted that in electronic equipment, black is the color-coded wiring for chassis-ground returns. However, in electrical power work, black wire is for high-side connections.

lets are wired in parallel to the 120-V line. One side has the white neutral wire. The other side connects to the red or black high side. Only one pair of connections is needed for the two sockets in the duplex outlet. However, extra screws are on the side if needed for wires to the next outlet in parallel.

Most outlets now have a third U-shaped or round opening in addition to the pair of blade receptacles. This third connection goes to a green screw on the inside frame of the outlet which is connected to the grounding wire. In the three-prong plug for appliances, the ground pin is connected to the metal case of the appliance with a green wire.

In addition, most receptacles now have one slot wider to fit this type of plug only one way. The wider blade is the side that should be connected to the white neutral. This safety feature means that the chassis ground of radio and television equipment will automatically be connected to the grounded side of the power line.

Single-Pole Switch. Figure 17-6 shows how the two terminals of the switch are wired in series, just in one side of the line. This must always be the high side. Note that black is into and red is out of the switch.

A lighted switch uses a small bulb across the switch terminals. When the switch is OFF or

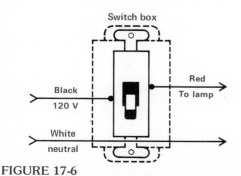

FIGURE 17-6

Wiring of single-pole single-throw switch.

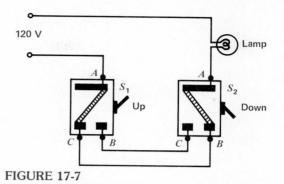

FIGURE 17-7

Wiring of three-way (single-pole double-throw) switches.

open, the 120 V across the open terminals lights the display. When the switch is ON or closed, there is no voltage across the switch itself. The switch light is off when the main light is turned on.

Three-way Switches. Two switches are used to turn on a light from two separate locations, as shown in Fig. 17-7. Each switch has three terminals. One is the common terminal A. This screw is usually copper, while the other two are brass. With the switch up, as shown for S_1, the common A is connected to terminal C. However, with the switch down, the common A is connected to terminal B, as shown for S_2. The B terminal of one switch is wired to the C terminal of the other switch.

The main terminal A on one switch is wired to one side of the lamp to be controlled, while the other side of the lamp returns to the 120-V ac line. The opposite side of the 120-V line is connected to the main terminal A on the other switch. In order to light the bulb, the switches must connect terminal A on one to A on the other. Then the lamp is across the 120-V line.

The lamp lights when one switch is up while the other is down, as shown in Fig. 17-7. Then A on S_1 from the power line connects to C

Chapter 17
The 60-Hz AC Power Line

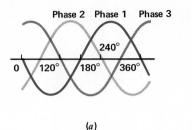
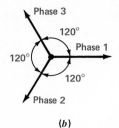

(a) (b)

FIGURE 17-8

Three-phase alternating voltage or current with 120° between each phase. (a) Sine waves. (b) Phasor diagram.

on S_1, to B on S_2, and to A on S_2 to the lamp. When both switches are up or down, one switch or the other is open and the bulb does not light.

*Practice Problems 17-4
(answers on page 347)
Answer true or false.*
(a) Figure 17-4 shows a three-wire, three-phase power line.
(b) Duplex outlets are in parallel with each other and the 120-V ac line.
(c) A light switch is in series with the lamp and the high side of the ac line.

17-5
THREE-PHASE POWER

In an alternator with a rotor field, if the stator armature has three windings equally spaced around the circle, the three armature windings will generate output voltages 120° out of phase with each other. The three-phase output is illustrated by the sine-wave voltages in Fig. 17-8a and the corresponding phasors in (b). The advantage of three-phase ac voltage is more efficient distribution of power. Also, ac induction motors are self-starting with three-phase alternating current. Finally, the ac ripple is easier to filter in dc power supplies.

Y Connections. In Fig. 17-9, the three windings are in the form of a Y, also called *Wye* or *star* connections. All three coils are joined at one end, with the opposite ends for the output terminals A, B, and C. Any pair of terminals is across two coils in series. By trigonometry, the voltage output across any two output terminals is $1.73 \times 120 = 208$ V.

Delta Connections. In Fig. 17-10, the three

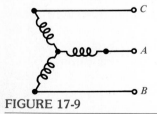

FIGURE 17-9

Y or wye connections for three-phase ac power.

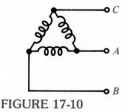

FIGURE 17-10

Delta or Δ connections for three-phase ac power.

windings are in the form of a Δ. Any pair of terminals is across one generator coil. However, the other coils are in a parallel leg. Therefore, the current capacity is increased by the factor 1.73.

Four-wire Power Line. In Fig. 17-11 the center point of the Y is used for a fourth line, which is the neutral wire. This way, power is available at 208 V three-phase or at 120 V single-phase. From terminal A, B, or C to the neutral line, the output is the 120 V across one coil. The 120-V single-phase power is for lighting circuits. Across any pair of terminals without the neutral, however, the output is 208 V for three-phase induction motors. Although illustrated here for the 120-V 60-Hz power line, it should be noted that three-phase connections can be used for other voltages and frequencies.

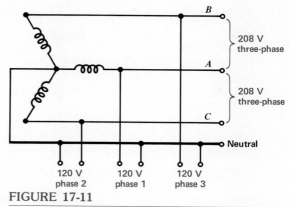

FIGURE 17-11

Y connections to four-wire line with neutral.

Practice Problems 17-5
(answers on page 347)
(a) What is the angle between three-phase voltages?
(b) In Fig. 17-9, how much is V_{AC} or V_{AB}?

Summary

1. The ac power line has a nominal voltage of 115 to 125 V and the exact frequency of 60 Hz.
2. For residential wiring, the three-wire single-phase Edison system is used to provide either 120 or 240 V.
3. In a motor, the rotating armature connects to the power line. The stator field coils provide the magnetic flux cut by the rotor. A dc motor has commutator segments contacted by graphite brushes for the external connections. An ac induction motor does not have brushes.
4. A fluorescent lamp has a phosphor coating on the inside wall that radiates light when the phosphor is excited by ultraviolet radiation from ionized mercury vapor.
5. In residential wiring, the 120-V duplex outlet boxes are wired in parallel. Light switches are in series with the lamp in the high side of the ac line. This side generally is red or black wire, while the grounded neutral is white.
6. In three-phase power, each phase is 120° out of phase. For the Y connections in Fig. 17-9, any pair of output terminals has $1.73 \times 120 = 208$ V.

Chapter 17
The 60-Hz AC Power Line

Self-Examination (Answers at back of book.)

1. An ac waveform that produces four cycles in the time it takes to produce one cycle of ac voltage from the power line has what frequency?
2. In Fig. 17-3, is the fluorescent bulb on or off when the starter switch S_1 is closed?
3. In Fig. 17-4, how much is the voltage from either A or B to the grounded neutral?
4. In Fig. 17-4, how much is the voltage between A and B?
5. Thermionic emission is produced by what electrodes in a fluorescent lamp?
6. Is a ballast inductance in series or parallel with a fluorescent lamp?
7. Are commutator segments used in a dc or ac motor?
8. What color is used for the grounded neutral in residential wiring?
9. In Fig. 17-11, how much is the voltage across two of the windings in series?
10. In Fig. 17-7, is the lamp on or off for the switch positions shown for S_1 and S_2?

Essay Questions

1. Define the following parts in the assembly of motors: (*a*) armature rotor; (*b*) field stator; (*c*) collector rings; (*d*) brushes; (*e*) commutator segments.
2. What is the function of a starting capacitor with an ac induction motor?
3. For fluorescent lighting give the function of (*a*) cathodes; (*b*) mercury vapor; (*c*) phosphor coating; (*d*) ballast.
4. What is the difference between *instant-start* and *rapid-start* fluorescent lamps?
5. Show a diagram of the three-wire single-phase ac power line to the service entrance in residential wiring.
6. Show a diagram of Y connections for three-phase ac power.
7. Show a diagram of Δ connections for three-phase ac power.
8. Show a schematic diagram for six duplex outlets wired to the ac power line.
9. Show a diagram of a single-pole switch wired to a duplex outlet.
10. Show a schematic diagram of three-way switches to light a bulb from two different locations.

Answers to Practice Problems

17-1	(a) T		17-3	(b) Cathode
	(b) T			(c) Open
	(c) T		17-4	(a) F
	(d) T			(b) T
17-2	(a) T			(c) T
	(b) T		17-5	(a) 120°
	(c) F			(b) 208 V
17-3	(a) Mercury vapor			

Review of Chapters 13 to 17

Summary

1. Iron, nickel, and cobalt are magnetic materials. Magnets have a north pole and south pole at opposite ends. Opposite poles attract; like poles repel.
2. A magnet has an invisible, external magnetic field. This magnetic flux is indicated by field lines. The direction of field lines outside the magnet is from north pole to south pole.
3. A permanent magnet is made of a hard magnetic material, such as alnico, to retain its magnetism indefinitely. Iron is a soft magnetic material which can be magnetized temporarily.
4. An electromagnet has an iron core that becomes magnetized when current flows in the coil winding.
5. Magnetic units are defined in Tables 13-1 and 14-2.
6. Continuous magnetization and demagnetization of an iron core by means of alternating current causes hysteresis losses, which increase with higher frequencies.
7. Ferrites are ceramic magnetic materials that are insulators.
8. Current in a conductor has an associated magnetic field with circular lines of force in a plane perpendicular to the wire. Their direction is counterclockwise when you look along the conductor in the direction of electron flow.
9. Motor action results from the net force of two fields that can aid or cancel. The direction of the resultant force is from the stronger field to the weaker.
10. The motion of magnetic flux cutting across a perpendicular conductor generates an induced emf. The amount of induced voltage increases with higher frequencies, more flux, and more turns of conductor.
11. Faraday's law of induced voltage is $v = N\, d\phi/dt$, where N is the turns and $d\phi/dt$ is the change in flux in webers per second.

12. Lenz' law states that an induced voltage must have the polarity that opposes the change causing the induction.
13. Alternating voltage varies in magnitude and reverses in direction. An ac voltage source produces alternating current.
14. One cycle includes the values between points having the same value and varying in the same direction. The cycle includes 360°, or 2π rad.
15. Frequency f equals the cycles per second (cps). One cps = 1 Hz.
16. Period T is time for one cycle. It equals $1/f$. When f is in cycles per second, T is in seconds.
17. Wavelength λ is the distance a wave travels in one cycle. $\lambda = v/f$.
18. The rms, or effective, value of a sine wave equals $0.707 \times$ peak value. Or the peak value equals $1.414 \times$ rms value. The average value equals $0.637 \times$ peak value.
19. Phase angle θ is the angular difference in time between corresponding values in the cycles for two sine waves of the same frequency.
20. Phasors, similar to vectors, indicate the amplitude and phase angle of alternating voltage or current. The length of the phasor is the amplitude while the angle is the phase.
21. The square wave and sawtooth wave are common examples of nonsinusoidal waveforms.
22. Harmonic frequencies are exact multiples of the fundamental frequency.
23. Dc motors generally use commutator segments with graphite brushes. Ac motors are usually the induction type without brushes.
24. Residential wiring generally uses three-wire single-phase power with the exact frequency of 60 Hz. Voltages available are 120 V to the grounded neutral or 240 V across the two high sides.
25. Three-phase ac power has three legs 120° out of phase. With Y connections, 208 V is available across any two legs.

Review Self-Examination (Answers at back of book.)

Choose (a), (b), (c), or (d).

1. Which of the following statements is true? (a) Alnico is commonly used for electromagnets. (b) Paper cannot affect magnetic flux because it is not a magnetic material. (c) Iron is generally used for permanent magnets. (d) Ferrites have lower permeability than air or vacuum.
2. Hysteresis losses (a) are caused by high-frequency alternating current

in a coil with an iron core; (b) generally increase with direct current in a coil; (c) are especially important with permanent magnets that have a steady magnetic field; (d) cannot be produced in an iron core, because it is a conductor.
3. A magnetic flux of 25,000 lines through an area of 5 cm² results in (a) 5 lines of flux; (b) 5000 Mx of flux; (c) flux density of 5000 G; (d) flux density corresponding to 25,000 A.
4. If 10 V is applied across a relay coil with 100 turns having 2 Ω of resistance, the total force producing magnetic flux in the circuit is (a) 10 Mx; (b) 50 G; (c) 100 Oe; (d) 500 ampere-turns.
5. The ac power-line voltage of 120 V rms has a peak value of (a) 100 V; (b) 170 V; (c) 240 V; (d) 338 V.
6. Which of the following can produce the most induced voltage? (a) 1-A direct current; (b) 50-A direct current; (c) 1-A 60-Hz alternating current; (d) 1-A 400-Hz alternating current.
7. Which of the following has the highest frequency? (a) $T = \frac{1}{1000}$ s; (b) $T = \frac{1}{60}$ s; (c) $T = 1$ s; (d) $T = 2$ s.
8. Two waves of the same frequency have opposite phase when the phase angle between them is (a) 0°; (b) 90°; (c) 360°; (d) π rad.
9. The 120-V 60-Hz power-line voltage is applied across a 120-Ω resistor. The current equals (a) 1 A, peak value; (b) 120 A, peak value; (c) 1 A, rms value; (d) 5 A, rms value.
10. When an alternating voltage reverses in polarity, the current it produces (a) reverses in direction; (b) has a steady dc value; (c) has a phase angle of 180°; (d) alternates at 1.4 times the frequency of the applied voltage.
11. In Fig. 17-4, the voltage from either A or B to the grounded neutral is (a) 240; (b) 208; (c) 170; (d) 120.
12. In Fig. 17-11, the voltage across any one coil is (a) 120 V, single-phase; (b) 120 V, three-phase; (c) 208 V, single-phase; (d) 208 V, three-phase.

References (Additional references at back of book.)

Books

Croft, T.: "Practical Electricity," 4th ed., McGraw-Hill Book Company, New York.

Morecock, E. M.: "Alternating-current Circuits," McGraw-Hill Book Company, New York.

Oppenheimer and Borchers: "Direct and Alternating Currents," McGraw-Hill Book Company, New York.
Siskind, C. S.: "Electrical Circuits," McGraw-Hill Book Company, New York.
Timbie, W. H.: "Elements of Electricity," John Wiley & Sons, Inc., New York.

Pamphlets and Handbooks

Magnetic Materials, *Allegheny Ludlum Steel Corp.*
National Electrical Code, *National Board of Fire Underwriters.*
Permanent Magnet Handbook, *Crucible Steel Company of America.*
Permanent-Magnet Materials and Their Selection, *Indiana Steel Products.*

Inductance

Chapter 18

Inductance is the ability of a conductor to produce induced voltage when the current varies. A long wire has more inductance than a short wire since more conductor length cut by magnetic flux produces more induced voltage. Similarly, a coil has more inductance than the equivalent length of straight wire because the coil concentrates magnetic flux. Components manufactured to have a definite value of inductance are just coils of wire, therefore, called *inductors*. Figure 18-1 shows typical inductors with their schematic symbols. In (*a*), the air-core rf choke is used to reduce the current for radio frequencies. The iron-core choke in (*b*) is for audio frequencies. Construction, operation, and uses of inductors are explained in the following topics:

18-1   Induction by Alternating Current
18-2   Self-inductance
18-3   Self-induced Voltage v_L
18-4   How v_L Opposes a Change in Current
18-5   Mutual Inductance
18-6   Transformers
18-7   Core Losses
18-8   Types of Cores
18-9   Variable Inductance
18-10 Inductances in Series or Parallel
18-11 Stray Inductance
18-12 Energy in Magnetic Field of Inductance
18-13 Troubles in Coils

18-1
INDUCTION BY ALTERNATING CURRENT

Induced voltage is the result of flux cutting across a conductor. This action can be produced by physical motion of either the magnetic field or the conductor. When the current in a conductor varies in amplitude, however, the variations of current and its associated magnetic field are equivalent to motion of the flux. As the current increases in value, the magnetic field expands outward from the conductor. When the current decreases, the field collapses into the conductor. As the field expands and collapses with changes of current, the flux is effectively in motion. Therefore, a varying current can produce induced voltage without the need for motion of the conductor.

Figure 18-2 illustrates the changes in magnetic field associated with a sine wave of alternating current. Since the alternating current varies in amplitude and reverses in direction, its associated magnetic field has the same variations. At point A, the current is zero and there is no flux. At B, the positive direction of current provides some field lines taken here in

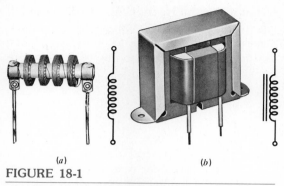

FIGURE 18-1
Typical inductors with their schematic symbols. (*a*) Air-core coil used as rf choke. Length is 2 in. (*b*) Iron-core coil used for 60 Hz. Height is 2 in.

field to expand and collapse again, but the directions are reversed. When the flux expands at points *F* and *G*, the field lines are clockwise, corresponding to current in the negative direction. From *G* to *H* and *I*, this clockwise field collapses into the wire.

The result of an expanding and collapsing field, then, is the same as a field in motion. This moving flux cuts across the conductor that is providing the current, producing induced voltage in the wire itself. Furthermore, any other conductor in the field, whether carrying current or not, also is cut by the varying flux and has induced voltage.

It is important to note that induction by a varying current results from the change in current, not the current value itself. The current must change to provide motion of the flux. A steady direct current of 1000 A, as an example of a large current, cannot produce any induced voltage as long as the current value is constant. A current of 1 μA changing to 2 μA, however, does induce voltage. Also, the faster the current changes, the higher the induced voltage because, when the flux moves at a higher speed, it can induce more voltage.

the counterclockwise direction. Point *C* has maximum current and maximum counterclockwise flux.

At *D* there is less flux than at *C*. Now the field is collapsing because of the reduced current. At *E*, with zero current, there is no magnetic flux. The field can be considered as having collapsed into the wire.

The next half-cycle of current allows the

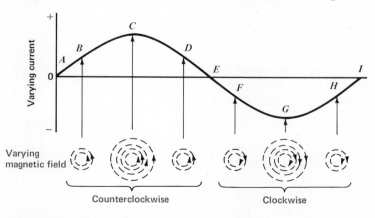

FIGURE 18-2
Magnetic field of an alternating current is effectively in motion as it expands and contracts with the current variations.

Chapter 18

Since inductance is a measure of induced voltage, the amount of inductance has an important effect in any circuit where the current changes. The inductance is an additional characteristic of the circuit besides its resistance. The characteristics of inductance are important in:

1. *Ac circuits.* Here the current is continuously changing and producing induced voltage. Lower frequencies of alternating current require more inductance to produce the same amount of induced voltage as a higher-frequency current.
2. *Dc circuits where the current changes in value.* It is not necessary for the current to reverse in direction. One example is a dc circuit being turned on or off. When the direct current is changing between zero and its steady value, the inductance affects the circuit at the time of switching. This effect with a sudden change is called the *transient response.* A steady direct current that does not change in value is not affected by inductance, however, because there can be no induced voltage without a change in current.

Practice Problems 18-1
(answers on page 378)
(a) In Fig. 18-1, which has more inductance, coil (a) or coil (b)?
(b) In Fig. 18-2, are the changes of current faster at time B or C?

18-2
SELF-INDUCTANCE

The ability of a conductor to induce voltage in itself when the current changes is its *self-inductance* or simply *inductance.* The symbol for inductance is L, for linkages of the magnetic flux, and its unit is the *henry* (H). This unit is named after Joseph Henry (1797–1878).

Definition of the Henry Unit. As illustrated in Fig. 18-3, one henry is the amount of inductance that allows one volt to be induced when the current changes at the rate of one ampere per second. The formula is

$$L = \frac{v_L}{di/dt} \qquad (18\text{-}1)$$

where v_L is in volts and di/dt is the current change in amperes per second.

Again the symbol d is used for *delta* (Δ) to indicate a small change. The factor di/dt for the current variation with respect to time really specifies how fast the current's associated magnetic flux is cutting the conductor to produce v_L.

Example 1. The current in an inductor changes from 12 to 16 A in 1 s. How much is the di/dt rate of current change in amperes per second?

Answer. The di is the difference between 16 and 12, or 4 A/s.

Example 2. The current in an inductor changes by 50 mA in 2 μs. How much is the di/dt rate of current change in amperes per second?

Answer.
$$\frac{di}{dt} = \frac{50 \times 10^{-3}}{2 \times 10^{-6}}$$
$$= 25 \times 10^3$$
$$= 25{,}000 \, \frac{\text{A}}{\text{s}}$$

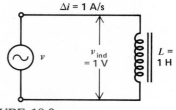

FIGURE 18-3

When a change of 1 A/s induces 1 V across L, its inductance equals 1 H.

Example 3. How much is the inductance of a coil that induces 40 V when its current changes at the rate of 4 A/s?

Answer. $L = \dfrac{v_L}{di/dt} = \dfrac{40}{4} = 10 \text{ H}$

Example 4. How much is the inductance of a coil that induces 1000 V when its current changes at the rate of 50 mA in 2 μs?

Answer. $L = \dfrac{v_L}{di/dt} = \dfrac{v_L \times dt}{di}$

$= \dfrac{1 \times 10^3 \times 2 \times 10^{-6}}{50 \times 10^{-3}}$

$= \dfrac{2 \times 10^{-3}}{50 \times 10^{-3}} = \dfrac{2}{50}$

$L = 0.04 \text{ H}$

Notice in Example 4 that the smaller inductance produces much more v_L than in Example 3. The very fast current change in Example 4 is equivalent to 25,000 A/s.

Inductance of Coils. In terms of physical construction, the inductance depends on how a coil is wound.[1] Note the following factors.

1. A greater number N of turns increases L because more voltage can be induced. Actually L increases in proportion to N^2. Double the number of turns in the same area and length increases the inductance four times.
2. More area A for each turn increases L. This means a coil with larger turns has more inductance. The L increases in direct proportion to A and as the square of the diameter of each turn.
3. L increases with the permeability of the core. For an air core μ_r is 1. With a magnetic core, L is increased by the μ_r factor as the magnetic flux is concentrated in the coil.
4. L decreases with more length for the same turns, as the magnetic field then is less concentrated.

These physical characteristics of a coil are illustrated in Fig. 18-4. For a long coil, where the length is at least ten times the diameter, the inductance can be calculated from the formula

$$L = \mu_r \times \dfrac{N^2 \times A}{l} \times 1.26 \times 10^{-6} \text{ H} \qquad (18\text{-}2)$$

where l is in meters and A is in square meters. The constant factor 1.26×10^{-6} is the absolute permeability of air or vacuum, in SI units, to calculate L in henrys.

For the air-core coil in Fig. 18-4,

$L = 1 \times \dfrac{10^4 \times 2 \times 10^{-4}}{0.2} \times 1.26 \times 10^{-6}$

$= 12.6 \times 10^{-6} \text{ H}$

$L = 12.6 \text{ μH}$

This value means that the coil can produce a self-induced voltage of 12.6 μV when its current changes at the rate of 1 A/s, as $v_L = L\,(di/dt)$. Furthermore, if the coil has an iron core with $\mu_r = 100$, then L will be 100 times greater.

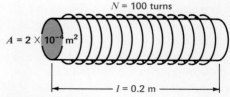

FIGURE 18-4
Physical factors for inductance of a coil. See text for calculating L.

---
[1] Methods of winding coils for a specific L are described in the *A.R.R.L. Handbook* and in *Bulletin 74* of the National Bureau of Standards.

Typical Coil Inductance Values. Air-core coils for rf applications have L values in millihenrys (mH) and microhenrys (μH). Note that

$$1 \text{ mH} = 1 \times 10^{-3} \text{ H}$$
$$1 \text{ }\mu\text{H} = 1 \times 10^{-6} \text{ H}$$

For example, an rf coil for the radio broadcast band of 540 to 1605 kHz has an inductance L of 250 μH or 0.250 mH. Iron-core inductors for the 60-Hz power line and for audio frequencies have inductance values of about 1 to 25 H.

Practice Problems 18-2
(answers on page 378)
(a) A coil induces 2 V with di/dt of 1 A/s. How much is L?
(b) A coil has L of 8 mH with 125 turns. If the number of turns is doubled, how much will L be?

18-3
SELF-INDUCED VOLTAGE v_L

The self-induced voltage across an inductance L produced by a change in current di/dt can be stated as

$$v_L = L \frac{di}{dt} \quad (18\text{-}3)$$

where v_L is in volts, L in henrys, and di/dt in amperes per second. This formula is just an inverted version of $L = v_L/(di/dt)$, giving the definition of inductance.

Actually both versions are based on Formula (15-1): $v = N(d\phi/dt)$ for magnetism. This gives the voltage in terms of how much magnetic flux is cut per second. When the magnetic flux associated with the current varies the same as i, then Formula (18-3) gives the same results for calculating induced voltage. Remember also that the induced voltage across the coil is actually the result of inducing electrons to move in the conductor, so that there is also an induced current.

Example 5. How much is the self-induced voltage across a 4-H inductance produced by a current change of 12 A/s?

Answer. $v_L = L \dfrac{di}{dt} = 4 \times 12$

$v_L = 48$ V

Example 6. The current through a 200-mH L changes from 0 to 100 mA in 2 μs. How much is v_L?

Answer. $v_L = L \dfrac{di}{dt}$

$= 200 \times 10^{-3} \times \dfrac{100 \times 10^{-3}}{2 \times 10^{-6}}$

$v_L = 10{,}000$ V

Note the high voltage induced in the 200-mH inductance because of the fast change in current.

The induced voltage is an actual voltage that can be measured, although v_L is produced only while the current is changing. When di/dt is present for only a short time, v_L is in the form of a voltage pulse. With a sine-wave current, which is always changing, v_L is a sinusoidal voltage 90° out of phase with i_L.

Practice Problems 18-3
(answers on page 378)
(a) L is 2 H and di/dt is 1 A/s. How much is v_L?
(b) For the same coil, di/dt is 100 A/s. How much is v_L?

18-4
HOW v_L OPPOSES A CHANGE IN CURRENT

By Lenz' law, the induced voltage must oppose the change of current that induces v_L. The polarity of v_L, therefore, depends on the direction of the current variation *di*. When *di* increases, v_L has the polarity that opposes the increase of current; when *di* decreases, v_L has the opposite polarity to oppose the decrease of current.

In both cases, the change of current is opposed by the induced voltage. Otherwise, v_L could increase to an unlimited amount without the need for adding any work. *Inductance, therefore, is the characteristic that opposes any change in current.* This is the reason why an induced voltage is often called a *counter emf* or *back emf*.

More details of applying Lenz' law to determine the polarity of v_L in a circuit are illustrated in Fig. 18-5. Note the directions carefully. In (*a*), the electron flow is into the top of the coil. This current is increasing. By Lenz' law v_L must have the polarity needed to oppose the increase. The induced voltage shown with the top side negative opposes the increase in current. The reason is that this polarity of v_L can produce current in the opposite direction, from minus to plus in the external circuit. Note that for this opposing current, v_L is the generator. This action tends to keep the current from increasing.

In (*b*), the source is still producing electron flow into the top of the coil but *i* is decreasing, because the source voltage is decreasing. By Lenz' law, v_L must have the polarity needed to oppose the decrease in current. The induced voltage shown with the top side positive now opposes the decrease. The reason is that this polarity of v_L can produce current in the same direction, tending to keep the current from decreasing.

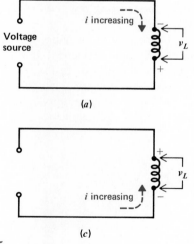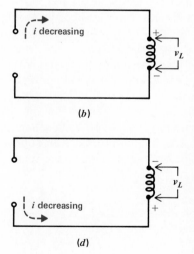

FIGURE 18-5

Determining the polarity of v_L that opposes the change in *i*. (*a*) Amount of *i* is increasing, and v_L produces an opposing current. (*b*) Amount of *i* is decreasing, and v_L produces an aiding current. (*c*) The *i* is increasing but in the opposite direction. (*d*) Same direction of *i* as in (*c*) but decreasing values.

In (c), the voltage source reverses polarity to produce current in the opposite direction, with electron flow into the bottom of the coil. This reversed direction of current is now increasing. The polarity of v_L must oppose the increase. As shown, now the bottom of the coil is made negative by v_L to produce current opposing the source current. Finally, in (d) the reversed current is decreasing. This decrease is opposed by the polarity shown for v_L to keep the current flowing in the same direction as the source current.

Notice that the polarity of v_L reverses for either a reversal of direction for i, or a reversal of change in di between increasing or decreasing values. When both the direction of the current and the direction of change are reversed, as in a comparison of (a) and (d), the polarity of v_L is the same.

Sometimes the formulas for induced voltage are written with a minus sign, in order to indicate the fact that v_L opposes the change, as specified by Lenz' law. However, the negative sign is omitted here so that the actual polarity of the self-induced voltage can be determined in typical circuits.

In summary, Lenz' law states that the reaction v_L opposes its cause, which is the change in i. When i is increasing, v_L produces an opposing current. When i is decreasing, v_L produces an aiding current.

Practice Problems 18-4
(answers on page 378)
Answer true or false. Refer to Fig. 18-5.
(a) In (a) and (b) the polarity of v_L is opposite.
(b) In (b) and (c) the polarity of v_L is the same.

18-5
MUTUAL INDUCTANCE

When the current in an inductor changes, the varying flux can cut across any other inductor nearby, producing induced voltage in both inductors. In Fig. 18-6, the coil L_1 is connected to a generator that produces varying current in the turns. The winding L_2 is not connected to L_1, but the turns are linked by the magnetic field. A varying current in L_1, therefore, induces voltage across L_1 and across L_2. If all the flux of the current in L_1 links all the turns of the coil L_2, each turn in L_2 will have the same amount of induced voltage as each turn in L_1. Furthermore, the induced voltage v_{L_2} can produce current in a load resistance connected across L_2.

When the induced voltage produces current in L_2, its varying magnetic field induces voltage in L_1. The two coils L_1 and L_2 have mutual inductance, therefore, because current in one can induce voltage in the other.

The unit of mutual inductance is the henry, and the symbol is L_M. *Two coils have L_M of one henry when a current change of one ampere per second in one coil induces one volt in the other coil.*

The schematic symbol for two coils with mutual inductance is shown in Fig. 18-7a for an air core, with an iron core in (b). Iron increases the mutual inductance, since it concentrates magnetic flux. Any magnetic lines that do not link the two coils result in *leakage flux*.

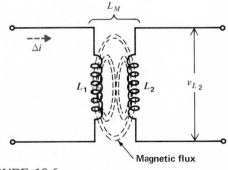

FIGURE 18-6

Mutual inductance L_M between L_1 and L_2 linked by magnetic flux.

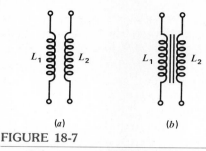

FIGURE 18-7

Schematic symbols for two coils with mutual inductance. (a) Air core. (b) Iron core.

Coefficient of Coupling. The fraction of total flux from one coil linking another coil is the coefficient of coupling k between the two coils. As examples, if all the flux of L_1 in Fig. 18-6 links L_2, then k equals 1, or unity coupling; if half the flux of one coil links the other, k equals 0.5. Specifically, the coefficient of coupling is

$$k = \frac{\text{flux linkages between } L_1 \text{ and } L_2}{\text{flux produced by } L_1}$$

There are no units for k, as it is just a ratio of two values of magnetic flux. The value of k is generally stated as a decimal fraction, like 0.5, rather than percent.

The coefficient of coupling is increased by placing the coils close together, possibly with one wound on top of the other, by placing them in parallel rather than perpendicular to each other, or by winding the coils on a common iron core. Several examples are shown in Fig. 18-8.

A high value of k, called *tight coupling*, allows the current in one coil to induce more voltage in the other coil. *Loose coupling*, with a low value of k, has the opposite effect. In the extreme case of zero coefficient of coupling, there is no mutual inductance. Two coils may be placed perpendicular to each other and far apart for essentially zero coupling when it is desired to minimize interaction between the coils.

Air-core coils wound on one form have values of k equal to 0.05 to 0.3, approximately, corresponding to 5 to 30 percent linkage. Coils on a common iron core can be considered to have practically unity coupling, with k equal to 1. As shown in Fig. 18-8c, for both windings L_1 and L_2 practically all the magnetic flux is in the common iron core.

Example 7. A coil L_1 produces 80 μWb of magnetic flux. Of this total flux, 60 μWb are linked with L_2. How much is k between L_1 and L_2?

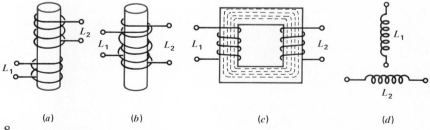

FIGURE 18-8

Examples of coupling between two coils linked by L_M. (a) L_1 and L_2 on paper or plastic form with air core; k is 0.1. (b) L_1 wound over L_2 for tighter coupling; k is 0.3. (c) L_1 and L_2 on the same iron core; k is 1. (d) Zero coupling between perpendicular air-core coils.

Answer. $k = \dfrac{60 \ \mu\text{Wb}}{80 \ \mu\text{Wb}}$

$k = 0.75$

Example 8. A 10-H inductance L_1 on an iron core produces 4 Wb of magnetic flux. Another coil L_2 is on the same core. How much is k between L_1 and L_2?

Answer. Unity or 1. All coils on a common iron core have practically perfect coupling.

Calculating L_M. The mutual inductance increases with higher values for the primary and secondary inductances and tighter coupling:

$$L_M = k \sqrt{L_1 \times L_2} \qquad \text{H} \qquad (18\text{-}4)$$

where L_1 and L_2 are the self-inductance values of the two coils, k is the coefficient of coupling, and L_M is the mutual inductance linking L_1 and L_2, in the same units as L_1 and L_2.

As an example, suppose that $L_1 = 2$ H and $L_2 = 8$ H, with both coils on an iron core for unity coupling. Then the mutual inductance is

$$L_M = 1 \sqrt{2 \times 8} = \sqrt{16} = 4 \text{ H}$$

The value of 4 H for L_M in this example means that when the current changes at the rate of 1 A/s in either coil, it will induce 4 V in the other coil.

Example 9. Two 400-mH coils L_1 and L_2 have a coefficient of coupling k equal to 0.2. Calculate L_M.

Answer. $L_M = k \sqrt{L_1 \times L_2}$
$= 0.2 \sqrt{400 \times 10^{-3} \times 400 \times 10^{-3}}$
$= 0.2 \times 400 \times 10^{-3}$
$= 80 \times 10^{-3}$
$L_M = 80$ mH

Example 10. If the above two coils had a mutual inductance L_M of 40 mH, how much would k be? (Note: Invert Formula (18-4) to find k.)

Answer. $k = \dfrac{L_M}{\sqrt{L_1 \times L_2}}$

$= \dfrac{40 \times 10^{-3}}{\sqrt{400 \times 10^{-3} \times 400 \times 10^{-3}}}$

$= \dfrac{40 \times 10^{-3}}{400 \times 10^{-3}}$

$k = 0.1$

Notice that the same two coils have one-half the mutual inductance L_M, because the coefficient of coupling k is 0.1 instead of 0.2.

Practice Problems 18-5
(answers on page 378)

(a) All the flux from the current in L_1 links L_2. How much is the coefficient of coupling k?
(b) L_M is 9 mH with k of 0.2. If k is doubled to 0.4, how much will L_M be?

18-6
TRANSFORMERS

The transformer is an important application of mutual inductance. As shown in Fig. 18-9, a transformer has the primary winding L_P connected to a voltage source that produces alternating current, while the secondary winding L_S is connected across the load resistance R_L. The purpose of the transformer is to transfer power from the primary, where the generator is connected, to the secondary, where the induced secondary voltage can produce current in the load resistance connected to L_S.

Although the primary and secondary are not connected to each other, power in the primary is coupled into the secondary by the magnetic field linking the two windings. The transformer is used to provide power for the load resistance R_L, instead of connecting R_L directly

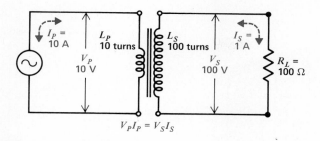

FIGURE 18-9
Iron-core transformer with 10:1 turns ratio. Primary current I_P induces secondary voltage V_S that produces current in the secondary load R_L.

across the generator, whenever the load requires an ac voltage higher or lower than the generator voltage. By having more or fewer turns in L_S, compared with L_P, the transformer can step up or step down the generator voltage to provide the required amount of secondary voltage. Typical transformers are shown in Figs. 18-10 and 18-11. It should be noted that a steady dc voltage cannot be stepped up or down by a transformer, because a steady current cannot produce induced voltage.

Turns Ratio. The ratio of the number of turns in the secondary to the number in the primary is the turns ratio of the transformer:

$$\text{Turns ratio} = \frac{N_S}{N_P} \qquad (18\text{-}5)$$

For example, 500 turns in the secondary and 50 turns in the primary provide a turns ratio of $^{500}\!/_{50}$, or 10:1.

Voltage Ratio. With unity coupling between primary and secondary, the voltage induced in each turn of the secondary is the same as the self-induced voltage of each turn in the primary. Therefore, the voltage ratio is in the same proportion as the turns ratio:

$$\frac{V_S}{V_P} = \frac{N_S}{N_P} \qquad (18\text{-}6)$$

When the secondary has more turns, the secondary voltage is higher and the primary voltage is stepped up. This principle is illustrated in Fig. 18-9 with a step-up ratio of $^{100}\!/_{10}$, or 10:1. When the secondary has fewer turns, the voltage is stepped down. In either case, the ratio is in terms of the primary voltage, which may be stepped up or down in the same proportion as the turns ratio.

These calculations apply only to iron-core transformers with unity coupling. Air-core

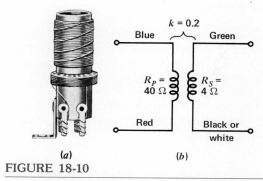

FIGURE 18-10
(a) Air-core rf transformer. Height is 2 in.
(b) Color code and dc resistance of windings.

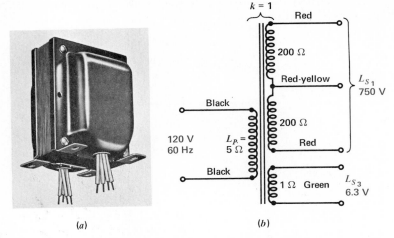

FIGURE 18-11

(a) Iron-core power transformer. Height is 5 in.
(b) Color code and dc resistance of windings.

transformers for rf circuits are generally tuned to resonance. In this case, the resonance factor is considered instead of the turns ratio.

Example 11. A power transformer has 100 turns for L_P and 600 turns for L_S. What is the turns ratio? How much is the secondary voltage V_S with a primary voltage V_P of 120 V?

Answer. The turns ratio is $600/100$, or $6:1$. Therefore, V_P is stepped up by the factor 6, making V_S equal to 6×120, or 720 V.

Example 12. A power transformer has 100 turns for L_P and 5 turns for L_S. What is the turns ratio? How much is the secondary voltage V_S with a primary voltage of 120 V?

Answer. The turns ratio is $5/100$, or $1:20$. V_P is stepped down by the factor $1/20$, therefore making V_S equal to $120/20$, or 6 V.

Secondary Current. By Ohm's law, the amount of secondary current equals the secondary voltage divided by the resistance in the secondary circuit. In Fig. 18-9, with a value of 100 Ω for R_L and negligible coil resistance assumed,

$$I_S = \frac{V_S}{R_L} = \frac{100 \text{ V}}{100 \text{ Ω}} = 1 \text{ A}$$

The power dissipated in R_L is $I_S{}^2 R_L$ or $V_S I_S$, which equals 100 W in this example.

The power used by the secondary load resistance is supplied by the generator in the primary. With current in the secondary winding, its magnetic field opposes the varying flux of the primary current. The generator must then produce more primary current to maintain the self-induced voltage across L_P and the secondary voltage developed in L_S by mutual induction. If the secondary current doubles, for instance, because the load resistance is reduced one-half, the primary current will also double in value to provide the required power for the secondary. Therefore, the effect of the secondary-load power on the generator is the same as though R_L were in the primary, except that in the secondary the voltage for R_L is stepped up or down by the turns ratio.

Current Ratio. With zero losses assumed for the transformer, the power in the secondary equals the power in the primary:

$$V_P I_P = V_S I_S \qquad (18\text{-}7)$$

or

$$\frac{I_P}{I_S} = \frac{V_S}{V_P} \qquad (18\text{-}8)$$

The current ratio is the inverse of the voltage ratio; that is, voltage step-up in the secondary means current step-down, and vice versa. The secondary does not generate power but only takes it from the primary. Therefore, the current step-up or step-down is in terms of the secondary current I_S, which is determined by the load resistance across the secondary voltage. These points are illustrated by the following two examples.

Example 13. A transformer with a 6:1 voltage step-up ratio has 720 V across 7200 Ω in the secondary. (a) How much is I_S? (b) How much is I_P?

Answer.

(a) $\quad I_S = \dfrac{V_S}{R_L} = \dfrac{720 \text{ V}}{7200 \text{ Ω}}$

$\quad I_S = 0.1 \text{ A}$

(b) With a turns and voltage ratio of 6:1, the current ratio is 1:6. Therefore,

$\quad I_S = \frac{1}{6} \times I_P$

or

$\quad I_P = 6 \times I_S = 6 \times 0.1$
$\quad I_P = 0.6 \text{ A}$

Example 14. A transformer with a 1:20 voltage step-down ratio has 6 V across 0.6 Ω in the secondary. (a) How much is I_S? (b) How much is I_P?

Answer.

(a) $\quad I_S = \dfrac{V_S}{R_L} = \dfrac{6 \text{ V}}{0.6 \text{ Ω}}$

$\quad I_S = 10 \text{ A}$

(b) $\quad I_P = \frac{1}{20} \times I_S$
$\quad = \frac{1}{20} \times 10$
$\quad I_P = 0.5 \text{ A}$

As an aid in these calculations, remember that the side with more voltage has less current. Both V and I are in the same proportion as the number of turns.

Total Secondary Power Equals Primary Power. Figure 18-12 illustrates a power transformer with two secondary windings L_1 and L_2. There can be one, two, or more secondary windings with unity coupling to the primary as long as all the windings are on the same iron core. Each secondary winding has induced voltage in proportion to its turns ratio with the primary winding, which is connected across the 120-V source.

The secondary winding L_1 has a turns ratio of 6:1, providing 720 V. The 7200-Ω load resistance R_1, across L_1, allows the 720 V to produce 0.1 A for I_1 in this secondary circuit. The power here is 720 V × 0.1 A, therefore, which equals 72 W.

The other secondary winding L_2 provides voltage step-down, with the ratio 1:20, resulting in 6 V for R_2. The 0.6-Ω load resistance in this circuit allows 10 A for I_2. Therefore, the power here is 6 V × 10 A, or 60 W. Since the windings have separate connections, each can have its individual values of voltage and current.

All the power used in the secondary circuits is supplied by the primary, however. In this

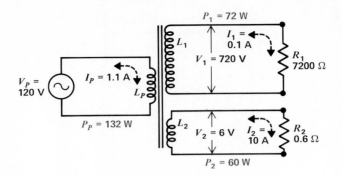

FIGURE 18-12
Total power used by the two secondary loads equals the power supplied by the source in the primary.

example, the total secondary power is 132 W, equal to 72 W for R_1 and 60 W for R_2. The power supplied by the 120-V source in the primary then is $72 + 60 = 132$ W.

The primary current I_P equals the primary power P_P divided by the primary voltage V_P. This is 132 W divided by 120 V, which equals 1.1 A for the primary current. The same value can be calculated as the sum of 0.6 A of primary current providing power for L_1, plus 0.5 A of primary current for L_2, resulting in the total of 1.1 A as the value of I_P.

This example shows how to analyze a loaded power transformer. The main idea is that the primary current depends on the secondary load. The calculations can be summarized as follows:

1. Calculate V_S from the turns ratio and V_P.
2. Use V_S to calculate $I_S = V_S/R_L$.
3. Use I_S to calculate $P_S = V_S \times I_S$.
4. Use P_S to find $P_P = P_S$.
5. Finally, I_P can be calculated as P_P/V_P.

With more than one secondary, calculate each I_S and P_S. Then add for the total secondary power, which equals the primary power.

Autotransformers. As illustrated in Fig. 18-13, an autotransformer consists of one continuous coil with a tapped connection such as terminal 2 between the ends at terminals 1 and 3. In (a) the autotransformer steps up the generator voltage. V_P between 1 and 2 is connected across part of the total turns, while V_S is induced across all the turns. With six times the turns for the secondary voltage, V_S is six times V_P.

In (b) the autotransformer steps down the primary voltage connected across the entire coil. Then the secondary voltage is taken across less than the total turns.

The winding that connects to the voltage source to supply power is the primary, while the secondary is across the load resistance R_L. The turns ratio and voltage ratio apply the same way as in a conventional transformer having an isolated secondary winding.

Autotransformers are used often because they are compact, efficient, and usually cost less with only one winding. However, the same wire size must be suitable for both the primary and secondary. Note that the autotransformer in Fig. 18-13 has only three leads, compared with four leads for the transformer in Fig. 18-9 with an isolated secondary.

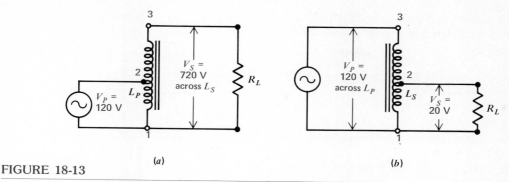

FIGURE 18-13
Autotransformer with tap at terminal 2 for 10 turns of 60-turn winding. (a) V_P between 1 and 2 stepped up across 1 and 3. (b) V_P between 1 and 3 stepped down across 1 and 2.

Isolation of the Secondary. In a transformer with a separate winding for L_S, as in Fig. 18-9, the secondary load is not connected directly to the ac power line in the primary. This isolation is an advantage in reducing the chance of electric shock. With an autotransformer, as in Fig. 18-13, the secondary is not isolated. Another advantage of an isolated secondary is the fact that any direct current in the primary is blocked from the secondary. Sometimes a transformer with a 1:1 turns ratio is used just for isolation from the ac power line.

Transformer Efficiency. Efficiency is defined as the ratio of power out to power in. Stated as a formula,

$$\text{Efficiency} = \frac{P_{\text{out}}}{P_{\text{in}}} \times 100\% \qquad (18\text{-}9)$$

For example, when the power out in watts equals one-half the power in, the efficiency is one-half, which equals 0.5×100 percent, or 50 percent. In a transformer, power out is secondary power, while power in is primary power.

Assuming zero losses in the transformer, power out equals power in and the efficiency is 100 percent. Power transformers actually, however, have an efficiency slightly less than 100 percent. The efficiency is approximately 80 to 90 percent for power transformers in receivers, with a power rating of 50 to 300 W. Transformers for higher power are more efficient because they require heavier wire, which has less resistance. In a transformer less than 100 percent efficient, the primary supplies more than the secondary power. The primary power missing from the output is dissipated as heat in the transformer.

Transformer Color Codes. The colors of the leads show the required connections in electronic circuits. For the rf transformer in Fig. 18-10, the leads are:

Blue—Output electrode of tube or transistor amplifier
Red—Dc supply voltage for this electrode
Green—Input electrode of next amplifier
Black or white—Return line of secondary winding

This system applies to all coupling transformers between amplifier stages, including iron-core transformers for audio circuits.

For the power transformer in Fig. 18-11, the primary is connected to the ac power line. The leads are:

Black—Primary leads without tap
Black with yellow—Tap on primary
Red—High voltage secondary to rectifier in power supply
Red with yellow—Tap on high-voltage secondary
Green with yellow—Low-voltage secondary for tube heaters

Practice Problems 18-6
(answers on page 378)
(a) A power transformer connected to the 120-V ac line has a turns ratio of 2:1. Calculate V_S.
(b) V_S is connected across a 2400-Ω R_L. Calculate I_S.

18-7
CORE LOSSES

The fact that the magnetic core can become warm, or even hot, shows that some of the energy supplied to the coil is used up in heat. The two main effects are eddy-current losses and hysteresis losses.

Eddy Currents. In any inductance with an iron core, alternating current induces voltage in the core itself. Since it is a conductor, the iron core has current produced by the induced voltage. This current is called an *eddy current* because it flows in a circular path through the cross section of the core, as illustrated in Fig. 18-14.

The eddy currents represent wasted power dissipated as heat in the core, equal to I^2R, where R is the resistance of the core. Note in Fig. 18-14 that the eddy-current flux opposes the coil flux, requiring more current in the coil to maintain its magnetic field. The higher the frequency of the alternating current in the inductance, the greater the eddy-current loss.

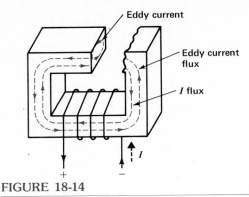

FIGURE 18-14

Cross-sectional view of iron core showing eddy currents.

Eddy currents can be induced in any conductor near a coil with alternating current, not only in its core. For instance, a coil has eddy-current losses in a metal cover. In fact, the technique of induction heating is an application of heat resulting from induced eddy currents.

Rf Shielding. The reason why a coil may have a metal cover, usually copper or aluminum, is to provide a shield against the varying flux of rf current. In this case, the shielding effect depends on using a good conductor for the eddy currents produced by the varying flux, rather than the magnetic materials used for shielding against static magnetic flux.

The shield cover not only isolates the coil from external varying magnetic fields, but also minimizes the effect of the coil's rf current for external circuits. The reason why the shield helps both ways is the same, as the induced eddy currents have a field that opposes the field that is inducing the current. It should be noted that the clearance between the sides of the coil and the metal should be equal to or greater than the coil radius, to minimize the effect of the shield in reducing the inductance.

Hysteresis Losses. Another factor with a mag-

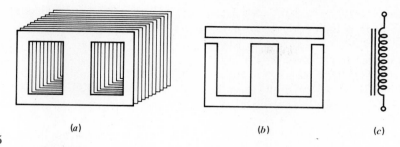

FIGURE 18-15

Laminated iron core. (a) Shell-type construction. (b) E- and I-shaped laminations. (c) Symbol for iron core.

netic core for rf coils is hysteresis losses, although these are not so great as eddy-current losses. The hysteresis losses result from the additional power needed to reverse the magnetic field in magnetic materials with rf alternating current.

Air-Core Coils. It should be noted that air has practically no losses from eddy currents or hysteresis. However, the inductance for small coils with an air core is limited to low values in the µH or mH range.

Practice Problems 18-7
(answers on page 378)
(a) Which has greater eddy-current losses, an iron core or an air core?
(b) Which produces more hysteresis losses, 60 Hz or 60 MHz?

18-8
TYPES OF CORES

In order to minimize losses while maintaining high flux density, the core can be made of laminated sheets insulated from each other, or insulated powdered-iron granules and ferrite materials can be used. These core types are illustrated in Figs. 18-15 and 18-16. The purpose is to reduce the amount of eddy currents.

Laminated Core. Figure 18-15a shows a shell-type core formed with a group of individual laminations. Each laminated section is insulated by a very thin coating of iron oxide and varnish. The insulating borders increase the resistance in the cross section of the core to reduce the eddy currents but allow a low-reluctance path for high flux density around the core. Transformers for audio frequencies and 60-Hz

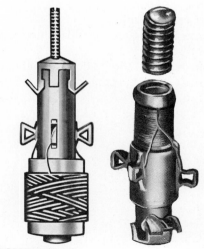

FIGURE 18-16

Rf coils with ferrite core. Width of coil is ½ in. (a) Variable L from 1 to 3 mH. (b) Tuning coil for 40 MHz.

Chapter 18
Inductance

power are generally made with a laminated iron core.

Powdered-Iron Core. To reduce eddy currents in the iron core of an inductance for radio frequencies, powdered iron is generally used. It consists of individual insulated granules pressed into one solid form called a *slug*.

Ferrite Core. The ferrites are synthetic ceramic materials that are ferromagnetic. They provide high values of flux density, like iron, but with the advantage of being insulators. Therefore, a ferrite core can be used for high frequencies with minimum eddy-current losses.

This core is usually a slug that can move in or out of the coil to vary L. In Fig. 18-16a, the screw at the top moves the core; in (b) the core has a hole to fit a plastic alignment tool for tuning the coil. Maximum L results with the slug in the coil.

Practice Problems 18-8
(answers on page 378)
Answer true or false.
(a) An iron core provides a coefficient of coupling k of 1.
(b) A laminated iron core reduces eddy-current losses.
(c) The ferrites have less eddy-current losses than iron.

18-9
VARIABLE INDUCTANCE

The inductance of a coil can be varied by one of the methods illustrated in Fig. 18-17. In (a), more or fewer turns can be used by connection to one of the taps on the coil. In (b), a slider contacts the coil to vary the number of turns used. These methods are for large coils. Note that the unused turns are short-circuited to prevent the tapped coil from acting as an autotransformer. The reason is that stepped-up voltage could cause arcing across the turns.

Figure 18-17c shows the schematic symbol for a coil with a slug of powdered iron or ferrite. The dotted lines indicate the core is not solid iron. The arrow shows the slug is variable. Usually, an arrow at the top means the adjustment is at the top of the coil. An arrow at the bottom, pointing down, shows the adjustment is at the bottom.

The symbol in (d) is a *variometer*, which is an arrangement for varying the position of one coil within the other. The total inductance of the series-aiding coils is minimum when they are perpendicular.

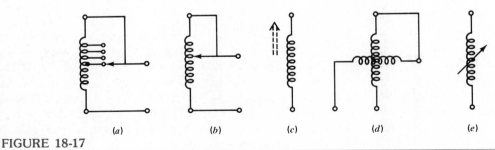

FIGURE 18-17
Methods of varying inductance. (a) Tapped coil. (b) Slider contact. (c) Adjustable slug. (d) Variometer. (e) Symbol for variable L.

For any method of varying L, the coil with an arrow in (e) can be used. However, an adjustable slug is usually shown as in (c).

A practical application of variable inductance is the *Variac* in Fig. 18-18. This unit is an autotransformer with a variable tap to change the turns ratio. The output voltage in the secondary can be varied from 0 to 140 V, with input from the 120-V 60-Hz power line. One use is to test equipment with voltage above or below the normal line voltage.

The Variac is plugged into the power line, and the equipment to be tested is plugged into the Variac. Note that the power rating of the Variac should be equal to or more than the power used by the equipment being tested.

Practice Problems 18-9
(*answers on page 378*)
Answer true or false.
(a) The Variac is an autotransformer with a variable tap for the primary.
(b) Figure 18-17c shows a ferrite core.

FIGURE 18-18

Variac rated at 300 W. Length is 5 in. (*General Radio Corp.*)

18-10
INDUCTANCES IN SERIES OR PARALLEL

As shown in Fig. 18-19, the total inductance of coils connected in series is the sum of the individual L values, as for series R. Since the series coils have the same current, the total induced voltage is a result of the total number of turns. Therefore, in series,

$$L_T = L_1 + L_2 + L_3 + \cdots + \text{etc.} \quad (18\text{-}10)$$

where L_T is in the same units of inductance as L_1, L_2, and L_3. This formula assumes no mutual induction between the coils.

Example 15. L_1 in Fig. 18-19 is 5 mH and L_2 is 10 mH. How much is L_T?

Answer.
$$L_T = 5 \text{ mH} + 10 \text{ mH}$$
$$L_T = 15 \text{ mH}$$

With coils connected in parallel, the total inductance is calculated from the reciprocal formula

$$\frac{1}{L_T} = \frac{1}{L_1} + \frac{1}{L_2} + \frac{1}{L_3} + \cdots + \text{etc.} \quad (18\text{-}11)$$

Again, no mutual induction is assumed, as illustrated in Fig. 18-20.

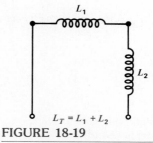

FIGURE 18-19

Inductances in series without mutual coupling.

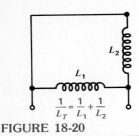

FIGURE 18-20
Inductances in parallel without mutual coupling.

Example 16. L_1 and L_2 in Fig. 18-20 are each 8 mH. How much is L_T?

Answer. $\dfrac{1}{L_T} = \dfrac{1}{8} + \dfrac{1}{8} = \dfrac{2}{8}$

$L_T = \dfrac{8}{2} = 4 \text{ mH}$

All the shortcuts for calculating parallel R can be used with parallel L, since both are based on the reciprocal formula. In this example L_T is $\frac{1}{2} \times 8 = 4$ mH.

Series Coils with L_M. This case depends on the amount of mutual coupling and on whether the coils are connected series-aiding or series-opposing. *Series-aiding* means that the common current produces the same direction of magnetic field for the two coils. The *series-opposing* connection results in opposite fields.

The coupling depends on the coil connections and direction of winding. Reversing either one reverses the field. In Fig. 18-21, L_1 and L_2 with the same direction of winding are connected series-aiding in (*a*). However, they are series-opposing in (*b*) because L_1 is connected to the opposite end of L_2.

To calculate the total inductance of two coils that are series-connected and have mutual inductance,

$$L_T = L_1 + L_2 \pm 2L_M \qquad (18\text{-}12)$$

The mutual inductance L_M is plus, increasing the total inductance, when the coils are series-aiding, or minus when they are series-opposing to reduce the total inductance.

Note the large dots just above the coils in Fig. 18-21. This method is generally used to indicate the sense of the windings without the need for showing the actual physical construction. Coils with dots at the same end have the same direction of winding. When current enters

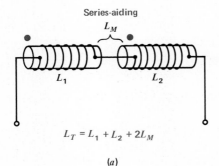

FIGURE 18-21
L_1 and L_2 in series but with mutual coupling L_M. (*a*) Aiding magnetic fields. (*b*) Opposing magnetic fields.

the dotted ends for two coils, their fields are aiding and L_M has the same sense as L.

How to Measure L_M. Formula (18-12) provides a method of determining the mutual inductance between two coils L_1 and L_2 of known inductance. First, the total inductance is measured for the series-aiding connection. Let this be L_{T_a}. Then the connections to one coil are reversed to measure the total inductance for the series-opposing coils. Let this be L_{T_o}. Then

$$L_M = \frac{L_{T_a} - L_{T_o}}{4} \qquad (18\text{-}13)$$

When the mutual inductance is known, the coefficient of coupling k can be calculated from the fact that $L_M = k\sqrt{L_1 L_2}$.

Example 17. Two series coils, each with an L of 250 µH, have a total inductance of 550 µH connected series-aiding and 450 µH series-opposing. (a) How much is the mutual inductance L_M between the two coils? (b) How much is the coupling coefficient k?

Answer.

(a) $L_M = \dfrac{L_{T_a} - L_{T_o}}{4} = \dfrac{550 - 450}{4} = \dfrac{100}{4}$

$L_M = 25 \ \mu H$

(b) $L_M = k\sqrt{L_1 L_2}$

or

$k = \dfrac{L_M}{\sqrt{L_1 L_2}} = \dfrac{25}{\sqrt{250 \times 250}} = \dfrac{25}{250} = \dfrac{1}{10}$

$k = 0.1$

Coils may also be in parallel with mutual coupling. However, the inverse relations with parallel connections and the question of aiding or opposing fields make this case complicated. Actually, it would hardly ever be used.

Practice Problems 18-10
(answers on page 378)

(a) A 500-µH coil and a 1-mH coil are in series without L_M. Calculate L_T.
(b) The same coils are in parallel without L_M. Calculate L_T.

18-11
STRAY INDUCTANCE

Although practical inductors are generally made as coils, all conductors have inductance. The amount of L is $v_L/(di/dt)$, as with any inductance producing induced voltage when the current changes. The inductance of any wiring not included in the conventional inductors can be considered stray inductance. In most cases, the stray inductance is very small, typical values being less than 1 µH. For high radio frequencies, though, even a small L can have an appreciable inductive effect.

One source of stray inductance is the connecting leads. A wire of 0.04 in diameter and 4 in long has an L of approximately 0.1 µH. At low frequencies, this inductance is negligible. However, consider the case of rf current where i varies from a 0- to 20-mA peak value in the short time of 0.025 µs for a quarter-cycle of a 10-MHz sine wave. Then v_L equals 80 mV, which is an appreciable inductive effect. This is one reason why the connecting leads must be very short in rf circuits.

As another example, wirewound resistors can have appreciable inductance when wound as a straight coil. This is why carbon resistors are preferred for minimum stray inductance in rf circuits. However, noninductive wirewound resistors can also be used. These are wound in such a way that adjacent turns have current in opposite directions, so that the magnetic fields oppose each other to cancel the inductance. Another application of this technique is twisting a pair of connecting leads to reduce the inductive effect.

Practice Problems 18-11
(answers on page 378)
Answer true or false.
(a) A straight wire 1 ft long can have L less than 1 μH.
(b) Carbon resistors have less L than wire-wound resistors.

18-12
ENERGY IN MAGNETIC FIELD OF INDUCTANCE

Magnetic flux associated with current in an inductance has electrical energy supplied by the voltage source producing the current. The energy is stored in the field, since it can do the work of producing induced voltage when the flux moves. The amount of electrical energy stored is

$$\text{Energy} = \mathcal{E} = \tfrac{1}{2}LI^2 \quad \text{J} \tag{18-14}$$

The factor of $\tfrac{1}{2}$ gives the average result of I in producing energy. With L in henrys and I in amperes, the energy is in watt-seconds, or joules. For a 10-H L with a 3-A I, the electrical energy stored in the magnetic field equals

$$\text{Energy} = \tfrac{1}{2}LI^2 = \frac{10 \times 9}{2} = 45 \text{ J}$$

This 45 J of energy is supplied by the voltage source that produces 3 A in the inductance. When the circuit is opened, the magnetic field collapses. The energy in the collapsing magnetic field is returned to the circuit in the form of induced voltage, which tends to keep the current flowing.

The entire 45 J is available for the work of inducing voltage, since no energy is dissipated by the magnetic field. With resistance in the circuit, however, the I^2R loss with induced current dissipates all the energy after a period of time.

Practice Problems 18-12
(answers on page 378)
(a) What is the unit of energy?
(b) Does a 4-H coil store double or one-half the energy of a 2-H coil?

18-13
TROUBLES IN COILS

The most common trouble in coils is an open winding. As illustrated in Fig. 18-22, an ohmmeter connected across the coil reads infinite resistance for the open circuit. It does not matter whether the coil has an air core or an iron core. Since the coil is open, it cannot conduct current and therefore has no inductance, because it cannot produce induced voltage. When the resistance is checked, the coil should be disconnected from the external circuit to eliminate any parallel paths that could affect the resistance readings.

DC Resistance of a Coil. A coil has dc resistance equal to the resistance of the wire used in the winding. The amount of resistance is less with heavier wire and fewer turns. For rf coils with inductance values up to several millihenrys, requiring 10 to 100 turns of fine wire, the dc resistance is 1 to 20 Ω, approximately. Inductors for 60 Hz and audio frequencies with several hundred turns may have resistance values of 10 to 500 Ω, depending on the wire size.

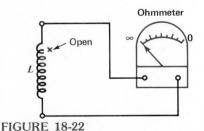

FIGURE 18-22

An open coil reads infinite ohms when its continuity is checked with an ohmmeter.

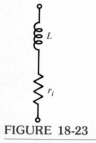

FIGURE 18-23

The internal dc resistance r_i is in series with the inductance of the coil.

As shown in Fig. 18-23, the dc resistance and inductance of a coil are in series, since the same current that induces voltage in the turns must overcome the resistance of the wire. Although resistance has no function in producing induced voltage, it is useful to know the dc coil resistance because if it is normal, usually the inductance can also be assumed to have its normal value.

Open Coil. An open winding has infinite resistance, as indicated by an ohmmeter reading. With a transformer having four leads or more, check the resistance across the two leads for the primary, across the two leads for the secondary, and across any other pairs of leads for additional secondary windings. For an autotransformer with three leads, check the resistance from one lead to each of the other two.

When the open circuit is inside the winding, it is usually not practical to repair the coil and the entire unit is replaced. In some cases, an open connection at the terminals can be resoldered.

Open Primary Winding. When the primary of a transformer is open, no primary current can flow and there is no voltage induced in any of the secondary windings.

Open Secondary Winding. When the secondary of a transformer is open, it cannot supply power to any load resistance across the open winding. Furthermore, with no current in the secondary, the primary current is also practically zero, as though the primary winding were open. The only primary current needed is the small magnetizing current to sustain the field producing induced voltage across the secondary without any load. If the transformer has several secondary windings, however, an open winding in one secondary does not affect transformer operation for the secondary circuits that are normal.

Short across Secondary Winding. In this case excessive primary current flows, as though it were short-circuited, often burning out the primary winding. The reason is that the large secondary current has a strong field that opposes the flux of the self-induced voltage across the primary, making it draw more current from the generator.

Practice Problems 18-13
(answers on page 378)
(a) The normal R of a coil is 18 Ω. How much will an ohmmeter read if the coil is open?
(b) The primary of a 3:1 step-up autotransformer is connected to the 120-V ac power line. How much will the secondary voltage be if the primary is open?

Summary

1. Varying current induces voltage in a conductor, since the expanding and collapsing field of the current is equivalent to flux in motion.

2. Lenz' law states that the induced voltage opposes the change in current causing the induction. Inductance, therefore, tends to keep the current from changing.
3. The ability of a conductor to produce induced voltage across itself when the current varies is its self-inductance, or inductance. The symbol is L, and the unit of inductance is the henry. One henry of inductance allows 1 V to be induced when the current changes at the rate of 1 A/s. For smaller units 1 mH = 1×10^{-3} H; 1 µH = 1×10^{-6} H.
4. To calculate the self-induced voltage, $v_L = L\,(di/dt)$, with v in volts, L in henrys, and di/dt in amperes per second.
5. Mutual inductance is the ability of varying current in one conductor to induce voltage in another conductor nearby. Its symbol is L_M, measured in henrys. $L_M = k\sqrt{L_1 L_2}$, where k is the coefficient of coupling.
6. A transformer consists of two or more windings with mutual inductance. The primary winding connects to the source voltage; the load resistance is connected across the secondary winding. A separate winding is an isolated secondary.
7. An autotransformer is a tapped coil, used to step up or step down the primary voltage. There are three leads with one connection common to both the primary and secondary.
8. A transformer with an iron core has essentially unit coupling. Therefore, the voltage ratio is the same as the turns ratio: $V_S/V_P = N_S/N_P$.
9. Assuming 100 percent efficiency for an iron-core power transformer, the power supplied to the primary equals the power used in the secondary.
10. Eddy currents are induced in the iron core of an inductance causing wasted power that heats the core. Eddy-current losses increase with higher frequencies of alternating current. To reduce eddy currents, the iron core is laminated with insulated sections. Powdered-iron and ferrite cores have minimum eddy-current losses for radio frequencies. Hysteresis losses also cause wasted power.
11. Assuming no mutual coupling, series inductances are added like series resistances. For parallel inductances, the total inductance is calculated by the reciprocal formula, as for parallel resistances.
12. The magnetic field of an inductance has stored energy $LI^2/2$. With I in amperes and L in henrys, the energy is in joules.
13. In addition to its inductance, a coil has dc resistance equal to the resistance of the wire in the coil. An open coil has infinitely high resistance.
14. An open primary in a transformer results in no induced voltage in any of the secondary windings.

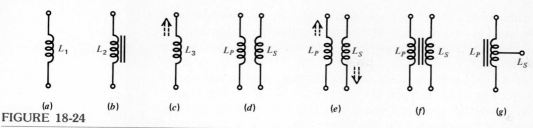

FIGURE 18-24
Summary of types of inductors. (a) Air-core coil. (b) Iron-core coil. (c) Adjustable ferrite core. (d) Air-core transformer. (e) Variable L_P and L_S. (f) Iron-core transformer. (g) Autotransformer.

15. Figure 18-24 summarizes the main types of inductors, or coils, with their schematic symbols.

Self-Examination (Answers at back of book.)

Choose (a), (b), (c), or (d).

1. Alternating current can induce voltage because alternating current has a (a) high peak value; (b) varying magnetic field; (c) stronger magnetic field than direct current; (d) constant magnetic field.
2. When current in a conductor increases, Lenz' law says that the self-induced voltage will (a) tend to increase the amount of current; (b) aid the applied voltage; (c) produce current opposite to the increasing current; (d) aid the increasing current.
3. A 5:1 voltage step-up transformer has 120 V across the primary and a 600-Ω resistance across the secondary. Assuming 100 percent efficiency, the primary current equals (a) ⅕ A; (b) 600 mA; (c) 5 A; (d) 10 A.
4. An iron-core transformer with an 8:1 step-up ratio has 120 V applied across the primary. The voltage across the secondary equals (a) 15 V; (b) 120 V; (c) 180 V; (d) 960 V.
5. With double the number of turns but the same length and area, the inductance is (a) the same; (b) double; (c) quadruple; (d) one-quarter.
6. Current changing from 4 to 6 A in 1 s induces 40 V in a coil. Its inductance equals (a) 40 mH; (b) 4 H; (c) 6 H; (d) 20 H.
7. A laminated iron core has reduced eddy-current losses because

(a) the laminations are stacked vertically; (b) the laminations are insulated from each other; (c) the magnetic flux is concentrated in the air gap of the core; (d) more wire can be used with less dc resistance in the coil.
8. Two 250-μH coils in series without mutual coupling have a total inductance of (a) 125 μH; (b) 250 μH; (c) 400 μH; (d) 500 μH.
9. The dc resistance of a coil made with 100 ft of No. 30 gage copper wire is approximately (a) less than 1 Ω; (b) 10.5 Ω; (c) 104 Ω; (d) more than 1 MΩ.
10. An open coil has (a) infinite resistance and zero inductance; (b) zero resistance and high inductance; (c) infinite resistance and normal inductance; (d) zero resistance and inductance.

Essay Questions

1. Define 1 H of self-inductance and 1 H of mutual inductance.
2. State Lenz' law in terms of induced voltage produced by varying current.
3. Refer to Fig. 18-5. Explain why the polarity of v_L is the same in (a) and (d).
4. Make a schematic diagram showing primary and secondary for an iron-core transformer with a 6:1 voltage step-up ratio: (a) using an autotransformer; (b) using a transformer with isolated secondary winding. Then (c) with 100 turns in the primary, how many turns are in the secondary for both cases?
5. Define the following: coefficient of coupling, transformer efficiency, stray inductance, and eddy-current losses.
6. Why are eddy-current losses reduced with the following cores: (a) laminated; (b) powdered iron; (c) ferrite?
7. Why is a good conductor used for an rf shield?
8. Show two methods of providing a variable inductance.
9. Derive the formula $L_M = (L_{T_a} - L_{T_o})/4$ from the fact that $L_{T_a} = L_1 + L_2 + 2L_M$ while $L_{T_o} = L_1 + L_2 - 2L_M$.
10. (a) Why will the primary of a power transformer have excessive current if the secondary is short-circuited? (b) Why is there no voltage across the secondary if the primary is open?
11. (a) Describe briefly how to check a coil for an open winding with an ohmmeter. What ohmmeter range should be used? (b) What leads will be checked on an autotransformer with one secondary and a transformer with two isolated secondary windings?
12. What is a Variac, and how is it used?

Problems (Answers to odd-numbered problems at back of book.)

1. Convert the following current changes to amperes per second: (a) zero to 4 A in 2 s; (b) zero to 50 mA in 5 μs; (c) 100 to 150 mA in 5 μs; (d) 150 to 100 mA in 5 μs.
2. Convert into henrys using powers of 10: (a) 250 μH; (b) 40 μH; (c) 40 mH; (d) 5 mH; (e) 0.005 H.
3. Calculate the values of v_L across a 5-mH inductance for each of the current variations in Prob. 1.
4. A coil produces a self-induced voltage of 50 mV when i varies at the rate of 25 mA/ms. How much is L?
5. A power transformer with an 8:1 turns ratio has 60 Hz 120 V across the primary. (a) What is the frequency of the secondary voltage? (b) How much is the secondary voltage? (c) With a load resistance of 10,000 Ω across the secondary, how much is the secondary current? Draw the schematic diagram showing primary and secondary circuits. (d) How much is the primary current? Assume 100 percent efficiency. (Note: 8:1 is the ratio of L_S to L_P.)
6. How much would the primary current be in a power transformer having a primary resistance of 5 Ω if it were connected by mistake to a 120-V dc line instead of the 120-V ac line?
7. For a 100-μH inductance L_1 and a 200-μH inductance L_2, calculate the following: (a) the total inductance L_T of L_1 and L_2 in series without mutual coupling; (b) the combined inductance of L_1 and L_2 in parallel without mutual coupling; (c) the L_T of L_1 and L_2 series-aiding, and series-opposing, with 10-μH mutual inductance; (d) the value of the coupling factor k.
8. Calculate the inductance L for the following long coils: (a) Air core, 20 turns, area 3.14 cm², length 25 cm; (b) same coil as (a) with ferrite core having a μ of 5000; (c) air core, 200 turns, area 3.14 cm², length 25 cm; (d) air core, 20 turns, area 3.14 cm², length 50 cm; (e) air core, 20 turns, diameter 4 cm, length 50 cm. (Note: 1 cm = 10^{-2} m, and 1 cm² = 10^{-4} m².)
9. Calculate the resistance of the following coils, using Table 10-1: (a) 400 turns, each using 3 in of No. 30 gage wire; (b) 40 turns, each using 3 in of No. 10 gage wire.
10. (a) Calculate the period T for one cycle of a 10-MHz sine wave. (b) How much is the time for one quarter-cycle? (c) If i increases from 0 to 20 mA in this time, how much is v_L across a 0.1-μH inductance?
11. Calculate the energy in joules stored in the magnetic field of a 300-mH L with a 90-mA I.

12. (a) A transformer delivers 400 W out with 500 W in. Calculate the efficiency in percent. (b) A transformer with 80 percent efficiency delivers 400 W total secondary power. Calculate the primary power.
13. A 20-mH L and a 40-mH L are connected series-aiding, with $k = 0.4$. Calculate L_T.
14. Calculate the inductance of the coil in Fig. 18-4 with an iron core having a μ_r of 100.

Answers to Practice Problems

18-1	(a) Coil (b) (b) Time B	18-8	(a) T (b) T (c) T
18-2	(a) $L = 2$ H (b) $L = 32$ mH	18-9	(a) T (b) T
18-3	(a) $v_L = 2$ V (b) $v_L = 200$ V	18-10	(a) $L_T = 1.5$ mH (b) $L_T = 0.33$ mH
18-4	(a) T (b) T	18-11	(a) T (b) T
18-5	(a) $k = 1$ (b) $L_M = 18$ mH	18-12	(a) Joule (b) Double
18-6	(a) $V_S = 240$ V (b) $I_S = 0.1$ A	18-13	(a) Infinite ohms (b) 0 V
18-7	(a) Iron core (b) 60 MHz		

Inductive Reactance

Chapter 19

When alternating current flows in an inductance L, the amount of current is much less than the resistance alone would allow. The reason is that the current variations induce a voltage across L that opposes the applied voltage. This additional opposition of an inductance to sine-wave alternating current is specified by the amount of its inductive reactance X_L. The X indicates reactance. It is an opposition to current, measured in ohms. The X_L is the ohms of opposition, therefore, that an inductance L has for sine-wave current.

The amount of X_L equals $2\pi fL$ ohms, with f in hertz and L in henrys. Note that the opposition in ohms of X_L increases for higher frequencies and more inductance.

The requirements for having X_L correspond to what is needed for producing induced voltage. There must be variations in current and its associated magnetic flux. For a steady direct current without any changes in current, the X_L is zero. However, with sine-wave alternating current, the X_L is the best way to analyze the effect of L. The topics explaining this important factor of X_L in sine-wave ac circuits are:

19-1 How X_L Reduces the Amount of I
19-2 $X_L = 2\pi fL$
19-3 Series or Parallel Inductive Reactances
19-4 Ohm's Law Applied to X_L
19-5 Applications of X_L for Different Frequencies.
19-6 Waveshape of v_L Induced by Sine-Wave Current.

19-1
HOW X_L REDUCES THE AMOUNT OF I

Figure 19-1 illustrates the effect of X_L in reducing the alternating current for a light bulb. The more ohms of X_L, the less current flows. When X_L reduces I to a very small value, the bulb cannot light.

In (a), there is no inductance, and the ac voltage source produces a 2.4-A current to light the bulb with full brilliance. This 2.4-A I results from 120 V applied across the 50-Ω R of the bulb's filament.

In (b), however, a coil is connected in series with the bulb. The coil has a dc resistance of only 1 Ω, which is negligible, but the reactance of the inductance is 1000 Ω. This X_L is a measure of the coil's reaction to sine-wave current in producing a self-induced voltage that opposes the applied voltage and reduces the current.

Chapter 19
Inductive Reactance

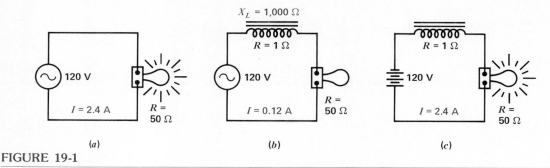

FIGURE 19-1
Illustrating the effect of X_L in reducing the amount of alternating current. (a) Bulb lights with 2.4 A. (b) Inserting an X_L of 1000 Ω reduces I to 0.12 A, and the bulb cannot light. (c) With direct current, the coil has no inductive reactance, and the bulb lights.

Now I is 120 V/1000 Ω, approximately, which equals 0.12 A. This I is not enough to light the bulb.

Although the dc resistance is only 1 Ω, the X_L of 1000 Ω for the coil limits the amount of alternating current to such a low value that the bulb cannot light. This X_L of 1000 Ω for a 60-Hz current can be obtained with an inductance L of approximately 2.65 H.

In (c), the coil is also in series with the bulb, but the applied battery voltage produces a steady value of direct current. Without any current variations, the coil cannot induce any voltage and, therefore, it has no reactance. The amount of direct current, then, is practically the same as though the dc voltage source were connected directly across the bulb, and it lights with full brilliance. In this case, the coil is only a length of wire, as there is no induced voltage without current variations. The dc resistance is the resistance of the wire in the coil.

In summary, we can make the following conclusions:

1. An inductance can have appreciable X_L in ac circuits, to reduce the amount of current. Furthermore, the higher the frequency of the alternating current, and the greater the inductance, the higher is the X_L opposition.
2. There is no X_L for steady direct current. In this case, the coil is just a resistance equal to the resistance of the wire.

These effects have almost unlimited applications in practical circuits. Consider how useful ohms of X_L can be for different kinds of current, compared with resistance, which always has the same ohms of opposition. One example is to use X_L where it is desired to have high ohms of opposition to alternating current but little opposition to direct current. Another example is to use X_L for more opposition to a high-frequency alternating current, compared with lower frequencies.

X_L Is an Inductive Effect. The reason why an inductance can have X_L to reduce the amount of alternating current is the fact that self-induced voltage is produced to oppose the applied voltage. In Fig. 19-2, V_L is the voltage across L, induced by the variations in sine-wave current produced by the applied voltage V_A.

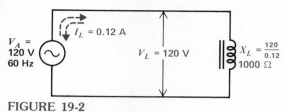

FIGURE 19-2
The inductive reactance X_L equals the V_L/I_L ratio in ohms.

The two voltages V_A and V_L are the same because they are in parallel. However, the current I_L is the amount that allows the self-induced voltage V_L to be equal to V_A. In this example, I is 0.12 A. This value of a 60-Hz current in the inductance produces a V_L of 120 V.

The Reactance Is a V/I Ratio. When we consider the V/I ratio for the ohms of opposition to the sine-wave current, this value is 120/0.12, which equals 1000 Ω. This 1000 Ω is what we call X_L, to indicate how much current can be produced by sine-wave voltage across an inductance. The ohms of X_L can be almost any amount, but the 1000 Ω here is a typical example.

The Effect of L and f on X_L. The X_L value depends on the amount of inductance and the frequency of the alternating current. If L in Fig. 19-2 were increased, it could induce the same 120 V for V_L with less current. Then the ratio of V_L/I_L would be greater, meaning more X_L for more inductance.

Also, if the frequency were increased in Fig. 19-2, the current variations would be faster with a higher frequency. Then the same L could produce the 120 V for V_L with less current. For this condition also, the V_L/I_L ratio would be greater because of the smaller current, indicating more X_L for a higher frequency.

Practice Problems 19-1
(answers on page 393)

(a) For the dc circuit in Fig. 19-1c, how much is X_L?
(b) For the ac circuit in Fig. 19-1b, how much is the V/I ratio for the ohms of X_L?

19-2
$X_L = 2\pi fL$

This formula includes the effects of frequency and inductance for calculating the reactance. The frequency is in hertz and L is in henrys for an X_L in ohms. As an example, we can calculate X_L for a 2.65-Hz L at the frequency of 60 Hz:

$$X_L = 2\pi fL \qquad (19\text{-}1)$$
$$= 6.28 \times 60 \times 2.65$$
$$X_L = 1000 \text{ Ω}$$

Note the following factors in the formula $X_L = 2\pi fL$

1. The constant factor 2π is always $2 \times 3.14 = 6.28$. It indicates the circular motion from which a sine wave is derived. Therefore, this formula applies only to sine-wave ac circuits. The 2π is actually 2π rad or 360° for a complete circle or cycle. Furthermore, $2\pi \times f$ is the angular velocity, in rad/s, for a rotating phasor corresponding to the sine-wave V or I of that particular frequency.
2. The frequency f is a time element. Higher frequency means the current varies at a faster rate. A faster current change can produce more self-induced voltage across a given amount of inductance.
3. The inductance L indicates the physical factors of the coil that determine how much voltage it can induce for a given current change.
4. X_L is in ohms, corresponding to a V_L/I_L ratio for sine-wave ac circuits, to determine

how much current L allows for a given applied voltage.

Stating X_L either as V_L/I_L or as $2\pi fL$ are two ways of specifying the same value of ohms. The $2\pi fL$ formula gives the effect of L and f on the X_L. The V_L/I_L ratio gives the result of $2\pi fL$ in reducing the amount of I.

The formula $2\pi fL$ shows that X_L is proportional to frequency. When f is doubled, for instance, X_L is doubled. This linear increase of inductive reactance with frequency is illustrated in Fig. 19-3.

The reactance formula also shows that X_L is proportional to the inductance. When the value of henrys for L is doubled, the ohms of X_L is also doubled. This linear increase of inductive reactance with frequency is illustrated in Fig. 19-4.

Example 1. How much is X_L of a 6-mH L at 41.67 kHz?

Answer. $X_L = 2\pi fL$
$= 6.28 \times 41.67 \times 10^3 \times 6 \times 10^{-3}$
$X_L = 1570 \; \Omega$

Example 2. Calculate the X_L of (a) a 10-H L at 60 Hz and (b) a 5-H L at 60 Hz.

Answer.

(a) For a 10-H L,
$X_L = 2\pi fL = 6.28 \times 60 \times 10$
$X_L = 3768 \; \Omega$

(b) For a 5-H L,
$X_L = \frac{1}{2} \times 3768$
$X_L = 1884 \; \Omega$

Example 3. Calculate the X_L of a 250-μH coil at (a) 1 MHz and (b) 10 MHz.

Answer.

(a) At 1 MHz,
$X_L = 2\pi fL = 6.28 \times 1 \times 10^6 \times 250 \times 10^{-6}$
$= 6.28 \times 250$
$X_L = 1570 \; \Omega$

(b) At 10 MHz,
$X_L = 10 \times 1570$
$X_L = 15{,}700 \; \Omega$

X_L increases with higher f

Frequency, Hz	$X_L = 2\pi fL$, Ω
0	0
100	200
200	400
300	600
400	800

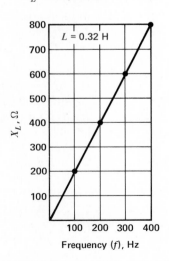

FIGURE 19-3

Linear increase of X_L with higher frequencies. L is constant at 0.32 H.

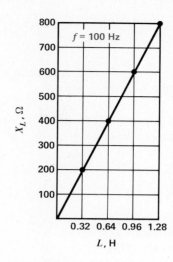

FIGURE 19-4

Linear increase of X_L with higher values of inductance. Frequency is constant at 100 Hz.

The last two examples illustrate the fact that X_L is proportional to frequency and inductance. In Example 2b, X_L is one-half the value in (a) because the inductance is one-half. In Example 3b, X_L is ten times more than in (a) because the frequency is ten times higher.

Finding L from X_L. Not only can X_L be calculated from f and L, but if any two factors are known, the third can be found. Very often X_L can be determined from voltage and current measurements. With the frequency known, L can be calculated as

$$L = \frac{X_L}{2\pi f} \qquad (19\text{-}2)$$

This formula just has the factors inverted from $X_L = 2\pi f L$. Use the basic units with ohms for X_L and hertz for f to calculate L in henrys.

Example 4. A coil with negligible resistance has 62.8 V across it with 0.01 A. How much is X_L?

Answer.
$$X_L = \frac{V_L}{I_L} = \frac{62.8 \text{ V}}{0.01 \text{ A}}$$
$$X_L = 6280 \; \Omega$$

Example 5. Calculate the L of the coil in Example 4 when the frequency is 1000 Hz.

Answer.
$$L = \frac{X_L}{2\pi f} = \frac{6280}{6.28 \times 1000} = \frac{6280}{6280}$$
$$L = 1 \text{ H}$$

Example 6. Calculate the inductance of a coil that has a 15,700-Ω X_L at 10 MHz.

Answer.
$$L = \frac{X_L}{2\pi f} = \frac{15{,}700}{6.28 \times 10 \times 10^6}$$
$$= \frac{15{,}700}{62.8} \times 10^{-6}$$
$$L = 250 \times 10^{-6}$$
$$L = 250 \; \mu\text{H}$$

Find f from X_L. For the third and final version of the inductive reactance formula,

Chapter 19
Inductive Reactance

$$f = \frac{X_L}{2\pi L} \quad (19\text{-}3)$$

Use the basic units of ohms for X_L and henrys for L to calculate the frequency in hertz.

Example 7. At what frequency will an inductance of 1 H have the reactance of 1000 Ω?

Answer. $f = \dfrac{X_L}{2\pi L} = \dfrac{1000}{6.28 \times 1} = 0.159 \times 10^3$

$f = 159$ Hz

Practice Problems 19-2
(answers on page 393)
(a) L is 1 H and f is 100 Hz. How much is X_L?
(b) L is 0.5 H and f is 100 Hz. How much is X_L?
(c) L is 1 H and f is 1000 Hz. How much is X_L?

19-3
SERIES OR PARALLEL INDUCTIVE REACTANCES

Since reactance is an opposition in ohms, inductive reactances in series or parallel are combined the same way as ohms of resistance. With series reactances the total reactance is the sum of the individual values, as shown in Fig. 19-5a. For example, the series reactances of 100 and 200 Ω add to equal 300 Ω of X_L across both inductances. Therefore, in series,

$$X_{L_T} = X_{L_1} + X_{L_2} + X_{L_3} + \cdots + \text{etc.} \quad (19\text{-}4)$$

For the case of parallel reactances, the combined reactance is calculated by the reciprocal formula. As shown in Fig. 19-5b, in parallel

$$\frac{1}{X_{L_T}} = \frac{1}{X_{L_1}} + \frac{1}{X_{L_2}} + \frac{1}{X_{L_3}} + \cdots + \text{etc.} \quad (19\text{-}5)$$

The combined parallel reactance will be less than the lowest branch reactance. Any short cuts for calculating parallel resistances also apply to the parallel reactances. For instance, the combined reactance of two equal reactances in parallel is one-half either reactance.

Practice Problems 19-3
(answers on page 393)
(a) X_L of 200 Ω is in series with a 300-Ω X_L. How much is the total X_{L_T}?
(b) X_L of 200 Ω is in parallel with a 300-Ω X_L. How much is the combined X_{L_T}?

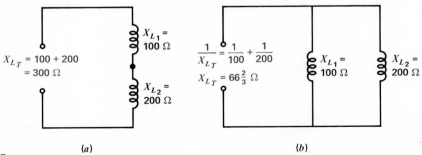

FIGURE 19-5
Combining inductive reactances. (a) X_{L_1} and X_{L_2} in series. (b) X_{L_1} and X_{L_2} in parallel.

19-4
OHM'S LAW APPLIED TO X_L

The amount of current in an ac circuit with just inductive reactance is equal to the applied voltage divided by X_L. Three examples are illustrated in Fig. 19-6. No dc resistance is indicated, since it is assumed to be practically zero for the coils shown. In (a), there is just one reactance of 100 Ω. Then I equals V/X_L, or 100 V/100 Ω, which is 1 A.

In (b), the total reactance is the sum of the two individual series reactances of 100 Ω each for a total of 200 Ω. The current, calculated as V/X_{L_T}, then equals 100 V/200 Ω, which is 0.5 A. This current is the same in both series reactances. Therefore, the voltage across each reactance equals its IX_L product. This is 0.5 A × 100 Ω, or 50 V across each X_L.

In (c) each parallel reactance has its individual branch current equal to the applied voltage divided by the branch reactance. Then each branch current equals 100 V/100 Ω, which is 1 A. The voltage is the same across both reactances, equal to the generator voltage, since they are all in parallel.

The total line current of 2 A is the sum of the two individual 1-A branch currents. With rms value for the applied voltage, all the calculated values of currents and voltage drops in Fig. 19-6 are also rms values.

Practice Problems 19-4
(answers on page 393)

(a) In Fig. 19-6b, how much is the I through both X_{L_1} and X_{L_2}?
(b) In Fig. 19-6c, how much is the V across both X_{L_1} and X_{L_2}?

19-5
APPLICATIONS OF X_L FOR DIFFERENT FREQUENCIES

The general use of inductance is to provide minimum reactance for relatively low frequencies but more for higher frequencies. In this way, the current in an ac circuit can be reduced for higher frequencies because of more X_L. There are many circuits where voltages of different frequencies are applied to produce current with different frequencies. Then, the general effect of X_L is to allow the most current for direct current and low frequencies, with less current for higher frequencies, as X_L increases.

Compare this frequency factor for ohms of X_L with ohms of resistance. The X_L increases with frequency, but R has the same effect in limiting direct current or alternating current of any frequency.

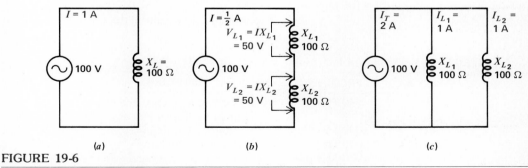

FIGURE 19-6
Circuit calculations with V, I, and ohms of X_L.
(a) One reactance. (b) Two series reactances.
(c) Two parallel reactances.

If 1000 Ω is taken as a suitable value of X_L for many applications, typical inductances can be calculated for different frequencies. These are listed in Table 19-1. At 60 Hz, for example, the inductance is 2.7 H for 1000 Ω of X_L. For this case, the inductance has practically no reactance for direct current or for very low frequencies below 60 Hz. Above 60 Hz, the inductive reactance is more than 1000 Ω.

Note that the smaller inductances at the bottom of the first column still have the same X_L of 1000 Ω as the frequency is increased. Typical rf coils, for instance, have an inductance value of the order of 100 to 300 μH. For the very high radio-frequency (VHF) range, only several microhenrys of inductance are needed for an X_L of 1000 Ω.

It is necessary to use smaller inductance values as the frequency is increased because a coil that is too large can have excessive losses at high frequencies. With iron-core coils, particularly, the hysteresis and eddy-current losses increase with frequency.

Practice Problems 19-5
(answers on page 393)
Refer to Table 19-1.
(a) Which frequency uses the smallest L for 1000 Ω of X_L?
(b) How much would X_L be for the 1.6-μH L at 200 MHz?

19-6 WAVESHAPE OF v_L INDUCED BY SINE-WAVE CURRENT

More details of inductive circuits can be analyzed by means of the waveshapes in Fig. 19-7, plotted for the calculated values in Table 19-2. The top curve shows a sine wave of current i_L flowing through a 6-mH inductance L. Since induced voltage depends on rate of change of current rather than the absolute value of i, the curve in (b) shows how much the current changes. In this curve the di/dt values are plotted for the current changes every 30° of the cycle. The bottom curve shows the actual induced voltage v_L. This v_L curve is similar to the di/dt curve because v_L equals the constant factor L multiplied by di/dt.

90° Phase Angle. The v_L curve at the bottom of Fig. 19-7 has its zero values when the i_L curve at the top is at maximum. This comparison shows that the curves are 90° out of phase. The v_L is a cosine wave of voltage for the sine wave of current i_L.

The 90° phase difference results from the fact that v_L depends on the di/dt rate of change, rather than i itself. More details of this 90° phase angle between v_L and i_L for inductance are explained in the next chapter.

Frequency. For each of the curves, the period

TABLE 19-1. Values of Inductance L for X_L of 1000 Ω

L*, (APPROX.)	FREQUENCY	REMARKS
2.7 H	60 Hz	Power-line frequency and low audio frequency
160 mH	1000 Hz	Medium audio frequency
16 mH	10,000 Hz	High audio frequency
160 μH	1000 kHz (rf)	In radio broadcast band
16 μH	10 MHz (HF)	In short-wave radio band
1.6 μH	100 MHz (VHF)	In FM broadcast band

*Calculated as $L = 1000/2\pi f$.

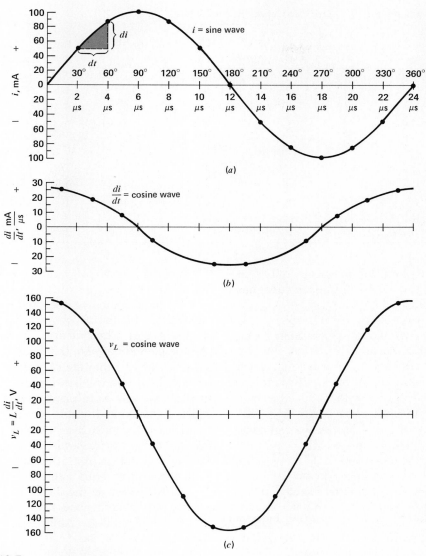

FIGURE 19-7.

Waveshapes of sine-wave current i and induced voltage v_L, plotted for values in Table 19-2.

TABLE 19-2. Values for $v_L = L(di/dt)$ Curves in Fig. 19-7

TIME		dt		di, mA	di/dt, mA/μs	L, mH	$v_L = L(di/dt)$, V
θ	μs	θ	μs				
30°	2	30°	2	50	25	6	150
60°	4	30°	2	36.6	18.3	6	109.8
90°	6	30°	2	13.4	6.7	6	40.2
120°	8	30°	2	−13.4	−6.7	6	−40.2
150°	10	30°	2	−36.6	−18.3	6	−109.8
180°	12	30°	2	−50	−25	6	−150
210°	14	30°	2	−50	−25	6	−150
240°	16	30°	2	−36.6	−18.3	6	−109.8
270°	18	30°	2	−13.4	−6.7	6	−40.2
300°	20	30°	2	13.4	6.7	6	40.2
330°	22	30°	2	36.6	18.3	6	109.8
360°	24	30°	2	50	25	6	150

T is 24 μs. Therefore, the frequency is $1/T$ or $1/24$ μs, which equals 41.67 kHz. Each curve has the same frequency.

Ohms of X_L. The ratio of v_L/i_L actually specifies the inductive reactance in ohms. For this comparison, we use the actual value of i_L, which has a peak value of 100 mA. The rate-of-change factor is included in the induced voltage v_L. Although the peak of v_L at 150 V is 90° before the peak of i_L at 100 mA, we can compare these two peak values. Then v_L/i_L is 150/0.1, which equals 1500 Ω.

This X_L is only an approximate value because v_L cannot be determined exactly for the large dt changes every 30°. If we used smaller intervals of time, the peak v_L would be 157 V. Then X_L would be 1570 Ω, the same as $2\pi fL$ Ω with a 6-mH L and a frequency of 41.67 kHz. This is the same X_L problem as Example 1 on page 382.

The Tabulated Values from 0 to 90°. The numerical values in Table 19-2 are calculated as follows: The i curve is a sine wave. This means it rises to one-half its peak value in 30°, to 0.866 of the peak in 60°, and the peak value is at 90°.

In the di/dt curve the changes in i are plotted. For the first 30° the di is 50 mA; the dt change is 2 μs. Then di/dt is 25 mA/μs. This point is plotted between 0 and 30° to indicate that 25 mA/μs is the rate of change of current for the 2-μs interval between 0 and 30°. If smaller intervals were used, the di/dt values could be determined more accurately.

During the next 2-μs interval from 30 to 60°, the current increases from 50 to 86.6 mA. The change of current during this time is 86.6 − 50, which equals 36.6 mA. The time is the same 2 μs for all the intervals. Then di/dt for the next plotted point is 36.6/2, or 18.3.

For the final 2-μs change before i reaches its peak at 100 mA, the di value is 100 − 86.6, or 13.4 mA, and the di/dt value is 6.7. All these values are listed in Table 19-2.

Notice that the di/dt curve in Fig. 19-7b has its peaks at the zero value of the i curve, while the peak i values correspond to zero on the di/dt curves. These conditions result because the sine wave of i has its sharpest slope at the zero values. The rate of change is greatest when the i curve is going through the zero axis. The i curve flattens near the peaks and has zero rate of change exactly at the peak. The curve must stop going up before it can come down. In summary, then, the di/dt curve and the i curve are 90° out of phase with each other.

The v_L curve follows the di/dt curve exactly as $v_L = L(di/dt)$. The phase of the v_L curve is exactly the same as the di/dt curve, 90° out of phase with the i curve. For the first plotted point,

$$v_L = L\frac{di}{dt} = 6 \times 10^{-3} \times \frac{50 \times 10^{-3}}{2 \times 10^{-6}} = 150 \text{ V}$$

The other v_L values are calculated the same way, multiplying the constant factor of 6 mH by the di/dt value for each 2-μs interval.

90 to 180°. In this quarter-cycle, the sine wave of i decreases from its peak of 100 mA at 90° to zero at 180°. This decrease is considered a negative value for di, as the slope is negative going downward. Physically, the decrease in current means its associated magnetic flux is collapsing, compared with the expanding flux as the current increases. The opposite motion of the collapsing flux must make v_L of opposite polarity, compared with the induced voltage polarity for increasing flux. This is why the di values are negative from 90 to 180°. The di/dt values are also negative, and the v_L values are negative.

180 to 270°. In this quarter-cycle, the current increases in the reverse direction. If the magnetic flux is considered counterclockwise around the conductor with $+i$ values, the flux is in the reversed clockwise direction with $-i$ values. Any induced voltage produced by expanding flux in one direction will have opposite polarity from voltage induced by expanding flux in the opposite direction. This is why the di values are considered negative from 180 to 270°, as in the second quarter-cycle, compared with the positive di values from 0 to 90°. Actually, increasing negative values and decreasing positive values are changing in the same direction. This is why v_L is negative for both the second and third quarter-cycles.

270 to 360°. In the last quarter-cycle, the negative i values are decreasing. Now the effect on polarity is like two negatives making a positive. The current and its magnetic flux have the negative direction. But the flux is collapsing, which induces opposite voltage from increasing flux. Therefore, the di values from 270 to 360° are positive, as are the di/dt values and the induced voltages v_L. Actually, the slope of the sine wave of current in the last 90° has the same positive upward direction as in the first 90°.

The same action is repeated for each cycle of sine-wave current. Then the current i_L and the induced voltage v_L are 90° out of phase. The reason is that v_L depends on di/dt, not on i alone.

The phase angle of 90° between V_L and I will always apply for any L with sine-wave current. Remember, though, that the specific com-

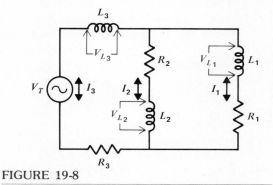

FIGURE 19-8

How 90° phase angle for a V_L applies in a circuit. I_1 lags V_{L_1} by 90°; I_2 lags V_{L_2} by 90°; I_3 lags V_{L_3} by 90°.

parison is only between the voltage across any one coil and the current flowing in its turns. To emphasize this important principle, Fig. 19-8 shows an ac circuit with a few coils and resistors. The phase angles in the circuit here are not to be figured now. However, for each L, the V_L is 90° out of phase with its I. The I lags V_L by 90°, or V_L leads I, because the peak of V_L is $\frac{1}{4}$ cycle earlier in time than the peak of I. For the examples in Fig. 19-8, I_1 lags V_{L_1} by 90°. Also, I_2 lags V_{L_2} by 90°. Finally, I_3, which is also I_T, lags V_{L_3} by 90°.

Practice Problems 19-6
(*answers on page 393*)
Refer to Fig. 19-7.
(a) At what angle does i have its maximum positive value?
(b) At what angle does v_L have its maximum positive value?
(c) What is the phase angle difference between the waveforms for i and v_L?

Summary

1. Inductive reactance, indicated X_L, is the opposition of an inductance to the flow of sine-wave alternating current.
2. X_L is measured in ohms because it limits the current to the value $I = V/X_L$. With V in volts and X_L in ohms, I is in amperes.
3. $X_L = 2\pi fL$. With f in hertz and L in henrys, X_L is in ohms.
4. With one constant L, its X_L increases proportionately with higher frequencies.
5. At one frequency, X_L increases proportionately with higher inductances.
6. With X_L and f known, the inductance $L = X_L/2\pi f$.
7. With X_L and L known, the frequency $f = X_L/2\pi L$.
8. The total X_L of reactances in series is the sum of the individual values, as for series resistances. Series reactances have the same current. The voltage across each inductive reactance is IX_L.
9. With parallel reactances, the total reactance is calculated by the reciprocal formula, as for parallel resistances. Each branch current is V/X_L. The total line current is the sum of the individual branch currents.

TABLE 19-3. Comparison of Inductance and Inductive Reactance

INDUCTANCE	INDUCTIVE REACTANCE
Symbol is L	Symbol is X_L
Measured in henry units	Measured in ohm units
Depends on construction of coil	Depends on frequency
$L = v_L/(di/dt)$, in H units	$X_L = v_L/i_L$ or $2\pi fL$, in Ω units

10. Table 19-3 summarizes the differences between L and X_L.
11. Table 19-4 compares X_L and R.

TABLE 19-4. Comparison of X_L and R

X_L	R
Ohm unit	Ohm unit
Increases for higher frequencies	Same for all frequencies
Phase angle is 90°	Phase angle is 0°

Self-Examination (Answers at back of book.)

Choose (a), (b), (c), or (d).

1. Inductive reactance is measured in ohms because it (a) reduces the amplitude of alternating current; (b) increases the amplitude of alternating current; (c) increases the amplitude of direct current; (d) has a back emf opposing a steady direct current.
2. Inductive reactance applies only to sine waves because it (a) increases with lower frequencies; (b) increases with lower inductance; (c) depends on the factor 2π; (d) decreases with higher frequencies.
3. An inductance has a reactance of 10,000 Ω at 10,000 Hz. At 20,000 Hz, its inductive reactance equals (a) 500 Ω; (b) 2000 Ω; (c) 20,000 Ω; (d) 32,000 Ω.
4. A 16-mH inductance has a reactance of 1000 Ω. If two of these are connected in series without any mutual coupling, their total reactance equals (a) 500 Ω; (b) 1000 Ω; (c) 1600 Ω; (d) 2000 Ω.
5. Two 5000-Ω inductive reactances in parallel have an equivalent reactance of (a) 2500 Ω; (b) 5000 Ω; (c) 10,000 Ω; (d) 50,000 Ω.
6. With 10 V applied across an inductive reactance of 100 Ω, the current equals (a) 10 µA; (b) 10 mA; (c) 100 mA; (d) 10 A.
7. A current of 100 mA through an inductive reactance of 100 Ω produces a voltage drop equal to (a) 1 V; (b) 6.28 V; (c) 10 V; (d) 100 V.
8. The inductance required for a 2000-Ω reactance at 20 MHz equals (a) 10 µH; (b) 15.9 µH; (c) 159 µH; (d) 320 µH.
9. A 160-µH inductance will have a 5000-Ω reactance at the frequency of (a) 5 kHz; (b) 200 kHz; (c) 1 MHz; (d) 5 MHz.
10. A coil has an inductive reactance of 1000 Ω. If its inductance is doubled and the frequency is doubled, then the inductive reactance will be (a) 1000 Ω; (b) 2000 Ω; (c) 4000 Ω; (d) 16,000 Ω.

Essay Questions

1. Explain briefly why X_L limits the amount of alternating current.
2. Give two differences and one similarity in comparing X_L and R.
3. Explain briefly why X_L increases with higher frequencies and more inductance.
4. Give two differences between inductance L of a coil and its inductive reactance X_L.
5. Referring to Fig. 19-7, why are waves a and b considered to be 90° out of phase, while waves b and c have the same phase?
6. Referring to Fig. 19-3, how does this graph show a linear proportion between X_L and frequency?
7. Referring to Fig. 19-4, how does this graph show a linear proportion between X_L and L?
8. Referring to Fig. 19-7, draw three similar curves, but for a sine wave of current with a period $T = 12$ μs for the full cycle. Use the same L of 6 mH. Compare the values of X_L obtained as $2\pi fL$ and v_L/i_L.
9. Referring to Fig. 19-3, tabulate the values of L that would be needed for each frequency listed but for an X_L of 2000 Ω. (Do not include 0 Hz.)
10. Calculate the inductance L needed for an X_L of 1000 Ω at the following five power frequencies: 50, 60, 120, 400, and 800 Hz. For 100 V applied, make a table comparing the amount of current, X_L, and L at the five different frequencies.
11. (a) Draw the circuit for a 40-Ω R across a 120-V 60-Hz source. (b) Draw the circuit for a 40-Ω X_L across a 120-V 60-Hz source. (c) Why is I equal to 3 A for both circuits? (d) Give two differences between the circuits.

Problems (Answers to odd-numbered problems at back of book.)

1. Calculate the X_L of a 0.5-H inductance at 100, 200, and 1000 Hz.
2. How much is the inductance for 628 Ω reactance at 100 Hz? 200 Hz? 1000 Hz? 500 kHz?
3. A coil with an X_L of 2000 Ω is connected across a 10-V ac generator. (a) Draw the schematic diagram. (b) Calculate the current. (c) How much is the voltage across the coil?
4. A 20-H coil has 10 V applied, with a frequency of 60 Hz. (a) Draw the schematic diagram. (b) How much is the inductive reactance of the coil? (c) Calculate the current. (d) What is the frequency of the current?

5. How much is the inductance of a coil with negligible resistance if the current is 0.1 A when connected across the 60-Hz 120-V power line?
6. Referring to Fig. 19-6, how much is the inductance of L_T, L_1, and L_2 if the frequency of the source voltage is 400 Hz?
7. How much is the inductance of a coil that has a reactance of 1000 Ω at 1000 Hz? How much will the reactance be for the same coil at 10 kHz?
8. How much is the reactance of a 10-μH inductance at 100 MHz?
9. A 1000-Ω X_{L_1} and a 4000-Ω X_{L_2} are in series across a 10-V 60-Hz source. Draw the schematic diagram and calculate the following: (a) total X_L; (b) current in X_{L_1} and in X_{L_2}; (c) voltage across X_{L_1} and across X_{L_2}; (d) inductance of L_1 and L_2.
10. The same 1000-Ω X_{L_1} and X_{L_2} are in parallel across the 10-V 60-Hz source. Draw the schematic diagram and calculate the following: branch currents in X_{L_1} and in X_{L_2}, total current in the generator, voltage across X_{L_1} and across X_{L_2}, inductance of L_1 and L_2.
11. At what frequencies will X_L be 2000 Ω for the following inductors: (a) 2 H; (b) 250 mH; (c) 800 μH; (d) 200 μH; (e) 20 μH?
12. A 6-mH L_1 is in series with an 8-mH L_2. The frequency is 40 kHz. (a) How much is L_T? (b) Calculate X_{L_T}. (c) Calculate X_{L_1} and X_{L_2} to see if their sum equals X_{L_T}.
13. Calculate X_L of a 1.2-mH coil at 216 kHz.
14. Calculate X_L of a 200-μH coil at 3.2 MHz.
15. Calculate X_L of a 2-H coil at 60 Hz.
16. How much is I when the X_L of Prob. 15 is connected to the 120-V 60-Hz power line?
17. A 250-mH inductor with negligible resistance is connected across a 10-V source. Tabulate the values of X_L and current in the circuit for alternating current at (a) 20 Hz; (b) 60 Hz; (c) 100 Hz; (d) 500 Hz; (e) 5000 Hz; (f) 15,000 Hz.
18. Do the same as in Prob. 17 for an 8-H inductor.

Answers to Practice Problems

19-1 (a) 0 Ω
 (b) 1000 Ω
19-2 (a) $X_L = 628$ Ω
 (b) $X_L = 314$ Ω
 (c) $X_L = 6280$ Ω
19-3 (a) $X_{L_T} = 500$ Ω
 $X_{L_T} = 120$ Ω
19-4 (a) 0.5 A
 (b) 100 V
19-5 (a) 100 MHz
 (b) 2000 Ω
19-6 (a) 90°
 (b) 0 or 360°
 (c) 90°

Inductive Circuits

Chapter 20

This unit analyzes circuits that combine inductive reactance X_L and resistance R. The main questions are: How do we combine the ohms of opposition, how much current flows, and what is the phase angle? In addition, the practical application of using a coil as a choke to reduce the current for a specific frequency is illustrated.

Finally, the general case of induced voltage across L is shown with nonsinusoidal current variations. Here, we compare the waveshapes of i_L and v_L instead of their phase. With nonsinusoidal waveforms, the circuit can be analyzed in terms of its L/R time constant instead of its reactance. Remember that X_L and its 90° phase angle apply only to sine waves. The topics are:

20-1 Sine-Wave i_L Lags v_L by 90°
20-2 X_L and R in Series
20-3 Impedance (Z)
20-4 X_L and R in Parallel
20-5 Q of a Coil
20-6 AF and RF Chokes
20-7 The General Case of Inductive Voltage
20-8 Calculating the L/R Time Constant

20-1
SINE-WAVE i_L LAGS v_L BY 90°

With sine-wave variations of current producing an induced voltage, the current lags its induced voltage by exactly 90°, as shown in Fig. 20-1. The inductive circuit in (a) has the current and voltage waveshapes shown in (b). The phasors in (c) show the 90° phase angle between i_L and v_L. Therefore, we can say that i_L lags v_L by 90°. Or, v_L leads i_L by 90°.

This 90° phase relationship between i_L and v_L is true in any sine-wave ac circuit, whether L is in series or parallel, and whether L is alone or combined with other components. We can always say that the voltage across any X_L is 90° out of phase with the current through it.

Why the Phase Angle is 90°. This results because v_L depends on the rate of change of i_L. As previously shown in Fig. 19-7 for a sine wave of i_L, the induced voltage is a cosine wave. In other words, v_L has the phase of di/dt, not the phase of i.

Why i_L Lags v_L. The 90° difference can be measured between any two points having the same value on the i_L and v_L waves. A convenient point is the positive peak value. Note that the i_L wave does not have its positive peak until 90° after the v_L wave. Therefore, i_L lags v_L by 90°. This 90° lag is in time. The time lag is one quarter-cycle, which is one-quarter of the time for a complete cycle.

Chapter 20
Inductive Circuits

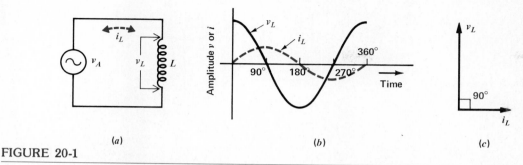

FIGURE 20-1
(a) Circuit with inductance L. (b) Sine wave of i_L lags v_L by 90°. (c) Phasor diagram.

Inductive Current Is the Same in a Series Circuit. The time delay and resultant phase angle for the current in an inductance apply only with respect to the voltage across the inductance. This condition does not change the fact that the current is the same in all parts of a series circuit. In Fig. 20-1a, the current in the generator, the connecting wires, and L must be the same because they are in series. At any instant, whatever the current value is at that time, it is the same in all the series components. The time lag is between current and voltage.

Inductive Voltage the Same across Parallel Branches. In Fig. 20-1a, the voltage across the generator and the voltage across L are the same because they are in parallel. There cannot be any lag or lead in time between these two parallel voltages. At any instant, whatever the voltage value is across the generator at that time, the voltage across L is the same. Considering the parallel voltage v_A or v_L, it is 90° out of phase with the current.

In this circuit the voltage across L is determined by the applied voltage, since they must be the same. The inductive effect here is to make the current have the values that produce $L(di/dt)$ equal to the parallel voltage.

The Frequency Is the Same for i_L and v_L. Although i_L lags v_L by 90°, both waves have the same frequency. The i_L wave reaches its peak values 90° later than the v_L wave, but the complete cycles of variations are repeated at the same rate. As an example, if the frequency of the sine wave v_L in Fig. 20-1b is 100 Hz, this is also the frequency for i_L.

Practice Problems 20-1
(answers on page 414)
Refer to Fig. 20-1.
(a) What is the phase between v_A and v_L?
(b) What is the phase between v_L and i_L?
(c) Does i_L lead or lag v_L?

20-2
X_L AND R IN SERIES

When a coil has series resistance, the current is limited by both X_L and R. This current I is the same in X_L and R since they are in series. Each has its own series voltage drop, equal to IR for the resistance and IX_L for the reactance.

Note the following points about a circuit that combines series X_L and R, as in Fig. 20-2:

1. The current is labeled I, rather than I_L, because I flows through all the series components.

Chapter 20
Inductive Circuits

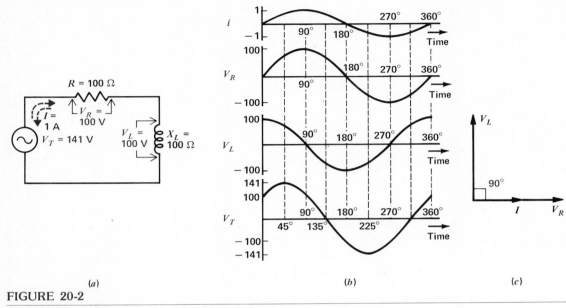

FIGURE 20-2
X_L and R in series. (a) Circuit. (b) Waveforms of current and voltages. (c) Phasor diagram.

2. The voltage across X_L, labeled V_L, can be considered an IX_L voltage drop, just as we use V_R for an IR voltage drop.
3. The current I through X_L must lag V_L by 90°, as this is the angle between current through an inductance and its self-induced voltage.
4. The current I through R and its IR voltage drop have the same phase. There is no reactance to sine-wave current in any resistance. Therefore, I and IR have the same phase, or this phase angle is 0°.

An example of such a circuit is shown in Fig. 20-2. R can be either the internal resistance of the coil or an external series resistance. The I and V values may be rms, peak, or instantaneous, as long as the same measure is applied to all. Peak values are used here for convenience in comparing the waveforms.

Phase Comparisons. Note the following:

1. V_L is 90° out of phase with I.
2. However, V_R has the same phase as I.
3. Therefore, V_L is also 90° out of phase with V_R.

Specifically, V_R lags V_L by 90°, just as the current I lags V_L. These phase relations are shown by the waveforms in Fig. 20-2b and the phasors in Fig. 20-3.

Combining V_R and V_L. As shown in Fig. 20-2b, when the V_R voltage wave is combined with the V_L voltage wave, the result is the voltage wave for the applied generator voltage V_T. The voltage drops must add to equal the applied voltage. The 100-V peak values for V_R and for V_L total 141 V, however, instead of 200 V, because of the 90° phase difference.

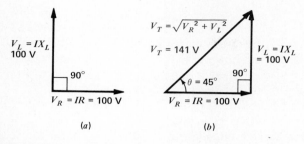

FIGURE 20-3
Addition of two voltages 90° out of phase. (a) Phasors at 90°. (b) Resultant of the two phasors is the hypotenuse of the right triangle.

Consider some instantaneous values to see why the 100-V peak V_R and 100-V peak V_L cannot be added arithmetically. When V_R is at its maximum of 100 V, for instance, V_L is at zero. The total for V_T then is 100 V. Similarly, with V_L at its maximum of 100 V, then V_R is zero and the total V_T is also 100 V.

Actually, V_T has its maximum value of 141 V at the time when V_L and V_R are each 70.7 V. When series voltage drops out of phase are combined, therefore, they cannot be added without taking the phase difference into account.

Phasor-Voltage Triangle. Instead of combining waveforms that are out of phase, we can add them more quickly by using their equivalent phasors, as shown in Fig. 20-3. The phasors in (a) just show the 90° angle without any addition. The method in (b) is to add the tail of one phasor to the arrowhead of the other, using the angle required to show their relative phase. V_R and V_L are at right angles because they are 90° out of phase. The sum of the phasors is a resultant phasor from the start of one to the end of the other. Since the V_R and V_L phasors form a right angle, the resultant phasor is the hypotenuse of a right triangle. The hypotenuse is the side opposite the 90° angle.

From the geometry of a right triangle, the pythagorean theorem states that the hypotenuse is equal to the square root of the sum of the squares of the sides. For the voltage triangle in Fig. 20-3b, therefore, the resultant is

$$V_T = \sqrt{V_R^2 + V_L^2} \qquad (20\text{-}1)$$

where V_T is the phasor sum of the two voltages V_R and V_L 90° out of phase.

This formula is for V_R and V_L when they are in series, since then they are 90° out of phase. All the voltages must be in the same units. When V_A is an rms value, V_R and V_L are also rms values.

In calculating the value of V_T, note that the terms V_R and V_L must each be squared before they are added to find the square root. For the example in Fig. 20-3,

$$V_T = \sqrt{100^2 + 100^2} = \sqrt{10{,}000 + 10{,}000}$$
$$= \sqrt{20{,}000}$$
$$V_T = 141 \text{ V}$$

Practice Problems 20-2
(answers on page 414)

(a) In a series circuit with X_L and R, what is the phase angle between I and V_R?
(b) What is the phase angle between V_R and V_L?

20-3 IMPEDANCE (Z)

A phasor triangle of R and X_L in series corresponds to the voltage triangle, as shown in Fig. 20-4. It is similar to the voltage triangle in Fig. 20-3, but the common factor I cancels because the current is the same in X_L and R. The resultant of the phasor addition of R and X_L is their total opposition in ohms, called impedance, with the symbol Z. The Z takes into account the 90° phase relation between R and X_L.

For the impedance triangle of a series circuit with reactance and resistance

$$Z = \sqrt{R^2 + X_L^2} \qquad (20\text{-}2)$$

With R and X_L in ohms, Z is also in ohms. For the example in Fig. 20-4

$$Z = \sqrt{100^2 + 100^2} = \sqrt{10{,}000 + 10{,}000}$$
$$= \sqrt{20{,}000}$$
$$Z = 141 \ \Omega$$

Note that the total impedance of 141 Ω divided into the applied voltage of 141 V results in 1 A of current in the series circuit. The IR voltage is 1×100, or 100 V; the IX_L voltage is also 1×100, or 100 V. The total of the series IR drops of 100 V each added by phasors

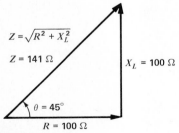

FIGURE 20-4
Phasor addition of R and X_L 90° out of phase in series circuit, to find the resultant impedance Z.

equals the applied voltage of 141 V. Finally, the applied voltage equals IZ, or 1×141, which is 141 V.

To summarize the similar phasor triangles for volts and ohms in a series circuit:

1. The phasor for R, IR, or V_R is at 0°.
2. The phasor for X_L, IX_L, or V_L is at 90°.
3. The phasor for Z, IZ, or V_T has the phase angle θ of the complete circuit.

Phase Angle with Series X_L. The angle between the generator voltage and its current is the phase angle of the circuit. Its symbol is θ (theta). In Fig. 20-3, the phase angle between V_T and IR is 45°. Since IR and I have the same phase, the angle is also 45° between V_T and I.

In the corresponding impedance triangle in Fig. 20-4, the angle between Z and R is also equal to the phase angle. Therefore, the phase angle can be calculated from the impedance triangle of a series circuit by the formula

$$\tan \theta_Z = \frac{X_L}{R} \qquad (20\text{-}3)$$

The tangent (tan) is a trigonometric function[1] of any angle, equal to the ratio of the opposite side to the adjacent side of a triangle. In this impedance triangle, X_L is the opposite side and R is the adjacent side of the angle. We use the subscript $_Z$ for θ to show θ_Z is found from the impedance triangle for a series circuit. To calculate this phase angle

$$\tan \theta_Z = \frac{X_L}{R} = \frac{100}{100} = 1$$

[1] Numerical trigonometry using the sine, cosine, or tangent functions for any angle is explained in B. Grob, "Mathematics Outline and Review Problems for Basic Electronics," McGraw-Hill Book Company, New York.

From the trigonometric table of sines, cosines, and tangents in Appendix E, the angle that has the tangent equal to 1 is 45°. Therefore, the phase angle is 45° in this example. The numerical values of the trigonometric functions can be found from a table, slide rule, or scientific calculator.

Note that the phase angle of 45° is halfway to 90° because R and X_L are equal.

Example 1. If a 30-Ω R and a 40-Ω X_L are in series with 100 V applied, find the following: Z, I, V_R, V_L, and θ_Z. What is the phase of V_L and V_R with respect to the phase of I? Prove that the sum of the series voltage drops equals the applied voltage V_T.

Answer.
$$Z = \sqrt{R^2 + X_L^2} = \sqrt{900 + 1600}$$
$$= \sqrt{2500}$$
$$Z = 50 \ \Omega$$
$$I = \frac{V_T}{Z} = \frac{100}{50} = 2 \ A$$
$$V_R = IR = 2 \times 30 = 60 \ V$$
$$V_L = IX_L = 2 \times 40 = 80 \ V$$
$$\tan \theta_Z = \frac{X_L}{R} = \frac{40}{30} = \frac{4}{3} = 1.33$$
$$\theta_Z = 53°$$

Therefore, I lags V_T by 53°. Furthermore, I and V_R have the same phase, and I lags V_L by 90°. Finally,

$$V_T = \sqrt{V_R^2 + V_L^2} = \sqrt{60^2 + 80^2}$$
$$= \sqrt{3600 + 6400} = \sqrt{10,000}$$
$$V_T = 100 \ V$$

Therefore, the sum of the voltage drops equals the applied voltage.

Series Combinations of X_L and R.

In a series circuit, the higher the value of X_L compared with R, the more inductive the circuit is. This means there is more voltage drop across the inductive reactance and the phase angle increases toward 90°. The series current lags the applied generator voltage. With all X_L and no R, the entire applied voltage is across X_L and θ_Z equals 90°.

Several combinations of X_L and R in series are listed in Table 20-1 with their resultant impedance and phase angle. Note that a ratio of 10:1 or more for X_L/R means that the circuit is practically all inductive. The phase angle of 84.3° is only slightly less than 90° for the ratio of 10:1, and the total impedance Z is approximately equal to X_L. The voltage drop across X_L in the series circuit will be practically equal to the applied voltage, with almost none across R.

At the opposite extreme, when R is ten times as large as X_L, the series circuit is mainly resistive. The phase angle of 5.7°, then, means the current has almost the same phase as the applied voltage, the total impedance Z is approximately equal to R, and the voltage drop across R is practically equal to the applied voltage, with almost none across X_L.

For the case when X_L and R equal each other, their resultant impedance Z is 1.41 times the value of either one. The phase angle then is 45°, halfway between 0° for resistance alone and 90° for inductive reactance alone.

Practice Problems 20-3
(answers on page 414)
(a) How much is Z_T for a 20-Ω R in series with a 20-Ω X_L?
(b) How much is V_T for 20 V across R and 20 V across X_L in series?
(c) What is the phase angle of this circuit?

TABLE 20-1. Series R and X_L Combinations

R, Ω	X_L, Ω	Z, Ω (APPROX.)	PHASE ANGLE θ_Z
1	10	$\sqrt{101} = 10$	84.3°
10	10	$\sqrt{200} = 14$	45°
10	1	$\sqrt{101} = 10$	5.7°

Note: θ_Z is the angle of Z_T with respect to the reference I in a series circuit.

20-4
X_L AND R IN PARALLEL

For parallel circuits with X_L and R, the 90° phase angle must be considered for each of the branch currents, instead of voltage drops in a series circuit. Remember that any series circuit has different voltage drops but with one common current. A parallel circuit has different branch currents but with one common voltage.

In the parallel circuit in Fig. 20-5, the applied voltage V_A is the same across X_L, R, and the generator, since they are all in parallel. There cannot be any phase difference between these voltages. Each branch, however, has its individual current. For the resistive branch, $I_R = V_A/R$; in the inductive branch, $I_L = V_A/X_L$.

The resistive branch current I_R has the same phase as the generator voltage V_A. The inductive branch current I_L lags V_A, however, because the current in an inductance lags the voltage across it by 90°.

The total line current, therefore, consists of I_R and I_L, which are 90° out of phase with each other. The vector sum of I_R and I_L equals the total line current I_T.

In Fig. 20-5b, the phasor sum of 10 A for I_R and 10 A for I_L is equal to 14 A. The branch currents are added by phasors here because they are the factors that are 90° out of phase in a parallel circuit. This method is similar to combining voltage drops 90° out of phase in a series circuit.

Phasor Current Triangle. Note that the phasor diagram in Fig. 20-5c has the applied voltage V_A of the generator as the reference phase. The reason is that V_A is the same throughout the parallel circuit.

The phasor for I_L is down, as compared with up for an X_L phasor. Here the parallel branch current I_L lags the parallel voltage reference V_A. In a series circuit the X_L voltage leads

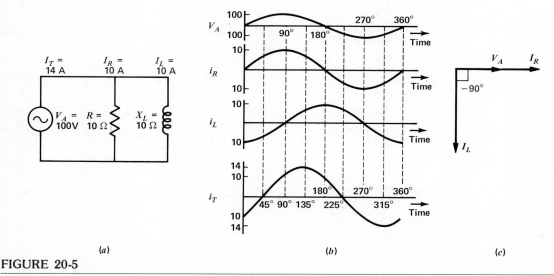

(a)   (b)   (c)

FIGURE 20-5
X_L and R in parallel. (a) Circuit. (b) Waveforms of applied voltage and branch currents. (c) Phasor diagram.

the series current reference I. For this reason the I_L phasor is shown with a negative 90° angle. The −90° means the current I_L lags the reference phasor V_A.

The phasor addition of the branch currents in a parallel circuit can be calculated by the phasor triangle for currents shown in Fig. 20-6. Peak values are used for convenience in this example, but when the applied voltage is an rms value, the calculated currents are also in rms values. To calculate the total line current, we have

$$I_T = \sqrt{I_R^2 + I_L^2} \qquad (20\text{-}4)$$

For the values in Fig. 20-6,

$$I_T = \sqrt{10^2 + 10^2} = \sqrt{100 + 100}$$
$$= \sqrt{200}$$
$$I_T = 14.14 \text{ A}$$

Impedance of X_L and R in Parallel. A practical approach to the problem of calculating the total impedance of X_L and R in parallel is to calculate the total line current I_T and divide this into the applied voltage:

$$Z_T = \frac{V_A}{I_T} \qquad (20\text{-}5)$$

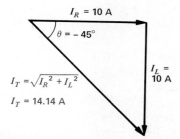

FIGURE 20-6

Phasor triangle of branch currents 90° out of phase in parallel circuit to find I_T.

For example, in Fig. 20-5, V_A is 100 V and the resultant I_T, obtained as the vector sum of the resistive and reactive branch currents, is equal to 14.14 A. Therefore,

$$Z_T = \frac{V_A}{I_T} = \frac{100 \text{ V}}{14.14 \text{ A}}$$
$$Z_T = 7.07 \text{ }\Omega$$

This impedance is the combined opposition in ohms across the generator, equal to the resistance of 10 Ω in parallel with the reactance of 10 Ω.

Note that the impedance for equal values of R and X_L in parallel is not one-half but equals 70.7 percent of either one. Still, the combined value of ohms must be less than the lowest ohms value in the parallel branches.

For the general case of calculating the impedance of X_L and R in parallel, any number can be assumed for the applied voltage because in the calculations for Z in terms of the branch currents the value of V_A cancels. A good value to assume for V_A is the value of either R or X_L, whichever is the higher number. This way there are no fractions smaller than one in calculation of the branch currents.

Example 2. What is the total Z of a 600-Ω R in parallel with a 300-Ω X_L? Assume 600 V for the applied voltage. Then

Answer.
$$I_R = \frac{600 \text{ V}}{600 \text{ }\Omega} = 1 \text{ A}$$
$$I_L = \frac{600 \text{ V}}{300 \text{ }\Omega} = 2 \text{ A}$$
$$I_T = \sqrt{I_R^2 + I_L^2} = \sqrt{1 + 4} = \sqrt{5}$$
$$I_T = 2.24 \text{ A}$$

Then, dividing the total line current into the assumed value of 600 V for the applied voltage gives

$$Z_T = \frac{V_A}{I_T} = \frac{600 \text{ V}}{2.24 \text{ A}}$$
$$Z_T = 268 \text{ }\Omega$$

The combined impedance of a 600-Ω R in parallel with a 300-Ω X_L is equal to 268 Ω, no matter how much the applied voltage is.

Phase Angle with Parallel X_L and R. In a parallel circuit, the phase angle is between the line current I_T and the common voltage V_A applied across all the branches. However, the resistive branch current I_R has the same phase as V_A. Therefore, the phase of I_R can be substituted for the phase of V_A. This is shown in Fig. 20-5c. The triangle of currents is in Fig. 20-6. To find θ_I from the branch currents, use the tangent formula

$$\tan \theta_I = -\frac{I_L}{I_R} \quad (20\text{-}6)$$

We use the subscript $_I$ for θ to show that θ_I is found from the triangle of branch currents in a parallel circuit. In Fig. 20-6, θ_I is $-45°$ because I_L and I_R are equal. Then $\tan \theta_I = -1$.

The negative sign is used for this current ratio because I_L is lagging at $-90°$, compared with I_R. The phase angle of $-45°$ here means that I_T lags I_R and V_A by 45°.

Note that the phasor triangle of branch currents gives θ_I as the angle of I_T with respect to the generator voltage V_A. This phase angle for I_T is with respect to the applied voltage as the reference at 0°. For the phasor triangle of voltages in a series circuit the phase angle θ_Z for Z_T and V_T is with respect to the series current as the reference at 0°.

Parallel Combinations of X_L and R. Several combinations of X_L and R in parallel are listed in Table 20-2. When X_L is ten times R, the parallel circuit is practically resistive because there is little inductive current in the line. The small value of I_L results from the high X_L. The total impedance of the parallel circuit is approximately equal to the resistance, then, since the high value of X_L in a parallel branch has little effect. The phase angle of $-5.7°$ is practically 0° because almost all the line current is resistive.

As X_L becomes smaller, it provides more inductive current in the main line. When X_L is $\frac{1}{10}$ R, practically all the line current is the I_L component. Then the parallel circuit is practically all inductive, with a total impedance practically equal to X_L. The phase angle of $-84.3°$ is almost $-90°$ because the line current is mostly inductive. Note that these conditions are opposite from the case of X_L and R in series.

When X_L and R are equal, their branch currents are equal and the phase angle is $-45°$. All these phase angles are negative for parallel I_L and I_R.

TABLE 20-2. Parallel Resistance and Inductance Combinations*

R, Ω	X_L, Ω	I_R, A	I_L, A	I_T, A (APPROX.)	$Z_T = V_A/I_T$, Ω	PHASE ANGLE θ_I
1	10	10	1	$\sqrt{101} = 10$	1	$-5.7°$
10	10	1	1	$\sqrt{2} = 1.4$	7.07	$-45°$
10	1	1	10	$\sqrt{101} = 10$	1	$-84.3°$

*$V_A = 10$ V. Note that θ_I is the angle of I_T with respect to the reference V_A in parallel circuits.

As additional comparisons between series and parallel circuits, remember that

1. The series voltage drops V_R and V_L have individual values that are 90° out of phase. Therefore, V_R and V_L are added by phasors to equal the applied voltage V_T. The phase angle θ_Z is between V_T and the common series current I. More series X_L allows more V_L to make the circuit more inductive with a larger positive phase angle for V_T with respect to I.
2. The parallel branch currents I_R and I_L have individual values that are 90° out of phase. Therefore, I_R and I_L are added by phasors to equal I_T, which is the main-line current. The negative phase angle $-\theta_I$ is between the line current I_T and the common parallel voltage V_A. Less parallel X_L allows more I_L to make the circuit more inductive with a larger negative phase angle for I_T with respect to V_A.

Practice Problems 20-4
(answers on page 414)
(a) How much is I_T for a branch current I_R of 2 A and I_L of 2 A?
(b) Find the phase angle θ_I.

20-5
Q OF A COIL

The ability of a coil to produce self-induced voltage is indicated by X_L, since it includes the factors of frequency and inductance. However, a coil has internal resistance equal to the resistance of the wire in the coil. This internal r_i of the coil reduces the current, which means less ability to produce induced voltage. Combining these two factors of X_L and r_i, the quality or merit of a coil is indicated by

$$Q = \frac{X_L}{r_i} = \frac{2\pi f L}{r_i} \qquad (20\text{-}7)$$

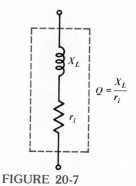

FIGURE 20-7
Q of a coil.

As shown in Fig. 20-7, the internal r_i is in series with X_L.

As an example, a coil with X_L of 500 Ω and r_i of 5 Ω has a Q of $500/5 = 100$. The Q is a numerical value without any units since the ohms cancel in the ratio of reactance to resistance. This Q of 100 means that the X_L of the coil is 100 times more than its r_i.

The Q of coils may range in value from less than 10 for a low-Q coil up to 1000 for a very high Q. Rf coils generally have a Q of about 30 to 300.

At low frequencies, r_i is just the dc resistance of the wire in the coil. However, for rf coils the losses increase with higher frequencies and the effective r_i increases. The increased resistance results from eddy currents and other losses.

Because of these losses, the Q of a coil does not increase without limit as X_L increases for higher frequencies. Generally, the Q can increase by a factor of about 2 for higher frequencies, within the range for which the coil is designed. The highest Q for rf coils generally results with an inductance value that provides an X_L of about 1000 Ω at the operating frequency.

More fundamentally, Q can be defined as

the ratio of reactive power in the inductance to the real power dissipated in the resistance. Then

$$Q = \frac{P_L}{P_{r_i}} = \frac{I^2 X_L}{I^2 r_i} = \frac{X_L}{r_i} = \frac{2\pi f L}{r_i}$$

which is the same as Formula (20-7).

Skin effect. Rf current tends to flow at the surface of a conductor, at very high frequencies, with little current in the solid core at the center. This skin effect results from the fact that current in the center of the wire encounters slightly more inductance because of the magnetic flux concentrated in the metal, compared with the edges where part of the flux is in air. For this reason, conductors for VHF currents are often made of hollow tubing. The skin effect increases the effective resistance, as a smaller cross-sectional area is used for the current path in the conductor.

AC Effective Resistance. When the power and current supplied to a coil are measured for rf applied voltage, the $I^2 R$ loss corresponds to a much higher resistance than the dc resistance measured with an ohmmeter. This higher resistance is the ac effective resistance R_e. Although a result of high-frequency alternating current, R_e is not a reactance. R_e is a resistive component because it draws in-phase current from the ac voltage source.

The factors that make the R_e of a coil more than its dc resistance include skin effect, eddy currents, and hysteresis losses. Air-core coils have low losses but are limited to small values of inductance.

For a magnetic core in rf coils, a powdered-iron or ferrite slug is generally used. In a powdered-iron slug, the granules of iron are insulated from each other to reduce eddy currents. Ferrite materials have small eddy-current

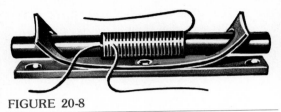

FIGURE 20-8

Ferrite coil antenna for radio receiver. Length is 4 in; inductance is 700 µH. (*J. W. Miller Co.*)

losses as they are insulators, although magnetic. A ferrite core is easily saturated. Therefore, its use must be limited to coils with low values of current. A common application is the ferrite-core antenna coil in Fig. 20-8.

To reduce the R_e for small rf coils, stranded wire can be made with separate strands insulated from each other and braided so that each strand is as much on the outer surface as all the other strands. This is called *litzendraht* or *litz wire*.

As an example of the total effect of ac losses, assume that an air-core rf coil of 50-µH inductance has a resistance of 1 Ω with the dc measurement of the battery in an ohmmeter. However, in an ac circuit with a 2-MHz current, the effective coil resistance R_e can increase to 12 Ω. The increased resistance reduces the Q of the coil.

Actually, the Q can be used to determine the effective ac resistance. Since Q is X_L/R_e, then R_e equals X_L/Q. For this 50-µH L at 2 MHz its X_L, equal to $2\pi f L$, is 628 Ω. The Q of the coil can be measured on a Q-meter, which operates on the principle of resonance. Let the measured Q be 50. Then $R_e = {}^{628}\!/_{50}$, equal to 12.6 Ω.

Example 3. An air-core coil has an X_L of 700 Ω and an R_e of 2 Ω. Calculate Q.

Answer. $Q = \dfrac{X_L}{R_e} = \dfrac{700}{2}$

$Q = 350$

Example 4. A 200-μH coil has a Q of 40 at 0.5 MHz. Find R_e.

Answer.
$$R_e = \frac{X_L}{Q} = \frac{2\pi fL}{Q}$$
$$= \frac{2\pi \times 0.5 \times 10^6 \times 200 \times 10^{-6}}{40}$$
$$= \frac{628}{40}$$
$$R_e = 15.7 \: \Omega$$

In general, the less the internal resistance for a coil, the higher is its Q.

Practice Problems 20-5
(answers on page 414)
(a) A 200-μH coil with an 8-Ω internal R_e has an X_L of 600 Ω. Calculate the Q.
(b) A coil with a Q of 50 has a 500-Ω X_L at 4 MHz. Calculate its internal R_e.

20-6
AF AND RF CHOKES

Inductance has the useful characteristic of providing more ohms of reactance at higher frequencies. Resistance has the same opposition at all frequencies and for direct current. These characteristics are applied to the circuit in Fig. 20-9, where X_L is much greater than R for the frequency of the ac source V_T. The result is that L has practically all the voltage drop in this series circuit with very little of the applied voltage across R.

The inductance L is used here as a *choke*. Therefore, a choke is an inductance in series with an external R to prevent the ac signal voltage from developing any appreciable output across R, at the frequency of the source.

The dividing line in calculations for a choke can be taken as X_L ten or more times the series R. Then the circuit is primarily inductive. Practically all the ac voltage drop is across L, with little across R. This case also results in θ of practically 90°, but the phase angle is not related to the action of X_L as a choke.

Figure 20-9b illustrates how a choke is used to prevent ac voltage in the input from developing voltage in the output for the next circuit. Note that the output here is V_R from point A to chassis ground. Practically all the ac input voltage is across X_L between points B and C. However, this voltage is not coupled out because neither B nor C is grounded.

The desired output across R could be di-

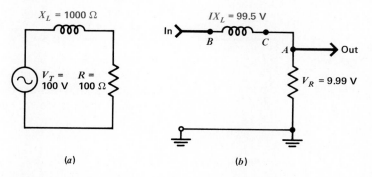

(a)   (b)

FIGURE 20-9
Coil used as a choke with X_L at least 10 × R. The V_L is practically all the applied voltage with almost none for V_R. (a) Equivalent circuit. (b) Actual circuit for input and output voltages.

Chapter 20
Inductive Circuits

TABLE 20-3. Typical Chokes for a Reactance of 1000 Ω*

F	L	REMARKS
100 Hz	1.6 H	Low audio frequency
1000 Hz	0.16 H	Medium audio frequency
10 kHz	16 mH	High audio frequency
1000 kHz	0.16 mH	Radio frequency
100 MHz	1.6 μH	Very high radio frequency

*For an X_L ten times a series R of 100 Ω.

rect current from the input side. Then X_L has no effect. A much lower frequency of ac voltage also could produce output voltage across R, as the same L would have too little X_L for a choke at the lower frequency.

Calculations for a Choke. Typical values for audio or radio frequencies can be calculated if we assume a series resistance of 100 Ω, as an example. Then X_L must be at least 1000 Ω. As listed in Table 20-3, at 100 Hz the relatively large inductance of 1.6 H provides 1000 Ω of X_L. Higher frequencies allow a smaller value of L for a choke with the same reactance. At 100 MHz, in the VHF range, the choke is only 1.6 μH.

Some typical chokes are shown in Fig. 20-10. The iron-core choke in (a) is for audio frequencies. The air-core choke in (b) is for radio frequencies. The rf choke in (c) has color coding, which is often used for small coils. The color values are the same as for resistors, but with L in microhenrys. As an example, a coil with yellow, red, and black stripes or dots is 42 μH.

Choosing a Choke for a Circuit. As an example of using these calculations, suppose that we have the problem of determining what kind of a coil to use as a choke for the following application. L is to be an rf choke in series with an external R of 300 Ω, with a current of 90 mA and a frequency of 0.2 MHz. Then X_L must be at least $10 \times 300 = 3000$ Ω. At f of 0.2 MHz,

$$L = \frac{X_L}{2\pi f} = \frac{30{,}000}{2\pi \times 2 \times 10^6} = \frac{30 \times 10^3}{12.56 \times 10^6}$$

$$= \frac{30}{12.56} \times 10^{-3}$$

$$L = 2.4 \text{ mH}$$

A typical commercial size easily available is 2.5 mH, with a current rating of 115 mA and an internal resistance of 20 Ω, similar to the rf choke in Fig. 20-10b. Note that the higher current rating is suitable. Also, the internal resistance is negligible compared with the external R. An inductance a little higher than the calculated value will provide more X_L, which is better for a choke.

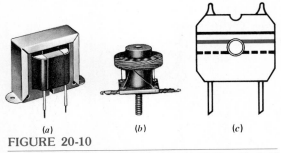

(a)   (b)   (c)

FIGURE 20-10

Typical chokes. (a) Choke for 60 Hz with 8-H inductance and r_i of 350 Ω. Width is 4 in. (b) Rf choke with 5-mH inductance and r_i of 50 Ω. Height is 1 in. (c) Small rf choke in plastic with leads for printed-circuit board. L is 42 μH. Width is ¾ in.

Practice Problems 20-6
(answers on page 414)

(a) How much is the minimum X_L for a choke in series with R of 80 Ω?
(b) If X_L is 800 Ω at 3 MHz, how much will X_L be at 6 MHz for the same coil?

20-7
THE GENERAL CASE OF INDUCTIVE VOLTAGE

The voltage across any inductance in any circuit is always equal to $L(di/dt)$. This formula gives the instantaneous values of v_L based on the self-induced voltage which is produced by a change in magnetic flux associated with a change in current.

A sine waveform of current i produces a cosine waveform for the induced voltage v_L, equal to $L(di/dt)$. This means v_L has the same waveform as i, but they are 90° out of phase for sine-wave variations.

The inductive voltage can be calculated as IX_L in sine-wave ac circuits. Since X_L is $2\pi fL$, the factors that determine the induced voltage are included in the frequency and inductance. Usually, it is more convenient to work with IX_L for the inductive voltage in sine-wave ac circuits, instead of $L(di/dt)$.

However, with a nonsinusoidal current waveform, the concept of reactance cannot be used. X_L applies only to sine waves. Then v_L must be calculated as $L(di/dt)$, which applies for any inductive voltage.

An example is illustrated in Fig. 20-11a for sawtooth current. This waveform is often used in the deflection circuits for the picture tube in television receivers. The sawtooth rise is a uniform or linear increase of current from zero to 90 mA in this example. The sharp drop in current is from 90 mA to zero. Note that the rise is relatively slow; it takes 90 μs. This is nine times longer than the fast drop in 10 μs.

The complete period of one cycle of this sawtooth wave is 100 μs. A cycle includes the rise of i to the peak value and its drop back to the starting value.

The Slope of i. The slope of any curve is a measure of how much it changes vertically for each horizontal unit. In Fig. 20-11a the increase of current has a constant slope. Here i increases 90 mA in 90 μs, or 10 mA for every 10 μs of time. Then di/dt is constant at 10 mA/10 μs for the entire rise time of the sawtooth waveform. Actually di/dt is the slope of the i curve. This is why the v_L waveform has a constant value of voltage during the linear rise of i.

The drop in i is also linear but much faster.

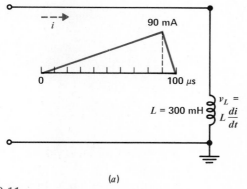
(a)

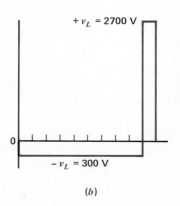
(b)

FIGURE 20-11

Rectangular waveshape of v_L produced by sawtooth current. (a) Waveform of i. (b) Induced voltage equal to $L(di/dt)$.

During this time, the slope is 90 mA/10 μs for di/dt.

The Polarity of v_L. In Fig. 20-11, apply Lenz' law to indicate that v_L opposes the change in current. With electron flow into the top of L, the v_L is negative to oppose an increase of current. This polarity opposes the direction of electron flow shown for the current i produced by the source. For the rise time, then, the induced voltage here is labeled $-v_L$.

During the drop of current, the induced voltage has opposite polarity, which is labeled $+v_L$. These voltage polarities are for the top of L with respect to chassis ground.

Calculations for v_L. The values of induced voltage across the 300-mH L are calculated as follows:

For the sawtooth rise:

$$-v_L = L\frac{di}{dt} = 300 \times 10^{-3} \times \frac{10 \times 10^{-3}}{10 \times 10^{-6}}$$

$$-v_L = 300 \text{ V}$$

For the sawtooth drop:

$$+v_L = L\frac{di}{dt} = 300 \times 10^{-3} \times \frac{90 \times 10^{-3}}{10 \times 10^{-6}}$$

$$+v_L = 2700 \text{ V}$$

The decrease in current produces nine times more voltage because the sharp drop in i is nine times faster than the relatively slow rise.

Remember that the di/dt factor can be very large, even with small currents, when the time is short. For instance, a current change of 1 mA in 1 μs is equivalent to the very high di/dt value of 1000 A/s.

In conclusion, it is important to note that v_L and i_L have different waveshapes with nonsinusoidal current. In this case, we compare the waveshapes instead of the phase angle in sine-wave circuits. Common examples of nonsinusoidal waveshapes for either v or i are the sawtooth waveform, square wave, and rectangular pulses.

Practice Problems 20-7
(answers on page 414)
Refer to Fig. 20-11.
(a) How much is di/dt in A/s for the sawtooth rise of i?
(b) How much is di/dt in A/s for the drop in i?

20-8 CALCULATING THE L/R TIME CONSTANT

With nonsinusoidal waveforms of i, the reaction of L to di/dt is the *transient response*, meaning a temporary result of a sudden change in i. The transient response of an inductive circuit is measured in terms of the ratio L/R, which is the time constant. In Fig. 20-12, L/R is the time for I to increase 63.2 percent when dc voltage is suddenly applied by the switch S.

To calculate the time constant

$$T = \frac{L}{R} \quad \text{s} \tag{20-8}$$

where T is the time constant and L the inductance in henry units, and R is in series with L. The R may be the internal coil resistance, an external resistance, or both in series. In Fig. 20-12

$$T = \frac{L}{R} = \frac{1 \text{ H}}{10 \text{ }\Omega} = 0.1 \text{ s}$$

Remember that if L were not present, the current in Fig. 20-12 would rise to $^{10}/_{10} = 1$ A instantly. Eventually, I will rise to 1 A, which is the steady-state value determined by V/R, and stay there. At the first instant when S is closed, though, the transient response of L opposes the

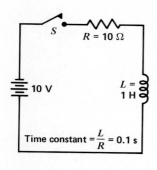
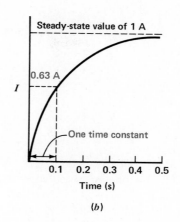

FIGURE 20-12
Transient response of inductive circuit. (a) Circuit. (b) Time for a 63 percent change in i is equal to L/R.

increase in I from zero. The time constant of 0.1 s here specifies that I will rise 63 percent or to 0.63 A in 0.1 s. I is the same in R and L, because they are in series.

The current will rise practically all the way to its steady-state value of 1 A in approximately 0.5 s, which is equal to five time constants. In other words, five time constants is the time for the inductance to complete its transient response and allow the steady-state value determined by R.

In general, the time constant always measures a 63 percent change. When i is increasing, it rises by 63 percent in one time constant. When i is decreasing, it drops by 63 percent in one time constant. For any part of the transient response curve, during one time constant, the value changes by 63 percent.

On the rise, the curve goes from zero to the steady-state value at the top in five time constants. For a decrease in i, the values drop from the steady-state value to zero in five time constants. More details of the L/R time constant and a comparison with the RC time constant for capacitive circuits are explained in Chap. 24.

Practice Problems 20-8
(answers on page 414)
(a) Calculate the time constant for L of 240 mH in series with 20 Ω.
(b) If the time constant in Fig. 20-12 were 12 ms, how long would it take for I to rise to 0.63 A?

Summary

1. In a sine-wave ac circuit, the current through an inductance lags 90° behind the voltage across the inductance because $v_L = L(di/dt)$. This fundamental fact is the basis of all the following relations.

Chapter 20
Inductive Circuits

TABLE 20-4. Comparison of R, X_L, and Z

R	$X_L = 2\pi f L$	$Z = \sqrt{R^2 + X_L^2}$
Ohms unit	Ohms unit	Ohms unit
IR voltage same phase as I	IX_L voltage leads I by 90°	IZ is applied voltage
Same for all frequencies	Increases at higher frequencies	Increases with X_L at higher frequencies

2. Therefore, inductive reactance X_L is a phasor quantity 90° out of phase with R. The phasor combination of X_L and R is their impedance Z.
3. These three types of opposition to current are compared in Table 20-4.
4. The phase angle θ is the angle between the applied voltage and its current.
5. The opposite characteristics for series and parallel circuits with X_L and R are summarized in Table 20-5.
6. The Q of a coil is X_L/r_i, where r_i is its internal resistance.
7. A choke is an inductance with X_L greater than the series R by a factor of 10 or more, for the purpose of providing practically all the ac applied voltage across L with little voltage across R.
8. In sine-wave circuits calculate V_L as IX_L. Then V_L has a phase angle 90° different from the current.
9. When the current is not a sine wave, figure $V_L = L(di/dt)$. Then the waveshape of V_L is different from the waveshape of current.
10. The time constant, equal to L/R, is the time in seconds for I to change by 63.2 percent, with L in henrys and R in ohms in series.

TABLE 20-5. Series and Parallel RL Circuits

X_L AND R IN SERIES	X_L AND R IN PARALLEL
I the same in X_L and R	V_A the same across X_L and R
$V_T = \sqrt{V_R^2 + V_L^2}$	$I_T = \sqrt{I_R^2 + I_L^2}$
$Z = \sqrt{R^2 + X_L^2}$	$Z = \dfrac{V_A}{I_T}$
V_L leads V_R by 90°	I_L lags I_R by 90°
$\tan \theta_Z = \dfrac{X_L}{R}$; θ increases as more X_L makes circuit inductive	$\tan \theta_I = -\dfrac{I_L}{I_R}$; $-\theta$ decreases as more X_L means less I_L

Self-Examination (Answers at back of book.)

Choose (a), (b), (c), or (d).

1. In a sine-wave ac circuit with inductive reactance, the (a) phase angle of the circuit is always 45°; (b) voltage across the inductance must be 90° out of phase with the applied voltage; (c) current through the inductance lags its induced voltage by 90°; (d) current through the inductance and voltage across it are 180° out of phase.
2. In a sine-wave ac circuit with X_L and R in series, the (a) voltages across R and X_L are in phase; (b) voltages across R and X_L are 180° out of phase; (c) voltage across R lags the voltage across X_L by 90°; (d) voltage across R leads the voltage across X_L by 90°.
3. In a sine-wave ac circuit with a 40-Ω R in series with a 30-Ω X_L, the total impedance Z equals (a) 30 Ω; (b) 40 Ω; (c) 50 Ω; (d) 70 Ω.
4. In a sine-wave ac circuit with a 90-Ω R in series with a 90-Ω X_L, phase angle θ equals (a) 0°; (b) 30°; (c) 45°; (d) 90°.
5. A 250-μH inductance is used as a choke at 10 MHz. At 12 MHz the choke (a) does not have enough inductance; (b) has more reactance; (c) has less reactance; (d) needs more turns.
6. The combined impedance of a 1000-Ω R in parallel with a 1000-Ω X_L equals (a) 500 Ω; (b) 707 Ω; (c) 1000 Ω; (d) 2000 Ω.
7. A coil with a 1000-Ω X_L at 3 MHz and 10-Ω internal resistance has a Q of (a) 3; (b) 10; (c) 100; (d) 1000.
8. In a sine-wave ac circuit with a resistive branch and inductive branch in parallel, the (a) voltage across the inductance leads the voltage across the resistance by 90°; (b) resistive branch current is 90° out of phase with the inductive branch current; (c) resistive and inductive branch currents have the same phase; (d) resistive and inductive branch currents are 180° out of phase.
9. With a 2-A I_R and a 2-A I_L in parallel branches, I_T is (a) 1 A; (b) 2 A; (c) 2.8 A; (d) 4 A.
10. In Fig. 20-11 di/dt for the drop in sawtooth current is (a) 90 mA/s; (b) 100 mA/s; (c) 100 A/s; (d) 9000 A/s.

Essay Questions

1. What characteristic of the current in an inductance determines the amount of induced voltage? State briefly why.
2. Draw a schematic diagram showing an inductance connected across

a sine-wave voltage source and indicate the current and voltage that are 90° out of phase.
3. Why does the voltage across a resistance have the same phase as the current through the resistance?
4. (a) Draw the sine waveforms for two voltages 90° out of phase, each with a peak value of 100 V. (b) Explain why their vector sum equals 141 V and not 200 V. (c) When will the sum of two 100-V drops in series equal 200 V?
5. (a) Define the phase angle of a sine-wave ac circuit. (b) State the formula for the phase angle in a circuit with X_L and R in series.
6. Define the following: (a) Q of a coil; (b) ac effective resistance; (c) rf choke; (d) sawtooth current.
7. Referring to Fig. 20-2, why do the waveshapes shown in (b) all have the same frequency?
8. Describe how to check the trouble of an open choke with an ohmmeter.
9. Redraw the circuit and graph in Fig. 20-11 for a sawtooth current with a peak of 30 mA.
10. Why is the R_e of a coil considered resistance rather than reactance?
11. Define the time constant of an inductive circuit.

Problems (Answers to odd-numbered problems at back of book.)

1. Draw the schematic diagram of a circuit with X_L and R in series across a 100-V source. Calculate Z, I, IR, IX_L, and θ, approximately, for the following values: (a) 100-Ω R, 1-Ω X_L; (b) 1-Ω R, 100-Ω X_L; (c) 50-Ω R, 50-Ω X_L.
2. Draw the schematic diagram of a circuit with X_L and R in parallel across a 100-V source. Calculate I_R, I_L, I_T, and Z for the following values: (a) 100-Ω R, 1-Ω X_L; (b) 1-Ω R, 100-Ω X_L; (c) 50-Ω R, 50-Ω X_L.
3. A coil has an inductance of 1 H and a 100-Ω internal resistance. (a) Draw the equivalent circuit of the coil showing its internal resistance in series with its inductance. (b) How much is the coil's inductive reactance at 60 Hz? (c) How much is the total impedance of the coil at 60 Hz? (d) How much current will flow when the coil is connected across a 100-V source with a frequency of 60 Hz? (e) How much is I with an f of 400 Hz?
4. Calculate the minimum inductance required for a choke in series with a resistance of 100 Ω when the frequency of the current is

5 kHz, 5 MHz, and 50 MHz. Do the same for the case where the series resistance is 10 Ω.

5. How much is the impedance Z of a coil that allows 0.3-A current when connected across a 120-V 60-Hz source? How much is the X_L of the coil if its resistance is 5 Ω? (Hint: $X_L^2 = Z^2 - R^2$.)
6. A 200-Ω R is in series with L across a 141-V 60-Hz generator V_T. The V_R is 100 V. Find L. (Hint: $V_L^2 = V_T^2 - V_R^2$.)
7. A 350-μH L has a Q of 35 at 1.5 MHz. Calculate the effective ac resistance R_e.
8. How much L is required to produce V_L equal to 6 kV when i_L drops from 300 mA to zero in 8 μs?
9. A 400-Ω R and 400-Ω X_L are in series with a 100-V 400-Hz source. Find Z, I, V_L, V_R, and θ_Z.
10. The same R and X_L of Prob. 9 are in parallel. Find I_R, I_L, I_T, Z, and θ_I.
11. The frequency is raised to 800 Hz for the parallel circuit in Prob. 10. Compare the values of I_R, I_L, and θ_I for the two frequencies of 400 and 800 Hz.
12. A 0.4-H L and a 180-Ω R are in series across a 120-V 60-Hz source. Find the current I and θ_Z.
13. An inductance L has 20 V across it, at 40 mA. The frequency is 5 kHz. Calculate X_L in ohms and L in henrys.
14. A 500-Ω R is in series with 300-Ω X_L. Find Z_T, I, and θ_Z. $V_T = 120$ V.
15. A 300-Ω R is in series with a 500-Ω X_L. Find Z_T, I, and θ_Z. Compare θ_Z here with Prob. 14, with the same 120 V applied.
16. A 500-Ω R is in parallel with 300-Ω X_L. Find I_T, Z_T, and θ_I. Compare θ_I here with θ_Z in Prob. 14, with the same 120 V applied.
17. The current shown in Fig. 20-13 flows through a 20-mH inductance. Show the corresponding waveform of induced voltage with values.

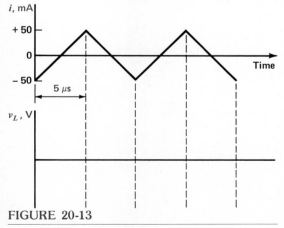

FIGURE 20-13

For Prob. 17.

Answers to Practice Problems

20-1 (a) $0°$
(b) $90°$
(c) lag

20-2 (a) $0°$
(b) $90°$

20-3 (a) $28.28\ \Omega$
(b) 28.28 V
(c) $\theta_Z = 45°$

20-4 (a) $I_T = 2.824$ A
(b) $\theta_I = -45°$

20-5 (a) $Q = 75$
(b) $R_e = 10\ \Omega$

20-6 (a) $X_L = 800\ \Omega$
(b) $X_L = 1600\ \Omega$

20-7 (a) $di/dt = 1000$ A/s
(b) $di/dt = 9000$ A/s

20-8 (a) 12 ms
(b) 12 ms

Review of Chapters 18 to 20

Summary

1. The ability of a conductor to produce induced voltage across itself when the current changes is its self-inductance, or inductance. The symbol is L, and the unit is the henry. One henry allows 1 V to be induced when the current changes at the rate of 1 A/s.
2. The polarity of the induced voltage always opposes the change in current that is causing the induced voltage. This is Lenz' law.
3. Mutual inductance is the ability of varying current in one coil to induce voltage in another coil nearby, without any connection between them. Its symbol is L_M and the unit is also the henry.
4. A transformer consists of two or more windings with mutual inductance. The primary connects to the source voltage, the secondary to the load. With an iron core, the voltage ratio between primary and secondary equals the turns ratio.
5. Efficiency of a transformer equals the ratio of power output from the secondary to power input to the primary, $\times$ 100 percent.
6. Eddy currents are induced in the iron core of an inductance, causing I^2R losses that increase with higher frequencies. Laminated-iron, powdered-iron, or ferrite cores have minimum eddy-current losses. Hysteresis also increases the losses.
7. Series inductances without mutual coupling add like series resistances; with parallel inductances, the combined inductance is calculated by the reciprocal formula.
8. Inductive reactance X_L equals $2\pi fL$ Ω, where f is in hertz and L in henrys. X_L increases with more inductance and higher frequencies.
9. A common application of X_L is an af or rf choke, which has high reactance for one group of frequencies but less reactance for lower frequencies.
10. X_L is a phasor quantity that has its current lagging 90° behind its induced voltage. In series circuits, R and X_L are added by phasors

because their voltage drops are 90° out of phase; in parallel circuits, the resistive and inductive branch currents are 90° out of phase.
11. Impedance Z, in ohms, is the total opposition of an ac circuit with resistance and reactance. For series circuits, $Z = \sqrt{R^2 + X_L^2}$ and $I = V_T/Z$. For parallel circuits, $I_T = \sqrt{I_R^2 + I_L^2}$ and $Z = V_A/I_T$.
12. The Q of a coil is X_L/r_i.
13. Energy stored by an inductance is $\frac{1}{2} LI^2$. With I in amperes, and L in henrys, the energy is in joules.
14. L/R is the time constant in seconds for I to change by 63.2 percent. The L is in henrys and R in ohms.
15. The voltage across L is always equal to $L(di/dt)$ for any waveshape of current.

Review Self-Examination (Answers at back of book.)

Choose (a), (b), (c), or (d).

1. A coil induces 200 mV when the current changes at the rate of 1 A/s. The inductance L is: (a) 1 mH; (b) 2 mH; (c) 200 mH; (d) 1000 mH.
2. Alternating current in an inductance produces maximum induced voltage when the current has its (a) maximum value; (b) maximum change in magnetic flux; (c) minimum change in magnetic flux; (d) rms value of 0.707 × peak.
3. An iron-core transformer connected to the 120-V 60-Hz power line has a turns ratio of 20:1. The voltage across the secondary equals (a) 20 V; (b) 60 V; (c) 120 V; (d) 2400 V.
4. Two 250-mH chokes in series have a total inductance of (a) 60 mH; (b) 125 mH; (c) 250 mH; (d) 500 mH.
5. Which of the following will have minimum eddy-current losses? (a) iron core; (b) laminated iron core; (c) powdered-iron core; (d) air core.
6. Which of the following will have maximum inductive reactance? (a) 2-H inductance at 60 Hz; (b) 2-mH inductance at 60 kHz; (c) 5-mH inductance at 60 kHz; (d) 5-mH inductance at 100 kHz.
7. A 100-Ω R is in series with 100 Ω of X_L. The total impedance Z equals (a) 70.7 Ω; (b) 100 Ω; (c) 141 Ω; (d) 200 Ω.
8. A 100-Ω R is in parallel with 100 Ω of X_L. The total impedance Z equals (a) 70.7 Ω; (b) 100 Ω; (c) 141 Ω; (d) 200 Ω.
9. If two waves have the frequency of 1000 Hz and one is at the maximum value when the other is at zero, the phase angle between them is (a) 0°; (b) 90°; (c) 180°; (d) 360°.

10. If an ohmmeter check on a 50-μH choke reads 3 Ω, the coil is probably (a) open; (b) defective; (c) normal; (d) partially open.

References (Additional references at back of book.)

Gillie, A. C.: "Electrical Principles of Electronics," McGraw-Hill Book Company, New York.

Oppenheimer, S. L., F. R. Hess and J. P. Borchers: "Direct and Alternating Currents," 2d ed., McGraw-Hill Book Company, New York.

Slurzberg, M. and W. Osterheld: "Essentials of Electricity and Electronics," 3d ed., McGraw-Hill Book Company, New York.

Capacitance

Chapter 21

Just as inductance is important for variations of current in a wire, capacitance is a similar but opposite characteristic that is important when the voltage changes across an insulator or dielectric. Specifically, capacitance is the ability of a dielectric to store electric charge. The unit is the farad, named after Michael Faraday.

The types of capacitances are named according to the dielectric. Most common are air, paper, mica, ceramic, and electrolytic capacitors. This chapter explains how capacitance becomes charged by a voltage source, and how the capacitor discharges. Commercial types of capacitors are shown in Figs. 21-4 to 21-10, with their color coding. Typical troubles of an open or short are explained, including the method of checking capacitors with an ohmmeter. The topics are

21-1 How Charge Is Stored in the Dielectric
21-2 Charging and Discharging a Capacitor
21-3 The Farad Unit of Capacitance
21-4 Typical Capacitors
21-5 Capacitor Color Coding
21-6 Parallel Capacitances
21-7 Series Capacitances
21-8 Stray Capacitive and Inductive Effects
21-9 Energy in Electrostatic Field of Capacitance
21-10 Troubles in Capacitors

21-1
HOW CHARGE IS STORED IN THE DIELECTRIC

It is possible for dielectric materials such as air or paper to hold an electric charge because free electrons cannot flow through an insulator. However, the charge must be applied by some source. In Fig. 21-1a, the battery can charge the capacitor shown. With the dielectric contacting the two conductors connected to the potential difference V, electrons from the voltage source accumulate on the side of the capacitor connected to the negative terminal of V. The opposite side of the capacitor connected to the positive terminal of V loses electrons.

As a result, the excess of electrons produces a negative charge on one side of the capacitor, while the opposite side has a positive charge. As an example, if 6.25×10^{18} electrons are accumulated, the negative charge equals 1 C. The charge on only one plate need be considered, as the number of electrons accumulated on one plate is exactly the same as the number taken from the opposite plate.

What the voltage source does is simply redistribute some electrons from one side of the capacitor to the other side. This process is charging the capacitor. The charging continues until the potential difference across the capacitor is equal to the applied voltage. Without any

Chapter 21
Capacitance

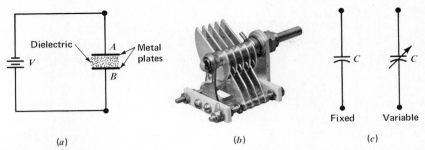

FIGURE 21-1

Capacitance stores charge in the dielectric between two conductors. (*a*) Structure. (*b*) Air-dielectric variable capacitor. Length is 2 in. (*c*) Schematic symbols for fixed and variable capacitances.

series resistance, the charging is instantaneous. Practically, however, there is always some series resistance. This charging current is transient, or temporary, as it flows only until the capacitor is charged to the applied voltage. Then there is no current in the circuit.

The result is a device for storing charge in the dielectric. Storage means that the charge remains even after the voltage source is disconnected. The measure of how much charge can be stored is the capacitance[1] C. More charge stored for a given amount of applied voltage means more capacitance. Components made to supply a specified amount of capacitance are called *capacitors*, or by their old name *condensers*.

Electrically, then, capacitance is the ability to store charge. Physically, a capacitance consists simply of two conductors separated by an insulator. For example, Fig. 21-1*b* shows a capacitor using air for the dielectric between the metal plates. There are many types with different dielectric materials, including paper, mica, and ceramics, but the schematic symbols shown in (*c*) apply to all capacitors.

Electric Field in the Dielectric. Any voltage has a field of electric lines of force between the opposite electric charges. The electric field corresponds to the magnetic lines of force of the magnetic field associated with electrical current.[2] What a capacitor does is concentrate the electric field in the dielectric between the plates. This concentration corresponds to a magnetic field concentrated in the turns of a coil. The only function of the capacitor plates and wire conductors is to connect the voltage source V across the dielectric. Then the electric field is concentrated in the capacitor, instead of being spread out in all directions.

Electrostatic Induction. The capacitor has opposite charges because of electrostatic induction by the electric field. Electrons that accumulate on the negative side of the capacitor provide electric lines of force that repel electrons from the opposite side. When this side

---

[1] The symbol F, for farad, has been recommended for capacitance, but C is used in this book.

[2] Electric and magnetic fields are compared in Fig. 14-6, Chap. 14.

loses electrons, it becomes positively charged. The opposite charges induced by an electric field correspond to the idea of opposite poles induced in magnetic materials by a magnetic field.

Practice Problems 21-1
(answers on page 441)
(a) In a capacitor, is the electric charge stored in the dielectric or in the metal plates?
(b) What is the unit of capacitance?

21-2
CHARGING AND DISCHARGING A CAPACITOR

These are the two main effects with capacitors. Applied voltage puts charge in the capacitor. The accumulation of charge results in a buildup of potential difference across the capacitor plates. When the capacitor voltage equals the applied voltage, there is no more charging. The charge remains in the capacitor, with or without the applied voltage connected.

The capacitor discharges when a conducting path is provided between the plates, without any applied voltage. Actually, it is only necessary that the capacitor voltage be more than the applied voltage. Then the capacitor can serve as voltage source, temporarily, to produce discharge current in the discharge path. The capacitor discharge continues until the capacitor voltage drops to zero or is equal to the applied voltage.

Applying the Charge. In Fig. 21-2a, the capacitor is neutral with no charge because it has not been connected to any source of applied voltage and there is no electrostatic field in the dielectric. Closing the switch in Fig. 21-2b, however, allows the negative battery terminal to repel free electrons in the conductor to plate A. At the same time, the positive terminal attracts free electrons from plate B. The side of the dielectric at plate A accumulates electrons because they cannot flow through the insulator, while plate B has an equal surplus of protons.

Remember that the opposite charges have an associated potential difference, which is the voltage across the capacitor. The charging process continues until the capacitor voltage equals the battery voltage, which is 10 V in this example. Then no further charging is possible

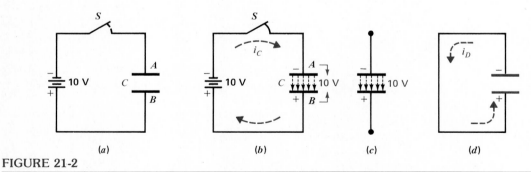

FIGURE 21-2

Storing charge in a capacitance. (a) Capacitor with no charge. (b) Battery charges capacitor to the applied voltage. (c) Stored charge remains in capacitor providing 10 V without the battery. (d) Discharging the capacitor.

because the applied voltage cannot make free electrons flow in the conductors.

Note that the potential difference across the charged capacitor is 10 V between plates A and B. There is no potential difference from each plate to its battery terminal, however, which is the reason why the capacitor stops charging.

Storing the Charge. The negative and positive charges on opposite plates have an associated electric field through the dielectric, as shown by the dotted lines in Fig. 21-2b and c. The direction of these electric lines of force is shown repelling electrons from plate B, making this side positive. It is the effect of electric lines of force through the dielectric that results in storage of the charge. The electric field distorts the molecular structure so that the dielectric is no longer neutral. The dielectric is actually stressed by the invisible force of the electric field. As evidence, the dielectric can be ruptured by a very intense field with high voltage across the capacitor.

The result of the electric field, then, is that the dielectric has charge supplied by the voltage source. Since the dielectric is an insulator that cannot conduct, the charge remains in the capacitor even after the voltage source is removed, as illustrated in Fig. 21-2c. You can now take this charged capacitor by itself out of the circuit, and it still has 10 V across the two terminals.

Discharging. The action of neutralizing the charge by connecting a conducting path across the dielectric is discharging the capacitor. As shown in Fig. 21-2d, the wire between plates A and B is a low-resistance path for discharge current. With the stored charge in the dielectric providing the potential difference, 10 V is available to produce discharge current. The negative plate repels electrons, which are attracted to the positive plate through the wire, until the positive and negative charges are neutralized. Then there is no net charge, the capacitor is completely discharged, the voltage across it equals zero, and there is no discharge current. Now the capacitor is in the same uncharged condition as in Fig. 21-2a. It can be charged again, however, by a source of applied voltage.

Nature of the Capacitance. A capacitor has the ability to store the amount of charge necessary to provide a potential difference equal to the charging voltage. If 100 V were applied in Fig. 21-2, the capacitor would charge to 100 V.

The capacitor charges to the applied voltage because, when the capacitor voltage is less, it takes on more charge. As soon as the capacitor voltage equals the applied voltage, no more charging current can flow. Note that any charge or discharge current flows through the conducting wires to the plates but not through the dielectric.

Charge and Discharge Currents. In Fig. 21-2b, i_C is in the opposite direction from i_D in d. In both cases the current is electron flow. However, i_C is charging current to the capacitor and i_D is discharge current from the capacitor. The charge and discharge currents must always be in opposite directions. In b, the negative plate of C accumulates electrons from the voltage source. In d, the charged capacitor serves as a voltage source to produce electron flow around the discharge path.

Practice Problems 21-2
(answers on page 441)
Refer to Fig. 21-2.
(a) If the applied voltage were 458 V, how much would the voltage be across C after it has charged?
(b) How much is the voltage across C after it is completely discharged?

21-3
THE FARAD UNIT OF CAPACITANCE

With more charging voltage, the electric field is stronger and more charge is stored in the dielectric. The amount of charge Q stored in the capacitance is therefore proportional to the applied voltage. Also, a larger capacitance can store more charge. These relations are summarized by the formula

$$Q = CV \quad \text{coulombs} \quad (21\text{-}1)$$

where Q is the charge stored in the dielectric in coulombs (C), and V is the voltage across the plates of the capacitor.

C is a physical constant, indicating the capacitance in terms of how much charge can be stored for a given amount of charging voltage. When one coulomb is stored in the dielectric with a potential difference of one volt, the capacitance is one *farad*.

Practical capacitors have sizes in millionths of a farad, or smaller. The reason is that typical capacitors store charge of microcoulombs or less. Therefore, the common units are

$$1 \text{ microfarad} = 1~\mu\text{F} = 1 \times 10^{-6} \text{ F}$$
$$1 \text{ micromicrofarad} = 1~\mu\mu\text{F} = 1 \times 10^{-12} \text{ F}$$
or
$$1 \text{ picofarad} = 1 \text{ pF} = 1 \times 10^{-12} \text{ F}$$

The picofarad and micromicrofarad units are the same but pF has become standard to eliminate confusion with microfarads.

Example 1. How much charge is stored in a 2-μF capacitor with 50 V across it?

Answer. $Q = CV = 2 \times 10^{-6} \times 50$
$Q = 100 \times 10^{-6}$ coulomb

Example 2. How much charge is stored in a 40-μF capacitor with 50 V across it?

Answer. $Q = CV = 40 \times 10^{-6} \times 50$
$Q = 2000 \times 10^{-6}$ coulomb

Note that the larger capacitor stores more charge for the same voltage, in accordance with the definition of capacitance as the ability to store charge.

The factors in $Q = CV$ can be inverted to

$$C = \frac{Q}{V} \quad (21\text{-}2)$$

or

$$V = \frac{Q}{C} \quad (21\text{-}3)$$

For all three formulas, the basic units are volts for V, coulombs for Q, and farads for C. The formula $C = Q/V$ actually defines one farad of capacitance as one coulomb of charge stored for one volt of potential difference.

Example 3. A constant current of 2 μA charges a capacitor for 20 s. How much charge is stored? Remember $I = Q/t$ or $Q = I \times t$.

Answer. $Q = I \times t = 2 \times 10^{-6} \times 20$
$Q = 40$ microcoulombs (μC)

Example 4. The voltage across the charged capacitor in Example 3 is 20 V. Calculate C.

Answer. $C = \dfrac{Q}{V} = \dfrac{40 \times 10^{-6}}{20} = 2 \times 10^{-6}$
$C = 2~\mu\text{F}$

Example 5. A constant current of 5 mA charges a 10-μF capacitor for 1 s. How much is the voltage across the capacitor?

Answer. Find the stored charge first:

$Q = I \times t = 5 \times 10^{-3} \times 1 = 5 \times 10^{-3}$ coulomb

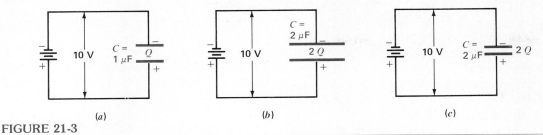

(a) (b) (c)

FIGURE 21-3
Increasing stored charge and capacitance by increasing plate area and decreasing distance between plates. (a) Capacitance of 1 µF. (b) 2-µF capacitance with twice the plate area and the same distance. (c) 2-µF capacitance with one-half the distance and the same plate area.

$$V = \frac{Q}{C} = \frac{5 \times 10^{-3}}{10 \times 10^{-6}} = \frac{5}{10} \times 10^3$$
$$V = 500 \text{ V}$$

Larger Plate Area Increases Capacitance. As illustrated in Fig. 21-3, when the area of each plate is doubled, the capacitance in (b) stores twice the charge of (a). The potential difference in both cases is still 10 V. This voltage produces a given strength of electric field. A larger plate area, however, means that more of the dielectric surface can contact each plate, allowing more lines of force through the dielectric between the plates and less flux leakage outside the dielectric. Then the field can store more charge in the dielectric. The result of larger plate area is more charge stored for the same applied voltage, which means the capacitance is larger.

Thinner Dielectric Increases Capacitance. As illustrated in Fig. 21-3c, when the distance between plates is reduced one-half, the capacitance stores twice the charge of Fig. 21-3a. The potential difference is still 10 V, but its electric field has greater flux density in the thinner dielectric. Then the field between opposite plates can store more charge in the dielectric. With less distance between the plates, the stored charge is greater for the same applied voltage, which means the capacitance is larger.

Dielectric Constant K_ϵ. This indicates the ability of an insulator to concentrate electric flux. Its numerical value is specified as the ratio of flux in the insulator compared with the flux in air or vacuum. The dielectric constant of air or vacuum is 1, since it is the reference.

Mica, for example, has an average dielectric constant of 6, meaning it can provide a density of electric flux six times as great as that of air or vacuum for the same applied voltage

TABLE 21-1. Dielectric Materials*

MATERIAL	DIELECTRIC CONSTANT K_ϵ	DIELECTRIC STRENGTH, V/mil
Air or vacuum	1	20
Ceramics	80–1200	600–1250
Glass	8	335–2000
Mica	3–8	600–1500
Oil	2–5	375
Paper	2–6	1250

*Exact values depend on the specific composition of different types.

and equal physical size. Insulators generally have a dielectric constant K_ϵ greater than 1, as listed in Table 21-1. Higher values of K_ϵ allow greater values of capacitance.

The dielectric constant for an insulator is actually its *relative permittivity,* with the symbol ϵ_r or K_ϵ, indicating the ability to concentrate electric flux. This factor corresponds to relative permeability, with the symbol μ_r or K_m, for magnetic flux. Both ϵ_r and μ_r are pure numbers without units, as they are just ratios.[1]

Dielectric Strength. Table 21-1 also lists breakdown-voltage ratings for typical dielectrics. Dielectric strength is the ability of a dielectric to withstand a potential difference without arcing across the insulator. This voltage rating is important because rupture of the insulator provides a conducting path through the dielectric. Then it cannot store charge, because the capacitor has been short-circuited. Since the breakdown voltage increases with greater thickness, capacitors for higher voltage ratings have more distance between the plates. This increased distance reduces the capacitance, however, all other factors remaining the same.

These physical factors for a parallel-plate capacitor are summarized by the formula

$$C = K_\epsilon \times \frac{A}{d} \times 8.85 \times 10^{-12} \text{ F} \qquad (21\text{-}4)$$

A is the area in square meters of either plate, and d is the distance in meters between plates. K_ϵ is the dielectric constant, or relative permittivity, as listed in Table 21-1. The constant factor 8.85×10^{-12} is the absolute permittivity

of air or vacuum, in SI, to calculate C in farads, which is an SI unit.

Example 6. Calculate C for two plates each with an area 2 m², separated by 1 cm, or 10^{-2} m, with a dielectric of air.

Answer. Substituting in Formula (21-4),

$$C = 1 \times \frac{2}{10^{-2}} \times 8.85 \times 10^{-12}$$
$$= 200 \times 8.85 \times 10^{-12}$$
$$= 1770 \times 10^{-12}$$
$$C = 1770 \text{ pF}$$

This value means the capacitor can store 1770×10^{-12} coulomb of charge with 1 V. Note the relatively small capacitance, in picofarad units, with the extremely large plates of 2 m², which is really the size of a table or a desk top.

If the dielectric used is paper with a dielectric constant of 6, then C will be six times greater. Also, if the spacing between plates is reduced by one-half to 0.5 cm, the capacitance will be doubled.

Practice Problems 21-3
(answers on page 441)
(a) A capacitor charged to 100 V has 1000 microcoulombs of charge. How much is C?
(b) A mica capacitor and ceramic capacitor have the same physical dimensions. Which has more C?

21-4
TYPICAL CAPACITORS

Commercial capacitors are generally classified according to the dielectric. Most common are air, mica, paper, and ceramic capacitors, plus the electrolytic type. Electrolytic capacitors use a molecular-thin oxide film as the dielectric,

---

[1] The absolute permittivity ϵ_o is 8.854×10^{-12} F/m, in SI units, for electric flux in air or vacuum. This value corresponds to an absolute permeability μ_o of $4\pi \times 10^{-7}$ H/m in SI units, for magnetic flux in air or vacuum.

TABLE 21-2. Types of Capacitors

DIELECTRIC	CONSTRUCTION	CAPACITANCE	BREAKDOWN, V
Air	Meshed plates	10–400 pF	400 (0.02-in air gap)
Ceramic	Tubular	0.5–1600 pF	500–20,000
	Disk	0.002–0.1 μF	
Electrolytic	Aluminum	5–1000 μF	10–450
	Tantalum	0.01–300 μF	6–50
Mica	Stacked sheets	10–5000 pF	500–20,000
Paper	Rolled foil	0.001–1 μF	200–1600

resulting in large capacitance values in little space. These types are compared in Table 21-2 and shown in Figs. 21-4 to 21-8.

There is no required polarity, since either side can be the more positive plate, except for electrolytic capacitors. These are marked to indicate which side must be positive to maintain the internal electrolytic action that produces the dielectric required to form the capacitance. It should be noted that the polarity of the charging source determines the polarity of the capacitor voltage.

Mica Capacitors. Thin mica sheets are stacked between tinfoil sections for the conducting plates to provide the required capacitance. Alternate strips of tinfoil are connected together and brought out as one terminal for one set of plates, while the opposite terminal connects to the other set of plates. The entire unit is generally in a molded Bakelite case. Mica capacitors are often used for small capacitance values of 50 to 500 pF; their length is $\frac{3}{4}$ in or less with about $\frac{1}{8}$-in thickness. Typical mica capacitors are shown in Fig. 21-4.

Paper Capacitors. In this construction, two rolls of tinfoil conductor separated by a tissue-paper insulator are rolled into a compact cylinder. Each outside lead connects to its roll of tinfoil as a plate. The entire cylinder is generally placed in a cardboard container coated with wax or encased in plastic. Paper capacitors are often used for medium capacitance values of 0.001 to 1.0 μF, approximately. The physical size for 0.05 μF is typically $1\frac{1}{2}$ in long with $\frac{1}{2}$-in diameter. Paper capacitors are shown in Fig. 21-5.

A black band at one end of a paper capacitor indicates the lead connected to the outside foil. This lead should be used for the ground or low-potential side of the circuit to take advantage of shielding by the outside foil. There is no required polarity, however, since the capacitance is the same no matter which side is

(a)

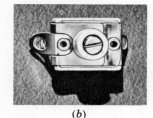

(b)

FIGURE 21-4

Mica capacitors, about $\frac{5}{8}$ in wide. (a) Fixed value, color-coded in picofarads. (b) Variable trimmer of 5 to 30 pF. (*El Menco*)

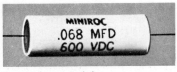

(a)

(b)

FIGURE 21-5

Paper capacitors (a) Tubular type 1 in long. C is 0.068 µF. (b) Encapsulated type with leads for printed-circuit board; length ¾ in; C is 430 pF.

grounded. It should also be noted that in the schematic symbol for C the curved line usually indicates the low-potential side of the capacitor.

Many capacitors of foil construction use a plastic film instead of tissue paper. Two types are Teflon[1] and Mylar[1] plastic film. These feature very high insulation resistance, of over 1000 MΩ, low losses, and longer service life without voltage breakdown, compared with paper capacitors. The plastic capacitors are available in sizes of 0.001 to 1.0 µF, like paper capacitors.

Ceramic Capacitors. The ceramic dielectric materials are made from earth fired under extreme heat. By use of titanium dioxide, or several types of silicates, very high values of dielectric constant K_ϵ can be obtained.

In the disk form, silver is fired onto both sides of the ceramic, to form the conductor plates. With a K_ϵ value of 1200, the disk ceramics feature capacitance values up to 0.01 µF in much less space than a paper capacitor.

For tubular ceramics, the hollow ceramic tube has a silver coating on the inside and outside surfaces. With values of 1 to 500 pF, these capacitors have the same applications as mica capacitors but are smaller. Typical ceramic capacitors are shown in Fig. 21-6.

Temperature Coefficient. Ceramic capacitors are often used for temperature compensation, to increase or decrease capacitance with a rise in temperature. The temperature coefficient is given in parts per million (ppm) per degree Celsius, with a reference of 25°C. As an example, a negative 750 ppm unit is stated as N750. A positive temperature coefficient of the same value would be stated as P750. Units that do not change in capacitance are labeled NPO.

Variable Capacitors. Figure 21-1b shows a variable air capacitor. In this construction, the fixed metal plates connected together form the *stator*. The movable plates connected together on the shaft form the *rotor*. Capacitance is varied by rotating the shaft to make the rotor plates mesh with the stator plates. They do not touch, however, since air is the dielectric. Full mesh is maximum capacitance. Moving the rotor completely out of mesh provides minimum capacitance.

A common application is the tuning capacitor in radio receivers. When you tune to different stations, the capacitance varies as the rotor moves in or out of mesh. Combined with an inductance, the variable capacitance then tunes the receiver to a different resonant frequency for each station. Usually two or three capacitor sections are *ganged* on one common shaft.

Electrolytic Capacitors. These capacitors are

---

[1] Du Pont trademarks.

Chapter 21
Capacitance 427

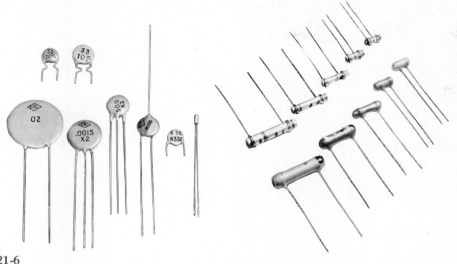

FIGURE 21-6
Ceramic capacitors shown actual size. (a) Disk type. (b) Tubular type. (*Centralab Division, Globe-Union Inc.*)

commonly used for capacitance values of 5 to 1000 μF because electrolytics provide the most capacitance in the smallest space with least cost. Figure 21-7 shows a typical electrolytic. The construction consists of two metal electrodes, usually aluminum, in an electrolyte of borax, phosphate, or carbonate. Between the two aluminum strips, absorbent gauze soaks up electrolyte to provide the required electrolysis.

When dc voltage is applied to form the capacitance during manufacture, the electrolytic action accumulates a molecular-thin layer of aluminum oxide at the junction between the positive aluminum electrode and the electrolyte. Since the oxide film is an insulator, there is capacitance between the positive aluminum electrode and the electrolyte in the gauze separator. The negative aluminum electrode simply provides a connection to the electrolyte.

With the extremely thin dielectric film, very large capacitance values can be obtained. The

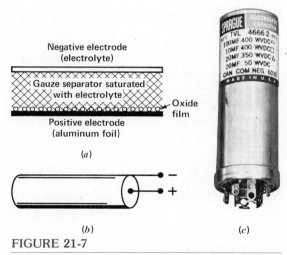

FIGURE 21-7
Construction of dry electrolytic capacitor. (a) Internal electrodes. (b) Foil rolled into cartridge. (c) Typical capacitor with multiple sections. Height is about 3 in. (*Sprague Electric Co.*)

area is increased by means of long strips of aluminum foil and gauze, which are rolled into a compact cylinder having very high capacitance. For example, an electrolytic capacitor the same size as a $0.1\text{-}\mu\text{F}$ paper capacitor, but rated at 10-V breakdown, may have $1000\ \mu\text{F}$ capacitance. Higher voltage ratings up to 450 V are often used in typical capacitance values of 8 to 80 μF.

Electrolytic capacitors must be connected so that the applied voltage maintains the positive electrode more positive than the negative terminal. Otherwise, the insulating oxide film is not formed and there is no capacitance.

Electrolytic capacitors are used in circuits that have a combination of dc voltage and ac voltage. The dc voltage maintains the polarity. A common application is for the electrolytic filter capacitors to eliminate the ac ripple in a dc power supply.

If the electrolytic is connected in opposite polarity, the reversed electrolysis forms gas and the capacitor becomes hot and may explode. This possibility applies only to electrolytic capacitors.

The disadvantage of electrolytics, in addition to the required polarization, is their relatively high leakage current, since the oxide film is not a perfect insulator. This leakage current through the dielectric is about 0.1 to 0.5 mA/μF of capacitance.

Nonpolarized electrolytic capacitors are also available for applications in ac circuits without any dc polarizing voltage. One use is for ac motors. A nonpolar electrolytic actually contains two capacitors, connected internally in series-opposing polarity.

Tantalum Capacitors. These are a new type of electrolytic, using tantalum instead of aluminum. Niobium is also used. The electrolyte may be wet or dry. They feature larger capacitance in a smaller size, longer shelf life, and less leakage current. Although the voltage ratings are lower than for aluminum electrolytics, tantalum capacitors are commonly used for low-voltage applications in transistor circuits. Typical capacitors are shown in Fig. 21-8.

Capacitance Tolerance. Ceramic disk capacitors for general applications usually have a tolerance of ± 20 percent. Paper capacitors usually have a tolerance of ± 10 percent. For closer tolerances, mica or ceramic tubular capacitors are used. These have tolerance values of ± 2 to 20 percent. Silver-plated mica capacitors are available with a tolerance of ± 1 percent.

The tolerance may be less on the minus side to make sure there is enough capacitance, particularly with electrolytic capacitors, which have a wide tolerance. For instance, a 20-μF electrolytic with a tolerance of -10 percent, $+50$ percent may have a capacitance of 18 to 30 μF. However, the exact capacitance value is not critical in most applications of capacitors for filtering, ac coupling, and bypassing.

Voltage Rating of Capacitors. This rating specifies the maximum potential difference that

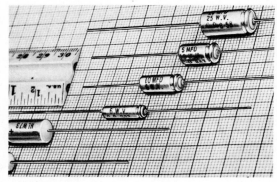

FIGURE 21-8

Low-voltage miniature electrolytic capacitors for transistor circuits. (*Cornell-Dubilier Electronics*)

can be applied across the plates without puncturing the dielectric. Usually the voltage rating is for temperatures up to about 60°C. Higher temperatures result in a lower voltage rating. Voltage ratings for general-purpose paper, mica, and ceramic capacitors are typically 200 to 500 V. Ceramic capacitors with ratings of 1 to 5 kV are also available.

Electrolytic capacitors are commonly used in 25-, 150-, and 450-V ratings. In addition, 6- and 10-V electrolytic capacitors are often used in transistor circuits. For applications where a lower voltage rating is permissible, more capacitance can be obtained in a smaller physical size.

The voltage ratings are for dc voltage applied. The breakdown rating is lower for ac voltage because of the internal heat produced by continuous charge and discharge.

The potential difference across the capacitor depends upon the applied voltage and is not necessarily equal to the voltage rating. A voltage rating higher than the potential difference applied across the capacitor provides a safety factor for long life in service. With electrolytic capacitors, however, the actual capacitor voltage should be close to the rated voltage to produce the oxide film that provides the specified capacitance.

Practice Problems 21-4
(answers on page 441)
(a) Which is the only capacitor type that has polarity?
(b) Which is physically smaller, a paper or ceramic capacitor, both with the same ratings?

21-5
CAPACITOR COLOR CODING

Mica and tubular ceramic capacitors are color-coded to indicate their capacitance value. Since coding is necessary only for very small sizes, the color-coded capacitance value is always in pF units. The colors used are the same as for resistor coding, from black for 0 up to white for 9.

Mica capacitors generally use the six-dot system shown in Fig. 21-9. Read the top row first from left to right, then the bottom row, in reverse order right to left. White for the first dot indicates the new EIA coding, but the capacitance value is read from the next three dots. As an example, if the colors are red, green, and brown for dots 2, 3, and 4, the capacitance is 250 pF. If the first dot is silver, it indicates a paper capacitor, but the capacitance is still read from dots 2, 3, and 4. Dot 5 specifies tolerance, while dot 6 gives the EIA class. There are seven classes from A to G specifying temperature coefficient, leakage resistance, and additional variable factors. Appendix G has more detailed information on the tolerance and class coding. Also listed are discontinued codes found on capacitors in old equipment.

For tubular ceramic capacitors, the system shown in Fig. 21-10 is used with color dots or bands. The wide color band specifying temperature coefficient indicates the left end, which is the side connected to the inner electrode. Capacitance is read from the next three colors, in either dots or stripes. For instance, brown, black, and brown for bands or dots 2, 3, and 4 means 100 pF.

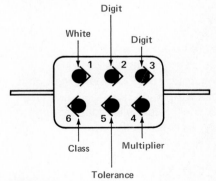

FIGURE 21-9

Six-dot color code for mica capacitors.

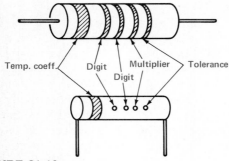

FIGURE 21-10
Color code for ceramic tubular capacitors.

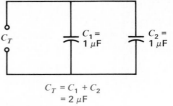

$$C_T = C_1 + C_2$$
$$= 2 \ \mu F$$

FIGURE 21-11
Capacitances in parallel.

Gray and white are used as decimal multipliers for very small values, with gray for 0.01 and white for 0.1. For instance, green, black, and white in dots 2, 3, and 4 means 50×0.1, or 5 pF. The color codes for tolerance and temperature coefficient of ceramic capacitors are listed in Appendix G.

In reading the color-coded capacitance value, keep in mind that mica capacitors generally range from 10 to 5000 pF. The small tubular ceramic capacitors are usually 0.5 to 1000 pF. With paper and ceramic disk capacitors, the capacitance and voltage rating is generally printed on the case. Where no voltage rating is specified, it is usually about 200 to 600 V. Electrolytic capacitors have the capacitance, voltage rating, and polarity printed on the case.

Practice Problems 21-5
(answers on page 441)
(a) How much is C for red, green, and black color stripes or dots?
(b) What letters are used to indicate a negative temperature coefficient?

21-6
PARALLEL CAPACITANCES

Connecting capacitances in parallel is equivalent to adding the plate areas. Therefore, the total capacitance is the sum of the individual capacitances. As illustrated in Fig. 21-11,

$$C_T = C_1 + C_2 + \cdots + \text{etc.} \qquad (21\text{-}5)$$

A 10-μF capacitor in parallel with a 5-μF capacitor, for example, provides a 15-μF capacitance for the parallel combination. The voltage is the same across the parallel capacitors. Note that adding parallel capacitances is opposite to the case of inductances in parallel, and resistances in parallel.

Practice Problems 21-6
(answers on page 441)
(a) How much is C_T for 0.01 μF in parallel with 0.02 μF?
(b) What C must be connected in parallel with 100 pF to make C_T of 250 pF?

21-7
SERIES CAPACITANCES

Connecting capacitances in series is equivalent to increasing the thickness of the dielectric. Therefore, the combined capacitance is less than the smallest individual value. As shown in Fig. 21-12, the combined equivalent capacitance is calculated by the reciprocal formula:

$$\frac{1}{C_T} = \frac{1}{C_1} + \frac{1}{C_2} + \cdots + \text{etc.} \qquad (21\text{-}6)$$

Any of the shortcut calculations for the reciprocal formula apply. For example, the

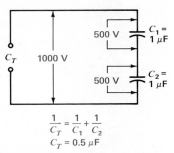

FIGURE 21-12

Capacitances in series.

charging current be 600 μA flowing for 1 s. The charge Q equals $I \times t$, then, or 600 μC. Both C_1 and C_2 have Q equal to 600 μC, as they are in the same series path for charging current.

Although the charge is the same in C_1 and C_2, they have different voltages because of different capacitance values. For each capacitor $V = Q/C$. The voltage V_1 across C_1 then is 600 μC/1 μF, which equals 600 V. For C_2, the voltage V_2 is 600 μC/2 μF, which equals 300 V.

Charging Current for Series Capacitances. The charging current is the same in all parts of the series path, including the junction between C_1 and C_2, even though this point is separated from the source voltage by two insulators. At the junction, the current is the resultant of electrons repelled by the negative plate of C_2 and attracted by the positive plate of C_1. The amount of current is how much would be produced by one capacitance of $\frac{2}{3}$ μF, which is the equivalent capacitance of C_1 and C_2 in series.

combined capacitance of two equal capacitances of 10 μF in series is 5 μF.

Capacitors are used in series to provide a higher voltage breakdown rating for the combination. For instance, each of three equal capacitances in series has one-third the applied voltage.

Voltage Division across Unequal Capacitances. In series, the voltage across each C is inversely proportional to its capacitance, as illustrated in Fig. 21-13. The smaller capacitance has the larger proportion of the applied voltage. The reason is that the series capacitances all have the same charge because they are in one current path. With equal charge, a smaller capacitance has a greater potential difference.

We can consider the amount of charge in the series capacitors in Fig. 21-13. Let the

Practice Problems 21-7
(answers on page 441)
(a) How much is C_T for two 0.2-μF capacitors in series?
(b) With 500 V applied across both, how much is V_C across each capacitor?
(c) How much is C_T for 100 pF in series with 50 pF?

21-8
STRAY CAPACITIVE AND INDUCTIVE EFFECTS

These two important characteristics can be evident in all circuits with all types of components. A capacitor has a small amount of inductance in the conductors. A coil has some capacitance between windings. A resistor has a small amount of inductance and capacitance. After all, a capacitance physically is simply an insulator between two points having a difference of

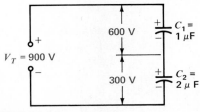

FIGURE 21-13

With series capacitances the smaller capacitance has more voltage for the same charge.

potential. An inductance is basically just a conductor carrying current.

Actually, though, these stray effects are usually quite small, compared with the concentrated or lumped values of capacitors or inductors. Typical values of stray capacitance may be 1 to 10 pF, while stray inductance is usually a fraction of 1 μH. For very high radio frequencies, however, when small values of L and C must be used, the stray effects become important. As another example, any wire cable has capacitance between the conductors.

Stray Circuit Capacitance. The wiring and the components in a circuit have capacitance to the metal chassis. This stray capacitance C_s is typically 5 to 10 pF. To reduce C_s, the wiring should be short, with the leads and components placed high off the chassis. Sometimes, for very high frequencies, the stray capacitance is included as part of the circuit design. Then changing the placement of components or wiring affects the circuit operation. Such critical *lead dress* is usually specified in the manufacturer's service notes.

Leakage Resistance of a Capacitor. Consider a capacitor charged by a dc voltage source. After the charging voltage is removed, a perfect capacitor would keep its charge indefinitely. After a long period of time, however, the charge will be neutralized by a small leakage current through the dielectric and across the insulated case between terminals, because there is no perfect insulator. For paper, ceramic, and mica capacitors, though, the leakage current is very slight or, inversely, the leakage resistance is very high. As shown in Fig. 21-14, the leakage resistance R_l is indicated by a high resistance in parallel with the capacitance C. For paper, ceramic, or mica capacitors R_l is 100 MΩ or more. However, electrolytic capacitors may have a leakage resistance of 0.5 MΩ or less.

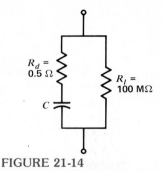

FIGURE 21-14

Equivalent circuit of a capacitor. R_l is leakage resistance; R_d is absorption loss dissipated in dielectric.

Absorption Losses in Capacitors. With ac voltage applied to a capacitor, the continuous charge, discharge, and reverse charging action cannot be followed instantaneously in the dielectric. This corresponds to hysteresis in magnetic materials. With high-frequency charging voltage for a capacitor, there may be a difference between the amount of ac voltage applied and the ac voltage stored in the dielectric. The difference can be considered *absorption loss* in the dielectric. With higher frequencies, the losses increase. In Fig. 21-14, the small value of 0.5 Ω for R_d indicates a typical value for paper capacitors. For ceramic and mica capacitors, the dielectric losses are even smaller. These losses need not be considered for electrolytic capacitors because they are generally not used for radio frequencies.

Power Factor of a Capacitor. The quality of a capacitor in terms of minimum loss is often indicated by its power factor, which states the fraction of input power dissipated as heat loss in the capacitor. The lower the numerical value of the power factor, the better is the quality of the capacitor. Since the losses are in the dielectric, the power factor of the capacitor is essentially the power factor of the dielectric, independent of capacitance value or voltage rating. At radio frequencies, approximate values of power factor are 0.000 for air or vacuum, 0.0004 for mica, about 0.01 for paper, and 0.0001 to 0.03 for ceramics.

The reciprocal of the power factor can be considered the Q of the capacitor, similar to the idea of Q of a coil. For instance, a power factor of 0.001 corresponds to a Q of 1000. A higher Q therefore means better quality for the capacitor.

Inductance of a Capacitor. Capacitors with a coiled construction, particularly paper and electrolytic capacitors, have some internal inductance. The larger the capacitor, the greater is its series inductance. Mica and ceramic capacitors have very little inductance, however, which is why they are generally used for radio frequencies.

For use above audio frequencies, the rolled-foil type of capacitor must have a noninductive construction. This means the start and finish of the foil winding must not be the terminals of the capacitor. Instead, the foil windings are offset. Then one terminal can contact all layers of one foil at one edge, while the opposite edge of the other foil contacts the second terminal. Most rolled-foil capacitors, including the paper and Mylar types, are constructed this way.

Distributed Capacitance of a Coil. As illustrated in Fig. 21-15, a coil has distributed capacitance C_d between turns. Note that each turn is a conductor separated from the next turn by an insulator, which is the definition of capacitance. Furthermore, the potential of each turn is different from the next, providing part of the total voltage as a potential difference to charge C_d. The result then is the equivalent circuit shown for an rf coil. L is the inductance and R_e its internal effective ac resistance in series with L, while the total distributed capacitance C_d for all the turns is across the entire coil.

Special methods for minimum C_d include *space-wound* coils, where the turns are spaced far apart; the honeycomb or *universal* winding, with the turns crossing each other at right angles; and the *bank winding*, with separate sections called *pies*. These windings are for rf coils. In audio and power transformers, a grounded conductor shield, called a *Faraday screen,* is often placed between windings to reduce capacitive coupling.

Reactive Effects in Resistors. As illustrated by the high-frequency equivalent circuit in Fig. 21-16, a resistor can include a small amount of inductance and capacitance. For carbon-composition resistors, the inductance is usually negligible. However, approximately 0.5 pF of capacitance across the ends may have an effect, particularly with large resistances used for high radio frequencies. Wirewound resistors definitely have enough inductance to be evident at radio frequencies. However, special resistors

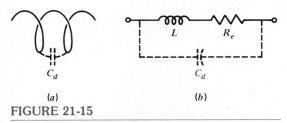

(a)　　　　(b)

FIGURE 21-15

Equivalent circuit of an rf coil. (a) Distributed capacitance C_d between turns of wire. (b) Equivalent circuit.

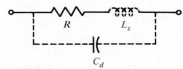

FIGURE 21-16

High-frequency equivalent circuit of a resistor.

are available with double windings in a non-inductive method based on cancellation of opposing magnetic fields.

Capacitance of an Open Circuit. An open switch or a break in a conducting wire has capacitance C_O across the open. The reason is that the open consists of an insulator between two conductors. With a voltage source in the circuit, C_O charges to the applied voltage. Because of the small C_O, in the order of picofarads, the capacitance charges to the source voltage in a short time. This charging of C_O is the reason why an open series circuit has the applied voltage across the open terminals. After a momentary flow of charging current, C_O charges to the applied voltage and stores the charge needed to maintain this voltage as long as the circuit is open.

Practice Problems 21-8
(answers on page 441)
Answer true or false.
(a) A two-wire cable has distributed C between the conductors.
(b) A coil has distributed C between the turns.
(c) The leakage resistance of ceramic capacitors is very high.

21-9
ENERGY IN ELECTROSTATIC FIELD OF CAPACITANCE

The electrostatic field of the charge stored in the dielectric has electrical energy supplied by the voltage source that charges C. This energy is stored in the dielectric. The proof is the fact that the capacitance can produce discharge current when the voltage source is removed. The electrical energy stored is

$$\text{Energy} = \varepsilon = \tfrac{1}{2}CV^2 \qquad \text{joules} \qquad (21\text{-}7)$$

where C is the capacitance in farads and V is the voltage across the capacitor. For example, a 1-μF capacitor charged to 400 V has stored energy equal to

$$\varepsilon = \tfrac{1}{2}CV^2 = \frac{1 \times 10^{-6} \times (4 \times 10^2)^2}{2}$$
$$= \frac{1 \times 10^{-6} \times (16 \times 10^4)}{2}$$
$$= 8 \times 10^{-2}$$
$$\varepsilon = 0.08 \text{ J}$$

This 0.08 J of energy is supplied by the voltage source that charges the capacitor to 400 V. When the charging circuit is opened, the stored energy remains as charge in the dielectric. With a closed path provided for discharge, the entire 0.08 J is available to produce discharge current. As the capacitor discharges, the energy is used in producing discharge current. When the capacitor is completely discharged, the stored energy is zero.

The stored energy is the reason why a charged capacitor can produce an electric shock, even when not connected into a circuit. When you touch the two leads of the charged capacitor, its voltage produces discharge current through your body. Stored energy greater than 1 J can be dangerous with a capacitor charged to a voltage high enough to produce an electric shock.

Practice Problems 21-9
(answers on page 441)
Answer true or false.
(a) The stored energy in C increases with more V.
(b) The stored energy decreases with less C.

21-10
TROUBLES IN CAPACITORS

Capacitors can become open or short-circuited. In either case, the capacitor is useless because it cannot store charge. A leaky capacitor is equivalent to a partial short-circuit where the dielectric gradually loses its insulating properties under the stress of applied voltage, lowering its resistance. A good capacitor has very high resistance of the order of megohms; a short-circuited capacitor has zero ohms resistance, or continuity; the resistance of a leaky capacitor is lower than normal.

Checking Capacitors with an Ohmmeter. A capacitor usually can be checked with an ohmmeter. The highest ohms range, such as $R \times 1$ MΩ, is preferable. Also, disconnect one side of the capacitor from the circuit to eliminate any parallel resistance paths that can lower the resistance. Keep your fingers off the connections, since the body resistance lowers the reading.

As illustrated in Fig. 21-17, the ohmmeter leads are connected across the capacitor. For a good capacitor, the meter pointer moves quickly toward the low-resistance side of the scale and then slowly recedes toward infinity. The reading when the pointer stops moving is the insulation resistance of the capacitor, which is normally very high. For paper, mica, and ceramic capacitors, the resistance can be 500 to 1000 MΩ, or more, which is practically infinite resistance. Electrolytic capacitors, however, have a lower normal resistance of about 0.5 MΩ. In all cases, discharge the capacitor before checking with the ohmmeter.

When the ohmmeter is initially connected, its battery charges the capacitor. This charging current is the reason the meter pointer moves away from infinity, since more current through the ohmmeter means less resistance. Maximum current flows at the first instant of charge. Then the charging current decreases as the capacitor voltage increases toward the applied voltage; therefore, the needle pointer slowly moves toward infinite resistance. Finally, the capacitor is completely charged to the ohmmeter battery voltage, the charging current is zero, and the ohmmeter reads just the small leakage current through the dielectric. This charging effect, called *capacitor action*, shows that the capacitor can store charge, indicating a normal capacitor. It should be noted that both the rise and fall of the meter readings are caused by charging. The capacitor discharges when the meter leads are reversed.

Ohmmeter Readings. Troubles in a capacitor are indicated as follows:

1. If an ohmmeter reading immediately goes practically to zero and stays there, the capacitor is short-circuited.
2. If the capacitor shows charging, but the final resistance reading is appreciably less than normal, the capacitor is leaky. Such capacitors are particularly troublesome in high-resistance circuits. When checking electrolytics, reverse the ohmmeter leads and take the higher of the two readings.
3. If the capacitor shows no charging action

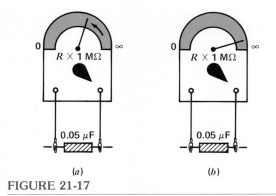

FIGURE 21-17

Checking a capacitor with an ohmmeter. R scale shown left to right, as on a VTVM. (*a*) Capacitor action as needle is moved by charging current. (*b*) Leakage-resistance reading after capacitor has charged.

but just reads very high resistance, it may be open. Some precautions must be remembered, however, since very high resistance is a normal condition for capacitors. Reverse the ohmmeter leads to discharge the capacitor, and check it again. In addition, remember that capacitance values of 100 pF, or less, normally have very little charging current for the low battery voltage of the ohmmeter.

Short-circuited Capacitors. In normal service, capacitors can become short-circuited because the dielectric deteriorates with age, usually over a period of years under the stress of charging voltage, especially with higher temperatures. This effect is more common with paper and electrolytic capacitors. The capacitor may become leaky gradually, indicating a partial short circuit, or the dielectric may be punctured, causing a short circuit.

Open Capacitors. In addition to the possibility of an open connection in any type of capacitor, electrolytics develop high resistance in the electrolyte with age, particularly at high temperatures. After service of a year or two, when the electrolyte dries up, the capacitor is partially open. Much of the capacitor action is gone, and the capacitor should be replaced.

Shelf Life. Except for electrolytics, capacitors do not deteriorate with age while stored, since there is no applied voltage. Electrolytic capacitors, however, like dry cells, should be used fresh from manufacture. The reason is the wet electrolyte.

Practice Problems 21-10
(answers on page 441)
(a) What is the ohmmeter reading for a shorted capacitor?
(b) Does capacitor action with an ohmmeter show the capacitor is good or bad?

Summary

1. A capacitor consists of two conductors separated by a dielectric insulator. Its ability to store charge is the capacitance C. Applying voltage to store charge is charging the capacitor; short-circuiting the two conductors of the capacitor to neutralize the charge is discharging the capacitor. Schematic symbols for C are summarized in Fig. 21-18.
2. The unit of capacitance is the farad. One farad of capacitance stores one coulomb of charge with one volt applied. Practical capacitors have much smaller capacitance values, from 1 pF to 1000 μF. One pF is 1×10^{-12} F; one μF is 1×10^{-6} F.
3. $Q = CV$, where Q is the charge in coulombs, C the capacitance in farads, and V the potential difference across the capacitor in volts.
4. Capacitance increases with larger plate area, and less distance between the plates.
5. The ratio of charge stored in different insulators to the charge stored in air is the dielectric constant K_ϵ of the material. Air or vacuum has a dielectric constant of 1.

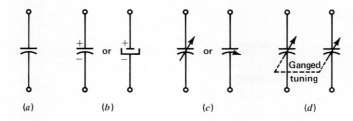

FIGURE 21-18

Schematic symbols for types of C. (a) Fixed-air, paper, mica, or ceramic. (b) Electrolytic, with polarity. (c) Variable. (d) Ganged on one shaft.

6. The most common types of commercial capacitors are air, paper, mica, ceramic, and electrolytic. Electrolytics are the only capacitors with polarity. The different types are compared in Table 21-2.
7. Mica and tubular ceramic capacitors are color-coded as shown in Figs. 21-9 and 21-10.
8. For parallel capacitors, $C_T = C_1 + C_2 + C_3 + \cdots +$ etc.
9. For series capacitors, $1/C_T = 1/C_1 + 1/C_2 + 1/C_3 + \cdots +$ etc.
10. The electric field of a capacitance has stored energy $CV^2/2$. With V in volts and C in farads, the energy is in joules.
11. When checked with an ohmmeter, a good capacitor shows charging current, and then the ohmmeter reading steadies at the insulation resistance. All types except electrolytics normally have a very high insulation resistance of 500 to 1000 MΩ. Electrolytics have more leakage current, with a typical resistance of 0.5 MΩ.
12. The main comparisons between the opposite characteristics of capacitance and inductance are summarized in Table 21-3.

TABLE 21-3. Comparison of Capacitance and Inductance

CAPACITANCE	INDUCTANCE
Symbol is C	Symbol is L
Farad unit	Henry unit
Stores charge Q	Conducts current I
Needs dielectric as insulator	Needs wire conductor
More plate area allows more C	More turns allow more L
Dielectric with higher K_ϵ or ϵ_r concentrates electric field for more C	Core with higher K_m or μ_r concentrates magnetic field for more L
$\frac{1}{C_T} = \frac{1}{C_1} + \frac{1}{C_2}$ in series	$L_T = L_1 + L_2$ in series
$C_T = C_1 + C_2$ in parallel	$\frac{1}{L_T} = \frac{1}{L_1} + \frac{1}{L_2}$ in parallel

Self-Examination (Answers at back of book.)

1. A capacitor consists of two (a) conductors separated by an insulator; (b) insulators separated by a conductor; (c) conductors alone; (d) insulators alone.
2. A capacitance of $0.02\ \mu F$ equals (a) 0.02×10^{-12} F; (b) 0.02×10^{-6} F; (c) 0.02×10^{6} F; (d) 200×10^{-12} F.
3. A $10\text{-}\mu F$ capacitance charged to 10 V has a stored charge equal to (a) $10\ \mu C$; (b) $100\ \mu C$; (c) $200\ \mu C$; (d) 1 C.
4. Capacitance increases with (a) larger plate area and greater distance between plates; (b) smaller plate area and less distance between plates; (c) larger plate area and less distance between plates; (d) higher values of applied voltage.
5. Which of the following statements is correct? (a) Air capacitors have a black band to indicate the outside foil. (b) Mica capacitors are available in capacitance values of 1 to 10 μF. (c) Electrolytic capacitors must be connected in the correct polarity. (d) Ceramic capacitors must be connected in the correct polarity.
6. Voltage applied across a ceramic dielectric produces an electrostatic field 100 times greater than in air. The dielectric constant K_ϵ of the ceramic equals (a) $33\frac{1}{3}$; (b) 50; (c) 100; (d) 10,000.
7. A six-dot mica capacitor color-coded white, red, green, brown, red, and yellow has the capacitance value of (a) 25 pF; (b) 124 pF; (c) 250 pF; (d) 925 pF.
8. The combination of two $0.02\text{-}\mu F$ 500-V capacitors in series has a capacitance and breakdown rating of (a) $0.01\ \mu F$, 500 V; (b) $0.01\ \mu F$, 1000 V; (c) $0.02\ \mu F$, 500 V; (d) $0.04\ \mu F$, 500 V.
9. The combination of two $0.02\text{-}\mu F$ 500-V capacitors in parallel has a capacitance and breakdown rating of (a) $0.01\ \mu F$, 1000 V; (b) $0.02\ \mu F$, 500 V; (c) $0.04\ \mu F$, 500 V; (d) $0.04\ \mu F$, 1000 V.
10. For a good $0.05\text{-}\mu F$ paper capacitor, the ohmmeter reading should (a) go quickly to 100 Ω, approximately, and remain there; (b) show low resistance momentarily and back off to a very high resistance; (c) show high resistance momentarily and then a very low resistance; (d) not move at all.

Essay Questions

1. Define capacitance with respect to physical structure and electrical function. Explain how a two-wire conductor has capacitance.
2. (a) What is meant by a dielectric material? (b) Name five common dielectric materials. (c) Define dielectric flux.

3. Explain briefly how to charge a capacitor. How is a charged capacitor discharged?
4. Define 1 F of capacitance. Convert the following into farads using powers of 10: (a) 50 pF; (b) 0.001 μF; (c) 0.047 μF; (d) 0.01 μF; (e) 10 μF.
5. State the effect on capacitance of (a) larger plate area; (b) thinner dielectric; (c) higher value of dielectric constant.
6. Give one reason for your choice of the type of capacitor to be used in the following applications: (a) 80-μF capacitance for a circuit where one side is positive and the applied voltage never exceeds 150 V; (b) 1.5-pF capacitance for an rf circuit where the required voltage rating is less than 500 V; (c) 0.05-μF capacitance for an audio circuit where the required voltage rating is less than 500 V.
7. (a) Give the capacitance value of six-dot mica capacitors color-coded as follows: (1) Black, red, green, brown, black, black. (2) White, green, black, black, green, brown. (3) White, gray, red, brown, silver, black. (b) Give the capacitance value of the tubular ceramic capacitors color-coded as follows: (4) Black, brown, black, black, brown. (5) Brown, gray, black, gray, black.
8. Draw a diagram showing the least number of 400-V 2-μF capacitors needed for a combination rated at 800 V with 2-μF total capacitance.
9. Given two identical uncharged capacitors. One is charged to 500 V and connected across the uncharged capacitor. Why will the voltage across both capacitors then be 250 V?
10. Describe briefly how you would check a 0.05-μF capacitor with an ohmmeter. State the ohmmeter indications for the case of the capacitor being good, short-circuited, or open.
11. Define the following: (a) stray circuit capacitance; (b) distributed capacitance of a coil; (c) leakage resistance of a capacitor; (d) power factor and Q of a capacitor.
12. Give two comparisons between the electric field in a capacitor and the magnetic field in a coil.
13. Give three types of troubles in capacitors.
14. When a capacitor discharges, why is its discharge current in the opposite direction from the charging current?

Problems (Answers to odd-numbered problems at back of book.)

1. How much charge in coulombs is in a 4-μF capacitor charged to 100 V?
2. A 4-μF capacitor has 400 μC of charge. (a) How much voltage is

across the capacitor? (b) How much is the voltage across an 8-μF capacitor with the same 400-μC charge?

3. A 2-μF capacitor is charged by a constant 3-μA charging current for 4 s. (a) How much charge is stored in the capacitor? (b) How much is the voltage across the capacitor?

4. A 1-μF capacitor C_1 and a 10-μF capacitor C_2 are in series with a constant 2-mA charging current. (a) After 5 s, how much charge is in C_1 and in C_2? (b) How much is the voltage across C_1 and across C_2?

5. Calculate C for a mica capacitor, with $K_\epsilon = 8$, a thickness of 0.02 cm, plates of 6 cm², and five sections in parallel. (Hint: 1 cm = 10^{-2} m and 1 cm² = 10^{-4} m²)

6. How much capacitance stores 6000 μC of charge with 150 V applied? The charge of how many electrons is stored? What type of capacitor is this most likely to be?

7. With 100 V across a capacitor, it stores 100 μC of charge. Then the applied voltage is doubled to 200 V. (a) How much is the voltage across the capacitor? (b) How much charge is stored? (c) How much is its capacitance?

8. Referring to the parallel capacitors in Fig. 21-11, calculate the charge Q_1 in C_1 and Q_2 in C_2 with 500 V. How much is the total charge Q_T in both capacitors? Calculate the total capacitance C_T as Q_T/V.

9. Calculate the energy in joules stored in (a) a 500-pF C charged to 10 kV; (b) a 1-μF C charged to 5 kV; (c) a 40-μF C charged to 400 V.

10. Three capacitors are in series. C_1 is 100 pF, C_2 is 100 pF, and C_3 is 50 pF. Calculate C_T.

11. Calculate C_T for the series-parallel combination of capacitors in Fig. 21-19a and b.

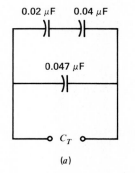
(a)

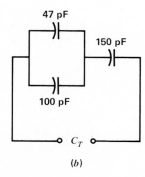
(b)

FIGURE 21-19
For Prob. 11.

Answers to Practice Problems

21-1	(a) Dielectric		21-6	(b) 150 pF
	(b) Farad		21-7	(a) 0.1 μF
21-2	(a) 458 V			(b) 250 V
	(b) 0 V			(c) 33.3 pF
21-3	(a) 10 μF		21-8	(a) T
	(b) Ceramic			(b) T
21-4	(a) Electrolytic			(c) T
	(b) Ceramic		21-9	(a) T
21-5	(a) 250 pF			(b) F
	(b) NPO		21-10	(a) 0 Ω
21-6	(a) 0.03 μF			(b) Good

Capacitive Reactance

Chapter 22

When a capacitor charges and discharges with varying voltage applied, alternating current can flow. Although there cannot be any current through the dielectric of the capacitor, its charge and discharge produces current in the circuit connected to the capacitor plates. How much alternating current flows with sine-wave voltage applied depends on the capacitive reactance X_C.

The amount of X_C is $1/(2\pi fC)$, with f in hertz and C in farads for X_C in ohms. The reactance of X_C is in ohm units, like X_L, but their effects are opposite in terms of frequency. While X_L is directly proportional to f, the X_C is inversely proportional to f. Because of the reciprocal relation in $X_C = 1/(2\pi fC)$, the ohms of X_C decrease for higher frequencies and more capacitance. The topics explaining these effects of X_C in sine-wave ac circuits are:

22-1 Alternating Current in a Capacitive Circuit
22-2 $X_C = 1/(2\pi fC)$
22-3 Series or Parallel Capacitive Reactances
22-4 Ohm's Law Applied to X_C
22-5 Applications of Capacitive Reactance
22-6 Sine-Wave Charge and Discharge Current

22-1 ALTERNATING CURRENT IN A CAPACITIVE CIRCUIT

The fact that current flows with ac voltage applied is demonstrated in Fig. 22-1, where the bulb lights in (a) and (b) because of the capacitor charge and discharge current. There is no current through the dielectric, which is an insulator. While the capacitor is being charged by increasing applied voltage, however, the charging current flows in one direction in the conductors to the plates. While the capacitor is discharging, when the applied voltage decreases, the discharge current flows in the reverse direction. With alternating voltage applied, the capacitor alternately charges and discharges.

First the capacitor is charged in one polarity, and then it discharges; next the capacitor is charged in the opposite polarity, and then it discharges again. The cycles of charge and discharge current provide alternating current in the circuit, at the same frequency as the applied voltage. This is the current that lights the bulb.

In Fig. 22-1a, the 4-μF capacitor provides enough alternating current to light the bulb brightly. In Fig. 22-1b, the 1-μF capacitor has less charge and discharge current because of the smaller capacitance, and the light is not so bright. Therefore, the smaller capacitor has

Chapter 22
Capacitive Reactance

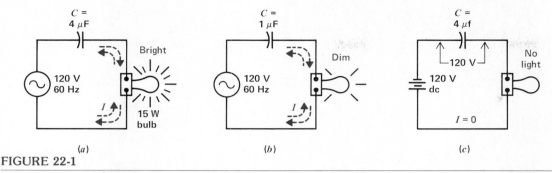

FIGURE 22-1
Current in a capacitive circuit. (a) The 4-μF capacitor allows enough 60-Hz current to light the bulb brightly. (b) Less current with smaller capacitor causes dim light. (c) Bulb cannot light with dc voltage applied.

more opposition to alternating current as less current flows with the same applied voltage; that is, it has more reactance for less capacitance.

In Fig. 22-1c, the steady dc voltage will charge the capacitor to 120 V. Because the applied voltage does not change, though, the capacitor will just stay charged. Since the potential difference of 120 V across the charged capacitor is a voltage drop opposing the applied voltage, no current can flow. Therefore, the bulb cannot light. The bulb may flicker on for an instant as charging current flows when voltage is applied, but this current is only temporary until the capacitor is charged. Then the capacitor has the applied voltage of 120 V, but there is zero voltage across the bulb.

As a result, the capacitor is said to *block* direct current or voltage. In other words, after the capacitor has been charged by a steady dc voltage, there is no current in the dc circuit. All the applied dc voltage is across the charged capacitor, with zero voltage across any series resistance.

In summary, then, this demonstration shows the following points:

1. Alternating current flows in a capacitive circuit with ac voltage applied.
2. A smaller capacitance allows less current, which means more X_C with more ohms of opposition.
3. Lower frequencies for the applied voltage result in less current and more X_C. With a steady dc voltage source, which corresponds to a frequency of zero, the opposition of the capacitor is infinite and there is no current. In this case the capacitor is effectively an open circuit.

These effects have almost unlimited applications in practical circuits because X_C depends on frequency. A very common use of a capacitor is to provide little opposition for ac voltage but to block any dc voltage. Another example is to use X_C for less opposition to a high-frequency alternating current, compared with lower frequencies.

Capacitive Current. The reason why a capacitor allows current to flow in an ac circuit is the alternate charge and discharge. If we insert an ammeter in the circuit, as shown in Fig. 22-2,

Chapter 22
Capacitive Reactance

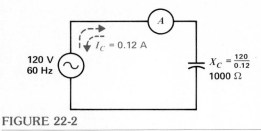

FIGURE 22-2
X_C is the ratio of V_C/I_C.

the ac meter will read the amount of charge and discharge current. In this example I_C is 0.12 A. This current is the same in the voltage source, the connecting leads, and the plates of the capacitor. However, there is no current in the dielectric between the plates of the capacitor.

Values for X_C. When we consider the ratio of V_C/I_C for the ohms of opposition to the sine-wave current, this value is $120/0.12$, which equals 1000 Ω. This 1000 Ω is what we call X_C, to indicate how much current can be produced by sine-wave voltage applied to a capacitor. In terms of current, $X_C = V_C/I_C$. In terms of frequency and capacitance, $X_C = 1/(2\pi fC)$.

The X_C value depends on the amount of capacitance and the frequency of the applied voltage. If C in Fig. 22-2 were increased, it could take on more charge for more charging current and then produce more discharge current. Then X_C is less for more capacitance. Also, if the frequency were increased in Fig. 22-2, the capacitor could charge and discharge faster to produce more current. This action also means V_C/I_C would be less, with more current for the same applied voltage. Therefore, X_C is less for higher frequencies. X_C can actually have almost any value, from practically zero to almost infinite ohms.

Practice Problems 22-1
(answers on page 456)
(a) Which has more reactance, a 0.1- or a 0.5-µF capacitor, at one frequency?
(b) Which allows more charge and discharge current, a 0.1- or a 0.5-µF capacitor?

22-2
$X_C = 1/(2\pi fC)$

This formula includes the effects of frequency and capacitance for calculating the ohms of reactance. The frequency is in hertz and C is in farads for X_C in ohms. As an example, we can calculate X_C for 2.65 µF and 60 Hz.

$$X_C = \frac{1}{2\pi fC} \quad (22\text{-}1)$$

$$= \frac{1}{2\pi \times 60 \times 2.65 \times 10^{-6}}$$

$$= \frac{0.159 \times 10^6}{60 \times 2.65} = \frac{159{,}000}{159}$$

$$X_C = 1000 \text{ Ω}$$

The constant factor 2π, equal to 6.28, indicates the circular motion from which a sine wave is derived. Therefore, the formula applies only to sine-wave circuits. To simplify calculations of X_C, the constant reciprocal $1/6.28$ can be taken as 0.159, approximately. Then

$$X_C = \frac{0.159}{fC} \quad (22\text{-}2)$$

Remember that C must be in farads for X_C in ohms. Although C values are usually microfarads (10^{-6}) or picofarads (10^{-12}), substitute the value of C in farads with the required negative power of 10.

Example 1. How much is X_C for (a) a 0.1-µF C at 1000 Hz? (b) a 1-µF C at the same frequency?

Answer.

(a) $X_C = \dfrac{0.159}{fC} = \dfrac{0.159 \times 10^6}{0.1 \times 1000} = \dfrac{0.159 \times 10^3}{0.1}$

$X_C = 1590 \text{ Ω}$

(b) At the same frequency, with ten times more C, X_C is 1590/10, which equals 159 Ω.

Note that X_C in (b) is one-tenth the X_C in (a) because C is ten times larger.

Example 2. How much is the X_C of a 100-pF C at (a) 1 MHz? (b) 10 MHz?

Answer.

(a) $X_C = \dfrac{0.159}{fC} = \dfrac{0.159}{1 \times 10^6 \times 100 \times 10^{-12}}$

$= \dfrac{0.159 \times 10^6}{100}$

$X_C = 1590$ Ω

(b) At ten times the frequency, X_C is 1590/10, which equals 159 Ω.

Note that X_C in (b) is one-tenth the X_C in (a) because f is ten times larger.

Example 3. How much is the X_C of a 240-pF C at 41.67 kHz?

Answer. $X_C = \dfrac{0.159}{fC}$

$= \dfrac{0.159}{41.67 \times 10^3 \times 240 \times 10^{-12}}$

$= \dfrac{0.159 \times 10^9}{41.67 \times 240}$

$X_C = 15{,}900$ Ω

X_C Is Inversely Proportional to Capacitance.

This statement means that X_C increases as the capacitance is reduced. In Fig. 22-3, when C is reduced by the factor of $\frac{1}{10}$, from 1.0 to 0.1 μF, then X_C increases ten times, from 1000 to 10,000 Ω. Also, decreasing C one-half, from 0.2 to 0.1 μF, doubles X_C, from 5000 to 10,000 Ω.

This inverse relation between C and X_C is illustrated by the graph in Fig. 22-3. Note that values of X_C increase downward on the graph, indicating negative reactance that is opposite from inductive reactance. With C increasing to the right, the decreasing values of X_C approach the zero axis of the graph.

X_C Is Inversely Proportional to Frequency.

Figure 22-4 illustrates the inverse relation between X_C and f. With f increasing to the right in the graph from 0.1 to 1 MHz, the negative value of X_C for the 159-pF capacitor decreases from 10,000 to 1000 Ω as the X_C curve comes closer to the zero axis.

The graphs are nonlinear because of the inverse relation between X_C and f or C. At one end, the curves approach infinitely high reactance for zero capacitance or zero frequency. At the other end, the curves approach zero reactance for infinitely high capacitance or frequency.

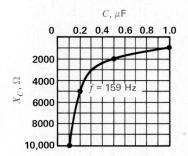

X_C Increases with Smaller C

$X_C = 1/(2\pi f C)$, Ω	C, μF
1000	1.0
2000	0.5
5000	0.2
10,000	0.1

For f = 159 Hz

FIGURE 22-3

Capacitive reactance X_C decreases with higher values of C.

Chapter 22
Capacitive Reactance

X_C Increases with Lower Frequencies

$X_C^* = 1/(2\pi fC)$, Ω   f, MHz

1000	1.0
2000	0.5
5000	0.2
10,000	0.1

*For $C = 159$ pF

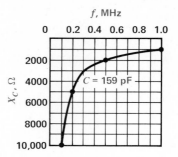

FIGURE 22-4

Capacitive reactance X_C decreases with higher frequencies.

In some applications, it is necessary to find the value of capacitance required for a desired value of X_C, at a specific frequency. For this case the reactance formula can be inverted:

$$C = \frac{0.159}{fX_C} \qquad (22\text{-}3)$$

Example 4. What capacitance is needed to have a 100-Ω X_C at 1 MHz?

Answer. $C = \dfrac{0.159}{fX_C} = \dfrac{0.159}{1 \times 10^6 \times 100}$

$= \dfrac{0.159 \times 10^{-6}}{1 \times 100}$

$= 0.001\ 59 \times 10^{-6}$ F

$C = 0.001\ 59\ \mu\text{F}$

Or, to find the frequency at which a given capacitance has a specified X_C, the reactance formula can be inverted to the form

$$f = \frac{0.159}{CX_C} \qquad (22\text{-}4)$$

Example 5. At what frequency will a 0.1-μF capacitor have an X_C equal to 1000 Ω?

Answer. $f = \dfrac{0.159}{CX_C} = \dfrac{0.159}{0.1 \times 10^{-6} \times 1000}$

$= \dfrac{0.159}{0.1 \times 10^{-6} \times 10^3}$

$= 0.159 \times 10^4$

$f = 1590$ Hz

In summary, Formula (22-2) gives X_C in terms of f and C; with Formula (22-3) we can calculate C when X_C and f are known; Formula (22-4) is used to find f with the values of C and X_C. The value of X_C can be measured as V_C/I_C.

Practice Problems 22-2
(answers on page 456)

The X_C is 400 Ω for a capacitor at 8 MHz.
(a) How much is X_C at 16 MHz?
(b) How much is X_C at 4 MHz?

22-3 SERIES OR PARALLEL CAPACITIVE REACTANCES

Because capacitive reactance is an opposition in ohms, series or parallel reactances are combined in the same way as resistances. As shown in Fig. 22-5a, series reactances of 100 and 200 Ω add to equal 300 Ω of X_{C_T}. The formula is

$$X_{C_T} = X_{C_1} + X_{C_2} + \cdots + \text{etc.} \qquad \text{in series} \qquad (22\text{-}5)$$

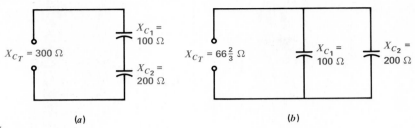

FIGURE 22-5
Reactances combine like resistances. (a) Addition of series reactances. (b) Two reactances in parallel equal their product over their sum.

For parallel reactances, the combined reactance is calculated by the reciprocal formula, as shown in Fig. 22-5b.

$$\frac{1}{X_{C_T}} = \frac{1}{X_{C_1}} + \frac{1}{X_{C_2}} + \cdots + \text{etc.} \quad \text{in parallel} \quad (22\text{-}6)$$

In Fig. 22-5b the parallel combination of 100 and 200 Ω is $66\frac{2}{3}$ Ω for X_{C_T}. The combined parallel reactance is smaller than the lowest branch reactance. Any short cuts for combining parallel resistances also apply to parallel reactances.

Combining reactances is opposite to the way capacitances are combined. The two procedures are equivalent, however, because capacitive reactance is inversely proportional to capacitance. The general case is that ohms of opposition add in series but combine by the reciprocal formula in parallel. This rule applies to resistances, to a combination of inductive reactances alone, or to capacitive reactances alone.

Practice Problems 22-3
(answers on page 456)
(a) How much is X_{C_T} for a 200-Ω X_{C_1} in series with a 300-Ω X_{C_2}?
(b) How much is X_{L_T} for a 200-Ω X_{L_1} in series with a 300-Ω X_{L_2}?

22-4
OHM'S LAW APPLIED TO X_C

The current in an ac circuit with X_C alone is equal to the applied voltage divided by the ohms of X_C. Three examples with X_C are illustrated in Fig. 22-6. In (a), there is just one reactance of 100 Ω. The current I then is equal to V/X_C, or 100 V/100 Ω, which is 1 A.

For the series circuit in (b), the total reactance, equal to the sum of the series reactances, is 300 Ω. Then the current is 100 V/300 Ω, which equals $\frac{1}{3}$ A. Furthermore, the voltage across each reactance is equal to its IX_C product. The sum of these series voltage drops equals the applied voltage.

For the parallel circuit in (c), each parallel reactance has its individual branch current, equal to the applied voltage divided by the branch reactance. The applied voltage is the same across both reactances, since they are all in parallel. In addition, the total line current of $1\frac{1}{2}$ A is equal to the sum of the individual branch currents of 1 and $\frac{1}{2}$ A each. With the applied voltage an rms value, all the calculated currents and voltage drops in Fig. 22-6 are also rms values.

Practice Problems 22-4
(answers on page 456)
(a) In Fig. 22-6b, how much is X_{C_T}?
(b) In Fig. 22-6c, how much is X_{C_T}?

Chapter 22

448 Capacitive Reactance

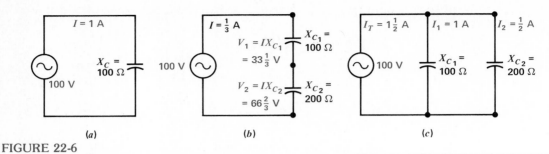

FIGURE 22-6
Circuit calculations with X_C. (a) $I = V/X_C$. (b) Sum of series voltage drops equals the applied voltage V_T. (c) Sum of parallel branch currents equals total line current I_T.

22-5 APPLICATIONS OF CAPACITIVE REACTANCE

The general use of X_C is to block direct current but provide low reactance for alternating current. In this way, a varying ac component can be separated from a steady direct current. Furthermore, a capacitor can have less reactance for alternating current of high frequencies, compared with lower frequencies.

Note the following differences in ohms of R, X_L, and X_C. Ohms of R remain the same for dc circuits or ac circuits. Ohms of reactance, however, either X_L or X_C, depend on the frequency. The effects of X_L and X_C are opposite, since X_L increases with frequency and X_C decreases with frequency.

If 100 Ω is taken as a desired value of X_C, capacitor values can be calculated for different frequencies, as listed in Table 22-1. The C values indicate typical capacitor sizes for different frequency applications.

*Practice Problems 22-5
(answers on page 456)*
A 20-μF C has 100 Ω of X_C at 60 Hz.
(a) How much is X_C at 120 Hz?
(b) How much is X_C at 6 Hz?

22-6 SINE-WAVE CHARGE AND DISCHARGE CURRENT

In Fig. 22-7 sine-wave voltage applied across a capacitor produces alternating charge and discharge current. The action is considered for each quarter-cycle. Note that the voltage v_C across the capacitor is the same as the applied voltage v_A at all times because they are in paral-

TABLE 22-1. Capacitance Values for a Reactance of 100 Ω

C (APPROX.)	FREQUENCY	REMARKS
27 μF	60 Hz	Power-line and low audio frequency
1.6 μF	1000 Hz	Medium audio frequency
0.16 μF	10,000 Hz	High audio frequency
1600 pF	1000 kHz (rf)	In AM radio broadcast band
160 pF	10 MHz (HF)	In short-wave radio band
16 pF	100 MHz (VHF)	In FM radio broadcast band

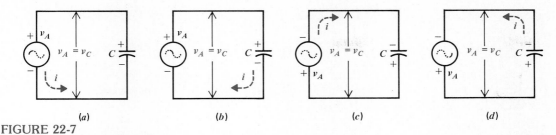

FIGURE 22-7
Capacitive charge and discharge current. (a) v_A increases positive to charge C. (b) C discharges as v_A decreases. (c) v_A increases negative to charge C in opposite polarity. (d) C discharges as reversed v_A decreases.

lel. The values of current i, however, depend on the charge and discharge of C. When v_A is increasing, it charges C to keep v_C at the same voltage as v_A; when v_A is decreasing, C discharges to maintain v_C at the same voltage as v_A. When v_A is not changing, there is no charge or discharge current.

During the first quarter-cycle in Fig. 22-7a, v_A is positive and increasing, charging C in the polarity shown. The electron flow is from the negative terminal of the source voltage, producing charging current in the direction indicated by the arrow for i. Next, when the applied voltage decreases during the second quarter-cycle, v_C also decreases by discharging. The discharge current is from the negative plate of C, through the source, and back to the positive plate. Note that the discharge current in (b) has the opposite direction from the charge current in (a).

For the third quarter-cycle in (c), the applied voltage v_A increases again but in the negative direction. Now C charges again but in reversed polarity. Here the charging current is in the opposite direction from the charge current in (a) but in the same direction as the discharge current in (b). Finally, the negative applied voltage decreases during the final quarter-cycle in (d). As a result, C discharges. This discharge current is opposite to the charge current in (c) but in the same direction as the charge current in (a).

For the sine wave of applied voltage, therefore, the capacitor provides a cycle of alternating charge and discharge current. Notice that capacitive current flows, for either charge or discharge, whenever the voltage changes, for either an increase or decrease. Also, i and v have the same frequency.

Calculating the Values of i_C. The greater the voltage change, the greater is the amount of capacitive current. Furthermore, a larger capacitor can allow more charge current when the applied voltage increases and produce more discharge current. Because of these factors the amount of capacitive current can be calculated as

$$i_C = C\frac{dv}{dt} \qquad (22\text{-}7)$$

where i is in amperes, with C in farads and dv/dt in volts per second. As an example, suppose that the voltage across a 240-pF capacitor changes by 25 V in 1 μs. The amount of capacitive current then is

$$i_C = C\frac{dv}{dt} = 240 \times 10^{-12} \times \frac{25}{1 \times 10^{-6}}$$
$$= 240 \times 25 \times 10^{-6} = 6000 \times 10^{-6}$$
$$= 6 \times 10^{-3} \text{ A}$$
$$i_C = 6 \text{ mA}$$

Notice how Formula (22-7) is similar to the capacitor charge formula $Q = CV$. When the voltage changes, this dv/dt factor produces a change in the charge Q. When the charge moves, this dq/dt change is the current i_C. Therefore, dq/dt or i_C is proportional to dv/dt. With the constant factor C, then, i_C becomes equal to $C(dv/dt)$.

The formula for capacitive current $i_C = C(dv/dt)$ corresponds to the formula for induced voltage $v_L = L(di/dt)$. In both cases there must be a change to have an effect. For inductance, v_L is induced when the current changes. For capacitance, i_C results when the voltage changes.

These formulas give the fundamental definitions for the amount of reactive effect for inductance or capacitance. Just as one henry is defined as the amount of inductance that produces one volt of v_L when the current changes at the rate of one ampere per second, one farad can also be defined as the amount of capacitance that produces one ampere of i_C when the voltage changes at the rate of one volt per second.

By means of Formula (22-7), then, i_C can be calculated to find the instantaneous value of charge or discharge current when the voltage changes across a capacitor.

Example 6. Calculate the instantaneous value of charging current i_C produced by a 6-μF C when its potential difference is increased by 50 V in 1 s.

Answer. $i_C = C\frac{dv}{dt} = 6 \times 10^{-6} \times \frac{50}{1}$
$i_C = 300$ μA

Example 7. Calculate i_C for the same C as in Example 6 where its potential difference is *decreased* by 50 V in 1 s.

Answer. For the same $C(dv/dt)$, i_C is the same 300 μA. However, this 300 μA is discharge current, which flows in the opposite direction from i_C on charge. If desired, the i_C for discharge current can be considered negative, or -300 μA.

Example 8. Calculate i_C produced by a 250-pF capacitor for a change of 50 V in 1 μs.

Answer. $i_C = C\frac{dv}{dt} = 250 \times 10^{-12} \times \frac{50}{1 \times 10^{-6}}$
$= 12{,}500 \times 10^{-6}$
$i_C = 12{,}500$ μA

Notice that more i_C is produced here, although C is smaller than in Example 6, because dv/dt is a much faster voltage change.

Waveshapes of v_C and i_C. More details of capacitive circuits can be analyzed by means of the waveshapes in Fig. 22-8, plotted for the calculated values in Table 22-2. The top curve shows a sine wave of voltage v_C across a 240-pF capacitance C. Since the capacitive current i_C depends on the rate of change of voltage, rather than the absolute value of v, the curve in (b) shows how much the voltage changes. In this curve, the dv/dt values are plotted for every 30° of the cycle.

The bottom curve shows the actual capacitive current i_C. This i_C curve is similar to the dv/dt curve because i_C equals the constant factor C multiplied by dv/dt.

All three curves are similar to the three curves shown in Fig. 19-7 for inductive circuits, but with the voltage and current curves inter-

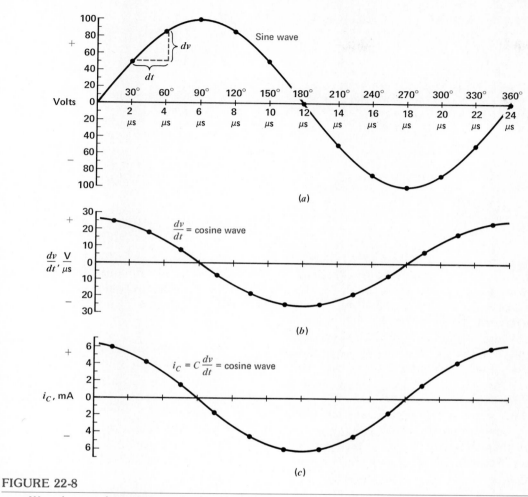

FIGURE 22-8

Waveshapes of sine-wave voltage with charge and discharge current i_C, plotted for values in Table 22-2.

changed. Both examples illustrate the effects of the rate of change in a sine wave.

90° Phase Angle. The i_C curve at the bottom in Fig. 22-8 has its zero values when the v_C curve at the top is at maximum. This comparison shows that the curves are 90° out of phase, as i_C is a cosine wave of current for the sine wave of voltage v_C. The 90° phase difference results from the fact that i_C depends on the dv/dt rate of change, rather than on v itself. More details of this 90° phase angle for capacitance are explained in the next chapter.

For each of the curves, the period T is 24 µs. Therefore, the frequency is $1/T$ or $1/24$ µs, which equals 41.67 kHz. Each curve has the same frequency, although there is a 90° phase difference between i and v.

Chapter 22
Capacitive Reactance

TABLE 22-2. Values for $i_C = C(dv/dt)$ Curves in Fig. 22-8

TIME		dt		dv, V	dv/dt, V/μS	C, pF	$i_C = C \times (dv/dt)$, mA
θ	μS	θ	μS				
30°	2	30°	2	50	25	240	6
60°	4	30°	2	36.6	18.3	240	4.4
90°	6	30°	2	13.4	6.7	240	1.6
120°	8	30°	2	−13.4	−6.7	240	−1.6
150°	10	30°	2	−36.6	−18.3	240	−4.4
180°	12	30°	2	−50	−25	240	−6
210°	14	30°	2	−50	−25	240	−6
240°	16	30°	2	−36.6	−18.3	240	−4.4
270°	18	30°	2	−13.4	−6.7	240	−1.6
300°	20	30°	2	13.4	6.7	240	1.6
330°	22	30°	2	36.6	18.3	240	4.4
360°	24	30°	2	50	25	240	6

Ohms of X_C. The ratio of v_C/i_C actually specifies the capacitive reactance, in ohms. For this comparison, we use the actual value of v_C, which has the peak of 100 V. The rate-of-change factor is included in i_C. Although the peak of i_C at 6 mA is 90° ahead of the peak of v_C at 100 V, we can compare these two peak values. Then v_C/i_C is $100/0.006$, which equals 16,667 Ω.

This X_C is only an approximate value because i_C cannot be determined exactly for the large dt changes every 30°. If we used smaller intervals of time, the peak i_C would be 6.28 mA with X_C then 15,900 Ω, the same as $1/(2\pi fC)$ with a 240-pF C and a frequency of 41.67 kHz. This is the same X_C problem as Example 3 on page 445.

Practice Problems 22-6
(answers on page 456)
Refer to the curves in Fig. 22-8.
(a) At what angle does v have its maximum positive value?
(b) At what angle does dv/dt have its maximum positive value?
(c) What is the phase angle difference between v_C and i_C?

Summary

1. Capacitive reactance, indicated by X_C, is the opposition of a capacitance to the flow of sine-wave alternating current.
2. X_C is measured in ohms because it limits the current to the value V/X_C. With V in volts and X_C in ohms, I is in amperes.
3. $X_C = 1/(2\pi fC)$. With f in hertz and C in farads, X_C is in ohms.
4. For one value of capacitance, X_C decreases with higher frequencies.
5. At one frequency, X_C decreases with higher values of capacitance.

Chapter 22
Capacitive Reactance

TABLE 22-3. Comparison of Capacitance and Capacitive Reactance

CAPACITANCE	CAPACITIVE REACTANCE
Symbol is C	Symbol is X_C
Measured in farad units	Measured in ohm units
Depends on construction of capacitor	Depends on frequency of sine-wave voltage
$C = i_C/(dv/dt)$ or Q/V	$X_C = v_C/i_C$ or $1/(2\pi fC)$

TABLE 22-4. Comparison of Inductive and Capacitive Reactances

X_L, Ω	X_C, Ω
Increases with more inductance	Decreases with more capacitance
Increases for higher frequencies	Decreases for higher frequencies
Allows more current for lower frequencies; passes direct current	Allows less current for lower frequencies; blocks direct current

6. With X_C and f known, the capacitance $C = 1/(2\pi f X_C)$.
7. With X_C and C known, the frequency $f = 1/(2\pi C X_C)$.
8. The total X_C of capacitive reactances in series equals the sum of the individual values, as for series resistances. The series reactances have the same current. The voltage across each reactance is IX_C.
9. With parallel capacitive reactances, the combined reactance is calculated by the reciprocal formula, as for parallel resistances. Each branch current is V/X_C. The total line current is the sum of the individual branch currents.
10. Table 22-3, above, summarizes the differences between C and X_C.
11. Table 22-4, above, compares the opposite reactances X_L and X_C.

Self-Examination (Answers at back of book.)

Choose (a), (b), (c), or (d).

1. Alternating current can flow in a capacitive circuit with ac voltage applied because (a) of the high peak value; (b) varying voltage produces charge and discharge current; (c) charging current flows when the voltage decreases; (d) discharge current flows when the voltage increases.
2. With higher frequencies, the amount of capacitive reactance (a) increases; (b) stays the same; (c) decreases; (d) increases only when the voltage increases.
3. At one frequency, larger capacitance results in (a) more reactance;

(b) the same reactance; (c) less reactance; (d) less reactance if the voltage amplitude decreases.
4. The capacitive reactance of a 0.1-μF capacitor at 1000 Hz equals (a) 1000 Ω; (b) 1600 Ω; (c) 2000 Ω; (d) 3200 Ω.
5. Two 1000-Ω X_C values in series have a total reactance of (a) 500 Ω; (b) 1000 Ω; (c) 1414 Ω; (d) 2000 Ω.
6. Two 1000-Ω X_C values in parallel have a combined reactance of (a) 500 Ω; (b) 707 Ω; (c) 1000 Ω; (d) 2000 Ω.
7. With 50-V rms applied across a 100-Ω X_C, the rms current in the circuit equals (a) 0.5 A; (b) 0.637 A; (c) 0.707 A; (d) 1.414 A.
8. With steady dc voltage from a battery applied to a capacitance, after it charges to the battery voltage, the current in the circuit (a) depends on the current rating of the battery; (b) is greater for larger values of capacitance; (c) is smaller for larger values of capacitance; (d) is zero for any capacitance value.
9. The capacitance needed for a 1000-Ω reactance at 2 MHz is (a) 2 pF; (b) 80 pF; (c) 1000 pF; (d) 2000 pF.
10. A 0.2-μF capacitance will have a reactance of 1000 Ω at the frequency of (a) 800 Hz; (b) 1 kHz; (c) 1 MHz; (d) 8 MHz.

Essay Questions

1. Why is capacitive reactance measured in ohms? State two differences between capacitance and capacitive reactance.
2. Referring to Fig. 22-1, explain briefly why the bulb lights in (a) but not in (c).
3. Explain briefly what is meant by two factors being inversely proportional. How does this apply to X_C and C? X_C and f?
4. In comparing X_L with X_C, give two differences and one similarity.
5. In comparing X_C and R, give two differences and one similarity.
6. Referring to Fig. 22-8, why are waves a and b considered to be 90° out of phase, while waves b and c have the same phase?
7. Referring to Fig. 22-3, how does this graph show an inverse relation between X_C and C?
8. Referring to Fig. 22-4, how does this graph show an inverse relation between X_C and f?
9. Referring to Fig. 22-8, draw three similar curves but for a sine wave of voltage with a period $T = 12$ μs for the full cycle. Use the same C of 240 pF. Compare the value of X_C obtained as $1/(2\pi fC)$ and v_C/i_C.
10. (a) What is the relation between charge q and current i? (b) How is this comparison similar to the relation between the two formulas $Q = CV$ and $i = C(dv/dt)$?

Problems (Answers to odd-numbered problems at back of book.)

1. Referring to Fig. 22-4, give the values of C needed for 2000 Ω of X_C at the four frequencies listed.
2. What size capacitance is needed for 100-Ω reactance at 100 kHz?
3. A capacitor with an X_C of 2000 Ω is connected across a 10-V 1000-Hz source. (a) Draw the schematic diagram. (b) How much is the current in the circuit? (c) What is the frequency of the current?
4. How much is the capacitance of a capacitor that draws 0.1 A from the 60-Hz 120-V power line?
5. A 1000-Ω X_{C_1} and a 4000-Ω X_{C_2} are in series across a 10-V source. (a) Draw the schematic diagram. (b) Calculate the current in the series circuit. (c) How much is the voltage across X_{C_1}? (d) How much is the voltage across X_{C_2}?
6. The 1000-Ω X_{C_1} and 4000-Ω X_{C_2} in Prob. 5 are in parallel across the 10-V source. (a) Draw the schematic diagram. (b) Calculate the branch current in X_{C_1}. (c) Calculate the branch current in X_{C_2}. (d) Calculate the total line current. (e) How much is the voltage across both reactances?
7. At what frequency will a 0.01-μF capacitor have a reactance of 10,000 Ω?
8. Four capacitive reactances of 100, 200, 300, and 400 Ω each are connected in series across a 40-V source. (a) Draw the schematic diagram. (b) How much is the total X_{C_T}? (c) Calculate I. (d) Calculate the voltages across each capacitance. (e) If the frequency of the applied voltage is 1600 kHz, calculate the required value of each capacitance.
9. Three equal capacitive reactances of 600 Ω each are in parallel. (a) How much is the equivalent combined reactance? (b) If the frequency of the applied voltage is 800 kHz, how much is the capacitance of each capacitor and how much is the equivalent combined capacitance of the three in parallel?
10. A 2-μF C is in series with a 4-μF C. The frequency is 5 kHz. (a) How much is C_T? (b) Calculate X_{C_T}. (c) Calculate X_{C_1} and X_{C_2} to see if their sum equals X_{C_T}.
11. A capacitor across the 120-V 60-Hz ac power line allows a 0.4-A current. (a) Calculate X_C and C. (b) What size C is needed to double the current?
12. A 0.01-μF capacitor is connected across a 10-V source. Tabulate the values of X_C and current in the circuit at 0 Hz (for steady dc voltage) and at 20 Hz, 60 Hz, 100 Hz, 500 Hz, 5 kHz, 10 kHz, and 455 kHz.
13. Calculate X_C for 470 pF at 2 MHz.

Chapter 22
Capacitive Reactance

14. What C is needed for the same X_C in Prob. 13 but at 500 Hz?
15. How much is I with 162 mV applied for the X_C in Probs. 13 and 14?
16. At what frequencies will X_C be 20,000 Ω for the following capacitors: (a) 2 μF; (b) 0.1 μF; (c) 0.05 μF; (d) 0.002 μF; (e) 250 pF; (f) 100 pF; (g) 47 pF?
17. What size C is needed to have X_C the same as the X_L of a 6-mH L at 100 kHz?
18. Refer to Fig. 22-9. (a) Which curve shows a direct proportion? (b) Which curve shows an inverse or reciprocal relation? (c) Which curve is for X_L? (d) Which curve is for X_C? (e) Determine the ohms values of reactance in both graphs for $L = 239$ μH and $C = 106$ pF.

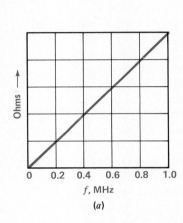
(a)

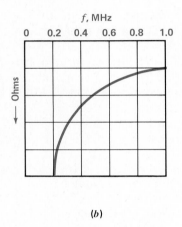
(b)

FIGURE 22-9
For Prob. 18.

Answers to Practice Problems

22-1	(a) 0.1 μF	22-4	(b) 66.7 Ω
	(b) 0.5 μF	22-5	(a) 50 Ω
22-2	(a) 200 Ω		(b) 1000 Ω
	(b) 800 Ω	22-6	(a) 90°
22-3	(a) 500 Ω		(b) 0 or 360°
	(b) 500 Ω		(c) 90°
22-4	(a) 300 Ω		

Capacitive Circuits

Chapter 23

This chapter analyzes circuits that combine capacitive reactance X_C and resistance R. The main questions are: How do we combine the ohms of opposition, how much current flows, and what is the phase angle? The method is similar to the procedures using impedance for inductive circuits, but remember that some important characteristics of X_C are opposite from X_L. In addition, methods of using series capacitances for a voltage divider are explained. Also, the practical application of a coupling capacitor shows how it is used to pass ac variations while blocking the steady dc value.

Finally, the general case of capacitive charge and discharge current is shown with nonsinusoidal voltage variations. With nonsinusoidal waveforms, the circuit can be analyzed in terms of its time constant, which is the product of $R \times C$. Remember that X_C and its 90° phase angle apply only to sine waves. The topics are:

23-1 Sine-Wave v_C Lags i_C by 90°
23-2 X_C and R in Series
23-3 X_C and R in Parallel
23-4 RF and AF Coupling Capacitors
23-5 Capacitive Voltage Dividers
23-6 The General Case of Capacitive Current i_C
23-7 Calculating the RC Time Constant

23-1
SINE-WAVE v_C LAGS i_C BY 90°

For a sine wave of applied voltage, the capacitor provides a cycle of alternating charge and discharge current, as shown in Fig. 23-1a. In (b), the waveshape of this charge and discharge current i_C is compared with the voltage v_C.

Note that the instantaneous value of i_C is zero when v_C is at its maximum value. At either its positive or negative peak, v_C is not changing. For one instant at both peaks, therefore, the voltage must have a static value before changing its direction. Then v is not changing and C is not charging or discharging. The result is zero current at this time.

Also note that i_C is maximum when v_C is zero. When v_C crosses the zero axis, i_C has its maximum value because then the voltage is changing most rapidly.

Therefore, i_C and v_C are 90° out of phase, since the maximum value of one corresponds to the zero value of the other; i_C leads v_C because i_C has its maximum value a quarter-cycle before the time that v_C reaches its peak. The phasors in Fig. 23-1c show i_C leading v_C by the counterclockwise angle of 90°. Here v_C is the horizontal

Chapter 23
Capacitive Circuits

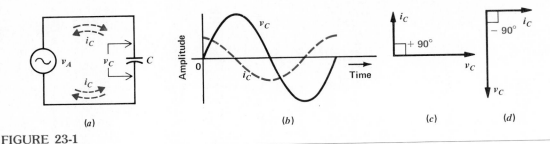

FIGURE 23-1
Capacitive current i_C leads v_C by 90°. (a) Circuit with sine-wave v_A across C. (b) Waveshapes of i_C 90° ahead of v_C in phase. (c) Phasor diagram of i_C leading v_C by counterclockwise angle of 90°. (d) Phasor diagram with i_C as reference to show v_C lagging i_C by clockwise angle of −90°.

phasor for the reference angle of 0°. In Fig. 23-1d, however, the current i_C is the horizontal phasor for reference. Since i_C must be 90° leading, v_C is shown lagging by the clockwise angle of −90°. In series circuits, the current i_C is the reference and then the voltage v_C can be considered to lag i_C by 90°.

The 90° phase angle results because i_C depends on the rate of change of v_C. As shown previously in Fig. 22-8 for a sine wave of v_C, the capacitive charge and discharge current is a cosine wave. This 90° phase between v_C and i_C is true in any sine-wave ac circuit, whether C is in series or parallel and whether C is alone or combined with other components. We can always say that for any X_C its current and voltage are 90° out of phase.

Capacitive Current the Same in Series Circuit. The leading phase angle of capacitive current is only with respect to the voltage across the capacitor, which does not change the fact that the current is the same in all parts of a series circuit. In Fig. 23-1a, for instance, the current in the generator, the connecting wires, and both plates of the capacitor must be the same because they are all in the same path.

Capacitive Voltage the Same across Parallel Branches. In Fig. 23-1a, the voltage is the same across the generator and C because they are in parallel. There cannot be any lag or lead in time between these two parallel voltages. At any instant, whatever the voltage value is across the generator at that time, the voltage across C is the same. With respect to the series current, however, both v_A and v_C are 90° out of phase with i_C.

The Frequency Is the Same for v_C and i_C. Although v_C lags i_C by 90°, both waves have the same frequency. For example, if the frequency of the sinewave v_C in Fig. 23-1b is 100 Hz, this is also the frequency of i_C.

Practice Problems 23-1
(answers on page 471)
Refer to Fig. 23-1.
(a) What is the phase between v_A and v_C?
(b) What is the phase between v_C and i_C?
(c) Does v_C lead or lag i_C?

23-2
X_C AND R IN SERIES
When resistance is in series with capacitive reactance (Fig. 23-2), both determine the current.

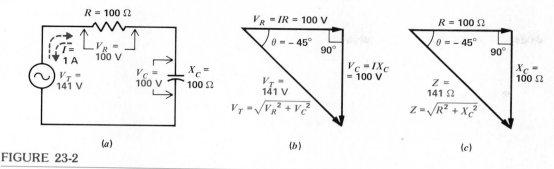

FIGURE 23-2
X_C and R in series. (a) Circuit. (b) Phasor triangle of voltages showing V_C lagging V_R by $-90°$. (c) Similar impedance triangle showing X_C lagging R by $-90°$.

I is the same in X_C and R since they are in series. Each has its own series voltage drop, equal to IR for the resistance and IX_C for the reactance.

If the capacitive reactance alone is considered, its voltage drop lags the series current I by $90°$. The IR voltage has the same phase as I, however, because resistance provides no phase shift. Therefore, R and X_C combined in series must be added by phasors because they are $90°$ out of phase with each other.

Phasor Addition of V_C and V_R. In Fig. 23-2b, the current phasor is shown horizontal, as the reference phase, because I is the same throughout the series circuit. The resistive voltage drop IR has the same phase as I. The capacitor voltage IX_C must be $90°$ clockwise from I and IR, as the capacitive voltage lags.

Note that the IX_C phasor is downward, exactly opposite from an IX_L phasor, because of the opposite phase angle. The phasor voltages V_R and V_C, being $90°$ out of phase, still form a right triangle. Therefore

$$V_T = \sqrt{V_R^2 + V_C^2} \qquad (23\text{-}1)$$

This formula applies just to series circuits because then V_C is $90°$ out of phase with V_R. All the voltages must be in the same units. When V_R and V_C are rms values, then V_T is an rms value.

In calculating the value of V_T, first square V_R and V_C, then add and take the square root. For the example in Fig. 23-2,

$$V_T = \sqrt{100^2 + 100^2} = \sqrt{10{,}000 + 10{,}000}$$
$$= \sqrt{20{,}000}$$
$$V_T = 141 \text{ V}$$

The two phasor voltages total 141 V instead of 200 V because the $90°$ phase means the peak value of one occurs when the other is at zero.

Phasor Addition of X_C and R. The voltage triangle in Fig. 23-2b corresponds to the impedance triangle in Fig. 23-2c because the common factor I can be canceled with the same current in X_C and R. Their phasor sum is the combined impedance

$$Z = \sqrt{R^2 + X_C^2} \qquad (23\text{-}2)$$

With R and X_C in ohms, Z is also in ohms. For the example in Fig. 23-2c,

$$Z = \sqrt{100^2 + 100^2}$$
$$= \sqrt{10{,}000 + 10{,}000}$$
$$= \sqrt{20{,}000}$$
$$Z = 141\ \Omega$$

Note that the total impedance of 141 Ω divided into the applied voltage of 141 V allows the current of 1 A in the series circuit. The IR voltage drop is 1×100, or 100 V; the IX_C voltage drop is also 1×100, or 100 V.

The phasor sum of the two series voltage drops of 100 V each equals the applied voltage of 141 V. Also, the applied voltage is equal to $I \times Z$, or 1×141, which is 141 V for V_T.

Phase Angle with Series X_C. As with inductive reactance, θ is the phase angle between the generator voltage and its series current. As shown in Fig. 23-2b and c, the θ can be calculated from the voltage or impedance triangle.

With series X_C, the phase angle is negative, clockwise from the zero reference angle of I, because the X_C voltage lags its current. To indicate the negative phase angle, therefore, this 90° phasor points downward from the horizontal reference, instead of upward as with series inductive reactance. To calculate the phase angle with series X_C and R,

$$\tan \theta = -\frac{X_C}{R} \qquad (23\text{-}3)$$

Using this formula for the circuit in Fig. 23-2c,

$$\tan \theta = -\frac{X_C}{R} = -\frac{100}{100} = -1$$
$$\theta = -45°$$

The negative sign means the angle is clockwise from zero, to indicate that V_T lags behind the leading I.

Series Combinations of X_C and R. In series, the higher the X_C compared with R, the more capacitive is the circuit. There is more voltage drop across the capacitive reactance, and the phase angle increases toward −90°. The series X_C always makes the current lead the applied voltage. With all X_C and no R, the entire applied voltage is across X_C, and θ equals −90°.

Several combinations of X_C and R in series are listed in Table 23-1, with their resultant impedance values and phase angle. Note that a ratio of 10:1, or more, for X_C/R means the circuit is practically all capacitive. The phase angle of −84.3° is almost −90°, and the total impedance Z is approximately equal to X_C. The voltage drop across X_C in the series circuit is then practically equal to the applied voltage, with almost none across R.

At the opposite extreme, when R is ten times more than X_C, the series circuit is mainly resistive. The phase angle of −5.7° then means the current has almost the same phase as the applied voltage; Z is approximately equal to R, and the voltage drop across R is practically equal to the applied voltage with almost none across X_C.

For the case when X_C and R equal each other, the resultant impedance Z is 1.41 times either one. The phase angle then is −45°, halfway between 0° for resistance alone and −90° for capacitive reactance alone.

TABLE 23-1. Series R and X_C Combinations

R, Ω	X_C, Ω	Z, Ω (APPROX.)	PHASE ANGLE θ_Z
1	10	$\sqrt{101} = 10$	−84.3°
10	10	$\sqrt{200} = 14$	−45°
10	1	$\sqrt{101} = 10$	−5.7°

Note: θ_Z is angle of Z_T or V_T with respect to the reference I in series circuits.

Practice Problems 23-2
(answers on page 471)
(a) How much is Z_T for a 20-Ω R in series with a 20-Ω X_C?
(b) How much is V_T for 20 V across R and 20 V across R in series?
(c) What is the phase angle of this circuit?

23-3
X_C AND R IN PARALLEL

Now the 90° phase angle for X_C must be with respect to branch currents instead of voltage drops in a series circuit. In the parallel circuit in Fig. 23-3a, the voltage is the same across X_C, R, and the generator, since they are all in parallel. There cannot be any phase difference between the parallel voltages.

Each branch, however, has its individual current. For the resistive branch, I_R is V_A/R; in the capacitive branch, $I_C = V_A/X_C$. These current phasors are shown in Fig. 23-3b.

Note that the phasor diagram has the generator voltage V_A as the reference phase because it is the same throughout the circuit. The resistive branch current I_R has the same phase as V_A, but the capacitive branch current I_C leads V_A by 90°.

The phasor for I_C is up, compared with down for an X_C phasor, because the parallel branch current I_C leads the reference V_A. This I_C phasor for a parallel branch current is opposite from an X_C phasor.

The total line current I_T then consists of I_R and I_C 90° out of phase with each other. The phasor sum of I_R and I_C equals I_T. As a result, the formula is

$$I_T = \sqrt{I_R^2 + I_C^2} \qquad (23\text{-}4)$$

In Fig. 23-2c, the phasor sum of 10 A for I_R and 10 A for I_C equals 14.14 A. The branch currents are added by phasors since they are the factors 90° out of phase in a parallel circuit, corresponding to the voltage drops 90° out of phase in a series circuit.

Impedance of X_C and R in Parallel. As usual, the impedance of a parallel circuit equals the applied voltage divided by the total line current: $Z = V_A/I_T$. In Fig. 23-3, for example,

$$Z = \frac{V_A}{I_T} = \frac{100}{14.14 \text{ A}} = 7.07 \text{ Ω}$$

Which is the opposition in ohms across the generator. This Z of 7.07 Ω is equal to the resistance of 10 Ω in parallel with the reactance of

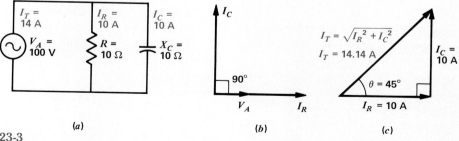

FIGURE 23-3
X_C and R in parallel. (a) Circuit. (b) Current phasors showing I_C leading V_A by 90°. (c) Phasor triangle of branch currents I_C and I_R to calculate total line current I_T.

10 Ω. Notice that the impedance of equal values of R and X_C is not one-half but equals 70.7 percent of either one.

Phase Angle in Parallel Circuits. In Fig. 23-3c, the phase angle θ is 45° because R and X_C are equal, resulting in equal branch currents. The phase angle is between the total current I_T and the generator voltage V_A. However, the phase of V_A is the same as the phase of I_R. Therefore θ is also between I_T and I_R.

Using the tangent formula to find θ from the current triangle in Fig. 23-3c gives

$$\tan \theta_I = \frac{I_C}{I_R} \qquad (23\text{-}5)$$

The phase angle is positive because the I_C phasor is upward, leading V_A by 90°. This direction is opposite from the lagging phasor of series X_C. The effect of X_C is no different, however. Only the reference is changed for the phase angle.

Note that the phasor triangle of branch currents for parallel circuits gives θ_I as the angle of I_T with respect to the generator voltage V_A. This phase angle for I_T is labeled θ_I with respect to the applied voltage. For the phasor triangle of voltages in a series circuit the phase angle for Z_T and V_T is labeled θ_Z with respect to the series current.

Parallel Combinations of X_C and R. In Table 23-2, when X_C is ten times R, the parallel circuit is practically resistive because there is little leading capacitive current in the main line. The small value of I_C results from the high reactance of shunt X_C. Then the total impedance of the parallel circuit is approximately equal to the resistance, since the high value of X_C in a parallel branch has little effect. The phase angle of 5.7° is practically 0° because almost all the line current is resistive.

As X_C becomes smaller, it provides more leading capacitive current in the main line. When X_C is $\frac{1}{10} R$, practically all the line current is the I_C component. Then, the parallel circuit is practically all capacitive, with a total impedance practically equal to X_C. The phase angle of 84.3° is almost 90° because the line current is mostly capacitive. Note that these conditions are opposite to the case of X_C and R in series. With X_C and R equal, their branch currents are equal and the phase angle is 45°.

As additional comparisons between series and parallel circuits, remember that

1. The series voltage drops V_R and V_C have individual values that are 90° out of phase. Therefore, V_R and V_C are added by phasors to equal the applied voltage V_T. The negative phase angle $-\theta_Z$ is between V_T and the common series current I. More series X_C allows more V_C to make the circuit more capacitive, with a larger negative phase angle for V_T with respect to I.
2. The parallel branch currents I_R and I_C have

TABLE 23-2. Parallel Resistance and Capacitance Combinations*

R, Ω	X_C, Ω	I_R, A	I_C, A	I_T, A (APPROX.)	Z_T, Ω (APPROX.)	PHASE ANGLE θ_I
1	10	10	1	$\sqrt{101} = 10$	1	5.7°
10	10	1	1	$\sqrt{2} = 1.4$	7.07	45°
10	1	1	10	$\sqrt{101} = 10$	1	84.3°

*$V_A = 10$ V. Note that θ_I is angle of I_T with respect to the reference V_A in parallel circuits.

individual values that are 90° out of phase. Therefore, I_R and I_C are added by phasors to equal I_T, which is the main-line current. The positive phase angle θ_I is between the line current I_T and the common parallel voltage V_A. Less parallel X_C allows more I_C to make the circuit more capacitive, with a larger positive phase angle for I_T with respect to V_A.

Practice Problems 23-3
(answers on page 471)
(a) How much is I_T for branch currents I_R of 2 A and I_C of 2 A?
(b) Find the phase angle θ_I.

23-4
RF AND AF COUPLING CAPACITORS

In Fig. 23-4, C_C is used in the application of a coupling capacitor. Its low reactance allows practically all the ac signal voltage of the generator to be developed across R. Very little of the ac voltage is across C_C.

The coupling capacitor is used for this application because at lower frequencies it provides more reactance, resulting in less ac voltage coupled across R and more across C_C. For dc voltage, all the voltage is across C with none across R, since the capacitor blocks direct current. As a result, the output signal voltage across R includes the desired higher frequencies but not direct current or very low frequencies. This application of C_C, therefore, is called *ac coupling.*

The dividing line for C_C to be a coupling capacitor at a specific frequency can be taken as X_C one-tenth or less of the series R. Then the series RC circuit is primarily resistive. Practically all the voltage drop of the ac generator is across R, with little across C. In addition, this case results in a phase angle of practically 0°.

Typical values of a coupling capacitor for audio or radio frequencies can be calculated if we assume a series resistance of 16,000 Ω. Then X_C must be 1600 Ω or less. Typical values for C_C are listed in Table 23-3. At 100 Hz, a coupling capacitor must be 1 μF to provide 1600 Ω of reactance. Higher frequencies allow a smaller value of C_C for a coupling capacitor having the same reactance. At 100 MHz in the VHF range the required capacitance is only 1 pF.

It should be noted that the C_C values are calculated for each frequency as a lower limit. At higher frequencies, the same size C_C will have less reactance than one-tenth of R, which improves the coupling.

Choosing a Coupling Capacitor for a Circuit. As an example of using these calculations, suppose that we have the problem of

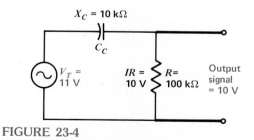

FIGURE 23-4
Series circuit for RC coupling. Small X_C compared with R allows practically all the applied voltage across R with little across C.

TABLE 23-3. Coupling Capacitors with a Reactance of 1600 Ω*

f	C_C	REMARKS
100 Hz	1 μF	Low audio frequencies
1000 Hz	0.1 μF	Medium audio frequencies
10 kHz	0.01 μF	High audio frequencies
1000 kHz	100 pF	Radio frequencies
100 MHz	1 pF	Very high frequencies

*For an X_C one-tenth of a series R of 16,000 Ω.

determining C_C for a transistorized audio amplifier. This application also illustrates the relatively large capacitance needed with low series resistance. C is to be a coupling capacitor for audio frequencies of 50 Hz and up, with a series R of 4000 Ω. Then the required X_C is 4000/10, or 400 Ω. To find C at 50 Hz

$$C = \frac{0.159}{f \times X_C} = \frac{0.159}{50 \times 400} = \frac{159{,}000 \times 10^{-6}}{20 \times 10^3}$$
$$= 7.95 \times 10^{-6}$$
$$C = 7.95 \ \mu F$$

A typical commercial size of low-voltage electrolytic readily available is 10 μF. The slightly higher capacitance value is better for coupling. The voltage rating can be 3 to 10 V, depending on the circuit, with a typical transistor supply voltage of 9 V. Although electrolytic capacitors have relatively high leakage current, they can be used for coupling capacitors in this application because of the low series resistance.

Practice Problems 23-4
(answers on page 471)
(a) X_C of a coupling capacitor is 70 Ω at 200 Hz. How much is its X_C at 400 Hz?
(b) From Table 23-3, what C would be needed for 1600 Ω of X_C at 50 MHz?

23-5
CAPACITIVE VOLTAGE DIVIDERS

When capacitors are connected in series across a voltage source, the series capacitors serve as a voltage divider. Each capacitor has part of the applied voltage, and the sum of all the series voltage drops equals the source voltage.

The amount of voltage across each is inversely proportional to its capacitance. For instance, with 2 μF in series with 1 μF, the smaller capacitor has double the voltage of the larger capacitor. Assuming 120 V applied, one-third of this, or 40 V, is across the 2 μF capacitor, with two-thirds, or 80 V, across the 1-μF capacitor.

The two series voltage drops of 40 and 80 V each add to equal the applied voltage of 120 V. The addition is just the arithmetic sum of the two voltages. It is only when voltages are out of phase with each other that the phasor addition becomes necessary.

AC Divider. With sine-wave alternating current, the voltage division between series capacitors can be calculated on the basis of reactance. In Fig. 23-5a, the total reactance is 120 Ω across the 120-V source. The current in the series circuit then is 1 A. This current is the same for X_{C_1} and X_{C_2} in series. Therefore, the

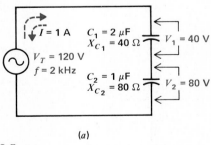
(a)

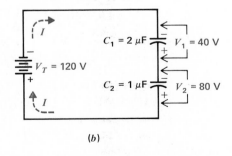
(b)

FIGURE 23-5
Series capacitors divide V_T inversely proportional to each C. (a) Ac divider. (b) Dc divider.

IX_C voltage across C_1 is 40 V, with 80 V across C_2.

The voltage division is proportional to the series reactances, as it is to series resistances. However, reactance is inversely proportional to capacitance. As a result, the smaller capacitance has more reactance and a greater part of the applied voltage.

DC Divider. In Fig. 23-5b, both C_1 and C_2 will be charged by the battery. The voltage across the series combination of C_1 and C_2 must equal V_T. When charging current flows, electrons repelled from the negative battery terminal accumulate on the negative plate of C_1, repelling electrons from its positive plate. These electrons flow through the conductor to the negative plate of C_2. With the positive battery terminal attracting electrons, the charging current from the positive plate of C_2 returns to the positive side of the dc source. Then C_1 and C_2 become charged in the polarity shown.

C_1 and C_2 are in the same series path for charging current. Therefore, both have the same amount of charge. However, the potential difference provided by the equal charges is inversely proportional to capacitance. The reason is that $Q = CV$, or $V = Q/C$. Therefore, the 1-μF capacitor has double the voltage of the 2-μF capacitor, with the same charge in both.

If you measure with a dc voltmeter across C_1, the meter reads 40 V. Across C_2 the dc voltage is 80 V. The measurement from the negative side of C_1 to the positive side of C_2 is the same as the applied battery voltage of 120 V.

If the meter is connected from the positive side of C_1 to the negative plate of C_2, however, the voltage is zero. These plates have the same potential because they are joined by a conductor of zero resistance.

The polarity marks at the junction between C_1 and C_2 indicate the voltage at this point with respect to the opposite plate for each capacitor. This junction is positive compared with the opposite plate of C_1 with a surplus of electrons. However, the same point is negative compared with the opposite plate of C_2, which has a deficiency of electrons.

In general, the following formula can be used for capacitances in series as a voltage divider:

$$V_C = \frac{C_T}{C} \times V_T \qquad (23\text{-}6)$$

Note that C_T is in the numerator, since it must be less than the smallest individual C with series capacitances. For the divider examples in Fig. 23-5a and b

$$V_1 = \frac{C_T}{C_1} \times 120 = \frac{2/3}{2} \times 120 = 40 \text{ V}$$

$$V_2 = \frac{C_T}{C_2} \times 120 = \frac{2/3}{1} \times 120 = 80 \text{ V}$$

This method applies to series capacitances as a divider for either dc or ac voltage, as long as there is no series resistance.

Practice Problems 23-5
(answers on page 471)
(a) C_1 of 10 pF and C_2 of 90 pF are across 20 kV. Calculate V_1 and V_2.
(b) In Fig. 23-5a, how much is X_{C_T}?

23-6
THE GENERAL CASE
OF CAPACITIVE CURRENT i_C

The capacitive charge and discharge current i_C is always equal to $C(dv/dt)$. A sine wave of voltage variations for v_C produces a cosine wave of current i. This means v_C and i_C have the same waveform, but they are 90° out of phase.

It is usually convenient to use X_C for calculations in sine-wave circuits. Since X_C is $1/(2\pi fC)$, the factors that determine the amount of charge and discharge current are included in f and C. Then I_C equals V_C/X_C. Or, if I_C is known, V_C can be calculated as $I_C \times V_C$.

With a nonsinusoidal waveform of voltage v_C, the concept of reactance cannot be used. X_C applies only to sine waves. Then i_C must be determined as $C(dv/dt)$. An example is illustrated in Fig. 23-6 to show the change of waveform here, instead of the change of phase angle in sine-wave circuits.

Note that the sawtooth waveform of voltage v_C corresponds to a rectangular waveform of current. The linear rise of the sawtooth wave produces a constant amount of charging current i_C because the rate of change is constant for the charging voltage. When the capacitor discharges, v_C drops sharply. Then discharge current is in the opposite direction from charge current. Also, the discharge current has a much larger value because of the faster rate of change in v_C.

An interesting feature of these capacitive waveshapes is the fact that they are the same as the inductive waveshapes shown before in Fig. 20-11, but with the current and voltage waveshapes interchanged. This comparison follows from the fact that both i_C and v_L depend on rate of change.

Practice Problems 23-6
(answers on page 471)
(a) In Fig. 23-6a, how much is dv/dt in V/s for the sawtooth rise from 0 to 90 V in 90 μs?
(b) How much is the charge current i_C, as $C(dv/dt)$ for this dv/dt?

23-7
CALCULATING THE RC TIME CONSTANT

With nonsinusoidal waveforms, the transient response of C is measured by its RC time constant. As a formula

$$T = R \times C \quad \text{s} \qquad (23\text{-}7)$$

where T is the time constant, C the capacitance in farads, and R the resistance in series with C for charge or discharge. As an example, for 1000 Ω of R in series with 4 μF of C

$$T = 1000 \times 4 \times 10^{-6}$$
$$= 4000 \times 10^{-6} = 4 \times 10^{-3}$$
$$T = 4 \text{ ms}$$

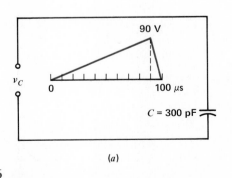 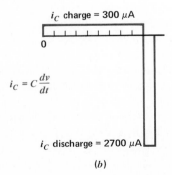

FIGURE 23-6
Waveshape of i_C equal to $C(dv/dt)$. (a) Sawtooth waveform of v_C. (b) Rectangular current waveform of i_C.

T is the time for the voltage across C to change by 63.2 percent. For instance, if the applied voltage is 100 V, then C will charge to 63.2 V in 4 ms for this example. The capacitor is completely charged in five time constants.

The RC time constant for capacitor voltage is similar to the L/R time constant for inductive current. In each case, the change is 63.2 percent in one time constant. This value is a property of the natural curve for a change in either v_C or i_L.

Actually, the RC time constant is much more important in its applications than the L/R time constant. The reason is that capacitors and resistors are small and economical to provide almost any value of RC time constant, without any coil problems. The applications of the RC time constant are explained in more detail in the next chapter, along with a comparison to the L/R time constant.

Practice Problems 23-7
(answers on page 471)
(a) How much is the RC time constant for 2 MΩ in series with 2 μF for charge?
(b) How much is the RC time constant for 100 Ω in series with 2 μF for discharge?

Summary

1. In a sine-wave ac circuit, the voltage across a capacitance lags its charge and discharge current by 90°.
2. Therefore, capacitive reactance X_C is a phasor quantity out of phase with its series resistance by $-90°$ because $i_C = C(dv/dt)$. This fundamental fact is the basis of all the following relations.
3. The vector combination of X_C and R in series is their impedance Z. These three types of ohms of opposition to current are compared in Table 23-4 below.
4. The opposite characteristics for series and parallel circuits with X_C and R are summarized in Table 23-5 on the next page.
5. In a comparison of capacitive and inductive circuits, I_L always lags V_L, but I_C leads V_C.
6. Two or more capacitors in series across a voltage source serve as a voltage divider. The smallest C has the largest part of the applied voltage.
7. A coupling capacitor has X_C less than its series resistance by the factor of $1/10$ or less, for the purpose of providing practically all the ac applied voltage across R with little across C.

TABLE 23-4. Comparison of R, X_C, and Z

R	$X_C = 1/(2\pi fC)$	$Z = \sqrt{R^2 + X_C^2}$
Ohms unit	Ohms unit	Ohms unit
IR voltage same phase as I	IX_C voltage lags I_C by 90°	IZ is the applied voltage
Same for all f	Decreases for higher f	Decreases with X_C

TABLE 23-5. Series and Parallel RC Circuits

X_C AND R IN SERIES	X_C AND R IN PARALLEL
I the same in X_C and R	V the same across X_C and R
$V_T = \sqrt{V_R^2 + V_C^2}$	$I_T = \sqrt{I_R^2 + I_C^2}$
$Z = \sqrt{R^2 + X_C^2}$	$Z = \dfrac{V}{I_T}$
V_C lags V_R by 90°	I_C leads I_R by 90°
$\tan \theta_Z = -\dfrac{X_C}{R}$; θ_Z increases as more X_C means more V_C	$\tan \theta_I = \dfrac{I_C}{I_R}$; θ_I decreases as more X_C means less I_C

8. In sine-wave circuits, $I_C = V_C/X_C$. Then I_C has a phase angle of 90° compared with V_C.
9. When the voltage is not a sine wave, $i_C = C(dv/dt)$. Then the waveshape of i_C is different from the voltage.
10. The RC time constant T in seconds is the product of C in farads and the series R in ohms. T is the time for a 63.2 percent change in V_C on charge or discharge.

Self-Examination (Answers at back of book.)

Choose (a), (b), (c), or (d).

1. In a capacitive circuit (a) a decrease in applied voltage makes a capacitor charge; (b) a steady value of applied voltage causes discharge; (c) an increase in applied voltage makes a capacitor discharge; (d) an increase in applied voltage makes a capacitor charge.
2. In a sine-wave ac circuit with X_C and R in series, the (a) phase angle of the circuit is 180° with high series resistance; (b) voltage across the capacitance must be 90° out of phase with its charge and discharge current; (c) voltage across the capacitance has the same phase as its charge and discharge current; (d) charge and discharge current of the capacitor must be 90° out of phase with the applied voltage.
3. When v_C across a 1-μF C drops from 43 to 42 V in 1 s, the discharge current i_C equals (a) 1 μA; (b) 42 μA; (c) 43 μA; (d) 43 A.
4. In a sine-wave ac circuit with R and C in parallel, (a) the voltage across C lags the voltage across R by 90°; (b) resistive I_R is 90° out of phase with I_C; (c) I_R and I_C have the same phase; (d) I_R and I_C are 180° out of phase.

5. In a sine-wave ac circuit with a 90-Ω R in series with a 90-Ω X_C, the phase angle equals (a) $-90°$; (b) $-45°$; (c) $0°$; (d) $90°$.
6. The combined impedance of a 1000-Ω R in parallel with a 1000-Ω X_C equals (a) 500 Ω; (b) 707 Ω; (c) 1000 Ω; (d) 2000 Ω.
7. With 100 V applied across two series capacitors of 5 μF each, the voltage across each capacitor will be (a) 5 V; (b) $33\frac{1}{3}$ V; (c) 50 V; (d) $66\frac{2}{3}$ V.
8. In a sine-wave ac circuit with X_C and R in series, the (a) voltages across R and X_C are in phase; (b) voltages across R and X_C are 180° out of phase; (c) voltage across R leads the voltage across X_C by 90°; (d) voltage across R lags the voltage across X_C by 90°.
9. A 0.01-μF capacitance in series with R is used as a coupling capacitor C_C for 1000 Hz. At 10,000 Hz: (a) C_C has too much reactance to be good for coupling; (b) C_C has less reactance, which improves the coupling; (c) C_C has the same reactance and coupling; (d) the voltage across R is reduced by one-tenth.
10. In an RC coupling circuit the phase angle is (a) 90°; (b) close to 0°; (c) $-90°$; (d) 180°.

Essay Questions

1. (a) Why does a capacitor charge when the applied voltage increases? (b) Why does the capacitor discharge when the applied voltage decreases?
2. A sine wave of voltage V is applied across a capacitor C. (a) Draw the schematic diagram. (b) Draw the sine waves of voltage and current out of phase by 90°. (c) Draw a vector diagram showing the phase angle of $-90°$ between V and I.
3. Why will a circuit with R and X_C in series be less capacitive as the frequency of the applied voltage is increased?
4. Define the following: coupling capacitor, sawtooth voltage, capacitive voltage divider.
5. Give two comparisons between RC circuits with sine-wave voltage applied and nonsinusoidal voltage applied.
6. Give three differences between RC circuits and RL circuits.
7. Compare the functions of a coupling capacitor with a choke coil, with two differences in their operation.
8. State two troubles possible in coupling capacitors and describe briefly how you would check with an ohmmeter.
9. Define the time constant of a capacitive circuit.

Chapter 23
Capacitive Circuits

Problems (Answers to odd-numbered problems at back of book.)

1. A 40-Ω R is in series with a 30-Ω X_C across a 100-V sine-wave ac source. (a) Draw the schematic diagram. (b) Calculate Z. (c) Calculate I. (d) Calculate the voltages across R and C. (e) What is the phase angle of the circuit?
2. A 40-Ω R and a 30-Ω X_C are in parallel across a 100-V sine-wave ac source. (a) Draw the schematic diagram. (b) Calculate each branch current. (c) How much is I_T? (d) Calculate Z. (e) What is the phase angle of the circuit? (f) Compare the phase of the voltage across R and X_C.
3. Draw the schematic diagram of a capacitor in series with a 1-MΩ resistance across a 10-V ac source. What size C is needed for equal voltages across R and X_C at frequencies of 100 Hz and 100 kHz?
4. Draw the schematic diagram of two capacitors C_1 and C_2 in series across 10,000 V. The C_1 is 900 pF and has 9000 V across it. (a) How much is the voltage across C_2? (b) How much is the capacitance of C_2?
5. In Fig. 23-2a, how much is C for the X_C value of 100 Ω at frequencies of 60 Hz, 1000 Hz, and 1 MHz?
6. A 1500-Ω R is in series with a 0.01-μF C across a 30-V source with a frequency of 8 kHz. Calculate X_C, Z_T, θ_Z, I, V_R, and V_C.
7. The same R and C as in Prob. 6 are in parallel. Calculate I_C, I_R, I_T, θ_I, Z, V_R, and V_C.
8. A 0.05-μF capacitor is in series with a 50,000-Ω R and a 10-V source. Tabulate the values of X_C, I, V_R, and V_C at the frequencies of 0 (for steady dc voltage), 20, 60, 100, 500, 5000, and 15,000 Hz.
9. A capacitive voltage divider has C_1 of 1 μF, C_2 of 2 μF, and C_3 of 4 μF in series across a 700-V source V_T. (a) Calculate V_1, V_2, and V_3 for a steady dc source. (b) Calculate V_1, V_2, and V_3 for an ac source with a frequency of 400 Hz.
10. (a) A 40-Ω X_C and a 30-Ω R are in series across a 120-V source. Calculate Z_T, I, and θ_Z. (b) The same X_C and R are in parallel. Calculate I_T, Z, and θ_I. (c) A 40-Ω X_L and a 30-Ω R are in series across a 120-V source. Calculate Z_T, I, and θ_Z. (d) The same X_L and R are in parallel. Calculate I_T, Z, and θ_I. [Note that capacitive reactance X_C is used in (a) and (b) compared with inductive reactance X_L in (c) and (d).]
11. Calculate the values for L and C in Prob. 10, with a frequency of 60 Hz for the source voltage.
12. A 500-Ω R is in series with 300-Ω X_C. Find Z_T, I, and θ_Z. $V_T = 120$ V.
13. A 300-Ω R is in series with a 500-Ω X_C. Find Z_T, I, and θ_Z. Compare θ_Z here with Prob. 12, with the same 120 V applied.

14. A 500-Ω R is parallel with a 300-Ω X_C. Find I_T, Z_T, and θ_I. Compare θ_I here with θ_Z in Prob. 12, with the same 120 V applied.
15. For the waveshape of capacitor voltage v_C in Fig. 23-7, show the corresponding charge and discharge current i_C, with values for a 200-pF capacitance. Compare these waveshapes with Prob. 17 in Chap. 20.

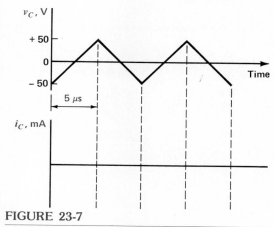

FIGURE 23-7
For Prob. 15.

Answers to Practice Problems

23-1 (a) 0°
     (b) 90°
     (c) Lag
23-2 (a) 28.28 Ω
     (b) 28.28 V
     (c) $\theta_Z = -45°$
23-3 (a) 2.828 A
     (b) $\theta_I = 45°$
23-4 (a) 35 Ω

23-4 (b) 2 pF
23-5 (a) $V_1 = 18$ kV
        $V_2 = 2$ kV
     (b) $X_{C_T} = 120$ Ω
23-6 (a) $dv/dt = 1 \times 10^6$ V/s
     (b) $i_C = 300$ μA
23-7 (a) 2 s
     (b) 200 μs

RC and L/R Time Constants

Chapter 24

Many applications of inductance are for sine-wave ac circuits, but any time the current changes, L has the effect of producing induced voltage. Examples of nonsinusoidal waveshapes include dc voltages that are switched on or off, square waves, sawtooth waves, and rectangular pulses. For capacitance, also, many applications are for sine waves, but any time the voltage changes, C produces charge or discharge current.

With nonsinusoidal voltage and current, the effect of L or C is to produce a change in waveshape. This effect can be analyzed by means of the time constant for RC and RL circuits. The time constant is the time for a change of 63.2 percent in the current through L or the voltage across C. More details are explained in the following topics:

24-1 Response of Resistance Alone
24-2 L/R Time Constant
24-3 High Voltage Produced by Opening RL Circuit
24-4 RC Time Constant
24-5 RC Charge and Discharge Curves
24-6 High Current Produced by Short-circuiting RC Circuit
24-7 RC Waveshapes
24-8 Long and Short Time Constants
24-9 Charge and Discharge with Short RC Time Constant
24-10 Long Time Constant for RC Coupling Circuit
24-11 Universal Time Constant Graph
24-12 Comparison of Reactance and Time Constant

24-1
RESPONSE OF RESISTANCE ALONE

In order to emphasize the special features of L or C, the circuit in Fig. 24-1 illustrates how ordinary a resistive circuit is. When the switch is closed, the battery supplies 10 V across the 10-Ω R and the resultant I is 1 A. The graph in (b) shows that I changes from 0 to 1 A instantly when the switch is closed. If the applied voltage is changed to 5 V, the current will change instantly to 0.5 A. If the switch is opened, I will immediately drop to zero.

Resistance has only opposition to current;

Chapter 24
RC and L/R Time Constants

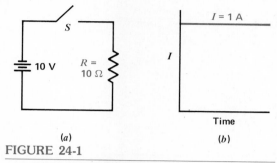

FIGURE 24-1

Response of circuit with R alone. When switch is closed, $I = 1$ A. (*a*) Circuit. (*b*) Graph of steady I.

there is no reaction to a change. The reason is that R has no concentrated magnetic field to oppose a change in I, like inductance, and no electric field to store charge that opposes a change in V, like capacitance.

Practice Problems 24-1
(*answers on page 493*)
Answer true or false.
(a) R does not produce induced voltage for a change in I.
(b) R does not produce charge or discharge current for a change in V.

24-2
L/R TIME CONSTANT

Consider the circuit in Fig. 24-2 where L is in series with R. When S is closed, the current changes as I increases from zero. Eventually, I will have the steady value of 1 A, equal to the battery voltage of 10 V divided by the circuit resistance of 10 Ω. While the current is building up from 0 to 1 A, however, I is changing and the inductance opposes the change. The action of the RL circuit during this time is its *transient response*, meaning a temporary condition existing only until the steady-state current of 1 A is reached. Similarly, when S is opened, the transient response of the RL circuit opposes the decay of current toward the steady-state value of zero.

The transient response is measured in terms of the ratio L/R, which is the time constant of an inductive circuit. To calculate the time constant

$$T = \frac{L}{R} \quad (24\text{-}1)$$

where T is the time constant in seconds and L the inductance in henrys. R is the ohms of resistance in series with L, being either the coil resistance, an external resistance, or both in series. In Fig. 24-2

$$T = \frac{L}{R} = \frac{1}{10} = 0.1 \text{ s}$$

Specifically, the time constant is a measure of how long it takes the current to change by 63.2 percent, or approximately 63 percent. In Fig. 24-2, the current increases from 0 to 0.63 A, which is 63 percent of the steady-state value, in the period of 0.1 s, which is the time constant. In the period of five time constants, the current is practically equal to its steady-state value of 1 A.

If the switch is opened now so that the current can decay to zero, I will decrease to 36.8 percent, or approximately 37 percent, of the steady-state value in one time constant. For the example in Fig. 24-2, I will decay from 1 to 0.37 A in one time constant. Note that the decrease to 0.37 A from 1 A is a change of 63 percent. The current decays practically to zero in five time constants.

The reason why L/R equals time can be illustrated as follows: Since induced voltage $V = L\,(di/dt)$, by transposing terms, L has the dimensions of $V \times T/I$. Dividing L by R results in $V \times T/IR$. As the IR and V factors cancel, T remains to indicate the dimension of time for the ratio L/R.

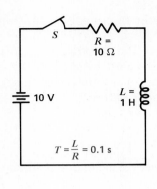
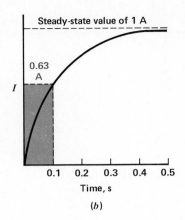

(a)   (b)

FIGURE 24-2
Transient response of circuit with R and L. When switch is closed, I rises from zero to the steady-state 1 A. (a) Circuit with time constant of 0.1 s. (b) Graph of I during five time constants.

Example 1. What is the time constant of a 20-H coil having 100 Ω of series resistance?

Answer. $T = \dfrac{L}{R} = \dfrac{20 \text{ H}}{100 \text{ Ω}}$

$T = 0.2$ s

Example 2. An applied dc voltage of 10 V will produce a steady-state current of 100 mA in the 100-Ω coil of Example 1. How much is the current after 0.2 s? After 1 s?

Answer. Since 0.2 s is one time constant, I then is 63 percent of 100 mA, which equals 63 mA. After five time constants, or 1 s, the current will reach its steady-state value of 100 mA and remain at this value as long as the applied voltage stays at 10 V.

Example 3. If a 1-MΩ R is added in series with the coil of Example 1, how much will the time constant be for the higher-resistance RL circuit?

Answer. $T = \dfrac{L}{R} = \dfrac{20 \text{ H}}{1{,}000{,}000}$

$= 20 \times 10^{-6}$ s

$T = 20$ μs   approximately

The L/R time constant becomes longer with larger values of L. More series R, however, makes the time constant shorter. Then the circuit is less inductive, with more series resistance.

Practice Problems 24-2
(answers on page 493)
(a) Calculate the time constant for 2 H in series with 100 Ω.
(b) Calculate the time constant for 2 H in series with 4000 Ω.

24-3
HIGH VOLTAGE PRODUCED BY OPENING RL CIRCUIT

When an inductive circuit is opened, the time constant for current decay becomes very short because L/R becomes smaller with the high resistance of the open circuit. Then the current drops toward zero much faster than the rise of current when the switch is closed. The result is a high value of self-induced voltage V_L across a coil whenever an *RL* circuit is opened. This high voltage can be much greater than the applied voltage.

There is no gain in energy, though, because the high voltage peak exists only for the short time the current is decreasing at a very fast rate at the start of the decay. Then, as I decays with a slower rate of change, the value of V_L is reduced. After the current has dropped to zero, there is no voltage across L.

This effect can be demonstrated by a neon bulb connected across the coil, as shown in Fig. 24-3. The neon bulb requires 90 V for ionization, at which time it glows. The source here is only 8 V, but when the switch is opened, the self-induced voltage is high enough to light the bulb for an instant. The sharp voltage pulse or spike is more than 90 V just after the switch is opened, when I drops very fast at the start of the decay in current.

Note that the 100-Ω R_1 is the internal resistance of the 2-H coil. This resistance is in series with L whether S is closed or open. The 4-kΩ R_2 across the switch is in the circuit only when S is opened, in order to have a specific resistance across the open switch. Since R_2 is much more than R_1, the L/R time constant is much shorter with the switch open.

Closing the Circuit. In Fig. 24-3a, the switch is closed to allow current in L and to store energy in the magnetic field. Since R_2 is short-circuited by the switch, the 100-Ω R_1 is the only resistance. The steady-state I is $V/R_1 = 8/100 = 0.08$ A. This value of I is reached after five time constants.

One time constant is $L/R = 2/100 = 0.02$ s. Five time constants equal $5 \times 0.02 = 0.1$ s. Therefore, I is 0.08 A after 0.1 s, or 100 ms. The energy stored in the magnetic field is 64×10^{-4} J, equal to $\frac{1}{2}LI^2$.

Opening the Circuit. When the switch is opened in Fig. 24-3b, R_2 is in series with L, making the total resistance 4100 Ω, or approximately 4 kΩ. The result is a much shorter time constant for current decay. Then L/R is $2/4000$, or 0.5 ms. The current decays practically to zero in five time constants, or 2.5 ms.

This rapid drop in current results in a magnetic field collapsing at a fast rate, inducing a high voltage across L. The peak v_L in this

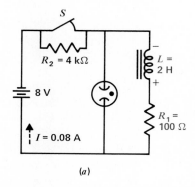
(a)

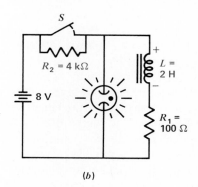
(b)

FIGURE 24-3

Demonstration of high voltage produced by opening inductive circuit. (a) With switch closed, 8 V applied cannot light the 90-V neon bulb. (b) Short time constant when S is opened results in a high v_L that lights bulb.

example is 320 V. Then v_L serves as the voltage source for the bulb connected across the coil. As a result, the neon bulb becomes ionized, and it lights for an instant.

Calculating the Peak of v_L. The value of 320 V for the peak induced voltage when S is opened in Fig. 24-3 can be determined as follows: With the switch closed, I is 0.08 A in all parts of the series circuit. The instant S is opened, R_2 is added in series with L and R_1. The energy stored in the magnetic field maintains I at 0.08 A for an instant before the current decays. With 0.08 A in the 4-kΩ R_2 its potential difference is $0.08 \times 4000 = 320$ V. The collapsing magnetic field induces this 320-V pulse to allow an I of 0.08 A at the instant the switch is opened.

The di/dt for v_L. The required rate of change in current is 160 A/s for the v_L of 320 V induced by the L of 2 H. Since $v_L = L\,(di/dt)$, this formula can be transposed to specify di/dt as equal to v_L/L. Then di/dt corresponds to 320 V/2 H, or 160 A/s. This value is the actual di/dt at the start of the decay in current when the switch is opened in Fig. 24-3b, as a result of the short time constant.[1]

Applications of Inductive Voltage Pulses. There are many uses of the high voltage generated by opening an inductive circuit. One example is the high voltage produced for the ignition system in an automobile. Here the circuit of the battery in series with a high-inductance spark coil is opened by the breaker points of the distributor to produce the high voltage needed for each spark plug. By opening an inductive circuit very rapidly, 10,000 V can easily be produced. Another important application is the high voltage of 10 to 24 kV for the anode of the picture tube in television receivers. One problem is that the high v_L produced when an inductive circuit is opened can cause arcing.

Practice Problems 24-3
(answers on page 493)
(a) Is the L/R time constant longer or shorter in Fig. 24-3 when S is opened?
(b) Which produces more v_L, a faster di/dt or a slower di/dt?

24-4
RC TIME CONSTANT

For capacitive circuits, the transient response is measured in terms of the product $R \times C$. To calculate the time constant

$$T = R \times C \qquad (24\text{-}2)$$

with R in ohms and C in farads, T is in seconds. In Fig. 24-4, for example, with an R of 3 MΩ and a C of 1 μF

$$T = 3 \times 10^6 \times 1 \times 10^{-6}$$
$$T = 3 \text{ s}$$

Note that the 10^6 for megohms and the 10^{-6} for microfarads cancel. Therefore, multiplying the units of M$\Omega \times \mu$F gives the RC product in seconds.

The reason why the RC product corresponds to time can be illustrated as follows: $C = Q/V$. The charge Q is the product of $I \times T$. The factor V is IR. Therefore, RC is equivalent to $(R \times Q)/V$, or $(R \times IT)/IR$. Since I and R cancel, T remains to indicate the dimension of time.

The Time Constant Indicates the Rate of Charge or Discharge. On charge, RC specifies the time it takes C to charge to 63 percent

---
[1] The di/dt value can be calculated from the slope at the start of decay, shown by the dashed line for curve b in Fig. 24-9.

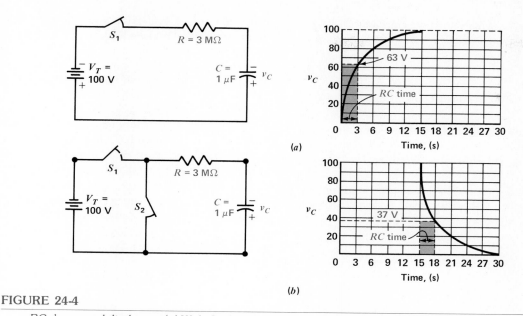

FIGURE 24-4

RC charge and discharge. (a) With S_1 closed, C charges through R to 63 percent of V_T in one time constant of 3 s. (b) With S_1 open and S_2 closed, C discharges through R to 37 percent of its initial voltage in one time constant of 3 s.

of the charging voltage. Similarly, on discharge, RC specifies the time it takes C to discharge 63 percent of the way down, to the value equal to 37 percent of the initial voltage across C at the start of discharge.

In Fig. 24-4a, for example, the time constant on charge is 3 s. Therefore, in 3 s, C charges to 63 percent of the 100 V applied, reaching 63 V in RC time. After five time constants, which is 15 s here, C is almost completely charged to the full 100 V applied. If C discharges after being charged to 100 V, then C will discharge down to 36.8 V or approximately 37 V in 3 s. After five time constants, C discharges down to zero.

A shorter time constant allows the capacitor to charge or discharge faster. If the RC product in Fig. 24-4 is 1 s, then C will charge to 63 V in 1 s instead of 3 s. Also, v_C will reach the full applied voltage of 100 V in 5 s instead of 15 s. Charging to the same voltage in less time means a faster charge.

On discharge also, the shorter time constant will allow C to discharge from 100 to 37 V in 1 s instead of 3 s. Also, v_C will be down to zero in 5 s instead of 15 s.

For the opposite case, a longer time constant means slower charge or discharge of the capacitor. More R or C results in a longer time constant.

RC Applications. Several examples are given here to illustrate how the time constant can be applied to RC circuits.

Example 4. What is the time constant of a 0.01-μF capacitor in series with a 1-MΩ resistance?

Answer.
$$T = R \times C$$
$$= 1 \times 10^6 \times 0.01 \times 10^{-6}$$
$$T = 0.01 \text{ s}$$

This is the time constant for charging or discharging, assuming the series resistance is the same for charge or discharge.

Example 5. With a dc voltage of 300 V applied, how much is the voltage across C in Example 4 after 0.01 s of charging? After 0.05 s? After 2 hours? After 2 days?

Answer. Since 0.01 s is one time constant, the voltage across C then is 63 percent of 300 V, which equals 189 V. After five time constants, or 0.05 s, C will be charged practically to the applied voltage of 300 V. After 2 hours or 2 days C will still be charged to 300 V if the applied voltage is still connected.

Example 6. If the capacitor in Example 5 is allowed to charge to 300 V and then discharged, how much is the capacitor voltage 0.01 s after the start of discharge? The series resistance is the same on discharge as on charge.

Answer. In one time constant C discharges to 37 percent of its initial voltage, or 0.37 $\times$ 300 V, which equals 111 V.

Example 7. If the capacitor in Example 5 is made to discharge after being charged to 200 V, how much will the voltage across C be 0.01 s later? The series resistance is the same on discharge as on charge.

Answer. In one time constant C discharges to 37 percent of its initial voltage, or 0.37 $\times$ 200, which equals 74 V.

This example shows that the capacitor can charge or discharge from any voltage value, not just after one RC time constant or five RC time constants.

Example 8. If a 1-MΩ resistance is added in series with the capacitor in Example 4, how much will the time constant be?

Answer. Now the series resistance is 2 MΩ. Therefore, RC is 2 $\times$ 0.01, or 0.02 s.

The RC time constant becomes longer with larger values of R and C. More capacitance means that the capacitor can store more charge. Therefore, it takes longer to store the charge needed to provide a potential difference equal to 63 percent of the applied voltage. More resistance reduces the charging current, requiring more time for charging the capacitor.

It should be noted that the RC time constant specifies just a rate. The actual amount of voltage across C depends upon the applied voltage as well as upon the RC time constant.

The capacitor takes on charge whenever its voltage is less than the applied voltage. The charging continues at the RC rate until either the capacitor is completely charged or the applied voltage decreases.

The capacitor discharges whenever its voltage is more than the applied voltage. The discharge continues at the RC rate until either the capacitor is completely discharged or the applied voltage increases.

To summarize these two important principles:

1. C charges when the net charging voltage is more than v_C.
2. C discharges when v_C is more than the net charging voltage.

The net charging voltage equals the difference between v_C and the applied voltage.

Practice Problems 24-4
(answers on page 493)
(a) How much is the RC time constant for 470 pF in series with 2 MΩ on charge?

(b) How much is the RC time constant for 470 pF in series with 1 kΩ on discharge?

24-5
RC CHARGE AND DISCHARGE CURVES

In Fig. 24-4, the RC charge curve has the rise shown because the charging is fastest at the start, then tapers off as C takes on additional charge at a slower rate. As C charges, its potential difference increases. Then the difference in voltage between V_T and v_C is reduced. Less potential difference reduces the current that puts the charge in C. The more C charges, the more slowly it takes on additional charge.

Similarly, on discharge, C loses its charge at a slower rate. At first, v_C has its highest value and can produce maximum discharge current. With the discharge continuing, v_C goes down and there is less discharge current. The more C discharges, the more slowly it can lose the remainder of its charge.

Charge and Discharge Current. There is often the question of how current can flow in a capacitive circuit with a battery as the dc source. The answer is that current flows anytime there is a change in voltage. When V_T is connected, the applied voltage changes from zero. Then charging current flows to charge C to the applied voltage. After v_C equals V_T, there is no net charging voltage and I is zero.

Similarly, C can produce discharge current anytime v_C is greater than V. When V_T is disconnected, v_C can discharge down to zero, producing discharge current in the opposite direction from the charging current. After v_C equals zero, there is no current.

Capacitance Opposes Voltage Changes Across Itself. This ability corresponds to the ability of inductance to oppose a change of current. In terms of the RC circuit, when the applied voltage increases, the voltage across the capacitance cannot increase until the charging current has stored enough charge in C. The increase in applied voltage is present across the resistance in series with C until the capacitor has charged to the higher applied voltage. When the applied voltage decreases, the voltage across the capacitor cannot go down immediately because the series resistance limits the discharge current.

The voltage across the capacitance in an RC circuit, therefore, cannot follow instantaneously the changes in applied voltage. As a result, the capacitance is able to oppose changes in voltage across itself. The instantaneous variations in V_T are present across the series resistance, however, since the series voltage drops must add to equal the applied voltage at all times.

Practice Problems 24-5
(answers on page 493)
(a) From the curve in Fig. 24-4a, how much is v_C after 3 s of charge?
(b) From the curve in Fig. 24-4b, how much is v_C after 3 s of discharge?

24-6
HIGH CURRENT PRODUCED BY SHORT-CIRCUITING RC CIRCUIT

Specifically, a capacitor can be charged slowly with a small charging current through a high resistance and then discharged fast through a low resistance to obtain a momentary surge, or pulse, of discharge current. This idea corresponds to the pulse of high voltage obtained by opening an inductive circuit.

The circuit in Fig. 24-5 illustrates the application of a battery-capacitor (BC) unit to fire a flash bulb for cameras. The flash bulb needs 5 A to ignite, but this is too much load current for the small 15-V battery that has a rating of

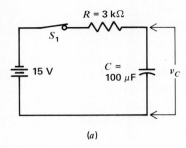 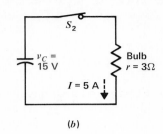

FIGURE 24-5
High current produced by discharging a charged capacitor through a low resistance. (a) When S_1 is closed, C charges to 15 V through 3 kΩ. (b) When S_2 is closed, v_C produces the peak discharge current of 5 A through 3 Ω.

30 mA for normal load current. Instead of using the bulb as a load for the battery, though, the 100-μF capacitor is charged by the battery through the 3-kΩ R in (a), and then the capacitor is discharged through the bulb in (b).

Charging the Capacitor. In Fig. 24-5a, S_1 is closed to charge C through the 3-kΩ R_1 without the bulb. The time constant of the RC charging circuit is 0.3 s.

After five time constants, or 1.5 s, C is charged to the 15 V of the battery. The peak charging current, at the first instant of charge, is V/R or 15 V/3 kΩ, which equals 5 mA. This value is an easy load current for the battery.

Discharging the Capacitor. In Fig. 24-5b, v_C is 15 V without the battery. Now S_2 is closed, and C discharges through the 3-Ω resistance of the bulb. The time constant for discharge with the lower R is $3 \times 100 \times 10^{-6}$, which equals 300 μs. At the first instant of discharge, when v_C is 15 V, the peak discharge current is 15/3, which equals 5 A. This current is enough to fire the bulb.

Energy Stored in C. When the 100-μF C is charged to 15 V by the battery, the energy stored in the electric field is $CV^2/2$, which equals 0.01 J, approximately. This energy is available to maintain v_C at 15 V for an instant when the switch is opened. The result is the 5-A I through the 3-Ω R at the start of the decay. Then v_C and i_C drop to zero in five time constants.

The dv/dt for i_C. The required rate of change in voltage is 0.05×10^6 V/s for the discharge current i_C of 5 A produced by the C of 100 μF. Since $i_C = C(dv/dt)$, this formula can be transposed to specify dv/dt as equal to i_C/C. Then dv/dt corresponds to 5 A/100 μF, or 0.05×10^6 V/s. This value is the actual dv/dt at the start of discharge when the switch is closed in Fig. 24-5b, as a result of the short time constant.[1]

Practice Problems 24-6
(answers on page 493)
(a) Is the RC time constant longer or shorter in Fig. 24-5b compared with (a)?

[1] See footnote on p. 476.

(b) Which produces more i_C, a faster dv/dt or a slower dv/dt?

24-7
RC WAVESHAPES

The voltage and current waveshapes in an RC circuit are shown in Fig. 24-6 for the case where a capacitor is allowed to charge through a resistance for RC time and then discharge through the same resistance for the same amount of time. It should be noted that this particular case is not typical of practical RC circuits, but the waveshapes show some useful details about the voltage and current. The RC time constant here equals 0.1 s.

Square Wave of Applied Voltage. The idea of closing S_1 to apply 100 V and then opening it to disconnect V_T at a regular rate corresponds to a square wave of applied voltage, as shown by the waveform a in Fig. 24-6. When S_1 is closed for charge, S_2 is open; when S_1 is open, S_2 is closed for discharge. Here the voltage is on for the RC time of 0.1 s and off for 0.1 s. The period of the square wave is 0.2 s, and f is 1/0.2 s or 5 Hz.

Capacitor Voltage v_C. As shown in Fig. 24-6b, the capacitor charges to 63 V, equal to 63 percent of the charging voltage, in the RC time of 0.1 s. Then the capacitor discharges because the applied V_T drops to zero. As a result, v_C drops to 37 percent of 63 V, or 23.3 V in RC time.

The next charge cycle begins with v_C at 23.3 V. The net charging voltage now is $100 - 23.3 = 76.7$ V. The capacitor voltage increases by 63 percent of 76.7 V, or 48.3 V. Adding 48.3 V to 23.3 V, then v_C rises to 71.6 V. On discharge, after 0.3 s, v_C drops to 37 percent of 71.6 V, or to 26.5 V.

Charge and Discharge Current. As shown in Fig. 24-6c, the current i has its positive peak at the start of charge and its negative peak at the start of discharge. On charge, i is calculated as the net charging voltage, which is $(V_T - v_C)/R$. On discharge, i always equals v_C/R.

At the start of charge, i is maximum because the net charging voltage is maximum before C charges. Similarly, the peak i for discharge occurs at the start when v_C is maximum before C discharges.

Note that i is actually an ac waveform around the zero axis, since the charge and discharge currents are in opposite directions. We are arbitrarily taking the charging current as positive values for i.

Resistor Voltage v_R. This waveshape in Fig. 24-6d follows the waveshape of current, as v_R is $i \times R$. Because of the opposite directions of charge and discharge current, the iR waveshape is an ac voltage.

Note that on charge v_R must always be equal to $V_T - v_C$ because of the series circuit. On discharge v_R has the same values as v_C because they are parallel, without V_T. Then S_2 is closed to connect R across C.

Why the i_C Waveshape Is Important. The v_C waveshape of capacitor voltage in Fig. 24-6 shows the charge and discharge directly, but the i_C waveshape is very interesting. First, the voltage waveshape across R is the same as the i_C waveshape. Also, whether C is charging or discharging, the i_C waveshape is really the same except for the reversed polarity. We can see the i_C waveshape as the voltage across R. It generally is better to connect an oscilloscope for voltage waveshapes across R, especially with one side grounded.

Finally, we can tell what v_C is from the v_R waveshape. The reason is that at any instant of

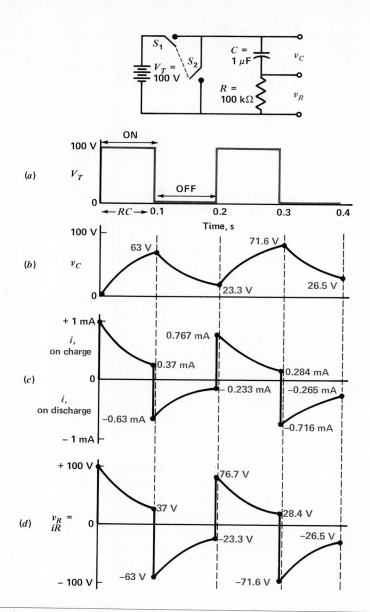

FIGURE 24-6

Charge and discharge of an RC circuit in RC time.

time V_T must equal the sum of v_R and v_C. Therefore v_C is equal to $V_T - v_R$, when V_T is charging C. For the case when C is discharging, there is no V_T. Then v_R is the same as v_C.

Practice Problems 24-7
(answers on page 493)
Refer to Fig. 24-6.
(a) When v_C is 63 V, how much is v_R?
(b) When v_R is 76.7 V, how much is v_C?

24-8 LONG AND SHORT TIME CONSTANTS

Useful waveshapes can be obtained by using RC circuits with the required time constant. In practical applications, RC circuits are used more than RL circuits because almost any value of an RC time constant can be obtained easily. With coils, the internal series resistance cannot be short-circuited and the distributed capacitance often causes resonance effects.

Long RC Time. Whether an RC time constant is long or short depends on the pulse width of the applied voltage. We can arbitrarily define a long time constant as at least five times longer than the time for applied voltage. As a result, C takes on very little charge. The time constant is too long for v_C to rise appreciably before the applied voltage drops to zero and C must discharge. On discharge also, with a long time constant, C discharges very little before the applied voltage rises to make C charge again.

Short RC Time. A short time constant can be defined as no more than one-fifth the time for applied voltage V_T. Then V_T is applied for a period of at least five time constants, allowing C to become completely charged. After C is charged, v_C remains at the value of V_T, while the voltage is applied. When V_T drops to zero, C discharges completely in five time constants and remains at zero while there is no applied voltage. On the next cycle, C charges and discharges completely again.

Differentiation. The voltage across R in an RC circuit is called differentiated output because v_R can change instantaneously. A short time constant is generally used for differentiating circuits to provide sharp pulses of v_R.

Integration. The voltage across C is called integrated output because it must accumulate over a period of time. A medium or long time constant is generally used for integrating circuits.

Practice Problems 24-8
(answers on page 493)
(a) V_T is on for 0.4 s and off for 0.4 s. RC is 6 ms for charge and discharge. Is this a long or short RC time constant?
(b) V_T is on for 2 μs and off for 2 μs. RC is 6 ms for charge and discharge. Is this a long or short RC time constant?

24-9 CHARGE AND DISCHARGE WITH SHORT RC TIME CONSTANT

Usually, the time constant is made much shorter or longer than the factor of 5, to obtain better waveshapes. In Fig. 24-7, RC is 0.1 ms. The frequency for the square wave is 25 Hz, with a period of 0.04 s, or 40 ms. One-half this period is the time V_T is applied. Therefore, the applied voltage is on for 20 ms and off for 20 ms. The RC time constant of 0.1 ms is shorter than the pulse width of 20 ms by a factor of $\frac{1}{200}$.

Square Wave of V_T Is Across C. The waveshape of v_C in Fig. 24-7b is essentially the same as the square wave of applied voltage. The reason is that the short time constant allows C to charge or discharge completely very soon after V_T is applied or removed. The charge or discharge time of five time constants is much less than the pulse width.

Sharp Pulses of i. The waveshape of i shows sharp peaks for the charge or discharge current. Each current peak is $V_T/R = 1$ mA, decaying to zero in five RC time constants. These pulses coincide with the leading and trailing edges of the square wave of V_T.

Actually, the pulses are much sharper than shown. They are not to scale horizontally in order to indicate the charge and discharge action. Also, v_C is actually a square wave like the applied voltage but with slightly rounded corners for the charge and discharge.

Note that the time axis of all the waveshapes is calibrated in seconds for the period of V_T, not in RC time constants.

Sharp Pulses of v_R. The waveshape of voltage across the resistor follows the current waveshape, as $v_R = iR$. Each current pulse of 1 mA across the 100-kΩ R results in a voltage pulse of 100 V.

More fundamentally, the peaks of v_R equal the applied voltage V_T before C charges. Then v_R drops to zero as v_C rises to the value of V_T.

On discharge, $v_R = v_C$, which is 100 V at the start of discharge. Then the pulse drops to zero in five time constants. The pulses of v_R in Fig. 24-7 are useful as timing pulses that match the edges of the square-wave applied voltage V_T. Either the positive or the negative pulses can be used.

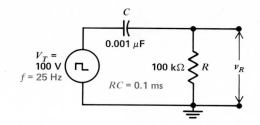

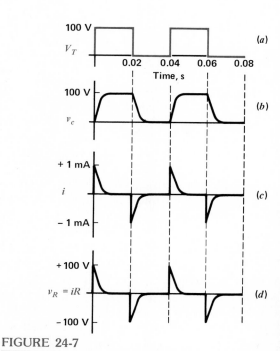

FIGURE 24-7

Charge and discharge of an RC circuit with a short time constant. Note that the waveshape of v_R in (d) has sharp voltage peaks for the leading and trailing edges of the square wave of applied voltage.

Practice Problems 24-9
(answers on page 493)
Refer to Fig. 24-7.
(a) Is the time constant here short or long?
(b) Is the square wave of applied voltage across C or R?

24-10
LONG TIME CONSTANT
FOR RC COUPLING CIRCUIT

The RC circuit in Fig. 24-8 is the same as in Fig. 24-7, but now the RC time constant is long because of the higher frequency of the applied

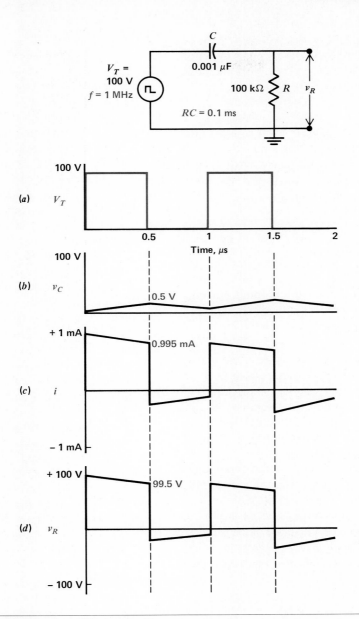

FIGURE 24-8

Charge and discharge of an RC circuit with a long time constant. Note that the waveshape of v_R in (d) has essentially the same waveform as the applied voltage.

voltage. Specifically, the RC time of 0.1 ms is 200 times longer than the 0.5-μs pulse width of V_T with a frequency of 1 MHz. Note that the time axis is calibrated in microseconds for the period of V_T, not in RC time constants.

Very Little of V_T is Across C. The waveshape of v_C in Fig. 24-8b shows very little voltage rise because of the long time constant. During the 0.5 μs that V_T is applied, C charges to only $\frac{1}{200}$ of the charging voltage. On discharge, also, v_C drops very little.

Square Wave of i. The waveshape of i stays close to the 1-mA peak at the start of charge. The reason is that v_C does not increase much, allowing V_T to maintain the charging current. On discharge, the reverse i for discharge current is very small because v_C is low.

Square Wave of V_T Is Across R. The waveshape of v_R is the same square wave as i, as $v_R = iR$. Actually, the waveshapes of i and v_R are essentially the same as the square-wave V_T applied. They are not shown to scale vertically in order to indicate the slight charge and discharge action.

Eventually, v_C will climb to the average dc value of 50 V, i will vary ±0.5 mA above and below zero, while v_R will vary ±50 V above and below zero. This application is an RC coupling circuit to block the average value of the varying dc voltage V_T as the capacitive voltage v_C, while v_R provides an ac voltage output having the same variations as V_T.

Practice Problems 24-10
(answers on page 493)
Refer to Fig. 24-8.
(a) Is the RC time constant here short or long?
(b) Is the square wave of applied voltage across R or C?

24-11
UNIVERSAL TIME CONSTANT GRAPH

We can determine transient voltage and current values for any amount of time, with the curves in Fig. 24-9. The rising curve a shows how v_C builds up as C charges in an RC circuit; the same curve applies to i_L, increasing in the inductance for an RL circuit. The decreasing curve b shows how v_C drops as C discharges or the decay of i_L in an inductance.

Note that the horizontal axis is in units of time constants rather than absolute time. Suppose that the time constant of an RC circuit is 5 μs. Therefore, one RC time unit = 5 μs, two RC units = 10 μs, three RC units = 15 μs, four RC units = 20 μs, and five RC units = 25 μs.

As an example, to find v_C after 10 μs of charging, we can take the value of curve a in Fig. 24-9 at two RC. This point is at 86 percent amplitude. Therefore, we can say that in this RC circuit with a time constant of 5 μs, v_C charges to 86 percent of the applied V_T, after 10 μs. Similarly, some important values that can be read from the curve are listed in Table 24-1.

If we consider curve a in Fig. 24-9 as an RC charge curve, v_C adds 63 percent of the net charging voltage for each additional unit of one time constant, although it may not appear so. For instance, in the second interval of RC time, v_C adds 63 percent of the net charging voltage, which is $0.37V_T$. Then 0.63×0.37 equals 0.23, which is added to 0.63 to give 0.86, or 86 percent, as the total charge from the start.

Slope at $t = 0$. The curves in Fig. 24-9 can be considered linear for the first 20 percent of change. In 0.1 time constant, for instance, the change in amplitude is 10 percent; in 0.2 time constant, the change is 20 percent. The dotted lines in Fig. 24-9 show that if this constant slope continued, the result would be 100 percent change in one time constant. This does not

Chapter 24
RC and L/R Time Constants

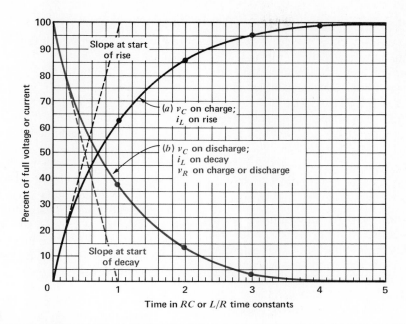

FIGURE 24-9
Universal time constant chart for RC or RL circuits. The rise or fall changes by 63 percent in one time constant.

happen, though, because the change is opposed by the energy stored in L and C. However, at the first instant of rise or decay, at $t = 0$, the change in v_C or i_L can be calculated from the dotted slope line.

Equation of the Decay Curve. The rising curve *a* in Fig. 24-9 may seem more interesting because it describes the buildup of v_C or i_L, but the decaying curve *b* is more useful. For RC circuits, curve *b* can be applied to

1. v_C on discharge
2. i and v_R on charge or discharge

If we use curve *b* for the voltage in RC circuits, the equation of this decay curve can be written as

$$v = V \times \varepsilon^{-t/RC} \qquad (24\text{-}3)$$

where *V* is the voltage at the start of decay and *v* is the reduced voltage after the time *t*. Specifically, *v* can be v_R on charge and discharge, or v_C only on discharge.

The constant ε is the base 2.718 for natural logarithms. The negative exponent $-t/RC$ in-

TABLE 24-1. Time Constant Factors

FACTOR	AMPLITUDE
0.2 time constant	20%
0.5 time constant	40%
0.7 time constant	50%
1 time constant	63%
2 time constants	86%
3 time constants	96%
4 time constants	98%
5 time constants	99%

dicates a declining exponential or logarithmic curve. The value of t/RC is the ratio of actual time of decline t to the RC time constant.

This equation can be converted to common logarithms for easier calculations. Since the natural base ε is 2.718, its logarithm to base 10 equals 0.434. Therefore, the equation becomes

$$v = \text{antilog}\left(\log V - 0.434 \times \frac{t}{RC}\right) \quad (24\text{-}4)$$

Calculations for v_R. As an example, let us calculate v_R dropping from 100 V, after RC time. Then the factor t/RC is 1. Substituting these values,

$$v_R = \text{antilog }(\log 100 - 0.434 \times 1)$$
$$= \text{antilog }(2 - 0.434) = \text{antilog } 1.566$$
$$v_R = 37 \text{ V}$$

All these logs are to base 10. The antilog of 1.566 is 37.

We can also use v_R to find v_C, which is $V_T - v_R$. Then $100 - 37 = 63$ V for v_C. These answers agree with the fact that in one time constant v_R drops 63 percent while v_C rises 63 percent. However, the formula can be used to calculate any decaying value on curve (b) in Fig. 24-9.

Calculations for t. Furthermore, Formula (24-4) can be transposed to find the time t for a specific voltage decay. Then

$$t = 2.33 RC \log \frac{V}{v} \quad (24\text{-}5)$$

where V is the higher voltage at the start and v is the lower voltage at the finish. The factor 2.33 is $1/0.434$.

As an example, let RC be 1 s. How long will it take for v_R to drop from 100 to 50 V? The required time for this decay is

$$t = 2.33 \times 1 \times \log \frac{100}{50}$$
$$= 2.33 \times 1 \times \log 2$$
$$= 2.33 \times 1 \times 0.3$$
$$t = 0.7 \text{ s} \quad \text{approximately}$$

This answer agrees with the fact that the time for a drop of 50 percent takes 0.7 time constant. However, Formula (24-5) can be used to calculate the time for any decay of v_C or v_R.

The formula cannot be used for a rise in v_C. However, if you convert this rise to an equivalent drop in v_R, the calculated time is the same for both cases.

Practice Problems 24-11
(answers on page 493)
Answer true or false for the universal curves in Fig. 24-9.
(a) Curve (a) applies to v_C on charge.
(b) Curve (b) applies to v_C on discharge.
(c) Curve (b) applies to v_R when C charges or discharges.

24-12
COMPARISON OF REACTANCE AND TIME CONSTANT
The formula for capacitive reactance includes the factor of time in terms of frequency as $X_C = 1/(2\pi fC)$. Therefore, X_C and the RC time constant are both measures of the reaction of C to a change in voltage. The reactance X_C is a special case but a very important one that applies only to sine waves. The RC time constant can be applied to any waveshape.

Phase Angle of Reactance. The capacitive charge and discharge current i_C is always equal to $C\,(dv/dt)$. A sine wave of voltage variations for v_C produces a cosine wave of current i_C. This means v_C and i_C are both sinusoids, but 90° out of phase.

In this case, it is usually more convenient to use X_C for calculations in sine-wave ac circuits to determine Z, I, and the phase angle θ. Then $I_C = V_C/X_C$. Moreover, if I_C is known, $V_C = I_C \times X_C$. The phase angle of the circuit depends on the amount of X_C compared with the resistance R.

Changes in Waveshape. With nonsinusoidal voltage applied, X_C cannot be used. Then i_C must be calculated as $C\,(dv/dt)$. In this comparison of i_C and v_C, their waveshapes can be different, instead of the change in phase angle for sine waves. The waveshapes of v_C and i_C depend on the RC time constant.

Coupling Capacitors. If we consider the application of a coupling capacitor, X_C must be one-tenth or less of its series R at the desired frequency. This condition is equivalent to having an RC time constant that is long compared with the period of one cycle. In terms of X_C, the C has little IX_C voltage, with practically all the applied voltage across the series R. In terms of a long RC time constant, C cannot take on much charge. Practically all the applied voltage is developed as $v_R = iR$ across the series resistance by the charge and discharge current. These comparisons are summarized in Table 24-2.

Inductive Circuits. Similar comparisons can be made between $X_L = 2\pi fL$ for sine waves and the L/R time constant. The voltage across any inductance is $v_L = L\,(di/dt)$. Sine-wave variations for i_L produce a cosine wave of voltage v_L, 90° out of phase.

In this case X_L can be used to determine Z, I, and the phase angle θ. Then $I_L = V_L/X_L$. Furthermore, if I_L is known, $V_L = I_L \times X_L$. The phase angle of the circuit depends on the amount of X_L compared with R.

With nonsinusoidal voltage, however, X_L cannot be used. Then v_L must be calculated as $L\,(di/dt)$. In this comparison, i_L and v_L can have different waveshapes, depending on the L/R time constant.

Choke Coils. For this application, the idea is to have almost all the applied ac voltage across L. The condition of X_L being at least ten times R corresponds to having a long time constant.

TABLE 24-2. Comparison of Reactance X_C and RC Time Constant

SINE-WAVE VOLTAGE	NONSINUSOIDAL VOLTAGE
Examples are 60-Hz power line, af signal voltage, rf signal voltage	Examples are dc circuit turned on and off, square waves, sawtooth waves
Reactance $X_C = \dfrac{1}{2\pi fC}$	Time constant $T = RC$
Larger C results in smaller reactance X_C	Larger C results in longer time constant
Higher frequency results in smaller X_C	Shorter pulse width corresponds to longer time constant
$I_C = \dfrac{V_C}{X_C}$	$i_C = C\dfrac{dv}{dt}$
X_C makes I_C and V_C 90° out of phase	Waveshape changes between i_C and v_C

The high value of X_L means practically all the applied ac voltage is across X_L as IX_L, with little IR voltage.

The long L/R time constant means i_L cannot rise appreciably, resulting in little v_R voltage across the resistor. The waveform for i_L and v_R in an inductive circuit corresponds to v_C in a capacitive circuit.

When Do We Use the Time Constant? In electronic circuits, the time constant is useful in analyzing the effect of L or C on the waveshape of nonsinusoidal voltages, particularly rectangular pulses. Another application is the transient response when a dc voltage is turned on or off. The 63 percent change in one time constant is a natural characteristic of v or i, where the magnitude of one is proportional to the rate of change of the other.

When Do We Use Reactance? The X_L and X_C are generally used for sine-wave V or I. We can determine Z, I, voltage drops, and phase angles. The phase angle of 90° is a natural characteristic of a cosine wave where its magnitude is proportional to the rate of change in a sine wave:

Practice Problems 24-12
(answers on page 493)
(a) Does an RC coupling circuit have a small or large X_C compared with R?
(b) Does an RC coupling circuit have a long or short time constant for the frequency of applied voltage?

Summary

1. The transient response of an inductive circuit with nonsinusoidal current is indicated by the time constant L/R. With L in henrys and R in ohms, T is the time in seconds for the current i_L to change by 63 percent. In five time constants, i_L reaches the steady value of V_T/R.
2. At the instant an inductive circuit is opened, high voltage is generated across L because of the fast current decay with a short time constant. The induced voltage $v_L = L\,(di/dt)$. The di is the change in i_L.
3. The transient response of a capacitive circuit with nonsinusoidal voltage is indicated by the time constant RC. With C in farads and R in ohms, T is the time in seconds for the voltage across the capacitor v_C to change by 63 percent. In five time constants, v_C reaches the steady value of V_T.
4. At the instant a charged capacitor is discharged through a low resistance, a high value of discharge current can be produced. The discharge current $i_C = C\,(dv/dt)$ can be large because of the fast discharge with a short time constant. The dv is the change in v_C.
5. The waveshapes of v_C and i_L correspond, as both rise relatively slowly to the steady-state value. This is integrated output.
6. Also i_C and v_L correspond, as they are the waveforms that can change instantaneously. This is differentiated output.

7. For both *RC* and *RL* circuits the resistor voltage $v_R = iR$.
8. A short time constant is one-fifth or less of the time for applied voltage.
9. A long time constant is greater than the time for applied voltage by a factor of 5 or more.
10. An *RC* circuit with a short time constant produces sharp voltage spikes for v_R at the leading and trailing edges of a square-wave applied voltage. The waveshape of applied voltage V_T is across the capacitor as v_C. See Fig. 24-7.
11. An *RC* circuit with a long time constant allows v_R to be essentially the same as the variations in applied voltage V_T, while the average dc value of V_T is blocked as v_C. See Fig. 24-8.
12. The universal rise and decay curves in Fig. 24-9 can be used for current or voltage in *RC* and *RL* circuits for any time up to five time constants.
13. The concept of reactance is useful for sine-wave ac circuits with *L* and *C*.
14. The time constant method is used with *L* or *C* to analyze non-sinusoidal waveforms.

Self-Examination (Answers at back of book.)

Choose (a), (b), (c), or (d).

1. A 250-μH *L* is in series with a 50-Ω *R*. The time constant is (a) 5 μs; (b) 25 μs; (c) 50 μs; (d) 250 μs.
2. If V_T is 500 mV in the preceding circuit, after 5 μs *I* rises to the value of (a) 3.7 mA; (b) 5 mA; (c) 6.3 mA; (d) 10 mA.
3. In the preceding circuit, *I* will have the steady-state value of 10 mA after (a) 5 μs; (b) 6.3 μs; (c) 10 μs; (d) 25 μs.
4. The arc across a switch when it opens an *RL* circuit is a result of the (a) long time constant; (b) large self-induced voltage across *L*; (c) low resistance of the open switch; (d) surge of resistance.
5. A 250-pF *C* is in series with a 1-MΩ *R*. The time constant is (a) 63 μs; (b) 100 μs; (c) 200 μs; (d) 250 μs.
6. If V_T is 100 V in the preceding circuit, after 250 μs, v_C rises to the value of (a) 37 V; (b) 50 V; (c) 63 V; (d) 100 V.
7. In the preceding circuit, v_C will have the steady-state value of 100 V after (a) 250 μs; (b) 630 μs; (c) 1000 μs or 1 ms; (d) 1.25 ms.
8. In the preceding circuit, after 3 hours v_C will be (a) zero; (b) 63 V; (c) 100 V; (d) 200 V.

9. For a square-wave applied voltage with the frequency of 500 Hz, a long time constant is (a) 1 ms; (b) 2 ms; (c) 3.7 ms; (d) 5 ms.
10. An RC circuit has a 2-μF C in series with a 1-MΩ R. The time of 6 s equals how many time constants? (a) one; (b) two; (c) three; (d) six.

Essay Questions

1. Give the formula, with units, for calculating the time constant of an RL circuit.
2. Give the formula, with units, for calculating the time constant of an RC circuit.
3. Redraw the RL circuit and graph in Fig. 24-2 for a 2-H L and a 100-Ω R.
4. Redraw the graphs in Fig. 24-4 to fit the circuit in Fig. 24-5 with a 100-μF C. Use a 3000-Ω R for charge but a 3-Ω R for discharge.
5. List two comparisons of RC and RL circuits for nonsinusoidal voltage.
6. List two comparisons between RC circuits with nonsinusoidal voltage and sine-wave voltage applied.
7. Define the following: (a) a long time constant; (b) a short time constant; (c) an RC differentiating circuit.
8. Redraw the horizontal time axis of the universal curve in Fig. 24-9, calibrated in absolute time units of milliseconds for an RC circuit with a time constant of 2.3 ms.
9. Redraw the circuit and graphs in Fig. 24-7 with everything the same except that R is 20 kΩ, making the RC time constant shorter.
10. Redraw the circuit and graphs in Fig. 24-8 with everything the same except that R is 500 kΩ, making the RC time constant longer.
11. Invert the equation $T = RC$, in two forms, to find R or C from the time constant.

Problems (Answers to odd-numbered problems at back of book.)

1. Calculate the time constant of the following inductive circuits: (a) L is 20 H and R is 400 Ω; (b) L is 20 μH and R is 400 Ω; (c) L is 50 mH and R is 50 Ω; (d) L is 40 μH and R is 2 Ω.
2. Calculate the time constant of the following capacitive circuits: (a) C is 0.001 μF and R is 1 MΩ; (b) C is 1 μF and R is 1000 Ω; (c) C is 0.05 μF and R is 250 kΩ; (d) C is 100 pF and R is 10 kΩ.

3. A 100-V source is in series with a 2-MΩ R and a 2-μF C. (a) How much time is required for v_C to be 63 V? (b) How much is v_C after 20 s?
4. The C in Prob. 3 is allowed to charge for 4 s and then made to discharge for 8 s. How much is v_C?
5. A 100-V source is applied in series with a 1-MΩ R and a 4-μF C that has already been charged to 63 V. How much is v_C after 4 s?
6. What value of R is needed with a 0.02-μF C for an RC time constant of 0.02 s? For 1 ms?
7. An RC circuit has a time constant of 1 ms. V_T applied is 200 V. How much is v_C on charge after 1.4 ms?
8. A 0.05-μF C charges through a 0.5-MΩ R but discharges through a 2-kΩ R. Calculate the time constants for charge and discharge. Why will the capacitor discharge faster than charge?
9. A 0.05-μF C is charged to 264 V. It discharges through a 40-kΩ R. How much is the time for v_C to discharge down to 132 V?
10. Referring to Fig. 24-6b, calculate the value of v_C on the next charge, starting from 26.5 V.
11. Use the slope line in Fig. 24-9b to calculate dv/dt at the start of the decay in v_C for the circuit in Fig. 24-5b.
12. Use the slope line in Fig. 24-9b to calculate di/dt at the start of the decay in i_L for the circuit in Fig. 24-3b. (Hint: You can ignore the steady 8 V and 100-Ω R_1 because they do not change the di/dt value.)

Answers to Practice Problems

24-1	(a) T	24-7	(b) $v_C = 23.3$ V
	(b) T	24-8	(a) Short
24-2	(a) 0.02 s		(b) Long
	(b) 0.5 ms	24-9	(a) Short
24-3	(a) Shorter		(b) Across C
	(b) Faster	24-10	(a) Long
24-4	(a) 940 μs		(b) Across R
	(b) 470 ns	24-11	(a) T
24-5	(a) 63.2 V		(b) T
	(b) 36.8 V		(c) T
24-6	(a) Shorter	24-12	(a) Small X_C
	(b) Faster		(b) Long time constant
24-7	(a) $v_R = 37$ V		

Review of Chapters 21 to 24

Summary

1. A capacitor, or condenser, consists of two conductors separated by an insulator, which is a dielectric material. With voltage applied to the conductors, charge is stored in the dielectric. One coulomb of charge stored with one volt applied corresponds to one farad of capacitance C. The common units of capacitance are microfarads ($\mu F = 10^{-6}$ F) or picofarads (pF $= 10^{-12}$ F).
2. Capacitance increases with plate area and larger values of dielectric constant but decreases with the distance between plates.
3. The most common types of capacitors are air, paper, mica, ceramic, and electrolytic. Electrolytics must be connected in the correct polarity. The color coding for mica and ceramic tubular capacitors is illustrated in Figs. 21-9 and 21-10.
4. The total capacitance of parallel capacitors is the sum of the individual values; the combined capacitance of series capacitors is found by the reciprocal formula. These rules are opposite from the formulas for resistors and inductors in series or parallel.
5. In checking with an ohmmeter, a good capacitor shows charging current and then the ohmmeter reads a very high value of ohms equal to the insulation resistance. A short-circuited capacitor reads zero ohms; an open capacitor does not show any charging current.
6. $X = 1/(2\pi fC)\ \Omega$, with f in hertz and C in farads. The higher the frequency and the greater the capacitance, the smaller X_C is.
7. A common application of X_C is in af or rf coupling capacitors, which have low reactance for one group of frequencies but more reactance for lower frequencies. This is just the opposite of an inductance used as a choke.
8. X_C is a phasor quantity where the voltage across the capacitor lags 90° behind its charge and discharge current. This phase angle of X_C is exactly opposite from the phase angle for X_L.

9. In series circuits, R and X_C are added by phasors because their voltage drops are 90° out of phase. Therefore, the total impedance Z equals $\sqrt{R^2 + X_C^2}$; the current I equals V_T/Z.
10. For parallel circuits, the resistive and capacitive branch currents are added by phasors: $I_T = \sqrt{I_R^2 + I_C^2}$; the impedance $Z = V_A/I_T$.
11. The time constant of a capacitive circuit equals $R \times C$. With R in ohms and C in farads, the RC product is the time in seconds for the voltage across C to change by 63 percent.
12. Capacitive charge or discharge current i_C is equal to $C(dv/dt)$ for any waveshape of v_C.
13. The time constant of an inductive circuit equals L/R. With L in henrys and R in ohms, L/R is the time in seconds for the current through L to change by 63 percent.
14. Induced voltage v_L is equal to $L(di/dt)$ for any waveshape of i_L.

Review Self-Examination (Answers at back of book.)

Answer true or false.

1. A capacitor can store charge because it has a dielectric between two conductors.
2. With 100 V applied, a 0.01-μF capacitor stores 1 μC of charge.
3. The smaller the capacitance, the higher the potential difference across it for a given amount of charge stored in the capacitor.
4. A 250-pF capacitance equals 250×10^{-12} F.
5. The thinner the dielectric, the more the capacitance and the lower the voltage breakdown rating for a capacitor.
6. Larger plate area increases the capacitance.
7. Capacitors in series provide less capacitance but a higher voltage breakdown rating for the combination.
8. Capacitors in parallel increase the total capacitance with the same voltage rating.
9. Two 0.01-μF capacitors in parallel have a total C of 0.005 μF.
10. A good 0.01-μF paper capacitor will show charging current and read 500 MΩ or more on an ohmmeter.
11. If the capacitance is doubled, the reactance is one-half.
12. If the frequency is doubled, the reactance is one-half.
13. The reactance of a 0.1-μF capacitor at 60 Hz is approximately 60 Ω.
14. In a series circuit, the voltage across X_C lags 90° behind the current.
15. The phase angle of a series circuit can be any angle between 0 and 90°, depending on the ratio of X_C to R.

16. In a parallel circuit, the voltage across X_C lags 90° behind its capacitive branch current.
17. In a parallel circuit of two resistances with 1 A in each branch, the total line current equals 1.414 A.
18. A 1000-Ω X_C in parallel with 1000-Ω R has a combined Z of 707 Ω.
19. A 1000-Ω X_C in series with 1000-Ω R has a total Z of 1414 Ω.
20. Neglecting its sign, the phase angle is 45° for both circuits in Probs. 18 and 19.
21. X_L and X_C are opposite reactances.
22. The total impedance of a 1-MΩ R in series with a 5-Ω X_C is approximately 1 MΩ with a phase angle of 0°.
23. The combined impedance of a 5-Ω R in shunt with a 1-MΩ X_C is approximately 5 Ω with a phase angle of 0°.
24. X_L and X_C change with frequency, but L and C do not depend on the frequency.
25. A long RC time constant corresponds to a large C and R.
26. When the RC time constant for discharge is calculated, R must be the resistance in the path for discharge current.
27. Resistance and impedance are both measured in ohms.
28. X_L and X_C are both measured in ohms.
29. Z can change with frequency because it includes reactance.
30. With 100 V applied, a 1-μF capacitor in series with a 1-MΩ R will have the transient value of $v_C = 63$ V in 1 s, while charging to the steady-state value of 100 V in 5 s.
31. A 1-μF capacitor charged to 2000 V has stored energy equal to 1 J.
32. A 2-H L is in series with a 1-kΩ R and a 100-V source. After 2 ms, the transient value of i_L is 63 mA, while rising to the steady-state value of 100 mA in 10 ms.
33. When the applied voltage increases, charging current can flow as the capacitor takes on additional charge.
34. When the applied voltage decreases, a charged capacitor can discharge because it has a higher potential difference than the source.
35. Capacitors in series have the same charge and discharge current.
36. Capacitors in parallel have the same voltage.
37. The phasor combination of a 30-Ω R in series with a 40-Ω X_C equals 70-Ω impedance.
38. A six-dot mica capacitor color-coded white, green, black, red, and yellow has the capacitance value of 500 pF.
39. Capacitive current can be considered leading current in a series circuit.
40. In a series circuit, the higher the value of X_C, the greater is its voltage drop compared with the IR drop.

References (Additional references at back of book.)

Boylestad, Robert: "Introductory Circuit Analysis," Charles E. Merrill, Inc., Englewood Cliffs, N.J.

Gillie, A. C.: "Electrical Principles of Electronics," McGraw-Hill Book Company, New York.

Jackson, H. W.: "Introduction to Electric Circuits," Prentice-Hall, Inc., Englewood Cliffs, N.J.

Slurzberg, M. and W. Osterheld: "Essentials of Electricity for Radio and Television," 3d ed., McGraw-Hill Book Company, New York.

Alternating-Current Circuits

Chapter 25

This chapter shows how to analyze sine-wave ac circuits that have R, X_L, and X_C. How do we combine these three types of ohms of opposition, how much current flows, and what is the phase angle? These questions are answered for both series and parallel circuits.

The problems are simplified by the fact that in series circuits X_L is at 90° and X_C at −90°, which are opposite phase angles. Then all of one reactance can be canceled by part of the other reactance, resulting in only a single net reactance. Similarly, in parallel circuits, I_L and I_C have opposite phase angles. These oppose each other for one net reactive line current.

Finally, the idea of how ac power and dc power can differ because of ac reactance is explained. Also, types of ac current meters are explained, including the wattmeter. The topics are:

25-1 AC Circuits with Resistance but no Reactance
25-2 Circuits with X_L Alone
25-3 Circuits with X_C Alone
25-4 Opposite Reactances Cancel
25-5 Series Reactance and Resistance
25-6 Parallel Reactance and Resistance
25-7 Series-Parallel Reactance and Resistance
25-8 Real Power
25-9 AC Meters
25-10 Wattmeters
25-11 Summary of Types of Ohms in AC Circuits
25-12 Summary of Types of Phasors in AC Circuits

25-1 AC CIRCUITS WITH RESISTANCE BUT NO REACTANCE

Combinations of series and parallel resistances are shown in Fig. 25-1. In both (a) and (b), all voltages and currents throughout the resistive circuit are in the same phase as the applied voltage. There is no reactance to cause a lead or lag in either current or voltage.

Series Resistances. For the circuit in Fig.

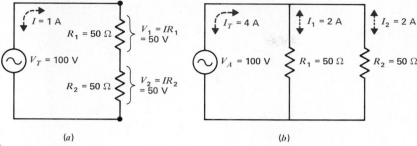

(a)   (b)

FIGURE 25-1

Ac circuits with resistance but no reactance. (a) R_1 and R_2 in series. (b) R_1 and R_2 in parallel.

25-1a, with two 50-Ω resistances in series across the 100-V source, the calculations are as follows:

$$R_T = R_1 + R_2 = 50 + 50 = 100 \; \Omega$$
$$I = \frac{V_T}{R_T} = \frac{100}{100} = 1 \; A$$
$$V_1 = IR_1 = 1 \times 50 = 50 \; V$$
$$V_2 = IR_2 = 1 \times 50 = 50 \; V$$

Note that the series resistances R_1 and R_2 serve as a voltage divider, as in dc circuits. Each R has one-half the applied voltage for one-half the total series resistance.

The voltage drops V_1 and V_2 are both in phase with the series current I, which is the common reference. Also I is in phase with the applied voltage V_T because there is no reactance.

Parallel Resistances. For the circuit in Fig. 25-1b, with two 50-Ω resistances in parallel across the 100-V source, the calculations are:

$$I_1 = \frac{V_A}{R_1} = \frac{100}{50} = 2 \; A$$
$$I_2 = \frac{V_A}{R_2} = \frac{100}{50} = 2 \; A$$
$$I_T = I_1 + I_2 = 2 + 2 = 4 \; A$$

With a total current of 4 A in the main line from the 100-V source, the combined parallel resistance is 25 Ω. This R_T equals 100 V/4 A for the two 50-Ω branches.

Each branch current has the same phase as the applied voltage. V_A is the reference because it is common to both branches.

Practice Problems 25-1
(answers on page 518)
Refer to Fig. 25-1.
(a) In (a), what is the phase angle between V_T and I?
(b) In (b), what is the phase angle between I_T and V_A?

25-2
CIRCUITS WITH X_L ALONE

The circuits with X_L in Figs. 25-2 and 25-3 correspond to the series and parallel circuits in Fig. 25-1, with ohms of X_L equal to the R values. Since the applied voltage is the same, the values of current correspond because ohms of X_L are just as effective as ohms of R in limiting the current or producing a voltage drop.

Although X_L is a phasor quantity with a 90° phase angle, all the ohms of opposition are the same kind of reactance in this example. Therefore, without any R or X_C, the series ohms

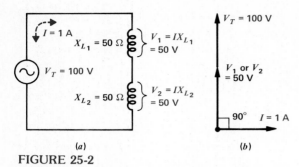

FIGURE 25-2

Series circuit with X_L alone. (a) Schematic diagram. (b) Phasor diagram.

of X_L can be combined directly. Similarly, the parallel I_L currents can be added.

X_L Values in Series. For Fig. 25-2a, the calculations are

$$X_{L_T} = X_{L_1} + X_{L_2} = 50 + 50 = 100 \ \Omega$$
$$I = \frac{V_T}{X_{L_T}} = \frac{100}{100} = 1 \ \text{A}$$
$$V_1 = IX_{L_1} = 1 \times 50 = 50 \ \text{V}$$
$$V_2 = IX_{L_2} = 1 \times 50 = 50 \ \text{V}$$

Note that the two series voltage drops of 50 V each add to equal the total applied voltage of 100 V.

With regard to the phase angle for the inductive reactance, the voltage across any X_L always leads the current through it by 90°. In Fig. 25-2b, I is the reference phasor because it is common to all the series components. Therefore, the voltage phasors for V_1 or V_2 across either reactance, or V_T across both reactances, are shown leading I by 90°.

I_L Values in Parallel. For Fig. 25-3a the calculations are

$$I_1 = \frac{V_A}{X_{L_1}} = \frac{100}{50} = 2 \ \text{A}$$
$$I_2 = \frac{V_A}{X_{L_2}} = \frac{100}{50} = 2 \ \text{A}$$
$$I_T = I_1 + I_2 = 2 + 2 = 4 \ \text{A}$$

These two branch currents can be added because they both have the same phase. This angle is 90° lagging the voltage reference phasor as shown in (b).

Since the voltage V_A is common to the branches, this voltage is across X_{L_1} and X_{L_2}. Therefore V_A is the reference phasor for parallel circuits.

Note that there is no fundamental change between Fig. 25-2b, which shows each X_L voltage leading its current by 90°, and Fig. 25-3b, showing each X_L current lagging its voltage by −90°. The phase angle between the inductive current and voltage is still the same 90°.

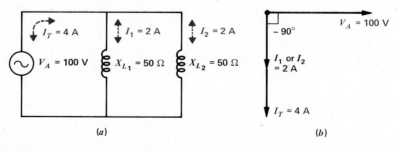

FIGURE 25-3

Parallel circuit with X_L alone. (a) Schematic diagram. (b) Phasor diagram.

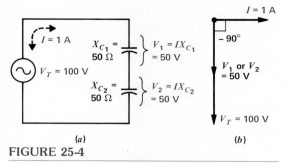

FIGURE 25-4

Series circuit with X_C alone. (a) Schematic diagram. (b) Phasor diagram.

Practice Problems 25-2
(answers on page 518)

(a) In Fig. 25-2, what is the phase angle of V_T with respect to I?
(b) In Fig. 25-3, what is the phase angle of I_T with respect to V_A?

25-3
CIRCUITS WITH X_C ALONE

Again, reactances are shown in Figs. 25-4 and 25-5 but with X_C values of 50 Ω. Since there is no R or X_L, the series ohms of X_C can be combined directly. Also the parallel I_C currents can be added.

X_C **Values in Series.** For Fig. 25-4a, the calculations for V_1 and V_2 are the same as before. These two series voltage drops of 50 V each add to equal the total applied voltage.

With regard to the phase angle for the capacitive reactance, the voltage across any X_C always lags its capacitive charge and discharge current I by 90°. For the series circuit in Fig. 25-4, I is the reference phasor. The capacitive current leads by 90°. Or, we can say that each voltage lags I by −90°.

I_C **Values in Parallel.** For Fig. 25-5, V_A is the reference phasor. The calculations for I_1 and I_2 are the same as before. However, now each of the capacitive branch currents or the I_T leads V_A by 90°.

Practice Problems 25-3
(answers on page 518)

(a) In Fig. 25-4, what is the phase angle of V_T with respect to I?
(b) In Fig. 25-5, what is the phase angle of I_T with respect to V_A?

25-4
OPPOSITE REACTANCES CANCEL

In a circuit with both X_L and X_C, the opposite phase angles enable one to cancel the effect of the other. For X_L and X_C in series, the net reactance is the difference between the two series reactances, resulting in less reactance

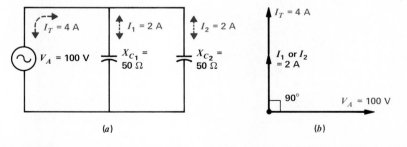

FIGURE 25-5

Parallel circuit with X_C alone. (a) Schematic diagram. (b) Phasor diagram.

than either one. In parallel circuits, the I_L and I_C branch currents cancel. The net line current then is the difference between the two branch currents, resulting in less total line current than either branch current.

X_L and X_C in Series. For the example in Fig. 25-6, the series combination of a 60-Ω X_L and a 40-Ω X_C in (a) and (b) is equivalent to the net reactance of the 20-Ω X_L shown in (c). Then, with 20 Ω as the net reactance across the 120-V source, the current is 6 A. This current lags the applied voltage V_T by 90° because the net reactance is inductive.

For the two series reactances in (a), the current is the same through both X_L and X_C. Therefore, the IX_L drop is 6 A × 60 Ω, or 360 V, and the IX_C drop is 6 A × 40 Ω, or 240 V.

Note that each individual reactive voltage drop can be more than the applied voltage. The sum of the series voltage drops still is 120 V, however, equal to the applied voltage. This results because the IX_L and IX_C voltages are opposite. The IX_L voltage leads the series current by 90°; the IX_C voltage lags the same current by 90°. Therefore, IX_L and IX_C are 180° out of phase with each other, which means they are of opposite polarity and cancel. Then the total voltage across the two in series is 360 V minus 240 V, which equals the applied voltage of 120 V.

If the values in Fig. 25-6 were reversed, with an X_C of 60 Ω and an X_L of 40 Ω, the net reactance would be a 20-Ω X_C. The current would be 6 A again, but with a lagging phase angle of −90° for the capacitive voltage. The IX_C voltage would then be larger at 360 V, with an IX_L value of 240 V, but the difference still equals the applied voltage of 120 V.

X_L and X_C in Parallel. In Fig. 25-7, the 60-Ω X_L and 40-Ω X_C are in parallel across the 120-V source. Then the 60-Ω X_L branch current I_L is 2 A, and the 40-Ω X_C branch current I_C is 3 A. The X_C branch has more current because its reactance is less than X_L.

In terms of phase angle, I_L lags the parallel voltage V_A by 90°, while I_C leads the same voltage by 90°. Therefore, the opposite reactive branch currents are 180° out of phase with each other and cancel. The net line current then is the difference between 3 A for I_C and 2 A for I_L, which equals the net value of 1 A. This re-

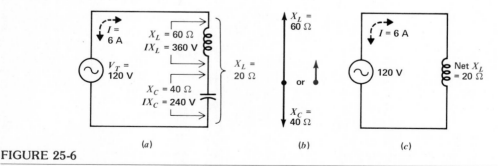

FIGURE 25-6

With X_L and X_C in series, their ohms of reactance cancel. (a) Series circuit. (b) Phasors for X_L and X_C with net resultant. (c) Equivalent circuit with net reactance of 20-Ω X_L.

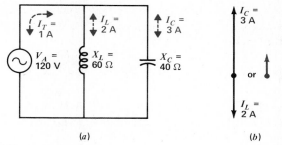

FIGURE 25-7
With X_L and X_C in parallel, their branch currents cancel. (a) Parallel circuit. (b) Phasors for branch currents I_C and I_L with net resultant. (c) Equivalent circuit with net line current of 1-A I_C.

sultant current leads V_A by 90° because it is capacitive current.

If the values in Fig. 25-7 were reversed, with an X_C of 60 Ω and an X_L of 40 Ω, I_L would be larger. I_L then equals 3 A, with an I_C of 2 A. The net line current is 1 A again but inductive, with a net X_L.

Practice Problems 25-4
(answers on page 518)
(a) In Fig. 25-6, how much is the net X_L?
(b) In Fig. 25-7, how much is the net I_C?

25-5
SERIES REACTANCE AND RESISTANCE

In this case, the resistive and reactive effects must be combined by phasors. For series circuits, the ohms of opposition are added to find Z. First add all the series resistances for one total R. Also combine all the series reactances, adding the same kind but subtracting opposites. The result is one net reactance, indicated X. It may be either capacitive or inductive, depending on which kind of reactance is larger. Then the total R and net X can be added by phasors to find the total ohms of opposition for the entire series circuit.

Magnitude of Z. After the total R and net reactance X are found, they can be combined by the formula

$$Z = \sqrt{R^2 + X^2} \qquad (25\text{-}1)$$

The circuit's total impedance Z is the phasor sum of the series resistance and reactance. Whether the net X is at +90° for X_L or −90° for X_C does not matter in calculating the magnitude of Z.

An example is illustrated in Fig. 25-8. Here the net series reactance in (b) is a 30-Ω X_C. This value is equal to a 60-Ω X_L subtracted from a 90-Ω X_C as shown in (a). The net 30-Ω X_C in (b) is in series with a 40-Ω R. Therefore

$$Z = \sqrt{R^2 + X^2} = \sqrt{(40)^2 + (30)^2}$$
$$= \sqrt{1600 + 900} = \sqrt{2500}$$
$$Z = 50 \text{ Ω}$$

I = V/Z. The current is 100 V/50 Ω in this example, or 2 A. This value is the magnitude, without considering the phase angle.

Chapter 25
Alternating-Current Circuits

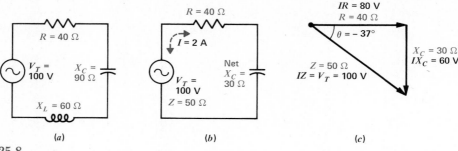

FIGURE 25-8

Impedance Z of series circuit. (a) R, X_L, and X_C in series. (b) Equivalent circuit with one net reactance. (c) Phasor diagram.

Series Voltage Drops. All the series components have the same 2-A current. Therefore, the individual drops in Fig. 25-8a are

$IR = 2 \times 40 = 80$ V
$IX_C = 2 \times 90 = 180$ V
$IX_L = 2 \times 60 = 120$ V

Since IX_C and IX_L are voltages of opposite polarity, the net reactive voltage is 180 minus 120 V, which equals 60 V. The phasor sum of IR at 80 V and the net reactive voltage IX of 60 V equals the applied voltage V of 100 V.

Phase Angle of Z. The phase angle of the series circuit is the angle whose tangent equals X/R. The angle is negative for X_C but positive for X_L.

In this example, X is the net reactance of 30 Ω for X_C and R is 40 Ω. Then $\tan \theta = -0.75$ and θ is $-37°$, approximately.

The negative angle for Z indicates lagging capacitive reactance for the series circuit. If the values of X_L and X_C were reversed, the phase angle would be $+37°$, instead of $-37°$, because of the net X_L. However, the magnitude of Z would still be the same.

More Series Components. How to combine any number of series resistances and reactances is illustrated by Fig. 25-9. Here the total series R of 40 Ω is the sum of 30 Ω for R_1 and 10 Ω for R_2. Note that the order of connection does not matter, since the current is the same in all series components.

The total series X_C is 90 Ω, equal to the sum of 70 Ω for X_{C_1} and 20 Ω for X_{C_2}. Similarly, the total series X_L is 60 Ω, equal to the sum of 30 Ω for X_{L_1} and 30 Ω for X_{L_2}.

The net reactance X equals 30 Ω, which is 90 Ω of X_C minus 60 Ω of X_L. Since X_C is larger than X_L, the net reactance is capacitive. The circuit in Fig. 25-9 is equivalent to Fig. 25-8,

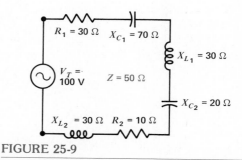

FIGURE 25-9

Series ac circuit with more components than Fig. 25-8, but the same values of Z, I, and θ.

therefore, since a 40-Ω R is in series with a net X_C of 30 Ω.

Practice Problems 25-5
(answers on page 518)
(a) In Fig. 25-8, how much is the net reactance?
(b) In Fig. 25-9, how much is the net reactance?

25-6 PARALLEL REACTANCE AND RESISTANCE

With parallel circuits, the branch currents for resistance and reactance are added by phasors. Then the total line current is found by the formula

$$I_T = \sqrt{I_R^2 + I_X^2} \qquad (25\text{-}2)$$

Calculating I_T. As an example, Fig. 25-10a shows a circuit with three branches. Since the voltage across all the parallel branches is the applied 100 V, the individual branch currents are

$$I_R = \frac{V_A}{R} = \frac{100 \text{ V}}{25 \text{ }\Omega} = 4 \text{ A}$$

$$I_L = \frac{V_A}{X_L} = \frac{100 \text{ V}}{25 \text{ }\Omega} = 4 \text{ A}$$

$$I_C = \frac{V_A}{X_C} = \frac{100 \text{ V}}{100 \text{ }\Omega} = 1 \text{ A}$$

The net reactive branch current I_X is 3 A, then, equal to the difference between the 4-A I_L and the 1-A I_C, as shown in (b).

The next step is to calculate I_T as the phasor sum of I_R and I_X. Then

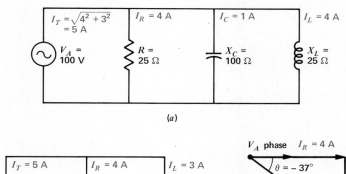

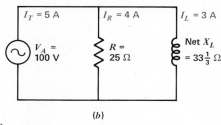

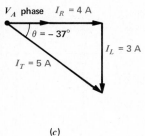

FIGURE 25-10
Total line current I_T of parallel circuit. (a) R, X_L, and X_C in parallel. (b) Equivalent circuit with net reactive branch current. (c) Phasor diagram.

$$I_T = \sqrt{I_R^2 + I_X^2} = \sqrt{4^2 + 3^2}$$
$$= \sqrt{16 + 9} = \sqrt{25}$$
$$I_T = 5 \text{ A}$$

The phasor diagram for I_T is shown in (c).

$Z_T = V_A/I_T$. This gives the total impedance of a parallel circuit. In this example, Z_T is 100 V/5 A, which equals 20 Ω. This value is the equivalent impedance of all three branches in parallel across the source.

Phase Angle. The phase angle of the parallel circuit is found from the branch currents. Now θ is the angle whose tangent equals I_X/I_R.

For this example, I_X is the net inductive current of the 3-A I_L. Also, I_R is 4 A. These phasors are shown in Fig. 25-10c. Then θ is a negative angle with the tangent of $-\tfrac{3}{4}$ or -0.75. This phase angle is $-37°$, approximately.

The negative angle for I_T indicates lagging inductive current. The value of $-37°$ is the phase angle of I_T with respect to the voltage reference V_A.

When Z_T is calculated as V_A/I_T for a parallel circuit, the phase angle of Z_T is the same value as for I_T but with opposite sign. In this example, Z_T is 20 Ω with a phase angle of $+37°$, for an I_T of 5 A with an angle of $-37°$.

We can consider that Z_T has the phase of the voltage source with respect to I_T.

More Parallel Branches. Figure 25-11 illustrates how any number of parallel resistances and reactances can be combined. The total resistive branch current I_R of 4 A is the sum of 2 A each for the R_1 branch and the R_2 branch. Note that the order of connection does not matter, since the parallel branch currents add in the main line. Effectively, two 50-Ω resistances in parallel are equivalent to one 25-Ω resistance.

Similarly, the total inductive branch current I_L is 4 A, equal to 3 A for I_{L_1} and 1 A for I_{L_2}. Also, the total capacitive branch current I_C is 1 A, equal to ½ A each for I_{C_1} and I_{C_2}.

The net reactive branch current I_X is 3 A, then, equal to a 4-A I_L minus a 1-A I_C. Since I_L is larger, the net current is inductive.

The circuit in Fig. 25-11 is equivalent to the circuit in Fig. 25-10, therefore. Both have a 4-A resistive current I_R and a 3-A net inductive current I_L. These values added by phasors make a total of 5 A for I_T in the main line.

Practice Problems 25-6
(answers on page 518)

(a) In Fig. 25-10, what is the net reactive branch current?
(b) In Fig. 25-11, what is the net reactive branch current?

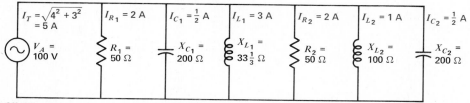

FIGURE 25-11

Parallel ac circuit with more components than Fig. 25-10, but with the same value of I_T, Z, and θ.

25-7
SERIES-PARALLEL REACTANCE AND RESISTANCE

Figure 25-12 shows how a series-parallel circuit can be reduced to a series circuit with just one reactance and one resistance. The method is straightforward as long as resistance and reactance are not combined in one parallel bank or series string.

Working backward toward the generator from the outside branch in Fig. 25-12a, we have an X_{L_1} and an X_{L_2} of 100 Ω each in series, which total 200 Ω. This string in (a) is equivalent to X_{L_5} in (b).

In the other branch, the net reactance of X_{L_3} and X_C is equal to 600 Ω minus 400 Ω This is equivalent to the 200 Ω of X_{L_4} in (b). The X_{L_4} and X_{L_5} of 200 Ω each in parallel are combined for an X_L of 100 Ω.

In (c), the 100-Ω X_L is in series with the 100-Ω $R_{1\text{-}2}$. This value is for R_1 and R_2 in parallel.

The phasor diagram for the equivalent circuit in (d) shows the total impedance Z of 141 Ω for a 100-Ω R in series with a 100-Ω X_L. With a 141-Ω impedance across the applied V_T of 100 V, the current in the generator is 0.7 A. The phase angle θ is 45° for this circuit.[1]

Practice Problems 25-7
(answers on page 518)
Refer to Fig. 25-12.
(a) How much is $X_{L_1} + X_{L_2}$?
(b) How much is $X_{L_3} - X_C$?
(c) How much is X_{L_4} in parallel with X_{L_5}?

25-8
REAL POWER

In an ac circuit with reactance, the current I supplied by the generator either leads or lags the generator voltage V. Then the product VI is not the real power produced by the generator, since the voltage may have a high value while the current is near zero, or vice versa. The real power, however, can always be calculated as I^2R, where R is the total resistive component of the circuit, because current and voltage have the same phase in a resistance. To find the corresponding value of power as VI, this product must be multiplied by the cosine of the phase angle θ. Then

$$\text{Real power} = I^2 R \qquad (25\text{-}3)$$

or

$$\text{Real power} = VI \cos \theta \qquad (25\text{-}4)$$

where V and I are in rms values, to calculate the real power, in watts. Multiplying VI by the cosine of the phase angle provides the resistive component for real power equal to I^2R.

For example, the ac circuit in Fig. 25-13 has 2 A through a 100-Ω R in series with the X_L of 173 Ω. Therefore

Real power = $I^2R = 4 \times 100$
Real power = 400 W

Furthermore, in this circuit the phase angle is 60° with a cosine of 0.5. The applied voltage is 400 V. Therefore

Real power = $VI \cos \theta = 400 \times 2 \times 0.5$
Real power = 400 W

In both examples, the real power is the same 400 W, because this is the amount of power supplied by the generator and dissipated in the resistance. Either formula can be used for calculating the real power, depending on which is more convenient.

---

[1] More complicated ac circuits with series-parallel impedances are analyzed with complex numbers, as explained in Chap. 26.

Chapter 25
Alternating-Current Circuits

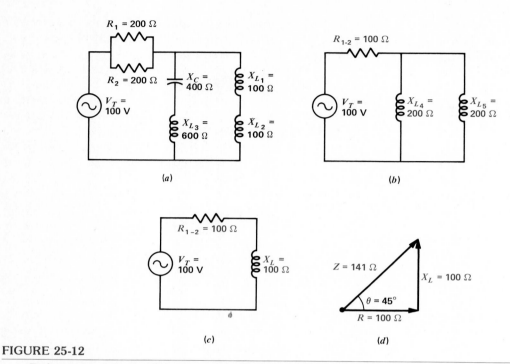

FIGURE 25-12

Reducing a series-parallel circuit with R, X_L, and X_C to a series circuit with one R and one X. (a) Actual circuit. (b) Simplified arrangement. (c) Series equivalent circuit. (d) Phasor diagram.

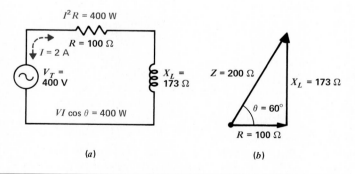

FIGURE 25-13

Real power in series circuit. (a) Schematic diagram. (b) Phasor diagram.

Real power can be considered as resistive power, which is dissipated as heat. A reactance does not dissipate power but stores energy in the electric or magnetic field.

Power Factor. Because it indicates the resistive component, $\cos \theta$ is the power factor of the circuit, converting the VI product to real power. For series circuits

$$\text{Power factor} = \cos \theta = \frac{R}{Z} \qquad (25\text{-}5)$$

or for parallel circuits

$$\text{Power factor} = \cos \theta = \frac{I_R}{I_T} \qquad (25\text{-}6)$$

In Fig. 25-13, as an example of a series circuit, we use R and Z for the calculations:

$$\text{Power factor} = \cos \theta = \frac{R}{Z} = \frac{100 \ \Omega}{200 \ \Omega} = 0.5$$

For the parallel circuit in Fig. 25-10, we use the resistive current I_R and the I_T:

$$\text{Power factor} = \cos \theta = \frac{I_R}{I_T} = \frac{4 \text{ A}}{5 \text{ A}} = 0.8$$

The power factor is not an angular measure but a numerical ratio, with a value between 0 and 1, equal to the cosine of the phase angle.

With all resistance and zero reactance, R and Z are the same for a series circuit, or I_R and I_T are the same for a parallel circuit, and the ratio is 1. Therefore, unity power factor means a resistive circuit. At the opposite extreme, all reactance with zero resistance makes the power factor zero, meaning that the circuit is all reactive.

Apparent Power. When V and I are out of phase because of reactance, the product of $V \times I$ is called apparent power. The unit is *voltamperes* (VA) instead of watts, since the watt is reserved for real power.

For the example in Fig. 25-13, with 400 V and the 2-A I, 60° out of phase, the apparent power is VI, or $400 \times 2 = 800$ VA. Note that apparent power is the VI product alone, without considering the power factor $\cos \theta$.

The power factor can be calculated as the ratio of real power to apparent power, as this ratio equals $\cos \theta$. As an example, in Fig. 25-13, the real power is 400 W, and the apparent power is 800 VA. The ratio of $^{400}/_{800}$ then is 0.5 for the power factor, the same as $\cos 60°$.

The VAR. This is an abbreviation for voltampere reactive. Specifically, VARs are voltamperes at the angle of 90°.

In general, for any phase angle θ between V and I, multiplying VI by $\sin \theta$ gives the vertical component at 90° for the value of the VARs. In Fig. 25-13, the $VI \sin 60° = 80 \times 0.866 = 692.8$ VAR.

Note that the factor $\sin \theta$ for the VARs gives the vertical or reactive component of the apparent power VI. However, multiplying VI by $\cos \theta$ as the power factor gives the horizontal or resistive component for the real power.

Correcting the Power Factor. In commercial use, the power factor should be close to unity for efficient distribution. However, the inductive load of motors may result in a power factor of 0.7, as an example, for the phase angle of 45°. To correct for this lagging inductive component of the current in the main line, a capacitor can be connected across the line to draw leading current from the source. The value of capacitance is calculated to take the same amount of voltamperes as the VARs of the load.

Chapter 25

510 Alternating-Current Circuits

Practice Problems 25-8
(*answers on page 518*)
(a) What is the unit for real power?
(b) What is the unit for reactive power?

25-9
AC METERS

The D'Arsonval moving-coil type of meter movement will not read if used in an ac circuit because the average value of an alternating current is zero. Since the two opposite polarities cancel, an alternating current cannot deflect the meter movement either up-scale or down-scale. An ac meter must produce deflection of the meter pointer up-scale regardless of polarity. This deflection is accomplished by one of the following three methods for ac meters.

1. *Thermal type.* In this method, the heating effect of the current, which is independent of polarity, is used to provide meter deflection. Two examples are the thermocouple type and the hot-wire meter.
2. *Electromagnetic type.* In this method, the relative magnetic polarity is maintained constant although the current reverses. Examples are the iron-vane meter, dynamometer, and wattmeter.
3. *Rectifier type.* The rectifier changes the ac input to dc output for the meter, which is usually a D'Arsonval movement. This type is the most common for ac voltmeters generally used for audio and radio frequencies.

All ac meters have scales calibrated in rms values, unless noted otherwise on the meter.

A thermocouple consists of two dissimilar metals, short-circuited at one end but open at the opposite side. Heat at the short-circuited junction produces a small dc voltage across the open ends, which are connected to a dc meter movement. In the hot-wire meter, current heats a wire to make it expand, and this motion is converted into meter deflection. Both types are used as ac meters for radio frequencies.

The iron-vane meter and dynamometer have very low sensitivity, compared with a D'Arsonval movement. They are used in power circuits, for either direct current or alternating current.

The rectifier type of ac voltmeter can use copper oxide, selenium, germanium, or a vacuum tube for the rectifier. A separate scale for a range of 10 V or less is generally necessary because of nonlinear rectification at low amplitudes. It should also be noted that the rectifier arrangement is not suitable for measuring current, because a current meter must have very low resistance. For this reason ac-dc multimeters generally measure ac voltages as well as dc voltages, but not alternating current.

Ac meters, other than the rectifier type, generally have a nonlinear scale that is crowded at the low end. This crowding of the scale readings results from the fact that the deflection is proportional to the square of the current.

Practice Problems 25-9
(*answers on page 518*)
Answer true or false.
(a) The iron-wave meter can read alternating current.
(b) The D'Arsonval meter movement is for direct current only.

25-10
WATTMETERS

The wattmeter uses fixed coils to indicate current in the circuit, while the movable coil indicates voltage (Fig. 25-14). The deflection then is proportional to power. Either dc power or real ac power can be read directly by the wattmeter.

In Fig. 25-14a, the coils L_{I_1} and L_{I_2} in series are the stationary coils serving as an ammeter to measure current. The two I termi-

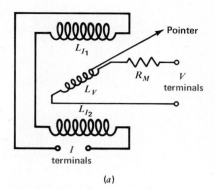 

FIGURE 25-14
Wattmeter. (*a*) Schematic of voltage and current coils. (*b*) 0- to 200-W meter. (*W. M. Welch Mfg. Co.*)

nals are connected in one side of the line in series with the load. The movable coil L_V and its multiplier resistance R_M are used as a voltmeter, with the V terminals connected across the line in parallel with the load. Then the current in the fixed coils is proportional to I, while the current in the movable coil is proportional to V. As a result, the deflection is proportional to the VI product, which is power.

Furthermore, it is the VI product for each instant of time that produces deflection. For instance, if the V value is high when the I value is low, for a phase angle close to 90°, there will be little deflection. The meter deflection is proportional to the watts of real power, therefore, regardless of the power factor in ac circuits. The wattmeter is commonly used to measure power from the 60-Hz power line. For radio frequencies, however, power is generally measured in terms of heat transfer.

Practice Problems 25-10
(*answers on page 518*)
(*a*) Does a wattmeter measure real or apparent power?
(*b*) In Fig. 25-14, does the movable coil measure V or I?

25-11
SUMMARY OF TYPES OF OHMS IN AC CIRCUITS

The differences in R, X_L, X_C, and Z are listed in Table 25-1, but the following general features should also be noted. Ohms of opposition limit the amount of current in dc circuits or ac circuits. Resistance R is the same for either case. However, ac circuits can have ohms of reactance because of the variations in alternating current or voltage. X_L is the reactance of an inductance with sine-wave changes in current. X_C is the reactance of a capacitor with sine-wave changes in voltage.

Both X_L and X_C are measured in ohms, like R, but reactance has a 90° phase angle, while the phase angle for resistance is 0°. A circuit with steady direct current cannot have any reactance.

Ohms of X_L or X_C are opposite, as X_L has a phase angle of $+90°$, while X_C has the angle of $-90°$. Any individual X_L or X_C always has a phase angle of exactly 90°.

Ohms of impedance Z result from the phasor combination of resistance and reactance. In fact, Z can be considered the general form of any ohms of opposition in ac circuits.

Chapter 25
Alternating-Current Circuits

TABLE 25-1. Types of Ohms in AC Circuits

	RESISTANCE R, Ω	INDUCTIVE REACTANCE X_L, Ω	CAPACITIVE REACTANCE X_C, Ω	IMPEDANCE Z, Ω
Definition	In-phase opposition to alternating or direct current	90° leading opposition to alternating current	90° lagging opposition to alternating current	Phasor combination of resistance and reactance $Z = \sqrt{R^2 + X^2}$
Effect of frequency	Same for all frequencies	Increases with higher frequencies	Decreases with higher frequencies	X_L component increases, but X_C decreases
Phase angle θ	0°	I_L lags V_L by 90°	V_C lags I_C by 90°	$\tan \theta = \pm \dfrac{X}{R}$ in series, or $\pm \dfrac{I_X}{I_R}$ in parallel

Z can have any phase angle, depending on the relative amounts of R and X. When Z consists mostly of R with little reactance, the phase angle of Z is close to 0°. With R and X equal, the phase angle of Z is 45°. Whether the angle is positive or negative depends on whether the net reactance is inductive or capacitive. When Z consists mainly of X with little R, the phase angle of Z is close to 90°.

The phase angle is θ_Z for Z or V_T with respect to the common I in a series circuit. With parallel branch currents, θ_I is for I_T in the main line with respect to the common voltage.

Practice Problems 25-11
(answers on page 518)
(a) Which of the following does not change with frequency: Z, X_L, X_C, or R?
(b) Which of the following has lagging current: R, X_L, or X_C?
(c) Which of the following has leading current: R, X_L, or X_C?

25-12
SUMMARY OF TYPES OF PHASORS IN AC CIRCUITS

The phasors for ohms, volts, and amperes are shown in Fig. 25-15. Note the following similarities and differences:

Series Components. In series circuits, ohms and voltage drops have similar phasors. The reason is the common I for all the series components. V_R or IR has the same phase as R; V_L or IX_C has the same phase as X_C; V_C or IX_C has the same phase as X_C.

Resistance. R, V_R, and I_R always have the same angle because there is no phase shift in a resistance. This applies to R in either a series or a parallel circuit.

Reactance. X_L and X_C are 90° phasors in opposite directions. X_L or V_L has the angle of +90° with an upward phasor. X_C or V_C has the angle of −90° with a downward phasor.

Reactive Branch Currents. The phasor of a parallel branch current is opposite from its reactance. Therefore, I_C is upward at +90°, opposite from X_C downward at −90°. Also, I_L is downward at −90°, opposite from X_L upward at +90°.

In short, I_C and I_L are opposite from each other, and both are opposite from their corresponding reactances.

Phase Angle θ_Z. The phasor resultant for ohms of reactance and resistance is the imped-

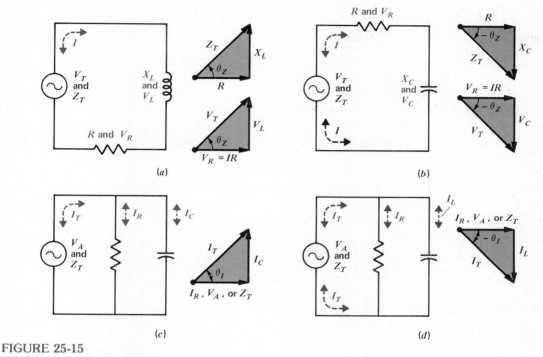

FIGURE 25-15

Summary of phasor relations in ac circuits. (a) X_L and R in series. (b) X_C and R in series. (c) Parallel branches with I_C and I_R. (d) Parallel branches with I_L and I_R.

ance Z. The phase angle θ for Z can be any angle between 0 and 90°. In a series circuit θ_Z for Z is the same as θ for V_T with respect to the common current I.

Phase Angle θ_I. The phasor resultant of branch currents is the total line current I_T. The phase angle of I_T can be any angle between 0 and 90°. In a parallel circuit, θ_I is the angle of I_T with respect to the applied voltage V_A.

The θ_I is the same value but of opposite sign from θ_Z for Z, which is the impedance of the combined parallel branches. The reason for the change of sign is that θ_I is for I with respect to V, but θ_Z is for V with respect to I.

Such phasor combinations are necessary in sine-wave ac circuits in order to take into account the effect of reactance. The phasors can be analyzed either graphically, as in Fig. 25-15, or by the shorter technique of complex numbers, with a j operator that corresponds to a 90° phasor. How to use complex numbers in the analysis of ac circuits is explained in the next chapter.

Practice Problems 25-12
(answers on page 518)

(a) Of the following three phasors, which two are 180° opposite: V_L, V_C, or V_R?
(b) Of the following three phasors, which two are in quadrature: I_R, I_T, or I_L?

Summary

1. In ac circuits with resistance alone, the circuit is analyzed the same way as for dc circuits, generally with rms ac values. Without any reactance, the phase angle is zero.
2. When capacitive reactances alone are combined, the X_C values are added in series and combined by the reciprocal formula in parallel, just like ohms of resistance. Similarly, ohms of X_L alone can be added in series or combined by the reciprocal formula in parallel, just like ohms of resistance.
3. Since X_C and X_L are opposite reactances, they cancel each other. In series, the ohms of X_C and X_L cancel. In parallel, the capacitive and inductive branch currents I_C and X_L cancel.
4. In ac circuits with R, X_L, and X_C, they can be reduced to one equivalent resistance and one net reactance.
5. In series, the total R and net X at 90° are combined as $Z = \sqrt{R^2 + X^2}$. The phase angle of the series R and X is the angle with tangent $\pm X/R$. First we calculate Z_T and then divide into V_T to find I.
6. For parallel branches, the total I_R and net reactive I_X at 90° are combined as $I_T = \sqrt{I_R^2 + I_X^2}$. The phase angle of the parallel R and X is the angle with tangent $\pm I_X/I_R$. First we calculate I_T and then divide into V_A to find Z_T.
7. R, X_L, X_C, and Z in ac circuits all are ohms of opposition. The differences with respect to frequency and phase angle are summarized in Table 25-1.
8. The phasor relations for resistance and reactance are summarized in Fig. 25-15.
9. In ac circuits with reactance, the real power in watts equals $I^2 R$, or $VI \cos \theta$, where θ is the phase angle. The real power is the power dissipated as heat in resistance. $\cos \theta$ is the power factor of the circuit.
10. The wattmeter measures real ac power or dc power.

Self-Examination (Answers at back of book.)

Choose (a), (b), (c), or (d).

1. In an ac circuit with resistance but no reactance (a) two 1000-Ω resistances in series total 1414 Ω; (b) two 1000-Ω resistances in series

total 2000 Ω; (c) two 1000-Ω resistances in parallel total 707 Ω; (d) a 1000-Ω R in series with a 400-Ω R totals 600 Ω.

2. An ac circuit has a 100-Ω X_{C_1}, a 50-Ω X_{C_2}, a 40-Ω X_{L_1}, and a 30-Ω X_{L_2}, all in series. The net reactance is equal to (a) an 80-Ω X_L; (b) a 200-Ω X_L; (c) an 80-Ω X_C; (d) a 220-Ω X_C.
3. An ac circuit has a 40-Ω R, a 90-Ω X_L, and a 60-Ω X_C, all in series. The impedance Z equals (a) 50 Ω; (b) 70.7 Ω; (c) 110 Ω; (d) 190 Ω.
4. An ac circuit has a 100-Ω R, a 100-Ω X_L, and a 100-Ω X_C, all in series. The impedance Z of the series combination is equal to (a) $33\frac{1}{3}$ Ω; (b) 70.7 Ω; (c) 100 Ω; (d) 300 Ω.
5. An ac circuit has a 100-Ω R, a 300-Ω X_L, and a 200-Ω X_C, all in series. The phase angle θ of the circuit equals (a) 0°; (b) 37°; (c) 45°; (d) 90°.
6. The power factor of an ac circuit equals (a) the cosine of the phase angle; (b) the tangent of the phase angle; (c) zero for a resistive circuit; (d) unity for a reactive circuit.
7. Which phasors in the following combinations are *not* in opposite directions? (a) X_L and X_C; (b) X_L and I_C; (c) I_L and I_C; (d) X_C and I_C.
8. In Fig. 25-8a, the voltage drop across X_L equals (a) 60 V; (b) $66\frac{2}{3}$ V; (c) 120 V; (d) 200 V.
9. In Fig. 25-10a, the combined impedance of the parallel circuit equals (a) 5 Ω; (b) 12.5 Ω; (c) 20 Ω; (d) 100 Ω.
10. The wattmeter (a) has voltage and current coils to measure real power; (b) has three connections, two of which are used at a time; (c) measures apparent power because the current is the same in the voltage and current coils; (d) can measure dc power but not 60-Hz ac power.

Essay Questions

1. Why can series or parallel resistances be combined in ac circuits the same way as in dc circuits?
2. (a) Why do X_L and X_C reactances in series cancel each other? (b) With X_L and X_C reactances in parallel, why do their branch currents cancel?
3. Give one difference in electrical characteristics comparing R and X_C, R and Z, X_C and C, X_L and L.
4. Name three types of ac meters.
5. Make a diagram showing a resistance R_1 in series with the load resistance R_L, with a wattmeter connected to measure the power in R_L.

Chapter 25
Alternating-Current Circuits

6. What is the difference between voltamperes and watts?
7. Make a phasor diagram for the circuit in Fig. 25-8a showing the phase of the voltage drops IR, IX_C, and IX_L with respect to the reference phase of the current I.
8. Explain briefly why the two opposite phasors at $+90°$ for X_L and $-90°$ for I_L both follow the principle that any self-induced voltage leads the current through the coil by $90°$.
9. Explain briefly why phasors for inductance are opposite from phasors for capacitance.
10. Why is it that a reactance phasor is always at exactly $90°$ but an impedance phasor can be less than $90°$?
11. Why must the impedance of a series circuit be more than either its reactance or resistance?
12. When resistance is increased in a series impedance, Z increases but θ decreases. Explain why.
13. Why must I_T in a parallel circuit be more than either I_R or I_X?
14. Why does the power factor of 0.707 correspond to a phase angle of $45°$?

Problems (Answers to odd-numbered problems at back of book.)

1. Refer to Fig. 25-1a. (a) Calculate the total real power supplied by the source. (b) Why is the phase angle zero? (c) What is the power factor of the circuit?
2. In a series ac circuit, 2 A flows through a 20-Ω R, a 40-Ω X_L, and a 60-Ω X_C. (a) Make a schematic diagram of the series circuit. (b) Calculate the voltage drop across each series component. (c) How much is the applied voltage? (d) Calculate the power factor of the circuit. (e) What is the phase angle θ?
3. A parallel circuit has the following five branches: three resistances of 30 Ω each; an X_L of 600 Ω; an X_C of 400 Ω. (a) Make a schematic diagram of the circuit. (b) If 100 V is applied, how much is the total line current? (c) What is the total impedance of the circuit? (d) What is the phase angle θ?
4. Referring to Fig. 25-8, assume that the frequency is doubled from 500 to 1000 Hz. Find X_L, X_C, Z, I, and θ for this higher frequency. Also, calculate L and C.
5. A series circuit has a 300-Ω R, a 500-Ω X_{C_1}, a 300-Ω X_{C_2}, an 800-Ω X_{L_1}, and a 400-Ω X_{L_2}, all in series with an applied voltage V of

400 V. (a) Draw the schematic diagram with all components. (b) Draw the equivalent circuit reduced to one resistance and one reactance. (c) Calculate Z_T, I, and the phase angle.

6. Do the same as Prob. 5 for a circuit with the same components in parallel across the voltage source.

7. A series circuit has a 600-Ω R, a 10-μH inductance L, and a 4-μF capacitance C, all in series with the 60-Hz 120-V power line as applied voltage. (a) Find the reactance of L and of C. (b) Calculate Z_T, I, and θ_Z.

8. Do the same as in Prob. 7 for the same circuit, but the 120-V source has the frequency of 10 MHz.

9. (a) Referring to the series circuit Fig. 25-6, what is the phase angle between the IX_L voltage of 360 V and the IX_C voltage of 240 V? (b) Draw the two sine waves for these voltages, showing their relative amplitudes and phase corresponding to the phasor diagram in Fig. 25-6b. Also show the resultant sine wave of voltage across the net X_L.

10. Do the same as in Prob. 9 for the currents in the parallel circuit of Fig. 25-7.

11. How much resistance dissipates 600 W of ac power, with 5-A rms current?

12. How much resistance must be inserted in series with a 1.9-H inductance to limit the current to 0.1 A from the 120-V 60-Hz power line?

13. How much resistance must be inserted in series with a 10-μF capacitance to provide a phase angle of $-45°$? The source is the 120-V 60-Hz power line.

14. With the same R as in Prob. 13, what value of C is necessary for the angle of $-45°$ at the frequency of 2 MHz?

15. How much capacitance, in microfarad units, must be inserted in series with a 200-Ω R for a current of 0.3 A, from the 120-V 60-Hz power line?

16. A parallel ac circuit has the following branch currents: $I_{R_1} = 4.2$ mA; $I_{R_2} = 2.4$ mA; $I_{L_1} = 7$ mA; $I_{L_2} = 1$ mA; $I_C = 6$ mA. Calculate I_T.

17. With 420 mV applied, an ac circuit has the following parallel branches: $R_1 = 100$ Ω; $R_2 = 175$ Ω; $X_{L_1} = 60$ Ω; $X_{L_2} = 420$ Ω; $X_C = 70$ Ω. Calculate I_T, θ_I, and Z_T.

18. The same components as in Prob. 17 are in series. Calculate Z_T, I, and θ_Z.

19. What R is needed in series with a 0.01-μF capacitor for a phase angle of $-64°$, with f of 800 Hz?

20. With the same R as in Prob. 19, what L will provide the phase angle of $64°$ at 800 Hz?

Answers to Practice Problems

25-1	(a) 0°	25-7	(b) 200 Ω
	(b) 0°		(c) 100 Ω
25-2	(a) 90°	25-8	(a) Watt
	(b) −90°		(b) Voltampere
25-3	(a) −90°	25-9	(a) T
	(b) 90°		(b) T
25-4	(a) 20 Ω	25-10	(a) Real power
	(b) 1 A		(b) V
25-5	(a) $X_C = 30$ Ω	25-11	(a) R
	(b) $X_C = 30$ Ω		(b) X_L
25-6	(a) $I_L = 3$ A		(c) X_C
	(b) $I_L = 3$ A	25-12	(a) V_L and V_C
25-7	(a) 200 Ω		(b) I_R and I_C

Complex Numbers for AC Circuits

Chapter 26

Complex numbers form a numerical system that includes the phase angle of a quantity, with its magnitude. Therefore, complex numbers are useful in ac circuits when the reactance of X_L or X_C makes it necessary to consider the phase angle.

Any type of ac circuit can be analyzed with complex numbers, but they are especially convenient for solving series-parallel circuits that have both resistance and reactance in one or more branches. Actually, the use of complex numbers is probably the best way to analyze ac circuits with series-parallel impedances. The topics here are:

26-1   Positive and Negative Numbers
26-2   The j Operator
26-3   Definition of a Complex Number
26-4   How Complex Numbers Are Applied to AC Circuits
26-5   Impedance in Complex Form
26-6   Operations with Complex Numbers
26-7   Magnitude and Angle of a Complex Number
26-8   Polar Form of Complex Numbers
26-9   Converting Polar to Rectangular Form
26-10   Complex Numbers in Series AC Circuits
26-11   Complex Numbers in Parallel AC Circuits
26-12   Combining Two Complex Branch Impedances
26-13   Combining Complex Branch Currents
26-14   Parallel Circuit with Three Complex Branches

26-1 POSITIVE AND NEGATIVE NUMBERS

Our common use of numbers as either positive or negative represents only two special cases. In their more general form, numbers have both quantity and phase angle. In Fig. 26-1, positive and negative numbers are shown as corresponding to the phase angles of 0 and 180°, respectively.

Chapter 26
Complex Numbers for AC Circuits

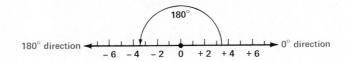

FIGURE 26-1

Positive and negative numbers.

For example, the numbers 2, 4, and 6 represent units along the horizontal or x axis, extending toward the right along the line of zero phase angle. Therefore, positive numbers really represent units having the phase angle of 0°. Or this phase angle corresponds to the factor of $+1$. To indicate 6 units with zero phase angle, then, 6 is multiplied by $+1$ as a factor for the positive number 6. The $+$ sign is often omitted, as it is assumed unless indicated otherwise.

In the opposite direction, negative numbers correspond to 180°. Or, this phase angle corresponds to the factor of -1. Actually, -6 represents the same quantity as 6 but rotated through the phase angle of 180°. The angle of rotation is the *operator* for the number. The operator for -1 is 180°; the operator for $+1$ is 0°.

Practice Problems 26-1
(*answers on page 539*)
(a) What is the angle for the number $+5$?
(b) What is the angle for the number -5?

26-2
THE j OPERATOR

The operator for a number can be any angle between 0 and 360°. Since the angle of 90° is important in ac circuits, the factor j is used to indicate 90°. See Fig. 26-2. Here, the number 5 means 5 units at 0°, the number -5 is at 180°, while $j5$ indicates the 90° angle.

The j is usually written before the number. The reason is that the j sign is a 90° operator, just as the $+$ sign is a 0° operator and the $-$ sign is a 180° operator. Any quantity at right angles to the zero axis, therefore, 90° counterclockwise, is on the $+j$ axis.

In mathematics, numbers on the horizontal axis are real numbers, including positive and negative values. Numbers on the j axis are called *imaginary numbers,* only because they are not on the real axis. Also, the abbreviation i is used in place of j. In electricity, however, j is used to avoid confusion with i as the symbol for current. Furthermore, there is nothing imaginary about electrical quantities on the j axis. An electric shock from $j500$ V is just as dangerous as 500 V positive or negative.

More features of the j operator are shown in Fig. 26-3. The angle of 180° corresponds to the j operation of 90° repeated twice. This angular rotation is indicated by the factor j^2. Note that the j operation multiplies itself, instead of adding.

Since j^2 means 180°, which corresponds to the factor of -1, we can say that j^2 is the same as -1. In short, the operator j^2 for a number means multiply by -1. For instance, $j^2 8$ is -8.

Furthermore, the angle of 270° is the same as $-90°$, which corresponds to the operator

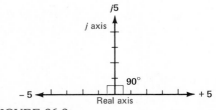

FIGURE 26-2

The j axis at 90° from real axis.

Chapter 26
Complex Numbers for AC Circuits

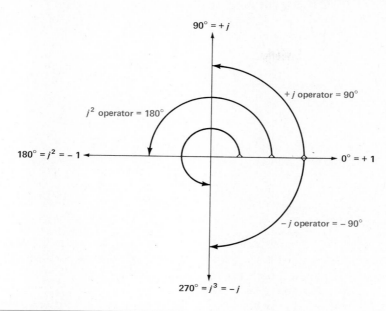

FIGURE 26-3
The *j* operator shows 90° rotation from real axis; −*j* operator is −90°; j^2 operator is 180° rotation back to real axis.

−*j*. These characteristics of the *j* operator are summarized as follows:

$0° = 1$
$90° = j$
$180° = j^2 = -1$
$270° = j^3 = j^2 \times j = -1 \times j = -j$
$360° = $ same as 0°

As examples, the number 4 or −4 represents 4 units on the real horizontal axis; *j*4 means 4 units with a leading phase angle of 90°; −*j*4 means 4 units with a lagging phase angle of −90°.

Practice Problems 26-2
(*answers on page 539*)
(a) What is the angle for the operator *j*?
(b) What is the angle for the operator −*j*?

26-3
DEFINITION OF A COMPLEX NUMBER

The combination of a real and imaginary term is a complex number. Usually, the real number is written first. As an example, 3 + *j*4 is a complex number including 3 units on the real axis added to 4 units 90° out of phase on the *j* axis. The name *complex number* just means that its terms must be added as phasors.

The phasors for several complex numbers are shown in Fig. 26-4. Note that the +*j* phasor is up for 90°; the −*j* phasor is down for −90°. The phasors are shown with the end of one joined to the start of the next, to be ready for addition. Graphically, the sum is the hypotenuse of the right triangle formed by the two phasors. Since a number like 3 + *j*4 specifies the phasors in rectangular coordinates, this system is the *rectangular form* of complex numbers.

Chapter 26
Complex Numbers for AC Circuits

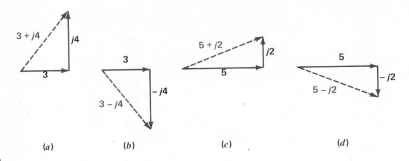

FIGURE 26-4
Phasors corresponding to real terms and j terms, in rectangular coordinates.

Be careful to distinguish a number like $j2$, where 2 is a coefficient, from j^2, where 2 is the exponent. The number $j2$ means 2 units up on the j axis of 90°. However, j^2 is the operator of -1, which is on the real axis in the negative direction.

Another comparison to note is between $j3$ and j^3. The number $j3$ is 3 units up on the j axis, while j^3 is the same as the $-j$ operator, which is down on the $-90°$ axis.

Also note that either the real term or j term can be the larger of the two. When the j term is larger, the angle is more than 45°; when the j term is smaller, the angle is less than 45°. If the j term and real term are equal, the angle is 45°.

Practice Problems 26-3
(*answers on page 539*)
Answer true or false.
(a) For $7 + j6$, the 6 is at 90° leading the 7.
(b) For $7 - j6$, the 6 is at 90° lagging the 7.

26-4
HOW COMPLEX NUMBERS ARE APPLIED TO AC CIRCUITS

The applications are just a question of using a real term for 0°, $+j$ for 90°, and $-j$ for $-90°$, to denote the phase angles. Specifically, Fig. 26-5 illustrates the following rules:

0° or a real number without any j operator is used for resistance R. For instance, 3 Ω of R is stated just as 3 Ω.

90° or $+j$ is used for inductive reactance X_L. For instance, a 4-Ω X_L is $j4$ Ω. This rule always applies to X_L, whether it is in series or parallel with R. The reason is the fact that X_L represents voltage across an inductance, which always leads the current through the inductance by 90°. The $+j$ is also used for V_L.

$-90°$ or $-j$ is used for capacitive reactance X_C. For instance, a 4-Ω X_C is $-j4$ Ω. This rule always applies to X_C, whether it is in series or

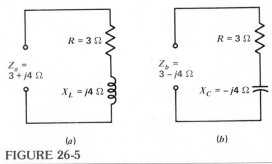

FIGURE 26-5
Rectangular form of complex numbers for impedances. (*a*) X_L is $+j$. (*b*) X_C is $-j$.

parallel with R. The reason is the fact that X_C represents voltage across a capacitor, which always lags the charge and discharge current of the capacitor by $-90°$. The $-j$ is also used for V_C.

With reactive branch currents, the sign for j is reversed, compared with reactive ohms, because of the opposite phase angle. As shown in Fig. 26-6a and b, $-j$ is used for inductive branch current I_L and $+j$ for capacitive branch current I_C.

Practice Problems 26-4
(answers on page 539)
(a) Write 3 kΩ of X_L with the j operator.
(b) Write 5 mA of I_L with the j operator.

26-5
IMPEDANCE IN COMPLEX FORM

The rectangular form of complex numbers is a convenient way to state the impedance of series resistance and reactance. In Fig. 26-5a, the impedance is $3 + j4$, as Z_a is the phasor sum of a 3-Ω R in series with $j4$ Ω for X_L. Similarly, Z_b is $3 - j4$ for a 3-Ω R in series with $-j4$ Ω for X_C. The minus sign results from adding the negative term for $-j$. More examples are:

For a 4-kΩ R and a 2-kΩ X_L in series, $Z_T = 4000 + j2000$

For a 3-kΩ R and a 9-kΩ X_C in series, $Z_T = 3000 - j9000$

For a zero R and a 7-Ω X_L in series, $Z_T = 0 + j7$

For a 12-Ω R and a zero reactance in series, $Z_T = 12 + j0$

Note the general form of stating $Z = R \pm jX$. If one term is zero, substitute 0 for this term, in order to keep Z in its general form. This procedure is not required, but there is usually less confusion when the same form is used for all types of Z.

The advantage of this method is that multiple impedances written as complex numbers can then be calculated as follows:

$$Z_T = Z_1 + Z_2 + Z_3 + \cdots + \text{etc.}$$
for series impedances

$$\frac{1}{Z_T} = \frac{1}{Z_1} + \frac{1}{Z_2} + \frac{1}{Z_3} + \cdots + \text{etc.}$$
for parallel impedances

or

$$Z_T = \frac{Z_1 \times Z_2}{Z_1 + Z_2} \quad \text{for two impedances in parallel}$$

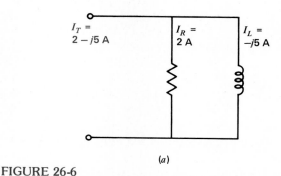 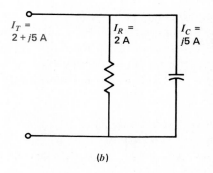

FIGURE 26-6

Rectangular form of complex numbers for branch currents. (a) I_L is $-j$. (b) I_C is $+j$.

Chapter 26
Complex Numbers for AC Circuits

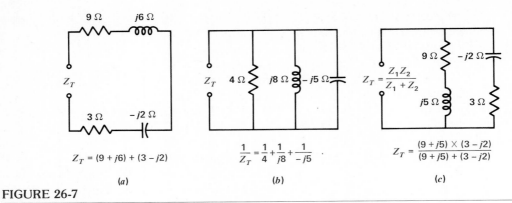

FIGURE 26-7
X_L is a $+j$ term and X_C is a $-j$ term, whether in series or parallel. (a) Series circuit. (b) Parallel branches. (c) Complex branch impedances Z_1 and Z_2 in parallel.

Examples are shown in Fig. 26-7. The circuit in (a) is just a series combination of resistances and reactances. Combining the real terms and j terms separately, $Z_T = 12 + j4$. The parallel circuit in (b) shows that X_L is $+j$ and X_C is $-j$ even though they are in parallel branches, as they are reactances, not currents.

So far, these types of circuits can be analyzed with or without complex numbers. For the series-parallel circuit in Fig. 26-7c, however, the notation of complex numbers is necessary to state the complex impedance Z_T, consisting of branches with reactance and resistance in one or more of the branches. Z_T is just stated here in its form as a complex impedance. In order to calculate Z_T, some of the rules described in the next section must be used for combining complex numbers.

Practice Problems 26-5
(answers on page 539)
Write the following impedances in complex form.
(a) X_L of 7 Ω in series with R of 4 Ω
(b) X_C of 7 Ω in series with zero R.

26-6
OPERATIONS WITH COMPLEX NUMBERS

Real numbers and j terms cannot be combined directly because they are 90° out of phase. The following rules apply:

For Addition or Subtraction. Add or subtract the real and j terms separately:

$$(9 + j5) + (3 + j2) = 9 + 3 + j5 + j2$$
$$= 12 + j7$$
$$(9 + j5) + (3 - j2) = 9 + 3 + j5 - j2$$
$$= 12 + j3$$
$$(9 + j5) + (3 - j8) = 9 + 3 + j5 - j8$$
$$= 12 - j3$$

The answer should be in the form of $R \pm jX$, where R is the algebraic sum of all the real or resistive terms and X is the algebraic sum of all the imaginary or reactive terms.

To Multiply or Divide a j Term by a Real Number. Just multiply or divide the numbers. The answer is still a j term. Note the algebraic signs in the following examples. If both factors have the same sign, either + or −, the answer is +; if one factor is negative, the answer is negative.

$$4 \times j3 = j12$$
$$j5 \times 6 = j30$$
$$j5 \times (-6) = -j30$$
$$-j5 \times 6 = -j30$$
$$-j5 \times (-6) = j30$$

$$j12 \div 4 = j3$$
$$j30 \div 6 = j5$$
$$-j30 \div (-6) = j5$$
$$-j30 \div 6 = -j5$$
$$j30 \div (-6) = -j5$$

To Multiply or Divide a Real Number by a Real Number. Just multiply or divide the real numbers, as in arithmetic, without any regard to j operation. The answer is still a real number.

To Multiply a j Term by a j Term. Multiply the numbers and the j coefficients to produce a j^2 term. The answer is a real term because j^2 is -1, which is on the real axis. Multiplying two j terms shifts the number 90° from the j axis to the real axis of 180°. As examples:

$$j4 \times j3 = j^2 12 = (-1)(12)$$
$$= -12$$
$$j4 \times (-j3) = -j^2 12 = -(-1)(12)$$
$$= 12$$

To Divide a j Term by a j Term. Divide the numbers and the j coefficients to produce a real number, as the j factors cancel. For instance:

$$j12 \div j4 = 3 \qquad -j12 \div j4 = -3$$
$$j30 \div j5 = 6 \qquad j30 \div (-j6) = -5$$
$$j15 \div j3 = 5 \qquad -j15 \div (-j3) = 5$$

To Multiply Complex Numbers. Follow the rules of algebra for multiplying two factors, each having two terms:

$$(9 + j5) \times (3 - j2) = 27 + j15 - j18 - j^2 10$$
$$= 27 - j3 + 10$$
$$= 37 - j3$$

Note that $-j^2 10$ equals $+10$ because the operator j^2 is -1 and $-j^2$ becomes $+1$.

To Divide Complex Numbers. This process becomes more involved because division of a real number by an imaginary number is not possible. Therefore, the denominator must first be converted to a real number without any j term.

Converting the denominator to a real number without any j term is called *rationalization* of the fraction. To do this, multiply both numerator and denominator by the *conjugate* of the denominator. Conjugate complex numbers have equal terms but opposite signs for the j term. For instance, $(1 + j2)$ has the conjugate $(1 - j2)$.

Rationalization is permissible because the value of a fraction is not changed when both numerator and denominator are multiplied by the same factor. This procedure is the same as multiplying by 1. In the following example of division with rationalization the denominator $(1 + j2)$ has the conjugate $(1 - j2)$:

$$\frac{4 - j1}{1 + j2} = \frac{4 - j1}{1 + j2} \times \left[\frac{(1 - j2)}{(1 - j2)}\right]$$
$$= \frac{4 - j8 - j1 + j^2 2}{1 - j^2 4}$$
$$= \frac{4 - j9 - 2}{1 + 4}$$
$$= \frac{2 - j9}{5}$$
$$= 0.4 - j1.8$$

Note that the product of a complex number and its conjugate always equals the sum of the squares of the numbers in each term. As another example, the product of $(2 + j3)$ and its conjugate $(2 - j3)$ must be $4 + 9$, which equals 13. Simple numerical examples of division and multiplication are given here because when the required calculations become too long, it is easier to divide and multiply complex numbers in polar form, as explained in Sec. 26-8.

*Practice Problems 26-6
(answers on page 539)*
(a) $(2 + j3) + (3 + j4) = ?$
(b) $(2 + j3) \times 2 = ?$

26-7
MAGNITUDE AND ANGLE OF A COMPLEX NUMBER

In electrical terms a complex impedance $(4 + j3)$ means 4 Ω of resistance and 3 Ω of inductive reactance with a leading phase angle of 90°. See Fig. 26-8a. The magnitude of the combined Z is the resultant, equal to $\sqrt{16 + 9} = \sqrt{25} = 5$ Ω. Finding the square root of the sum of the squares is vector or phasor addition of two terms in quadrature, 90° out of phase.

The phase angle[1] of the resultant is the angle whose tangent is $\frac{3}{4}$ or 0.75. This angle equals 37°. Therefore, $4 + j3 = 5\underline{/37°}$.

When calculating the tangent ratio note the *j* term is the numerator and the real term is the denominator because the tangent of the phase angle is the ratio of the opposite side to the adjacent side. With a negative *j* term, the tangent is negative, which means a negative phase angle.

Note the following definitions: $(4 + j3)$ is the complex number in rectangular coordinates. The real term is 4. The imaginary term is

[1] Appendix E, Trigonometric Functions, explains the sine, cosine, and tangent of an angle. See also B. Grob, "Problems Workbook in Technical Mathematics for Basic Electronics," McGraw-Hill Book Company, New York.

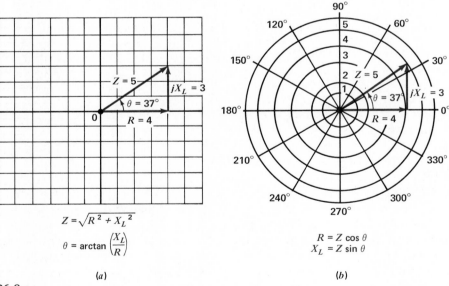

FIGURE 26-8

Magnitude and angle of a complex number. (*a*) Rectangular form. (*b*) Polar form.

$j3$. The resultant 5 is the magnitude, absolute value, or modulus of the complex number. Its phase angle or argument is 37°. The resultant value by itself can be written as $|5|$, with vertical lines to indicate it is the magnitude without the phase angle. The magnitude is the value a meter would read. For instance, with a current of $5\underline{/37°}$ A in a circuit, an ammeter reads 5 A. As additional examples:

$2 + j4 = \sqrt{4 + 16}$ (arctan 2) $= 4.47\underline{/63°}$
$4 + j2 = \sqrt{16 + 4}$ (arctan 0.5) $= 4.47\underline{/26.5°}$
$8 + j6 = \sqrt{64 + 36}$ (arctan 0.75) $= 10\underline{/37°}$
$8 - j6 = \sqrt{64 + 36}$ (arctan -0.75)
$\qquad = 10\underline{/-37°}$
$4 + j4 = \sqrt{16 + 16}$ (arctan 1) $= 5.66\underline{/45°}$
$4 - j4 = \sqrt{16 + 16}$ (arctan -1)
$\qquad = 5.66\underline{/-45°}$

Note that arctan 2, for example, means the angle with a tangent equal to 2. This can also be indicated as $\tan^{-1} 2$. In either case, the angle is specified as having 2 for its tangent, and the angle is 63.4°.

Practice Problems 26-7
(answers on page 539)
(a) For the complex impedance $10 + j10 \, \Omega$, calculate the magnitude.
(b) Calculate the phase angle.

26-8
POLAR FORM OF COMPLEX NUMBERS

Calculating the magnitude and phase angle of a complex number is actually converting to an angular form in polar coordinates. As shown in Fig. 26-8, the rectangular form $4 + j3$ is equal to $5\underline{/37°}$ in polar form. In polar coordinates, the distance out from the center is the magnitude of the vector Z. Its phase angle θ is counterclockwise from the 0° axis.

To convert any complex number to polar form:

1. Find the magnitude by phasor addition of the j term and real term.
2. Find the angle whose tangent is the j term divided by the real term. As examples:

$2 + j4 = 4.47\underline{/63°}$
$4 + j2 = 4.47\underline{/26.5°}$
$8 + j6 = 10\underline{/37°}$
$8 - j6 = 10\underline{/-37°}$
$4 + j4 = 5.66\underline{/45°}$
$4 - j4 = 5.66\underline{/-45°}$

These examples are the same as those given for finding the magnitude and phase angle of a complex number.

The magnitude in polar form must be more than either term in rectangular form, but less than the arithmetic sum of the two terms. For instance, in $8 + j6 = 10\underline{/37°}$ the magnitude of 10 is more than 8 or 6 but less than their sum of 14.

Applied to ac circuits with resistance for the real term and reactance for the j term, then, the polar form of a complex number states the resultant impedance and its phase angle. Note the following cases for an impedance where either the resistance or reactance is reduced to zero.

$0 + j5 = 5\underline{/90°}$ $\qquad$ $5 + j0 = 5\underline{/0°}$
$0 - j5 = 5\underline{/-90°}$

The polar form is much more convenient for multiplying or dividing complex numbers. The reason is that multiplication in polar form is reduced to addition of the angles, and the angles are just subtracted for division in polar form. The following rules apply:

For Multiplication. Multiply the magnitudes but add the angles algebraically:

$$24\underline{/40°} \times 2\underline{/30°} = 48\underline{/+70°}$$
$$24\underline{/40°} \times [-2\underline{/30°}] = -48\underline{/+70°}$$
$$12\underline{/-20°} \times 3\underline{/-50°} = 36\underline{/-70°}$$
$$12\underline{/-20°} \times 4\underline{/5°} = 48\underline{/-15°}$$

When you multiply by a real number, just multiply the magnitudes:

$$4 \times 2\underline{/30°} = 8\underline{/30°}$$
$$4 \times 2\underline{/-30°} = 8\underline{/-30°}$$
$$-4 \times 2\underline{/30°} = -8\underline{/30°}$$
$$-4 \times [-2\underline{/30°}] = 8\underline{/30°}$$

This rule follows from the fact that a real number has an angle of 0°. When you add 0° to any angle, the sum equals the same angle.

For Division. Divide the magnitudes but subtract the angles algebraically:

$$24\underline{/40°} \div 2\underline{/30°} = 12\underline{/40° - 30°}$$
$$= 12\underline{/10°}$$

$$12\underline{/20°} \div 3\underline{/50°} = 4\underline{/20° - 50°}$$
$$= 4\underline{/-30°}$$

$$12\underline{/-20°} \div 4\underline{/50°} = 3\underline{/-20° - 50°}$$
$$= 3\underline{/-70°}$$

To divide by a real number, just divide the magnitudes:

$$12\underline{/30°} \div 2 = 6\underline{/30°}$$
$$12\underline{/-30°} \div 2 = 6\underline{/-30°}$$

This rule is also a special case that follows from the fact that a real number has a phase angle of 0°. When you subtract 0° from any angle, the remainder equals the same angle.

For the opposite case, however, when you divide a real number by a complex number, the angle of the denominator changes its sign in the answer in the numerator. This rule still follows the procedure of subtracting angles for division, since a real number has a phase angle of 0°. As examples:

$$\frac{10}{5\underline{/30°}} = \frac{10\underline{/0°}}{5\underline{/30°}} = 2\underline{/0° - 30°} = 2\underline{/-30°}$$

$$\frac{10}{5\underline{/-30°}} = \frac{10\underline{/0°}}{5\underline{/-30°}}$$
$$= 2\underline{/0° - [-30°]} = 2\underline{/+30°}$$

Stated another way, we can say that the reciprocal of an angle is the same angle but with opposite sign. Note that this operation is similar to working with powers of 10. Angles and powers of 10 follow the general rules of exponents.

Practice Problems 26-8
(answers on page 539)
(a) $6\underline{/20°} \times 2\underline{/30°} = ?$
(b) $6\underline{/20°} \div 2\underline{/30°} = ?$

26-9
CONVERTING POLAR TO RECTANGULAR FORM

Complex numbers in polar form are convenient for multiplication and division, but they cannot be added or subtracted. The reason is that changing the angle corresponds to the operation of multiplying or dividing. When complex numbers in polar form are to be added or subtracted, therefore, they must be converted back into rectangular form.

Consider the impedance $Z\underline{/\theta}$ in polar form. Its value is the hypotenuse of a right triangle with sides formed by the real term and j term in rectangular coordinates. See Fig. 26-9. Therefore, the polar form can be converted to rectan-

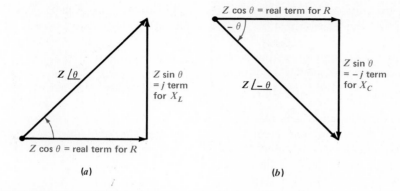

FIGURE 26-9
Converting polar form of Z/θ to rectangular $R + jX$. (a) Positive angle θ in first quadrant has $+j$ term. (b) Negative angle $-\theta$ in fourth quadrant has $-j$ term.

gular form by finding the horizontal and vertical sides of the right triangle. Specifically:

Real term for $R = Z \cos \theta$
j term for $X = Z \sin \theta$

In Fig. 26-9a, assume that Z/θ in polar form is $5/37°$. The sine of $37°$ is 0.6 and its cosine is 0.8.

To convert to rectangular form:

$R = Z \cos \theta = 5 \times 0.8 = 4$
$X = Z \sin \theta = 5 \times 0.6 = 3$

Therefore,

$5/37° = 4 + j3$

This example is the same as the illustration in Fig. 26-8. The $+$ sign for the j term means it is X_L, not X_C.

In Fig. 26-9b, the values are the same, but the j term is negative when θ is negative. The negative angle has a negative j term because the opposite side is in the fourth quadrant, where the sine is negative. However, the real term is still positive because the cosine is positive.

Note that R for $\cos \theta$ is the horizontal phasor, which is an adjacent side of the angle. The X for $\sin \theta$ is the vertical phasor, which is opposite the angle. The $+X$ is X_L; the $-X$ is X_C. You can ignore the sign of θ in calculating $\sin \theta$ and $\cos \theta$ because the values are the same up to $+90°$ or down to $-90°$.

These rules apply for angles in the first or fourth quadrant, from 0 to $90°$ or from 0 to $-90°$. As examples:

$14.14/45° = 10 + j10$
$14.14/-45° = 10 - j10$
$10/90° = 0 + j10$
$10/-90° = 0 - j10$
$100/30° = 86.6 + j50$
$100/-30° = 86.6 - j50$
$100/60° = 50 + j86.6$
$100/-60° = 50 - j86.6$

When going from one form to the other, keep in mind whether the angle is smaller or greater than $45°$ and if the j term is smaller or larger than the real term. For angles between 0 and $45°$, the opposite side, which is the j term, must be smaller than the real term. For angles between 45 and 90° the j term must be larger than the real term.

To summarize how complex numbers are used in ac circuits in rectangular and polar form:

1. For addition or subtraction, complex numbers must be in rectangular form. This procedure applies to the addition of impedances in a series circuit. If the series impedances are in rectangular form, just combine all the real terms and j terms separately. If the series impedances are in polar form, they must be converted to rectangular form to be added.
2. For multiplication and division, complex numbers are generally used in polar form because the calculations are faster. If the complex number is in rectangular form, convert to polar form. With the complex number available in both forms, then you can quickly add or subtract in rectangular form and multiply or divide in polar form. Sample problems showing how to apply these methods in the analysis of ac circuits are illustrated in the following sections.

Practice Problems 26-9
(*answers on page 539*)
Convert to rectangular form.
(a) $14.14 \underline{/45°}$.
(b) $14.14 \underline{/-45°}$.

26-10
COMPLEX NUMBERS IN SERIES AC CIRCUITS

Refer to the diagram in Fig. 26-10. Although a circuit like this with only series resistances and reactances can be solved just by phasors, the complex numbers show more details of the phase angles.

Z_T in Rectangular Form. The total Z_T in (a) is the sum of the impedances:

$$Z_T = 2 + j4 + 4 - j12 = 6 - j8$$

The total series impedance then is $6 - j8$. Actually, this amounts to adding all the series resistances for the real term and finding the algebraic sum of all the series reactances for the j term.

Z_T in Polar Form. We can convert Z_T from rectangular to polar form as follows:

$$Z_T = 6 - j8 = \sqrt{36 + 64} \underline{/\arctan -8/6}$$
$$= \sqrt{100} \underline{/\arctan -1.33}$$
$$Z_T = 10 \underline{/-53°} \ \Omega$$

The angle of $-53°$ for Z_T means this is the phase angle of the circuit. Or the applied voltage and the current are $53°$ out of phase.

Calculating I. The reason for the polar form is to divide Z_T into the applied voltage V_T to calculate the current I. See Fig. 26.10b. Note that the V_T of 20 V is a real number without any j term. Therefore, the applied voltage is $20 \underline{/0°}$. This angle of $0°$ for V_T makes it the reference phase for the following calculations. We can find the current as

$$I = \frac{V_T}{Z_T} = \frac{20 \underline{/0°}}{10 \underline{/-53°}} = 2 \underline{/0° - (-53°)}$$
$$I = 2 \underline{/53°} \ A$$

Note that Z_T has the negative angle of $-53°$ but the sign changes to $+53°$ for I because of the division into a quantity with the angle of $0°$. In general, the reciprocal of an angle in polar form is the same angle with opposite sign.

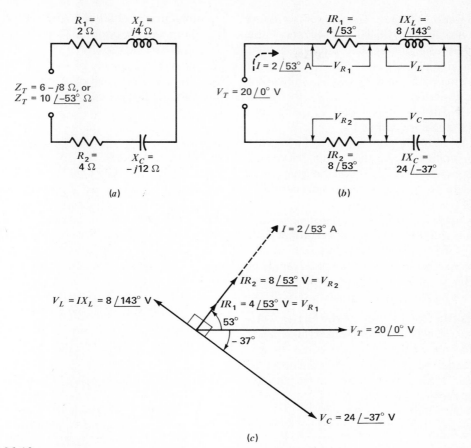

FIGURE 26-10
Complex numbers applied to series ac circuits. See text for analysis. (a) Circuit with series impedances. (b) Current and voltages. (c) Phasor diagram of current and voltages.

Phase Angle of the Circuit. The fact that I has the angle of $+53°$ means it leads V_T. The positive angle for I shows the series circuit is capacitive, with leading current. This angle is more than $45°$ because the net reactance is more than the total resistance, resulting in a tangent function greater than 1.

Finding Each IR Drop. To calculate the voltage drops around the circuit, each resistance or reactance can be multiplied by I:

$$V_{R_1} = IR_1 = 2\underline{/53°} \times 2\underline{/0°} = 4\underline{/53°} \text{ V}$$
$$V_L = IX_L = 2\underline{/53°} \times 4\underline{/90°} = 8\underline{/143°} \text{ V}$$
$$V_C = IX_C = 2\underline{/53°} \times 12\underline{/-90°}$$
$$= 24\underline{/-37°} \text{ V}$$
$$V_{R_2} = IR_2 = 2\underline{/53°} \times 4\underline{/0°} = 8\underline{/53°} \text{ V}$$

Phase of Each Voltage. The phasors for these voltages are in Fig. 26-10c. They show the phase angles using the applied voltage V_T as the zero reference phase.

The angle of 53° for V_{R_1} and V_{R_2} shows that the voltage across a resistance has the same phase as I. These voltages lead V_T by 53° because of the leading current.

For V_C, its angle of −37° means it lags the generator voltage V_T by this much. However, this voltage across X_C still lags the current by 90°, which is the difference between 53° and −37°.

The angle of 143° for V_L in the second quadrant is still 90° leading the current at 53°, as 143° − 53° = 90°. With respect to the generator voltage V_T, though, the phase angle of V_L is 143°.

V_T Equals the Phasor Sum of the Series Voltage Drops. If we want to add the voltage drops around the circuit to see if they equal the applied voltage, each V must be converted to rectangular form. Then these values can be added. In rectangular form then

$$
\begin{aligned}
V_{R_1} &= 4\underline{/53°} = & 2.408 &+ j3.196 \text{ V} \\
V_L &= 8\underline{/143°} = & -6.392 &+ j4.816 \text{ V} \\
V_C &= 24\underline{/-37°} = & 19.176 &- j14.448 \text{ V} \\
V_{R_2} &= 8\underline{/53°} = & 4.816 &+ j6.392 \text{ V} \\
& \text{Total } V = & 20.008 &- j0.044 \text{ V}
\end{aligned}
$$

or

$V_T = 20\underline{/0°}$ V   approximately

Note that for $8\underline{/143°}$ in the second quadrant, the cosine is negative for a negative real term but the sine is positive for a positive j term.

Practice Problems 26-10
(answers on page 539)
Refer to Fig. 26-10.
(a) What is the phase of I to V_T?
(b) What is the phase of V_L to V_T?
(c) What is the phase of V_L to V_R?

26-11
COMPLEX NUMBERS IN PARALLEL AC CIRCUITS

A useful application here is converting a parallel circuit to an equivalent series circuit. See Fig. 26-11, with a 10-Ω X_L in parallel with a 10-Ω R. In complex notation, R is $10 + j0$ while X_L is $0 + j10$. Their combined parallel impedance Z_T equals the product over the sum. Or

$$Z_T = \frac{(10 + j0) \times (0 + j10)}{(10 + j0) + (0 + j10)}$$

$$= \frac{10 \times j10}{10 + j10} = \frac{j100}{10 + j10}$$

Converting to polar form for division,

$$Z_T = \frac{j100}{10 + j10} = \frac{100\underline{/90°}}{14.14\underline{/45°}} = 7.07\underline{/45°}$$

Converting the Z_T of $7.07\underline{/45°}$ into rectangular form to see its resistive and reactive components,

Real term = 7.07 cos 45°
         = 7.07 × 0.707 = 5
j term = 7.07 sin 45°
        = 7.07 × 0.707 = 5

Therefore,

$Z_T = 7.07\underline{/45°}$ in polar form
$Z_T = 5 + j5$ in rectangular form

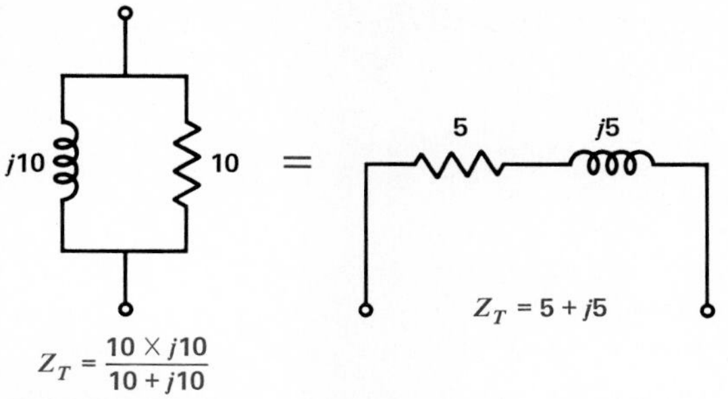

tion. An inductive branch has susceptance $-jB$, while a capacitive branch has $+jB$.

With parallel branches of conductance and susceptance the total admittance $Y_T = G \pm jB$. For the two branches in Fig. 26-11, as an example, $Y_T = 0.1 - j0.1$ S in rectangular form. In polar form,

$$Y_T = 0.14\underline{/-45°}\ \text{S}$$

This numerical value is the same as I_T with 1 V applied across Z_T of $7.07\underline{/45°}\ \Omega$.

FIGURE 26-11
Complex numbers applied to a parallel ac circuit to convert a parallel bank to an equivalent series impedance.

$Z_T = \dfrac{10 \times j10}{10 + j10}$

The rectangular form of Z_T means that 5-Ω R in series with 5-Ω X_L is the equivalent of 10-Ω R in parallel with 10-Ω X_L, as shown in Fig. 26-11.

Admittance Y and Susceptance B. In parallel circuits, it is usually easier to add branch currents than to combine reciprocal impedances. For this reason, branch conductance G is often used instead of branch resistance, where $G = 1/R$. Similarly, reciprocal terms can be defined for complex impedances. The two main types are *admittance Y*, which is the reciprocal of impedance, and *susceptance B*, which is the reciprocal of reactance. These reciprocals can be summarized as follows:

$$\text{Conductance} = G = \frac{1}{R}\ \ \text{S}$$

$$\text{Susceptance} = B = \frac{1}{\pm X}\ \ \text{S}$$

$$\text{Admittance} = Y = \frac{1}{Z}\ \ \text{S}$$

The phase angle for B or Y is the same as current. Therefore, the sign is opposite from the angle of X or Z because of the reciprocal rela-

Practice Problems 26-11
(answers on page 539)
(a) Z of $3 + j4\ \Omega$ is parallel with R of $2\ \Omega$. State Z_T in rectangular form.
(b) Do the same as in (a) for X_C instead of X_L.

26-12
COMBINING TWO COMPLEX BRANCH IMPEDANCES

A common application is a circuit with two branches Z_1 and Z_2, where each is a complex impedance with both reactance and resistance. See Fig. 26-12. A circuit like this can be solved only graphically or by complex numbers. Actu-

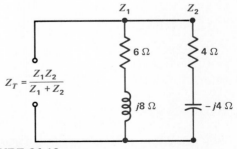

FIGURE 26-12
Finding Z_T for any two complex impedances Z_1 and Z_2 in parallel. See text for solution.

ally, using complex numbers is the shortest method.

The procedure here is to find Z_T as the product divided by the sum for Z_1 and Z_2. A good way to start is to state each branch impedance in both rectangular and polar forms. Then Z_1 and Z_2 are ready for addition, multiplication, and division. The solution of this circuit follows:

$$Z_1 = 6 + j8 = 10\underline{/53°}$$
$$Z_2 = 4 - j4 = 5.66\underline{/-45°}$$

The combined impedance

$$Z_T = \frac{Z_1 \times Z_2}{Z_1 + Z_2}$$

Use the polar form of Z_1 and Z_2 to multiply, but add in rectangular form:

$$Z_T = \frac{10\underline{/53°} \times 5.66\underline{/-45°}}{6 + j8 + 4 - j4} = \frac{56.6\underline{/8°}}{10 + j4}$$

Converting the denominator to polar form for easier division,

$$10 + j4 = 10.8\underline{/22°}$$

Then

$$Z_T = \frac{56.6\underline{/8°}}{10.8\underline{/22°}}$$

Therefore,

$$Z_T = 5.24 \, \Omega\underline{/-14°}$$

We can convert Z_T into rectangular form. The R component is $5.24 \times \cos(-14°)$ or $5.24 \times 0.97 = 5.08$. Note that $\cos\theta$ is positive in the first and fourth quadrants. The j component equals $5.24 \times \sin(-14°)$ or $5.24 \times (-0.242) = -1.27$. In rectangular form, then,

$$Z_T = 5.08 - j1.27$$

Therefore, this series-parallel circuit combination is equivalent to $5.08 \, \Omega$ of R in series with $1.27 \, \Omega$ of X_C. This problem can also be done in rectangular form by rationalizing the fraction for Z_T.

Practice Problems 26-12
(*answers on page 539*)
Refer to Fig. 26-12.
(a) Add $(6 + j8) + (4 - j4)$ for the sum of Z_1 and Z_2.
(b) Multiply $10\underline{/53°} \times 5.66\underline{/-45°}$ for the product of $Z_1 \times Z_2$.

26-13 COMBINING COMPLEX BRANCH CURRENTS

An example with two branches is shown in Fig. 26-13, to find I_T. The branch currents can just be added in rectangular form for the total I_T of parallel branches. This method corresponds to adding series impedances in rectangular form to find Z_T. The rectangular form is necessary for adding phasors.

Adding the branch currents in Fig. 26-13

$$I_T = I_1 + I_2$$
$$= (6 + j6) + (3 - j4)$$
$$I_T = 9 + j2 \text{ A}$$

Note that I_1 has $+j$ for the $+90°$ of capacitive current, while I_2 has $-j$ for inductive current. These current phasors have the opposite signs from their reactance phasors.

In polar form the I_T of $9 + j2$ A is calculated as the phasor sum of the branch currents.

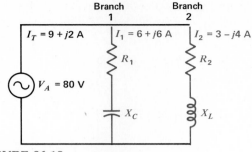
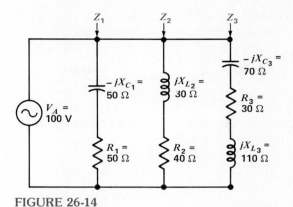

FIGURE 26-13
Finding I_T for two complex branch currents in parallel.

$$I_T = \sqrt{9^2 + 2^2} = \sqrt{85} = 9.22 \text{ A}$$
$$\tan \theta = 2/9 = 0.22$$
$$\theta = 12.53°$$

Therefore, I_T is $9 + j2$ A in rectangular form or $9.22\underline{/12.53°}$ A in polar form. The complex currents for any number of branches can be added in rectangular form.

Practice Problems 26-13
(answers on page 539)
(a) Find I_T in rectangular form for I_1 of $0 + j2$ A and I_2 of $4 + j3$ A.
(b) Find I_T in rectangular form for I_1 of $6 + j7$ A and I_2 of $3 - j9$ A.

26-14
PARALLEL CIRCUIT WITH THREE COMPLEX BRANCHES

Because the circuit in Fig. 26-14 has more than two complex impedances in parallel, the method of branch currents is used. There will be several conversions between rectangular and polar form, since addition must be in rectangular form, but division is easier in polar form. The sequence of calculations is:

FIGURE 26-14
Finding Z_T for any three complex impedances in parallel. See text for solution by means of complex branch currents.

1. Convert each branch impedance to polar form. This is necessary for dividing into the applied voltage V_A to calculate the individual branch currents. If V_A is not given, any convenient value can be assumed. Note that V_A has a phase angle of $0°$ because it is the reference.
2. Convert the individual branch currents from polar to rectangular form so that they can be added for the total line current. This step is necessary because the resistive and reactive components must be added separately.
3. Convert the total line current from rectangular to polar form for dividing into the applied voltage to calculate Z_T.
4. The total impedance can remain in polar form with its magnitude and phase angle, or can be converted to rectangular form for its resistive and reactive components.

These steps are used in the following calculations to solve the circuit in Fig. 26-14. All the values are in A, V, or Ω units.

Branch Impedances. Each Z is converted from rectangular form to polar form

$Z_1 = 50 - j50 = 70.7\underline{/-45°}$
$Z_2 = 40 + j30 = 50\underline{/+37°}$
$Z_3 = 30 + j40 = 50\underline{/+53°}$

Branch Currents. Each I is calculated as V_A divided by Z in polar form

$I_1 = \dfrac{V_A}{Z_1} = \dfrac{100}{70.7\underline{/-45°}} = 1.414\underline{/+45°}$

$\phantom{I_1 = \dfrac{V_A}{Z_1}} = 1 + j1$

$I_2 = \dfrac{V_A}{Z_2} = \dfrac{100}{50\underline{/37°}} = 2.00\underline{/-37°}$

$\phantom{I_2 = \dfrac{V_A}{Z_2}} = 1.6 - j1.2$

$I_3 = \dfrac{V_A}{Z_3} = \dfrac{100}{50\underline{/53°}} = 2.00\underline{/-53°}$

$\phantom{I_3 = \dfrac{V_A}{Z_3}} = 1.2 - j1.6$

The polar form of each I is converted to rectangular form, for addition of the branch currents.

Total Line Current. In rectangular form

$I_T = I_1 + I_2 + I_3$
$ = 1 + 1.6 + 1.2 + j1 - j1.2 - j1.6$
$I_T = 3.8 - j1.8$

Converting $3.8 - j1.8$ into polar form
$I_T = 4.2\underline{/-25.4°}$

Total Impedance. In polar form,

$Z_T = \dfrac{V_A}{I_T}$

$ = \dfrac{100}{4.2\underline{/-25.4°}}$

$Z_T = 23.8\underline{/+25.4°}$

Converting $23.8\underline{/+25.4°}$ into rectangular form,
$Z_T = 21.5 + j10.2$

Therefore, the complex ac circuit in Fig. 26-14 is equivalent to the combination of 21.5 Ω of R in series with 10.2 Ω of X_L.

This problem can also be done by combining Z_1 and Z_2 in parallel as $Z_1 Z_2 / (Z_1 + Z_2)$. Then combine this value with Z_3 in parallel to find the total Z_T of the three branches.

Practice Problems 26-14
(answers on page 539)
Refer to Fig. 26-14.
(a) State Z_2 in rectangular form for branch 2.
(b) State Z_2 in polar form.
(c) Find I_2.

Summary

1. In complex numbers, resistance R is a real term and reactance is a j term. Thus, an 8-Ω R is 8; an 8-Ω X_L is $j8$; an 8-Ω X_C is $-j8$. The general form of a complex impedance with series resistance and reactance then is $Z = R \pm jX$, in rectangular form.
2. The same notation can be used for series voltages where $V = V_R \pm jV_X$.
3. For branch currents $I_T = I_R \pm jI_X$, but the reactive branch currents have signs opposite from impedances. Capacitive branch current is jI_C, while inductive branch current is $-jI_L$.

4. The complex branch currents are added in rectangular form for any number of branches to find I_T.
5. To convert from rectangular to polar form: $R \pm jX = Z/\theta$. The magnitude of Z is $\sqrt{R^2 + X^2}$. Also, θ is the angle with $\tan = X/R$.
6. To convert from polar to rectangular form, $Z/\theta = R \pm jX$, where R is $Z \cos \theta$ and the j term is $Z \sin \theta$. A positive angle has a positive j term; a negative angle has a negative j term. Also, the angle is more than 45° for a j term larger than the real term; the angle is less than 45° for a j term smaller than the real term.
7. The rectangular form must be used for addition or subtraction of complex numbers.
8. The polar form is usually more convenient in multiplying and dividing complex numbers. For multiplication, multiply the magnitudes and add the angles; for division, divide the magnitudes and subtract the angles.
9. To find the total impedance Z_T of a series circuit, add all the resistances for the real term and find the algebraic sum of the reactances for the j term. The result is $Z_T = R \pm jX$. Then convert Z_T to polar form for dividing into the applied voltage to calculate the current.
10. To find the total impedance Z_T of two complex branch impedances Z_1 and Z_2 in parallel, Z_T can be calculated as $Z_1 Z_2 / (Z_1 + Z_2)$.

Self-Examination (Answers at back of book.)

Match the values in the column at the left with those at the right.

1. $24 + j5 + 16 + j10$
2. $24 - j5 + 16 - j10$
3. $j12 \times 4$
4. $j12 \times j4$
5. $j12 \div j3$
6. $(4 + j2) \times (4 - j2)$
7. 1200 Ω of R + 800 Ω of X_C
8. 5 A of I_R + 7 A of I_C
9. 90 V of V_R + 60 V of V_L
10. $14/28° \times /22°$
11. $14/28° \div 2/22°$
12. $15/42° \times 3/0°$
13. $6/-75° \times 4/30°$
14. $50/45° \div 5/-45°$
15. $60/-80° \div 5/5°$
16. Admittance Y

(a) $14/50°$
(b) $7/6°$
(c) $12/-85°$
(d) $40 + j15$
(e) $90 + j60$ V
(f) $45/42°$
(g) $24/-45°$
(h) 4
(i) $10/90°$
(j) -48
(k) $5 + j7$ A
(l) 20
(m) $40 - j15$
(n) $j48$
(o) $1200 - j800$ Ω
(p) $1/Z$

Chapter 26
Complex Numbers for AC Circuits

Problems (Answers to odd-numbered problems at back of book.)

1. State Z in rectangular form for the following series circuits: (a) 4-Ω R and 3-Ω X_C; (b) 4-Ω R and 3-Ω X_L; (c) 3-Ω R and 6-Ω X_L; (d) 3-Ω R and 3-Ω X_C.
2. Draw the schematic diagram for the impedances in Prob. 1.
3. Convert the following impedances to polar form: (a) $4 - j3$; (b) $4 + j3$; (c) $3 + j$; (d) $3 - j3$.
4. Convert the following impedances to rectangular form: (a) $5/\!-\!27°$; (b) $5/27°$; (c) $6.71/63.4°$; (d) $4.24/\!-\!45°$.
5. Find the total Z_T in rectangular form for the following three series impedances: (a) $12/10°$; (b) $25/15°$; (c) $34/26°$.
6. Multiply the following, in polar form: (a) $45/24° \times 10/54°$; (b) $45/\!-\!24° \times 10/54°$; (c) $18/\!-\!64° \times 4/14°$; (d) $18/\!-\!64° \times 4/\!-\!14°$.
7. Divide the following, in polar form: (a) $45/24° \div 10/10°$; (b) $45/24° \div 10/\!-\!10°$; (c) $500/\!-\!72° \div 5/12°$; (d) $500/\!-\!72° \div 5/\!-\!12°$.
8. Match the four phasor diagrams in Fig. 26-4a, b, c, and d with the four circuits in Fig. 26-5 and 26-6.
9. Find Z_T in polar form for the series circuit in Fig. 26-7a.
10. Find Z_T in polar form for the series-parallel circuit in Fig. 26-7c.
11. Solve the circuit in Fig. 26-12 to find Z_T in rectangular form by rationalization.
12. Solve the circuit in Fig. 26-12 to find Z_T in polar form, using the method of branch currents. Assume an applied voltage of 56.6 V.
13. Show the equivalent series circuit of Fig. 26-12.
14. Solve the circuit in Fig. 26-14 to find Z_T in polar form, without using branch currents. (Find the Z of two branches in parallel; then combine this Z with the third branch Z.)
15. Show the equivalent series circuit of Fig. 26-14.
16. Refer to Fig. 26-13. (a) Find Z_1 and Z_2 for the two branch currents given. (b) Calculate the values needed for R_1, R_2, X_C, and X_L for these impedances. (c) What are the L and C values for a frequency of 60 Hz?
17. Solve the series ac circuit in Fig. 25-8 in the previous chapter by the use of complex numbers. Find Z/θ, I/θ, and each V/θ. Prove that the sum of the complex voltage drops around the circuit equals the applied voltage V_T. Make a phasor diagram showing all phase angles with respect to V_T.
18. The following components are in series: $L = 100$ μH, $C = 20$ pF,

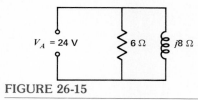

FIGURE 26-15
For Prob. 20.

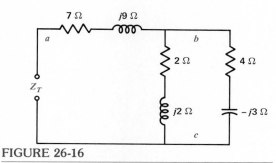

FIGURE 26-16
For Prob. 21.

$R = 2000\ \Omega$. At the frequency of 2 MHz calculate X_L, X_C, Z_T, I, θ, V_R, V_L, and V_C. The applied $V_T = 80$ V.

19. Solve the same circuit as in Prob. 18 for the frequency of 4 MHz. Give three effects of the higher frequency.
20. In Fig. 26-15, show that $Z_T = 4.8\ \Omega$ and $\theta = 36.9°$ by (a) the method of branch currents; (b) calculating Z_T as $Z_1 Z_2/(Z_1 + Z_2)$.
21. In Fig. 26-16, find $Z_T/\underline{\theta}$ by calculating Z_{bc} of the parallel bank and combining with the series Z_{ab}.

Answers to Practice Problems

26-1	(a) 0°	26-5	(a) $4 + j7$
	(b) 180°		(b) $0 - j7$
26-2	(a) 90°	26-6	(a) $5 + j7$
	(b) −90 or 270°		(b) $4 + j6$
26-3	(a) T	26-7	(a) 14.14 Ω
	(b) T		(b) 45°
26-4	(a) $j3$ kΩ	26-8	(a) $12/\underline{50°}$
	(b) $-j5$ mA		(b) $3/\underline{-10°}$

Chapter 26
Complex Numbers for AC Circuits

26-9   (a) $10 + j10$
        (b) $10 - j10$

26-10   (a) $53°$
        (b) $143°$
        (c) $90°$

26-11   (a) $(6 + j8) / (5 + j4)$
        (b) $(6 - j8) / (5 - j8)$

26-12   (a) $10 + j4$
        (b) $56.6 \underline{/8°}$

26-13   (a) $4 + j5$ A
        (b) $9 - j2$ A

26-14   (a) $40 + j30$
        (b) $50 \underline{/37°}$ Ω
        (c) $2 \underline{/-37°}$ A

Resonance

This chapter explains how X_L and X_C can be combined to favor one particular frequency, the resonant frequency to which the *LC* circuit is tuned. The resonance effect occurs when the inductive and capacitive reactances are equal.

The main application of resonance is in rf circuits for tuning to an ac signal of the desired frequency. All examples of tuning in radio and television receivers, transmitters, and electronics equipment in general are applications of resonance. The topics are:

Chapter 27

27-1   The Resonance Effect
27-2   Series Resonance
27-3   Parallel Resonance
27-4   The Resonant Frequency $f_r = 1/(2\pi\sqrt{LC})$
27-5   Q Magnification Factor of Resonant Circuit
27-6   Bandwidth of Resonant Circuit
27-7   Tuning
27-8   Mistuning
27-9   Analysis of Parallel Resonant Circuits
27-10   Damping of Parallel Resonant Circuits
27-11   Choosing *L* and *C* for a Resonant Circuit

27-1 THE RESONANCE EFFECT

Inductive reactance increases as the frequency is increased, but capacitive reactance decreases with higher frequencies. Because of these opposite characteristics, for any *LC* combination there must be a frequency at which the X_L equals the X_C, as one increases while the other decreases. This case of equal and opposite reactances is called *resonance*, and the ac circuit is then a *resonant circuit*.

Any *LC* circuit can be resonant. It all depends on the frequency. At the resonant frequency, an *LC* combination provides the resonance effect. Off the resonant frequency, either below or above, the *LC* combination is just another ac circuit.

The frequency at which the opposite reactances are equal is the *resonant frequency*. This frequency can be calculated as $f_r = 1/(2\pi\sqrt{LC})$ where *L* is the inductance in henrys, *C* is the capacitance in farads, and f_r is the resonant frequency in hertz that makes $X_L = X_C$.

In general, we can say that large values of *L* and *C* provide a relatively low resonant frequency. Smaller values of *L* and *C* allow higher values for f_r. The resonance effect is most useful for radio frequencies, where the required values of microhenrys for *L* and picofarads for *C* are easily obtained.

The most common application of resonance in rf circuits is called *tuning*. In this use, the *LC* circuit provides maximum voltage output at the resonant frequency, compared with the amount of output at any other frequency either below or above resonance. This idea is illustrated in Fig. 27-1*a*, where the *LC* circuit

Chapter 27
Resonance

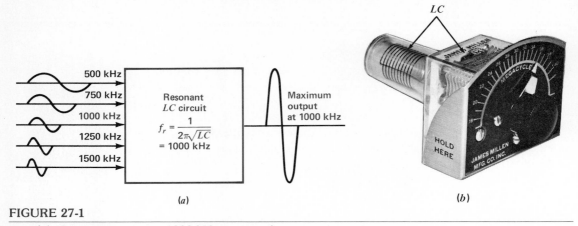

FIGURE 27-1
(a) Circuit resonant at 1000 kHz to provide maximum output at this f_r. (b) Wavemeter as an example of tuning an *LC* circuit to resonance. (*James Millen Mfg. Co. Inc.*)

resonant at 1000 kHz magnifies the effect of this particular frequency. The result is maximum output at 1000 kHz, compared with lower or higher frequencies.

For the wavemeter in (b), note that the capacitance *C* can be varied to provide resonance at different frequencies. The wavemeter can be tuned to any one frequency in a range depending on the *LC* combination.

All examples of tuning in radio and television are applications of resonance. When you tune a radio to one station, the *LC* circuits are tuned to resonance for that particular carrier frequency. Also, when you tune a television receiver to a particular channel, the *LC* circuits are tuned to resonance for that station. There are almost unlimited uses for resonance in ac circuits.

Practice Problems 27-1
(*answers on page 569*)
Refer to Fig. 27-1.
(a) Give the resonant frequency.
(b) Give the frequency that has maximum output.

27-2
SERIES RESONANCE

In the series ac circuit in Fig. 27-2a, when the frequency of the applied voltage is 1000 kHz, the reactance of the 239-μH inductance equals 1500 Ω. At the same frequency, the reactance of the 106-pF capacitance also is 1500 Ω. Therefore, this *LC* combination is resonant at 1000 kHz. This is f_r, because the inductive reactance and capacitive reactance are equal at this frequency.

In a series ac circuit, inductive reactance leads by 90°, compared with the zero reference angle of the resistance, while capacitive reactance lags by 90°. Therefore, X_L and X_C are 180° out of phase. The opposite reactances cancel each other completely when they are equal.

Figure 27-2b shows X_L and X_C equal, resulting in a net reactance of zero ohms. The only opposition to current then is the coil resistance r_S, which is the limit on how low the series resistance in the circuit can be. With zero reactance and just the low value of series resistance, the generator voltage produces the greatest

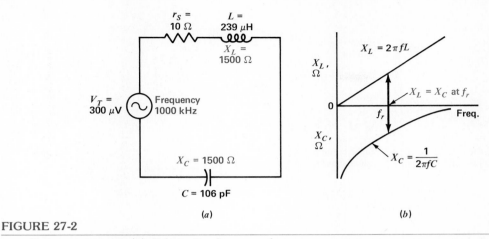

FIGURE 27-2
Series resonance. (a) Schematic diagram of series r_S, L, and C. (b) X_C and X_L are equal and opposite at the resonant frequency f_r.

amount of current in the series LC circuit at the resonant frequency. The series resistance should be as small as possible for a sharp increase in current at resonance.

Maximum Current at Series Resonance. The main characteristic of series resonance is the resonant rise of current to its maximum value of V_T/r_S at the resonant frequency. For the circuit in Fig. 27-2a, the maximum current at series resonance is 30 µA, equal to 300 µV/10 Ω. At any other frequency either below or above the resonant frequency, there is less current in the circuit.

This resonant rise of current to 30 µA at 1000 kHz is illustrated in Fig. 27-3. In (a), the amount of current is shown as the amplitude of individual cycles of the alternating current produced in the circuit by the ac generator voltage. Whether the amplitude of one ac cycle is considered in terms of peak, rms, or average value, the amount of current is greatest at the resonant frequency. In (b), the current amplitudes are plotted on a graph for frequencies at and near the resonant frequency, producing a typical *response curve* for a series resonant circuit. The response curve in (b) can be considered as an outline of the increasing and decreasing amplitudes for the individual cycles shown in (a).

The response curve of the series resonant circuit shows that the current is small below resonance, rises to its maximum value at the resonant frequency, and then drops off to small values also above resonance. To prove this fact, Table 27-1 lists the calculated values of impedance and current in the circuit of Fig. 27-2 at the resonant frequency of 1000 kHz, and at two frequencies below and two frequencies above resonance.

Below resonance, at 600 kHz, X_C is more than X_L and there is appreciable net reactance, which limits the current to a relatively low value. At the higher frequency of 800 kHz, X_C decreases and X_L increases, making the two reactances closer to the same value. The net reactance is then smaller, allowing more current.

At the resonant frequency, X_L and X_C are equal, the net reactance is zero, and the current has its maximum value equal to V_T/r_S.

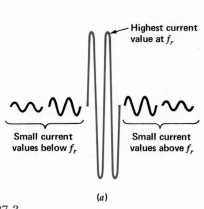
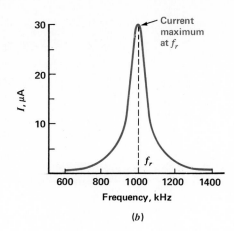

FIGURE 27-3
Graphs showing maximum current at resonance, for the series circuit in Fig. 27-2. (a) Amplitudes of individual cycles. (b) Response curve showing amount of I below and above resonance. Values of I in Table 27-1.

Above resonance at 1200 and 1400 kHz, X_L is greater than X_C, providing net reactance that limits the current to much smaller values than at resonance.

In summary:

1. Below the resonant frequency, X_L is small, but X_C has high values that limit the amount of current.
2. Above the resonant frequency, X_C is small, but X_L has high values that limit the amount of current.
3. At the resonant frequency, X_L equals X_C, and they cancel to allow maximum current.

Minimum Impedance at Series Resonance. Since the reactances cancel at the resonant frequency, the impedance of the series circuit is minimum, equal to just the low value of series resistance. This minimum impedance at reso-

TABLE 27-1. Series-Resonance Calculations for the Circuit in Fig. 27-2*

FREQUENCY, kHz	$X_L =$ $2\pi fL$, Ω	$X_C =$ $1/(2\pi fC)$, Ω	NET REACTANCE, Ω		Z_T, Ω†	$I =$ V_T/Z_T, μA†	$V_L =$ IX_L, μV	$V_C =$ IX_C, μV
			$X_C - X_L$	$X_L - X_C$				
600	900	2500	1600		1600	0.19	171	475
800	1200	1875	675		675	0.44	528	825
$f_r \rightarrow$ 1000	1500	1500	0	0	10	30	45,000	45,000
1200	1800	1250		550	550	0.55	990	688
1400	2100	1070		1030	1030	0.29	609	310

*$L = 239$ μH, $C = 106$ pF, $V_T = 300$ μV, $r_s = 10$ Ω.
†Z_T and I calculated without r_s when its resistance is very small compared with the net X_L or X_C. Z_T and I are resistive at f_r.

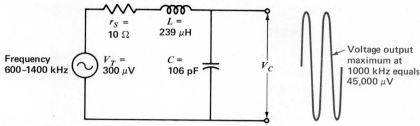

FIGURE 27-4
Series circuit selects frequency by producing maximum IX_C voltage output across C at resonance.

nance is resistive, resulting in zero phase angle. At resonance, therefore, the resonant current is in phase with the generator voltage.

Resonant Rise in Voltage Across Series L or C. The maximum current in a series LC circuit at resonance is useful because it produces maximum voltage across either X_L or X_C at the resonant frequency. As a result, the series resonant circuit can select one frequency by providing much more voltage output at the resonant frequency, compared with frequencies above and below resonance. Figure 27-4 illustrates the resonant rise in voltage across the capacitance in a series ac circuit. At the resonant frequency of 1000 kHz, the voltage across C rises to the value of 45,000 μV, while the input voltage is only 300 μV.

In Table 27-1, the voltage across C is calculated as IX_C and across L as IX_L. Below the resonant frequency, X_C has a higher value than at resonance, but the current is small. Similarly, above the resonant frequency, X_L is higher than at resonance, but the current has a low value because of the inductive reactance. At resonance, although X_L and X_C cancel each other to allow maximum current, each reactance by itself has an appreciable value. Since the current is the same in all parts of a series circuit, the maximum current at resonance produces maximum voltage IX_C across C and an equal IX_L voltage across L for the resonant frequency.

Although the voltage across X_C and X_L is reactive, it is an actual voltage that can be measured. In Fig. 27-5, the voltage drops around the series resonant circuit are

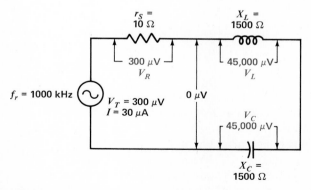

FIGURE 27-5
Voltage drops around series resonant circuit.

45,000 µV across C and 45,000 µV across L, with 300 µV across r_S. The voltage across the resistance is equal to the generator voltage and has the same phase.

Across the series combination of both L and C, the voltage is zero because the two series voltage drops are equal and opposite. In order to use the resonant rise of voltage, therefore, the output must be connected across either L or C alone. We can consider the V_L and V_C voltages as similar to the idea of two batteries connected in series opposition. Together, the net resultant is zero for the equal and opposite voltages, but each battery still has its own potential difference.

In summary, for a series resonant circuit the main characteristics are:

1. The current I is maximum at the resonant frequency f_r.
2. I is in phase with the generator voltage, or the phase angle of the circuit is 0°.
3. The voltage is maximum across either L or C alone.
4. The impedance is minimum at f_r, equal only to the low r_S.

Practice Problems 27-2
(answers on page 569)
Answer true or false, for series resonance.
(a) X_L and X_C are maximum.
(b) X_L and X_C are equal.
(c) I is maximum.

27-3
PARALLEL RESONANCE

With L and C in parallel as shown in Fig. 27-6, when X_L equals X_C, the reactive branch currents are equal and opposite at resonance. Then they cancel each other to produce minimum current in the main line. Since the line current is minimum, the impedance is maximum. These relations are based on r_S being very small compared with X_L at resonance. In this case, the branch currents are practically equal when X_L and X_C are equal.

Minimum Line Current at Parallel Resonance. To show how the current in the main line dips to its minimum value when the parallel LC circuit is resonant, Table 27-2 lists the values of branch currents and the total line current for the circuit in Fig. 27-6.

With L and C the same as in the series circuit of Fig. 27-2, X_L and X_C have the same values at the same frequencies. Since L, C, and the generator are in parallel, the voltage applied across the branches equals the generator voltage of 300 µV. Therefore, each reactive branch current is calculated as 300 µV divided by the reactance of the branch.

The values in the top row of Table 27-2 are obtained as follows: at 600 kHz the capacitive branch current equals 300 µV/2500 Ω, or 0.12 µA. The inductive branch current at this frequency is 300 µV/900 Ω, or 0.33 µA. Since this is a parallel ac circuit, the capacitive current leads by 90° while the inductive current lags by 90°, compared with the reference angle of the generator voltage, which is applied across the parallel branches. Therefore, the opposite currents are 180° out of phase, canceling each other in the main line. The net current in the line, then, is the difference between 0.33 and 0.12, which equals 0.21 µA.

Following this procedure, the calculations show that as the frequency is increased toward resonance, the capacitive branch current increases because of the lower value of X_C, while the inductive branch current decreases with higher values of X_L. As a result, there is less net line current as the two branch currents become more nearly equal.

At the resonant frequency of 1000 kHz, both reactances are 1500 Ω, and the reactive branch currents are both 0.20 µA, canceling each other completely.

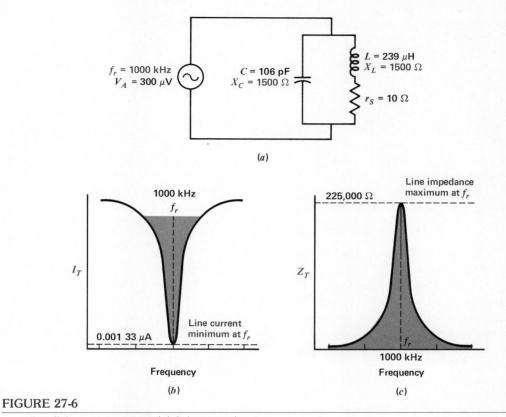

FIGURE 27-6
Parallel resonant circuit. (a) Schematic diagram of L and C in parallel branches. (b) Response curve of I_T shows line current dips to minimum at f_r. (c) Response curve of Z_T shows it rises to maximum at f_r.

Above the resonant frequency, there is more current in the capacitive branch than in the inductive branch, and the net line current increases above its minimum value at resonance.

The dip in I_T to its minimum value at f_r is shown by the graph in Fig. 27-6b. At parallel resonance, I_T is minimum and Z_T is maximum.

The in-phase current due to r_S in the inductive branch can be ignored off resonance because it is so small compared with the reactive line current. At the resonant frequency when the reactive currents cancel, however, the resistive component is the entire line current. Its value at resonance equals 0.001 33 µA in this example. This small resistive current is the minimum value of the line current at parallel resonance.

Maximum Line Impedance at Parallel Resonance. The minimum line current resulting from parallel resonance is useful because it corresponds to maximum impedance in the line

TABLE 27-2. Parallel-Resonance Calculations for the Circuit in Fig. 27-6*

FREQUENCY, kHz	$X_C = 1/(2\pi fC)$, Ω	$X_L = 2\pi fL$, Ω	$I_C = V/X_C$, μA	$I_L = V/X_L$, μA†	NET REACTIVE LINE CURRENT, μA		I_T, μA†	$Z_T = V_A/I_T$, Ω†
					$I_L - I_C$	$I_C - I_L$		
600	2500	900	0.12	0.33	0.21		0.21	1400
800	1875	1200	0.16	0.25	0.09		0.09	3333
$f_r \to$ 1000	1500	1500	0.20	0.20	0	0	0.001 33	225,000‡
1200	1250	1800	0.24	0.17		0.08	0.08	3800
1400	1070	2100	0.28	0.14		0.14	0.14	2143

*$L = 239$ μH, $C = 106$ pF, $V_A = 300$ μV.
†I_L, I_T, and Z_T calculated approximately without r_S when its resistive component of the line current is very small compared with I_L.
‡At resonance, Z_T calculated by Formula (27-8). Z_T and I_T are resistive at f_r.

across the generator. Therefore, an impedance that has a high value for just one frequency but a low impedance for other frequencies, either below or above resonance, can be obtained by using a parallel LC circuit resonant at the desired frequency. This is another method of selecting one frequency by resonance. The response curve in Fig. 27-6c shows how the impedance rises to maximum for parallel resonance.

The main application of parallel resonance is the use of an LC tuned circuit as the load impedance Z_L in the output circuit of rf amplifiers. Because of the high impedance, then, the gain of the amplifier is maximum at f_r. The voltage gain of an amplifier is directly proportional to Z_L. The advantage of a resonant LC circuit is that Z is maximum only for an ac signal at the resonant frequency. Also, L has practically no dc resistance, which means practically no dc voltage drop.

Referring to Table 27-2, the total impedance of the parallel ac circuit is calculated as the generator voltage divided by the total line current. At 600 kHz, for example, Z_T equals 300 $\mu V/0.21$ μA, or 1400 Ω. At 800 kHz, the impedance is higher because there is less line current.

At the resonant frequency of 1000 kHz, the line current is at its minimum value of 0.001 33 μA. Then the impedance is maximum, equal to 300 $\mu V/0.001 33$ μA, or 225,000 Ω.

Above 1000 kHz, the line current increases, and the impedance decreases from its maximum value.

The idea of how the line current can have a very low value even though the reactive branch currents are appreciable is illustrated in Fig. 27-7. In (a), the resistive component of the total line current is shown as though it were a separate branch drawing an amount of resistive current from the generator in the main line equal to the current resulting from the coil resistance. Each reactive branch current has its value equal to the generator voltage divided by the reactance. Since they are equal and of opposite phase, however, in any part of the circuit where both reactive currents are present, the net amount of electron flow in one direction at any instant of time corresponds to zero current. The graph in (b) shows how equal and opposite currents for I_L and I_C cancel.

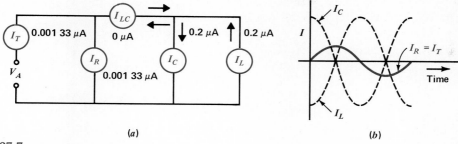

FIGURE 27-7
Distribution of currents in parallel circuit at resonance. Resistive current shown as an equivalent branch for I_R. (a) Circuit with branch currents for R, L, and C. (b) Graph of equal and opposite reactive currents I_L and I_C.

If a meter is inserted in series with the main line to indicate total line current I_T, it dips sharply to the minimum value of line current at the resonant frequency. With minimum current in the line, the impedance across the line is maximum at the resonant frequency. The maximum impedance at parallel resonance corresponds to a high value of resistance, without reactance, since the line current is then resistive with zero phase angle.

In summary, for a parallel resonant circuit, the main characteristics are:

1. The line current I_T is minimum at the resonant frequency.
2. I_T is in phase with the generator voltage V_A, or the phase angle of the circuit is 0°.
3. The impedance Z_T, equal to V_A/I_T, is maximum at f_r because of the minimum I_T.

The LC Tank Circuit. It should be noted that the individual branch currents are appreciable at resonance, although I_T is minimum. For the example in Table 27-2, at f_r either the I_L or I_C equals 0.2 µA. This current is greater than the I_C values below f_r or the I_L values above f_r.

The branch currents cancel in the main line because I_C is at 90° with respect to the source V_A while I_L is at −90°, making them opposite with respect to each other.

However, inside the LC circuit, I_L and I_C do not cancel because they are in separate branches. Then I_L and I_C provide a circulating current in the LC circuit, which equals 0.2 µA in this example. For this reason, a parallel resonant LC circuit is often called a *tank circuit*.

Because of the energy stored by L and C, the circulating tank current can provide full sine waves of current and voltage output when the input is only a pulse. The sine-wave output is always at the natural resonant frequency of the LC tank circuit. This ability of the LC circuit to supply complete sine waves is called the *flywheel effect*.

Practice Problems 27-3
(answers on page 569)
Answer true or false, for parallel resonance.
(a) I_L and I_C are maximum.
(b) I_L and I_C are equal.
(c) I_T is minimum.

27-4
THE RESONANT FREQUENCY
$f_r = 1/(2\pi\sqrt{LC})$

This formula is derived from $X_L = X_C$. Using f_r to indicate the resonant frequency in the formulas for X_L and X_C, we have

$$2\pi f_r L = \frac{1}{2\pi f_r C}$$

Inverting the factor f_r gives

$$2\pi L (f_r)^2 = \frac{1}{2\pi C}$$

Inverting the factor $2\pi L$ gives

$$f_r^2 = \frac{1}{(2\pi)^2 LC}$$

The square root of both sides is then

$$f_r = \frac{1}{2\pi\sqrt{LC}} \qquad (27\text{-}1)$$

With units of henrys for L and farads for C, the resonant frequency f_r is in hertz. Since $1/2\pi$ is a numerical value equal to $1/6.28$, or 0.159, a more convenient form for calculations is

$$f_r = \frac{0.159}{\sqrt{LC}} \quad \text{Hz} \qquad (27\text{-}2)$$

For example, to find the resonant frequency of the LC combination in Fig. 27-2, the values of 239×10^{-6} H and 106×10^{-12} F are substituted for L and C:

$$f_r = \frac{0.159}{\sqrt{239 \times 10^{-6} \times 106 \times 10^{-12}}}$$

$$= \frac{0.159}{\sqrt{253 \times 10^{-16}}}$$

Taking the square root of the denominator,

$$f_r = \frac{0.159}{15.9 \times 10^{-8}} = \frac{0.159}{0.159 \times 10^{-6}}$$

$$= \frac{0.159 \times 10^6}{0.159}$$

$$f_r = 1 \times 10^6 \text{ Hz} = 1 \text{ MHz} = 1000 \text{ kHz}$$

For any LC circuit, series or parallel, $f_r = 1/(2\pi\sqrt{LC})$ is the resonant frequency that makes the inductive and capacitive reactances equal.

How f_r Varies with L or C. It is important to note that higher values of L or C result in lower values of f_r. Also, an LC circuit can be resonant at any frequency from a few Hz to many MHz, depending upon the inductance and capacitance values.

As examples, an LC combination with the relatively large values of an 8-H inductance and a 20-μF capacitance is resonant at the low audio frequency of 12.6 Hz; a small inductance of 2 μH will resonate with the small capacitance of 3 pF at the high radio frequency of 64.9 MHz. These examples are solved in the next two problems for more practice with the resonant frequency formula. Such calculations are often used in practical applications of tuned circuits. Probably the most important feature of any LC combination is its resonant frequency.

Example 1. Calculate the resonant frequency for an 8-H L and a 20-μF C.

Answer.
$$f_r = \frac{1}{2\pi\sqrt{LC}} = \frac{0.159}{\sqrt{8 \times 20 \times 10^{-6}}}$$

$$= \frac{0.159 \times 10^3}{\sqrt{160}} = \frac{159}{12.65}$$

$$f_r = 12.6 \text{ Hz}$$

Example 2. Calculate the resonant frequency for a 2-μH L and a 3-pF C.

Answer.
$$f_r = \frac{1}{2\pi\sqrt{LC}}$$
$$= \frac{0.159}{\sqrt{2 \times 10^{-6} \times 3 \times 10^{-12}}}$$
$$= \frac{0.159}{\sqrt{6 \times 10^{-18}}} = \frac{0.159 \times 10^9}{\sqrt{6}}$$
$$= \frac{159 \times 10^6}{\sqrt{6}} = \frac{159}{2.45} \times 10^6$$
$$f_r = 64.9 \text{ MHz}$$

More specifically, f_r decreases inversely as the square root of L or C. For instance, if L or C is quadrupled, f_r is reduced by one-half. Suppose f_r is 6 MHz with an LC combination. If L or C is made four times larger, then f_r will be reduced to 3 MHz. Or to take the opposite case, to double f_r, the value of L or C must be reduced by one-fourth, or both reduced by one-half.

LC Product Determines f_r. There are any number of LC combinations that can be resonant at one frequency. Table 27-3 lists five possible combinations of L and C resonant at 1000 kHz. The resonant frequency is the same because when either L or C is decreased by the factor of 10 or 2, the other is increased by the same factor, resulting in a constant value for the LC product.

The reactance at resonance changes with different combinations of L and C, but in all five cases X_L and X_C are equal to each other at 1000 kHz. This is the resonant frequency determined by the value of the LC product in $f_r = 1/(2\pi\sqrt{LC})$.

Measuring L or C by Resonance. Of the three factors L, C, and f_r in the resonant-frequency formula, any one can be calculated when the other two are known. The resonant frequency of the LC combination can be found experimentally by determining the frequency that produces the resonant response in an LC combination. With a known value of either L or C, and the resonant frequency determined, the third factor can be calculated. This method is commonly used for measuring inductance or capacitance. A test instrument for this purpose is the Q meter, which also measures the Q of a coil.

Calculating C from f_r. The C can be taken out of the square root sign or radical in the resonance formula, as follows:

$$f_r = \frac{1}{2\pi\sqrt{LC}}$$

Squaring both sides to eliminate the radical gives

$$f_r^2 = \frac{1}{(2\pi)^2 LC}$$

Inverting C and f_r^2 gives

$$C = \frac{1}{(2\pi)^2 f_r^2 L} = \frac{1}{4\pi^2 f_r^2 L} = \frac{0.0254}{f_r^2 L} \quad (27\text{-}3)$$

TABLE 27-3. *LC* Combinations Resonant at 1000 kHz

L, μH	C, pF	LC PRODUCT	X_L, Ω AT 1000 kHz	X_C, Ω AT 1000 kHz
23.9	1060	25,334	150	150
119.5	212	25,334	750	750
239	106	25,334	1500	1500
478	53	25,334	3000	3000
2390	10.6	25,334	15,000	15,000

With f_r in hertz, the units are farads for C and henrys for L.

The constant factor 0.0254 in the numerator is the reciprocal of 39.44 for $4\pi^2$ in the denominator. These numbers remain the same for any values of f_r, L, and C.

Calculating L from f_r. Similarly, the resonance formula can be transposed to find L. Then

$$L = \frac{1}{(2\pi)^2 f_r^2 C} = \frac{1}{4\pi^2 f_r^2 C} = \frac{0.0254}{f_r^2 C} \quad (27\text{-}4)$$

With Formula (27-4), L is determined by its f_r with a known value of C. Similarly, C is determined from Formula (27-3) by its f_r with a known value of L.

Example 3. What value of C resonates with a 239-μH L at 1000 kHz?

Answer.
$$C = \frac{0.0254}{f_r^2 L}$$
$$= \frac{0.0254}{(1 \times 10^6)^2 \times 239 \times 10^{-6}}$$
$$= \frac{0.0254}{1 \times 10^{12} \times 239 \times 10^{-6}}$$
$$= \frac{0.0254}{239 \times 10^6} = \frac{0.0254}{239} \times 10^{-6}$$
$$= \frac{25{,}400}{239} \times 10^{-12}$$
$$= 106 \times 10^{-12} \text{ F}$$
$$C = 106 \text{ pF}$$

Example 4. What value of L resonates with a 106-pF C at 1000 kHz?

Answer.
$$L = \frac{0.0254}{f_r^2 C}$$
$$= \frac{0.0254}{1 \times 10^{12} \times 106 \times 10^{-12}}$$
$$= \frac{0.0254}{106}$$
$$= \frac{25{,}400}{106} \times 10^{-6} = 239 \times 10^{-6} \text{ H}$$
$$L = 239 \; \mu\text{H}$$

These values are from the LC circuit illustrated in Fig. 27-2 for series resonance and Fig. 27-6 for parallel resonance.

Practice Problems 27-4
(answers on page 569)
(a) To increase f_r, must the C be more or less?
(b) C is increased from 100 to 400 pF. L must be decreased from 800 μH to what value for the same f_r?

27-5
Q MAGNIFICATION FACTOR OF RESONANT CIRCUIT

The quality, or *figure of merit*, of the resonant circuit, in sharpness of resonance, is indicated by the factor Q. In general, the higher the ratio of the reactance at resonance to the series resistance, the higher is the Q and the sharper the resonance effect.

Q of Series Circuit. In a series resonant circuit we can calculate Q from the following formula:

$$Q = \frac{X_L}{r_S} \quad (27\text{-}5)$$

where Q is the figure of merit, X_L is the inductive reactance at the resonant frequency, and r_S is the resistance in series with X_L. For the series resonant circuit in Fig. 27-2,

$$Q = \frac{1500 \; \Omega}{10 \; \Omega} = 150$$

Q is a numerical factor without any units, because it is a ratio of reactance to resistance and the ohms cancel. Since the series resistance limits the amount of current at resonance, the lower the resistance, the sharper is the increase to maximum current at the resonant frequency, and the higher the Q. Also, a higher value of reactance at resonance allows the maximum current to produce a higher value of voltage output.

The Q has the same value if calculated with X_C instead of X_L, since they are equal at resonance. However, the Q of the circuit is generally considered in terms of X_L, because usually the coil has the series resistance of the circuit. In this case, the Q of the coil and the Q of the series resonant circuit are the same. If extra resistance is added, the Q of the circuit will be less than the Q of the coil. The highest possible Q for the circuit is the Q of the coil.

The value of 150 can be considered as a high Q. Typical values are 50 to 250, approximately. Less than 10 is a low Q; more than 300 is a very high Q.

Higher L/C Ratio Can Provide Higher Q. As shown before in Table 27-3, different LC combinations can be resonant at the same frequency. However, the amount of reactance at resonance is different. More X_L can be obtained with a higher L and smaller C for resonance, although X_L and X_C must be equal at the resonant frequency. Therefore, both X_L and X_C are higher with a higher L/C ratio for resonance.

More X_L can allow a higher Q if the ac resistance does not increase as much as the reactance. With typical rf coils, an approximate rule is that maximum Q can be obtained when X_L is about 1000 Ω.

Q Rise in Voltage Across Series L or C. The Q of the resonant circuit can be considered a magnification factor that determines how much the voltage across L or C is increased by the resonant rise of current in a series circuit. Specifically, the voltage output at series resonance is Q times the generator voltage:

$$V_L = V_C = Q \times V_{gen} \qquad (27\text{-}6)$$

In Fig. 27-4, for example, the generator voltage is 300 μV and Q is 150. The resonant rise of voltage across either L or C then equals 300 μV × 150, or 45,000 μV. Note that this is the same value calculated in Table 27-1 for V_C or V_L at resonance.

How to Measure Q in a Series Resonant Circuit. The fundamental nature of Q for a series resonant circuit is seen from the fact that the Q can be determined experimentally by measuring the Q rise in voltage across either L or C and comparing this voltage with the generator voltage. As a formula,

$$Q = \frac{V_{out}}{V_{in}} \qquad (27\text{-}7)$$

where V_{out} is the ac voltage measured across the coil or capacitor and V_{in} is the generator voltage.

Referring to Fig. 27-5, suppose that you measure with an ac voltmeter across L or C and this voltage equals 45,000 μV at the resonant frequency. Also, measure the generator input of 300 μV. Then

$$Q = \frac{V_{out}}{V_{in}} = \frac{45{,}000 \; \mu V}{300 \; \mu V}$$
$$Q = 150$$

This method is better than the X_L/r_s formula for determining Q because r_s is the ac resistance of the coil, which is not so easily measured. Remember that the coil's ac resistance can be more than double the dc resistance measured with an ohmmeter. In fact, measuring Q with Formula (27-7) makes it possible to calculate the ac resistance. These points are illustrated in the following examples.

Example 5. A series circuit resonant at 0.4 MHz develops 100 mV across a 250-μH L with a 2-mV input. Calculate Q.

Answer. $\quad Q = \dfrac{V_{out}}{V_{in}} = \dfrac{100 \; mV}{2 \; mV}$
$\qquad\qquad Q = 50$

Example 6. How much is the ac resistance of the coil in the preceding example?

Answer. The Q of the coil is 50. We need to know the reactance of this 250-μH coil at the frequency of 0.4 MHz. Then,

$$X_L = 2\pi f L = 2\pi \times 0.4 \times 10^6 \times 250 \times 10^{-6}$$
$$= 2\pi \times 100$$
$$X_L = 628 \; \Omega$$

Also,

$$Q = \frac{X_L}{r_S} \quad \text{or} \quad r_S = \frac{X_L}{Q}$$
$$= \frac{628 \; \Omega}{50}$$
$$r_S = 12.56 \; \Omega$$

Q of Parallel Circuit. In a parallel resonant circuit, where r_S is very small compared with X_L, the Q also equals X_L/r_S. Note that r_S is still the resistance of the coil in series with X_L (see Fig. 27-8). The Q of the coil determines the Q of the parallel circuit here because it is less than the Q of the capacitive branch. Capacitors used in tuned circuits generally have a very high Q because of their low losses. In Fig. 27-8, the Q is 1500 Ω/10 Ω, or 150, the same as the series resonant circuit with the same values.

This example assumes that the generator resistance is very high and that there is no other resistance branch shunting the tuned circuit. Then the Q of the parallel resonant circuit is the same as the Q of the coil. Actually, shunt resistance can lower the Q of a parallel resonant circuit, as analyzed in Sec. 27-10.

Q Rise in Impedance Across Parallel Resonant Circuit. For parallel resonance, the Q magnification factor determines by how much the impedance across the parallel LC circuit is increased because of the minimum line current.

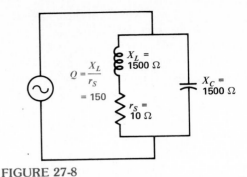

FIGURE 27-8

Q of a parallel resonant circuit in terms of X_L and its series resistance r_S.

Specifically, the impedance across the parallel resonant circuit is Q times the inductive reactance at the resonant frequency:

$$Z_T = Q \times X_L \tag{27-8}$$

Referring back to the parallel resonant circuit in Fig. 27-6, as an example, X_L is 1500 Ω and Q is 150. The result is a rise of impedance to the maximum value of 150 $\times$ 1500 Ω, or 225,000 Ω, at the resonant frequency.

Since the line current equals V_A/Z_T, the minimum value of line current then is 300 μV/225,000 Ω, which equals 0.001 33 μA.

At f_r the minimum line current is $1/Q$ of either branch current. In Fig. 27-7, I_L or I_C is 0.2 μA and Q is 150. Therefore, I_T is 0.2/150, or 0.001 33 μA, which is the same answer as V_A/Z_T. Or, stated another way, the circulating tank current is Q times the minimum I_T.

How to Measure Z_T of a Parallel Resonant Circuit. Formula (27-8) is also useful in its inverted version as $Q = Z_T/X_L$. We can measure Z_T by the method illustrated in Fig. 27-9. Then Q can be calculated.

To measure Z_T, first tune the LC circuit to resonance. Then adjust R_1 to the resistance that makes its ac voltage equal to the ac voltage

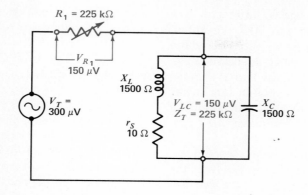

FIGURE 27-9

Adjust R_1 to make V_{R_1} equal V_{LC}. Then $Z_T = R_1$.

across the tuned circuit. With equal voltages, Z_T must be the same as R_1.

For the example here, which corresponds to the parallel resonance shown in Figs. 27-6 and 27-8, Z_T is 225,000 Ω. Therefore, Q equals Z_T/X_L or 225,000/1500, which equals 150.

Example 7. In Fig. 27-9, assume that with 4-V ac input signal for V_T, the voltage across R_1 is 2 V when R_1 is 225 kΩ. Determine Z_T and Q.

Answer. Because they divide V_T equally, Z_T is 225 kΩ, the same as R_1. The amount of input voltage does not matter, as the voltage division determines the relative proportions between R_1 and Z_T. With 225 kΩ for Z_T and 1.5 kΩ for X_L, the Q is $225/1.5$, or Q = 150.

Example 8. A parallel LC circuit tuned to 200 kHz with a 350-μH L has a measured Z_T of 17,600 Ω. Calculate Q.

Answer. First, calculate X_L as $2\pi fL$ at f_r:

$$X_L = 2\pi \times 200 \times 10^3 \times 350 \times 10^{-6} = 440 \text{ Ω}$$

Then,

$$Q = \frac{Z_T}{X_L} = \frac{17,600}{440}$$

$$Q = 40$$

Practice Problems 27-5
(answers on page 569)

(a) In a series resonant circuit, V_L is 300 mV with input of 3 mV. Calculate Q.
(b) In a parallel resonant circuit, X_L is 500 Ω. With a Q of 50, calculate Z_T.

27-6 BANDWIDTH OF RESONANT CIRCUIT

When we say an LC circuit is resonant at one frequency, this is true for the maximum resonance effect. However, other frequencies close to f_r also are effective. For series resonance, frequencies just below and above f_r produce increased current, but a little less than the value at resonance. Similarly, for parallel resonance, frequencies close to f_r can provide a high impedance, although a little less than the maximum Z_T.

Therefore, any resonant frequency has an associated band of frequencies that provide resonance effects. How wide the band is depends on the Q of the resonant circuit. Actually, it is practically impossible to have an LC circuit with a resonant effect at only one frequency. The width of the resonant band of frequencies centered around f_r is called the *bandwidth* of the tuned circuit.

Measurement of Bandwidth. The group of frequencies with a response 70.7 percent of maximum, or more, is generally considered the bandwidth of the tuned circuit, as shown in Fig. 27-10b. The resonant response here is increasing current for the series circuit in (a). Therefore, the bandwidth is measured between the two frequencies, f_1 and f_2, producing 70.7 percent of the maximum current at f_r.

For a parallel circuit, the resonant response is increasing impedance Z_T. Then the bandwidth is measured between the two frequencies allowing 70.7 percent of the maximum Z_T at f_r.

The bandwidth indicated on the response curve in Fig. 27-10b equals 20 kHz. This value is the difference between f_2 at 60 kHz and f_1 at 40 kHz, both with 70.7 percent response.

Compared with the maximum current of 100 mA for f_r at 50 kHz, f_1 below resonance and f_2 above resonance each allow a rise to 70.7 mA. All frequencies in this band 20 kHz wide allow 70.7 mA, or more, as the resonant response in this example.

Bandwidth Equals f_r/Q. Sharp resonance with high Q means narrow bandwidth. The lower the Q, the broader is the resonant response and the greater the bandwidth.

Also, the higher the resonant frequency, the greater is the range of frequency values included in the bandwidth for a given sharpness of resonance. Therefore, the bandwidth of a resonant circuit depends on the factors f_r and Q. The formula is

$$f_2 - f_1 = \Delta f = \frac{f_r}{Q} \qquad (27\text{-}9)$$

where Δf is the total bandwidth in the same units as the resonant frequency f_r. The bandwidth Δf can also be abbreviated BW.

For example, a series circuit resonant at 800 kHz with a Q of 100 has a bandwidth of

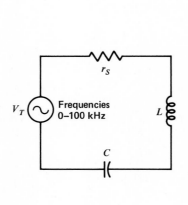

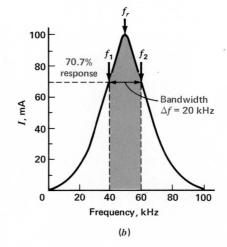

FIGURE 27-10

Bandwidth of a tuned *LC* circuit. (a) Series circuit with input of 0 to 100 kHz. (b) Response curve with bandwidth Δf equal to 20 kHz between f_1 and f_2.

$800/100$, or 8 kHz. Then the I is 70.7 percent of maximum, or more, for all frequencies for a band 8 kHz wide. This frequency band is centered around 800 kHz, from 796 to 804 kHz.

With a parallel resonant circuit having a Q higher than 10, Formula (27-9) also can be used for calculating the bandwidth of frequencies which provide 70.7 percent or more of the maximum Z_T. However, the formula cannot be used for parallel resonant circuits with low Q, as the resonance curve then becomes unsymmetrical.

High Q Means Narrow Bandwidth. The effect for different values of Q is illustrated in Fig. 27-11. Note that a higher Q for the same resonant frequency results in less bandwidth. The slope is sharper for the sides or *skirts* of the response curve, in addition to its greater amplitude.

The Edge Frequencies. Either f_1 or f_2 is separated from f_r by one-half of the total bandwidth. For the top curve in Fig. 27-11, as an example, with a Q of 80, Δf is ±5 kHz centered around 800 kHz for f_r. To determine the edge frequencies:

$$f_1 = f_r - \frac{\Delta f}{2} = 800 - 5 = 795 \text{ kHz}$$

$$f_2 = f_r + \frac{\Delta f}{2} = 800 + 5 = 805 \text{ kHz}$$

These examples assume the resonance curve is symmetrical. This is true for a high-Q parallel resonant circuit and a series resonant circuit with any Q.

Example 9. An LC circuit resonant at 2000 kHz has a Q of 100. Find the total bandwidth Δf and the edge frequencies f_1 and f_2.

Answer. $\quad \Delta f = \dfrac{f_r}{Q} = \dfrac{2000 \text{ kHz}}{100} = 20 \text{ kHz}$

$$f_1 = f_r - \frac{\Delta f}{2} = 2000 - 10$$
$$= 1990 \text{ kHz}$$
$$f_2 = f_r + \frac{\Delta f}{2} = 2000 + 10$$
$$= 2010 \text{ kHz}$$

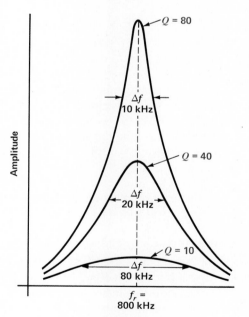

FIGURE 27-11

Higher Q provides a sharper resonant response. Amplitude is I for series resonance or Z_T for parallel resonance. Bandwidth at half-power frequencies is Δf.

Example 10. Do the same as in Example 9 for an f_r equal to 6000 kHz and the same Q of 100.

Answer. $\quad f = \dfrac{f_r}{Q} = \dfrac{6000 \text{ kHz}}{100} = 60 \text{ kHz}$

$$f_1 = 6000 - 30 = 5970 \text{ kHz}$$
$$f_2 = 6000 + 30 = 6030 \text{ kHz}$$

Notice that Δf is three times as wide for the same Q because f_r is three times higher.

Half-Power Points. It is simply for convenience in calculations that the bandwidth is defined between the two frequencies having 70.7 percent response. At each of these frequencies, the net capacitive or inductive reactance equals the resistance. Then the total impedance of the series reactance and resistance is 1.4 times greater than R. With this much more impedance, the current is reduced to $1/1.414$, or 0.707, of its maximum value.

Furthermore, the relative current or voltage value of 70.7 percent corresponds to 50 percent in power, since power is I^2R or V^2/R and the square of 0.707 equals 0.50. Therefore, the bandwidth between frequencies having 70.7 percent response in current or voltage is also the bandwidth in terms of half-power points. Formula (27-9) is derived for Δf between the points with 70.7 percent response on the resonance curve.

Measuring Bandwidth to Calculate Q. The half-power frequencies f_1 and f_2 can be determined experimentally. For series resonance, find the two frequencies at which the current is 70.7 percent of maximum I. Or, for parallel resonance, find the two frequencies that make the impedance 70.7 percent of the maximum Z_T. The following method uses the circuit in Fig. 27-9 for measuring Z_T, but with different values to determine its bandwidth and Q.

1. Tune the circuit to resonance and determine its maximum Z_T at f_r. In this example, assume that Z_T is 10,000 Ω at the resonant frequency of 200 kHz.
2. Keep the same amount of input voltage, but change its frequency slightly below f_r to determine the frequency f_1 which results in a Z_1 equal to 70.7 percent of Z_T. The required value here is $0.707 \times 10{,}000$, or 7070 Ω, for Z_1 at f_1. Assume this frequency f_1 is determined to be 195 kHz.
3. Similarly, find the frequency f_2 above f_r that results in the impedance Z_2 of 7070 Ω. Assume f_2 is 205 kHz.
4. The total bandwidth between the half-power frequencies equals $f_2 - f_1$ or $205 - 195$. Then $\Delta f = 10$ kHz.
5. Then $Q = f_r/\Delta f = 200 \text{ kHz}/10 \text{ kHz}$, or the $Q = 20$.

Practice Problems 27-6
(answers on page 569)
(a) An LC circuit with f_r of 10 MHz has a Q of 40. Calculate the half-power bandwidth.
(b) For an f_r of 500 kHz and bandwidth Δf of 10 kHz, calculate Q.

27-7
TUNING

This means obtaining resonances at different frequencies by varying either L or C. As illustrated in Fig. 27-12, the variable capacitance C can be adjusted to tune the series LC circuit to resonance at any one of the five different frequencies. Each of the voltages V_1 to V_5 indicates an ac input with a specific frequency. Which one is selected for maximum output is determined by the resonant frequency of the LC circuit.

When C is set to 424 pF, for example, the resonant frequency of the LC circuit is 500 kHz for f_{r_1}. The input voltage that has the frequency of 500 kHz then produces a resonant rise of current which results in maximum output voltage across C. At other frequencies, such as 707 kHz, the voltage output is less than the input. With C at 424 pF, therefore, the LC tuned to 500 kHz selects this frequency by providing much more voltage output compared with other frequencies.

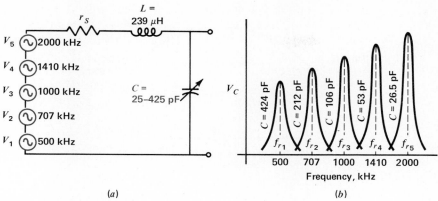

FIGURE 27-12
Tuning a series circuit. (*a*) Circuit with input voltages at different frequencies. (*b*) Resonant responses at different frequencies when *C* is varied. Relative amplitudes not to scale.

Suppose that we want maximum output for the ac input voltage that has the frequency of 707 kHz. Then *C* is set at 212 pF to make the *LC* circuit resonant at 707 kHz for f_{r_2}. Similarly, the tuned circuit can resonate at a different frequency for each input voltage. In this way, the *LC* circuit is tuned to select the desired frequency.

The variable capacitance *C* can be set at the values listed in Table 27-4 to tune the *LC* circuit to different frequencies. Only five frequencies are listed here, but any one capacitance value between 26.5 and 424 pF can tune the 239-μH coil to resonance at any frequency in the range of 500 to 2000 kHz. It should be noted that a parallel resonant circuit also can be tuned by varying *C* or *L*.

Tuning Ratio. When an *LC* circuit is tuned, the change in resonant frequency is inversely proportional to the square root of the change in *L* or *C*. Referring to Table 27-4, notice that when *C* is decreased by one-fourth, from 424 to 106 pF, the resonant frequency doubles from 500 to 1000 kHz. Or the frequency is increased by the factor $1/\sqrt{1/4}$, which equals 2.

Suppose that we want to tune through the whole frequency range of 500 to 2000 kHz. This is a tuning ratio of 4:1 for the highest frequency to the lowest frequency. Then the capacitance must be varied from 424 to 26.5 pF, a 16:1 capacitance ratio.

Radio Tuning Dial. Figure 27-13 illustrates a common application of resonant circuits in tuning a receiver to the carrier frequency of a desired station in the band. The tuning is done by the air capacitor *C*, which can be varied from

TABLE 27-4. Tuning *LC* Circuit by Varying *C*

L, μH	C, pF	f_r, kHz
239	424	500
239	212	707
239	106	1000
239	53	1410
239	26.5	2000

Chapter 27
Resonance

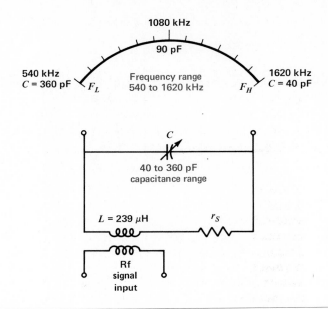

FIGURE 27-13
Application of tuning *LC* circuit through the AM radio band of 540 to 1620 kHz.

360 pF with the plates completely in mesh to 40 pF out of mesh. The fixed plates form the *stator,* while the *rotor* has the plates that move in and out.

Note that the lowest frequency F_L at 540 kHz is tuned in with the highest C at 360 pF. Resonance at the highest frequency F_H at 1620 kHz results with the lowest C at 40 pF.

The capacitance range of 40 to 360 pF tunes through the frequency range from 1620 kHz down to 540 kHz. F_L is one-third F_H because the maximum C is nine times the minimum C. The tuning dial, in kHz, usually omits the last zero to save space.

The same idea applies to tuning through the commercial FM broadcast band of 88 to 108 MHz, with smaller values of L and C. Also, television receivers are tuned to a specific broadcast channel by resonance at the desired frequencies.

Practice Problems 27-7
(*answers on page 569*)
(a) When a tuning capacitor is completely in mesh, is the station tuned in the highest or lowest frequency in the band?
(b) A tuning ratio of 2:1 in frequency requires what ratio of variable L or C?

27-8
MISTUNING

For example, suppose that a series *LC* circuit is tuned to 1000 kHz but the frequency of the input voltage is 17 kHz, completely off resonance. The circuit could provide a Q rise in output voltage for current having the frequency of 1000 kHz, but there is no input voltage and therefore no current at this frequency.

The input voltage produces current that has the frequency of 17 kHz. This frequency cannot produce a resonant rise in current, however, because the current is limited by the net reactance. When the frequency of the input voltage and the resonant frequency of the *LC* circuit are not the same, therefore, the mistuned circuit has very little output compared with the *Q* rise in voltage at resonance.

Similarly, when a parallel circuit is mistuned, it does not have a high value of impedance. Furthermore, the net reactance off resonance makes the *LC* circuit either inductive or capacitive.

Series Circuit off Resonance. When the frequency of the input voltage is lower than the resonant frequency of a series *LC* circuit, the capacitive reactance is greater than the inductive reactance. As a result, there is more voltage across the capacitive reactance than across the inductive reactance. The series *LC* circuit is capacitive below resonance, therefore, with capacitive current leading the generator voltage.

Above the resonant frequency, the inductive reactance is greater than the capacitive reactance. As a result, the circuit is inductive above resonance, with inductive current that lags the generator voltage. In both cases, there is much less output voltage than at resonance.

Parallel Circuit off Resonance. With a parallel *LC* circuit, the smaller amount of inductive reactance below resonance results in more inductive branch current than capacitive branch current. The net line current is inductive, therefore, making the parallel *LC* circuit inductive below resonance, as the line current lags the generator voltage.

Above the resonant frequency, the net line current is capacitive because of the higher value of capacitive branch current. Then the parallel *LC* circuit is capacitive, with line current leading the generator voltage. In both cases the total impedance of the parallel circuit is much less than the maximum impedance at resonance. Note that the capacitive and inductive effects off resonance are opposite for series and parallel *LC* circuits.

Practice Problems 27-8
(answers on page 569)
(a) Is a series resonant circuit inductive or capacitive below resonance?
(b) Is a parallel resonant circuit inductive or capacitive below resonance?

27-9 ANALYSIS OF PARALLEL RESONANT CIRCUITS

Parallel resonance is more complex than series resonance because the reactive branch currents are not exactly equal when X_L equals X_C. The reason is that the coil has its series resistance r_S in the X_L branch, while the capacitor has only X_C in its branch.

For high-*Q* circuits, we consider r_S to be negligible. In low-*Q* circuits, however, the inductive branch must be analyzed as a complex impedance with X_L and r_S in series. This impedance is in parallel with X_C, as shown in Fig. 27-14. The total impedance Z_T can then be calculated by using complex numbers, as explained in Chap. 26.

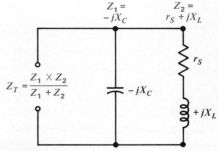

FIGURE 27-14

General method of calculating Z_T for a parallel resonant circuit as $Z_1 \times Z_2/(Z_1 + Z_2)$ with complex numbers.

High-Q Circuit. We can apply the general method in Fig. 27-14 to the parallel resonant circuit shown before in Fig. 27-6 to see if Z_T is 225,000 Ω. In this example, X_L and X_C are 1500 Ω and r_S is 10 Ω. The calculations are

$$Z_T = \frac{Z_1 \times Z_2}{Z_1 + Z_2} = \frac{-j1500 \times (j1500 + 10)}{-j1500 + j1500 + 10}$$
$$= \frac{-j^2 2.25 \times 10^6 - j15,000}{10}$$
$$= -j^2 2.25 \times 10^5 - j1500$$
$$= 225,000 - j1500$$
$$Z_T = 225,000 \underline{/0°} \; \Omega$$

Note that $-j^2$ is $+1$. Also, the reactive $j1500$ Ω is negligible compared with the resistive 225,000 Ω. This answer for Z_T is the same as $Q \times X_L$, or $150 \times 15,000$, because of the high Q with negligibly small r_S.

Low-Q Circuit. We can consider a Q less than 10 as low. For the same circuit in Fig. 27-6, if r_S is 300 Ω with an X_L of 1500 Ω, the Q will be 1500/300, which equals 5. For this case of appreciable r_S, the branch currents cannot be equal when X_L and X_C are equal because then the inductive branch will have more impedance and less current.

With a low-Q circuit Z_T must be calculated in terms of the branch impedances. For this example, the calculations are simpler with all impedances stated in kilohms:

$$Z_T = \frac{Z_1 \times Z_2}{Z_1 + Z_2} = \frac{-j1.5 \times (j1.5 + 0.3)}{-j1.5 + j1.5 + 0.3}$$
$$= \frac{-j^2 2.25 - j0.45}{0.3} = -j^2 7.5 - j1.5$$
$$= 7.5 - j1.5 = 7.65 \underline{/-11.3°} \; \text{k}\Omega$$
$$Z_T = 7650 \underline{/-11.3°} \; \Omega$$

The phase angle θ is not zero because the reactive branch currents are unequal, even though X_L and X_C are equal. The appreciable value of r_S in the X_L branch makes this branch current smaller than I_C in the X_C branch.

Criteria for Parallel Resonance. The frequency f_r that makes $X_L = X_C$ is always $1/(2\pi\sqrt{LC})$. However, for low-Q circuits f_r does not necessarily provide the desired resonance effect. The three main criteria for parallel resonance are

1. Zero phase angle and unity power factor.
2. Maximum impedance Z_T and minimum line current.
3. $X_L = X_C$. This is the resonance at $f_r = 1/(2\pi\sqrt{LC})$.

These three effects do not occur at the same frequency in parallel circuits that have a low Q. The condition for unity power factor is often called *antiresonance* in a parallel LC circuit to distinguish it from the case of equal X_L and X_C.

It should be noted that when Q is 10 or higher, though, the parallel branch currents are practically equal when $X_L = X_C$. Then at $f_r = 1/(2\pi\sqrt{LC})$, the line current is minimum with zero phase angle, and the impedance is maximum.

For a series resonant circuit there are no parallel branches to consider. Therefore, the current is maximum at exactly f_r, whether the Q is high or low.

Practice Problems 27-9
(answers on page 569)
(a) Is the Q of 8 a high or low value?
(b) With this Q, will the I_L be more or less than I_C in the parallel branches when $X_L = X_C$?

27-10
DAMPING OF PARALLEL RESONANT CIRCUITS

In Fig. 27-15a, the shunt R_P across L and C is a damping resistance because it lowers the Q of the tuned circuit. R_P may represent the resistance of the external source driving the parallel resonant circuit, or R_P can be an actual resistor added for lower Q and greater bandwidth. Using the parallel R_P to reduce Q is better than increasing the series resistance r_S because the resonant response is more symmetrical with shunt damping.

The effect of varying the parallel R_P is opposite from the series r_S. A lower value of R_P lowers the Q and reduces the sharpness of resonance. Remember that less resistance in a parallel branch allows more current. This resistive branch current cannot be canceled at resonance by the reactive currents. Therefore, the resonant dip to minimum line current is less sharp with more resistive line current. Specifically, when Q is determined by parallel resistance

$$Q = \frac{R_P}{X_L} \qquad (27\text{-}10)$$

This relation with shunt R_P is the reciprocal of the Q formula with series r_S. Reducing R_P decreases Q, but reducing r_S increases Q. The damping can be done by series r_S, parallel R_P, or both.

Parallel R_P Without r_S. In Fig. 27-15a, Q is determined only by the R_P, as no series r_S is

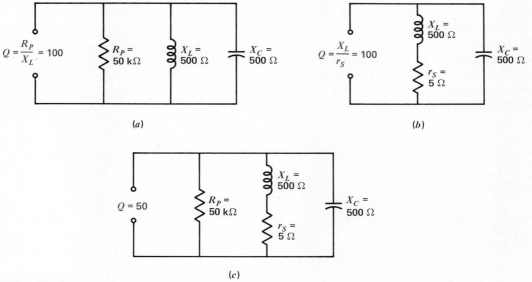

FIGURE 27-15

Q of parallel resonant circuit in terms of coil resistance r_S and parallel damping resistor R_P. (a) Parallel R_P but negligible r_S. (b) Series r_S but no R_P branch. (c) Both R_P and r_S. See Formula (27-11) for calculating Q.

shown. We can consider that r_S is zero or very small. Then the Q of the coil is infinite or high enough to be greater than the damped Q of the tuned circuit, by a factor of 10 or more. The Q of the damped resonant circuit here is $R_P/X_L = 50{,}000/500 = 100$.

Series r_S Without R_P. In Fig. 27-15b, Q is determined only by the coil resistance r_S, as no shunt damping resistance is used. Then $Q = X_L/r_S = 500/5 = 100$. This value is the Q of the coil, which is also the Q of the parallel resonant circuit without shunt damping.

Conversion of r_S or R_P. Q is 100 for the circuits in both (a) and (b) because the 50,000-Ω R_P is equivalent to the 5-Ω r_S as a damping resistance. One value can be converted to the other. The $r_S = X_L^2/R_P$. Also, the $R_P = X_L^2/r_S$. In this example, r_S is $250{,}000/50{,}000 = 5\ \Omega$, or R_P is $250{,}000/5 = 50{,}000\ \Omega$.

Damping with Both r_S and R_P. Figure 27-15c shows the general case of damping where both r_S and R_P must be considered. Then the Q of the circuit can be calculated as

$$Q = \frac{X_L}{r_S + X_L^2/R_P} \qquad (27\text{-}11)$$

For the values in Fig. 27-15c

$$Q = \frac{500}{5 + 250{,}000/50{,}000} = \frac{500}{5 + 5} = \frac{500}{10}$$
$$Q = 50$$

The Q is lower here compared with (a) or (b) because this circuit has both series and shunt damping.

It should be noted that for an r_S of zero, Formula (27-11) can be inverted and simplified to $Q = R_P/X_L$. This is the same as Formula (27-10) for shunt damping alone.

For the opposite case of R_P being infinite for an open parallel path, Formula (27-11) reduces to X_L/r_S. This is the same as Formula (27-5) without shunt damping.

Practice Problems 27-10
(*answers on page 569*)
(a) A parallel resonant circuit has an X_L of $1000\ \Omega$ and an r_S of $20\ \Omega$, without any shunt damping. Calculate Q.
(b) A parallel resonant circuit has an X_L of $1000\ \Omega$, negligible r_S, and shunt R_P of $50\ k\Omega$. Calculate Q.
(c) How much is Z_T at f_r for the circuits in both (a) and (b)?

27-11
CHOOSING L AND C FOR A RESONANT CIRCUIT

The following example illustrates how resonance is really just an application of X_L and X_C. Suppose that we have the problem of determining the inductance and capacitance for a circuit to be resonant at 159 kHz. First, we need a known value for either L or C, in order to calculate the other. Which one to choose depends on the application. In some cases, particularly at very high frequencies, C must be the minimum possible value, which might be about 10 pF. At medium frequencies, though, we can choose L for the general case where an X_L of $1000\ \Omega$ is desirable and can be obtained. Then the inductance of the required L, equal to $X_L/2\pi f$, is 0.001 H or 1 mH, for the inductive reactance of $1000\ \Omega$.

For resonance at 159 kHz with a 1-mH L, the required C is $0.001\ \mu F$ or 1000 pF. This value of C can be calculated for an X_C of $1000\ \Omega$, equal to X_L at the f_r of 159 kHz, or from Formula (27-3). In either case, if you substitute 1×10^{-9} F for C and 1×10^{-3} H for L in the resonant frequency formula, f_r will be 159 kHz.

This combination is resonant at 159 kHz whether L and C are in series or parallel. In series, the resonant effect is to produce maximum current and maximum voltage across L or C at 159 kHz. In parallel, the resonant effect at 159 kHz is minimum line current and maximum impedance across the generator.

If we assume the 1-mH coil used for L has an internal resistance of 20 Ω, the Q of the coil is 50. This value is also the Q of the series resonant circuit. If there is no shunt damping resistance across the parallel LC circuit, its Q is also 50. With a Q of 50 the bandwidth of the resonant circuit is 159 kHz/50, which equals 3.18 kHz for Δf.

Practice Problems 27-11
(answers on page 569)
(a) What is f_r for 1000 pF of C and 1 mH of L?
(b) What is f_r for 250 pF of C and 1 mH of L?

Summary

Series and parallel resonance are compared in Table 27-5. The main difference is that series resonance produces maximum current and very low impedance at f_r, but with parallel resonance the line current is minimum to provide a very high impedance. Remember that these formulas for parallel resonance are very close approximations that can be used for circuits with a Q higher than 10. For series resonance, the formulas apply whether the Q is high or low.

TABLE 27-5. Comparison of Series and Parallel Resonance

SERIES RESONANCE	PARALLEL RESONANCE (HIGH Q)
$f_r = \dfrac{1}{2\pi \sqrt{LC}}$	$f_r = \dfrac{1}{2\pi \sqrt{LC}}$
I maximum at f_r with θ of 0°	I_T minimum at f_r with θ of 0°
Impedance Z minimum at f_r	Impedance Z maximum at f_r
$Q = X_L/r_S$, or	$Q = X_L/r_S$, or
$Q = V_{out}/V_{in}$	$Q = Z_{max}/X_L$
Q rise in voltage $= Q \times V_{gen}$	Q rise in impedance $= Q \times X_L$
Bandwidth $\Delta f = f_r/Q$	Bandwidth $\Delta f = f_r/Q$
Capacitive below f_r, but inductive above f_r	Inductive below f_r, but capacitive above f_r
Needs low-resistance source for low r_S, high Q, and sharp tuning	Needs high-resistance source for high R_P, high Q, and sharp tuning
Source is inside LC circuit	Source is outside LC circuit

Self-Examination (Answers at back of book.)

Choose (a), (b), (c), or (d).

1. For a series or parallel *LC* circuit, resonance occurs when (a) X_L is ten times X_C or more; (b) X_C is ten times X_L or more; (c) $X_L = X_C$; (d) the phase angle of the circuit is 90°.
2. When either *L* or *C* is increased, the resonant frequency of the *LC* circuit (a) increases; (b) decreases; (c) remains the same; (d) is determined by the shunt resistance.
3. The resonant frequency of an *LC* circuit is 1000 kHz. If *L* is doubled but *C* is reduced to one-eighth of its original value, the resonant frequency then is (a) 250 kHz; (b) 500 kHz; (c) 1000 kHz; (d) 2000 kHz.
4. A coil has a 1000-Ω X_L and a 5-Ω internal resistance. Its *Q* equals (a) 0.005; (b) 5; (c) 200; (d) 1000.
5. In a parallel *LC* circuit, at the resonant frequency, the (a) line current is maximum; (b) inductive branch current is minimum; (c) total impedance is minimum; (d) total impedance is maximum.
6. At resonance, the phase angle equals (a) 0°; (b) 90°; (c) 180°; (d) 270°.
7. In a series *LC* circuit, at the resonant frequency, the (a) current is minimum; (b) voltage across *C* is minimum; (c) impedance is maximum; (d) current is maximum.
8. A series *LC* circuit has a *Q* of 100 at resonance. When 5 mV is applied at the resonant frequency, the voltage across *C* equals (a) 5 mV; (b) 20 mV; (c) 100 mV; (d) 500 mV.
9. An *LC* circuit resonant at 1000 kHz has a *Q* of 100. The bandwidth between half-power points equals (a) 10 kHz between 995 and 1005 kHz; (b) 10 kHz between 1000 and 1010 kHz; (c) 5 kHz between 995 and 1000 kHz; (d) 200 kHz between 900 and 1100 kHz.
10. In a low-*Q* parallel resonant circuit, when $X_L = X_C$ (a) I_L equals I_C; (b) I_L is less than I_C; (c) I_L is more than I_C; (d) the phase angle is 0°.

Essay Questions

1. (a) State two characteristics of series resonance. (b) With a microammeter measuring current in the series *LC* circuit of Fig. 27-2, describe the meter readings for the different frequencies from 600 to 1400 kHz.

2. (a) State two characteristics of parallel resonance. (b) With a microammeter measuring current in the main line for the parallel LC circuit in Fig. 27-6a, describe the meter readings for the different frequencies from 600 to 1400 kHz.
3. State the Q formula for the following LC circuits: (a) series resonant; (b) parallel resonant, with series resistance r_S in the inductive branch; (c) parallel resonant, with zero series resistance but shunt R_P.
4. Explain briefly why a parallel LC circuit is inductive but a series LC circuit is capacitive below f_r.
5. What is the effect on Q and bandwidth of a parallel resonant circuit if its shunt damping resistance is decreased from 50,000 to 10,000 Ω?
6. Describe briefly how you would use an ac meter to measure the bandwidth of a series resonant circuit for calculating its Q.
7. Why is a low-resistance generator good for a high Q in series resonance, while a high-resistance generator is needed for a high Q in parallel resonance?
8. Referring to Fig. 27-13, why is it that the middle frequency of 1080 kHz does not correspond to the middle capacitance value of 200 pF?
9. (a) Give three criteria for parallel resonance. (b) Why is the antiresonant frequency f_a different from f_r with a low-Q circuit? (c) Why are they the same for a high-Q circuit?
10. Show how Formula (27-11) reduces to R_P/X_L when r_S is zero.
11. (a) Specify the edge frequencies f_1 and f_2 for each of the three response curves in Fig. 27-11. (b) Why does lower Q allow more bandwidth?
12. (a) Why does maximum Z_T for a parallel resonant circuit correspond to minimum line current? (b) Why does zero phase angle for a resonant circuit correspond to unity power factor?

Problems (Answers to odd-numbered problems at back of book.)

1. Find f_r for a series circuit with a 10-μF C, a 16-H L, and a 5-Ω r_S.
2. Find f_r for a parallel circuit with a 2-μF C, a 2-H L, and a 5-Ω r_S.
3. Find f_r for L of 80 μH and C of 120 pF.
4. The f_r is 1.6 MHz with L of 40 μH. Find C.
5. In a series resonant circuit, X_L is 1500 Ω and the internal coil resistance is 15 Ω. At the resonant frequency: (a) How much is the Q of the circuit? (b) How much is X_C? (c) With a generator voltage of 15 mV, how much is I? (d) How much is the voltage across X_C?

6. In a parallel resonant circuit, X_L is 1200 Ω, the resistance of the coil is practically zero, but there is a 36,000-Ω resistance across the LC circuit. At the resonant frequency: (a) How much is the Q of the circuit? (b) How much is X_C? (c) With a generator voltage of 12 mV and zero resistance in the main line, how much is the main-line current? (d) How much is the voltage across L, C, and R? (e) How much is the impedance across the main line?
7. What value of L is necessary with a C of 100 pF for series resonance at 1 MHz? At 4 MHz?
8. Calculate the C needed with a 350-μH L for a 200-kHz f_r.
9. Calculate the lowest and highest values of C needed with 0.1-μH L to tune through the commercial FM broadcast band of 88 to 108 MHz.
10. (a) At what frequency will a 200-μH coil with a 20-Ω r_S have a 1000-Ω X_L? (b) What size C is needed for a 1000-Ω X_C at this frequency? (c) What is f_r for this LC combination? (d) How much is the Q of the coil?
11. Draw the schematic diagram of a parallel resonant circuit with the L, C, and r_S of Prob. 10. Let the applied voltage be 5 V. Calculate the values of main-line current I_T, Z_T, and θ at: (a) the resonant frequency f_r; (b) 0.1 MHz below f_r; (c) 0.1 MHz above f_r.
12. For the series resonant circuit in Fig. 27-16: (a) How much is X_L? (b) Calculate L in millihenrys and C in microfarads. (c) Calculate the Q and bandwidth. (d) Calculate V_C and V_L. (e) If L is doubled and C is one-half, what is f_r? Calculate Q and bandwidth for this case. (f) If the original values of both L and C are doubled, calculate f_r.
13. Redraw Fig. 27-16 as a parallel resonant circuit with the original values and r_S in series with the coil. (a) What are the Q and bandwidth? (b) Calculate Z_T at f_r.
14. For the same circuit as Prob. 13, let r_S increase to 500 Ω. (a) What is the Q now? (b) Calculate Z_T at f_r, one-half f_r, and twice f_r. (c) What value of R_P would be used for the same Q if r_S were zero?

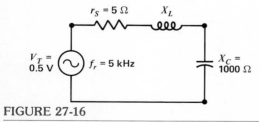

FIGURE 27-16

For Prob. 12.

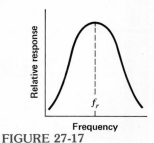

FIGURE 27-17
For Prob. 18.

15. For the series circuit in Fig. 27-16: (a) Tabulate the I and θ values every kilohertz from 2 to 9 kHz. (b) Draw the response curve showing I vs. frequency.
16. For the values in Fig. 27-16 connected for parallel resonance: (a) Calculate Z_T and θ every kilohertz from 2 to 9 kHz. (b) Draw the response curve showing Z_T vs. frequency.
17. A series resonant circuit produces 240 mV across L with a 2-mV input. (a) How much is the Q of the coil? (b) Calculate r_S, if L is 5 mH and f_r is 0.3 MHz. (c) How much C is needed for this f_r?
18. Refer to the relative response curve in Fig. 27-17. (a) For $f_r = 10.7$ MHz and $Q = 50$, determine the bandwidth Δf and the edge frequencies f_1 and f_2. (b) Do the same for a lower f_r of 456 kHz. Q is still 50.

Answers to Practice Problems

27-1	(a) 1000 kHz	27-6	(b) $Q = 50$
	(b) 1000 kHz	27-7	(a) Lowest
27-2	(a) F		(b) 1:4
	(b) T	27-8	(a) Capacitive
	(c) T		(b) Inductive
27-3	(a) F	27-9	(a) Low
	(b) T		(b) Less
	(c) T	27-10	(a) $Q = 50$
27-4	(a) Less		(b) $Q = 50$
	(b) 200 μH		(c) $Z_T = 50$ kΩ
27-5	(a) $Q = 100$	27-11	(a) $f_r = 159$ kHz
	(b) $Z_T = 25$ kΩ		(b) $f_r = 318$ kHz
27-6	(a) $\Delta f = 0.25$ MHz		

Filters

Chapter 28

A filter separates different components that are mixed together. For instance, a mechanical filter can separate particles from liquid, or small particles from large particles. An electrical filter can separate different frequency components.

Generally, inductors and capacitors are used for filtering because of their opposite frequency characteristics. X_L increases but X_C decreases with higher frequencies. In addition, their filtering action depends on whether L and C are in series or parallel with the load. The most common filtering applications are separating audio from radio frequencies, or vice versa, and separating ac variations from the average dc level. More details are explained in the following topics:

28-1 Examples of Filtering
28-2 Direct Current Combined with Alternating Current
28-3 Transformer Coupling
28-4 Capacitive Coupling
28-5 Bypass Capacitors
28-6 Filter Circuits
28-7 Low-pass Filters
28-8 High-pass Filters
28-9 Resonant Filters
28-10 Interference Filters

28-1
EXAMPLES OF FILTERING

Electronic circuits often have currents of different frequencies corresponding to voltages of different frequencies. The reason is that a source produces current with the same frequency as the applied voltage. As examples, the ac signal input to an audio circuit can have high and low audio frequencies; an rf circuit can have a wide range of radio frequencies in its input; the audio detector in a radio has both radio frequencies and audio frequencies in the output. Finally, the rectifier in a power supply produces dc output with an ac ripple superimposed on the average dc level.

In such applications where the current has different frequency components, it is usually necessary either to favor or to reject one frequency or a band of frequencies. Then an electrical filter is used to separate higher or lower frequencies.

The electrical filter can pass the higher-frequency component to the load resistance, which is the case of a high-pass filter, or a low-pass filter can be used to favor the lower frequencies. In Fig. 28-1a, the high-pass filter allows 10 kHz to produce output, while rejecting or attenuating the lower frequency of 100 Hz. In (b), the filtering action is reversed to pass the lower frequency of 100 Hz, while attenuating 10 kHz. These examples are for high and low audio frequencies.

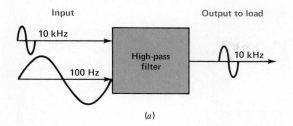

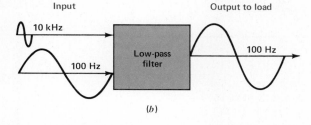

FIGURE 28-1

Function of electrical filters. (*a*) High-pass filter couples higher frequencies to the load. (*b*) Low-pass filter couples lower frequencies to the load.

For the case of audio mixed with radio frequencies, a low-pass filter allows the audio frequencies in the output. Or, a high-pass filter allows the radio frequencies to be passed to the load.

Practice Problems 28-1
(*answers on page 592*)
A high-pass filter will pass which of the following:
(*a*) 10 or 500 kHz.
(*b*) 60 Hz or a steady dc level.

28-2
DIRECT CURRENT COMBINED WITH ALTERNATING CURRENT

Current that varies in amplitude but does not reverse in polarity is considered *pulsating* or *fluctuating* direct current. It is not a steady direct current because its value fluctuates. However, it is not alternating current because the polarity remains the same, either positive or negative. The same idea applies to voltages.

Figure 28-2 illustrates how a circuit can have pulsating direct current or voltage. Here, the steady dc voltage of the battery V_B is in series with the ac voltage V_A. Since the two series generators add, the voltage across R_L is the sum of the two applied voltages, as shown by the waveshape of v_R in (*b*).

If values are taken at opposite peaks of the ac variation, when V_A is at +10 V, it adds to the +20 V of the battery to provide +30 V across R_L; when the ac voltage is −10 V, it bucks the battery voltage of +20 V to provide +10 V across R_L. When the ac voltage is at zero, the voltage across R_L equals the battery voltage of +20 V.

The combined voltage v_R then consists of the ac variations fluctuating above and below the battery voltage as the axis, instead of the zero axis for ac voltage. The result is a pulsating

Chapter 28
Filters

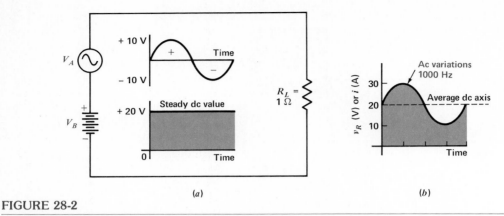

FIGURE 28-2
Pulsating direct current and voltage. (*a*) Circuit. (*b*) Graph of voltage across R_L equal to V_B plus V_A. Frequency of ac voltage is 1000 Hz.

dc voltage, since it is fluctuating but always has positive polarity with respect to zero.

The pulsating direct current i through R_L has the same waveform, fluctuating above and below the steady dc level of 20 A. The i and v values are the same because R is 1 Ω.

Another example is illustrated in Fig. 28-3. If the 100-Ω R_L is connected across the 120-V 60-Hz ac power line in (*a*), the current in R_L will be V/R_L. This alternating current is a sine wave, with an rms value of 1.2 A.

Also, if you connect the same R_L across the 200-V dc source in (*b*), instead of using the ac source, the steady direct current in R_L will be $200/100$, or 2 A. The battery source voltage and its current are considered steady dc values because there are no variations.

However, suppose that the ac source V_A

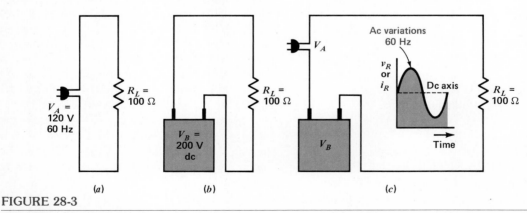

FIGURE 28-3
An example of providing fluctuating dc voltage across R_L. (*a*) Ac source alone. (*b*) Dc source alone. (*c*) Ac source in series with dc source.

and dc source V_B are connected in series with R_L, as in (c). What will happen to the current and voltage for R_L? Will V_A or V_B supply the current? The answer is that both sources will. Each voltage source produces current as though the other were not there, assuming the sources have negligibly small internal impedance. The result then is the fluctuating dc voltage or current shown, with the ac variations of V_A superimposed on the average dc level of V_B.

DC and AC Components. The pulsating dc voltage v_R in Fig. 28-2 is just the original ac voltage V_A with its axis shifted to a dc level by the battery voltage V_B. In effect, a dc component has been inserted into the ac variations.

Referring back to Fig. 28-2, if you measure across R_L with a dc voltmeter, it will read the dc level of 20 V. An ac voltmeter will read the rms value of the variations, which is 7.07 V.

It is convenient, therefore, to consider the pulsating or fluctuating voltage and current in two parts. One is the steady dc component, which is the axis or average level of the variations; the other is the ac component, consisting of the variations above and below the dc axis. Here the dc level for V_T is +20 V, while the ac component equals 10 V peak or 7.07 V rms value.

It should be noted that with respect to the dc level the fluctuations represent alternating voltage or current that actually reverses in polarity. For example, the change of v_R from +20 to +10 V is just a decrease in positive voltage compared with zero. However, compared with the dc level of +20 V, the value of +10 V is 10 V more negative than the axis.

Typical Examples of DC Level with AC Component. As a common application, electron tubes and transistors always have fluctuating dc voltage or current when used for amplifying an ac signal. The tube or transistor amplifier needs steady dc voltages to operate. The signal input is an ac variation, usually with a dc axis to establish the desired operating level. The amplified output is also an ac variation superimposed on a dc supply voltage that supplies the required power output. Therefore, the input and output circuits have fluctuating dc voltage.

The examples in Fig. 28-4 illustrate two possibilities, in terms of polarities with respect to chassis ground. In (a), the waveform is always positive, as in the previous examples. This example could apply to collector voltage on an NPN transistor amplifier. Note the specific values. The average dc axis is the steady dc level. The positive peak equals the dc level plus the

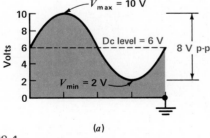
(a)

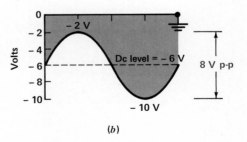
(b)

FIGURE 28-4

Typical examples of dc voltage with ac component. (a) Positive dc values. (b) Negative dc values.

peak ac value. The minimum point equals the dc level minus the peak ac value. The peak-to-peak value of the ac component and its rms value are the same as for the ac signal alone. However, it is better to subtract the minimum from the maximum for the peak-to-peak value, in case the waveform is unsymmetrical.

In (b) all the values are negative. This example could apply to control-grid voltage in a vacuum-tube amplifier. Notice that here the positive peak of the ac component subtracts from the dc level because of the opposite polarities. Now the negative peak adds to the negative dc level to provide a maximum point of negative voltage.

Separating the AC Component. In many applications, the circuit has pulsating dc voltage, but only the ac component is desired. Then the ac component can be passed to the load, while the steady dc component is blocked, either with transformer coupling or with capacitive coupling. A transformer with a separate secondary winding isolates or blocks steady direct current in the primary. A capacitor isolates or blocks a steady dc voltage.

Practice Problems 28-2
(answers on page 592)
For the fluctuating dc waveform in Fig. 28-4a, specify the following voltages:
(a) Average dc level.
(b) Maximum and minimum values.
(c) Peak-to-peak of ac component.
(d) Peak and rms of ac component.

28-3
TRANSFORMER COUPLING

Remember that a transformer produces induced secondary voltage just for variations in primary current. With pulsating direct current in the primary, the secondary has output voltage only for the ac variations, therefore. The dc component in the primary has no effect in the secondary.

In Fig. 28-5, the pulsating dc voltage in the primary produces pulsating primary current. The dc axis corresponds to a steady value of primary current that has a constant magnetic field, but only when the field changes can secondary voltage be induced. Therefore, only the fluctuations in the primary can produce output in the secondary. Since there is no output for

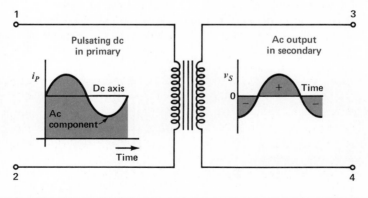

FIGURE 28-5
Transformer coupling blocks dc component. With pulsating direct current in primary, the output voltage in secondary has ac component only.

the steady primary current, this dc level corresponds to the zero level for the ac output in the secondary.

When the primary current increases above the steady level, this increase produces one polarity for the secondary voltage as the field expands; when the primary current decreases below the steady level, the secondary voltage has reverse polarity as the field contracts. The result in the secondary is an ac variation having opposite polarities with respect to the zero level.

The phase of the ac secondary voltage may be as shown or 180° opposite, depending on the connections and direction of the windings. Also, the ac secondary output may be more or less than the ac component in the primary, depending on the turns ratio. This ability to isolate the steady dc component in the primary while providing ac output in the secondary applies to all transformers with a separate secondary winding, whether iron-core or air-core.

Practice Problems 28-3
(*answers on page 592*)
(a) Is transformer coupling an example of a high-pass or low-pass filter?
(b) In Fig. 28-5, what is the level of v_S for the average dc level of i_P?

28-4
CAPACITIVE COUPLING

This method is probably the most common type of coupling in amplifier circuits. The coupling means connecting the output of one circuit to the input of the next. The requirements are to include all frequencies in the desired signal, while rejecting undesired components. Usually, the dc component must be blocked from the input to ac amplifiers.

In Fig. 28-6, the pulsating dc voltage across input terminals 1 and 2 is applied to the RC coupling circuit. C_C will charge to the steady dc level, which is the average charging voltage. The steady dc component is blocked, therefore, since it cannot produce voltage across R. However, the ac component is developed across R, between the output terminals 3 and 4. Note that the zero axis of the ac voltage output corresponds to the average level of the pulsating dc voltage input.

The DC Component Across C. The voltage across C_C is the steady dc component of the input voltage because the variations of the ac component are symmetrical above and below the average level. Furthermore, the series resistance is the same for charge and discharge. As a result, any increase in charging voltage above the average level is counteracted by an equal discharge below the average.

In Fig. 28-6, for example, when v_{in} increases from 20 to 30 V, this effect on charging C_C is nullified by the discharge when v_{in} decreases from 20 to 10 V. At all times, however, v_{in} has a positive value that charges C_C in the polarity shown.

The net result is that only the average level is effective in charging C_C, since the variations from the axis neutralize each other. After a period of time, depending on the RC time constant, C_C will charge to the average value of the pulsating dc voltage applied, which is 20 V here.

The AC Component Across R. Although C_C is charged to the average dc level, when the pulsating input voltage varies above and below this level, the charge and discharge current produces IR voltage corresponding to the fluctuations of the input. When v_{in} increases above the average level, C_C takes on charge, producing charging current through R. Even though the charging current may be too small to affect the voltage across C_C appreciably, the IR drop across a large value of resistance can be practically equal to the ac component of the input

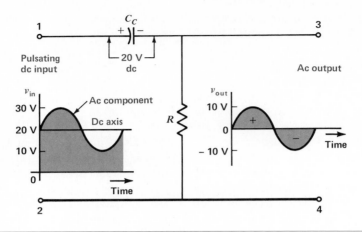

FIGURE 28-6

RC coupling blocks dc component. With pulsating dc voltage applied, output voltage across *R* has ac component only.

voltage. In summary, a long *RC* time constant is needed for good coupling. (See Fig. 24-8 on p. 485.)

If the polarity is considered, in Fig. 28-6, the charging current produced for an increase of v_{in} produces electron flow from the low side of *R* to the top, adding electrons to the negative side of C_C. The voltage at the top of *R* is then positive with respect to the line below.

When v_{in} decreases below the average level, *C* loses charge. The discharge current then is in the opposite direction through *R*. The result is negative polarity for the ac voltage output across *R*.

When the input voltage is at its average level, there is no charge or discharge current, resulting in zero voltage across *R*. The zero level in the ac voltage across *R* corresponds to the average level of the pulsating dc voltage applied to the *RC* circuit.

The end result is that with positive pulsating dc voltage applied, the values above the average produce the positive half-cycle of the ac voltage across *R*; the values below the average produce the negative half-cycle. Only this ac voltage across *R* is coupled to the next circuit.

It is important to note that there is practically no phase shift. This rule applies to all *RC* coupling circuits, since *R* must be ten or more times X_C. Then the reactance is negligible compared with the series resistance, and the phase angle of less than 5.7° is practically zero.

Voltages Around the *RC* Coupling Circuit. If you measure the pulsating dc voltage input voltage across points 1 and 2 in Fig. 28-6 with a dc voltmeter, it will read the average level of 20 V. A voltmeter that reads only ac values across these same points will read the fluctuating ac component, equal to 7 V rms.

Across points 1 and 3, a dc voltmeter reads the steady dc value of 20 V across C_C. An ac voltmeter across 1 and 3 reads zero.

However, an ac voltmeter across the output *R* between points 3 and 4 will read the ac voltage of 7 V rms; a dc voltmeter across *R* reads zero.

Typical Coupling Capacitors. Common values of rf and of coupling capacitors for different sizes of series *R* are listed in Table 28-1. In all cases the coupling capacitor blocks the steady

TABLE 28-1. Typical AF and RF Coupling Capacitors*

FREQUENCY	VALUES OF C_C			FREQUENCY BAND
	$R = 1.6$ kΩ	$R = 16$ kΩ	$R = 160$ kΩ	
100 Hz	10 μF	1 μF	0.1 μF	Low af
1000 Hz	1 μF	0.1 μf	0.01 μF	Medium af
10 kHz	0.1 μF	0.01 μF	0.001 μF	High af
100 kHz	0.01 μF	0.001 μF	100 pF	Low rf
1 MHz	0.001 μF	100 pF	10 pF	Medium rf
10 MHz	100 pF	10 pF	1 pF	High rf
100 MHz	10 pF	1 pF	0.1 pF	Vhf

*For coupling circuit in Fig. 28-6; $X_{C_C} = \frac{1}{10} R$.

dc component of the input voltage, while the ac component is passed to the resistance.

The size of C_C required depends on the frequency of the ac component. At each frequency, the values of capacitance in the horizontal row have an X_C equal to one-tenth the resistance value for each column. Typical audio coupling capacitors, then, are about 0.1 to 10 μF, depending on the lowest audio frequency to be coupled and the size of the series resistance. Similarly, typical rf coupling capacitors are about 1 to 100 pF.

Values of C_C more than about 1 μF are usually electrolytic capacitors, which must be connected in the correct polarity. These can be very small, many being $\frac{1}{2}$ in long, with a low voltage rating of 3 to 25 V for transistor circuits. Also, the leakage current of electrolytic capacitors is not a serious problem in this application because of the low voltage and small series resistance for transistor coupling circuits.

Practice Problems 28-4
(answers on page 592)

(a) In Fig. 28-6, what is the voltage level for the v_{out} across R corresponding to the average dc level of v_{in}?
(b) Which of the following is a typical audio coupling capacitor with a 1-kΩ R: 1 pF, 0.001 μF, or 5 μF?

28-5
BYPASS CAPACITORS

A bypass is a path around a component. In circuits, the bypass is a parallel or shunt path. Capacitors are often used in parallel with resistance, to bypass the ac component of a pulsating dc voltage. The result, then, is a steady dc voltage across the RC parallel combination, if the bypass capacitance is large enough to have little reactance for the lowest frequency of the ac variations.

As illustrated in Fig. 28-7, the capacitance C_1 in parallel with R_1 is an ac bypass capacitor for R_1. For any frequency at which X_{C_1} is one-tenth of R_1, or less, the ac component is bypassed around R_1 through the low reactance in the shunt path. The result is practically zero ac voltage across the bypass capacitor because of its low reactance.

Since the voltage is the same across R_1 and C_1 because they are in parallel, there is also no ac voltage across R_1 for the frequency at which C_1 is a bypass capacitor. We can say that R is bypassed for the frequency at which X_C is one-tenth of R. The bypassing also applies to higher

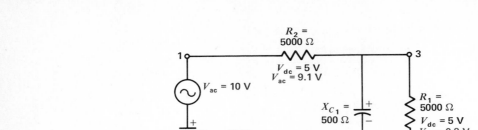

FIGURE 28-7
Bypass capacitor C_1 short-circuits R_1 for ac component of pulsating dc input voltage.

frequencies where X_C is less than one-tenth of R. Then the ac voltage across the bypass capacitor is even closer to zero because of its lower reactance.

Bypassing the AC Component of a Pulsating DC Voltage. The voltages in Fig. 28-7 are calculated by considering the effect of C_1 separately for V_{dc} and for V_{ac}. For direct current, C_1 is practically an open circuit. Then its reactance is so high compared with the 5000-Ω R_1 that X_{C_1} can be ignored as a parallel branch. Therefore, R_1 can be considered as a voltage divider in series with R_2. Since R_1 and R_2 are equal, each has 5 V, equal to one-half V_{dc}. Although this dc voltage division depends on R_1 and R_2, the dc voltage across C_1 is the same 5 V as across its parallel R_1.

For the ac component of the applied voltage, however, the bypass capacitor has very low reactance. In fact, X_{C_1} must be one-tenth of R_1, or less. Then the 5000-Ω R_1 is so high compared with the low value of X_{C_1} that R_1 can be ignored as a parallel branch. Therefore, the 500-Ω X_{C_1} can be considered as a voltage divider in series with R_2.

With an X_{C_1} of 500 Ω, this value in series with the 5000-Ω R_2 allows approximately one-eleventh of V_{ac} to be developed across C_1. This ac voltage, equal to 0.9 V here, is the same across R_1 and C_1 in parallel. The remainder of the ac applied voltage, equal to approximately 9.1 V, is across R_2. In summary, then, the bypass capacitor provides an ac short circuit across its shunt resistance, so that little or no ac voltage can be developed, without affecting the dc voltages.

Measuring voltages around the circuit in Fig. 28-7, a dc voltmeter reads 5 V across R_1 and 5 V across R_2. An ac voltmeter across R_2 reads 9.1 V, which is almost all the ac input voltage. Across the bypass capacitor C_1 the ac voltage is only 0.9 V.

In Table 28-2, typical sizes for rf and af bypass capacitors are listed. The values of C have been calculated at different frequencies for an X_C one-tenth the shunt resistance given in each column. Note that smaller values of R require larger values of C for bypassing. Also, when X_C equals one-tenth of R at one frequency, X_C will be even less for higher frequencies, improving the bypassing action. Therefore, the size of bypass capacitors should be considered on the basis of the lowest frequency to be bypassed.

It should be noted that the applications of coupling and bypassing for C are really the same, except that C_C is in series with R and the

TABLE 28-2. Typical AF and RF Bypass Capacitors*

FREQUENCY	VALUES OF C			FREQUENCY BAND
	$R = 16$ kΩ	$R = 1.6$ kΩ	$R = 160$ Ω	
100 Hz	1 μF	10 μF	100 μF	Low af
1000 Hz	0.1 μF	1 μF	10 μF	Medium af
10 kHz	0.01 μF	0.1 μF	1 μF	High af
100 kHz	0.001 μF	0.01 μF	0.1 μF	Low rf
1 MHz	100 pF	0.001 μF	0.01 μF	Medium rf
10 MHz	10 pF	100 pF	0.001 μF	High rf
100 MHz	1 pF	10 pF	100 pF	Vhf

*For RC bypass circuit in Fig. 28-7; $X_{C_1} = \frac{1}{10} R$.

bypass C is in parallel with R. In both cases X_C must be one-tenth or less of R. Then C_C couples the ac signal to R. Or the shunt bypass short-circuits R for the ac signal.

Bypassing Radio Frequencies but Not Audio Frequencies. See Fig. 28-8. At the audio frequency of 1000 Hz, C_1 has a reactance of 1.6 MΩ. This reactance is so much higher than R_1 that the impedance of the parallel combination is essentially equal to the 16,000 Ω of R_1. Then R_1 and R_2 serve as a voltage divider for the applied af voltage of 10 V. Each of the equal resistances has one-half the applied voltage,
equal to 5 V across R_2 and 5 V across R_1. This 5 V at 1000 Hz is also present across C_1, since it is in parallel with R_1.

For the rf voltage at 1 MHz, however, the reactance of the bypass capacitor is only 1600 Ω. This is one-tenth of R_1. Then X_{C_1} and R_1 in parallel have a combined impedance equal to approximately 1600 Ω.

Now, with a 1600-Ω impedance for the R_1C_1 bank in series with the 16,000 Ω of R_2, the voltage across R_1 and C_1 is one-eleventh the applied rf voltage. Then there is 0.9 V across the lower impedance of R_1 and C_1, with 9.1 V across the larger resistance of R_2. As a result,

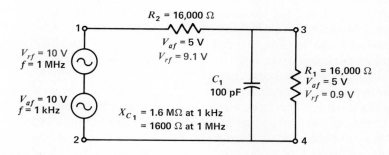

FIGURE 28-8

C_1 bypasses R_1 for radio frequencies but not audio frequencies.

the rf component of the applied voltage can be considered bypassed. C_1 is the rf bypass capacitor across R_1.

Practice Problems 28-5
(answers on page 592)
(a) In Fig. 28-8, is C_1 an af or rf bypass?
(b) Which of the following is a typical audio bypass capacitor across a 1-kΩ R: 1 pF, 0.001 μF, or 5 μF?

28-6
FILTER CIRCUITS

In terms of their function, filters can be classified as either low-pass or high-pass. A low-pass filter allows the lower-frequency components of the applied voltage to develop output voltage across the load resistance, while the higher frequency components are attenuated or reduced in the output. A high-pass filter does the opposite, allowing the higher-frequency components of the applied voltage to develop voltage across the output load resistance.

The case of an RC coupling circuit is an example of a high-pass filter because the ac component of the input voltage is developed across R while the dc voltage is blocked by the series capacitor. Furthermore, with higher frequencies in the ac component, more ac voltage is coupled. For the opposite case, a bypass capacitor is an example of a low-pass filter. The higher frequencies are bypassed, but the lower the frequency, the less the bypassing action. Then lower frequencies can develop output voltage across the shunt bypass capacitor.

In order to make the filtering more selective in terms of which frequencies are passed to produce output voltage across the load, filter circuits generally combine inductance and capacitance. Since inductive reactance increases with higher frequencies, while capacitive reactance decreases, the two opposite effects improve the filtering action.

With combinations of L and C, filters are named to correspond to the circuit configuration. Most common types are the L, T, and π, shown in Figs. 28-9 to 28-11. Any one of the three can function as either a low-pass or a high-pass filter.

For either low-pass or high-pass filters with L and C the reactance X_L must increase with higher frequencies, while X_C decreases. The frequency characteristics of X_L and X_C cannot be changed. However, the circuit connections are opposite to reverse the filtering action.

In general, high-pass filters use:

1. Coupling capacitance C in series with the load. Then X_C can be low for high frequencies to be passed to R_L, while low frequencies are blocked.
2. Choke inductance L in parallel across R_L. Then the shunt X_L can be high for high frequencies to prevent a short circuit across R_L, while low frequencies are bypassed.

The opposite characteristics for low-pass filters are:

1. Inductance L in series with the load. The high X_L for high frequencies can serve as a choke, while low frequencies can be passed to R_L.
2. Bypass capacitance C in parallel across R_L. Then high frequencies are bypassed by a small X_C, while low frequencies are not affected by the shunt path.

Practice Problems 28-6
(answers on page 592)
(a) Is series C for high-pass or low-pass filtering?
(b) Is parallel C for high-pass or low-pass filtering?

28-7
LOW-PASS FILTERS

Figure 28-9 illustrates low-pass circuits from the case of a single filter element with a shunt bypass capacitor in (a) or a series choke in (b), to the more elaborate combinations of an L-type filter in (c), a T type in (d), and a π type in (e) and (f). With an applied input voltage having different frequency components, the low-pass filter action results in maximum low-frequency voltage across R_L, while most of the high-frequency voltage is developed across the series choke or resistance.

In (a), the shunt capacitor C bypasses R_L for high frequencies. In (b), the choke L acts as a voltage divider in series with R_L. Since L has maximum reactance for the highest frequencies, this component of the input voltage is developed across L, with little across R_L. For lower frequencies, L has low reactance, and most of the input voltage can be developed across R_L.

In (c), the use of both the series choke and bypass capacitor improves the filtering by providing sharper cutoff between the low frequencies that can develop voltage across R_L and the higher frequencies stopped from the load by producing maximum voltage across L. Similarly, the T-type circuit in (d) and the π-type circuits in (e) and (f) improve filtering.

Using the series resistance in (f), instead of a choke, provides an economical π filter needing less space.

The ability to reduce the amplitude of undesired frequencies is the *attenuation* of the filter. The frequency at which the attenuation reduces the output to 70.7 percent response is the *cutoff frequency*.

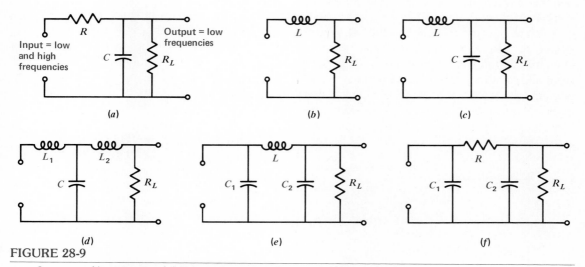

FIGURE 28-9
Low-pass filter circuits. (a) Bypass capacitor C in parallel with R_L. (b) Choke L in series with R_L. (c) The inverted-L type with choke and bypass. (d) T type with two chokes and one bypass. (e) π type with one choke and two bypass capacitors. (f) π type with series resistor instead of choke.

Passband and Stop Band. As illustrated in Fig. 28-10, a low-pass filter attenuates frequencies above the cutoff frequency of 15 kHz in this example. Any component of the input voltage having a frequency lower than 15 kHz can produce output voltage across the load. These frequencies are in the *passband*. Frequencies of 15 kHz or more are in the *stop band*. The sharpness of filtering between the passband and the stop band depends on the type of circuit. In general, the more L and C components, the sharper the response of the filter can be. Therefore, π and T types are better filters than the L type and the bypass or choke alone.

The response curve in Fig. 28-10 is illustrated for the application of a low-pass filter attenuating rf voltages while passing audio frequencies to the load. This is necessary where the input voltage has rf and af components but only the audio voltage is desired for the af circuits that follow the filter.

A good example is filtering the audio output of the detector circuit in a radio receiver, after the rf-modulated carrier signal has been rectified. Another common application of low-pass filtering is where the steady dc component of pulsating dc input must be separated from the higher frequency 60-Hz ac component, as in the pulsating dc output of a power rectifier.

Circuit Variations. The choice between the T-type filter with a series input choke and the π type with a shunt input capacitor depends upon the internal resistance of the generator supplying input voltage to the filter. A low-resistance generator needs the T filter so that the choke can provide a high series impedance for the bypass capacitor. Otherwise, the bypass must have extremely large values to short-circuit the low-resistance generator for high frequencies.

The π filter is more suitable with a high-resistance generator where the input capacitor can be effective as a bypass. For the same reasons, the L filter can have the shunt bypass either in the input for a high-resistance generator or across the output for a low-resistance generator.

For all the filter circuits, the series choke can be connected either in the high side of the line, as in Fig. 28-9, or in series in the opposite side of the line, without any effect on the filtering action. Also, the series components can be connected in both sides of the line for a *balanced filter* circuit.

Practice Problems 28-7
(answers on page 592)
(a) Which diagrams in Fig. 28-9 show a π-type filter?
(b) Does the response curve in Fig. 28-10 show low-pass or high-pass filtering?

28-8
HIGH-PASS FILTERS

As illustrated in Fig. 28-11, the high-pass filter passes to the load all frequencies higher than the cutoff frequency, while lower frequencies cannot develop appreciable voltage across the load. The graph in (a) shows the response of a high-pass filter with a stop band of 0 to 50 Hz.

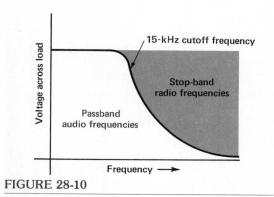

FIGURE 28-10

Response of low-pass filter with cutoff at 15 kHz, passing audio voltage but attenuating radio frequencies.

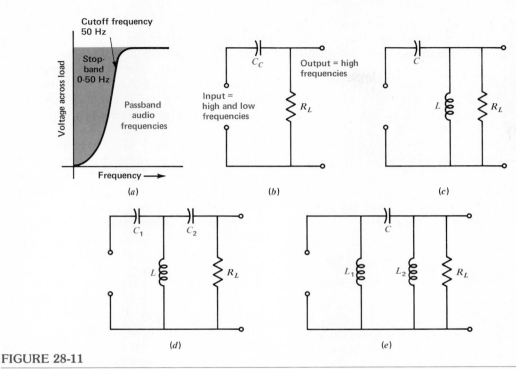

FIGURE 28-11

High-pass filters. (a) Response curve for af filter cutting off at 50 Hz. (b) RC coupling circuit. (c) Inverted L. (d) T type. (e) π type.

Above the cutoff frequency of 50 Hz, the higher audio frequencies in the passband can produce af voltage across the output load resistance.

The high-pass filtering action results from using C_C as a coupling capacitor in series with the load, as in (b). The L, T, and π types use the inductance for a high-reactance choke across the line. In this way the higher-frequency components of the input voltage can develop very little voltage across the series capacitance, allowing most of this voltage to be produced across R_L. The inductance across the line has higher reactance with increasing frequencies, allowing the shunt impedance to be no lower than the value of R_L.

For low frequencies, however, R_L is effectively short-circuited by the low inductive reactance across the line. Also C_C has high reactance and develops most of the voltage at low frequencies, stopping these frequencies from developing voltage across the load.

Bandpass Filtering. A high-pass filter can be combined with a low-pass filter. Then the net result is to pass the band of frequencies that are not stopped by either circuit. Such a bandpass response is shown in Fig. 28-12 for audio frequencies. In this example, the only frequencies passed by both filters are 50 to 15,000 Hz. It should be noted, though, that filters for a spe-

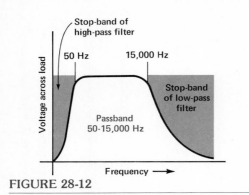

FIGURE 28-12 Bandpass response curve for audio frequencies.

cific band of frequencies are most often used at radio frequencies, as another application of resonant circuits.

Constant-k Filter. If we consider an L type, as a basic example, the values of inductance and capacitance can be designed to make the product of X_L and X_C constant at all frequencies. The purpose is to have the filter present a constant impedance at the input and output terminals. The constant-k filter can be high-pass or low-pass.

The m-derived Filter. This is a modified form of the constant-k filter. The design is based on the ratio of the filter cutoff frequency to the frequency of infinite attenuation. This ratio determines the m factor, which is generally between 0.8 and 1.25. The m-derived filter also can be high-pass or low-pass. The advantage is very sharp cutoff. Details on the design of filters can be found in most electronics handbooks.

Practice Problems 28-8
(answers on page 592)
Refer to Fig. 28-11.
(a) Which diagram shows a T-type filter?
(b) Does the response curve in (a) show high-pass or low-pass filtering?

28-9
RESONANT FILTERS

Tuned circuits provide a convenient method of filtering a band of radio frequencies because relatively small values of L and C are necessary for resonance. A tuned circuit provides filtering action by means of its maximum response at the resonant frequency.

The width of the band of frequencies affected by resonance depends on the Q of the tuned circuit, a higher Q providing narrower bandwidth. Because resonance is effective for a band of frequencies below and above f_r, resonant filters are called *band-stop* or *bandpass* filters. Series or parallel LC circuits can be used for either function, depending on the connections with respect to R_L.

Series Resonance Filters. A series resonant circuit has maximum current and minimum impedance at the resonant frequency. Connected in series with R_L, as in Fig. 28-13a, the series-tuned LC circuit allows frequencies at and near resonance to produce maximum output across R_L. Therefore, this is a case of band-pass filtering. When the series LC circuit is connected across R_L as in (b), however, the resonant circuit provides a low-impedance shunt path that short-circuits R_L. Then there is minimum output. This action corresponds to a shunt bypass capacitor, but the resonant circuit is more selective, short-circuiting R_L just for frequencies at and near resonance. For the bandwidth of the tuned circuit, therefore, the series resonant circuit in shunt with R_L provides band-stop filtering.

Parallel Resonance Filters. A parallel resonant circuit has maximum impedance at the resonant frequency. Connected in series with R_L, as in Fig. 28-14a, the parallel-tuned LC circuit provides maximum impedance in series with R_L, at and near the resonant frequency.

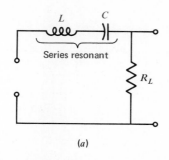
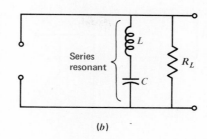

FIGURE 28-13

Filtering action of series resonant circuit. (a) Bandpass in series with R_L. (b) Band stop in shunt with R_L.

Then these frequencies produce maximum voltage across the LC circuit but minimum output voltage across R_L. This is a band-stop filter, therefore, for the bandwidth of the tuned circuit.

The parallel LC circuit connected across R_L, however, as in (b), provides a bandpass filter. At resonance, the high impedance of the parallel LC circuit allows R_L to develop its output voltage. Below resonance, R_L is short-circuited by the low reactance of L; above resonance, R_L is short-circuited by the low reactance of C. For frequencies at the near resonance, though, R_L is shunted by a high impedance, resulting in maximum output voltage.

L-type Resonant Filter. Series and parallel resonant circuits can be combined in L, T, or π sections to improve the filtering. Figure 28-15 illustrates the L-type filter, with band-stop filtering for the circuit arrangement in (a) but bandpass filtering in (b). The circuit in (a) is a band-stop filter because the parallel resonant circuit is in series with the load, while the series resonant

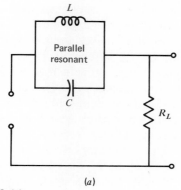
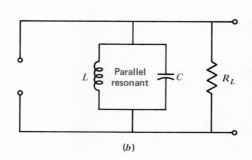

FIGURE 28-14

Filtering action of parallel resonant circuit. (a) Band stop in series with R_L. (b) Bandpass in shunt with R_L.

Chapter 28

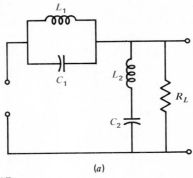 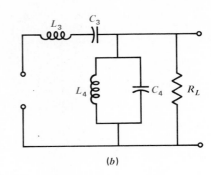

FIGURE 28-15

Inverted-L filter with resonant circuits. (*a*) Band stop. (*b*) Bandpass.

circuit is in shunt with the load. In (*b*) the bandpass filtering results from connecting the series resonant circuit in series with the load, while the parallel resonant circuit is across the load.

Practice Problems 28-9
(answers on page 592)
Answer true or false.
(a) A parallel resonant LC circuit in series with the load is a band-stop filter.
(b) A series resonant LC circuit in series with the load is a bandpass filter.

28-10
INTERFERENCE FILTERS

Voltage or current not at the desired frequency represents interference. Usually, such interference can be eliminated by a filter. Some typical applications are (1) low-pass filter to eliminate rf interference from the 60-Hz power-line input to a receiver, (2) high-pass filter to eliminate rf interference from the signal picked up by a television receiving antenna, and (3) resonant filter to eliminate an interfering radio frequency from the desired rf signal. The resonant band-stop filter is called a *wavetrap*.

Power-Line Filter. Although the power line is a source of 60-Hz voltage, it is also a conductor for interfering rf currents produced by motors, fluorescent lighting circuits, and rf equipment. When a receiver is connected to the power line, the rf interference can produce noise and whistles in the receiver output. To minimize this interference, the filter shown in Fig. 28-16 can be used. The filter is plugged into the wall outlet for 60-Hz power, while the receiver is plugged into the filter. An rf bypass capacitor across the line with two series rf chokes forms a low-pass balanced L-type filter. Using a choke in each side of the line makes the circuit balanced to ground.

The chokes provide high impedance for interfering rf current but not for 60 Hz, isolating the receiver input connections from rf interference in the power line. Also, the bypass capacitor short-circuits the receiver input for radio frequencies but not for 60 Hz. The unit then is a low-pass filter for 60-Hz power applied to the receiver while rejecting higher frequencies. The current rating means the filter can be used for equipment that draws 3 A or less from the power line without excessive heat in the chokes.

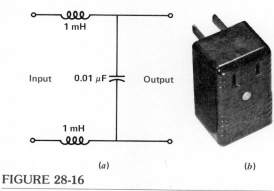

FIGURE 28-16
Power-line filter to pass 60 Hz but filter out radio frequencies. (a) Circuit of balanced L-type low-pass filter. (b) Filter unit, rated 3 A, 120 V. (*P. R. Mallory and Co., Inc.*)

Television Antenna Filter. When a television receiver has interference in the picture resulting from radio frequencies below the television broadcast band, picked up by the receiving antenna, this rf interference can be reduced by the high-pass filter shown in Fig. 28-17. The filter attenuates frequencies below 54 MHz, which is the lowest frequency for channel 2.

At frequencies lower than 54 MHz the series capacitances provide increasing reactance with a larger voltage drop, while the shunt inductances have less reactance and short-circuit the load. Higher frequencies are passed to the load as the series capacitive reactance decreases and the shunt inductive reactance increases.

Connections to the filter unit are made at the receiver end of the line from the antenna. Either end of the filter is connected to the antenna terminals on the receiver, with the opposite end connected to the antenna line.

Practice Problems 28-10
(*answers on page 592*)
Answer true or false.
(a) A wavetrap is a band-stop filter.
(b) The television antenna filter in Fig. 28-17 is a high-pass filter with series capacitors.

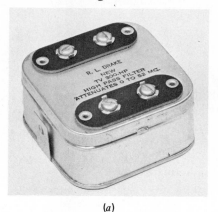

FIGURE 28-17
Television antenna filter to pass channel frequencies above 54 MHz but attenuate lower frequencies. (a) Filter unit. (b) Underside of cover showing series capacitors and shunt chokes. (*R. L. Drake Co.*)

Summary

1. A filter separates high and low frequencies. With input of different frequencies, the high-pass filter allows the higher frequencies to produce output voltage across the load; a low-pass filter provides output voltage for the lower frequencies.
2. Pulsating or fluctuating direct current varies in amplitude but does not reverse its direction. Similarly, a pulsating or fluctuating dc voltage varies in amplitude but maintains one polarity, either positive or negative.
3. The pulsating direct current or voltage consists of a steady dc level, equal to the average value, and an ac component that reverses in polarity with respect to the average level. The dc and ac components can be separated by filters.
4. An RC coupling circuit is effectively a high-pass filter for pulsating direct current. C_C blocks the steady dc voltage but passes the ac component.
5. A transformer with an isolated secondary winding also is effectively a high-pass filter. With pulsating direct current in the primary, only the ac component produces output voltage in the secondary.
6. A bypass capacitor in parallel with R provides a low-pass filter.
7. Combinations of L, C, and R can be arranged as L, T, or π filters for more selective filtering. All three arrangements can be used for either low-pass or high-pass action.
8. In high-pass filters, the capacitance must be in series with the load as a coupling capacitor, with shunt R or L across the line.
9. For low-pass filters, the capacitance is across the line as a bypass capacitor, while R or L then must be in series with the load.
10. A bandpass or band-stop filter has in effect two cutoff frequencies. The bandpass filter passes to the load those frequencies in the band between the cutoff frequencies, while attenuating all other frequencies higher and lower than the passband. A band-stop filter does the opposite, attenuating the band between the cutoff frequencies, while passing to the load all other frequencies higher and lower than the stop band.
11. Resonant circuits are generally used for bandpass or band-stop filtering with radio frequencies.
12. For bandpass filtering, the series resonant LC circuit must be in series with the load, for minimum series opposition, while the high impedance of parallel resonance is across the load.

Chapter 28
Filters **589**

13. For band-stop filtering, the circuit is reversed, with the parallel resonant LC circuit in series with the load, while the series resonant circuit is in shunt across the load.
14. A wavetrap is an application of the resonant band-stop filter.

Self-Examination (Answers at back of book.)

1. With input frequencies from direct current up to 15 kHz, a high-pass filter allows the most output voltage to be developed across the load resistance for which of the following frequencies? (*a*) Direct current; (*b*) 15 Hz; (*c*) 150 Hz; (*d*) 15,000 Hz.
2. With input frequencies from direct current up to 15 kHz a low-pass filter allows the most output voltage to be developed across the load resistance for which of the following frequencies? (*a*) Direct current; (*b*) 15 Hz; (*c*) 150 Hz; (*d*) 15,000 Hz.
3. An $R_C C_C$ coupling circuit is a high-pass filter for pulsating dc voltage because: (*a*) C_C has high reactance for high frequencies; (*b*) C_C blocks dc voltage; (*c*) C_C has low reactance for low frequencies; (*d*) R_C has minimum opposition for low frequencies.
4. A transformer with an isolated secondary winding is a high-pass filter for pulsating direct primary current because: (*a*) the steady primary current has no magnetic field; (*b*) the ac component of the primary current has the strongest field; (*c*) only variations in primary current can induce secondary voltage; (*d*) the secondary voltage is maximum for steady direct current in the primary.
5. Which of the following is a low-pass filter? (*a*) L type with series C and shunt L; (*b*) π type with series C and shunt L; (*c*) T type with series C and shunt L; (*d*) L type with series L and shunt C.
6. A bypass capacitor C_b across R_b provides low-pass filtering because: (*a*) current in the C_b branch is maximum for low frequencies; (*b*) voltage across C_b is minimum for high frequencies; (*c*) voltage across C_b is minimum for low frequencies; (*d*) voltage across R_b is minimum for low frequencies.
7. An ac voltmeter across C_C in Fig. 28-6 reads (*a*) practically zero; (*b*) 7.07 V; (*c*) 10 V; (*d*) 20 V.
8. Which of the following L-type filters is the best band-stop filter? (*a*) Series resonant LC circuit in series with the load and parallel resonant LC circuit in shunt; (*b*) parallel resonant LC circuit in series with the load and series resonant LC circuit in shunt; (*c*) series

resonant *LC* circuits in series and in parallel with the load; (*d*) parallel resonant *LC* circuits in series and in parallel with the load.
9. A 455-kHz wavetrap is a resonant *LC* circuit tuned to 455 kHz and connected as a (*a*) band-stop filter for frequencies at and near 455 kHz; (*b*) bandpass filter for frequencies at and near 455 kHz; (*c*) band-stop filter for frequencies from direct current up to 455 kHz; (*d*) bandpass filter for frequencies from 455 kHz up to 300 MHz.
10. A power-line filter for rejecting rf interference has (*a*) rf coupling capacitors in series with the power line; (*b*) rf chokes in shunt across the power line; (*c*) 60-Hz chokes in series with the power line; (*d*) rf bypass capacitors in shunt across the power line.

Essay Questions

1. What is the function of an electrical filter?
2. Give two examples where the voltage has different frequency components.
3. (*a*) What is meant by pulsating direct current or voltage? (*b*) What are the two components of a pulsating dc voltage? (*c*) How can you measure the value of each of the two components?
4. Define the function of the following filters in terms of output voltage across the load resistance: (*a*) High-pass filter. Why is an $R_c C_c$ coupling circuit an example? (*b*) Low-pass filter. Why is an $R_b C_b$ bypass circuit an example? (*c*) Bandpass filter. How does it differ from a coupling circuit? (*d*) Band-stop filter. How does it differ from a bandpass filter?
5. Draw circuit diagrams for the following filter types. No values are necessary. (*a*) T-type high-pass and T-type low-pass; (*b*) π-type low-pass, balanced with a filter reactance in both sides of the line.
6. Draw the circuit diagrams for L-type bandpass and L-type band-stop filters. How do these two circuits differ from each other?
7. Draw the response curve for each of the following filters: (*a*) low-pass cutting off at 20,000 Hz; (*b*) high-pass cutting off at 20 Hz; (*c*) bandpass for 20 to 20,000 Hz; (*d*) bandpass for 450 to 460 kHz.
8. Give one similarity and one difference in comparing a coupling capacitor and a bypass capacitor.
9. Give two differences between a low-pass filter and a high-pass filter.
10. Explain briefly why the power-line filter in Fig. 28-16 passes 60-Hz alternating current but not 1-MHz rf current.

Problems (Answers to odd-numbered problems at back of book.)

1. Refer to the RC coupling circuit in Fig. 28-6, with R equal to 160,000 Ω. (a) Calculate the required value for C_C at 1000 Hz. (b) How much is the steady dc voltage across C_C and across R? (c) How much is the ac voltage across C_C and across R?
2. Refer to the R_1C_1 bypass circuit in Fig. 28-8. (a) Why is 1 MHz bypassed but not 1 kHz? (b) If C_1 were doubled in capacitance, what is the lowest frequency that could be bypassed, maintaining a 10:1 ratio of R to X_C?
3. Calculate the C_C needed to couple frequencies of 50 to 15,000 Hz with a 500-kΩ R.
4. Show the fluctuating plate current i_b of a vacuum tube that has an average dc axis of 24 mA and a square-wave ac component with a 10-mA peak value. Label the dc axis, maximum and minimum positive values, and the peak-to-peak alternating current.
5. Show the fluctuating grid voltage v_C of a vacuum tube that has an average dc axis of -8 V and a sine-wave ac component with 3 V peak value. Label the dc axis, maximum and minimum negative values, and peak-to-peak ac voltage.
6. Do the same as in Prob. 4 for the fluctuating collector current i_C of an NPN transistor.
7. (a) Draw an inverted L-type band-stop filter used as a wavetrap for 455 kHz. (b) Give the inductance necessary with a 80-pF C.
8. (a) Referring to Fig. 28-6, calculate the value of C_C necessary for coupling 50 Hz when R is 500 kΩ. (b) Referring to Fig. 28-7, calculate the value of C_1 necessary for bypassing R_1 at 50 Hz.
9. Referring to the audio tone-control switch in Fig. 28-18, calculate the required capacitance values for the following: (a) C_1 to bypass R_1 at

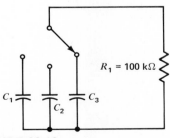

FIGURE 28-18
For Prob. 9.

10,000 Hz; (b) C_2 to bypass R_1 at 5000 Hz; (c) C_3 to bypass R_1 at 2000 Hz.

10. Referring to the *RC* low-pass filter in Fig. 28-9a, draw the schematic diagram with values of 75 kΩ for R, 0.001 μF for C, and 10 MΩ for R_L. (a) For 10 V input, calculate the values of V_C at 1, 2, 5, 10, and 15 kHz. (b) Draw the response curve of the filter, plotting V_C vs. frequency.

Answers to Practice Problems

28-1	(a) 500 kHz	28-5	(b) 5 μF
	(b) 60 Hz	28-6	(a) High-pass
28-2	(a) 6 V		(b) Low-pass
	(b) 10 and 2 V	28-7	(a) e and f
	(c) 8 V		(b) Low-pass
	(d) 4 and 2.8 V	28-8	(a) d
28-3	(a) High pass		(b) High-pass
	(b) 0 V	28-9	(a) T
28-4	(a) 0 V		(b) T
	(b) 5 μF	28-10	(a) T
28-5	(a) Rf		(b) T

Review of Chapters 25 to 28

Summary

1. X_C and X_L are opposite reactances. In series, the ohms of X_C and X_L cancel; in parallel, the capacitive and inductive branch currents cancel.
2. As a result, circuits with R, X_C, and X_L can be reduced to one net reactance and one equivalent resistance. In series circuits, the net reactance is added with the total resistance by phasors: $Z = \sqrt{R^2 + X^2}$; then $I = V/Z$. In parallel circuits, the net reactive branch current is added with the total resistive branch current by phasors: $I_T = \sqrt{I_R^2 + I_X^2}$; then $Z = V/I_T$.
3. The ohms of R, X_C, X_L, and Z in ac circuits are compared in Table 25-1.
4. In ac circuits with reactance, the real power in watts equals I^2R, or $VI \cos \theta$, where θ is the phase angle of the circuit and $\cos \theta$ is the power factor.
5. Ac meter scales are generally calibrated in rms values.
6. The wattmeter uses the dynamometer ac meter movement to read V and I at the same time, measuring watts of real power.
7. Since X_L and X_C cancel in series, when they are equal, the net reactance is zero. In parallel, the net reactive branch current is zero. The specific frequency that makes $X_L = X_C$ is the resonant frequency $f_r = 1/(2\pi \sqrt{LC})$.
8. Larger values of L and C mean lower resonant frequencies, as f_r is inversely proportional to the square root of L or C. If the value of L or C is quadrupled, for instance, f_r will decrease by one-half.
9. For a series resonant LC circuit, the current is maximum, since the opposition is just the low resistance of the conductors; the voltage drop across each reactance is maximum, but they are equal and opposite; the phase angle is zero. The reactive voltage at resonance is Q times greater than the applied voltage.

10. For a parallel resonant LC circuit, the impedance is maximum with minimum line current, since the reactive branch currents cancel. The impedance at resonance is Q times the X_L value, but it is resistive with a phase angle of zero.
11. The Q of the resonant circuit equals X_L/r_s for resistance in series with X_L, or R_p/X_L for resistance in parallel with X_L.
12. The bandwidth between half-power points is f_r/Q.
13. A filter uses inductance and capacitance to separate high or low frequencies. A low-pass filter allows low frequencies to develop output voltage across the load; a high-pass filter does the same for high frequencies. Series inductance or shunt capacitance provides low-pass filtering; series capacitance or shunt inductance provides high-pass filtering.
14. A fluctuating or pulsating direct current or voltage varies in amplitude but with one direction or polarity. It is equivalent to an ac component varying in opposite directions around the average-value axis. The axis, which is just the arithmetical average of all the values in one cycle, is the steady dc value.
15. An RC coupling circuit is effectively a high-pass filter for pulsating dc voltage, passing the ac component but blocking the dc component.
16. A transformer with an isolated secondary is a high-pass filter for pulsating direct current, allowing ac variations in the secondary but no output for the dc level of primary current.
17. A bypass capacitor in parallel with R is effectively a low-pass filter, since its low reactance reduces the voltage across R for high frequencies.
18. The main types of filter circuits are π type, L type, and T type. These can be high-pass or low-pass, depending on how the L and C components are connected.
19. Resonant circuits can be used as bandpass or band-stop filters. For bandpass filtering, series resonant circuits are in series with the load or parallel resonant circuits are across the load. For band-stop filtering, parallel resonant circuits are in series with the load or series resonant circuits are across the load.
20. A wavetrap is an application of a resonant band-stop filter.

Review Self-Examination (Answers at back of book.)

Fill in the numerical answer.

1. A sine-wave ac circuit with 100-Ω R_1 and 200-Ω R_2 in series has R_T of _____ Ω.
2. With 100-Ω X_{L_T} in series with 200-Ω X_{L_T}, the total X_C is _____ Ω.

3. With 100-Ω X_{L_T} in series with 200-Ω X_{L_T} the X_{L_T} is _____ Ω.
4. Two X_C branches of 500 Ω each in parallel have X_{L_T} of _____ Ω.
5. Two X_L branches of 500 Ω each in parallel have X_{L_T} of _____ Ω.
6. With 500-Ω X_L in series with 300-Ω X_C, the net X_L is _____ Ω.
7. With 500-Ω X_C in series with 300-Ω X_L, the net X_C is _____ Ω.
8. With 10-Ω X_L in series with 10-Ω R, the total Z_T is _____ Ω.
9. With 10-Ω X_C in series with 10-Ω R, the total Z_T is _____ Ω.
10. With 14 V applied across 14-Ω Z_T, the I is _____ A.
11. With 10-Ω X_L and 10-Ω R in series, θ is _____ degrees.
12. With 10-Ω X_C and 10-Ω R in series, θ is _____ degrees.
13. With 10-Ω X_L and 10-Ω R in parallel across 10 V, the amount of each branch I is _____ A.
14. In question 13, the total line current I_T equals _____ A.
15. In questions 13 and 14, Z_T of the parallel branches equals _____ Ω.
16. With 120 V, an I of 10 A, and θ of 60°, a wattmeter reads _____ W.
17. A 60-Hz ac voltage with a peak value of 500 V will be read by an iron-vane meter as _____ V.
18. L of 10 H and C of 40 μF has f_r of _____ Hz.
19. L of 100 μH and C of 400 pF has f_r of _____ MHz.
20. In question 19, if C is reduced to 100 pF, f_r increases to _____ MHz.
21. In question 20, if L is increased to 400 μH, the f_r decreases to _____ MHz.
22. In a series resonant circuit with 10 mV applied across a 1-Ω R, a 1000-Ω X_L, and a 1000-Ω X_C, at resonance the current is _____ mA.
23. In question 22, the voltage across X_C equals _____ V.
24. In a parallel resonant circuit with a 1-Ω r_S in series with a 1000-Ω X_L in one branch and a 1000-Ω X_C in the other branch, with 10 mV applied, the voltage across X_C equals _____ mV.
25. In question 24, Z of the parallel resonant circuit equals _____ MΩ.
26. An LC circuit resonant at 500 kHz has a Q of 100. Its total bandwidth between half-power points equals _____ kHz.
27. In question 26, if the Q is reduced to 10 by a shunt R_P, the bandwidth becomes _____ kHz.
28. A coupling capacitor for 40 to 15,000 Hz in series with a 0.5-MΩ resistor has the capacitance of _____ μF.
29. A bypass capacitor for 40 to 15,000 Hz in shunt with a 1000-Ω R has the capacitance of _____ μF.
30. A pulsating dc voltage varying in a symmetrical sine wave between 100 and 200 V has the average value of _____ V.

Answer true or false.

31. In an ac circuit with X_C and R in series, if f is increased, the current will increase.

32. In an ac circuit with X_L and R in series, if f is increased, the current will be reduced.
33. The voltampere is a unit of apparent power.
34. A circuit with a phase angle of 90° does not dissipate any real power.
35. Effective value and rms value of a sine-wave ac voltage are the same.
36. In a capacitive voltage divider, the smallest C has the most voltage.
37. A steady direct current in the primary of a transformer cannot produce any ac output voltage in the secondary.
38. A π-type filter with shunt capacitances is a low-pass filter.
39. An L-type filter with a parallel resonant LC circuit in series with the load is a band-stop filter.
40. With pulsating dc voltage applied across an RC coupling circuit, the average dc level of the applied voltage is across C.

References (Additional references at back of book.)

Gillie, A. C.: *"Electrical Principles of Electronics,"* McGraw-Hill Book Company, New York.

Siskind, C. S.: *"Electricity, Direct and Alternating Current,"* 2d ed., McGraw-Hill Book Company, New York.

Turner, R. P.: *"Basic Electronic Test Instruments,"* Holt, Rinehart and Winston, Inc., New York.

Weick, C. B.: "Principles of Electronic Technology," McGraw-Hill Book Company, New York.

Zeines, B.: "Electric Circuit Analysis," Prentice-Hall, Inc., Englewood Cliffs, N.J.

Vacuum Tubes

Chapter 29

This chapter describes the construction of tubes, the different types of tubes, and their applications as amplifiers or as rectifiers. An amplifier circuit increases the amplitude of a desired signal. For instance, the amplifier can have 1-mV input and 100-mV output for a voltage gain of 100. A rectifier converts alternating current to direct current. The ac input to the rectifier has positive and negative polarities, but the dc output has only one polarity, either positive or negative.

Similarly, semiconductor diodes and transistors are used for rectifiers and amplifiers, as described in Chap. 30. Separate diodes and transistors are *discrete components*, but many of them can be combined in an integrated circuit (IC) unit, as described in Chap. 31. In general, then, the functions of amplifiers, oscillators, rectifiers, and electronic switching can be performed by tubes, transistors, or integrated circuits.

The analysis of tubes and semiconductor devices starts the subject of *active components* for electronic circuits. Essentially, the active components are tubes and semiconductors for amplification or rectification. The *passive components* R, L, and C are still necessary. However, when they are combined with tubes or semiconductor devices, the result is almost unlimited possibilities for using electronic circuits in many functions. The topics here are:

29-1   Rectifiers, Amplifiers, and Oscillators
29-2   Construction of Tubes
29-3   Diodes
29-4   Plate Current
29-5   Diode Rectifier Circuit
29-6   Triodes
29-7   How a Triode Amplifies the Control-Grid Voltage
29-8   Triode Characteristics
29-9   Tube Parameters
29-10 Tetrodes
29-11 Pentodes
29-12 Tube Types
29-13 The Cathode-Ray Tube (CRT)
29-14 Troubles in Vacuum Tubes

Chapter 29

29-1 RECTIFIERS, AMPLIFIERS, AND OSCILLATORS

Most electronic circuits can be divided in these three groups. A rectifier changes its ac input to dc output. An amplifier circuit amplifies its input signal. An oscillator circuit is a special case of the amplifier, but the oscillator generates ac output from its dc power supply, without any ac input signal.

Amplifiers. The block diagram in Fig. 29-1 illustrates an amplifier for audio signal. In general, the term *signal* is used for the variations in voltage or current corresponding to the desired information. The ac input of 0.2 V here could be the audio signal from a phonograph. By means of a tube or transistor, with its associated passive components, this audio amplifier circuit provides an output of 8 V. Then this signal can be used to drive a power amplifier stage that provides enough current to drive a loudspeaker. Amplifiers connected with the output driving the input of the next stage are *cascaded stages*.

For the voltage amplifier in Fig. 29-1, the ac output signal of 8 V is forty times more than the input signal of 0.2 V. Note that the output signal is shown with reversed polarity, as many amplifier circuits invert the phase of the signal by 180°.

For the passive components, R is always useful in providing a desired IR voltage, for either direct current or alternating current. C is commonly used for capacitive coupling to pass the ac signal while blocking any dc component. Or a transformer can be used for coupling. C can also be used as a bypass capacitor. L can be used as an audio or rf choke, and for transformers. LC resonant circuits are generally used in rf amplifiers.

It should be noted that nonsinusoidal ac voltages can be amplified. In addition, dc amplifiers are used.

Although the amplifier in Fig. 29-1 is amplifying ac signal, the circuit requires a steady dc supply voltage. The reason is that transistors or tubes need dc voltage for the electrodes in order to conduct any current at all. In general, any amplifier operates by having the input signal variations control a larger amount of dc power in the output circuit. As a result, amplifier circuits have currents and voltages that include ac variations on an average dc axis.

Oscillators. The block diagram in Fig. 29-2 illustrates an rf feedback oscillator. There is no ac input signal, but the oscillator circuit generates rf output at the natural resonant frequency of the LC circuit tuned to 1 MHz.

The only difference between this circuit and an rf amplifier for a 1-MHz signal is that the oscillator has positive feedback, from the output to the input. This polarity reinforces the signal variations of the input circuit. The positive feedback can be obtained by two phase reversals of 180°.

Although a tuned rf oscillator circuit is shown here for a 1-MHz sine-wave output, oscillator circuits also can be used to generate pulses, square waves, or sawtooth waves. The

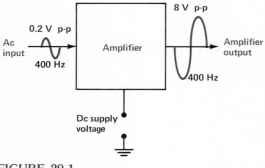

FIGURE 29-1

Function of an amplifier circuit. The ac signal input of 0.2 V is amplified to 8 V p-p.

Chapter 29
Vacuum Tubes

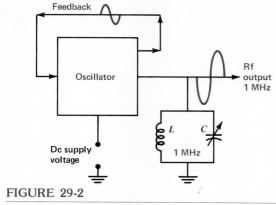

FIGURE 29-2
An rf feedback oscillator circuit, tuned to 1 MHz as the resonant frequency.

frequency can be almost any value, from low audio frequencies to extremely high radio frequencies.

Rectifiers. The block diagram in Fig. 29-3 illustrates a rectifier, changing the 60-Hz ac input voltage to fluctuating dc output voltage. The diode allows current in only one direction, for one polarity of applied voltage. Therefore, current flows in the output circuit only during the half-cycles of the ac input voltage that turn the diode on. This circuit is a half-wave rectifier.

Although not shown here, the output circuit usually has shunt filter capacitors, with a series choke or resistor, to remove the fluctuations in the dc output. The entire circuit then is a dc power supply.

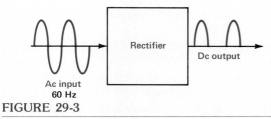

FIGURE 29-3
A rectifier circuit converts its ac input to pulsating dc output.

The function of the power supply could be to provide the dc supply voltage needed for the amplifier and oscillator in Figs. 29-1 and 29-2. Diode semiconductors or tubes can be used as the rectifiers, but semiconductor diodes are most common because of their high efficiency and small size. Although 60 Hz is illustrated here for the ac power line, it should be noted that the ac input to the rectifier can have any frequency.

Practice Problems 29-1
(*answers on page 623*)
(a) In Fig. 29-1, how much is the voltage gain for the audio signal?
(b) In Fig. 29-2, what is the resonant frequency of the *LC* circuit for the tuned rf oscillator?

29-2
CONSTRUCTION OF TUBES

As illustrated in Fig. 29-4, a vacuum tube has a glass or metal envelope enclosing metal electrodes in a vacuum. The electrodes make it

FIGURE 29-4
Typical amplifier tube. Height is $1\frac{1}{2}$ in.

possible to have electrons flow through the evacuated space inside the tube. The electrode that emits electrons is the *cathode.* Generally, the cathode is heated by a wire filament, resulting in thermionic emission of electrons. The electrode that collects the emitted electrons is the *anode,* or the *plate.* In general, an anode is a positive electrode while the cathode is negative with respect to the anode.

The plate has a positive potential applied with respect to the cathode, so that the emitted electrons are attracted to provide plate current. Between the cathode and plate the tube can also have a wire mesh serving as a control-grid electrode to increase or decrease electron flow to the plate. The current flow is almost instantaneous, a typical transit time being 0.001 μs for electrons emitted from the cathode to reach the plate.

The vacuum tube is fundamentally a low-current device, as the electron flow is limited by the amount of thermionic emission from the cathode. For conventional sizes, typical values of plate current are less than 1 to about 100 mA. Compared with transistors, most vacuum tubes have much higher internal resistance because of the relatively small plate current.

In their historical development, the first tubes were diodes, with a cathode and anode.

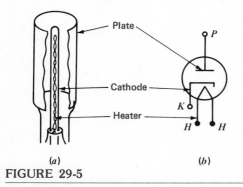

FIGURE 29-5

Diode vacuum tube. (*a*) Construction. (*b*) Schematic symbol.

Then triodes were invented, making use of the grid to control plate current. These were followed by tetrodes and pentodes. The tetrode has four electrodes, including two grids; the pentode has five electrodes with three grids. They all need the cathode to emit electrons and the plate to collect the emitted electrons. These are the main types of vacuum tubes. The source for the filament or heater is often called *A voltage,* with *B voltage* used for the plate and *C voltage* for the grid.

Practice Problems 29-2
(*answers on page 623*)
(*a*) Which electrode emits electrons?
(*b*) What is the polarity of plate voltage, with respect to cathode?

29-3
DIODES

A diode has just two electrodes: cathode and plate, as shown in Fig. 29-5. Note the schematic symbol. The heater is not counted as an electrode because it is merely an incandescent filament to heat the cathode electrically. If any other method were convenient for heating the cathode to its emission temperature, it would serve just as well.

Note that the plate surrounds the cathode. Therefore, electrons emitted from the surface of the cathode sleeve can be attracted to the metal anode to provide plate current. The plate is usually made of iron, nickel, or molybdenum. Larger plate area is used for tubes that conduct larger values of plate current.

Thermionic Emission. When a metal is heated, electrons in the atoms are accelerated in their random motion because of the added heat energy. With enough heat to make the metal glow as an incandescent filament, some internal electrons gain enough velocity to break away from the surface of the metal. In a sense, the electrons can be considered as "boiled off" the surface. Now the metal is a cathode that supplies electrons by thermionic emission.

Cathode Construction. The two types are illustrated schematically in Fig. 29-6. In (a), the filament-cathode is heated directly by electrical current, serving as an incandescent filament that produces thermionic emission. In (b), the cathode is heated indirectly by a separate heater. The directly heated type is usually called a *filament*, while the filament for the indirectly heated cathode is a *heater*.

Filament Current. Power for heating the filament is obtained by applying the rated filament voltage so that the correct amount of current can flow to heat the filament. The ratings are given in the manufacturer's tube manual. For instance, many tubes are rated at 6.3 V, 0.3 A for the heater; that is, 6.3 V applied will produce the rated heater current of 0.3 A. Most receiver tubes use an indirectly heated cathode, with a 60-Hz ac heater current.

The Vacuum. After the tube is assembled, air is exhausted from the envelope. There are several reasons why the vacuum is necessary. First, the heated filament would oxidize in air and burn. Also, the cathode emits more electrons in a vacuum. Finally, when it is desired that only the electrons emitted from the cathode travel to the plate, without ionizing air molecules, the electrodes must be in a vacuum.

When enough air has been pumped out, the envelope is sealed off. To improve the vacuum further, the entire assembly is then heated to force out any additional gas molecules in the metal electrodes. At this time, a magnesium *getter* compound, mounted on a small disk inside, vaporizes because of the heating. The resultant chemical action of the vaporized magnesium, with the gases released from the metal, removes the final traces of gas in the tube. After the tube cools off, the vaporized getter condenses on the inside of the envelope, forming the silvery film usually seen in glass tubes.

*Practice Problems 29-3
(answers on page 623)*
(a) Thermionic emission is from which electrode?
(b) Which electrode collects electrons?

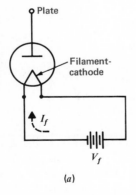 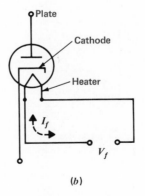

FIGURE 29-6
(a) Filament-cathode heated with direct current. (b) Indirectly heated cathode with ac power for heater.

29-4
PLATE CURRENT

The plate is made positive with respect to the cathode in order to attract the emitted electrons. Furthermore, the plate has a continuous path for electron flow returning to the cathode in the external circuit through the plate supply.

Space Charge. This is a cloud of electrons near the cathode, produced by thermionic emission. Since electrons are negative, they are attracted to the anode with its positive accelerating voltage.

Plate-Current Characteristics. If a milliammeter is connected in series in the plate circuit, as shown in Fig. 29-7a, the meter can read the values of plate current I_b produced by different amounts of plate voltage V_b. Remember that V_b must be positive. The results are tabulated and plotted on the graph in (b).

The procedure is simply to vary the potentiometer R to obtain the desired value of V_b and then read the corresponding I_b. A tube with an indirectly heated cathode is shown here, in order to concentrate on the plate current. The heater circuit is omitted because it operates at rated voltage and is not varied for this experiment.

The resulting graph is a plate characteristic curve, plotting the amount of I_b for a corresponding V_b. For example, the curve shows that a plate voltage of 100 V allows a plate current of 8 mA. Such a curve is a voltampere characteristic of the diode.

Practice Problems 29-4
(answers on page 623)
(a) In Fig. 29-7a, does the milliammeter read I_b, V_b, or heater current?
(b) In Fig. 29-7b, for V_b of 50 V, how much is I_b?

29-5
DIODE RECTIFIER CIRCUIT

The fact that the diode allows current to flow in only one direction means that it is useful as a rectifier to change alternating current to direct current. Such a diode rectifier circuit is shown in Fig. 29-8.

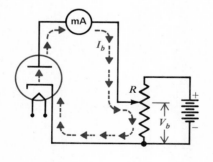

(a)

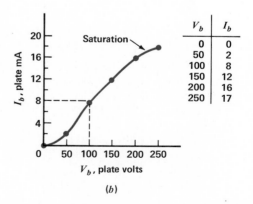

(b)

FIGURE 29-7

Measuring I_b for different plate voltages V_b to obtain the diode plate characteristic curve. (a) Circuit. (b) Graph.

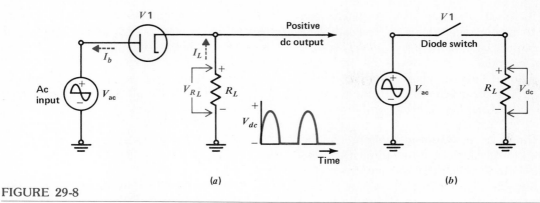

FIGURE 29-8

Diode tube in a half-wave rectifier circuit. (a) Ac input to V1 rectified to pulsating dc output. (b) Equivalent circuit with V1 as a switch.

Load Resistance. The plate current becomes useful when it is made to flow through an external load R_L outside the tube. The purpose of using R_L is to have a load that can develop an IR_L voltage drop outside the tube that depends on the amount of plate current flowing within the tube. Then the IR_L drop is output voltage that can be coupled to another circuit. For the diode rectifier circuit in Fig. 29-8, therefore, R_L is connected in series with the diode V1 and the ac input voltage.

Half-Wave Rectifier. The voltage applied to the diode plate-cathode circuit through R_L is the ac input voltage. There is no dc supply voltage for the tube now because the ac input voltage makes the diode plate positive every half-cycle. When the plate is driven positive, plate current flows through the tube and the input voltage source, returning to the cathode through R_L. I_b is I_L through the load. For the half-cycle when the input voltage makes the diode plate negative, there is no plate current and no output voltage across R_L.

The result is half-cycles, or half-waves, of current through R_L in the direction shown. Current cannot flow in the opposite direction because I flows only from cathode to plate in the tube.

The output voltage across R_L equals IR_L. This output is a dc voltage, therefore, because it has just one polarity. The circuit is a half-wave rectifier because it produces a half-wave of rectified output for each cycle of the ac input.

Ripple Frequency. The fluctuating component is the ripple in the dc output. In this circuit, the ripple is half-wave pulsations at the same frequency as the ac input. Therefore, the ripple frequency of the half-wave rectifier for the 60-Hz ac power-line voltage is 60 Hz. The ripple can be removed, however, by using series filter chokes and shunt filter capacitors.

DC Polarity. The rectified output has the polarity shown in Fig. 29-8 because R_L is in the cathode side of the diode circuit. Then I_b flowing in the plate circuit must return from the low side of R_L, through its resistance, and back to the cathode. Since the electron flow is from minus to plus, the cathode side of R_L must be the positive side of the rectified output voltage.

The positive polarity of V_L is only with respect to the opposite side of R_L. The cathode is still less positive than the plate voltage supplied by the ac input when it drives the diode into conduction. Otherwise, there could be no plate current. The plate voltage is just the ac input, however, without any rectification.

The Diode as a One-way Switch. In Fig. 29-9b, the rectifier circuit is redrawn to show the diode as a switch that lets current flow in only one direction. When the diode plate is positive, plate current flows. Then the tube is on, or the switch V1 is closed. With the diode plate negative, the diode is an open circuit, since its resistance is practically infinite without plate current. Then the switch V1 is open.

As a result, we can consider the rectification in terms of the switching action. Only when V_{ac} is positive does the ac input voltage become connected to R_L through V1. Now we can also see why V_{dc} across R_L must be positive at the cathode end. This terminal is connected to the positive side of V_{ac} by the closed switch when V1 conducts. Therefore, the only time any output can be produced is when the cathode side of R_L is positive.

As a one-way switch, the diode tube and a semiconductor diode really have the same function. The anode, or positive side of the semiconductor, corresponds to the plate of the tube. In either case, positive voltage applied to the anode makes the diode conduct. More details of rectifiers, including full-wave circuits, are explained with semiconductor diodes in Chap. 30. Practically all rectifier circuits now use solid-state devices.

Practice Problems 29-5
(answers on page 623)
Refer to Fig. 29-8.
(a) Does diode V1 conduct when the ac input voltage makes the plate positive or negative?
(b) Is the cathode side of the dc output voltage positive or negative?

29-6
TRIODES

As illustrated in Fig. 29-9, a cathode and a plate are needed to provide plate current like a diode, but in addition the triode has the control-grid electrode. The grid is a fine metal wire, usually nickel, molybdenum, or iron, wrapped around two supports placed in the space between the cathode and the plate. All electrons attracted to the plate from the cathode go through the openings in the grid. The grid is connected to a base pin, however, so that it can have voltage applied to determine the amount of electrons that travel from cathode to plate. Now the plate current depends on two factors: the plate voltage and the control-grid voltage.

C− Voltage. The potential applied to the control grid is usually a small negative voltage with respect to the cathode, as shown in Fig. 29-10. This voltage completes the alphabet, with A voltage for the heater, B+ voltage for the plate, and C− for the control grid. The grid voltage is usually made negative so that it cannot attract electrons. Then there will be no grid current. The function of the control grid is not

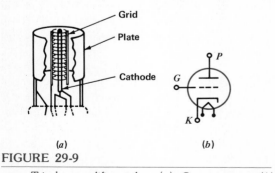

FIGURE 29-9

Triode amplifier tube. (a) Construction. (b) Schematic symbol.

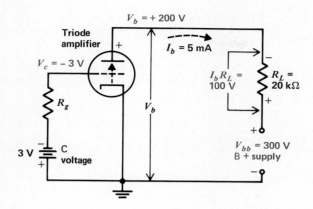

FIGURE 29-10

Triode amplifier circuit. B+ voltage is for plate and C− voltage for grid.

to provide current but to have its voltage control the plate current.

The effect of the grid voltage on plate current can be summarized as follows:

1. Less negative grid voltage increases the plate current.
2. More negative grid voltage decreases the plate current.
3. If the grid voltage is made negative enough, there will be no plate current even with the plate positive.

Grid-Cutoff Voltage. This is the amount of negative grid voltage required to cut off the plate current with a specified amount of positive voltage on the plate. For some tubes, the grid-cutoff voltage may be as little as −1 V; for others, the grid-cutoff voltage may be −60 V, as typical examples. The grid-cutoff voltage is a characteristic of the tube construction, but also depends on the amount of plate voltage.

Triode Circuit. Plate current flows in Fig. 29-10 because the plate has its positive voltage and the control-grid potential is less than the grid-cutoff voltage. The electron flow inside the tube is from the cathode, through the spaces between the grid wire, to the anode.

In the external plate circuit, the electrons flow through the 20-kΩ plate-load resistor R_L and the B supply, returning to the cathode through the chassis. Assuming a 5-mA plate current value, the $I_b R_L$ drop is $0.005 \times 20,000 = 100$ V.

The plate-to-cathode circuit of the tube and R_L are in series as a voltage divider across the 300-V supply V_{bb}. Therefore, the plate-to-cathode voltage is 300 V minus the 100-V drop across R_L, or V_b is $300 - 100 = 200$ V. As a formula

$$V_b = V_{bb} - I_b R_L \qquad (29\text{-}1)$$

This formula is used to determine the actual plate-to-cathode voltage.

Example 1. Calculate V_b for a 20-mA I_b with a 2000-Ω R_L and a plate supply of 250 V.

Answer.
$$\begin{aligned}
V_b &= V_{bb} - I_b R_L \\
&= 250 - (20 \times 10^{-3} \times 2 \times 10^3) \\
&= 250 - 40 \\
V_b &= 210 \text{ V}
\end{aligned}$$

TABLE 29-1. Vacuum-Tube Letter Symbols

V_{bb} = dc plate supply voltage, equal to B+
V_b = average dc plate-cathode voltage
v_b = instantaneous value of the fluctuating dc plate voltage with signal variations
v_p = ac component of the fluctuating dc plate voltage. This is the amplified ac signal output
V_{cc} = control-grid dc bias supply voltage
V_c = average control-grid dc voltage. This equals V_{cc} when there is no control-grid current
v_c = instantaneous value of the fluctuating dc grid voltage with signal variations
v_g = ac signal input to the control grid

Control-Grid Bias. A bias voltage is a constant voltage that is used to obtain the desired method of operation. In Fig. 29-10 the V_c of -3 V is a negative dc grid bias. Its function is to maintain the average grid voltage negative, even with an ac signal input. The ac signal swings the grid voltage in a positive direction, but the negative dc bias is more than the positive peak of the ac signal.

Symbols for Tube Voltages. In analyzing vacuum-tube amplifier circuits, we must distinguish between plate and grid voltages and between plate-to-cathode voltage and the B+ supply. Furthermore, when the ac signal is amplified, the resultant grid and plate voltages have fluctuating dc waveforms with an average dc value and an ac component. The symbols in Table 29-1 are generally used.

The same system of nomenclature also applies to the plate current, with I_b for average dc, i_b for instantaneous dc, and i_p for ac values. These symbols are used for diodes, triodes, tetrodes, or pentodes.

Practice Problems 29-6
(answers on page 623)
(a) Does negative v_c increase or decrease i_b?
(b) How much is v_b with a B+ of 180 V and V_{R_L} of 60 V?

29-7
HOW A TRIODE AMPLIFIES THE CONTROL-GRID VOLTAGE

The main advantage in having the control grid is the fact that it enables the vacuum tube to amplify a small voltage at the grid to provide much more voltage at the plate. The amplification results from the ability of the control-grid voltage to vary the plate current.

Consider the numerical examples listed in Table 29-2. For the circuit in Fig. 29-10, let the average plate current I_b be 5 mA with -3 V for the control-grid bias voltage V_c. These values are listed in the middle row of Table 29-2. With a 3-mA I_b, the plate voltage equals 200 V. The 300-V B+ voltage drops to 200 V for the plate because of the 100-V IR drop across the 20,000-Ω R_L.

TABLE 29-2. Amplification in the Plate Circuit*

	v_c, V	i_b, mA	$i_b R_L$, V	$v_b = 300 - i_b R_L$, V
	-2	6	120	180
Average level →	-3	5	100	200
	-4	4	80	220

*R_L = 20 kΩ and B+ or V_{bb} = 300 V.

Now suppose the grid voltage v_c is reduced from -3 to -2 V. The grid, being less negative, allows more plate current. For the case of i_b increasing to 6 mA, then, the voltage drop across R_L increases to 120 V. The voltage for v_b is less, however, falling to 180 V as the remainder of the 300-V B+ supply minus the $i_b R_L$ drop.

For the third case, in the bottom row, when v_c is made more negative, from -3 to -4 V, less plate current flows. Then i_b is reduced to 4 mA. This change results in a smaller $i_b R_L$ drop of 80 V. The resulting plate voltage, equal to 300 V minus 80 V, rises to 220 V.

Note the effect of the change as v_c varies ± 1 V, above and below the middle value of -3 V. Similarly, i_b varies ± 1 mA around the 5-mA middle value.

So far, all that the control-grid voltage has done is vary the plate current. In terms of plate voltage, however, v_b varies ± 20 V above and below the middle value of 200 V. Here is where the voltage amplification is accomplished.

The changes in plate current through the plate-load resistance allow the plate voltage to vary by a much larger amount than the grid-voltage variation. In this example, the grid voltage of ± 1 V has been multiplied by a factor of 20 to produce a corresponding plate voltage variation of ± 20 V.

For voltage amplification, the requirements are that (1) the grid voltage must vary the plate current and (2) the plate circuit must have a plate-load resistance that can vary the plate voltage as the plate current varies.

Phase Inversion. If we consider the variations in Table 29-2 as peak values of a sine-wave ac signal, the resultant waveforms are shown in Fig. 29-11. In (a) the grid voltage v_c is a fluctuating dc voltage with variations above and below the bias axis of -3 V. These variations in v_c produce the variations of i_b in waveform (b). The plate current varies above and below the

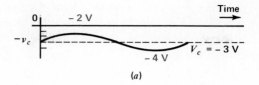

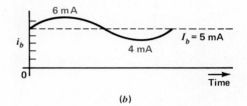

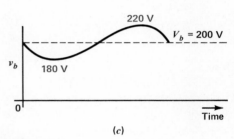

FIGURE 29-11
Ladder diagram of waveshapes for Table 29-2. (a) Sine-wave grid voltage v_c. (b) Plate current i_b. (c) Plate voltage v_b equal to $V_{bb} - i_b R_L$.

average dc axis of 5 mA. The voltage across R_L has similar variations, as $v_L = i_b R_L$.

However, the plate voltage v_b shown in (c) has polarity opposite from i_b and v_L. The reason is that v_b is the difference between V_{bb} and v_L. As v_L increases, the value of positive dc plate voltage v_b decreases. Therefore, the variations in v_b have opposite polarity from the variations in v_c. Or the input signal is inverted by $180°$ in the amplified output.

Voltage Gain. The symbol is A_V. As a formula

$$A_V = \frac{V_{\text{out}}}{V_{\text{in}}} \tag{29-2}$$

These are ac signal voltages, in rms, peak, or peak-to-peak values, as long as the same measure is used for both. A_V has no units since it is a ratio of two voltages. For this example, A_V is 20 V/1 V = 20.

Practice Problems 29-7
(answers on page 623)
Refer to Fig. 29-11.
(a) How much is the average negative dc grid bias voltage v_c?
(b) How much is the ac input signal voltage to the grid?
(c) How much is the average dc plate voltage V_b?
(d) How much is the amplified ac output voltage?

29-8
TRIODE CHARACTERISTICS

The plate current is not linear for all values of electrode voltages. Therefore, it is necessary to show tube characteristics by curves. The way that plate current i_b is affected by both plate voltage v_b and control-grid voltage v_c is shown by the characteristic curves in Fig. 29-12.

One characteristic is shown separately in (a), with -2 V for v_c. With this grid voltage, the triode curve shows that 80 V on the plate allows slightly less than 4-mA i_b; or a 120-V v_b produces 8 mA, as examples. The v_b values are plate-to-cathode voltages, not the supply voltage.

The manufacturer's tube manual shows a family of plate characteristics for different values of grid voltage. Typical curves are shown in (b). Note that the curve marked -2 V for v_c is the same characteristic shown in (a).

The family of characteristics, however, has more information for typical values of negative grid voltage. From left to right, the individual curves show plate characteristics for v_c values of 0, -2, -4, -6, and more negative voltages up to -24 V. The family of curves shows all the plate-current characteristics of the tube.

For any fixed value of v_c, just read the one

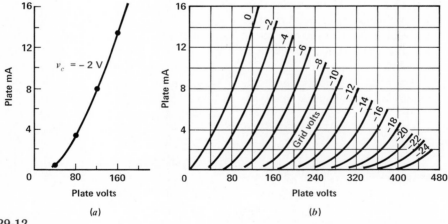

FIGURE 29-12

Triode characteristic curves. (a) Graph of plate characteristics v_b and i_b for one value of grid voltage. (b) Family of plate characteristics for different v_c values.

curve to determine i_b for a specified v_b. For instance, with -4 V for v_c, this curve intersects the vertical line for a 160-V v_b at the horizontal line for an 8-mA i_b. Therefore, the grid potential of -4 V allows 8 mA of i_b with 160 V for v_b. As another example, -8 V for v_c with the same 160 V for v_b allows only 1 mA plate current. The cutoff grid voltage resulting in zero plate current with 160 V at the plate is approximately -10 V.

Practice Problems 29-8
(answers on page 623)
Refer to Fig. 29-12.
(a) In (a), how much is i_b with 160 V at the plate?
(b) In (b), how much is i_b with 160 V at the plate and a grid bias of -4 V?

29-9
TUBE PARAMETERS

A parameter is a measurement that describes a particular characteristic. For amplifier tubes, the main characteristics are the amplification factor μ, or mu, the internal plate-cathode resistance r_p, and the grid-plate transconductance g_m. These parameters are for the tube itself, not the amplifier circuit.

Amplification Factor. This ratio compares the effectiveness of the control-grid voltage v_c to the plate voltage v_b in changing the plate current i_b.

$$\mu = \frac{\Delta v_b}{\Delta v_c} \quad \text{for the same } i_b \quad (29\text{-}3)$$

There are no units for μ because it is a ratio of two voltages and the volt units cancel.

Plate Resistance. This R is the equivalent resistance of the internal plate-to-cathode circuit. Since plate voltage is applied across the tube and plate current is flowing through it, the tube has a resistance equal to its v/i ratio.

$$r_p = \frac{\Delta v_b}{\Delta i_b} \quad \text{for the same } v_c \quad (29\text{-}4)$$

This value is considered the ac or small-signal r_p, since it is calculated from small changes. The dc or large-signal values, indicated as R_b, can be calculated simply as V_b/I_b.

Transconductance. This factor states how effective the grid voltage is in changing the plate current.

$$g_m = \frac{\Delta i_b}{\Delta v_c} \quad \text{for the same } v_b \quad (29\text{-}5)$$

In Fig. 29-12b, let the grid voltage change from -4 to -2 V for Δv_c of 2 V. Reading these two curves on the same vertical line of 160 V for v_b, the i_b is increased from 8 to 13 mA. This Δi_b is 5 mA. Then to calculate this transconductance,

$$g_m = \frac{\Delta i_b}{\Delta v_c} = \frac{0.005 \text{ A}}{2 \text{ V}}$$
$$= 0.0025 \text{ S} = 2500 \times 10^{-6} \text{ S}$$
$$g_m = 2500 \text{ }\mu\text{S}$$

The unit for g_m is the siemens for conductance, since any i/v ratio is the reciprocal of ohms.

Since vacuum tubes are low-current devices, the transconductance will be a fraction of 1 S. It is more convenient, therefore, to specify g_m in microsiemens. Typical values for receiver tubes are 2000 to 14,000 μS.

Practice Problems 29-9
(answers on page 623)
(a) A grid voltage change of ± 2 V swings the plate current ± 24 mA. Calculate the g_m of the tube.

(b) A tube has 400 mA of I_b with 300 V for V_b. Calculate the dc plate resistance R_b.

29-10
TETRODES

This construction is similar to the triode, with cathode, control grid, and plate, but there is an additional electrode called the screen grid between the control grid and the plate. See Fig. 29-13.

The control grid is grid No. 1, placed close to the cathode so that it still functions to control the space charge. The screen grid is grid No. 2, closer to the plate. It is not used to control the plate current but has steady positive dc voltage to help accelerate electrons to be collected by the plate. The path for plate current inside the tube is from the cathode, through the control grid, and through the spaces in the screen grid to be collected by the plate.

Since the screen grid is positive, it will collect some electrons. These provide screen-grid current that returns to the cathode through the screen-grid circuit. The screen-grid current is *waste current*, however, since it is not used in the output circuit.

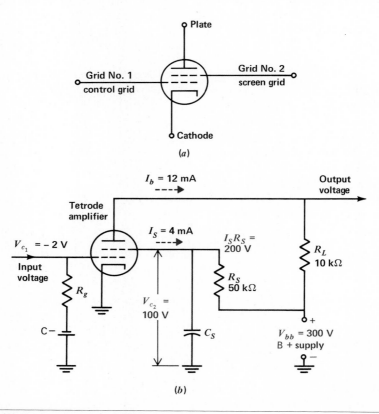

FIGURE 29-13
The tetrode. (a) Schematic symbol. (b) Circuit with positive dc voltage for the screen grid, in addition to plate voltage.

Screen-Grid Voltage. The positive screen-grid potential V_{c_2} can be obtained from the same B+ supply that provides plate voltage. As shown in Fig. 29-13b, the resistor R_S, called the *screen-dropping resistor*, is in series with the B+ voltage. Then the screen current returning through the B+ supply to the cathode flows through R_S. The $I_S R_S$ voltage drop allows the screen-grid voltage to be less than the B+ voltage by the amount of voltage across the screen-dropping resistor.

In Fig. 29-13b, the $I_S R_S$ voltage drop of 4 mA through 50 kΩ equals 200 V across R_S. The remainder for the screen-grid-to-cathode circuit then is $300 - 200 = 100$ V. As a formula,

$$V_{c_2} = V_{bb} - I_S R_S \qquad (29\text{-}6)$$

Example 2. Calculate V_{c_2} with a 20-kΩ R_S, an 8-mA I_S, and 250 V for the supply voltage V_{bb}.

Answer.
$$\begin{aligned} V_{c_2} &= V_{bb} - I_S R_S \\ &= 250 - (0.008 \times 20{,}000) \\ &= 250 - 160 \\ V_{c_2} &= 90 \text{ V} \end{aligned}$$

Screen Bypass Capacitor. In order to make sure that the voltage remains fixed at its steady dc value, the screen-grid resistor is bypassed by C_S. It returns to the cathode through the chassis ground connections. The capacitance of C_S must be large enough to have reactance low enough to be one-tenth or less of R_S. The reactance is figured at the lowest frequency of the ac signal being amplified. Because of the bypass C_S, there should be practically zero ac signal voltage at the screen grid.

Grid-Plate Capacitance. The screen grid also reduces the capacitance between the control grid and plate inside the tube. This capacitance is noted as C_{gp} in Fig. 29-14. The reason for the capacitance is simply the fact that the control grid and plate are two metal conductors separated by an insulator, which is the vacuum in the tube.

As Fig. 29-14b shows, the typical value of 4 pF for a triode is reduced by a factor of 1/1000 to the low value of 0.004 pF for C_{gp} in the equivalent tube with a screen grid. The reason for the reduced value of C_{gp} is that the screen grid acts as a shield between the plate and control grid for an ac signal.

Secondary Emission. Metals have the property of releasing electrons when the surface is bombarded by incident electrons. No heat is

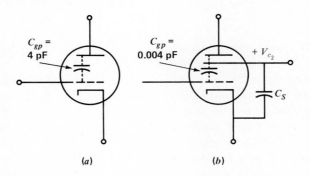

FIGURE 29-14

Interelectrode capacitance between plate and control grid. (a) Triode. (b) Screen-grid tube.

necessary. The requirement is high positive voltage to provide a strong accelerating field so that the incident electrons can strike at high velocity. The electrons released are then called *secondary electrons,* and the process is *secondary emission.*

In a vacuum tube, the metal plate is bombarded by the electrons attracted from the cathode. Therefore, the plate has secondary emission. In a diode or triode, though, secondary electrons are no problem because any secondary electrons near the plate are collected by the positive anode.

In a tetrode, however, the positive screen grid can attract secondary electrons. This effect occurs when the plate voltage drops below the screen voltage. Then the plate current is reduced. For this reason, tetrodes are not commonly used in amplifier circuits. Generally, when a screen-grid tube is desired for an amplifier, a pentode is used.

Practice Problems 29-10
(*answers on page 623*)
Refer to Fig. 29-13b.
(a) How much is the dc voltage from screen grid to cathode?
(b) How much is the ac signal voltage from screen grid to cathode?

29-11
PENTODES

As shown in Fig. 29-15, the pentode has the same construction as a tetrode but with the addition of a *suppressor grid* in the space between the screen grid and the plate. Since it is a screen-grid tube, the pentode has the advantages of little C_{gp} and the operating characteristics of a tetrode. However, there is negligible secondary emission from the plate because of the suppressor grid.

In Fig. 29-15a, the first grid is the control grid, grid No. 2 is the screen grid, and grid No. 3 is the suppressor grid. The suppressor grid is not used for input or output signal but has a fixed potential, usually equal to the cathode voltage. In most pentodes, the suppressor is connected internally to the cathode. When the suppressor has its own external pin connection, it is connected either to the cathode pin or to chassis ground.

Since the suppressor is close to the plate but has the cathode potential that is negative with respect to the plate, any secondary electrons emitted are repelled back to the plate. The plate can still attract electrons from the cathode, however, to provide plate current.

Referring to the pentode circuit in Fig. 29-15b, the screen-grid voltage is operated at a fixed positive dc voltage, as for a tetrode. Just as in triodes, the input voltage to be amplified is connected to the control-grid circuit, which has its required negative C− bias voltage. The amplified output signal voltage in the plate circuit results from plate current flowing through the plate-load resistor R_L.

Current Paths. In a circuit like Fig. 29-15b, it is important to keep in mind the different currents for each electrode. Let us assume I_b is 12 mA and I_{c_2} is 3 mA. This means a 12-mA I_b flows in the plate circuit through R_L. Also, a 3-mA I_{c_2} flows in the screen-grid circuit through R_S. In the B+ supply, both I_b and I_{c_2} add to result in 15 mA.

This 15 mA is actually the space current in the tube. The reason is that the negative side of the B+ supply is returned to the cathode. If you insert a milliammeter in the cathode circuit, it will read 15 mA.

Furthermore, if the control grid goes positive, it can attract electrons to produce grid current I_{c_1}. Then I_{c_1} flows through R_g and the C− supply back to the cathode.

In summary, the cathode current is the total space current, equal to the sum of all the

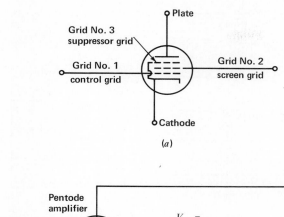

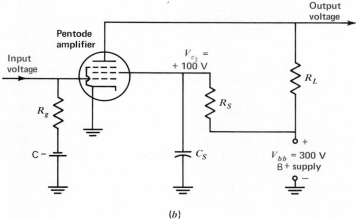

FIGURE 29-15

The pentode. (a) Schematic symbol, with suppressor grid tied internally to cathode. (b) Amplifier circuit, with input signal to control grid and output from plate.

individual electrode currents. As a formula, $I_K = I_b + I_{c_2} + I_{c_1}$. If we assume 40, 8, and 1 mA for these values, as an example, $I_K = 40 + 8 + 1 = 49$ mA. Usually, though, I_{c_1} is zero.

Pentode Characteristics. As for triodes, the manufacturer's tube manual shows plate characteristic curves of pentodes for different values of control-grid voltage. For the v_{c_1} value of -1 V in Fig. 29-16, as an example, the curve shows that plate current increases from approximately 5 mA with 100 V for v_b to 5.5 mA with 400 V.

The very slow rise of the curve means that the plate current increases very little with increased plate voltage in a pentode. This characteristic is a result of the screen-grid voltage, which is constant. In terms of the plate resistance of the tube, its r_p is very high because i_b changes very little with a change in v_b.

Remote Cutoff Pentodes. As an example, a pentode with a sharp cutoff of -3 V for V_{c_1} may

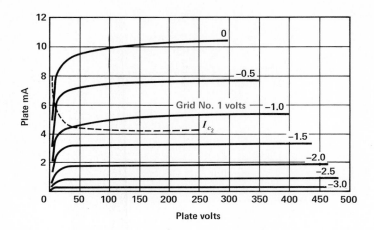

FIGURE 29-16

Plate characteristic curves for pentode tube.

correspond to a similar remote cutoff tube that cuts off plate current with -30 V at the control grid. The remote cutoff pentodes are also called *supercontrol* or *variable-μ* tubes. In these tubes, the control grid has its turns closely spaced at both ends but wider spacing at the center. Then higher negative values of V_{c_1} are required for the wider spaces of the grid to cut off electron flow. These pentodes are used in amplifier circuits where the grid bias is varied over a wide range of voltages.

Practice Problems 29-11
(*answers on page 623*)
Answer true or false.
(a) Grid 1 is the control grid. It usually has negative dc bias and ac signal input.
(b) Grid 2 is the screen grid. It usually has positive dc voltage and zero ac signal.
(c) Grid 3 is the suppressor grid. It usually is at cathode potential.

29-12
TUBE TYPES

The first digits in the type number for amplifier tubes indicate the heater voltage. For example, the 6AU6 heater voltage is 6.3 V, and the 17JT6 uses 17 V. The 6.3-V tubes are for parallel heaters. Other heater voltages such as 3, 5, 17, 25, 35, and 50 V are for series circuits, where the heater voltages add to equal the ac power-line voltage of 120 V.

For the cathode-ray tube (CRT), however, the first digits give the screen size. The heater voltage is usually 6.3 V.

Pin Connections. Four types of base diagrams or socket connections are shown in Fig. 29-17. The pins are numbered clockwise, looking at the socket or the tube base from the bottom. The top view of the socket is opposite. Wider spacing between the end pins makes the tube fit the socket only one way.

For small amplifier tubes, the seven-pin socket in (a) and nine-pin *novar* socket in (b) are most common. Larger tubes use the twelve-pin *duodecar* base in (c). The eight-pin octal socket is seldom used for tubes anymore.

The novar base comes in two sizes: small-button and large-button with a wider glass bulb. The duodecar base also comes in two sizes.

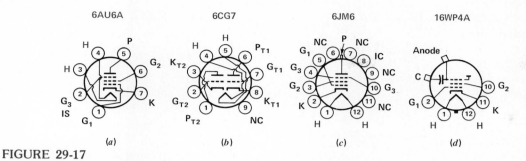

FIGURE 29-17
Tube sockets with pin connections. Bottom view of socket or base of tube. For picture tube in (d), the C indicates external conductive coating; anode button is a recessed cavity.

The notation IS indicates an internal shield; NC means no internal connection. It should be noted that some tubes have a top cap for the plate connection, generally for high voltage. See Fig. 29-17c.

Heater Pins. You can check continuity with an ohmmeter to find an open heater. The heater pins are 3 and 4 for the seven-pin base or 4 and 5 for the nine-pin base. On the duodecar base, the end pins 1 and 12 across the wide space are for the heater. For a CRT also, the heater pins are usually the end pins.

Multipurpose Tubes. To save space, one envelope often includes two or more tube functions. As examples, the 6FQ7 is a twin-triode and the 6AN8 is a triode-pentode. A *compactron* usually combines three or four functions in one envelope.

Beam Power Tubes. These tubes operate as pentodes, but instead of a suppressor grid, the tube has internal beam-forming plates tied to the cathode. The feature of this construction is low screen current, which is important for power tubes. In this application, the screen grid can be at approximately the same potential as the plate because of the small waste current in the screen-grid circuit. Typical tubes are the 6JM6, 6V6, 35L6, and 50C5.

Tube Shields. Glass tubes often have an external metal shield which slips over the tube and contacts spring clips connected to the chassis. Tubes that have two or more sections in one envelope usually have an internal shield between sections. A shield minimizes interaction between tubes amplifying high frequencies.

High-Power Tubes. In transmitters, larger tubes are generally used to produce the required amount of output power. Typical ratings are several hundred watts up to kilowatts. The larger tubes may have a water-cooled anode.

Phototubes. These have a photocathode which emits photoelectrons with incident light. The electrons are collected by a positive anode to provide photoelectric current proportional to the light. Such PE cells can be used in many light-control circuits. Semiconductor devices are also used for photoelectric effects, but phototubes are preferred where heat is a problem, as in oil-burner control circuits.

Practice Problems 29-12
(*answers on page 623*)
(a) What is the heater voltage for the 21JZ6 amplifier tube?
(b) What are the heater pins on this duodecar base?

29-13
THE CATHODE-RAY TUBE (CRT)

As illustrated in Fig. 29-18, this tube consists of an electron gun, deflection plates, and a fluorescent screen inside the evacuated glass envelope. Although the cathode, control grid and anode electrodes are constructed as cylinders, their function is the same as in conventional vacuum tubes. The cathode is heated to emit electrons, and the control grid controls the flow of electrons attracted by the positive potential of the anodes. High voltages are used, on the order of 2 to 80 kV for the last anode. The entire cylinder assembly can be considered an *electron gun,* producing electrons formed into a narrow beam attracted to the screen.

The inside surface of the front glass faceplate is coated with a fluorescent material that emits light when bombarded by electrons. Green and white are two common colors of illumination that can be produced by the screen, depending on its chemical composition. A green screen is generally used for the CRT in oscilloscopes, because green phosphors are most efficient.

When the electron beam hits the screen, it produces a spot of light visible through the glass. In order to deflect the electron beam and move the position of the light spot on the screen, deflection voltage can be applied to the deflection plates. A pair of horizontal deflection plates provides the potential difference needed to move the electron beam left or right. Similarly, the pair of vertical deflection plates can move the beam up or down. This type of CRT is used in an oscilloscope (Fig. 29-19). This test instrument shows the waveform of voltage applied to its input terminals. An example is the sine-wave photograph shown in Fig. 16-1.

Similarly, the picture reproduced on the screen of the picture tube in a television receiver is another application of the CRT. However, picture tubes use magnetic deflection because of the larger screen. In this case, a yoke with deflection coils is slipped over the neck of the tube, against the wide bell.

In CRT type numbers, the first digits give the screen size in inches for the diameter of a round screen or the diagonal of a rectangular screen. The phosphor numbers at the end of the type number are generally P1 for a green

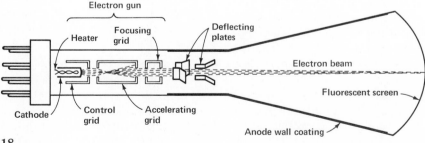

FIGURE 29-18

Cathode-ray tube (CRT) using electrostatic deflection and focusing. Screen diameter is generally 3 or 5 in for oscilloscopes.

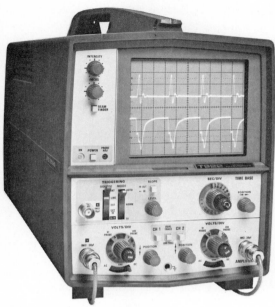

FIGURE 29-19

Oscilloscope with 8 × 10 cm green screen. This unit is a dual-trace "scope," with two input channels. (*Tektronix*)

screen, P4 for a white screen, and P22 for a red, green, and blue screen in color picture tubes. For instance, the 5BP1 CRT for oscilloscopes has a 5-in green screen. For oscilloscopes, the screen size is often given in centimeters.

Practice Problems 29-13
(*answers on page 623*)
(a) In a CRT, which electrode has the highest positive voltage?
(b) What is the phosphor number for a green screen?

29-14
TROUBLES IN VACUUM TUBES

The most common troubles in vacuum tubes are an open heater or weak emission from the cathode. Sometimes there may be an internal short circuit between electrodes, or the tube may be microphonic. Also, tubes may have excessive cathode-to-heater leakage, which causes hum. In all cases, the defective tube must be replaced.

Open Filament. There is no current, and the tube is cold. In glass tubes usually you can see that the filament is not lit. The possibility of an open can be checked by the ohmmeter, with power off. Just place the ohmmeter leads across the two filament pins and check continuity.

Typical resistance for the cold heater is 1 to 50 Ω, depending on the tube. A low resistance reading indicates that there is no open and that the filament is normal. Note that the cold resistance of the filament is much less than the hot resistance, by a factor of about $\frac{1}{10}$. For example, a tube rated for a 0.3-A heater current at 6.3 V has a hot resistance in normal operation of 6.3 V/0.3 A, which equals 21 Ω. The normal resistance of the cold heater, however, as checked by an ohmmeter, is about 2 Ω.

Microphonics. A tube with electrodes that are not perfectly rigid is called *microphonic* because the electrodes can act like a microphone if the tube vibrates. In a receiver, when a microphonic tube is tapped, it produces a hollow ringing noise that dies out slowly. At loud volume settings, the sound can make the microphonic tube vibrate, and it causes a sustained howl.

Cathode-to-Heater Leakage. If the insulation resistance between cathode and heater is not high enough, this leakage resistance can cause hum in the receiver. With 60-Hz alternating current for the heater, the hum frequency is 60 Hz.

Tube Noise. Although it is not necessarily a defect, tubes do produce a small amount of noise. In a receiver the tube noise produces a continuous hissing or "frying" sound. This noise can often be heard between stations when the volume is all the way up.

Much of the tube noise results from the random flow of electrons forming the plate current and is called *shot effect*. The resultant noise voltage is in the order of microvolts, but for tubes with a small amount of signal voltage input that will be amplified to a great extent, the noise generated in the first stages of amplification may be significant.

In general, the more grids there are, the more noise the tube will generate because of the *partition effect* of the grids on the space current, causing a more random electron flow to the plate. Triodes are therefore less noisy than equivalent multigrid types. As a result, where low tube noise is important, triodes may be used in preference to pentodes.

Tube Tester. This instrument provides operating voltages for a tube inserted in its proper socket, and conduction is indicated by a meter on the panel that reads "good" or "bad." There are two types. In the emission tester, only static plate current is indicated. In the mutual-conductance type, a small ac voltage is applied to the grid, and the change in plate current indicates whether the tube is good or bad. This type, which is preferable, can have the meter scale calibrated to read g_m in microsiemens.

The scale of a tube checker has a question mark at the center range to indicate a tube of doubtful quality. For power tubes, it is usually better to replace a questionable tube.

Practice Problems 29-14
(*answers on page 623*)
(a) How much is the resistance on an open heater?
(b) Which can produce more tube noise: triodes or pentodes?

Summary

1. The main types of electronic circuits for vacuum tubes and semiconductors are rectifiers, amplifiers, and oscillators, as illustrated in Figs. 29-1 to 29-3.
2. Diodes have a heated cathode and anode. The cathode emits electrons by thermionic emission; the cloud of emitted electrons near the cathode is the space charge. The anode attracts the electrons when it is positive with respect to cathode. The plate current can flow in only one direction, from cathode to plate. The more positive the plate voltage v_b, the more plate current i_b, up to saturation.
3. In a filament cathode, the heated filament emits electrons and serves as the cathode. With an indirectly heated cathode, a separate heater makes the insulated cathode emit electrons.
4. The diode is useful as a rectifier to change alternating to direct current. One diode is used as a half-wave rectifier; two diodes can be used in a full-wave rectifier circuit.
5. In a triode, the control grid determines how many electrons from the space charge are attracted by the positive plate to provide plate

TABLE 29-3. Types of Vacuum Tubes

TUBE	SYMBOL*	ELECTRODES	CHARACTERISTICS	APPLICATIONS
Diode		Plate Cathode	When plate is positive, conducts current only from cathode to plate	One diode for half-wave rectifier; two diodes for full-wave rectifier
Triode		Plate Control grid Cathode	Negative grid controls plate current; has large C_{gp}	Triode amplifier with external R_L, for af amplifiers; low tube noise
Tetrode		Plate Screen grid Control grid Cathode	Screen grid reduces C_{gp}; screen must have +dc voltage and bypass C to cathode for ac signal	Tetrode little used because of secondary emission from plate
Pentode		Plate Suppressor Screen grid Control grid Cathode	Suppressor grid eliminates secondary emission; screen grid provides high μ and high r_p	For af or rf amplifiers; sharp cutoff except variable-μ tubes

*Tubes shown with indirectly heated cathode.

current. The more negative the control grid voltage v_c is, the less the plate current. The negative v_c value that results in zero i_b is the grid-cutoff voltage. The fixed negative dc voltage for V_c is the grid bias.

6. The function of the triode is to amplify a grid-voltage variation. Amplification requires a steady positive dc plate voltage to allow plate current, with an external R_L in series between the plate and B+. Plate current flowing through R_L can produce changes in plate voltage much greater than the changes in grid voltage.
7. The three main characteristics for triode, tetrode, and pentode amplifier tubes are the amplification factor μ equal to $\Delta v_b / \Delta v_c$, plate resistance r_p equal to $\Delta v_b / \Delta i_b$, and transconductance g_m equal to $\Delta i_b / \Delta v_c$.
8. Tetrodes have the cathode, control-grid, and plate structure of a triode, plus the screen grid between the control grid and plate. The

screen grid must have a positive dc voltage, bypassed to the cathode at the ac signal frequency. The screen grid reduces C_{gp}.

9. Pentodes have the cathode, control grid, screen grid, and plate plus the suppressor grid between the plate and screen grid. The suppressor grid eliminates the effect of secondary emission from the plate. The suppressor grid is generally connected internally to the cathode.
10. Table 29-3 summarizes the main points about pentodes, tetrodes, triodes, and diodes. The nomenclature for electrode voltages is given in Table 29-1 on page 606.
11. In type numbers for amplifier tubes, the first number gives the heater voltage, the letters indicate the type, and the final number indicates pins actually used.
12. The most common tube troubles are low emission or an open heater.

Self-Examination (Answers at back of book.)

Choose (a), (b), (c), or (d).

1. In a diode the plate current increases when the (a) plate voltage is made more positive; (b) plate voltage is made less positive; (c) control-grid voltage is made less negative; (d) saturation plate current is reached.
2. With an indirectly heated cathode, the heater voltage (a) is applied to the cathode; (b) is separate from the cathode circuit; (c) must be a steady dc voltage; (d) is equal to the C− bias voltage.
3. In a triode, tetrode, or pentode, the plate current increases when the (a) control-grid voltage is made more negative; (b) control-grid voltage is made less negative; (c) plate voltage is made less positive; (d) screen-grid voltage is made less positive.
4. The screen grid in a pentode (a) makes the plate current more dependent on plate voltage; (b) has the same potential as the cathode; (c) decreases the grid-plate capacitance, as compared with a triode; (d) eliminates the problem of secondary emission from the plate.
5. The suppressor grid in a pentode (a) reduces the grid-plate capacitance, as compared with a triode; (b) eliminates the problem of secondary emission from the plate; (c) usually has a positive dc potential less than the plate voltage; (d) is usually connected internally to the control grid.
6. With 250 V for B+ and 10 mA for I_b through a 20-kΩ R_L, the plate-cathode voltage V_b equals (a) 10; (b) 20; (c) 50; (d) 250.

7. In a half-wave rectifier circuit, the diode conducts when the (a) plate is made positive by the positive alternation of the ac input voltage; (b) plate is made negative by the negative alternation of the ac input voltage; (c) cathode is made positive by the positive alternation of the ac input voltage; (d) ac input voltage is at its average value of zero.
8. In a pentode, when v_c varies ± 2 V, the plate current varies ± 10 mA. The grid-plate transconductance g_m in microsiemens equals (a) 500 μS; (b) 1000 μS; (c) 2000 μS; (d) 5000 μS.
9. The heater power required for the 12BA6 remote-cutoff pentode is (a) 7 V at 0.15 A; (b) 12.6 V at 0.15 A; (c) 6.3 V at 0.3 A; (d) 22.6 V at 0.3 A.
10. With an 8-mA I_b, a 2-mA I_{c_2}, and a 500-μA I_{c_1}, the total cathode current I_K equals (a) 2 mA; (b) 8 mA; (c) 10.5 mA; (d) 500 mA.

Essay Questions

1. Give the functions for rectifier, amplifier, and oscillator circuits.
2. Draw the schematic symbols for diode, triode, and pentode vacuum tubes, labeling all electrodes.
3. Show the schematic symbols for a filament cathode and an indirectly heated cathode.
4. Define the following in terms of vacuum-tube operation: (a) saturation plate current; (b) space charge; (c) grid-plate capacitance.
5. Draw the schematic diagram of a triode amplifier circuit with a plate-load R_L. Give the function of (a) A voltage; (b) B+ voltage; (c) C− voltage; (d) R_L.
6. Draw the schematic diagram of a pentode amplifier with a plate load resistance R_L. (a) What is the function of the screen-grid resistor R_S? (b) What is the function of the screen bypass capacitor C_S? (c) What is the path of plate current? (d) What is the path of screen current? (e) Why is there no control-grid current?
7. Define μ, r_p, and g_m for a triode or pentode amplifier tube.
8. Define the following symbols: V_{bb}, V_b, v_b, V_{c_2}, and V_{c_1}.
9. Referring to Fig. 29-15, why is a bypass capacitor used for R_S in the screen-grid circuit but not for R_L in the plate circuit?
10. How would you test for an open heater with an ohmmeter?
11. Draw the schematic diagram of a half-wave rectifier circuit.
12. Refer to Fig. 29-8. (a) Why is the dc output positive? (b) Why is the ripple frequency 60 Hz?

Problems (Answers to odd-numbered problems at back of book.)

1. (a) Calculate V_{c_2}, with a 4-mA I_{c_2}, a 25-kΩ R_S, and 250 V for B+. (b) Calculate V_b, with a 12-mA I_b, a 5-kΩ R_L, and 250 V for B+.
2. Referring to the triode plate characteristic curves in Fig. 29-12, how much is the plate current i_b for the following? (a) v_b is 120 V and v_c is zero volts; (b) same v_b, but v_c is -2 V; (c) same v_b, but v_c is -6 V; (d) v_c is -6 V, but v_b is 200 V; (e) v_b is 400 V, and v_c is -18 V. (f) How much negative grid voltage v_c is required for zero i_b with 400 V at the plate?
3. Referring to the pentode plate characteristic curves in Fig. 29-16, how much is i_b with -2.5 V for v_c when v_b has values of 100, 200, 250, and 300 V? Why do these values indicate that plate current is relatively independent of plate voltage in a screen-grid tube?
4. Referring to the family of plate characteristics for the pentode in Fig. 29-16: (a) Tabulate the values of i_b, with a constant v_b at 250 V, for v_c values in 0.5-V steps from 0 to 3.5 V. (b) Draw a graph showing i_b on the vertical axis and v_c on the horizontal axis. (This graph is in the second quadrant for the negative values of v_c.)
5. Referring to Fig. 29-10, with v_b constant at 200 V, calculate g_m for a v_c change of 0.5 V from an average of -1 V.
6. (a) Calculate R_S to drop V_{bb} of 300 V to 90 V for V_{c_2}, with an I_{c_2} of 4 mA. (b) Calculate C_S needed to bypass R_S at 50 Hz.
7. Calculate R_L for V_b of 40 V with 300 V for V_{bb} and 2 mA of I_b.
8. How much is the hot resistance of the heater in the 38HK7 diode-beam power tube, with heater current of 450 mA?

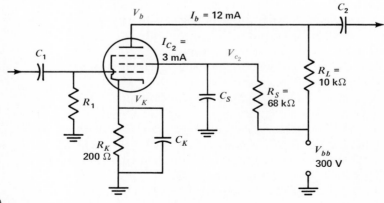

FIGURE 29-20

For Probs. 13 and 14.

9. Calculate the voltage gain of a triode amplifier with 300-mV input signal and 12-V output signal.
10. In Fig. 29-15: (a) Calculate the total cathode current I_K if $I_b = 12$ mA, $I_{c_2} = 4$ mA, and $I_{c_1} = 0$. (b) Calculate I_b if $I_K = 18$ mA, $I_{c_1} = 1$ mA, and $I_{c_2} = 4$ mA.
11. (a) An amplifier has a 50-μV ac input signal and a 600-μV output. How much is the voltage gain A_V? (b) If the gain were 24, how much would the ac output voltage be?
12. Calculate the C needed to bypass a 200-Ω cathode resistance at the lowest audio frequency of 50 Hz.
13. For the pentode amplifier circuit in Fig. 29-20, calculate V_b, V_{c_2}, I_K, and V_K.
14. In Fig. 29-20, how much would R_K be for V_K of 1.5 V?

Answers to Practice Problems

29-1 (a) 40
      (b) 1 MHz
29-2 (a) Cathode
      (b) Positive
29-3 (a) Cathode
      (b) Plate or anode
29-4 (a) I_b
      (b) 2 mA
29-5 (a) Positive
      (b) Positive
29-6 (a) Decreases
      (b) 120 V
29-7 (a) −3 V
      (b) ±1 V
      (c) 200 V
      (d) ±20 V
29-8 (a) 13 mA
      (b) 8 mA
29-9 (a) 12,000 μS
      (b) 750 Ω
29-10 (a) 100 V
      (b) Zero
29-11 (a) T
      (b) T
      (c) T
29-12 (a) 21 V
      (b) Pins 1 and 12
29-13 (a) Anode
      (b) P1
29-14 (a) Infinity
      (b) Triodes

Semiconductor Diodes and Transistors

Chapter 30

This chapter discusses how the semiconductor elements germanium and silicon are used for diode rectifiers and transistor amplifiers. Because of their small size and efficient operation, these semiconductor devices have replaced vacuum tubes in most electronic circuits. Typical examples are shown in Fig. 30-1. The transistor was invented in 1948 by J. Bardeen and W. H. Brattain of the Bell Telephone Laboratories.

The name *transistor* is derived from "transfer resistor," indicating a solid-state device. Instead of depending on thermal emission, the solid semiconductor has free charges that are controlled by the input voltage to provide rectification or amplification in the output. Since there is no heated filament, the operation of semiconductor devices is instantaneous, without any warmup time. The topics here are:

30-1   Types of Semiconductor Devices
30-2   Characteristics of Semiconductors
30-3   N-type and P-type Doping
30-4   Current in Semiconductors
30-5   The PN Junction
30-6   Diode Rectifier Circuits
30-7   Transistors
30-8   Transistor Amplifier Circuits
30-9   The CE Amplifier Circuit
30-10 Collector Characteristic Curves
30-11 Load-Line Analysis
30-12 Bias Stabilization
30-13 Field-Effect Transistor (FET)
30-14 Silicon Controlled Rectifier (SCR)
30-15 Types of Transistors
30-16 Special-Purpose Diodes
30-17 Transistor Troubles

Chapter 30
Semiconductor Diodes and Transistors

FIGURE 30-1

Typical transistors. Height is $\frac{1}{4}$–$\frac{1}{2}$ in without leads.

30-1
TYPES OF SEMICONDUCTOR DEVICES

A semiconductor is a material which has a resistance in between that of a conductor and an insulator. The special feature of the semiconductor elements, however, is that the atomic structure allows their conductivity to be increased by adding impurity elements, a process called *doping*. The purpose is to increase the number of free charges that can be moved by an external applied voltage. When the number of free electrons is increased, the doped semiconductor is negative or N type; reducing the number of free electrons makes the material P type. Both germanium (Ge) and silicon (Si) can be used with either P-type or N-type doping, but silicon semiconductor devices are most common.

Where two opposite types meet, the result is a PN or NP junction. The junction is a continuous structure of the solid semiconductor, but with opposite free charges on opposite sides. A small internal contact potential across the junction keeps the opposite charges separate. It is the junction that is useful for practical semiconductor devices, because the junction voltage controls the flow of current.

A PN junction by itself is a semiconductor diode with two electrodes. The PN diode can be used as a rectifier because it conducts current in one direction.

With either a P or N between two opposite types, the result is a PNP or NPN triode transistor. Now there are three electrodes with two junctions. These electrodes are the *emitter*, the *base* in the middle, and the *collector*. The emitter supplies charges through its junction with the base to provide current to the collector through its junction with the base. For a typical transistor amplifier, the base-emitter junction voltage in the input circuit controls the current through the collector junction for the output circuit.

The field-effect transistor (FET) is also an amplifier, with the same functions as PNP and NPN transistors. However, the FET is constructed to operate with only one polarity of charge carriers, either P or N. The FET has a restricted channel that can be electrically expanded or reduced to control the charge carriers for current. The FET electrodes are *source, gate,* and *drain,* corresponding to emitter, base, and collector in junction transistors. These electrodes are summarized in Table 30-1 and compared with vacuum tubes.

An integrated circuit (IC) combines transistors and diodes with resistors and capacitors on one silicon chip. More details of IC units are described in Chap. 31.

The silicon controlled rectifier (SCR) or thyristor is a power rectifier that has a gate electrode to control the start of conduction. This function is useful for control circuits in industrial electronics.

Finally, there are many types of semiconductor diodes that have useful applications besides rectification, because of the special characteristics of a PN junction. These include capacitance that varies with the junction voltage and photoelectric effects.

Chapter 30
Semiconductor Diodes and Transistors

TABLE 30-1. Electrodes in Transistors and Vacuum Tubes

FUNCTION	VACUUM TUBE	JUNCTION TRANSISTOR	FIELD-EFFECT TRANSISTOR
Source of charges	Cathode	Emitter	Source
Control of charges	Control grid	Base	Gate
Collector of charges	Plate	Collector	Drain

Practice Problems 30-1
(*answers on page 668*)
(a) What electrode in a junction transistor corresponds to the cathode in a tube?
(b) What electrode in the FET corresponds to the collector in a junction transistor?

30-2
CHARACTERISTICS OF SEMICONDUCTORS

The main features of pure semiconductors without doping are: (1) R more than metal conductors but less than insulators; (2) negative temperature coefficient α as R decreases with heat; and most important, (3) electron valence of ± 4. This valence means the atom has four electrons in the outside ring, halfway to the goal of 8 electrons.

As an example of atomic structure, Si has the atomic number 14. This atom has 14 protons in the nucleus balanced by 14 electrons in the outer shells. The electrons are distributed in rings of 2, 8, and 4.

Covalent Bonds. Such atoms do not usually gain or lose the 4 valence electrons but share them with neighboring atoms. The result is a stable configuration with 8 outer electrons. This union of atoms sharing the valence electrons is called a *covalent bond*.[1]

Resistivity. With filled covalent bonds, a semiconductor does not have the individual free electrons present in metal conductors. This factor is why semiconductors have more resistance than metals. As a comparison, approximate values of resistivity ρ in $\Omega \cdot$ cm units are: 2×10^{-6} for copper, 50 for Ge, 50×10^3 for Si, and 2×10^{12} for mica.

Crystal Structure. Si and Ge have atoms in a regular network or lattice of cubes to form a crystal solid. A crystal has a definite geometrical form of the internal atoms. A diamond is an example of the crystalline structure for pure carbon. When a crystal is broken into smaller segments, each has the same structure as the original crystal.

If we visualize atoms of Si interconnected by covalent bonds, as in Fig. 30-2, the result is a lattice of atoms to illustrate the crystalline structure. Because of this crystal lattice of covalent

---

[1] More details of atomic structure and electron valence are explained in Chap. 1. Covalent bonds between silicon atoms are illustrated in Fig. 10-15 in Chap. 10.

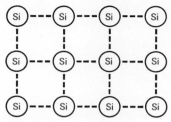

FIGURE 30-2
Crystal lattice structure of a semiconductor, with covalent bonds between Si atoms.

bonds, it is possible to add impurity atoms for doping. The purpose is to alter the electrical characteristics of the semiconductor.

Intrinsic Semiconductors. A pure semiconductor without any doping is called an intrinsic semiconductor. Silicon itself is an element in most common rocks. Sand is silicon dioxide. The element Si was discovered in 1823 and Ge in 1886. Germanium is recovered from the ash of certain coals. The oxides of both Ge and Si are reduced chemically to produce the elements with almost 100 percent purity. Figure 30-3 shows a solid bar of silicon and the slices or disks that are doped for use in semiconductor devices.

(a) (b)

FIGURE 30-3
Forms of pure silicon. (a) Solid bar. (b) Wafer disks. Diameter $1\frac{1}{2}$ in. (*Dow Corning*)

Extrinsic Semiconductors. The doping results in an extrinsic semiconductor, meaning it is not in the natural form. The difference is that an intrinsic semiconductor has only covalent bonds for all the atoms, but an extrinsic semiconductor also has free charges as a result of the doping.

Temperature Coefficient α. Intrinsic semiconductors have a negative α, since R decreases with heat. The reason is in the covalent bond structure. However, an extrinsic semiconductor acts more like a metal conductor because both have free charges that can be moved easily by applied voltage. Therefore extrinsic semiconductors, like metal conductors, have a positive α, since R increases with heat.

Practice Problems 30-2
(answers on page 668)
(a) What is the electron valence for Si and Ge?
(b) Which has the lowest resistivity ρ: copper, paper, or intrinsic silicon?

30-3 N-TYPE AND P-TYPE DOPING

The elements used for doping generally have an electron valence of either 5 or 3. As a result, the doped semiconductor has either an excess or shortage of electrons in the covalent bond structure. A semiconductor with excess electrons is an N type, while a P type has a shortage of electrons. The doping elements are listed in Table 30-2.

Free Electron Charges in N-Type Semiconductor. The doping elements arsenic, antimony, and phosphorus have a valence of 5. For each of these atoms there are 5 electrons in the outermost ring. In a covalent bond with Ge or Si atoms having 4 valence electrons, each impurity atom provides an extra electron.

TABLE 30-2. Doping Elements and Semiconductors

ELEMENT	SYMBOL	ATOMIC NUMBER	VALENCE ELECTRONS	APPLICATIONS
Antimony	Sb	51	5	*Donor* impurity elements. Give electrons to form N-type semiconductor. As and Sb used for Ge; phosphorus for Si
Arsenic	As	33	5	
Phosphorus	P	15	5	
Germanium	Ge	32	4	*Intrinsic* semiconductors used in pure crystal form to be doped with impurity elements. Doped semiconductors are *extrinsic*
Silicon	Si	14	4	
Aluminum	Al	13	3	*Acceptor* impurity elements. Take electrons to form P-type semiconductor. Ga and In used for Ge; Al and B for Si
Boron	B	5	3	
Gallium	Ga	31	3	
Indium	In	49	3	

This idea is illustrated in Fig. 30-4, where the crystal lattice of Si atoms includes one phosphorus atom. Four of the five valence electrons of the impurity element become part of the covalent bond structure. However, the 1 extra electron can be considered a free charge because it is not needed for a covalent bond.

Free Hole Charges in P-Type Semiconductor. The doping elements aluminum, boron, gallium, and indium have a valence of 3. For each of these atoms there are 3 electrons in the outermost ring. In a covalent bond with Ge or Si atoms, there are 7 electrons instead of 8 for each bond with an impurity element. The 1 missing electron in such a covalent bond can be considered as a free positive charge called a *hole charge*. Figure 30-5 illustrates a hole charge in the crystal lattice of doped silicon.

A hole has the same amount of positive charge as a proton, equal to an electron but with opposite polarity. However, a hole charge is not a proton. The proton is a stable charge in the nucleus that is not free to move. A hole is a positive charge outside the nucleus present only in semiconductors because of unfilled covalent bonds.

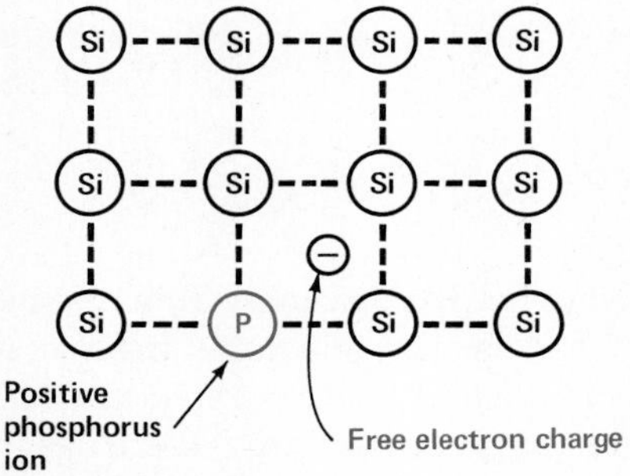

FIGURE 30-4

Crystal lattice structure of Si doped with phosphorus. The covalent bonds have one free electron for each phosphorus atom.

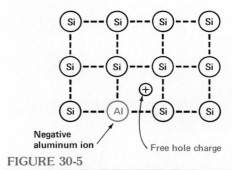

FIGURE 30-5

Crystal lattice structure of Si doped with aluminum. The covalent bonds have one free positive hole charge for each Al atom.

Hole Current. The idea of hole charges moving to provide hole current is illustrated in Fig. 30-6. In (a), along the top row, a hole charge is shown at point 1, along with some filled covalent bonds. Suppose that a valence electron from the filled bond at point 2 moves to point 1. As shown in (b), the bond at point 1 becomes filled and there is a hole charge at point 2. Similarly, an electron can move from point 3 to point 2 to fill this bond. With this sequence, the hole charge is moving from point 1 to point 6 to provide hole current, from left to right here.

To produce this hole current, voltage could be applied across the semiconductor with the positive terminal at point 1. The direction of hole current is the same as conventional current, opposite from electron flow. All symbols for current in semiconductors are shown for the direction of hole current.

It should be noted, though, that hole current flows only in P-type semiconductors. The current is electron flow in N-type semiconductors and all wire conductors.

Fixed Ion Charges. The free charges in a doped semiconductor are balanced by ions of the impurity element. For example, in Fig. 30-4, the phosphorus atom with a valence of 5 supplies 1 free electron that can easily move through the crystal. As a result, the nucleus of the phosphorus atom has 1 extra proton.

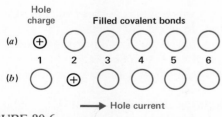

FIGURE 30-6

Hole charge moving in a P-type semiconductor to provide hole current.

Any atom with unbalanced charges is an ion. The phosphorus atom in Fig. 30-4 is a positive ion, therefore.

In Fig. 30-5, the aluminum atom effectively has an *extra* electron as a part of a covalent bond structure, with 8 electrons whenever the free hole charge moves away from the impurity atom. Therefore, the aluminum atom becomes a negative ion.

The ions are fixed charges that are not easily moved in the solid crystal. Since the ions are impurity atoms, they are present only in the doped semiconductor. The fixed charges of the ions are important because they provide an internal contact potential across a PN junction.

It should be noted that the doping really does not add or subtract charges. The semiconductor is still neutral, with equal positive and negative charges. However, the doping redistributes the valence electrons so that more free charges are available.

Practice Problems 30-3
(*answers on page 668*)
(a) Is the doped semiconductor illustrated in Fig. 30-4 N type or P type?
(b) What is the type in Fig. 30-5?

30-4
CURRENT IN SEMICONDUCTORS

With doping, an N-type semiconductor now has a large supply of free electrons, a result of the added impurity atoms. Then the electrons become the dominant or *majority charges*. Still there are *minority charges* of holes. Similarly, a P-type semiconductor has majority hole charges, but there are minority electron charges.

When the majority charges are made to move in the semiconductor by an applied voltage, the result is a relatively large amount of *forward current* or *easy current*. This easy current is a flow of electrons as the majority carriers in N-type semiconductors, or hole current in P-type semiconductors. The amount of *I* is in the range of milliamperes and amperes.

When minority charges move, this current is in the reverse direction, compared with the forward current of the majority charges. The reason is simply that the polarity of minority charges is always opposite from the majority charges.

This very small current of minority charges is called *reverse current* or *leakage current*. It is in the order of microamperes.

Furthermore, the reverse current increases with higher temperatures, as more minority charges are produced by an increase of thermal energy. This increase in the reverse current of minority charges is the reason why temperature is very important in the operation of bipolar transistors.

Drift Current. This term is used for charges in motion that are forced to move by the electric field of a potential difference, which is the usual method of producing current. For instance, the electron flow produced by applied voltage is a drift current. Hole charges in a P-type semiconductor can also provide a drift current with an applied voltage. However, hole current is in the opposite direction from electron flow. The drift current in semiconductors, therefore, can be electron flow or hole current. Also, it can be a motion of majority charges for forward current or minority charges for reverse current.

Diffusion Current. This results from a difference in concentration of the charge carriers in adjacent areas of the crystal lattice in a solid semiconductor. An applied voltage is not necessary. The difference in charge concentration in the solid results in motion.

The charge carriers have a random motion of their own from heat energy, but the motion is restricted to the solid material. The tendency is for charges to move from areas of high concentration to areas of lower concentration. This idea is similar to a drop of ink diffusing through a glass of water.

Either electrons or hole charges can diffuse through a solid semiconductor when one side has a high concentration of majority carriers. In a transistor, the majority charges into the base from the emitter junction diffuse through the base to the collector junction.

Practice Problems 30-4
(answers on page 668)
(a) Are electrons majority or minority charge carriers in N-type semiconductors?
(b) Are hole charges majority or minority carriers in N-type semiconductors?

30-5
THE PN JUNCTION

Figure 30-7 illustrates a magnified view of the junction to analyze the effect of the P and N semiconductors. Actually, the regular structure of the single crystal is the same through the P and N bulk materials and their junction.

Some electrons in the N side are attracted to the P side, while at the same time an equal number of hole charges move in the opposite direction. These charges produce neutral electron-hole pairs. However, this effect at the junction uncovers the charged ions of the impurity atoms. These ions provide an internal barrier potential V_b across the two sides of the junction. V_b has the polarity to prevent any more free electrons or hole charges from crossing the junction. In effect, V_b maintains the free electrons in the N semiconductor and the hole charges in the P semiconductor, to prevent the opposite sides from becoming neutralized.

The Internal Barrier Potential (V_b). Although it is an internal contact potential that cannot be measured directly, the effect can be overcome by 0.3 V for a Ge junction or 0.7 V for Si. The barrier voltage is more for Si because its lower

Chapter 30
Semiconductor Diodes and Transistors

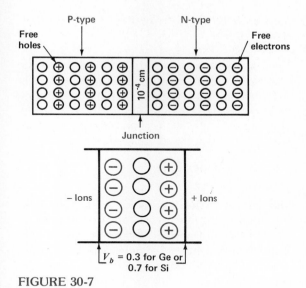

FIGURE 30-7

PN junction with depletion zone magnified to show how ion charges produce internal contact potential V_b at the barrier.

atomic number allows more stability in the covalent bonds.

The V_b is a characteristic of the element. Therefore these values apply to all PN junctions for semiconductor diodes and transistors of any size or power rating. V_b is 0.7 V for any junction in all semiconductors made of silicon; V_b is 0.3 V for any junction made of germanium.

The barrier voltage is what makes the junction useful because the effect of V_b can be controlled by an external voltage. With forward voltage applied in the polarity to cancel V_b, forward current flows. V_b can be neutralized either completely or partially. Reverse voltage is applied in the polarity that does not cancel V_b.

Depletion Zone. Because of its neutral electron-hole pairs, the junction area is the depletion zone, meaning it has no free charges. However, it still has the ion charges anchored in position to produce V_b.

Effect of Temperature on V_b. The values of 0.3 V for Ge and 0.7 V for Si are at normal room temperature of 25°C. However, V_b decreases at higher temperatures. The reason is more minority charge carriers. Specifically, V_b decreases 2.5 mV/°C rise in temperature, for both Ge and Si.

Forward Voltage. See Fig. 30-8a. The for-

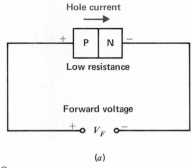
(a)

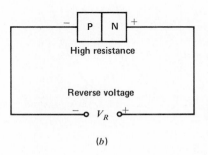
(b)

FIGURE 30-8

(a) Polarity for applying forward voltage V_F across PN junction. Note direction of hole current in the forward direction from P to N. (b) Polarity for reverse voltage V_R. Then there is no forward current.

ward voltage is applied by wire conductors to the P and N electrodes for the bulk materials. Such a connection without any barrier potential is called an *ohmic contact*. Then the external voltage is applied through the bulk materials to the PN junction. Forward current flows as the forward voltage neutralizes V_b. The required polarity is

$+V_F$ to the P electrode
$-V_F$ to the N electrode   forward

In other words, the polarity of V_F corresponds to the semiconductor types.

This polarity of the external V_F can neutralize the internal V_b. The reason is that $+V_F$ at the P electrode repels hole charges to the side of the junction that has negative ion charges for V_b. Then the negative ions are neutralized. Also, $-V_F$ repels electrons to the junction at the positive side of V_b to neutralize the positive ions.

Reverse Voltage. In Fig. 30-8b, the reverse voltage V_R attracts minority charges away from the junction. Then the ion charges at the junction remain intact to maintain V_b. The polarity of V_R is reversed from V_F. Specifically, the required polarity is

$-V_R$ to the P electrode
$+V_R$ to the N electrode   reverse

Voltampere Characteristic. Figure 30-9 summarizes the effects of forward or reverse voltages for Ge and Si. Separate graphs are shown because the forward characteristics in (a) are for V_F in tenths of a volt and high currents in the ampere range, while the reverse characteristics in (b) are for higher voltages with very small reverse currents.

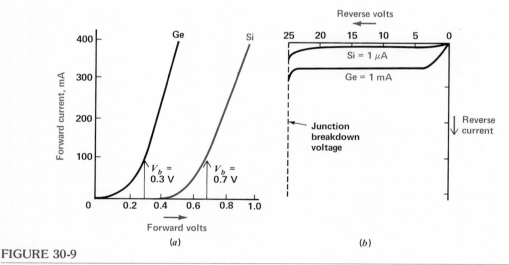

FIGURE 30-9

Voltampere characteristic curves of Ge and Si junctions. (a) Small forward voltage and high current. (b) High reverse voltage and small reverse leakage current.

Forward Current. As shown in Fig. 30-9a, for Si with V_b of 0.7 V, forward current flows when V_F approaches 0.5 V. With 0.7 V or more for V_b, forward current increases sharply to the maximum value at saturation. Then the external V_F completely neutralizes the internal V_b.

A middle value of V_F for forward current is 0.6 V for a Si junction. Similarly, for a Ge junction the range of V_F for forward current is 0.1 to 0.3 V. Typical values of forward current are 0.2 to 20 A.

Reverse Current. With reverse voltage, only a small reverse current of minority charges can flow, as shown in Fig. 30-9b. The separate curves indicate typical values of 1 mA for Ge and 1 μA for Si. Note the advantage of Si with practically zero reverse current.

This current has the symbol I_{CO}, indicating a small cutoff current. The junction is practically an open circuit with reverse voltage, compared to a short circuit with forward voltage.

I_{CO} is in the opposite direction from the forward current. Since I_{CO} consists of minority charges, the reverse current increases with temperature. For every 10°C rise, the I_{CO} doubles approximately for both Ge and Si.

Junction Breakdown. The reverse current is constant at a very small value until V_R reaches the junction breakdown voltage, shown as 25 V in Fig. 30-9b. Then a relatively large reverse current can flow. Most important, the reverse voltage across the junction is constant at the breakdown value. This effect is used for voltage regulator diodes.

For transistor amplifiers and rectifier diodes, though, the forward characteristics in Fig. 30-9a are used. Then the forward current can be controlled to provide amplification and rectification.

Practice Problems 30-5
(answers on page 668)

(a) V_F of 0.6 V is applied to an Si junction. Will forward current flow?
(b) Does V_b increase or decrease with temperature?
(c) Does I_{CO} increase or decrease with temperature?

30-6
DIODE RECTIFIER CIRCUITS

The standard symbol for a semiconductor diode is an arrow and bar showing the direction of hole current (Fig. 30-10). Therefore the arrow is the P side and the bar is the N side. Positive voltage applied to the P arrow makes the diode conduct since this side is the anode, while the N bar is the cathode.

The arrow and bar are generally marked on the diode. If not, a dot or band at one end indicates the cathode side. Or a + mark at the cathode shows this is the terminal for positive dc voltage output.

The only numbering system is the letter N for semiconductors and the prefix 1 for diodes with one junction. As an example, the diode in Fig. 30-12 is type number 1N3754. The last digits indicate a specific type. In schematic diagrams, the semiconductor diodes are labelled X, Y, or CR for crystal rectifier.

However, general-purpose silicon diode rectifiers for receivers often do not have a type number. They usually have maximum ratings of 1 A forward current and 200 to 400 V reverse breakdown voltage. The forward voltage drop across a silicon diode is approximately 1 V, which is very low compared with about 18 V for a tube.

Half-Wave Rectifier. In Fig. 30-11, the input voltage V_{ac} is applied to the diode in series with the output load resistor R_L. For the positive half-cycles of ac input, the P side of the diode is

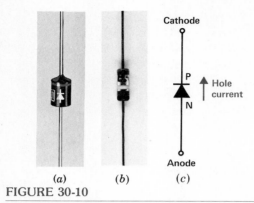

FIGURE 30-10

Semiconductor diode rectifiers. Length about ½ in without leads. (*a*) Silicon power diode rated at 1 A. Arrow shows hole current. (*b*) Germanium detector diode. Band at cathode end. (*c*) Schematic symbol showing direction of hole current.

positive. This is the polarity for forward current. Then the diode conducts. V_{ac} then produces current through R_L, providing fluctuating dc voltage output V_L across R_L.

The polarity of V_L is positive at the top of R_L in Fig. 30-11 because this end is connected to the positive terminal of V_{ac} through the conducting diode. On the negative half-cycle of ac input, the P side of the diode is negative. This polarity provides reverse voltage, and the diode cannot conduct. Then there is no output.

Although not a steady dc value, the fluctuating output V_L is a dc voltage because it has only one polarity. The fluctuating component is the ac ripple in the dc output. With 60 Hz ac voltage from the power line, the ripple frequency is 60 Hz. In a dc power supply, a filter is used to reduce the amplitude of ac ripple.

Operating Characteristics. The graph in Fig. 30-12a shows how much dc output voltage can be produced across R_L for the half-wave power supply in (*b*). Three curves are shown for different values of C because the filter capacitor at the cathode of the rectifier has a big effect on the operating characteristics.

Besides reducing the ac ripple, C can charge to the peak value of the ac input voltage, which is close to 170 V. How much C discharges between positive peaks depends on R_L. Lower values of R_L, corresponding to more load current I_L, allow more discharge from C, resulting in less dc output voltage.

To take one example from the curve for a 50-μF C, the dc load current I_L of 100 mA allows 145 V for V_L. The dc output voltage is more than the rms ac input because C charges to the peak value. The 5.6-Ω R_S is a *surge-*

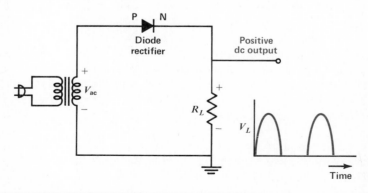

FIGURE 30-11

Half-wave rectifier circuit using silicon diode.

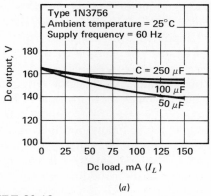
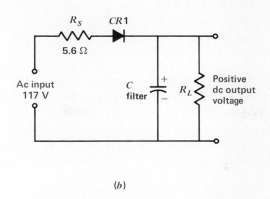

FIGURE 30-12

Operating characteristics of half-wave power supply with filter C and load R_L. (a) Graph of dc output voltage vs. load current. (b) Circuit.

limiting resistor. It protects the diode against excessive charging current at the start of operation when C charges from zero.

Inverted Diode Rectifier. If the connections to CR1 are reversed in Fig. 30-12, the dc output with respect to chassis ground will have negative polarity. Then the ac input is applied to the cathode. Driving the cathode negative is the same as making the anode positive. The inverted diode then conducts on the negative half-cycle of ac input. Also, the dc output voltage across R_L in the anode circuit then is negative.

Full-Wave Rectifier. In Fig. 30-13, both alternations of the ac input produce dc output. Two diodes are necessary. The diode Y1 conducts for one alternation when its anode is driven positive, while Y2 rests as its anode is negative.

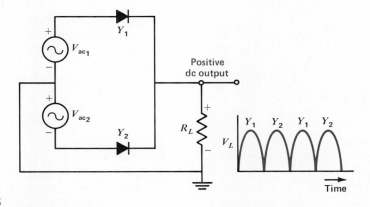

FIGURE 30-13

Full-wave rectifier circuit with two diodes.

On the next alternation, the ac input voltage reverses in polarity and Y2 conducts without Y1. The ac input supplies equal and opposite voltages, usually with a center-tapped secondary winding in the power transformer.

Although the diodes conduct on opposite half-cycles, notice that in both cases the polarity is the same for the dc output across R_L. For either Y1 or Y2 the dc load R_L is in the cathode return circuit. The top of R_L is connected to a cathode for positive dc output from both diodes.

The ripple frequency for the full-wave rectifier is double the frequency of the ac input, as each half-cycle produces a fluctuation of dc output voltage. With the 60-Hz power line as an ac input, the ripple frequency in the dc output is 120 Hz.

Power-Supply Hum. In a receiver, insufficient filtering of the ac ripple can cause hum to be heard from the loudspeaker. Excessive hum is often the result of an open or leaky filter capacitor. The hum frequency is 60 Hz for a half-wave rectifier or 120 Hz with a full-wave rectifier, with an ac input of 60 Hz. A higher ripple frequency means better filtering as the same shunt C has one-half reactance for double the frequency.

Power Supplies. The following types of rectifier circuits are often used for a dc power supply operating from the 60-Hz ac power line.

1. *Half-wave rectifier.* One diode for a simple and economical circuit. The ripple is 60 Hz.
2. *Full-wave rectifier.* Two diodes can supply double the load current. The ripple is 120 Hz.
3. *Full-wave bridge rectifier.* Four diodes are used for full-wave operation without a center tap in the ac input. The ripple is 120 Hz.
4. *Voltage doubler.* Two diodes provide a series circuit for dc output voltage double the ac input voltage. Either a half-wave or full-wave circuit can be used.

Detector Circuits. A detector is just a low-power diode rectifier for a few volts of ac signal. For the audio detector in AM radio receivers, as an example, the ac input is modulated rf signal. This ac voltage must be rectified to filter out the rf carrier wave and extract the variations of amplitude in the audio modulation. Generally, a detector uses a very small Ge or Si diode as a half-wave rectifier, as shown in Fig. 30-10b.

Checking Diodes with an Ohmmeter. The diode should have at least 100 times more R in the reverse direction, compared with the forward resistance. Just connect the ohmmeter across the diode in one polarity, and then reverse the leads for the opposite polarity. A silicon diode has practically infinite R in the reverse direction.

When R is very high in both directions, the diode is open. When R is very low in both directions, the diode is shorted. If parallel paths confuse the readings, disconnect one side of the diode from the circuit.

Practice Problems 30-6
(answers on page 668)
(a) What is the ripple frequency of a full-wave power supply?
(b) Referring to the graph in Fig. 30-12a, how much is V across R_L with 100 μF for C and 150 mA for I_L?

30-7
TRANSISTORS

As shown in Fig. 30-14, the transistor consists of a PN junction and an NP junction, by making either a P or N semiconductor between opposite types. The purpose is to have the first section supply charges, either holes or electrons, to

Chapter 30
Semiconductor Diodes and Transistors

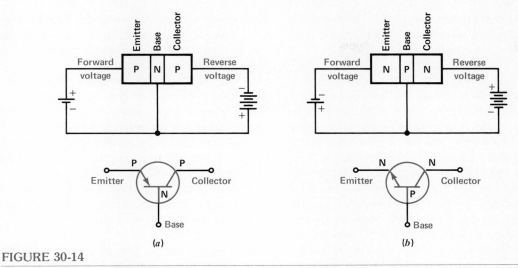

FIGURE 30-14

Junction transistors illustrating emitter, base, and collector. (*a*) PNP. (*b*) NPN.

be collected by the third section, through the middle section. The electrode that supplies charges is the *emitter;* the electrode at the opposite end to collect the charges is the *collector.* The *base* in the middle forms two junctions between emitter and collector, to control the collector current.

Emitter. The emitter-base junction is biased with forward voltage. Typical values are 0.2 V for Ge or 0.6 V for Si.

As shown for the PNP transistor in Fig. 30-14*a*, the P emitter supplies hole charges to its junction with the base. This direction is indicated by the emitter arrow for forward hole current in the schematic symbol. The arrow pointed into the base shows a PN junction between emitter and base, corresponding to the symbol for a PN diode.

For the NPN transistor in (*b*), the emitter supplies electrons to the base. Therefore, the symbol for the N emitter shows the arrow out from the base, opposite to the direction of electron flow.

In the schematic symbols, only the emitter has an arrow. The arrow pointing into the base means a PNP transistor; the arrow out from the base means an NPN transistor. Practically all small transistors for audio and rf amplifiers are NPN, made of silicon, with a typical forward bias of 0.6 V between base and emitter.

Collector. Its function is to remove charges from the junction with the base. In Fig. 30-14*a*, the PNP transistor has a P collector receiving hole charges. For the NPN transistor in (*b*), the N collector receives electrons.

The collector-base junction always has reverse voltage. Typical values are 4 to 100 V. This polarity means no majority charges can flow from collector to base. However, in the opposite direction, from base to collector, the collector voltage attracts the charges in the base supplied by the emitter.

Base. The base in the middle separates the emitter and collector. The base-emitter junction is forward-biased. As a result, the resistance is very low for the emitter circuit. The base-collector junction is reverse-biased, providing a much higher resistance in the collector circuit.

Collector Current. The final requirement for transistor action is to have the collector current controlled by the emitter-base circuit. The emitter has heavy doping to supply majority charges. However, the base has only light doping and is very thin, so that its charges can move to the collector junction. The collector voltage is relatively high. Because of these factors, practically all the charges supplied by the emitter to the base are made to flow in the collector circuit. Typically, 98 to 99 percent or more of the emitter charges provide collector current (I_C). The remaining 1 to 2 percent or less becomes base current (I_B).

Consider the currents for an NPN transistor. The N emitter supplies electrons to the P base. Here the electrons are minority charges. Because of light doping in the base, though, very few of the electrons can recombine with hole charges. Any recombination of charges in the base provides the very small I_B returning from base to emitter.

There is a heavy concentration of free electron charges in the base at the emitter junction, because of the forward bias. As a result, almost all the electrons diffuse through the thin base to the collector junction.

The N collector has reverse voltage of positive polarity. For electrons moving from the base, however, the positive collector voltage attracts these free charges. As a result, the electrons diffused from the emitter side of the base move into the collector to provide a drift current of electrons for I_C in the collector circuit. For a PNP transistor, all voltage polarities are reversed and I_C at the collector electrode consists of hole charges.

Electrode Currents. As an example, in Fig. 30-15, the emitter supplies 10 mA of forward current. This is I_E. Of this, 9.8 mA is injected into the collector circuit. This is I_C. Only 0.2 mA or 200 µA of I_B flows through the base terminal to return to the emitter. As a formula

$$I_E = I_C + I_B \tag{30-1}$$

Note that I_E in Fig. 30-15 is marked negative, only to indicate its direction is opposite from I_C and I_B. It is standard practice to consider

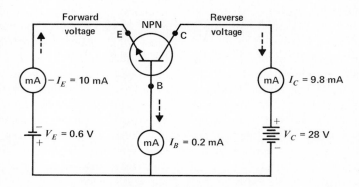

FIGURE 30-15

Electrode currents I_E, I_B, and I_C shown for NPN transistor. Dotted arrows show direction of electron flow for I in external circuit.

hole current into a semiconductor as the positive direction of I. Since I_E is electron flow into the transistor, this I is in the negative direction.

Algebraically, the values are -10 mA $+$ 9.8 mA $+ 0.2$ mA $= 0$. In practical terms, the formula states that the collector and base currents must add to equal the emitter current, which is the source.

Example 1. How much is I_E with 100 mA for I_C and 800 µA or 0.8 mA for I_B?

Answer. $\quad I_E = I_C + I_B = 100 + 0.8$
$\quad\quad\quad\quad I_E = 100.8$ mA

Example 2. With 4 A for I_E and 3.8 A for I_C, how much is I_B?

Answer. Transposing Formula (30-1),

$I_B = I_E - I_C$
$\quad\; = 4 - 3.8$
$I_B = 0.2$ A or 200 mA

For power transistors, I_E and I_C are generally in amperes.

The Base Current Controls the Collector Current. When I_B is increased by more forward voltage, this means more majority charges are in the base to be injected into the collector. Therefore, increasing I_B means more I_C. For the opposite effect with less forward voltage and less I_B, the collector current is reduced.

Practice Problems 30-7
(answers on page 668)
(a) Does the base-emitter junction have forward or reverse bias?
(b) Does the collector-emitter junction have forward or reverse bias?
(c) I_C is 1 mA. I_B is 5 µA. How much is I_E?
(d) An Si transistor has 0.1-V forward bias. Is it conducting or cut off?

30-8
TRANSISTOR AMPLIFIER CIRCUITS

Since a PNP or NPN transistor has only three electrodes, one must be a common connection for two pairs of terminals for input and output signals. The general case of three electrodes, with one common, is illustrated in Fig. 30-16a. Specifically, the three possibilities for amplifier circuits are: common base (CB) in (b), common emitter (CE) in (c), and common collector (CC) in (d). The circuits are shown for NPN transistors, but they are the same for PNP transistors with all polarities reversed. Although the common electrode is shown grounded here, it need not be connected to chassis ground. The main features of these circuits are compared in Table 30-3.

For all these amplifier circuits, the collector has reverse voltage and the emitter-base junction requires forward bias. It is labeled either V_{BE} or V_{EB}. This average forward voltage is called *bias voltage*, and the corresponding current is *bias current*. The bias sets the average dc values in the amplifier. Then ac signal varies the values around the bias level.

The common-emitter circuit is the one generally used for transistor amplifiers because it has the most gain with cascaded stages. However, each type of circuit has special features.

Common-Base (CB) Circuit. In Fig. 30-16b, the input voltage is applied to the emitter, with respect to the grounded base. The amplified output is taken from the collector. R_L is in series with the collector supply V_{CC}. The positive side of V_{CC} supplies reverse voltage for the N collector.

In the CB circuit the emitter input has low resistance r_i because I_E is high. The output resistance r_o for the collector is high. Typical values for a small-signal transistor with I_E of 1.5 mA are 20 Ω for r_i and 1 MΩ for r_o. These

Chapter 30
Semiconductor Diodes and Transistors

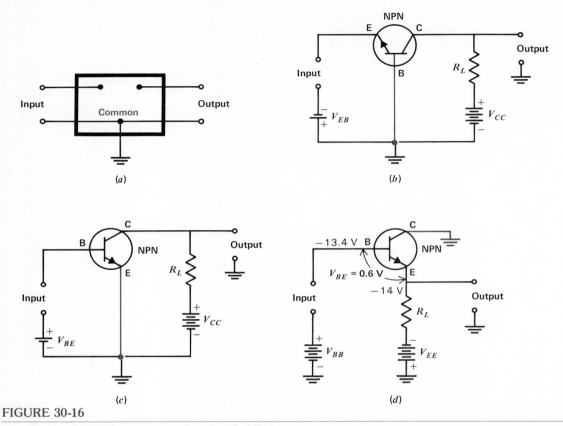

FIGURE 30-16

Types of amplifier circuits, shown with NPN transistor. All polarities are reversed for a PNP transistor. (*a*) General case of a common terminal for two pairs of connections. (*b*) Common base (CB). (*c*) Common emitter (CE). (*d*) Common collector (CC) or emitter-follower.

TABLE 30-3. Comparison of Transistor Circuits

CHARACTERISTIC	COMMON BASE (CB)	COMMON EMITTER (CE)	COMMON COLLECTOR (CC)
Signal into	Emitter	Base	Base
Signal out of	Collector	Collector	Emitter
Advantage	Stability	High gain	High r_i
Phase inversion	No	Yes	No
Similar vacuum-tube circuit	Grounded grid	Grounded cathode	Grounded plate (cathode-follower)

values are for the internal resistance of each electrode to the common base.

The CB circuit is seldom used. It has no current gain from input to output because I_C must be less than I_E. The voltage gain can be high, but the output is shunted by the low input resistance of the next stage. The only advantage is that the CB circuit has the best stability with an increase in temperature. The reason is that reverse leakage current from collector to base is not amplified in the CB circuit.

Common-Emitter (CE) Circuit. See Fig. 30-16c. Input voltage is applied to the base instead of the emitter, which is now the grounded electrode. Note that the emitter is shown at the bottom of the schematic symbol. The input circuit here involves I_B instead of I_E. As a result the r_i for the CE circuit is much higher than for the CB circuit. A typical value is 1000 Ω for r_i.

The output voltage is still taken from the collector with its R_L. The V_{CC} is positive for reverse bias on the N collector. A typical value for r_o in the collector output circuit is 50 kΩ.

In the input circuit, the forward bias V_{BE} is applied to the base instead of the emitter. Note the polarity. Positive V_{BE} to the P base corresponds to negative V_{EB} at the negative emitter. Both are forward-bias voltages, with the polarity the same as the N or P electrode.

Furthermore, the positive V_{BE} at the base for forward bias uses the same voltage polarity as positive V_{CC} for reverse voltage at the N collector. This feature allows the practical convenience of using one voltage supply for both forward bias in the input and reverse bias in the output. As an example, the CE circuit in Fig. 30-17 uses R_1 to drop the collector supply of 12 V to 0.6 V for the base.

The CE circuit has current gain because I_C is much larger than I_B. The voltage gain is the same as for the CB circuit. With a higher r_i, however, the CE circuit can be used in cascaded amplifiers where the collector output of one stage drives the base input of the next. The CE circuit is the amplifier generally used for transistors because it has the best combination of voltage gain and current gain. The disadvantage is that reverse leakage current is amplified in the CE circuit, but bias stabilization methods can be used.

Only the CE amplifier inverts the polarity of signal voltage. This phase inversion of 180° is neither good nor bad, but just a result of the circuit connections. When the base input signal increases the forward voltage, the collector voltage of the same polarity decreases because of the voltage drop across R_L.

Common-Collector (CC) Circuit. See Fig. 30-16d. Signal input is applied to the base, as in the CE circuit. However, the collector is grounded, instead of the emitter. Therefore, the emitter has R_L for output signal.

Now there are two questions for the electrode voltages: How to apply reverse voltage for the grounded collector and forward bias for the base. Note that the emitter is at −14 V with respect to chassis ground. The collector is at chassis ground. In effect, the collector is connected to the positive side of the emitter supply voltage. This makes the N collector positive with respect to the emitter, as reverse voltage for V_{CE}.

For forward voltage, the base must be more positive than the emitter. V_E in this example is −14 V. Therefore V_B is made −13.4 V. The actual bias V_{BE} then is 0.6 V.

In the CC circuit, the input circuit has high r_i. A typical value is 150 kΩ. The output in the emitter circuit has low r_o of about 80 Ω. Note that for the CC circuit, the input resistance is high and output resistance is low, compared with low r_i and high r_o for the other circuits.

There is no voltage gain in the CC circuit because the output signal voltage across R_L in the emitter circuit provides negative feedback to

Chapter 30
Semiconductor Diodes and Transistors

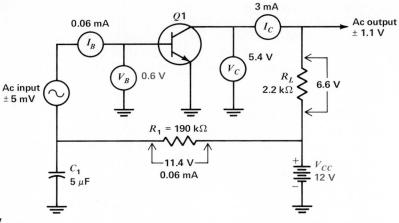

FIGURE 30-17
Common-emitter amplifier circuit. NPN takes $+V_{CC}$ for reverse voltage and $+V_B$ for forward bias. V_B is 0.6 V for Si.

the base input circuit. However, there is appreciable current gain.

Emitter-Follower. This is the name generally used for the CC circuit, because it corresponds to the cathode-follower circuit with vacuum tubes. The output signal follows the polarity of the input signal. These circuits are often used for impedance matching, from a high-impedance source to a low-impedance load.

Darlington Pair. This circuit consists of two emitter-followers in cascade. The two stages are usually packaged in one unit, with dc coupling internally and just three external leads. The package provides higher input resistance and more current gain than just one stage.

Practice Problems 30-8
(answers on page 668)
(a) Which circuit has input signal to the base and output from the collector?
(b) Which circuit has the most gain?
(c) Which circuit has the highest input resistance?

30-9
THE CE AMPLIFIER CIRCUIT

Typical values of V_{BE} are in tenths of a volt for a junction transistor. The required bias at the base for a class A amplifier is 0.6 V for Si or 0.2 V for Ge. Furthermore, the maximum ac input signal without overload distortion is ± 0.1 V. These values are summarized in Table 30-4. Note that 0.1 V is 100 mV.

TABLE 30-4. Input Voltages V_{BE} at 25°C

	CUT-IN VOLTAGE	SATURATION VOLTAGE	ACTIVE REGION	AVERAGE BIAS VOLTAGE
Ge	0.1	0.3	0.1–0.3	0.2
Si	0.5	0.7	0.5–0.7	0.6

Without any forward bias a junction transistor is cut off by its own internal barrier potential. The cut-in voltage in the first column of Table 30-4 is the lowest V_{BE} that allows appreciable I_C. This V_{BE} for any I_C at all is only 0.1 V less than the average forward bias.

The saturation voltage in the second column of Table 30-4 is the highest V_{BE} that allows it to produce proportional changes in I_C. At saturation, the maximum I_C does not increase with an increase of forward voltage. The saturation V_{BE} is only 0.1 V more than the average forward bias.

Circuit Components. The transistor amplifier itself is usually labelled Q, as for Q1 in Fig. 30-17. R_L is the collector load. It is in series with the positive V_{CC} of 12 V for reverse collector voltage on the NPN transistor.

One supply voltage is used for both collector and base in the CE circuit, as positive base bias and positive collector voltage are needed. However, the collector voltage is too high for base bias. Therefore, R_1 is connected in series as a voltage-dropping resistor for the base.

The required forward-bias voltage for Si is 0.6 V. The corresponding bias current for this transistor is taken as 60 μA or 0.06 mA, as an example. Therefore, the voltage drop across R_1 is $0.06 \text{ mA} \times 190 \text{ k}\Omega = 11.4 \text{ V}$. Then $12 - 11.4 = 0.6$ V remains for V_{BE} as forward voltage for the base.

The bypass C_1 allows the ac input signal voltage to vary the base current without the series resistance of R_1. Then very small changes of input voltage can produce appreciable changes in base current. We are assuming an ac input of ±5 mV for the base input signal, or 10 mV p-p.

Signal Variations. The input signal of base current i_B is shown in Fig. 30-18a. This waveform shows i_B varies by 10 μA above the 60-μA

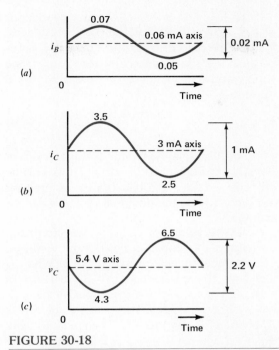

FIGURE 30-18

Ladder diagram of signal waveforms for circuit in Fig. 30-17. (a) Base current i_B. (b) Collector current i_C. (c) Output voltage $V_C = V_{CC} - i_C R_L$.

bias axis, up to the peak of 70 μA. On the down side, i_B decreases by 10 μA, from 60 to 50 μA. Positive signal voltage in the forward direction increases i_B, while negative signal voltage reduces i_B. The peak-to-peak signal in i_B then is $70 - 50 = 20$ μA, or 0.02 mA.

The variations in i_B cause corresponding variations in i_C, as shown in waveform (b). Let the current transfer ratio be 50, meaning this is the ratio of collector output current to base input current. Then the i_B variations of ±10 μA swing i_C by $50 \times 10 = 500$ μA or 0.5 mA.

The average dc level for I_C is taken as 3 mA for a small-signal low-power transistor. Then the ac signal swing of i_C is ±0.5 mA above and below the axis of 3 mA. In i_C, the peak-to-peak signal is $3.5 - 2.5 = 1$ mA.

The signal changes in i_C produce variations in the voltage drop $i_C R_L$ across the collector load for the output circuit. As a result, V_C varies because it is the difference between V_{CC} of the supply and the voltage drop across R_L. As a formula,

$$V_C = V_{CC} - i_C R_L \tag{30-2}$$

For example, with an average level of 3-mA I_C through the 2.2-kΩ R_L, this voltage drop is $0.003 \times 2200 = 6.6$ V. Subtracting 6.6 V from 12 V, the difference is 5.4 V for the average V_C.

The variations of v_C are in waveform (c). This shows the amplified signal output voltage. The average dc level or axis is 5.4 V. When i_B increases, the v_C decreases to 4.3 V because of a larger voltage drop across R_L. On the next half-cycle i_B decreases. Then less voltage across R_L allows v_C to rise to 5.4 V. Then peak-to-peak signal voltage is $6.5 - 4.3 = 2.2$ V for v_C at the collector. This amplified output voltage is 180° out of phase with the signal input voltage at the base.

The basis for Formula (30-2) is just the fact that R_L and the collector-emitter circuit of the transistor are in series with each other as a voltage divider from the high side of V_{CC} to chassis ground. The equivalent circuit is shown in Fig. 30-19. R_L is the external collector load. R_Q is the internal resistance of the transistor conducting current from emitter to collector. In this example, Q1 is conducting 3 mA with 5.4 V for V_{CE}. The equivalent R_Q is $5.4/0.003 = 1800$ Ω. Note that V_Q is the same as V_C.

Gain. To calculate the voltage amplification, or gain, with p-p values for Fig. 30-17,

$$A_V = \frac{V_{out}}{V_{in}} \tag{30-3}$$

$$A_V = \frac{2.2 \text{ V}}{0.01 \text{ V}} = 220$$

The current amplification is

$$A_I = \frac{i_{out}}{i_{in}} \tag{30-4}$$

$$A_I = \frac{i_C}{i_B} = \frac{1 \text{ mA p-p}}{0.02 \text{ mA p-p}} = 50$$

The power amplification is

$$A_P = A_V \times A_I \tag{30-5}$$
$$A_P = 220 \times 50 = 11{,}000$$

Since the CE circuit has both voltage and current amplification, the power gain is high. Power gain is desirable because it means the voltage output can drive a low-impedance circuit without too much loss of voltage amplification.

Cascaded Stages. As shown in Fig. 30-20, the amplifiers $Q1$, $Q2$, $Q3$, and $Q4$ are in cascade. With CE amplifiers, the collector output of one stage drives the base input of the next. It is important to realize that the voltage amplifica-

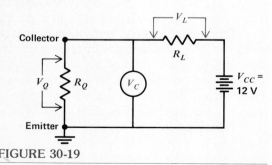

FIGURE 30-19

Equivalent voltage divider circuit for V_C in series with V_L.

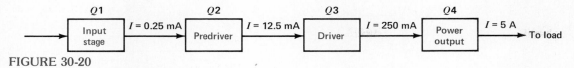

FIGURE 30-20
Amplifier stages in cascade.

tion need not build up the signal to a level greater than ±0.1 V. This is the maximum signal swing without distortion for junction transistors. However, the cascaded stages provide enough signal current to drive the base of the output stage Q4.

As an example, suppose that Q4 is a power output stage to drive a loudspeaker as the load that needs 5 A. With an average level of 5 A for I_C in Q4, its I_B would be in the order of 250 mA or 5 A/20, with a current transfer ratio of 20. This I_B of 250 mA can be supplied by the collector output of the driver-stage Q3. Similarly, Q3 with an I_C of 250 mA would have I_B of 12.5 mA with a current ratio of 20. Also, with a current ratio of 50 for Q2, its I_B would be 12.5 mA/50 = 0.25 mA, or 250 µA. This drive for base current in Q3 can be provided by the collector current of the input stage Q1.

Practice Problems 30-9
(answers on page 668)
(a) In Fig. 30-17, how much is the base-bias voltage V_{BE}?
(b) In Fig. 30-19, how much is V_Q with V_L of 5.5 V?
(c) In Fig. 30-18, how much is the p-p signal output in i_c?

30-10
COLLECTOR CHARACTERISTIC CURVES

The transistor is a nonlinear device, since the collector current does not vary in direct proportion to changes in either V_C or V_B. Therefore, it is necessary to show the characteristics by curves. Figure 30-21 illustrates a circuit to vary the voltages experimentally in order to determine the values of I_C.

The main idea here is to change V_{CE} to determine I_C in the output. However, I_C also depends on how much I_B is in the input, for the CE circuit. The results for one value of I_B at 50 µA are shown by the graph in (b). Finally, the values for all typical operating conditions are shown in (c). This is a family of collector characteristic curves for different values of base current. The curves are for I_B values, not V_B, because the base current determines I_C. Furthermore, I_B is not linear with respect to V_B.

Such curves are provided by the manufacturer, in a transistor manual, or in application notes. It should be noted that for a CB circuit, the collector curves would be shown for different values of emitter current.

The curve in (b) shows that with a 50-µA I_B, the collector current is about 3 mA. Note that I_B is in microamperes here while I_C is in milliamperes. In general, I_C must be larger than I_B, as I_C is almost all the emitter current, but I_B is only 1 or 2 percent of I_E.

The collector current increases very little as the reverse voltage V_{CE} is increased to 16 V. The reason is that I_C is limited by how much emitter current and base current are permitted by the forward voltage V_{BE} in the input circuit.

The results for different values of I_B are shown in the family of collector characteristic curves in (c). Each curve can be considered as a single graph of I_C vs. V_{CE} for a specific value of base current I_B.

As an example of how to read the family of

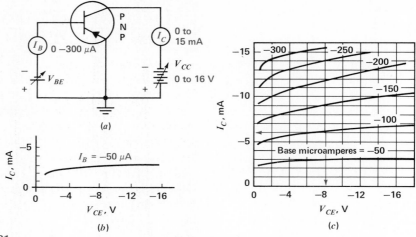

FIGURE 30-21

Collector characteristic curves. These show a separate graph for each I_B value in the CE circuit. (*a*) Circuit to determine I_C vs. V_C with a fixed I_B. (*b*) Typical collector characteristic for I_B of 50 μA. (*c*) Family of collector curves for different values of I_B.

curves, assume I_B is 100 μA for the second curve up from the bottom in (*c*). Then for a V_{CE} of 8 V, the value of I_C is 6 mA, as shown by the arrows on the graph.

Beta Characteristic. The ratio of collector current to base current is the β (beta) characteristic of the transistor. As a formula,

$$\beta = \frac{I_C}{I_B} \qquad (30\text{-}6)$$

The beta characteristic is the current transfer ratio for a common-emitter circuit. There are no units, since this is a ratio of two currents. Values for β are 10 to 300. For a small-signal low-power transistor 200 is typical, while power transistors have β of 20 to 30 as a typical value.

Example 3. Calculate β from Fig. 30-21*c* for a 6-mA I_C and a 100-μA I_B with V_{CE} at 8 V.

Answer. The 100-μA I_B is 0.1 mA. Then

$$\beta = \frac{I_C}{I_B} = \frac{6 \text{ mA}}{0.1 \text{ mA}}$$
$$\beta = 60$$

Example 4. Calculate I_C for a 1-mA I_B and a β of 50.

Answer. I_C is $\beta \times I_B$. Also, $I_B = I_C/\beta$. Then

$$I_C = \beta \times I_B$$
$$= 50 \times 1$$
$$I_C = 50 \text{ mA}$$

Alpha Characteristic. The α (alpha) is the current transfer ratio for a common-base circuit. Then the input current is emitter current, although the output is still collector current. As a formula,

$$\alpha = \frac{I_C}{I_E} \qquad (30\text{-}7)$$

Values of α are 0.97 to 0.99, with 0.98 as typical. This means 98 percent of the emitter current is injected into the collector. The α must be less than 1, as I_C cannot be more than I_E.

Example 5. Calculate α for 6.1-mA I_E, 6-mA I_C, and 0.1-mA I_B.

Answer. These values show that I_E is the total of I_C and I_B, but the base current is not used for calculating alpha.

$$\alpha = \frac{I_C}{I_E} = \frac{6 \text{ mA}}{6.1 \text{ mA}}$$
$$\alpha = 0.984$$

These values for α and β are static dc values, without ac signal input. The ac dynamic values are calculated for small changes in the currents. It should be noted that α and β are related as $\beta = \alpha/(1 - \alpha)$.

Cutoff Frequency. The frequency at which the value of α or β drops to 0.707 of its value at 1 kHz is the cutoff frequency of the transistor. Internal capacitances and diffusion time in the base cause the drop at higher frequencies. For small-signal transistors in rf circuits, the cutoff frequency is typically 300 MHz. Typical values of collector-base capacitance are 5 to 10 pF for small transistors and up to 900 pF for power transistors.

Collector Cutoff Current. If the emitter were open, there would be no forward current into the base. Then the collector junction with the base is the same as a diode with reverse bias provided by the collector voltage. The normal I_C of majority carriers is zero, but the diode has its reverse saturation current I_{CO}. In a transistor, I_{CO} is a reverse current of minority carriers in the same direction as the collector current of majority carriers. The problem with I_{CO} is that the output current does not drop completely to zero when the input circuit is cut off.

I_{CO} is very sensitive to temperature because it consists of minority carriers. As in diodes, the reverse saturation current doubles approximately for every 10°C rise in temperature, for both Ge and Si. However, I_{CO} for Si is about $\frac{1}{1000}$ of the value for Ge. As an example, for a 5-W Si transistor with I_C of 5 A, the I_{CO} is only 10 µA at 25°C. However, at 150°C, I_{CO} rises to several milliamperes.

Saturation. The collector is saturated, with maximum I_C, when V_C drops below the base voltage. Then the collector junction becomes forward-biased, instead of having its normal reverse voltage. The low V_C can result from the voltage drop across R_L.

Practice Problems 30-10
(answers on page 668)
(a) In Fig. 30-21c, how much is I_C for 4 V of V_{CE} and 300 µA of I_B?
(b) How much is β when a change of 10 µA in i_B changes i_C by 3 mA?

30-11
LOAD-LINE ANALYSIS

The collector characteristic curves show electrode voltages and currents for the transistor itself, without a load in the output circuit. Actually an external load impedance is necessary to provide amplified output voltage. A typical circuit is shown in Fig. 30-22a with a 5-Ω R_L.

Although the transistor is nonlinear, R_L has a linear voltampere characteristic. To see the effect of R_L on the collector voltage and current, the straight-line characteristic of R_L is superimposed on the collector characteristic curves, as in (b). This graphical analysis with the load line

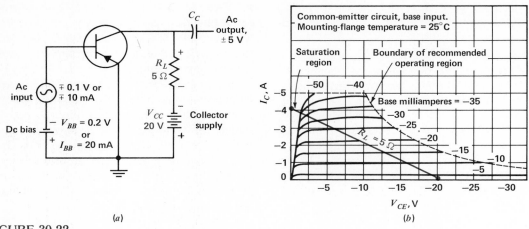

FIGURE 30-22

Load-line analysis for circuit with 5-Ω R_L and 20-V collector supply. (a) CE circuit. (b) Construction of load line between points at 20 V for V_C and $^{20}/_5 = 4$ A for I_C.

of R_L can be used to determine specific values. The details of the load-line intercepts with the collector characteristics are shown separately in Fig. 30-23.

The CE circuit in Fig. 30-22a uses a PNP germanium power transistor, with amperes of collector current. In the output circuit, the collector supply voltage V_{CC} is 20 V. In the input circuit, a V_{BB} of 0.2 V provides the forward bias for a 20-mA base current. It should be noted that currents and voltages are negative for the base and collector of a PNP transistor, but only

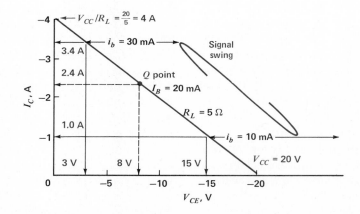

FIGURE 30-23

Details of load line for Fig. 30-22. See text for this graphical analysis.

the values are considered here. I_C and I_B are in the negative direction because hole charges are moving out of the transistor.

Constructing the Load Line. All values of collector current and voltage with a specified R_L are on the load line of R_L, drawn on the family of collector characteristics. To construct the load line, we need only the values of R_L and the supply voltage V_{CC}, which determine the two end points. One point is at V_{CC}, equal to 20 V on the horizontal axis where I_C is zero. This is one operating point because the collector voltage equals V_{CC} when there is no I_C and no voltage drop across R_L.

The opposite point is at the extreme value of collector current where V_C would be zero with the voltage drop across R_L equal to the supply voltage V_{CC}. This end of the load line is at $I_C = V_{CC}/R_L$ on the vertical axis where V_C is zero. This point on the vertical I_C axis is at $20/5 = 4$ A.

The straight line drawn between 4 A on the vertical axis and 20 V on the horizontal axis is the load line for the 5-Ω R_L with a 20-V supply. For any value of i_C, the corresponding v_C must be on the load line, which takes into account the $i_C R_L$ voltage drop.

Q Point. Where the load line intersects the collector curve for the base-bias current of 20 mA in this example is the Q point or quiescent point. This point specifies the static dc values without any ac signal input. The operating point of 20 mA for I_B is chosen here because it is a middle value between saturation and cutoff of the collector current.

As shown in Fig. 30-23, the curve for a 20-mA I_B intersects the load line where the value of V_C is 8 V and I_C equals 2.4 A. As a check on these values, V_C and the voltage drop $I_C R_L$ must add to equal the 20-V supply. $I_C R_L$ is $2.4 \times 5 = 12$ V. Adding this 12 V to the 8 V of V_C, the sum then equals the 20 V of V_{CC}. These are average dc values without signal, or *quiescent* values.

AC Signal Swing. With input signal to the base, the ac drive changes the base current up to the peak of 30 mA and down to the minimum of 10 mA. These values of i_B are two curves up and down from the Q point of 20 mA. The intercepts with the load line are shown in Fig. 30-23.

The peak i_B of 30 mA produces a peak collector current of 3.4 A; the minimum i_B of 10 mA produces a minimum i_C of 1.0 A. These values are read by projecting the points of intersection over to the vertical axis of I_C. As a result, the peak-to-peak swing of collector current is $3.4 - 1.0 = 2.4$ A. Then A_I is $2.4/0.02 = 120$, for the current gain.

We can read the minimum and peak values of v_C by projecting the points of intersection down to the axis of collector voltage. These values are 3 and 15 V. Note that the minimum v_C corresponds to the maximum i_C, because of the voltage drop across R_L. The peak-to-peak swing in v_C then is $15 - 3 = 12$ V, for v_{out}. With a 0.2-V p-p input, then A_V is $12/0.2 = 60$. The power gain is $60 \times 120 = 7200$.

Class of Operation. The method of operation for the amplifier in Fig. 30-22 is class A, meaning that collector output current flows for the full 360° of the input signal, without any part of the signal being cut off. In class B operation, output current flows for 180°, or one-half the input cycle. For class C operation, output current flows for less than 180°, usually 120°.

The class of operation is determined by the Q point set by the bias of the input circuit and by the amount of ac signal. Class A operation is the most inefficient but has the least distortion. Class C operation has the most distortion but the highest efficiency. Class B operation is between A and C.

With one amplifier in a stage for audio signal, the operation must be class A. However, two audio amplifiers can operate class B in a push-pull circuit where each supplies opposite halves of the signal. Class C operation is used in tuned rf power amplifiers, where the *LC* circuit can supply full sine waves of output.

Power Dissipation. The quiescent dc power dissipated at the collector junction is $V_C \times I_C$. For the example in Fig. 30-22,

$$P = V_C \times I_C$$
$$= 8 \text{ V} \times 2.4 \text{ A}$$
$$P = 19.2 \text{ W}$$

Typical power ratings range from 75 to 300 mW for small transistors up to 5 to 100 W for power transistors, rated at 25°C.

At higher temperatures up to a maximum of 150 to 200°C, the maximum permissible power dissipation must be derated. In many cases, the power rating is reduced 50 percent for operation at one-half the maximum temperature.

Letter Symbols for Transistors. Because of the combination of an ac component on a dc axis, it is important to distinguish between the different currents and voltages. In general, the capital letters *V* and *I* are used for average dc values. Double subscripts, as in V_{CC}, indicate the supply voltage that does not change. The small letters *v* and *i* indicate values that vary with time. These symbols are summarized in Table 30-5.

Note that in the symbol I_{CBO} for leakage current the system is to use the letter O for the open electrode in the subscript. This also applied to BV_{CBO} for breakdown voltage.

In the symbol h_{fe} the *h* stands for *hybrid parameters*. These are combinations of voltage and current ratios in the forward or reverse direction used in analyzing transistors as two-port networks. The *f* subscript is for a forward characteristic from the base input to collector output, while the *e* indicates the common-emitter circuit.

Practice Problems 30-11
(*answers on page 668*)
Answer true or false.
(a) In Fig. 20-23, when I_C is 4 A, then V_C is zero.
(b) The power dissipation with 5 V for V_C and 5 A for I_C is 5 W.
(c) The h_{fe} of 75 corresponds to β of 75.

TABLE 30-5. Letter Symbols for Transistors

SYMBOL	DEFINITION	NOTES
V_{CC}	Collector supply voltage	Same system for collector currents; also for base or emitter voltages and currents. Also applies to drain, gate, and source of field-effect transistors
V_C	Average dc voltage	
v_c	Ac component	
v_C	Instantaneous value	
V_c	Rms value of ac component	
I_{CBO}	Collector cutoff current, emitter open	Reverse leakage current
BV_{CBO}	Breakdown voltage, collector to base, emitter open	Ambient temperature (T_A) is 25°C
h_{fe}	Small-signal foward-current transfer ratio (CE)	Same as ac beta (CE)

30-12
BIAS STABILIZATION

The power dissipated as heat at the collector junction raises the junction temperature. This increases the collector leakage current, which is I_{CBO}. In the CE circuit, more I_{CBO} increases the forward-bias current. The reason is that the reverse current of minority charges in I_{CBO} is in the same direction as the forward current of majority charges from the emitter. The collector current increases, which raises the junction temperature. Since the action is cumulative, excessive I_C and heat can destroy the collector junction, resulting in an internal short circuit to the base. This effect is called *thermal runaway*.

Methods of preventing thermal runaway include operating the transistor well within its maximum power ratings and stabilizing the bias against any increase in the average collector current. Also, Si transistors have very little leakage current.

Self-Bias in the Emitter Circuit. In Fig. 30-24, the emitter bias V_E of 0.8 V results from the voltage drop $I_E R_E$. This voltage is self-bias because V_E depends on the transistor's own emitter current.

However, note that V_E is positive at the N emitter of the NPN transistor. This is opposite from the polarity for forward bias. Therefore, a voltage divider is used in the base circuit to provide the required positive bias voltage at the P base. Here the $R_1 R_2$ divider from the 9-V supply line provides 1.4 V positive at the base, which is the forward polarity. The net bias voltage from base to emitter then is V_{BE}, equal to $1.4 - 0.8 = 0.6$ V.

Stabilizing Effect of R_E. The purpose of stabilization is to prevent I_C from increasing with more leakage current. The self-bias of V_E in the emitter circuit accomplishes this effect, as it opposes the forward bias in the base circuit.

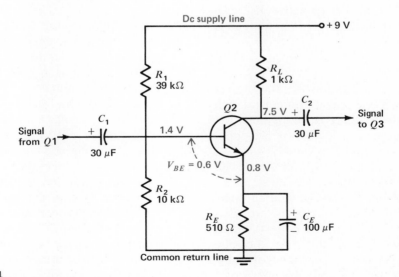

FIGURE 30-24
Typical audio amplifier circuit with emitter self-bias for stabilization and $R_1 R_2$ voltage divider for base bias.

Any increase of I_C also increases I_E. The increase in reverse V_E, though, reduces the forward bias V_{BE} to decrease I_C. The result can be a steady value of average I_C that does not increase with temperature. A higher R_E improves the stabilization, as the circuit becomes more like a CB stage, which has the best stability.

Temperature-compensating Components. Additional techniques for bias stability include *compensating diodes, thermistors,* and *sensistors*. These components can be used in the bias network for the base to maintain a stable value for the average I_C. The forward characteristic of a compensating diode is used to control the bias current in the base circuit. A thermistor decreases its R with higher temperatures; a sensistor increases its R.

Practice Problems 30-12
(answers on page 668)
Refer to Fig. 30-24.
(a) How much is the voltage across R_2?
(b) When V_E increases to 0.87 V, how much is the bias V_{BE}?

30-13
FIELD-EFFECT TRANSISTOR (FET)

This device is a semiconductor amplifier with the same function as junction transistors, but the FET construction provides a very high input resistance, in megohms. Also, the FET can take several volts for the input circuit, compared with tenths of a volt for junction transistors.

As illustrated in Fig. 30-25, the FET operation depends on controlling current through a semiconductor channel of one polarity. An N channel is shown here, but a P channel can be used instead. The bulk or substrate material is neutral or lightly doped silicon. This only serves as a platform on which the other electrodes are diffused.

When voltage is applied between the drain and source at opposite ends of the channel, the current through the channel is controlled by the gate. The terms anode and cathode are not used here because the channel can be either N or P, but current always flows from the source to the drain.

The FET is a unipolar device, as the charge carriers in the channel have only one polarity. A conventional PNP or NPN transistor is bipolar with both majority and minority charge carriers.

The FET has very high input resistance from gate to source. Also, the FET is less sensitive to temperature, x-rays, and cosmic radiation, which can produce minority carriers in bipolar transistors. The disadvantages are less gain for a given bandwidth and smaller power ratings, compared with bipolar transistors.

In summary, the electrodes of the FET correspond to the emitter, base, and collector in a junction transistor, as follows:

1. *Source.* This is the terminal where the charge carriers enter the channel bar to provide current through the channel. Source current is I_S. The source corresponds to the emitter.
2. *Drain.* This is the terminal where current leaves the channel. Drain current is I_D. The drain corresponds to the collector.
3. *Gate.* This electrode controls the conductance of the channel between the source and drain. Input signal voltage is generally applied to the gate. The gate voltage is V_G. The gate corresponds to the base, but the gate voltage controls the electric field in the channel, while the base controls the current in a bipolar transistor.
4. *Substrate or bulk.* This electrode is usually connected to the source, with both grounded in a common-source circuit.

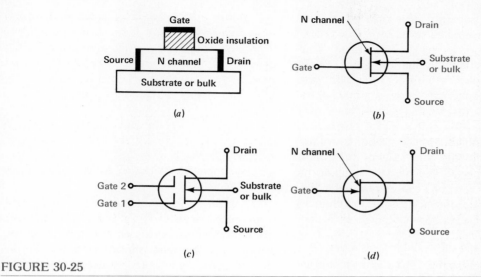

FIGURE 30-25

Field-effect transistor. (*a*) Construction with insulated gate. (*b*) Symbol for IGFET. (*c*) With double gate. (*d*) Symbol for JFET with PN junction at gate.

The arrow for hole charges into the channel indicates an N channel. The source and drain have no polarity, as they are just ohmic contacts. All the electrodes are listed in Table 30-1 on page 626.

Function of the Gate. Input signal is generally applied to the gate, with amplified output from the drain. In the input circuit, the gate and channel act like two plates of a capacitor. A charge of one polarity on the gate induces an equal and opposite charge in the channel. As a result, the conductivity of the channel can be increased or decreased by the gate voltage. With an N channel, positive voltage at the gate induces negative charges in the channel to allow more electron flow from source to drain.

JFET. In the junction-gate type, or JFET, the gate and channel form a conventional PN junction. However, this junction has high input resistance because of reverse bias.

IGFET. The insulated gate consists of a metal electrode separated from the channel by a thin layer of glassy silicon dioxide. Voltage applied to the gate can induce charges in the channel to control the drain current, but there is no PN junction. The insulated gate has very high input resistance independent of the polarity of the gate voltage. Also, the characteristics are not sensitive to temperature. This type is a metal-oxide semiconductor field-effect transistor (MOSFET) or an IGFET.

The insulated gate is essentially an electrostatic device with very high resistance. A typical r_{in} is 15 MΩ. An IGFET not connected in a circuit may require a shorting ring on the leads to protect against a buildup of static charge. The IGFET has many uses for low-power amplifiers and in counting circuits. This method of construction is also applied in IC units.

FET Types. When the channel can conduct current from source to drain only with gate voltage applied, the transistor is an *enhancement type.* This means the gate voltage must increase the charge carriers in the channel to allow any drain current to flow. When the channel conducts with zero gate voltage, the transistor is a *depletion type.* In schematic symbols the depletion type has a solid channel bar, as in Fig. 30-25. For the enhancement type, the channel line is shown broken by the D, B, and S electrodes.

FET Amplifier Circuit. The common-source (CS) circuit in Fig. 30-26 is similar to a common-emitter (CE) amplifier. This FET is an N-channel depletion type. In the input circuit the gate has zero bias. Then the ac input swings the gate voltage ± 0.5 V, or 1 V p-p. As a result, the channel conductance changes to vary the drain current I_D.

In the output circuit, the varying I_D produces corresponding changes in the voltage drop across the 2-kΩ R_L. Note the drain voltage v_D is the difference between the supply V_{DD} and v_{RL}, just as in a collector circuit. The result is voltage amplification. In this example, v_{out} is 10 V p-p. Then A_V is $^{10}/_1 = 10$.

FET Amplifier Circuits. As with all amplifiers, any one electrode can be common for the input

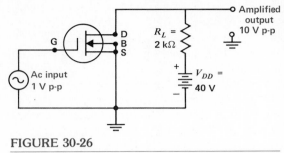

FIGURE 30-26

Common-source circuit, corresponding to CE amplifier.

and output signals. The three possibilities are common source, corresponding to the CE amplifier; common gate, similar to the CB amplifier; and common drain or source-follower, like the emitter-follower. These are summarized in Table 30-6, with a comparison to vacuum tubes.

Drain Characteristic Curves. As shown in Fig. 30-27, the characteristics of the FET are shown by a family of curves of drain current vs. drain voltage for different gate voltages. These curves correspond to a family of collector characteristics for different base currents.

In the drain characteristics, note that the curve for zero gate voltage is in the middle, approximately. With the Q point at $V_{GS} = 0$ V, the ac signal input can swing the gate voltage positive and negative. It should be noted, however, that a small negative bias on the gate can be used.

TABLE 30-6. Amplifier Circuits for Transistors and Vacuum Tubes

VACUUM TUBE	JUNCTION TRANSISTOR	FIELD-EFFECT TRANSISTOR
Grounded cathode	Common emitter	Common source
Grounded grid	Common base	Common gate
Cathode-follower	Emitter-follower	Source-follower

The load line for a specific R_L and V_{DD} can be drawn on the FET characteristic curves, using the same method shown for the collector curves in Fig. 30-22. For the FET, the bottom end of the load line is where $V_D = V_{DD}$. The top end on the vertical axis is where $I_D = V_{DD}/R_L$.

The gate circuit can accomodate a wide range of input voltages, compared to bipolar transistors. Because of the very high gate resistance, there is practically no gate current. Therefore, the IGFET takes voltage drive for the ac input signal, like grid signal for a vacuum tube.

Transconductance g_m. The mutual transconductance between the input voltage to the gate and the output current from the drain is

$$g_m = \frac{\Delta i_D}{\Delta v_G} \quad \text{with } V_D \text{ constant} \quad (30\text{-}8)$$

In Fig. 30-27, assume V_D is constant at 10 V. Then a change in V_{GS} of 1 V, from 0 to -1 V, decreases i_D by 5 mA, from 15 to 10 mA. Therefore,

$$g_m = \frac{5\text{ mA}}{1\text{ V}} = 5 \times 10^{-3}\text{ S} = 5000\ \mu\text{S}$$

Typical values of g_m are 1000 to 10,000 μS.

Practice Problems 30-13
(answers on page 668)
(a) Which electrode in the FET corresponds to the collector in a junction transistor?
(b) In the common-source circuit, which electrode has the signal input?
(c) The i_D changes by 10 mA when v_G changes by 1 V. How much is g_m?

30-14
SILICON CONTROLLED RECTIFIER (SCR)

As shown in Fig. 30-28, the SCR is a four-layer device used as a silicon rectifier, but conduction is controlled by a gate electrode. When forward voltage is applied between cathode and anode, in series with an external load, no appreciable load current can flow until the barrier voltage at the gate-cathode junction is overcome. This internal reverse bias is about 0.7 V.

Forward voltage is applied for one-half cycle of the ac input between the anode and cathode. However, note that the cathode-anode voltage is distributed over four layers, reducing

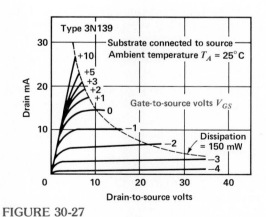

FIGURE 30-27

Family of drain characteristic curves. (From RCA transistor manual)

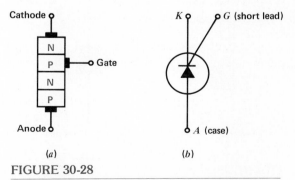

FIGURE 30-28

Silicon controlled rectifier. (a) Four-layer construction. (b) Symbol.

the amount of forward voltage at the gate. But with a high enough forward voltage applied, the SCR can conduct. This value is the *forward-breakover voltage.*

When the SCR is conducting, this is the ON state. The SCR has a very low resistance then, serving as a silicon diode rectifier. In the OFF state, though, the SCR is a very high resistance, practically an open circuit.

The Gate Trigger. The breakover voltage to turn on the SCR is controlled by the gate. A small positive voltage here neutralizes the internal reverse bias of the gate to allow conduction in the main circuit for the load.

In normal operation, the forward voltage from the anode circuit is below the breakover voltage but the SCR is switched on by the gate. Once the SCR is turned on by the gate signal, the load current is independent of the gate voltage or current. The SCR remains on, conducting its *holding current,* until the main circuit reduces the forward voltage below the value required to sustain conduction. Then when the SCR is off, the gate signal can turn it on again.

Thyristors. This is the general name for gate-controlled rectifiers like the SCR. The characteristics are similar to the thyratron gas tube, where the grid voltage starts conduction. The main types are the SCR, triac, and unijunction transistor. The SCR and triac are also in the class of *four-layer diodes,* because of the construction with alternate P and N layers. They are used as a gate-controlled switch (GCS).

Triac. As shown in Fig. 30-29, the construction enables the triac to conduct for either polarity of load voltage in the main circuit. Also, the gate can have either positive or negative forward voltage. In effect, the triac is a bidirectional SCR. There is no anode or cathode in the triac, as current can flow in either direction between main terminals 1 and 2.

Diac. This is also bidirectional between main terminals 1 and 2, but it does not have a gate. There are just three layers.

Unijunction Transistor (UJT). As shown in Fig. 30-30, the UJT has an emitter and two connections to the base, without a collector terminal. The UJT is not used as a transistor amplifier, but is used in switching and timing applications. In effect, the emitter serves as a gate to control the resistive voltage divider between $B1$ and $B2$.

The PUT. This abbreviation is for a programmable unijunction transistor. However, the construction is a four-layer device with a gate, like the SCR. A resistive voltage divider in the gate circuit determines the firing voltage, though, similar to the idea of the divided base resistances in the UJT.

Practice Problems 30-14
(answers on page 668)
Answer true or false.
(a) The SCR is used as a gate-controlled switch.

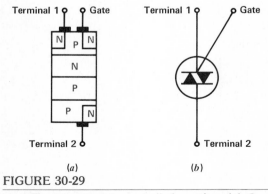

FIGURE 30-29

The triac, a gate-controlled rectifier. (*a*) Construction. (*b*) Symbol.

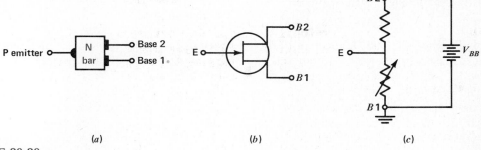

(a) (b) (c)

FIGURE 30-30

The unijunction transistor (UJT). (a) Construction. (b) Symbol. (c) Equivalent circuit for double-base connections.

(b) The triac is bidirectional.
(c) The UJT is a four-layer device like the triac.

30-15
TYPES OF TRANSISTORS

NPN and PNP transistors are the bipolar junction type. The FET is unipolar, with either an N or P channel. Most junction transistors are NPN, using Si because of its lower leakage current and higher temperature ratings. Also, Si is generally used for the FET and integrated circuits. Ge is used for some power transistors because of its lower internal voltage drop.

Construction. The first transistors were the point-contact type, with two thin wires or "whiskers" welded on a semiconductor block to form two junctions. A later type was the alloy junction, where the doping material was fused as dots on opposite sides of a P or N base. The advantage was a bigger area for the junction. However, transistors and integrated circuits now generally use the planar construction illustrated in Fig. 30-31. In this method the junctions are parallel planes.

The starting material is a uniform P or N silicon crystal. Then impurities can be diffused in the vapor state onto the crystal block. Vertical penetration is controlled by temperature and time. The surface area depends on masking and the use of photochemical techniques for etching specific areas. Specific techniques are used for *mesa transistors* and *epitaxial transistors,* for the purpose of providing low saturation voltage for the collector. This feature is important for the switching transistors used in digital circuits.

In the planar construction, the base thickness is reduced to only 5×10^{-6} m. The size of

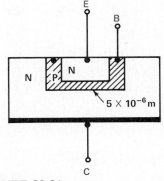

FIGURE 30-31

Diffused planar construction generally used for transistors and integrated circuits.

Chapter 30
Semiconductor Diodes and Transistors

the whole transistor chip is only about $\frac{1}{8}$ in^2 for medium power ratings.

Transistor Packages. After fabrication of the individual transistor chips, wires are bonded to the electrodes and connected to the pins. The entire unit is sealed, either in epoxy plastic or in a metal case with inert filler. This *encapsulation* is necessary for mechanical ruggedness and to prevent moisture from contaminating the semiconductor. Three common types of transistor packages are shown in Fig. 30-32. The types in (a) and (b) are generally soldered into the printed-circuit board, instead of using a socket. The reason is that oxidation at a socket connection can affect the very small base-emitter voltage.

Heat Sink. Heat is a problem with power transistors. Although there is no heater, a power transistor in normal operation can be hot enough to burn your fingers. The failure rate can be ten times greater for a temperature rise of 25°C above normal. It is necessary, therefore, to use a heat sink, which is a metal structure to radiate heat, mainly away from the collector junction.

In power transistors of 10 to 25 W, the collector is usually connected internally to the steel or aluminum case, as shown in Fig. 30-32c. In (b), the collector is tied internally to the copper mounting tab, in a medium-power transistor. For both types, the transistor uses the metal chassis for heat dissipation.

If the collector must be insulated from chassis ground, a thin mica insulating spacer is used. The spacer is generally covered with a silicon grease for better heat transfer. For power transistors of 25 W or more, a separate heat sink with large radiating fins is generally used.

Type Numbers. The letter N indicates a semiconductor, with a numerical prefix for the number of junctions. Diodes are numbered 1N, junction transistors are 2N, and an FET with one gate is 3N. The digits that follow are for specific types registered with the Electronic Industries Association (EIA). An example is the 2N5102 junction transistor. However, foreign transistor types not registered with the EIA are

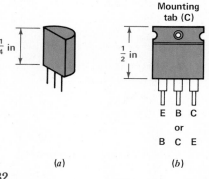
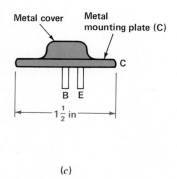

FIGURE 30-32

Transistor packages. (a) Plastic encapsulation for small-signal types. (b) Plastic package with metal tab as heat sink for medium-power transistors, the SCR, and triacs. (c) Metal can for high-power transistors.

often labelled 2SA for PNP and 2SC for NPN types.

Practice Problems 30-15
(*answers on page 668*)
Answer true or false.
(a) The FET is unipolar, while junction transistors are bipolar.
(b) The metal power tab on an NPN transistor is usually connected internally to the base.
(c) Most small-signal transistors are NPN, silicon, in a plastic package.
(d) A heat sink is used to radiate heat away from the collector junction.

30-16
SPECIAL-PURPOSE DIODES

A semiconductor diode is just a PN junction. Since current flows only one way, the main application is rectification. However, the junction has additional properties that are useful. With reverse bias, the junction has capacitance which can be controlled by an external voltage. Also, the PN junction is photosensitive. Furthermore, the reverse breakdown voltage can be used for voltage-reference diodes. The schematic symbols for these diodes are shown in Fig. 30-33.

Tunnel Diodes. These are also called *Esaki diodes,* named after the man who discovered that heavy doping can cause a tunneling effect of charge carriers through the depletion zone at the junction. One important feature of the tunnel diode is its *negative resistance* for a specific range of forward voltage. In a negative resistance, the current decreases for an increase of applied voltage. Because of this characteristic, the tunnel diode can operate as an amplifier or oscillator. Essentially, this diode is a low-power device for microwave frequencies with relative freedom from radiation effects.

Compensating Diodes. These are used for bias stabilization in transistor circuits.

Varistor. As shown in Fig. 30-33, two junctions of opposite polarities are used. The varistor is often used as a shunt for the collector to protect against excessive voltage spikes, either positive or negative.

Also, a single diode rectifier can be used in series with the collector. This function is to isolate the collector from voltage spikes that have forward polarity.

Zener[1] **Diodes.** These are also called *voltage-reference diodes.* They are silicon diodes designed for a specific reverse breakdown voltage. When the *avalanche current* flows, the voltage across the diode remains constant. This char-

[1] Named after C. A. Zener, who analyzed voltage breakdown of insulators.

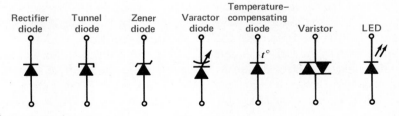

FIGURE 30-33

Schematic symbols for types of semiconductor diodes.

acteristic makes the diode useful as a voltage regulator, instead of gas tubes. Multiple diodes in series increase the voltage rating.

Varactor Diodes. The barrier voltage at a reverse-biased junction enables the junction to serve as a capacitance because of the separated charges in the depletion zone. Most important, the amount of junction capacitance can be controlled by the reverse voltage. The C changes as the depletion zone is made wider or narrower by changes in V. Diodes made for this use as a *voltage-sensitive capacitor* are *varactors* or *varicaps*. A common application is electronic tuning for television receivers.

Photoconductive Diodes. These are made of a photosensitive material, such as cadmium sulfide, where the resistance decreases with more light. The applications include many light-control devices.

Light-emitting Diodes (LED). The PN junction can radiate light, as energy is released by the recombination of charges. For greater efficiency, special compounds of gallium (Ga) are used.

The light is emitted when forward voltage is applied. Typical ratings are 1.2 V with a forward current of 20 mA. Many of these diodes are connected in series to provide a display for a number or letter. The radiation is in the wavelengths for red, green, and yellow light.

Figure 30-34 shows the seven segments of a display that are used to form the digits 0 to 9, as they would be used for an electronic calculator. Each segment containing the LED units can be turned on or off by digital control circuits to form the desired digit. For instance, when all segments are on, the digit is 8. If only the center one is off, the digit is 0.

The same idea of segments for a digit is used with a liquid-crystal display (LCD) as with

FIGURE 30-34

Seven segments for a numerical display, either LED or LCD.

an LED display. The liquid-crystal display depends on ambient light for its energy. Only the orientation of the molecules is controlled by applied voltage to turn the display on or off. As a result, the LCD display requires much less load current than the LED display.

Practice Problems 30-16
(answers on page 668)
Is forward or reverse voltage used on the following:
(a) Zener diode.
(b) Varactor.
(c) LED.

30-17
TRANSISTOR TROUBLES

Failures generally result from an open weld at the wire leads to the semiconductor, a short circuit caused by momentary overloads, and circuit failures that cause transistor overheating. In most cases a defective transistor is internally short-circuited or open, and simple tests will reveal the trouble.

Some problems, like an increase in leakage, a drop in breakdown voltage, or excessive noise are more difficult to detect, and direct substitution may be the easiest way to localize the fault. Transistor testers are available to check the transistor in or out of the circuit, for an open circuit, short circuit, leakage, and beta.

However, open and short circuits can be checked with a multimeter.

In-Circuit Tests. These tests are very helpful because transistors are usually soldered into place. Voltage measurements can determine if the junctions are intact and if the transistor is conducting properly.

Check Forward Bias. Measure the base-to-emitter voltage by putting the voltmeter leads directly across these terminals. V_{BE} should be about 0.2 V for Ge and 0.6 V for Si transistors. Referring back to Fig. 30-24, as an example, a VOM from base to emitter should read 0.6 V for V_{BE}. If a VTVM is used, measure each voltage to the common line and subtract the readings.

If the reading is zero for V_{BE}, the base-emitter junction is short-circuited. If V_{BE} is 0.8 V or higher, the base-emitter junction is probably open. One word of caution here. These voltage readings apply to class A amplifiers. In pulse circuits, it is normal for a reverse bias of a few volts to appear between base and emitter, in order to cut off I_C until the input pulse drives the transistor into conduction.

Checking $I_C R_L$ Voltage. To check for correct collector current, put the voltmeter between the collector and the supply voltage to read the voltage drop across the load. In Fig. 30-24, a VOM across the 1-kΩ R_L should read 1.5 V. Then divide this reading by the dc resistance in the collector circuit to calculate the current I_C. For this example, the normal I_C value is 1.5 V/1 kΩ = 1.5 mA.

If there is no voltage drop across R_L, then I_C must be zero. Then V_C at the collector will have the same value as the supply voltage V_{CC}. For the opposite trouble, excessive I_C can cause excessive voltage across R_L, resulting in zero or very low V_C.

If I_C is zero or very low, the transistor may be open. However, check for opens in the emitter circuit before making a replacement.

If excessive I_C is flowing, short-circuit the base-to-emitter voltage and repeat the test. Under these conditions only the small leakage current should flow. If I_C is still high, the transistor collector is probably short-circuited.

Measuring I_C. Calculation of collector current may be difficult if the collector circuit contains little resistance or an unknown value, such as a transformer primary. In this case current measurements may be better. You can open the collector circuit easily by cutting the foil of the printed-circuit board with a razor blade. Then put the leads of your VOM, set to read milliamperes or amperes, across the cut. You can bridge the cut with solder when the test is finished.

Checking the Emitter Circuit. You can also check current by measuring the voltage drop across the emitter resistance and dividing by the value of the resistor. However, be careful in your analysis. This voltage will read almost normal even if the emitter resistor is open. The reason is that the voltmeter resistance then completes the emitter circuit. So check the value of the emitter resistor first.

Resistance Tests. Resistance measurements in transistor circuits are often misleading, as the internal battery of the ohmmeter may forward-bias a junction and provide a parallel path for current. Know the voltage polarity at your ohmmeter leads and make sure all junctions are reverse-biased for ohmmeter readings. Or, best of all, use an ohmmeter with low-power ohms, which does not have enough voltage for forward bias on transistors.

Out-of-Circuit Tests. Ohmmeter checks on a transistor are useful for finding opens or short circuits. Avoid the $R \times 1$ scale when checking low-power transistors. On this scale the current in the leads is usually more than 100 mA when the leads are short-circuited, even though the current through the meter movement may be less. Check the resistance between base and collector and reverse the leads. Do the same for base and emitter. You should get a very high reading when the junction is reverse-biased and a very low reading when the junction is forward-biased. These tests really check each transistor junction as a diode.

The actual reading in ohms depends upon your ohmmeter, the range selected, and the type of transistor. For conclusive tests, make a comparison with another transistor of the same type that is known to be good. It is helpful to realize that the reverse resistance of a silicon junction is usually infinite.

Replacing Transistors. Small-signal and medium-power transistors are usually soldered on the printed-circuit board. However, the transistor can be removed easily, without damage to the PC board, by using desoldering techniques to suck out the old solder. The trick is to remove all the solder from the terminals so that the leads can be straightened to lift out the transistor. Desoldering methods for PC boards are described in Appendix H.

Practice Problems 30-17
(*answers on page 668*)

(a) When the emitter junction is shorted to the base, how much is V_{BE}?
(b) When the collector terminal is open, how much is I_C?
(c) When the base has no forward bias, how much is I_C?

Summary

The main types of semiconductor devices are listed in Table 30-7. Examples of pin connections are shown in Fig. 30-35. More details appear in transistor manuals, replacement guides, and cross-reference guides published by manufacturers.

Self-Examination (Answers at back of book.)

Answer true or false.

1. Silicon and germanium are semiconductors with a valence of 4.
2. Phosphorus with an atomic number 15 has 5 valence electrons.
3. Doping with phosphorus makes silicon N type, with majority electron charges and minority hole charges.
4. Hole current is the movement of positive charges in the opposite direction from electron flow.
5. Hole current into an electrode is considered the positive direction of current.

Chapter 30
Semiconductor Diodes and Transistors

TABLE 30-7. Semiconductor Devices

SYMBOL	TYPE	SYMBOL	TYPE
(PNP transistor symbol; B, C, E)	PNP transistor; needs negative V_C	(Triac symbol; G, MT_1, MT_2)	Triac thyristor; MT is main terminal; uses either polarity of MT voltage
(NPN transistor symbol; B, C, E)	NPN transistor; needs positive V_C	(Diac symbol)	Diac trigger diode; uses either polarity of voltage
(JFET symbol; G, D, S)	Junction field-effect transistor (JFET); N channel	(Diode symbol; A, K)	Diode rectifier; arrow shows hole current
(IGFET symbol; G, D, Sub, S)	Field-effect transistor; N channel, depletion IGFET	(Tunnel diode symbol)	Tunnel diode or Esaki diode
(IGFET symbol; G, D, Sub, S)	Field-effect transistor; P channel, enhancement IGFET	(Zener diode symbol)	Zener diode or voltage reference diode
(Dual-gate IGFET symbol; G_1, G_2, D, Sub, S)	IGFET with two gates	(Varactor symbol)	Capacitive diode or varactor
(SCR symbol; A, G, K)	Silicon controlled rectifier (SCR); gate G is shorter lead	(Temperature-dependent diode symbol, $t°$)	Temperature-dependent diode
(UJT symbol; E, B_1, B_2)	Unijunction transistor (UJT); thyristor with B_1 and B_2 as gates for N bar	(Photosensitive diode symbol)	Photosensitive diode; light in
		(LED symbol)	Photoemissive diode; light out (LED)

6. Atoms of the impurity element in a doped semiconductor provide fixed ion charges.
7. A drift current requires a difference of potential, but a diffusion current requires only a difference in charge density.
8. The internal barrier potential at a PN junction for silicon is approximately 9 V.

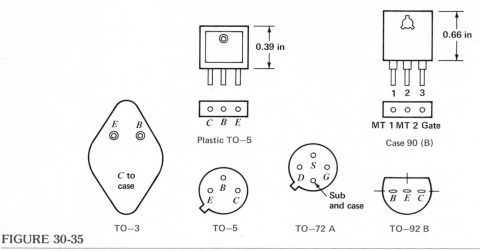

FIGURE 30-35

Typical transistor pin connections, bottom view. The TO is "transistor outline."

9. The collector circuit in a transistor amplifier always has reverse voltage, while the base-emitter circuit has forward bias.
10. For an NPN transistor, negative voltage is needed at the collector.
11. For an NPN transistor, positive voltage at the base, with respect to emitter, provides forward bias.
12. The common-emitter amplifier circuit is used most often because it has the best combination of voltage and current gain.
13. The arrow for current in semiconductors indicates the direction of electron flow.
14. The internal barrier potential V_b decreases at higher temperatures.
15. The semiconductor type number 2N34 indicates a diode with two electrodes.
16. When a change of i_B from 30 to 40 mA changes i_C from 500 to 900 mA, the beta characteristic equals 40.
17. When $I_C = 99$ mA and $I_B = 1$ mA, then I_E is 100 mA and the alpha characteristic equals 0.99.
18. With an alpha of 0.98 and $I_E = 50$ mA, then $I_C = 52$ mA.
19. With $I_C = 100$ mA and a beta of 50, then $I_B = 2$ mA.
20. For the transistor in question 19, $I_E = 52$ mA.
21. The dc output voltage from a power supply increases with higher values of filter capacitance but decreases with more load current.
22. With an ac input from the 60-Hz power line, the ripple frequency is 60 Hz in the dc output of either a half-wave or full-wave rectifier.

23. Bias stabilization is often used with transistor amplifiers to prevent thermal runaway.
24. In Fig. 30-17 the average I_E is 3.06 mA.
25. The FET is a unipolar transistor with high input resistance.
26. When a 1-V increase in gate voltage changes the drain current 10 mA in an FET, its g_m equals 10,000 μS.
27. An ac input signal of ± 1 V can be applied to the gate of an FET, but this is too much input voltage for any bipolar transistor amplifier.
28. The SCR is a silicon diode rectifier with a gate-control electrode.
29. Without forward bias for the input, I_C in the output is zero.
30. The triac is a bidirectional gate-controlled rectifier.
31. The LED emits light.
32. The capacitive diode uses reverse voltage.

Essay Questions

1. Show the atomic structure of silicon with atomic number 14, indicating the four valence electrons.
2. Define the following: (a) doping; (b) N-type silicon; (c) P-type silicon; (d) PN junction; (e) internal barrier potential; (f) ion charge; (g) depletion zone.
3. (a) Give two characteristics of semiconductors. (b) Name four types of semiconductor devices.
4. Compare the following: (a) electron and hole charges; (b) electron current and hole current.
5. Define the positive direction of current in semiconductors.
6. Compare the following: (a) majority and minority carriers; (b) drift current and diffusion current; (c) free charges and ion charges; (d) intrinsic and extrinsic semiconductors.
7. For a PN junction, show a battery applying forward voltage and reverse voltage.
8. Show the schematic symbol for PNP and NPN transistors, indicating emitter, base, and collector.
9. Give one feature of the CB, CE, and CC circuits, in addition to the information on which terminal is common.
10. Why is the I_{CO} for a PN junction called reverse saturation current?
11. Why is the reverse leakage current I_{CBO} in a transistor essentially the same as I_{CO} in a diode?
12. Why does the value of I_{CO} or I_{CBO} increase with temperature?
13. Draw the complete schematic diagram of a common-emitter ampli-

fier, using an NPN transistor with an external R_L. Indicate where input voltage is applied and where output voltage is obtained. Give typical values for the forward bias in the input circuit and collector voltage in the output. Assume a silicon transistor.
14. What is meant by thermal runaway in a transistor?
15. Show two methods of forward bias for the CE circuit.
16. Show the complete schematic diagram of a full-wave rectifier circuit using silicon diodes for the 120-V power line with a step-down transformer for 40 V to each diode.
17. Define class A, B, and C operations for amplifier circuits.
18. Why is a transistor cut off with zero bias in the emitter-base circuit?
19. Give the functions of source, gate, and channel in the FET and compare these electrodes to a bipolar transistor.
20. Define JFET, MOSFET, IGFET, junction gate, insulated gate, N channel, and P channel for field-effect transistors.
21. Show the schematic symbols for an N-channel FET, depletion type and enhancement type.
22. Draw the diagram of an FET in a common-source amplifier circuit, with 30 V for V_{DD} and a 1500-Ω R_L.
23. Define transconductance g_m for an FET.
24. Define SCR, thyristor, triac, and UJT.
25. List three types of semiconductor diodes other than silicon power rectifiers.
26. What is meant by a four-layer device?
27. Describe briefly how you could measure I_C as 2.4 A in Fig. 30-22a, without opening the collector circuit.
28. Give two troubles that can cause zero I_C.

Problems (Answers to odd-numbered problems at back of book.)

1. V_B is 18.4 V. V_E is 17.8 V. How much is V_{BE}?
2. $I_B = 500$ μA and $I_C = 22$ mA. Calculate I_E. (b) $I_E = 5$ A and $I_B = 80$ mA. Calculate I_C. (c) $I_E = 41$ mA and $I_C = 40$ mA. Calculate the I_B.
3. For the half-wave rectifier in Fig. 30-12, how much is the dc output voltage with (a) 250-μF filter capacitance and 50-mA load current I_L? (b) 50-μF C and 150-mA I_L?
4. Referring to the CE circuit in Fig. 30-17, list the values of V_C, I_C, V_B, I_B, and the total emitter current I_E.

5. Referring to the CE circuit with bias stabilization in Fig. 30-24, give the values with polarity for V_C, V_{CE}, V_B, V_{BE}, and V_E.
6. In Fig. 30-24, determine I_C, I_E, and I_B.
7. From the values in Prob. 6, calculate the dc values of α and β for the transistor in Fig. 30-24.
8. A transistor amplifier has $V_{CC} = 50$ V, $V_C = 10$ V, and an R_L of 200 Ω. Calculate I_C.
9. Calculate the quiescent dc power dissipated at the collector for the transistor in Prob. 8.
10. Refer to the collector characteristic curves in Fig. 30-21b. (a) Give the values of I_C for the different base currents of -100, -150, and -200 μA, with a constant collector-emitter voltage V_{CE} of -12 V. (b) Give the values of I_C for different collector-emitter voltages V_{CE} of -4, -8, and -12 V, with a constant base current of -150 μA.
11. An FET has -2-V gate voltage for 14-mA I_D. When V_G changes to -2.4 V, I_D decreases to 11 mA. Calculate g_m in microsiemens.
12. Given that $I_C = 9$ mA and $I_B = 150$ μA, how much is I_E? Calculate α and β for this transistor.
13. Given that I_B is 40 μA and β is 70, calculate I_C.
14. (a) Given that $I_E = 20$ mA and α is 0.98, calculate I_C. (b) How much is the base current I_B? (c) Calculate β for this transistor.
15. In a common-emitter circuit, i_B variations of ± 100 μA swing i_C by ± 5 mA, with an R_L of 2000 Ω. (a) How much is the current gain? (b) Calculate the peak-to-peak ac output voltage across R_L. (c) If the input voltage is ± 40 mV, how much is the voltage gain?
16. Determine the forward bias V_{BE} and reverse collector voltage V_{CE} and draw the circuit diagram with a PNP transistor for the following electrode voltages to ground: (a) Common-emitter circuit with $V_{CC} = -6$ V, $V_{R_L} = 2$ V, $V_C = -1$ V, and $V_E = 0.8$ V. (b) Common-collector circuit (emitter-follower) with $V_C = 0$ V, $V_{EE} = 6$ V, $V_E = 3.2$ V, and $V_B = 3$ V.
17. Refer to the collector characteristic curves in Fig. 30-21c. Draw the load line for a 1.6-kΩ R_L with 16 V for V_{CC}. Assume a base-bias current I_b of 100 μA and an ac signal swing of ± 50 μA. Determine the peak-to-peak i_C and v_C. Calculate A_I, A_V, and the power gain.
18. Refer to the drain characteristic curves for the FET in Fig. 30-27. Draw the load line for a 1.5-kΩ R_L with 30 V for V_{DD}. Assume a gate bias voltage of -2 V and an ac signal swing of ± 1 V. Draw a ladder diagram showing waveforms with peak values for v_G, i_D, and v_D. (A ladder diagram means that the waveforms are in line vertically with respect to time, as in Fig. 30-18.)
19. For each of the circuits in Fig. 30-36, on the next page, indicate whether the diode and transistors are cut off or conducting.

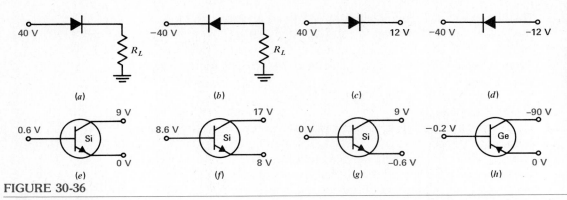

FIGURE 30-36

For Prob. 19.

Answers to Practice Problems

30-1   (a) Emitter
       (b) Drain
30-2   (a) ±4
       (b) Silicon
30-3   (a) N type
       (b) P type
30-4   (a) Majority
       (b) Minority
30-5   (a) Yes
       (b) Decreases
       (c) Increases
30-6   (a) 120 Hz
       (b) 152 V (approx.)
30-7   (a) Forward
       (b) Reverse
       (c) $I_E = 1005\ \mu A$
       (d) Cut off
30-8   (a) CE
       (b) CE
       (c) CC
30-9   (a) 0.6 V
       (b) 6.5 V
       (c) 1 mA

30-10   (a) 14.8 mA
        (b) 300
30-11   (a) T
        (b) F
        (c) T
30-12   (a) 1.4 V
        (b) 0.53 V
30-13   (a) Drain
        (b) Gate
        (c) 10,000 μS
30-14   (a) T
        (b) T
        (c) F
30-15   (a) T
        (b) F
        (c) T
        (d) T
30-16   (a) Reverse
        (b) Reverse
        (c) Forward
30-17   (a) Zero
        (b) Zero
        (c) Zero

Review of Chapters 29 and 30

Summary

1. In vacuum tubes the cathode is heated to emit electrons. The anode or plate collects these electrons to provide plate current. The main tube types are diodes, triodes, tetrodes, and pentodes, as listed in Table 29-3. Semiconductor devices are listed in Table 30-7.
2. Diodes are used as rectifiers to change alternating current to direct current. Current can flow through the diode only when the anode is positive.
3. Triodes, tetrodes, and pentodes are used for amplifiers since the control-grid voltage controls the plate current. Making the control-grid voltage more negative decreases the plate current; making it less negative increases the plate current.
4. Transistors are amplifiers making use of controlled current flow in solid semiconductors such as germanium and silicon. No heater is necessary.
5. N-type germanium or silicon is doped with an impurity that makes it negative with a surplus of free electrons; P type has a surplus of positive hole charges.
6. The transition between P and N semiconductors is a PN junction. The junction has an internal barrier voltage of 0.7 V for silicon or 0.3 V for germanium that blocks forward current. However, forward voltage of the same polarity as the electrodes allows current in the forward direction. Reverse voltage of opposite polarity prevents forward current.
7. A current of hole charges is in the direction of conventional current, opposite from electron flow. Hole current into the semiconductor is considered the positive direction of I.
8. A PNP transistor has a wafer of an N-type semiconductor forming two junctions with two P-type semiconductors. An NPN transistor is the reverse. Both types are used as amplifiers. The semiconductor in

the center is the base. The emitter at one end has forward bias to supply charges to its junction with the base; the collector at the other end has reverse bias and receives charges from its junction with the base.
9. In the schematic symbol the arrowhead on the emitter indicates the direction of hole current into the base for a P emitter or out from the base for an N emitter.
10. The main types of transistor circuits are common base (CB), common emitter (CE), and common collector (CC). The CE circuit is generally used for amplifiers because it has the most gain.
11. The two main characteristics of a transistor are its α (alpha), which is the ratio of collector current to emitter current, and β (beta), which is the ratio of collector current to base current.
12. Silicon and germanium diodes are generally used in rectifier circuits.
13. A half-wave rectifier uses one diode. With 60-Hz ac input, the ripple frequency in the dc output is 60 Hz. A full-wave rectifier uses two diodes to rectify both halves of the ac input cycle; the ripple frequency is 120 Hz.
14. The SCR is a silicon power rectifier with a gate electrode to control the start of current between anode and cathode.
15. The field-effect transistor (FET) has a gate electrode to control the electric field in the channel between source and drain. The gate corresponds to the base; the source supplies charges like an emitter; the drain corresponds to a collector receiving the charges. Input signal voltage is applied to the gate, and amplified output is taken from the drain, in the common-source (CS) circuit.
16. The main characteristic of the FET is its transconductance g_m equal to $\Delta i_D / \Delta v_G$.

Review Self-Examination (Answers at back of book.)

Answer true or false.

1. A thermionic cathode is heated to emit electrons.
2. The anode has a positive potential with respect to the cathode to attract electrons.
3. Plate current can flow in only one direction.
4. An open heater results in zero plate current.
5. An N-type semiconductor has free electrons, while a P type has hole charges.
6. The emitter is always forward-biased, while the collector has reverse voltage.

7. The schematic symbol with an arrowhead into the base indicates an NPN transistor.
8. Forward bias is applied to an NPN transistor by + voltage to the emitter and − voltage to the base.
9. The β characteristic of transistors compares collector current to base current.
10. Typical values of β for transistors are 0.98 to 0.99.
11. In the NPN transistor, collector current is hole charges supplied by the emitter.
12. The + mark or a band on silicon diode power rectifiers indicates where positive dc output voltage is obtained.
13. Positive voltage applied to the collector of an NPN transistor is the polarity for reverse voltage.
14. Minority charge carriers have opposite polarity from majority carriers.
15. I_{CO} in a diode and I_{CBO} in a transistor consist of majority carriers.
16. A typical value of input signal for a PNP transistor is about 8 V.
17. Bias stabilization is used to prevent thermal runaway.
18. The FET has a very high input resistance.
19. The gate electrode in the FET corresponds to the collector in a bipolar transistor.
20. The SCR is a silicon rectifier with a gate electrode to control when current flows from cathode to anode.
21. Typical forward bias on a silicon NPN transistor is 0.6 V.
22. The Zener diode is used as a constant-voltage source.

References (Additional references at back of book.)

Books

Bell, D. A.: "Fundamentals of Electronic Devices," Reston Publishing Company, Inc., Reston, Va.

Cutler, P.: "Semiconductor Circuit Analysis," McGraw-Hill Book Company, New York.

Fitchen, F. C.: "Transistor Circuit Analysis and Design," D. Van Nostrand Company, Inc., Princeton, N.J.

Hibberd, R. B.: "Integrated Circuits," McGraw-Hill Book Company, New York.

Kiver, M.: "Transistors," McGraw-Hill Book Company, New York.

Millman, J. and C. Halkias: "Electronic Devices and Circuits," McGraw-Hill Book Company, New York.

"Radio Amateur's Handbook," American Radio Relay League, Newington, Conn.

Ristenbatt, M. and R. Riddle: "Transistor Physics and Circuits," Prentice-Hall, Inc., Englewood Cliffs, N.J.

Sams, H. W.: "Transistor Substitution Handbook," Howard W. Sams & Co., Inc., Indianapolis, Ind.

Sowa, W. A.: "Active Devices for Electronics," Rinehart Press, San Francisco, Calif.

Pamphlets, Catalogues, and Manuals

Application Notes, Motorola Semiconductor Products Inc., Phoenix, Ariz.

Application Notes, Texas Instruments Inc., Dallas, Tex.

GE Transistor Manual, General Electric Co., Semiconductor Products, Syracuse, N.Y.

RCA Receiving Tube Manual, RCA Electronic Components and Devices, Harrison, N.J.

RCA Solid-State Devices Manual, RCA Solid-State Division, Somerville, N.J.

RCA Transistor Manual, RCA Electronic Components and Devices, Harrison, N.J.

Semiconductor Cross-reference Guide, Motorola Semiconductor Products Inc., Phoenix, Ariz.

Solid-State Replacement Guide, RCA Electronic Components and Devices, Harrison, N.J.

Integrated Circuits

Chapter 31

Transistor chips are small, but when they are packaged as discrete transistors with separate resistors and capacitors, the circuits are relatively large. Actually, the techniques of making diffused planar transistors can be combined with printed circuits to integrate all the components on one chip. The result is an integrated circuit (IC). The integration methods can be applied to bipolar transistors or MOSFET types. Typical IC units are shown in Fig. 31-1. An IC unit in a package 1 in long by $\frac{3}{8}$ in can contain dozens of transistor circuits.

Common uses are in electronic calculators, watches, and computers for digital applications. In addition, audio amplifiers, rf amplifiers, and the color circuits in television receivers generally use IC units, as applications of linear amplifiers. The power rating for IC units is generally 100 mW to 5 W. More details are in the following topics:

31-1 Types of Integrated Circuits
31-2 Production of Integrated Circuits
31-3 Integrated Components
31-4 The Differential Amplifier
31-5 Linear IC Applications

31-1 TYPES OF INTEGRATED CIRCUITS

The example in Fig. 31-2 illustrates a transistor, resistor, and capacitor integrated on a single silicon wafer. Inductors are not generally integrated because they take too much space. Transistors and diodes are the easiest components to integrate. Resistors are not precise, but the circuits are designed to depend on the ratios of resistances rather than absolute values. The range of R values is 100 to 25,000 Ω, approximately. The range of C values is approximately 3 to 30 F. External components are connected to the IC unit for larger values of R and C, with any L required for tuning.

For the construction in Fig. 31-2, the diffused layers at the left form an NPN transistor. Note that the collector has a reverse NP junction with the wafer to isolate the transistor from the silicon substrate or platform. In the middle section, R between the two ohmic contacts is determined by the length and width of the strip. C at the right is the capacitance across the reverse-biased junction of the N material to the P wafer.

Monolithic IC. In this type, all the components are formed as part of a single P-type or N-type wafer. This is the construction shown in Fig. 31-2.

Chapter 31
Integrated Circuits

FIGURE 31-1
IC packages. Round IC unit is TO-5, with diameter of ½ in. The 16-pin dual inline package (DIP) is ¾ in long. Flat pack at right is 1 in.

Thin-film IC. In this type, the substrate is ceramic or glass, which is an insulator. All the components are evaporated on this insulating platform. The advantage is good isolation between the integrated components.

Thick-film IC. In this type, R and C are formed on the substrate, but the transistors are added as discrete chips.

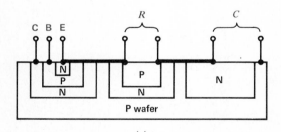

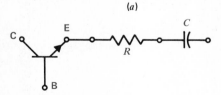

FIGURE 31-2
Integrated circuit on silicon wafer. (a) Construction with NPN transistor, R, and C. (b) Schematic diagram.

Hybrid IC. This type combines monolithic and thin-film units on a single ceramic platform. Discrete transistors may also be included, where high power rating is needed.

Digital and Linear IC Units. In terms of their function, IC chips are classified in either of these two broad groups. Digital circuits are pulse circuits. The digital IC chips contain an array of pulse switching circuits to form logic functions, as explained in Chap. 32, Digital Electronics.

Linear IC chips generally contain several amplifier circuits for either audio or rf signals. A linear amplifier operates on the linear portion of its forward transfer characteristic curve, resulting in an amplified duplicate of the input signal.

LSI and MSI Chips. Digital IC units contain many circuits of the same type, repeating a basic logic circuit, such as the gate function. The degree of complexity is indicated by LSI for large-scale integration, including over 100 gates, and MSI for medium-scale integration, with 10 to 100 gates.

Practice Problems 31-1
(answers on page 684)
Answer true or false.
(a) Digital IC units contain gate circuits for pulse waveforms.

(b) The unit for an audio amplifier would be a linear IC.
(c) Figure 32-2 illustrates the hybrid IC construction.

31-2 PRODUCTION OF INTEGRATED CIRCUITS

For the monolithic type, IC chips are made by the batch of 25 to 100 chips from one silicon wafer. Figure 31-3 illustrates the steps in construction, from the wafer slices of a silicon ingot in (a), to the scribed chips on the wafer in (b), and the diffused components in (c) and (d).

The individual silicon wafers are about 10 mil thick, with a diameter of 1.5 to 2 in. First, each wafer is polished to a mirror finish by acid etching. Then a layer of silicon dioxide (SiO_2) is formed on the wafer. This thin glass insulator protects the silicon surface and serves as a barrier to the doping for semiconductor junctions. The oxide coating is opened in a window pattern by photochemical techniques, to allow doping where desired. A sequence of oxidation and diffusion operations results in layers of the doped semiconductors. The oxidation forms silicon dioxide as a barrier and insulator. Then the diffusion forms the junctions.

Metallic paths are needed to interconnect the components in each IC chip. These are essentially aluminum printed wiring. They terminate at the edges of the chip, where tiny

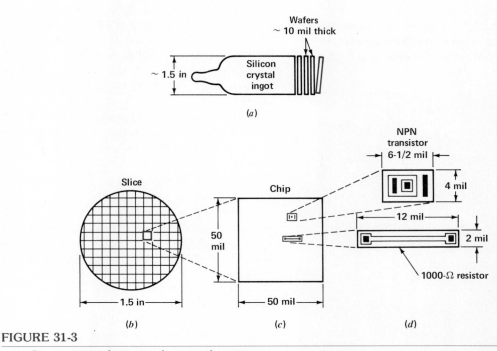

FIGURE 31-3

Construction of integrated-circuit chips starting from silicon ingot in (a) to slice in (b) and IC chip in (c), with magnified view of diffused components in (d).

wires are bonded for connection to the external leads.

Automatic equipment with test probes is used to check the aluminum electrode pads. Each complete chip is tested sequentially by a large number of needle-tipped probes. This wafer probing on many electrodes for all the chips is controlled and monitored by computers.

Since the complete silicon wafer has the physical properties of a thin sheet of glass, the chips are separated by methods similar to cutting glass. A fine diamond point is used to scribe the wafer into the chips (Fig. 31-3b). All the chips on one wafer are the same. The slice is held firmly and then stressed to break along the scribe lines. This division of the slice into chips is called "dicing."

Chips that are defective in probe testing are sorted out. Also, microscopic examination shows chips with imperfections due to dicing.

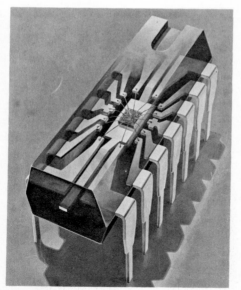

FIGURE 31-4
IC chip in DIP unit. Length of package is 1 in.

Then each chip is mounted in its package, which is generally a plastic case. The chip is essentially glued with epoxy. The terminal leads on the package are connected to the internal electrodes on the chip with 1.5-mil wire, aluminum or gold. Connections are made either by ultrasonic bonding or thermocompression bonding. The ultrasonic energy vibrates the bonding tip to produce heat. The final IC package is shown in Fig. 31-4. Note that most of the space in the package is needed for the connecting leads and pins.

The IC package is generally in the round TO-5 case, ceramic flat pack, or dual inline form (DIP). The TO-5 style may be a glass-metal package, hermetically sealed in dry nitrogen, or a plastic case. The flat pack and DIP packages may be ceramic or plastic. Ceramic packages have a better hermetic seal than plastic.

Typical pin connections for IC units are shown in Fig. 31-5. The round TO-5 package generally has 8, 10, or 12 leads. The dual inline and flat pack may have 8, 14, 16, 24, or 28 pins. Corresponding sockets are available for all the IC packages, but they are often soldered on the printed-circuit board.

Practice Problems 31-2
(*answers on page 684*)
Answer true or false.
(a) Dicing is the process of breaking the silicon slice into 25 to 100 chips.
(b) Silicon dioxide (SiO_2) is an insulator, like glass.

31-3
INTEGRATED COMPONENTS
These include transistors, diodes, resistors, and capacitors. Isolation between adjacent components is provided by a reverse-biased PN junction. The isolation R can be 100 MΩ or more. The advantages of integrated components are

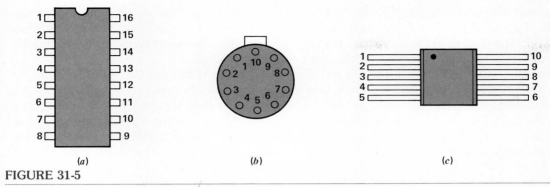

FIGURE 31-5

Pin connections for IC packages. Dot or notch at pin 1. (a) 16-pin DIP in TO-116 case. Top view. (b) 10-pin TO-5 case. Bottom view. (c) 10-pin ceramic flat pack in TO-81 case. Top view.

miniaturization and matched characteristics for the components with respect to operating values and temperature gradient.

Integrated Bipolar Transistors. Most common is the NPN construction shown before in Fig. 31-2. Note that the P substrate forms a reverse diode with the P collector. Also, the P substrate could be the collector for a PNP transistor with the adjacent N and P layers. Then the combination would form a complementary pair of NPN and PNP transistors.

Integrated Diodes. An N-type diffusion into the P substrate can produce diode junctions where they are needed on the chip. However, this is essentially the same as the emitter-base junction on a transistor. Therefore, the diode function can be provided by joining the transistor collector to base, serving as the anode (Fig. 31-6). The collector should not be left open, to prevent parasitic circuits with other components on the chip. Furthermore, the diode characteristics are matched to the transistors in the chip, as they are all made the same way. These diodes can be used for transistor bias in the base circuit, temperature compensation, and isolating the incorrect polarity of voltage for the collector on a separate transistor.

The collector can also be connected to the emitter for diode operation. This way both junctions are used in parallel for the diode.

Integrated Zener Diodes. By using reverse voltage on the emitter-base junction, it can serve as a zener diode. The zener voltage is typically 7 V. Zener diodes are used for voltage regulation, to stabilize the supply voltage against changes with different load currents.

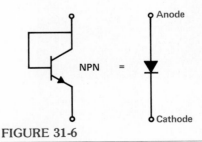

FIGURE 31-6

Triode transistor connected as a diode.

Integrated Resistors. A common method uses a strip of lightly doped P material, as shown in Fig. 31-7. The N layer below the P layer will be reverse-biased for isolation from the substrate. The silicon dioxide provides insulation from the next layer. External connections to the resistor are made with the aluminum pads.

This type of monolithic R has a resistivity of about 200 Ω for the entire chip. However, the ratio of length to width of the R strip can provide values of 100 to 25,000 Ω. Higher values of 10 to 50 kΩ can be obtained by special doping techniques to reduce the cross-sectional area of the R strip. This type of resistor is a *pinch resistor*.

The tolerance on monolithic resistors is ±30 to ±50 percent, which is poor for precise values. Also, the temperature coefficient of resistivity is high. However, the resistance ratios on the same chip can be controlled within 2 to 3 percent. Therefore, the circuits are designed to depend on a resistance ratio, rather than the absolute value of R.

Integrated Capacitors. The two types are the junction type illustrated for C in Fig. 31-2 and the MOS type in Fig. 31-8. In the junction type, the capacitance is across the depletion zone of a PN junction with reverse bias. Typical values are 0.1 to 0.4 pF per square mil, with reverse bias of 5 V. If an area of 100 square mil is used out of the 2500 square mil on a chip, the C then will be 10 to 40 pF.

For the MOS capacitor in Fig. 31-8, the N^+ layer forms the bottom plate, the SiO_2 insulating layer is the dielectric, and the aluminum metallization serves as the opposite plate. The C value of 3 to 30 pF depends just on the dielectric and plate area. The + sign for N or P indicates heavy doping.

COS/MOS or C/MOS Transistors. This abbreviation stands for complementary symmetry with P-channel and N-channel MOSFET or IGFET units, generally of the enhancement type. Figure 31-9 illustrates a COS/MOS inverter circuit in digital logic. The function of the inverter is to change the conditions in a train of pulse signals from high to low amplitude, or vice versa.

In Fig. 31-9a, the FET at the right uses the N substrate for the N channel. The gate, drain, and source electrodes are shown with heavy lines. The P^+ areas are used as guard bands to separate the different electrodes from each other. For the FET at the left, the P well forms the P channel. In this FET, the gate, drain, and source electrodes are separated from each other by N^+ guard bands.

In Fig. 31-9b, the inverter circuit makes use of complementary symmetry. The substrate

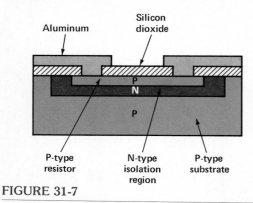

FIGURE 31-7

Cross section of P-type diffused resistor in IC chip. (From *RCA Manual IC-42*)

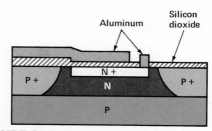

FIGURE 31-8

Cross section of MOS type of monolithic capacitor. (From *RCA Manual IC-42*)

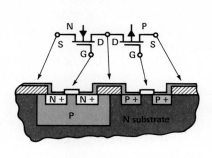
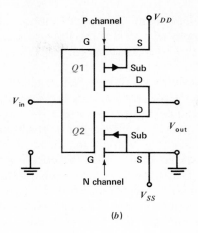

(a) (b)

FIGURE 31-9
(a) Cross section of COS/MOS diffused transistors on IC chip. (b) Inverter circuit for construction in (a). (From *RCA Manual SC-16*)

of the P channel is at $+V_{DD}$, and the N channel is at ground or 0 V. When the input voltage is zero, the P-channel FET is conducting but the N-channel FET is off. Then the output voltage is close to V_{DD}. When the input has a positive voltage pulse, however, conditions are reversed. Then the P-channel FET is off and the N-channel FET is on. These conditions make the output voltage close to 0 V. The net result is that V_{out} is high when V_{in} is low and V_{out} is low when V_{in} is high, which is the inverting function.

Supply Voltage. Typical values are 15 to 40 V for linear IC units and 3 to 15 V for digital IC units, with monolithic construction. Usually, positive polarity is required, as for collector voltage on NPN transistors.

Practice Problems 31-3
(answers on page 684)
Answer true or false.
(a) A typical value for an integrated monolithic C is 80 μF.
(b) To use an NPN transistor as a diode, join the collector and base.

31-4
THE DIFFERENTIAL AMPLIFIER

This circuit is commonly used in linear IC units for a broad variety of functions. The schematic diagram of the basic amplifier circuit is shown in Fig. 31-10. Q1 and Q2 are NPN transistors in a *differential pair*. They are dc coupled between the two emitters.

The stages are balanced with equal values of R_L from one V_{CC} supply. In the emitter circuit, R_1 can adjust the emitter-base bias for equal values of I_C in Q1 and Q2. R2 is used to increase the resistance in the emitter circuit. Its effect is to provide a constant-current source to stabilize the bias current. Furthermore, the common resistance for both emitter circuits provides feedback.

In many cases, the bias circuit in the emitter branch uses a separate transistor conducting saturation current. This method has the advantages of constant bias current and elimination of the need for negative bias voltage.

There are several possible methods for input and output signals, as follows:

Chapter 31
Integrated Circuits

1. *Differential output.* This signal voltage between points 1 and 2 is the difference between the collector voltages for Q1 and Q2.
2. *Differential input.* Two signal voltages of opposite polarities are connected to the base input terminals. This example of opposite voltages is also called a *push-pull* signal.
3. *Single-ended input.* Either input terminal can be used alone for input signal, with the other input terminal grounded.
4. *Single-ended output.* Signal voltage is taken from the collector of either Q1 or Q2 alone, at point 1 or 2. With differential input, the single-ended outputs at 1 or 2 have opposite polarities.

Amplification of a Differential Signal. We can take an example of differential input and output voltages, assuming a gain of 100 for each stage. Suppose the input V_1 to the Q1 base is +1 mV. This signal is amplified and inverted by Q1 for −100 mV at point A. Also, the Q2 base has input V of −1 mV, of opposite polarity from V_1. This amplified signal output from Q2 at B then is +100 mV.

Each stage operates as a CE amplifier. The emitter voltage is constant for both Q1 and Q2, as the opposite base signals cancel in the common-emitter branch.

The differential output is the difference between the signal voltages at 1 and 2. This difference in the output is 100 mV − (−100 mV) = 200 mV. The difference voltage for input signal is 2 mV. Therefore, the voltage gain of the differential amplifier is $^{200}\!/_2 = 100$, the same as for a single stage. The reason for the gain with differential operation is the fact that the collector voltages vary in opposite directions.

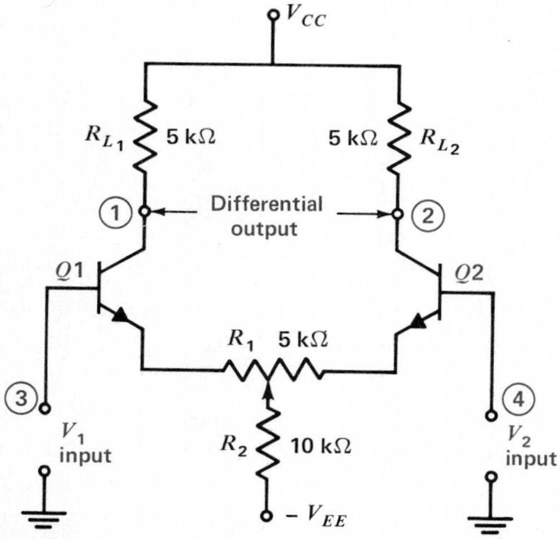

Common-mode Rejection. The advantage of the differential amplifier is that it has no gain for input signals applied to Q1 and Q2 in the same phase, which is the *common mode*. The reason is that the difference voltage between 1 and 2 equals zero when they are at the same potential.

As an example, any variations in the supply voltage would be applied in the common mode and, therefore, rejected. In short, the circuit amplifies differential signals but rejects common-mode signals. The degree of common-mode rejection depends on the balance of the two stages.

Offset Voltage. This is the amount of differential voltage output with no input. Ideally, the offset should be zero, but the stages are not perfectly balanced either in I_C or in input signals. Temperature variations and pickup of stray signals may cause offset.

FIGURE 31-10
Basic circuit of differential amplifier.

Operational Amplifiers. In its fundamental form, the operational amplifier uses a pair of dc-coupled stages, with differential input but single-ended output. However, many combinations of feedback can be used to provide a circuit with the advantages of high gain, wide bandwidth for af or rf signals, high input impedance, and low output impedance. The applications include amplifier, comparator, follower, differentiator, integrator, adder, subtractor, or phase shifter.

Practice Problems 31-4
(answers on page 684)
Refer to Fig. 31-10.
(a) Which terminals are used for differential output?
(b) Which terminals are used for differential input?

31-5
LINEAR IC APPLICATIONS

In general, linear IC units are for linear amplifiers and analog circuits. An analog signal has continuous variations, as in a sine wave, corresponding to the desired information. Pulse or digital circuits operate between the two discrete states of on or off.

A listing of linear IC applications includes the following:

Audio amplifiers. Usually in pairs, for stereo.
Rf amplifiers and oscillators. For AM and FM radio or television receivers.
IF amplifiers. For AM, FM, or television receivers, usually with envelope detector. IF is the intermediate frequency of the carrier signal before it is detected.
3.58-MHz color circuits. Subcarrier regenerator, amplifiers, and demodulators for color television receivers.
Operational amplifiers. For wide-band amplification.

Voltage regulators. To stabilize a dc supply voltage.
Choppers. To chop a dc waveform into ac segments for easier amplification, and then convert back into the dc form. The chopper is used for amplifying steady dc or very low frequency variations.
Diode and transistor arrays. Have separate connections, but the integrated unit provides matched characteristics.

In addition, specialized IC chips are available for electronic calculators, electronic watches, the video games units for television receivers, and the frequency synthesizer used in CB radios instead of separate crystals.

An example of a linear IC unit containing dual audio amplifiers is shown in Fig. 31-11 on the next page. Digital IC applications are explained in Chap. 32, Digital Electronics.

In Fig. 31-11a, the DIP unit has 14 pins, numbered from the notch, top view. The unit can deliver up to 2 W per audio channel into 8-Ω or 16-Ω loudspeakers. Supply voltage is 10 to 26 V. The total supply current is 15 mA without signal up to a maximum load current of 500 mA.

In Fig. 31-11b note the triangle symbol for the IC amplifiers. Each half is for one audio channel. The numbers on the amplifier refer to the pin numbers in (a). Components outside the dashed box are external parts that must be added to the IC unit. Note the 100-kΩ feedback resistor between the output and pin 7 or 8 for each amplifier.

Practice Problems 31-5
(answers on page 684)
(a) In Fig. 31-11a, which two pins are used for the audio output signals?
(b) In Fig. 31-11b, what size C is used for coupling to the loudspeakers?

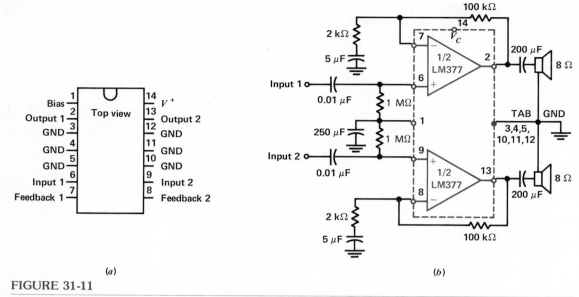

FIGURE 31-11
IC unit for stereo audio amplifier. (a) Dual inline package. (b) Typical circuit application. Components outside dotted box are not in IC amplifier. (*National Semiconductor LM377*)

Summary

The following terms and definitions summarize the main features of integrated circuits.

Bipolar. NPN and PNP transistors.
Chip. Tiny semiconductor containing integrated circuits.
Common mode. Signals in the same phase.
Complementary symmetry. Symmetry between P channel and N channel or PNP and NPN transistors.
COS/MOS or C/MOS. Complementary symmetry with MOSFET devices.
Dicing. Breaking the semiconductor slice into chips.
Differential amplifier. High-gain amplifier with provision for differential input and output.
Differential input. An input with opposite polarities, or input in push-pull.
Differential output. Difference between two outputs.
Digital IC. Circuit which uses gate circuits for digital logic with pulse signals.

DIP. Double inline pins for IC package.
Gate. On-off circuit for digital applications.
Hybrid IC. Circuit which combines monolithic construction and separate transistors.
IC. Integrated circuit with semiconductors, R, and C on one chip.
Inverter. Digital circuit used to reverse high or low condition of pulse.
Linear IC. IC unit with linear amplifiers for af and rf applications.
LSI. Large-scale integration for digital IC chips with 100 or more gates.
Microelectronics. Type of electronics in which integrated circuits are used to miniaturize electronic equipment.
Monolithic IC. IC units with all semiconductor components formed on one substrate.
MOS. Metal-oxide semiconductor type of FET. Same as IGFET.
MSI. Medium-scale integration for digital IC chips with less than 100 gates.
Offset. Output of differential amplifier without any input.
Operational amplifier. High-gain amplifier with feedback network that determines the characteristics.
Substrate. Platform on which the IC components are formed.
Thick-film IC. Circuit in which R and C are formed on glass or ceramic substrate, but transistors are added as discrete chips.
Thin-film IC. Circuit in which R, C, and transistors are deposited on glass or ceramic substrate.

Self-Examination (Answers at back of book.)

Answer true or false.

1. Figure 32-2 illustrates construction of a monolithic linear IC.
2. Typical values of L integrated on a chip are 10 to 100 mH.
3. Typical values of C integrated on a chip are 3 to 30 pF.
4. An NPN transistor can be used as a diode by connecting collector to base or emitter.
5. A typical dc supply voltage for IC units is 3 to 40 V.
6. Transistors on a monolithic linear IC unit are generally NPN on a silicon substrate.
7. Linear and digital IC units usually have similar applications.
8. In the differential amplifier in Fig. 31-10, R_{L_1} and R_{L_2} are the collector load resistors.
9. In Fig. 31-10, the voltage across R_{L_2} provides differential output.
10. For the stereo audio amplifiers in Fig. 31-11b, the input coupling capacitors are 0.01 μF.

Essay Questions

1. Compare the applications of linear and digital IC units.
2. Compare the construction for monolithic, thin-film, and hybrid IC units.
3. From a manual of integrated circuits, list the circuits on two types of linear IC units and two digital IC units.
4. Compare the packaging and pin connections for TO-5 and DIP IC units.
5. For the differential amplifier in Fig. 31-10, give three ways to supply input signal and three ways to take output signal.
6. For a differential amplifier, what is meant by common-mode rejection and by offset voltage?
7. Give five types of components that are usually integrated in monolithic construction.
8. What is meant by complementary symmetry?

Answers to Practice Problems

31-1	(a) T	31-3	(b) T
	(b) T	31-4	(a) 1 and 2
	(c) F		(b) 3 and 4
31-2	(a) T	31-5	(a) 2 and 13
	(b) T		(b) 200 μF
31-3	(a) F		

Digital Electronics

Chapter 32

Electronic circuits can be divided into two broad groups, analog and digital. In analog circuits, the voltage and current waveforms are similar to the signal variations. A digital signal, however, is a group of pulses with the same level but either on or off. This is also called a *binary* signal because it has just two bits of information.

A common use of digital circuits is in numerical counting. These applications include the important fields of computers, electronic calculators, digital clocks, and test equipment. A digital frequency meter is shown in Fig. 32-1. A comparison between analog and digital equipment can be seen from the analog VTVM in Fig. 7-22 and digital VTVM in Fig. 7-23, described in Chap. 7.

Furthermore, digital signals have an advantage over analog signals in reducing the effects of noise. For this reason, digital circuits are used in communications electronics, in addition to numerical counting and digital control. Some basic features of digital electronics are described in the following topics:

32-1   Binary Numbers
32-2   Binary Arithmetic
32-3   Symbolic Logic for Switching Circuits
32-4   Truth Tables
32-5   Basic Logic Circuits
32-6   Diode Gate Circuits
32-7   Diode-Transistor Logic (DTL)
32-8   Transistor-Transistor Logic (TTL)
32-9   Multivibrator (MV) Circuits
32-10 Flip-flop Circuits
32-11 Counters

32-1 BINARY NUMBERS

Only the digits 0 and 1 are used, representing the two levels of a binary signal. Examples of binary numbers are 001, 010, and 111. Any number of places is possible, each with a 1 or 0.

Figure 32-2 illustrates how the two levels of a pulse waveform correspond to the binary digits 1 and 0.

In general, the digits for number systems are 0, 1, 2, 3, 4, 5, 6, 7, 8, and 9. How many digits are used is the *base* or *radix*. As an exam-

Chapter 32
Digital Electronics

FIGURE 32-1

Digital frequency meter. (*Hewlett Packard*)

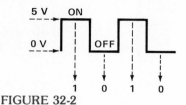

FIGURE 32-2

Specifying binary digits 1 and 0 for ON and OFF conditions of voltage pulses.

ple, binary numbers use base two, as only the two digits 0 and 1 are used. The base can be indicated by a subscript, as in $(111)_2$.

Our conventional decimal system for counting uses all ten digits from 0 to 9. Therefore, the base or radix is ten for decimal numbers.

The highest digit for any place is one less than the base. In decimal numbers, 9 is the highest digit, while 1 is the highest digit in binary numbers, for counting in successive places.

How Each Place Is Used for Different Counts. To count numbers more than the base itself, the complete number must have an order of places for larger counts. The order is indicated by the *decimal point* in decimal numbers or the *binary point* in binary numbers.

Figure 32-3 illustrates the decimal orders. For numbers greater than 1, the first place to the left of the decimal point is for the count of the digits alone, from 0 to 9. The second place is for counts of tens. Next is the place for counts of 10^2, or hundreds, and then for 10^3, or thousands. The counts for successive places increase in multiples of 10.

For decimal fractions less than 1, the places are to the right of the decimal point. These values decrease in multiples of $\frac{1}{10}$ or 0.1.

The same method is used with binary numbers. However, the first place to the left of the point can be only 0 or 1. As usual, higher counts are carried over to the next place. The second place is in counts of two. Successive places to the left are for counts of 2^2, or fours; 2^3, or eights; etc.

Places to the right of the binary point can be used for fractions less than 1, as in decimal numbers. However, the values decrease in multiples of $\frac{1}{2}$ instead of $\frac{1}{10}$.

For the binary example in Fig. 32-4, $(111)_2$ is $4 + 2 + 1 = (7)_{10}$. The binary number 111 is actually 111.0, but the point is usually omitted, as in decimal numbers.

As another example, $(1100)_2$ is $8 + 4 + 0 + 0 = (12)_{10}$. Binary numbers

$10^3 =$ THOUSANDS	$10^2 =$ HUNDREDS	$10^1 =$ TENS	DIGITS 0 TO 9	QUANTITY
0	1	1	1	$(111)_{10}$

FIGURE 32-3

Order of places in a decimal number.

$2^3 =$ EIGHTS	$2^2 =$ FOURS	$2^1 =$ TWOS	DIGITS 0 OR 1	QUANTITY
0	1	1	1	$(7)_{10}$

FIGURE 32-4

Order of places in a binary number.

need more places for counting because the base is smaller.

The general method of counting in number systems with a different base or radix is illustrated in Fig. 32-5. Included with 10 and 2 is the octal radix of 8, which is also used in digital circuits.

Converting Binary to Decimal. Just make the count for each place and add the values. As an example, the seven places in $(1010101)_2$ show the count

$$64 + 0 + 16 + 0 + 4 + 0 + 1 = (85)_{10}.$$

Converting Decimal to Binary. There are several methods, but a common system is as follows:

1. Divide the decimal number successively by 2, until the quotient is zero.
2. Note whether each remainder is 0 or 1.
3. The remainders of 0 and 1 are the digits in the binary number.

The places of the digits 0 or 1 in the binary number are to the left of the binary point, in the same order as the divisions. Continue the divisions until the last quotient is zero.

As an example, to convert $(13)_{10}$ to binary form:

$$13 \div 2 = 6 \quad \text{with remainder of } 1$$
$$6 \div 2 = 3 \quad \text{with remainder of } 0$$
$$3 \div 2 = 1 \quad \text{with remainder of } 1$$
$$1 \div 2 = 0 \quad \text{with remainder of } 1$$

The remainders, in the order of the divisions, are 1, 0, 1, and 1. In the correct places starting from the binary point, they form $(1101)_2$. As a check on this example, $(1101)_2$ equals $8 + 4 + 0 + 1 = 13$.

Shifting the Place Point. In decimal numbers moving the point one place to the right multiplies the number by 10. For instance, 10×53.7 and 537.0 are equal. Moving the point one place to the left divides by 10.

In binary numbers, shifting the point one place multiplies or divides by 2. For instance,

$(BASE)^3$	$(BASE)^2$	BASE	DIGITS	BASE OR RADIX
1000	100	Ten	0 to 9	10
8	4	Two	0 or 1	2
512	64	Eight	0 to 7	8

FIGURE 32-5

Order of places for numbers with base 10, 2, or 8.

$(111.0)_2$ is $(7)_{10}$, but $(1110.0)_2$ is $(14)_{10}$. The 7 is doubled to 14 by moving the binary point one place to the right.

For division by 2, the point is moved one place to the left. For instance, $(111.0)_2$ is $(7)_{10}$, but $(11.10)_2$ is $(3.5)_{10}$. The binary places in 11.1 have the values of $2 + 1 + \frac{1}{2}$ which equals $3\frac{1}{2}$ or 3.5. The 3.5 is one-half of 7.

Practice Problems 32-1
(answers on page 709)
(a) Convert $(1111)_2$ to decimal form.
(b) Convert 10 in decimal form to a binary number.
(c) How much is $(000)_2$ as a decimal number?

32-2
BINARY ARITHMETIC

Only the digits 0 and 1 can be used in each place. For binary addition, all possible sums of 0 and 1 are

$0 + 0 = 0$ $1 + 0 = 1$
$0 + 1 = 1$ $1 + 1 = (10)_2$

The last sum is not decimal number ten but has 0 and 1 in the two binary places because of the carry of 1.

We cannot have the digit 2 as a bit. However, 1 in the second binary place has a weight of two. The first binary place is 0, so that the sum will not be three. The general rule is that for binary $1 + 1$ the sum is 0 with a carry of 1 to the next higher order.

To add numbers with more than one digit, start next to the binary point. As an example,

$$\begin{array}{r}(110)_2 \\ +(111)_2 \\ \hline 1101 \end{array}$$

In the third column from the point, the binary addition of $1 + 1 +$ carry of 1 results in the sum of 1 in this column and carry of 1 to the fourth column.

As a check on the answer, this problem in decimal numbers is $6 + 7 = 13$.

The only two possible combinations with a carry are

(a) $1 + 1 =$ sum of 0 with carry of 1. This is binary 10, which equals decimal two.
(b) $1 + 1 +$ carry of $1 =$ sum of 1 with carry of 1. This is binary 11, which equals decimal three.

For binary multiplication, all possible products of 0 and 1 are

$0 \times 0 = 0$ $1 \times 0 = 0$
$0 \times 1 = 0$ $1 \times 1 = 1$

These values are the same as in decimal multiplication. For long hand multiplication with more digits, the partial products are added.

More details on binary numbers for computer mathematics can be found in the books listed in the bibliography at the end of this chapter. Binary addition is probably the most important operation for digital counting circuits. Actually, multiplication is a short cut method for adding large quantities. Furthermore, subtraction can be done by methods of adding complements of the numbers and dropping the highest place. Finally, division can be reduced to a system of subtraction.

Practice Problems 32-2
(answers on page 709)
(a) Add $(101)_2 + (010)_2$.
(b) Add $(111)_2 + (111)_2$.

32-3
SYMBOLIC LOGIC FOR SWITCHING CIRCUITS

This system specifies the results for different combinations of two conditions: ON or OFF, open or closed, and 1 or 0. For instance, when two switches are in series to light a bulb, it is reasonable to expect that both switches must be on for the bulb to be on. The relationship between the input and output is a logic function. The system is symbolic logic because symbols such as A or B and 0 or 1 are used to represent the switch conditions.

As a historical note, symbolic logic is also called *Boolean algebra,* named after George Boole, an English mathematician (1815–1864). He started the system of putting ordinary statements into symbols so that they could be analyzed mathematically. These concepts were not applied to electronics until Claude Shannon of Bell Telephone Laboratories used logic functions to simplify switching circuits for the telephone system. Today, all digital electronic circuits make use of symbolic logic. The basic logic functions are AND and OR, with their inverted or negative functions NAND and NOR.

The AND Function. This is demonstrated by the two switches in series in Fig. 32-6a. Only if both switches are closed will the bulb be able to light. If either S_1 or S_2 or both switches are off, the light cannot go on.

Consider both switches as one gate circuit. This AND gate has two inputs, one for each switch. The output is current for the bulb. Using the symbol 1 for the ON position and 0 for the OFF position of each switch, the possible combinations are as follows:

S_1	0	1	0	1
S_2	0	0	1	1
LIGHT	OFF	OFF	OFF	ON

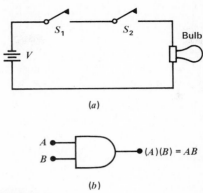

FIGURE 32-6

The AND gate. (a) Equivalent circuit. (b) Symbol as a logic function.

Only when both inputs are at the 1 position can the light go on.

The symbol in Fig. 32-6b indicates that the AND gate performs a logic multiplication. With inputs of A and B, the output is $(A)(B)$, $A \times B$, or AB. In symbolic logic or Boolean algebra, the AND function does not mean addition in the sense of arithmetic. Instead, we can think of multiplication signs as indicating the AND function.

The logic block symbol represents the switching circuits. Actually, diodes and transistors are used, pulsed to conduct maximum current in the ON position. The shape of the symbol indicates the AND function. This means all the inputs must be ON to have output. More than two inputs may be required for the one output, but still all the inputs must be ON.

The OR Function. This is demonstrated by the two switches in parallel in Fig. 32-7a. Now either S_1 or S_2 can be closed to light the bulb. Also, the bulb lights with both S_1 and S_2 ON. The only condition to turn the bulb off is with both S_1 and S_2 OFF.

Chapter 32
Digital Electronics

FIGURE 32-8

Symbol for exclusive OR gate with output function circled to show exclusive operation.

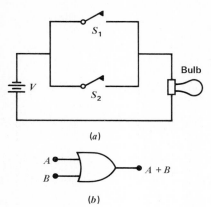

FIGURE 32-7

The OR gate. (*a*) Equivalent circuit. (*b*) Symbol as a logic function.

Using 1 for ON and 0 for OFF, the possible switch combinations are:

S_1	0	1	0	1
S_2	0	0	1	1
LIGHT	OFF	ON	ON	ON

The symbol in Fig. 32-7*b* indicates that the OR gate performs a logic addition. With inputs of *A* and *B*, the output is $A + B$. However, the + sign does not mean addition as in arithmetic. In symbolic logic we can think of the + sign as indicating the OR function. This means all inputs must be OFF to turn off the output.

The symbol in Fig. 32-8 represents the *exclusive* OR function. Figure 32-7 is an *inclusive* OR function. The difference is that the exclusive OR function has the output OFF when the inputs are either all OFF or all ON. Note that the output of $A + B$ is circled in Fig. 32-8 to indicate the exclusive OR function. An OR function is inclusive, however, unless it is specified as exclusive.

Inversion or Negation. An inverter circuit has its output signal of opposite polarity from the input. For instance, with positive input voltage to the base of an NPN transistor in a CE amplifier, the amplified collector output voltage is negative.

The inversion function is also called *negation*. As an example, assume input *A* can be 0 or 1. When *A* is at 0, its inversion is 1, or 1 can be inverted to 0. The inversion also applies to the ON and OFF states.

The logic symbol for inversion or negation is a bar over the function to indicate the opposite state. For instance, $\bar{A}$ means "not *A*." The schematic symbol generally used for negation is a small circle. In Fig. 32-9*a*, the *A* input to the amplifier with a circle is $\bar{A}$ in the output.

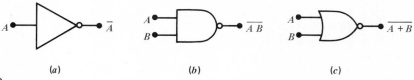

FIGURE 32-9

Use of small circle in symbol to show inverting or negating function. (*a*) Input *A* is not *A* in output. Triangle is symbol for amplifier. (*b*) NAND gate. (*c*) NOR gate, inclusive.

Chapter 32
Digital Electronics

TABLE 32-1.

GATE	INPUTS	OUTPUT	BOOLEAN FUNCTION
AND	All ON (HIGH)	ON (HIGH)	$x = AB$
OR	All OFF (LOW)	OFF (LOW)	$x = A + B$
NAND	All ON (HIGH)	OFF (LOW)	$x = \overline{AB}$
NOR	All OFF (LOW)	ON (HIGH)	$x = \overline{A + B}$

The NAND Function. This is an AND gate but with inverted output. In Fig. 32-9b, the circle in the output of the AND symbol shows inversion. The NAND output function is $\overline{AB}$, or not AB, instead of AB. In terms of the circuit, the NAND gate must have both inputs ON to have the output in the OFF state.

The NOR Function. As shown in Fig. 32-9c, this is an inclusive OR gate but with inverted output. The NOR output function $\overline{A + B}$ is not (A or B), instead of having A or B. In terms of the circuit, the NOR gate must have all inputs OFF to have the output in the ON state. These functions are summarized in Table 32-1 for the AND, OR, NAND, and NOR gates.

Application of AND Gate. Figure 32-10 shows two groups of input pulses to an AND gate, with the output signal. The AND gate has output only when all inputs are ON, or at the high level. Therefore, the output is high only at the time intervals marked 1 and 5. At these times, input A is high and input B is high. At all other times, when only one input is high, the output is low.

Practice Problems 32-3
(answers on page 709)
Answer true or false.
(a) The AND gate corresponds to two switches in series.
(b) The OR gate corresponds to two switches in parallel.
(c) The NAND gate is an inverted OR gate.

32-4
TRUTH TABLES

These are listings of all possible combinations of variables, such as input signals to a logic gate, with the resulting output for each combination. As an example, Table 32-2 shows the truth table for an AND function with two inputs. The table is generally in binary notation, with 1 for ON or HIGH and 0 for OFF or LOW. The third vertical column here lists the truth for the combinations

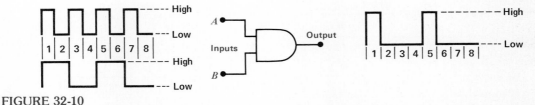

FIGURE 32-10

Pulse output of AND gate is at high level only when both inputs are high.

TABLE 32-2. Truth Table for AND Function

INPUTS		OUTPUT
A	B	(A)(B)
0	0	$0 \times 0 = 0$
0	1	$0 \times 1 = 0$
1	0	$1 \times 0 = 0$
1	1	$1 \times 1 = 1$

TABLE 32-3. Truth Table for OR Function

INPUTS		OUTPUT
C	D	C + D
0	0	0 or 0 = 0
0	1	0 or 1 = 1
1	0	1 or 0 = 1
1	1	1 or 1 = 1

in each horizontal row. The AND function has the output of 1 only when both inputs are 1.

All possible combinations must be listed in the truth table. For two variables with two input symbols, the combinations are 2^2 or 4. With three variables there are $2^3 = 8$ combinations. In general, the combinations equal 2^N, where N is the number of inputs.

The truth table for a two-input OR function is shown in Table 32-3. When either input is 1, or both are 1, the output is 1.

One application of truth tables is to show that certain logic functions are identical. This means the same results can be produced with a different circuit. Table 32-4 shows that the logic function $(\bar{A})(\bar{B})$ is identical to $\overline{A + B}$.

As a start for the proof, the vertical columns I and II list the four possible binary combinations of the inputs A and B. In columns III and IV the inverted forms for $\bar{A}$ with $\bar{B}$ are shown by changing 1 to 0 or 0 to 1. Column V lists the results of multiplying (ANDing) $\bar{A}$ by $\bar{B}$.

Column VI goes back to the original values to add (OR) $A + B$. Then this is negated in column VII.

If we compare this last column for $\overline{A + B}$ with column V for $(\bar{A})(\bar{B})$, the values in corresponding rows are the same. This result of $(\bar{A})(\bar{B}) = \overline{A + B}$ is known as De Morgan's identity, which is one of the laws of Boolean algebra.

Practice Problems 32-4
(answers on page 709)
(a) What is the negation of binary 1?
(b) What is the negation of binary 0?
(c) The truth table for an OR function with four inputs has how many binary combinations?

TABLE 32-4. Truth Table for Identity $(\bar{A})(\bar{B}) = \overline{A + B}$

I	II	III	IV	V	VI	VII
A	B	$\bar{A}$	$\bar{B}$	$(\bar{A})(\bar{B})$	A + B	$\overline{A + B}$
0	0	1	1	$1 \times 1 = 1$	0 or 0 = 0	1
0	1	1	0	$1 \times 0 = 0$	0 or 1 = 1	0
1	0	0	1	$0 \times 1 = 0$	1 or 0 = 1	0
1	1	0	0	$0 \times 0 = 0$	1 or 1 = 1	0

32-5
BASIC LOGIC CIRCUITS

The circuit components are resistors, diodes, and transistors, including the junction type and the FET. A diode is switched on by applying forward voltage. The conducting diode is practically a short circuit. It has little resistance and an internal voltage drop of less than 1 V.

Applying reverse voltage to a conducting diode cuts it off. Then the diode is an open circuit with very high resistance.

Similarly, a transistor is switched on by applying enough forward bias to produce saturation. Then the transistor output voltage is very low, with the collector saturation voltage less than 1 V.

When the transistor is cut off, without any output current, the output voltage is high. This value is equal to the supply voltage, as there is no IR drop across R_L.

A transistor in the CE circuit is also an inverter. It negates a logic function from the input at the base to the collector output.

Logic circuits generally combine all these components in integrated circuits. Complete IC chips are available for most logic functions.

Positive and Negative Logic. It is necessary to define the binary 1 and 0 with respect to positive and negative voltage polarities. In positive logic the 1 represents the presence of a positive voltage with respect to chassis ground (Fig. 32-11a). In general, the 1 state is high, with voltage more positive than the 0 state.

Note that with negative voltages, ground level can be the 1 state. Then ground is more positive than a negative potential (Fig. 32-11b).

Negative logic uses the opposite polarity of voltage for the 1 state. Since positive logic is more common, however, all the logic circuits here are shown with 1 for the more positive state. The amount of supply voltage is generally 5, 6, or 12 V.

Types of Logic Circuits. Based on the components, the following abbreviations are common:

TTL is transistor-transistor logic.
DTL is diode-transistor logic.
RTL is resistor-transistor logic.

In addition, diode circuits are used for the AND gate and OR gate. Followed by a CE stage these functions become inverted to the NAND gate and NOR gate.

Fan-in and Fan-out. The number of inputs that can be connected to the logic circuit is its fan-in factor. The maximum number of circuits that can be driven by the output is the fan-out factor. This effectively specifies how many parallel loads can be connected without too much current in the driver circuit.

Practice Problems 32-5
(answers on page 709)

(a) In positive logic is ground potential 0 or 1 with respect to −5 V?
(b) Is the collector voltage high or low with saturation current?

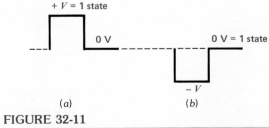

FIGURE 32-11

Positive logic with binary 1 for more positive voltage. (a) $+V$ to ground. (b) Ground to $-V$.

32-6
DIODE GATE CIRCUITS

The two main types of diode logic circuits are the AND gate illustrated in Fig. 32-12 and the OR gate in Fig. 32-13. A fan-in of only two inputs is illustrated here, but more can be used with a diode for each input.

Diode AND Gate. In Fig. 32-12, $D1$ and $D2$ are normally conducting to be ON. Note the $R_1 R_2$ voltage divider from the 5-V supply. The positive voltage across R_2 is applied to the anode for both diodes. $D1$ and $D2$ are conducting, therefore, without any input voltage at the cathode. Then V_{out} is at the 0 level. Point C is effectively shorted to ground by the conducting diodes.

However, input of $+3$ V at the cathode of $D1$ or $D2$ cuts off the diode. Remember that positive voltage at the cathode is equivalent to negative voltage at the anode.

This circuit is an AND gate because both inputs must be HIGH to have the output HIGH. Input to either $D1$ or $D2$ cuts off one diode, but the other conducting diode keeps V_{out} LOW.

Only when $+3$ V is applied to the cathode for $D1$ and $D2$ can both diodes be cut off. Then the diode circuit is open and point C has $+2$ V across R_2. The $+2$ V in the output is the HIGH level or 1 state.

Using positive logic for the binary notation:

V_1	V_2	V_{out}
0	0	0
0	1	0
1	0	0
1	1	1

The inputs are at binary 1 for $+3$ V. In the output, binary 1 corresponds to approximately $+2$ V. The output is at 1 only when both inputs are at 1. This is the AND function.

Diode OR Gate. The opposite of the AND gate is an inclusive OR gate, which needs all inputs to be LOW to have the output LOW. Figure 32-13 shows an OR gate using diodes. Note that $D1$ and $D2$ are inverted here, and negative supply voltage is used, opposite from the AND gate in Fig. 32-12. These two changes convert the AND gate to an OR gate using positive logic for both.

In Fig. 32-13, the negative voltage across R_2 from the -5 V of the supply is applied to the cathode for $D1$ and $D2$. Since this corresponds to positive voltage at the anode, both diodes are

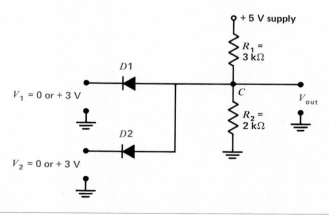

FIGURE 32-12
Circuit for AND gate, using diodes.

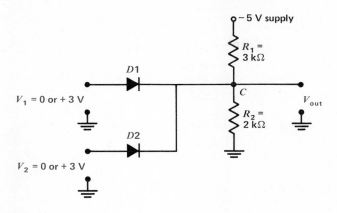

FIGURE 32-13
Circuit for OR gate, using diodes.

conducting, with zero input voltage at the anodes. Then point C in the output is close to ground potential at the 0 state.

When +3 V is applied to the anode for either D1 or D2, this voltage is passed to point C by a conducting diode. Then C rises to +3 V, approximately, for the 1 state. This positive voltage at both cathodes cuts off the other diode that has 0 V input at the anode.

When +3 V is applied to both D1 and D2, both diodes conduct. Then the output at C is also at approximately 3 V for the 1 state.

The binary combinations are:

V_1	V_2	V_{out}
0	0	0
0	1	1
1	0	1
1	1	1

Only when both inputs are at 0 is the output at 0. When either input is 1 or both inputs are 1, the output is 1. This is the OR function.

Practice Problems 32-6
(*answers on page 709*)

(a) For the AND gate, when the output is 1, what are both inputs?

(b) For an OR gate, when the output is 0, what are both inputs?

32-7
DIODE-TRANSISTOR LOGIC (DTL)

These circuits use diode gates with a CE transistor amplifier as an inverter stage. The DTL combination then is a NAND or NOR circuit.

NAND Gate. In Fig. 32-14, D1 and D2 with R_A form an AND gate. The diodes are biased on by positive anode voltage from R_A to the supply V_A. Then point C is close to ground potential. Both inputs V_1 and V_2 must be high with positive voltage at the cathode to cut off the diodes to allow point C to rise to V_A. When only one diode is cut off, the other conducting diode keeps the gate output low, close to ground potential.

R_1 in the coupling circuit to $Q1$ isolates the loading effect of the low v_{in} from base to emitter. C_1 is a speed-up capacitor across R_1. Fast variations in gate voltage can be coupled through C_1.

The output voltage of the diode AND gate is the input voltage V_B to the base of the NPN transistor $Q1$. Without any input voltage from the gate, $Q1$ is cut off by negative base voltage,

Chapter 32
Digital Electronics

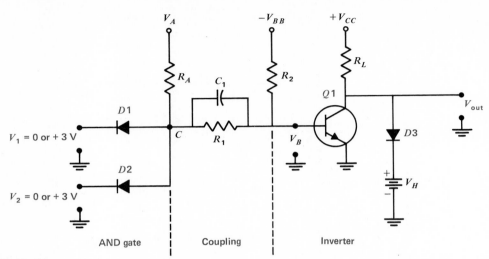

FIGURE 32-14

NAND gate, using diodes and transistors in DTL circuit.

from $-V_{BB}$ through R_2. Then the collector voltage is high at $+V_{CC}$ without any collector current. V_{out}, therefore, is high at the 1 state.

However, positive drive at the base makes Q1 conduct. Then the collector voltage is low because of the voltage drop across R_L. As a result, V_{out} drops to the 0 state.

D3 in the collector circuit is a diode clamp. It holds the high state of V_C at a value no greater than the clamping level V_H. Any V_C greater than V_H makes D3 conduct. Then the collector is effectively connected to V_H.

The truth table for the NAND gate in Fig. 32-14 shows the following conditions:

V_1	V_2	V_B	V_{out}
0	0	0	1
0	1	0	1
1	0	0	1
1	1	1	0

Note that V_B is the output of the diode AND gate. Both inputs must be at 1 for 1 at the input to Q1. However, these conditions are negated for V_{out}. When the input of Q1 is low, the output is high. Only when the gate output is high does the Q1 output become low.

NOR Gate. In Fig. 32-15, the diodes of the OR gate are reversed from the AND gate in Fig. 32-14, and Q1 is a PNP transistor instead of NPN. Both diodes are biased on by negative voltage at the cathode, from R_A to the supply $-V_A$. Without any input, therefore, the diodes are conducting to make point C close to ground potential. R_A and $-V_A$ have the values needed to keep the diodes on at all times in this circuit.

The inputs V_1 and V_2 in Fig. 32-15 are negative pulses of -3 V. With both diodes passing the input voltage to point C, the output voltage becomes close to -3 V.

Note that with positive logic ground potential is the 1 state, compared with -3 V for the 0 state. Without any input voltage to the diodes, point C is at ground potential through the conducting diodes. This is the 1 state. When both

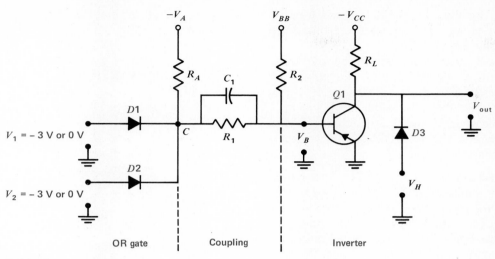

FIGURE 32-15

NOR gate, using diodes and transistors in DTL circuit.

diodes pass -3 V to point C, it is at the 0 state. This -3 V is the negative base drive V_B for $Q3$. When only one diode has -3 V input, the short circuit through the other conducting diode without input can ground point C for the 1 state.

The PNP transistor is held cut off by positive base voltage through R_2 from V_{BB}. When the diode gate supplies negative drive of -3 V for V_B, then $Q1$ conducts. The V_{out} at the collector is low when $Q1$ conducts but high when it is cut off.

A tabulation of the input voltages to the diodes and V_{out} from $Q1$ is as follows:

V_1	V_2	V_{out}
-3	-3	0
-3	0	-3
0	-3	-3
0	0	-3

The corresponding truth table in binary notation is as follows:

V_1	V_2	V_{out}
0	0	1
0	1	0
1	0	0
1	1	0

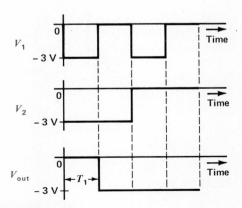

FIGURE 32-16

Pulse waveforms for NOR circuit in Fig. 32-15.

V_{out} for the NOR gate is inverted from an inclusive OR gate. The output is at 1 only when both inputs are at 0.

These results are shown by the pulse waveforms in Fig. 32-16. All the pulses are negative to -3 V, which is the 0 state. Note that V_2 has double the pulse width of V_1.

Only when V_1 and V_2 are both at -3 V for the 0 state is V_{out} at the 1 state. This occurs during the time marked T_1.

Resistor-Transistor Logic (RTL). These circuits are similar to DTL gates. However, the diodes are replaced by resistors. In a voltage divider circuit, the values of R determine what part of the applied voltage and input pulses are passed to the transistor inverter stage. The RTL circuits can be used for NAND and NOR gates.

Practice Problems 32-7
(answers on page 709)

(a) In Fig. 32-14, is point C at 1 or 0 when V_{out} is at 0?

(b) In Fig. 32-15, is point C at 1 or 0 when V_{out} is at 1?

32-8
TRANSISTOR-TRANSISTOR LOGIC (TTL)

See the circuit with four NPN transistors in Fig. 32-17. Note that $Q1$ has two emitter connections for inputs 1 and 2. Each emitter-base junction serves as a diode. All the transistors are dc-coupled from the dual inputs for $Q1$ to the output terminal between $Q3$ and $Q4$.

We consider just the inputs and outputs for the NAND gate in Fig. 32-17. If either or both inputs are grounded at the 0 state, the output is 3.3 V for the 1 state. For the opposite case, both inputs at 1 make the output 0. Remember that the NAND function is the negation of the AND function.

These circuits are generally on an IC chip. The required supply voltage is $+5$ V for the NPN stages. The output voltage, when it is high

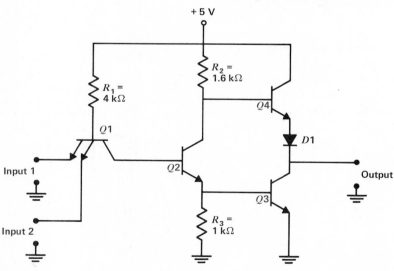

FIGURE 32-17

Circuit for NAND gate, using transistor-transistor logic (TTL). Note multiple emitters in $Q1$. Dc coupling is used for the complete circuit on an IC chip.

Chapter 32
Digital Electronics

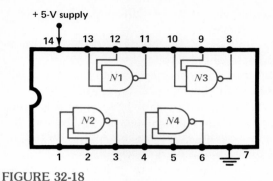

FIGURE 32-18

A quad IC chip with four NAND gates, using TTL circuits. Each gate has two inputs. Top view of 14 pins, numbered counterclockwise from notch at left.

at the 1 state, is 3.3 V. The low output, or 0 state, is approximately 0 V. At the low-voltage state, which has high current, the unit can deliver or *sink* 16 mA as load current. The response time is rated at 10 ns or 0.01 µs for switching from one state to the other.

A typical IC chip with four (QUAD) NAND units is shown in Fig. 32-18. Each NAND gate is an independent unit with two inputs and one output. However, all use the +5-V supply that must be connected to pin 14 and the common ground at pin 7. These gates can be interconnected to perform almost any logic function.

Combining NAND Gates for the OR Function.
Figure 32-19 shows how three NAND gates can

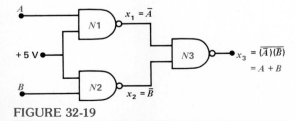

FIGURE 32-19

NAND gates N1, N2 and N3 combined to provide OR function for output.

be used for an OR gate. The NAND gates N1 and N2 are used as inverters for each input. Both outputs are connected as dual inputs to N3, which serves as a standard NAND gate.

Tracing the signals as logic functions, the input to N1 is A, and the output is $\bar{A}$. Also, the input to N2 is B, and the output is $\bar{B}$. With these two inputs to the NAND gate N3, its output is the negation of $(\bar{A})(\bar{B})$. The negation twice in the AND function makes the result identical to the OR function. This identity can be proven from Table 32-4, where the negation of $\overline{AB}$ is the same as $A + B$.

COS/MOS Digital IC Units. For TTL logic circuits, the IGFET or MOSFET construction can be used instead of bipolar transistors. A common application uses complementary symmetry, with P-channel and N-channel field-effect transistors. Series and parallel combinations provide the NAND and NOR functions. This arrangement is COS/MOS, which is complementary symmetry with MOSFET units having an insulated gate.

Figure 32-20 illustrates the logic diagram for a COS/MOS integrated circuit for two pairs of NOR gates. The IC chip has terminals for two inputs and one output for each gate, the drain supply voltage V_{DD} of 3 to 15 V, and the common source terminal, usually connected to chassis ground. All P-unit substrates are con-

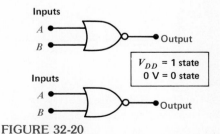

FIGURE 32-20

Logic diagram for COS/MOS IC chip with a pair of two-input NOR gates.

nected internally to V_{DD}; all N-unit substrates are connected to V_{SS}. When the output goes high at the 1 level, it is connected to V_{DD}. The 0 level is 0 V.

For each NOR gate, the output is LOW, or not HIGH, when either or both inputs are at the 1 state. Also, both inputs must be at 0, for output at the 1 level. This is the negation of the OR function, which has the output at 1 when either or both inputs are at 1.

Practice Problems 32-8
(answers on page 709)
(a) For the NAND gate in Fig. 32-17 with inputs of A and B, what is the logic function in the output?
(b) Using the IC chip in Fig. 32-18 for the circuit in Fig. 32-19, which pin of N3 has the output of $A + B$?

32-9
MULTIVIBRATOR (MV) CIRCUITS

The multivibrator is a pulse generator, producing square-wave or rectangular-wave output. The free-running MV produces output without any signal input. Two amplifiers are used, with the output of one stage driving the input of the other stage.

In Fig. 32-21, the collector of Q1 drives the base of Q2 through the $R_1 C_1$ coupling circuit. Also, the collector output of Q2 has feedback to the base of Q2, through $R_2 C_2$.

Since each stage inverts the polarity of its input signal, the feedback is positive, in the same polarity as the original input. Therefore, oscillations can be produced.

The oscillations are in the ON-OFF conditions for each stage. When the circuit is first turned on, one stage must conduct slightly more than the other. This unbalance is amplified in the two stages to produce maximum current in the conducting stage.

The sharp rise in I_C produces a sharp drop in V_C. As a result, this negative-going voltage in the conducting stage cuts off the other stage. How long it remains cut off depends on the RC time constant in the coupling circuit. When the cutoff stage starts conducting, it cuts off the other stage.

The rate at which the stages are cut off is the oscillator frequency. One cycle includes the cutoff time for both stages.

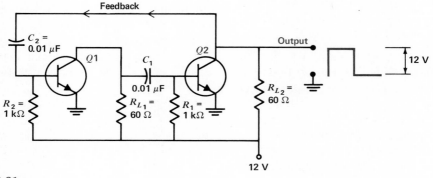

FIGURE 32-21

Basic multivibrator (MV) circuit. This type is collector-coupled, free-running, or astable, and has symmetrical pulses in the output. Frequency is approximately 70 kHz.

R_{L_1} and R_{L_2} are the collector load resistors. C_1 with R_1 couples V_C of Q1 to the base of Q2. The opposite coupling circuit is R_2C_2. In addition, R_1 and R_2 provide forward bias from the 12-V supply.

In the output, the voltage is a train of symmetrical square waves, or unsymmetrical rectangular waves. The output voltage can be taken from the collector of either Q1 or Q2 but with inverted polarities, as one collector voltage is high when the other is low. The symmetry depends on equal values for the components in the two stages.

Types of MV Circuits. Based on the method of feedback, we can have:

Collector-coupled MV. The Q1 collector drives the Q2 base, and the Q2 collector drives the Q1 base, as in Fig. 32-21.
Emitter-coupled MV. The Q1 collector drives the Q2 base, but Q2 has feedback to Q1 only through a common emitter-resistor for both stages.

Multivibrators are also classified according to stability. A stable stage will remain in the OFF condition until it is triggered into conduction by external pulses. The three classes are:

Astable MV. Neither stage is stable. One cuts off the other at the MV repetition rate. This type is simply a *free-running multivibrator.* It is an oscillator circuit to generate pulses at the MV frequency.
Bistable MV. This circuit has both stages in the stable condition, with one ON and the other OFF. An input pulse can reverse the states.
Monostable MV. This circuit has one stage stable. It can be used as a circuit to delay pulses.

Specifically, the bistable MV is a flip-flop circuit, which is commonly used for logic functions. In addition, the *Schmitt trigger* is a bistable MV used to square off sine waves to produce pulse signals.

Clock Generator. This circuit is a free-running MV to generate pulses accurately at a specific frequency. The clock generator controls the timing of the logic-gates, to synchronize all the operations in a digital system. A frequency divider circuit can be used with the clock to divide the output in multiples of $\frac{1}{2}$.

Practice Problems 32-9
(*answers on page 709*)
Refer to Fig. 32-21.
(a) Which coupling circuit determines the cutoff time for Q2?
(b) When V_C for Q2 is high at 12 V, how much is V_C for Q1?

32-10
FLIP-FLOP CIRCUITS

In terms of multivibrators, the bistable trigger circuits are flip-flops. This means each stage can change abruptly between cutoff and conduction. The circuit can stay in either state indefinitely in the absence of a trigger pulse.

The Bistable Characteristic. When the flip-flop is in either state, it remains that way until the next input pulse arrives. This condition is one stable state. Then an input pulse makes the circuit flip to the opposite condition. This is also a stable state. It remains that way until another pulse makes the circuit flop back to its original state. Each change from the previous state is like the operation of a toggle switch. Furthermore, the flip-flop has memory, as it remains in one state until toggled to the opposite state.

SET and RESET. We can define the flipped state with an input pulse as the SET position. The

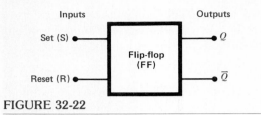

FIGURE 32-22

General form of basic flip-flop (FF) circuit with reset-set (RS) inputs.

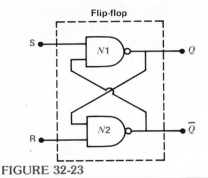

FIGURE 32-23

Using two NAND gates for basic reset-set FF.

SET pulse reverses the states from the condition without any input. Then a RESET pulse flops the states to put the flip-flop back into its original condition.

General Form of a Flip-flop. In Fig. 32-22, the flip-flop is illustrated just as a bistable device with input and output terminals. The input terminals are labeled S and R for set and reset pulses. These are two different input connections used for the bistable device. The output terminals are labelled Q and its negation $\bar{Q}$. These are two different output connections, arranged to make $\bar{Q}$ always the opposite of Q. As an example, if the voltage at Q is high, the voltage at $\bar{Q}$ must be low.

To make the states more general, if Q is at binary 1, then $\bar{Q}$ must be at 0. Or, when Q is at 0, then $\bar{Q}$ is at 1. These states for a reset-set (RS) flip-flop can be summarized as follows:

INPUT		OUTPUT	
S	R	Q	$\bar{Q}$
1	0	1	0
0	1	0	1

In short, a SET pulse makes the Q output high for binary 1. A RESET pulse puts the $\bar{Q}$ output at 1, which is the opposite state. Different circuits can be used, but such a binary table specifies the bistable characteristics of any flip-flop as a logic element.

Reset-Set Flip-flop. This type with just two inputs and two outputs is a basic RS flip-flop. It is a bistable device that can be in either the SET or RESET condition. A limition on the RS flip-flop is that both inputs should not be the same. Also, it has no input terminal for clock pulses.

Using Logic Gates for RS Flip-Flop. Note that the truth table for the flip-flop is similar to a logic function. Therefore, a combination of logic gates can provide the same results. Figure 32-23 illustrates how two NAND gates are interconnected to form an RS flip-flop. The gates are cross-coupled, with the output of one feeding the input of the other. Two NOR gates can be used the same way to form a flip-flop.

The basic requirement is that the outputs Q and $\bar{Q}$ be opposite. In Fig. 32-23 with NAND gates, Q is high when S is low. However, with two NOR gates, Q is high when S is high.

Clocked RS Flip-Flop. Figure 32-24 shows a reset-set FF with an additional input terminal for clock pulses. This flip-flop cannot change states unless a clock pulse is present. It is often necessary to clock the FF for counting circuits. The clocked type is a synchronous FF, as the operation is timed by the clock pulses.

D-type FF. In Fig. 32-25, only one input terminal is used, in addition to the clock terminal. The purpose is to avoid the indeterminate con-

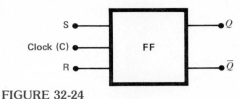

FIGURE 32-24
Clocked RS flip-flop.

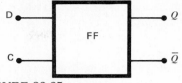

FIGURE 32-25
D-type flip-flop. D input is for both S and R. Clock input is C.

dition of both R and S at the same level in the basic RS flip-flop. The D input is passed directly to one of the inputs but inverted internally for the opposite gate.

JK-type FF. This is the most common type because it has no ambiguous states and can be clocked. One form is illustrated in Fig. 32-26a, with a photo in (b). Some of the main features in the operation of the type 7476 are:

1. Information in the form of a high or low voltage can be taken from either the Q or $\bar{Q}$ terminal in the output.
2. A high or low pulse can be fed into any of the following inputs: S, R, J, K, and clock.
3. When either S or R, but not both, goes either high or low, this input reverses the outputs. This action is independent of the other inputs.

4. For the specific case of both S and C high, the state of the flip-flop depends on the clock and the level of the J and K inputs.
5. When the clock input is low, the J and K inputs can toggle the output.

For the specific conditions of S and R high and the clock low, the J and K inputs have the following effects:

1. The state of J and K both low or both high does not toggle the output.
2. J high and K low makes the Q output high and $\bar{Q}$ low.
3. J low and K high makes the Q output low and $\bar{Q}$ high.

In summary, either the S or R input can toggle the output, as long as both are not the

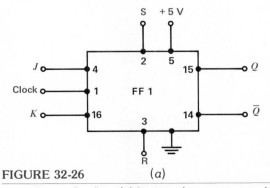

FIGURE 32-26

JK-type flip-flop. (a) Input and output terminals for one FF on type 7476. (b) Photo of IC unit with two separate JK flip-flops. Length is $\frac{3}{4}$ in.

same. The condition of both S and R low is not permitted. The condition of both S and R high allows the J, K, or clock inputs to toggle the outputs.

In terms of the clock pulses, every two changes result in one cycle of change in the FF output. Therefore, the circuit is a divide-by-two or binary flip-flop.

Practice Problems 32-10
(*answers on page 709*)

Answer true or false, for flip-flops.
(a) It is a bistable circuit.
(b) Either a set or clear pulse input can toggle the output.
(c) The Q and $\bar{Q}$ outputs cannot both be high at the same time.
(d) The JK type has an input terminal for clock pulses.

32-11
COUNTERS

The basic binary counter is a chain of flip-flops. Each divides by two. Successive flip-flops divide in multiples of two. The division is in the frequency of the clock-pulse input. This frequency division is equivalent to counting the clock pulses in multiples of 2, 4, 8, 16, etc. The count of 16 can be modified to 10 for a decade counter, by skipping six counts.

The counter is also a *register*. A counting register makes the count for each order of places. A *shift register* can shift the count between chains of flip-flops for each binary place in the count.

Figure 32-27 shows a divide-by-eight counter. This is a *ripple counter* because the flip-flops are in cascade. The J and K inputs of all the flip-flops are tied together and connected to +5 V. This input makes each FF ready to toggle. When the set and clear are both high, the FF will toggle each time the clock input goes low.

Note that the clock pulses are fed to the clock input of the first flip-flop FF-A. However, the clock input of FF-B is from the Q output of FF-A. Also, the Q output of FF-B feeds the clock input of FF-C. The divide-by-eight output of the counter is taken from the Q output of FF-C, as the final flip-flop.

This circuit is a binary counter that divides by two in three successive steps. For every eight changes of the clock, the Q output of FF-A changes four times, the Q output of FF-B changes two times, and the Q output of FF-C changes only one time. This final output, then, is the clock input divided by eight.

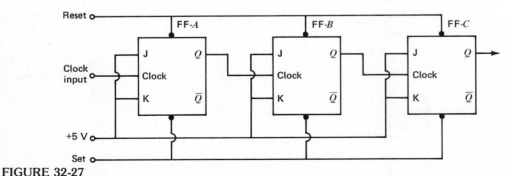

FIGURE 32-27

A divide-by-eight ripple-type counter, using three flip-flops.

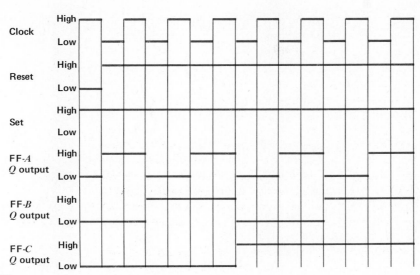

FIGURE 32-28
Pulse waveforms for the counter in Fig. 32-27.

Figure 32-28 shows waveforms for the divide-by-eight ripple counter. Note that each successive flip-flop changes only when the preceding Q output goes from a high to a low.

A disadvantage of the cascade counter is its relatively slow speed. The reason is that each FF must wait for the previous FF to change state. When a faster counter is desired, the synchronous type is used. In a synchronous counter, the clock pulses are fed directly to each FF.

For a decade counter, four flip-flops are used to obtain the maximum count of 16. However, a NAND gate and NOR gate are added. The NAND gate is connected in such a way that six of the counts will not be used. The AND gate is used for steering the reset pulses for the flip-flops.

Practice Problems 32-11
(*answers on page 709*)
(a) In Fig. 32-27 how much is the division factor at the Q output of FF-B?
(b) In Fig. 32-28, how many pulses, both high and low, are shown for the Q output of FF-C, for the 16 clock pulses?

Summary

Table 32-5 on the next page shows the basic building blocks for digital circuits. In addition, the terminology in digital electronics is summarized in the following alphabetical list:

AND *gate.* Logic function comparable to series switches.
Baud. Unit of frequency in bits or bytes per second, in transmission of digital signals.

Chapter 32
Digital Electronics

TABLE 32-5. Symbols for Digital Circuits

FUNCTION	SYMBOL
AND	A, B → AB
OR	A, B → $A+B$
Exclusive OR	A, B → $\overline{(A+B)}$
Inverter	A → $\overline{A}$
NAND	A, B → $\overline{AB}$
NOR	A, B → $\overline{A+B}$
Set-reset, flip-flop	R, S → FF → Q, $\overline{Q}$

Binary. Based on a scale of 2, using the two digits 1 and 0.

BCD. Binary-coded decimal. Each digit in a decimal number is processed in its binary equivalent.

Bit. Abbreviation for binary digit.

Byte. A sequence of binary digits used as a unit, but usually shorter than a word.

CLEAR INPUT. Same as RESET for flip-flop circuits.

Clock. Usually a multivibrator to generate pulses at an exact frequency to synchronize the logic circuits.

C/MOS. Complementary symmetry in terms of P channel and N channel.

Complement. A binary opposite, changing between 0 and 1.

Counter circuit. Usually a group of flip-flops. The output has one pulse for a predetermined number of input pulses.

CTL. Complementary transistor logic with NPN and PNP transistors.

DCTL. Direct-coupled transistor logic.

Decimal. Based on a scale of 10.

DTL. Diode-transistor logic.

D-type flip-flop. Has a D input terminal for clock pulses.

Digit. One character in a number. Binary numbers have the two digits 0 and 1. Decimal numbers have ten digits from 0 to 9.

ECL. Emitter-coupled logic.

Exclusive OR *Gate.* A logic function in which the output is ON only when the two inputs are not in the same state.

Fan-in. The number of inputs connected to a logic gate.

Fan-out. The maximum number of output circuits that can be used as a load.

FET. Field-effect transistor.

Flip-flop (*FF*). Bistable circuit that can store one bit of information. The outputs at the Q and $\bar{Q}$ terminals must be opposite.

Gate. Logic circuit with two or more inputs designed to give one output only for specific combinations of input pulses.

Inverter. Negates input to provide complement in output. For example, 1 becomes inverted or negated to 0, and 0 becomes 1. Also a NOT function.

JK flip-flop. Has two control inputs at J and K terminals, with clock input, for Q and $\bar{Q}$ output.

Logic. A form of mathematics based on two-state truth tables. Electronic logic circuits use gates and flip-flops to perform logic functions.

Memory. Also called *storage.* The bistable FF is an example of temporary memory. Magnetic tapes or discs and punched cards are examples of permanent storage.

Microprocessor. IC unit with digital circuits forming subsystem of a computer.

MOSFET. Metal-oxide semiconductor field-effect transistor. Has insulated-gate electrode.

NAND *gate.* Logic function in which both inputs must be high for output to be low. Is negation of AND gate.

NOR *gate.* Logic function in which output is low when either of the inputs, but not both, is high. Is negation of inclusive OR gate.

Octal. Based on a scale of eight.

OR *gate.* Logic function comparable to parallel switches.

Radix. The base for a number system. Radix is 2 for binary numbers and 10 for decimal numbers.

RAM. Random access memory.

RCTL. Resistor-capacitor transistor logic.

Read. Acquire information for use or storage.

ROM. Read or write memory.

RTL. Resistor-transistor logic.

Readout. Visual display of information.

RESET INPUT. Same as CLEAR for flip-flop circuits. Flops the bistable circuit back to its initial stage.

RS flip-flop. Has set and reset input terminals for Q and $\bar{Q}$ output.

SET INPUT. For flip-flop circuits. Flips the bistable circuit to the opposite condition from its initial state.

Shift register. Can shift the count by one order of places.

Storage. Also called *memory.* Holds data to be retrieved at a later time.
Truth table. Lists all possible states of input and output for a logic function.
Word. An ordered set of characters that is the normal unit of information for a computer operation.
Write. Introduce information into storage.

Self-Examination (Answers at back of book.)

1. What is the radix for binary numbers?
2. The digit 0 with carry of 1 is the sum for what binary addition?
3. Which logic gate is similar to the function of two series switches?
4. Which logic gate is similar to the function of two parallel switches?
5. Which logic function has the output low only when both inputs are high?
6. In any flip-flop, when the $\bar{Q}$ output is at 1, what is the state of the Q terminal?
7. In positive logic, is a pulse at -3 V to chassis ground the 0 or 1 level?
8. How many inputs can be supplied to a logic gate with a fan-in factor of four?
9. Which circuit is used for a clock generator, a free-running MV or *JK* flip-flop?
10. What type of circuit is used for counters?
11. What are the possible binary states for the Q and $\bar{Q}$ output terminals of a flip-flop?
12. How many FF circuits are needed to divide by 16?

Essay Questions

1. Define base, radix, and carry for number systems.
2. Define bistable, astable, and free-running for multivibrators.
3. Define fan-in and fan-out factors for a logic gate.
4. Give one advantage and one disadvantage in comparing binary and decimal numbers.
5. Give one advantage and one disadvantage in comparing analog and digital circuits.
6. What is meant by negation or inversion?
7. Compare the following pairs of logic gates: (*a*) AND with OR; (*b*) AND with NAND; (*c*) OR with NOR.

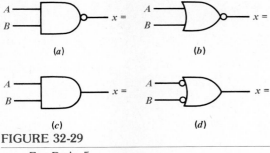

FIGURE 32-29

For Prob. 5.

8. Show the truth table for the NOR gate in Fig. 32-15.
9. Use a truth table to show that $(\bar{A})(\bar{B}) = \overline{A + B}$.
10. Show the logic symbols for AND, OR, NAND, and NOR gates.
11. Describe briefly two types of multivibrator circuits.
12. Describe briefly two types of flip-flop circuits.

Problems (Answers to odd-numbered problems at back of book.)

1. Convert the following decimal numbers to their binary equivalents: (a) 39; (b) 62.
2. Convert $(187)_{10}$ to binary form.
3. Convert $(1110)_2$ to a decimal number.
4. Add $(16)_{10}$ and $(9)_{10}$ in binary form and check that the sum equals $(25)_{10}$.
5. Give the output equation for each of the logic symbols in Fig. 32-29.

Answers to Practice Problems

32-1	(a) 15	32-4	(a) 0
	(b) 1010		(b) 1
	(c) Zero		(c) 16
32-2	(a) $(111)_2$	32-5	(a) 1
	(b) $(1110)_2$		(b) Low
32-3	(a) T	32-6	(a) 1
	(b) T		(b) 0
	(c) F	32-7	(a) 1

32-7	(b) 0		32-10	(b) T
32-8	(a) $\overline{AB}$			(c) T
	(b) pin 8			(d) T
32-9	(a) C_1 and R_1		32-11	(a) 4
	(b) 0 V			(b) 1
32-10	(a) T			

Review of Chapters 31 and 32

Summary

1. Integrated circuits combine semiconductor devices R and C on one silicon chip. The advantages are small space and reliability.
2. Linear IC units are commonly used for af and rf amplifiers.
3. Digital IC units are commonly used for gates, flip-flops, and additional types of logic circuits.
4. The terminology of IC units is described in the Summary for Chap. 31.
5. Digital electronics uses binary numbers with only the digits 0 and 1 to represent the two states of a pulse signal.
6. The basic gates for digital logic are AND, NAND, OR, and NOR, as listed in Table 32-5.
7. The flip-flop is a bistable circuit that can be changed from one state to the opposite by input pulses.
8. The terminology for digital electronics is described in the Summary for Chap. 32.

Review Self-Examination (Answers at back of book.)

1. Is a typical capacitance on an IC unit 20 pF or 20 µF?
2. Is an N channel or a P channel used for an FET or bipolar transistor?
3. Does a typical DIP package have 18 or 4 pins?
4. Is a unit with quad audio amplifiers a linear or digital IC?
5. What is the binary equivalent of decimal 7?
6. How much is $1 + 1$ in the binary system?
7. What logic gate has an output of AB with an input of A and B?

8. In the flip-flop, when Q output is at the HIGH level, what is the level of $\bar{Q}$?
9. Which flip-flop uses clock pulses: *RS* or *JK*?
10. Can a clock generator use a bistable MV or the astable MV circuit?

References (Additional references at back of book.)

Deboo, G. J. and C. N. Burrous: "Integrated Circuits and Semiconductor Devices," McGraw-Hill Book Company, New York.

Ilardi, F.: "Computer Circuit Analysis—Theory and Applications," Prentice-Hall, Inc., Englewood Cliffs, N.J.

RCA Solid-State Devices Manual, RCA Solid-State Division, Sommerville, N.J.

Bibliography

Mathematics

Cooke, N. M. and H. Adams: "Basic Mathematics for Electronics," McGraw-Hill Book Company, New York.
Heineman, E. R.: "Plane Trigonometry with Tables," McGraw-Hill Book Company, New York.
Lippin, G.: "Circuit Problems and Solutions," Hayden Book Company, Inc., New York.
Rice, H. S., and R. M. Knight: "Technical Mathematics," McGraw-Hill Book Company, New York.
Roberts and Stockton: "Elements of Mathematics," Addison-Wesley Publishing Company, Inc., Reading, Mass.
Stein, P.: "Graphical Analysis," Hayden Book Company, Inc., New York.

Radio Communications and Electronics

Chute, G. M.: "Electronics in Industry," McGraw-Hill Book Company, New York.
DeFrance, J. J.: "General Electronic Circuits," Holt, Rinehart and Winston, Inc., New York.
Grob, B.: "Basic Television," McGraw-Hill Book Company, New York.
Hayt, W., and J. Kemmerly: "Engineering Circuit Analysis," McGraw-Hill Book Company, New York.
Kaufman, M.: "Radio Operator's License Q and A Manual," Hayden Book Company, Inc., New York.
Lurch, E. N.: "Fundamentals of Electronics," John Wiley & Sons, Inc., New York.
Prensky, S. D.: "Electronic Instrumentation," Prentice-Hall, Inc., Englewood Cliffs, N.J.
"Radio Amateur's Handbook," American Radio Relay League, Newington, Conn.
Richter, H.: "Practical Electrical Wiring," McGraw-Hill Book Company, New York.
Shrader, R. L.: "Electronic Communication," McGraw-Hill Book Company, New York.

Slurzberg, M., and W. Osterheld: "Essentials of Radio-Electronics," McGraw-Hill Book Company, New York.
Wilson, J. A. and Glass, D.: "Study Guide for CET Examinations," Howard W. Sams & Co., Inc., Indianapolis, Ind.
Zbar, P. B.: "Electronic Instruments and Measurements," McGraw-Hill Book Company, New York.

Transistors and Tubes

Bell, D. A.: "Fundamentals of Electronics Devices," Reston Publishing Co., Reston, Virginia.
Cutler, P.: "Semiconductor Circuit Analysis," McGraw-Hill Book Company, New York.
GE Transistor Manual, General Electric Co., Semiconductor Products, Syracuse, N.Y.
HEP (Hobbyist-Experimenter) Semiconductor Cross-Reference Guide, Motorola Semiconductors, Phoenix, Ariz.
Kiver, M. S.: "Transistors," McGraw-Hill Book Company, New York.
Millman, J., and C. Halkias: "Electronic Devices and Circuits," McGraw-Hill Book Company, New York.
RCA Receiving Tube Manual, RCA Electronic Components and Devices, Harrison, N.J.
RCA Silicon Power Circuits Manual, RCA Electronic Components and Devices, Harrison, N.J.
RCA Solid-State Devices Manual, RCA Solid-State Division, Sommerville, N.J.
Tomer, R. B.: "Semiconductor Handbook," Howard W. Sams & Co., Inc., Indianapolis, Ind.

Digital Electronics and Integrated Circuits

Deboo, G. J. and C. N. Burrous: "Integrated Circuits and Semiconductor Devices," McGraw-Hill Book Company, New York.
Ilardi, F.: "Computer Circuit Analysis—Theory and Applications," Prentice-Hall, Inc., Englewood Cliffs, N.J.
Jung, W. G.: "IC Op-Amp Cookbook," Howard W. Sams & Co., Inc., Indianapolis, Ind.

Malvino, A. P., and D. Leach: "Digital Principles and Applications," McGraw-Hill Book Company, New York.
Millman, J., and C. Halkias: "Analog and Digital Integrated Circuits," McGraw-Hill Book Company, New York.
Rosenthal, M. P.: "Understanding Integrated Circuits," Hayden Book Company, Inc., New York.

Manufacturers' Periodical Publications

Aerovox Research Worker, Aerovox Corp., New Bedford, Mass.
Amphenol Engineering News, Amphenol Corp., Broadview, Ill.
Capacitor, Cornell-Dubilier Electric Corp., Newark, N.J.
International Rectifier News, International Rectifier Corp., El Segundo, Calif.
RCA Service News, RCA Service Co., Camden, N.J.
Sylvania News, GTE Sylvania Electric Products, Inc., Batavia, N.Y.
Tekscope, Tektronix, Inc., Beaverton, Oreg.

Magazines

Electronics, McGraw-Hill Publications, New York.
Electronic Service Dealer, Ledger Publishing Co., Waltham, Mass.
Electronic Servicing, Howard W. Sams & Co., Inc., Indianapolis, Ind.
Electronic Technician/Dealer, Harcourt Brace Jovanovich Publications, New York.
QST, American Radio Relay League, West Hartford, Conn.
Radio-Electronics, Gernsback Publications, Inc., New York.
Spectrum, Institute of Electrical and Electronic Engineers, New York.

Sources for Films and Filmstrips

McGraw-Hill Films, New York.
National Audiovisual Center, General Services Administration, Washington, D.C. 20409.

Public relations departments of electronics companies, including:
    Fairchild Semiconductor Division, Mountain View, Calif.
    New York Telephone Company, New York.
    Tektronix, Inc., Beaverton, Oreg.
    Western Electric Company, New York.

Industry Standards

Electronic Industries Association (EIA), Washington, D.C.
Institute of Electrical and Electronics Engineers (IEEE), New York.
International Society of Certified Electronic Technicians (ISCET), Indianapolis, Ind.

Electronic Frequency Spectrum

Appendix A

FREQUENCY OR WAVELENGTH*	NAME	APPLICATIONS
0 Hz	Steady direct current or voltage	Dc motors, solenoids, relays, electrode voltages for tubes and transistors
16–16,000 Hz	Audio frequencies	60-Hz power, ac motors, audio amplifiers, microphones, loudspeakers, phonographs, tape recorders, high-fidelity equipment, public address systems, and intercoms
16–30 kHz	Ultrasonic frequencies or very low radio frequencies	Sound waves for ultrasonic cleaning, vibration testing, thickness gaging, flow detection, and sonar; electromagnetic waves for induction heating
30 kHz–30,000 MHz	Radio frequencies (see Appendix B)	Radio communications and broadcasting, including television, radio navigation, radio astronomy, industrial, medical, scientific, and military radio
30,000–300,000 MHz or 1–0.1 cm	Extra-high frequencies	Experimental, weather radar, amateur, government
300,000–7600 Å	Infrared light rays	Heating, infrared photography
7600–3900 Å	Visible light rays	Color, illumination, photography
3900–320 Å	Ultraviolet rays	Sterilizing, deodorizing, medical
320–0.1 Å	X-rays	Thickness gages, inspection, medical
0.1–0.006 Å	Gamma rays	Radiation detection; more penetrating than hardest x-rays
Shortest of all electromagnetic waves	Cosmic rays	Exist in outer space; can penetrate 70 m of water or 1 m of lead

*Frequency and wavelength are inversely proportional to each other. The higher the frequency, the shorter the wavelength, and vice versa. Af and rf waves are generally considered in terms of frequency because the wavelength is so long. The exception is microwaves, which are often designated by wavelength because their frequencies are so high. Light waves, x-rays, and gamma rays are also generally considered in wavelength because their frequencies are so high. The units of wavelength are the micrometer, equal to 10^{-6} m; the nanometer, equal to 10^{-9} m; and the angstrom Å, equal to 10^{-10} m.

Appendix A
Electronic Frequency Spectrum

The four main categories of electromagnetic radiation and their frequencies can be summarized as follows:

1. Radio frequency waves from 30 kHz to 300,000 MHz.
2. Heat waves or infrared rays from 1×10^{13} to 2.5×10^{14} Hz. Infrared means below the frequency of visible red light.
3. Visible light frequencies from about 2.5×10^{14} Hz for red up to 8×10^{14} Hz for blue and violet.
4. Ionizing radiation such as ultraviolet rays, x-rays, gamma rays, and cosmic rays from about 8×10^{14} Hz for ultraviolet light to above 5×10^{20} Hz for cosmic rays. Ultraviolet means above the frequencies of blue and violet visible light.

FCC Frequency Allocations from 30 kHz to 300,000 MHz

Appendix B

The main categories of radio frequencies can be summarized as follows:

1. Very-low frequencies (VLF) below 30 kHz.
2. Low frequencies (LF) 30 to 300 kHz.
3. Medium frequencies (MF) 0.3 to 3 MHz.
4. High frequencies (HF) 3 to 30 MHz.
5. Very-high frequencies (VHF) 30 to 300 MHz.
6. Ultra-high frequencies (UHF) 300 to 3000 MHz.
7. Super-high frequencies (SHF) 3000 to 30,000 MHz.
8. Extra-high frequencies (EHF) 30,000 to 300,000 MHz.

Details of the frequencies assigned by the Federal Communications Commission for specific services are listed on the next page.

Appendix B

FCC Frequency Allocations from 30 kHz to 300,000 MHz

BAND	ALLOCATION	REMARKS
30–535 kHz	Includes maritime communications and navigation, aeronautical radio navigation	Low and medium radio frequencies
535–1605 kHz	Standard radio broadcast band	AM broadcasting
1605 kHz–30 MHz	Includes amateur radio, loran, government radio, international shortwave broadcast, fixed and mobile communications, radio navigation, industrial, scientific and medical, and the CB radio band	Amateur bands 3.5–4.0 MHz and 28–29.7 MHz; industrial, scientific, and medical band 26.95–27.54 MHz; citizen's band class D for voice is 26.965–27.405 MHz in forty 10-kHz channels
30–50 MHz	Government and nongovernment, fixed and mobile	Includes police, fire, forestry, highway, and railroad services; VHF band starts at 30 MHz
50–54 MHz	Amateur	6-m band
54–72 MHz	Television broadcast channels 2 to 4	Also fixed and mobile services
72–76 MHz	Government and nongovernment services	Aeronautical marker beacon on 75 MHz
76–88 MHz	Television broadcast channels 5 and 6	Also fixed and mobile services
88–108 MHz	FM broadcast	Also available for facsimile broadcast; 88–92 MHz educational FM broadcast
108–122 MHz	Aeronautical navigation	Localizers, radio range, and airport control
122–174 MHz	Government and nongovernment, fixed and mobile, amateur broadcast	144–148 MHz amateur band
174–216 MHz	Television broadcast channels 7 to 13	Also fixed and mobile services
216–470 MHz	Amateur, government and nongovernment, fixed and mobile, aeronautical navigation, citizens' radio	Radio altimeter, glide path, and meteorological equipment; citizens' radio band 462.5–465 MHz; civil aviation 225–400 MHz; UHF band starts at 300 MHz
470–890 MHz	Television broadcasting	UHF television broadcast channels 14 to 83
890–3000 MHz	Aeronautical radio navigation, amateur broadcast, studio-transmitter relay, government and nongovernment, fixed and mobile	Radar bands 1300–1600 MHz
3000–30,000 MHz	Government and nongovernment, fixed and mobile, amateur broadcast, radio navigation	Super-high frequencies (SHF); 8400–8500 MHz satellite communications
30,000–300,000 MHz	Experimental, government, amateur	Extra-high frequencies (EHF)

Alphabetical Listing of the Chemical Elements

Appendix C

ELEMENT	SYMBOL	ATOMIC NUMBER	YEAR OF DISCOVERY	REMARKS*
Actinium	Ac	89	1899	
Aluminum	Al	13	1825	Metal conductor
Antimony	Sb	51	Ancient	
Argon	Ar	18	1894	Inert gas
Arsenic	As	33	1649	
Barium	Ba	56	1808	Ends 8-electron O shell
Beryllium	Be	4	1798	
Bismuth	Bi	83	1753	
Boron	B	5	1808	
Bromine	Br	35	1826	
Cadmium	Cd	48	1817	
Calcium	Ca	20	1808	Ends 8-electron M shell
Carbon	C	6	Ancient	Semiconductor
Cerium	Ce	58	1803	Starts rare-earth series 58 to 71
Cesium	Cs	55	1860	Photosensitive
Chlorine	Cl	17	1774	Active gas
Chromium	Cr	24	1798	
Cobalt	Co	27	1735	
Copper	Cu	29	Ancient	Metal conductor
Fluorine	F	9	1771	Active gas
Gallium	Ga	31	1875	
Germanium	Ge	32	1886	Semiconductor

*Rare-earth elements with atomic numbers 58 to 71 and 90 to 103 are omitted.

Appendix C
Alphabetical Listing of the Chemical Elements

ELEMENT	SYMBOL	ATOMIC NUMBER	YEAR OF DISCOVERY	REMARKS
Gold	Au	79	Ancient	Metal conductor
Hafnium	Hf	72	1923	Starts 32-electron N shell
Helium	He	2	1895	Inert gas
Hydrogen	H	1	1766	Active gas
Indium	In	49	1863	
Iodine	I	53	1811	
Iridium	Ir	77	1804	
Iron	Fe	26	Ancient	Magnetic
Krypton	Kr	36	1898	Inert gas
Lanthanum	La	57	1839	
Lead	Pb	82	Ancient	
Lithium	Li	3	1817	
Magnesium	Mg	12	1755	
Manganese	Mn	25	1774	
Mercury	Hg	80	Ancient	
Molybdenum	Mo	42	1781	
Neon	Ne	10	1898	Inert gas
Nickel	Ni	28	1751	
Niobium	Nb	41	1801	
Nitrogen	N	7	1772	
Osmium	Os	76	1804	
Oxygen	O	8	1774	
Palladium	Pd	46	1803	N shell has 18 electrons
Phosphorus	P	15	1669	
Platinum	Pt	78	1735	
Polonium	Po	84	1898	
Potassium	K	19	1807	
Radium	Ra	88	1898	
Radon	Rn	86	1900	Inert gas
Rhenium	Re	75	1925	
Rhodium	Rh	45	1803	
Rubidium	Rb	37	1861	
Ruthenium	Ru	44	1844	
Scandium	Sc	21	1879	
Selenium	Se	34	1818	Photosensitive
Silicon	Si	14	1823	Semiconductor
Silver	Ag	47	Ancient	Metal conductor
Sodium	Na	11	1807	
Strontium	Sr	38	1790	Ends 8-electron N shell
Sulfur	S	16	Ancient	
Tantalum	Ta	73	1802	
Technetium	Tc	43	1937	
Tellurium	Te	52	1783	

ELEMENT	SYMBOL	ATOMIC NUMBER	YEAR OF DISCOVERY	REMARKS
Thallium	Tl	81	1861	
Thorium	Th	90	1829	Starts rare-earth series 90 to 103
Tin	Sn	50	Ancient	
Titanium	Ti	22	1791	
Tungsten	W	74	1783	
Uranium	U	92	1789	
Vanadium	V	23	1831	
Xenon	Xe	54	1898	Inert gas
Yttrium	Y	39	1843	
Zinc	Zn	30	1746	
Zirconium	Zr	40	1789	

Physics Units

Appendix D

All the units are based on the fundamental dimensions of length, mass, and time. These are considered basic quantities, compared with derived quantities such as area, force, velocity, and acceleration, which are only different combinations of length, mass, and time. Each of the basic dimensions has units in the English system and in the decimal or metric system, as listed in Table D-1.

SYSTEMS OF UNITS

The cgs system is an abbreviation for its basic units of centimeters, grams, and seconds. The mks system based on meters, kilograms, and seconds provides larger units which are closer to practical values, since the kilogram is 1000 g and the meter is 100 cm. The SI system is based on and very similar to the mks system.

In many cases, it is necessary to convert between English and metric units. Then the following conversions can be used:

LENGTH

1 meter = 39.37 inches
 1 inch = 2.54 centimeters

MASS

1 kilogram = 2.2 pounds
 1 gram = 0.03527 ounce

The basic unit of time is the second in all systems.

TABLE D-1. Units for Basic Dimensions

DIMENSION	ENGLISH UNIT	METRIC UNIT	
		CGS	MKS AND SI
Length (L)	foot	centimeter	meter
Mass (M)	slug*	gram	kilogram
Time (T)	second	second	second

*1 slug is the mass of a 1-lb weight.

MASS
The dimension of mass is often considered similar to the weight of an object. However, weight is actually the force due to the acceleration of gravity. To define mass more specifically, it is necessary to use Newton's second law of motion: $F = Ma$. This can be transposed to $M = F/a$, which states that the mass is defined by how much force is necessary for a given amount of acceleration.

DERIVED QUANTITIES
A simple example is velocity, as a combination of the basic dimensions of length and time. The units can be feet per second, meters per second, or centimeters per second. Still, basically velocity is just L/T, meaning it is the time rate of change of length. Sometimes speed and velocity are used interchangeably. However, velocity is a vector quantity that has direction, while speed is a scalar quantity without direction. Another derived quantity is acceleration, the time rate of change of velocity. In the mks and SI systems the unit is meters per second per second, or m/s^2. The basic dimensions are L/T^2, which results from L/T for velocity, divided by T. Additional derived quantities are force, work or energy, and power.

FORCE
Newton's law of acceleration is used to derive the units of force. With the formula $F = Ma$ in the cgs system, one *dyne* is the force needed for an acceleration of one centimeter per second per second with a mass of one gram. In the mks and SI systems, one *newton* is the force needed for an acceleration of one meter per second per second with a mass of one kilogram. One newton equals 10^5 dyne. To convert to English units, 1 N = 0.225 lb.

WORK AND ENERGY
Work W is the product of force F times the distance s through which the force acts. As a formula, $W = Fs$. For example, if you lift a 20-lb weight through a distance of 2 ft, the work equals 40 ft · lb.

In the cgs system, F is in dyne-centimeters, which is an *erg*. A larger cgs unit, also used in SI, is the *joule*, equal to 10^7 ergs.

In the mks system, F is in newton-meters. This unit is the same as 10^7 ergs. Or 1 N · m equals 1 J. The joule unit of work is named after James P. Joule (1818–1889), an important English physicist.

Energy is the ability to do work. Kinetic energy is due to the motion of a mass, as when you throw a ball. Potential energy is stored energy, as in a coiled spring. The units for both kinetic and potential energy are the same as for work.

POWER

This is the time rate of doing work, or $P = W/T$. The practical unit in the metric system is the joule per second, equal to 1 *watt*. In the English system the unit is foot-pounds per second. For a larger unit, 550 ft · lb/s equal 1 hp.

TEMPERATURE SCALES

The Celsius scale, formerly known as the centigrade scale, invented by A. Celsius, has 100 divisions between 0° for the freezing point of water and 100° for the boiling point. The Fahrenheit scale, invented by G. D. Fahrenheit, is still used for weather observations and general purposes. On this scale, the freezing point is 32° while the boiling point is 212°, with 180 divisions between. To convert from one scale to the other,

$$T_C = \frac{5}{9}(T_F - 32°) \quad \text{and} \quad T_F = \frac{9}{5}T_C + 32°$$

The Kelvin or absolute temperature scale was devised by Lord Kelvin. On this scale, the zero point is absolute zero, 273° below 0°C. At 0 K any material loses all its thermal energy. The divisions of the °C scale and the K scale are the same. (Note: The SI unit for temperature is K for Kelvin, without the degree symbol.)

To convert from °C to K, just add 273°. Thus, 0°C equals 273 K. To convert from K to °C, subtract 273°. Then 0 K equals -273°C.

Average room temperature is generally considered about 20 to 25°C. This equals 68 to 77°F. On the absolute scale, the corresponding temperatures are 293 to 298 K.

UNITS OF HEAT ENERGY

In the cgs system, one *calorie* is the amount of heat needed to raise the temperature of one gram of water by one degree Celsius. In mks units, the amount of heat to raise the temperature of one kilogram of water by one degree Celsius is equal to one kilocalorie. The SI system uses the joule as a unit of heat as well as a unit of work. In English units, the *British thermal unit* (*Btu*) is the amount of heat needed to raise the temperature of one pound of water by one degree Fahrenheit. The Btu is the larger unit, as 1 Btu equals 252 calories. The fact that heat is a form of energy can be seen from the use of the joule; 1 calorie is equivalent to 4.19 J. One joule equals 0.24 calorie. Table D-2 summarizes all the units described here.

TABLE D-2. Summary of Physics Units

QUANTITY	ENGLISH (FPS) UNITS	CGS UNITS	MKS AND SI UNITS	DIMENSIONS
Length	foot	centimeter	meter	L
Mass	slug	gram	kilogram	M
Time	second	second	second	T
Velocity	feet per second	centimeters per second	meters per second	L/T
Acceleration	feet per second per second	centimeters per second per second	meters per second per second	L/T^2
Force	pound	dyne	newton	$Ma = ML/T^2$
Work and energy	foot-pound*	erg = dyne-centimeters 10^7 ergs = 1 joule	joule = newton-meter	force $\times L$
Power	foot-pounds per second; 1 hp = 500 ft · lb/s	ergs per second; 10^7 ergs/s = 1 W	watt = joule per second	work/T
Heat	Btu	calorie	kilocalorie (mks); joule (SI)	1 calorie = 4.19 joules

*Foot-pound is a unit of work, while pound-foot is used for angular torque.

For more details on physics, chemistry, the periodic table, and atomic structure the following references can be helpful:

Chemical Education Material Study: "Chemistry, and Experimental Science," W. H. Freeman and Company, San Francisco, Calif.
Dull, Metcalfe, and Williams: "Modern Physics," Holt, Rinehart and Winston, Inc., New York.
Ewing-Meyer: "Chemistry: A Survey of Principles," John Wiley & Sons, Inc., New York.
Holton and Roller: "Foundations of Modern Physical Science," Addison-Wesley Publishing Company, Inc., Reading, Mass.
"Key to Welch Periodic Chart of the Atoms," Sargent Welch Scientific Co., Chicago, Ill.
Semat, H.: "Fundamentals of Physics," Holt, Rinehart and Winston, Inc., New York.
White, Manning, and Weber: "Practical Physics," McGraw-Hill Book Company, New York.

Trigonometric Functions

Appendix E

The six functions of an angle θ are sine, cosine, tangent, cotangent, secant, and cosecant. All are numerical ratios, comparing the sides formed by the angle θ in a right triangle, as in Fig. E-1. Any one function specifies the angle.

The values for $\sin \theta$, $\cos \theta$, and $\tan \theta$ are listed in Table E-1. These functions are defined in terms of the right triangle, as follows:

$$\sin \theta = \frac{\text{opposite side}}{\text{hypotenuse}} = \frac{a}{c}$$

$$\cos \theta = \frac{\text{adjacent side}}{\text{hypotenuse}} = \frac{b}{c}$$

$$\tan \theta = \frac{\text{opposite side}}{\text{adjacent side}} = \frac{a}{b}$$

These trigonometric functions are commonly used in ac circuits. Tan θ is especially useful for finding the phase angle θ because the tangent function uses only the sides without the hypotenuse.

It may be of interest to note that the other three trigonometric functions are reciprocals. These are: $\cot \theta = 1/\tan \theta$; $\sec \theta = 1/\cos \theta$; $\csc \theta = 1/\sin \theta$.

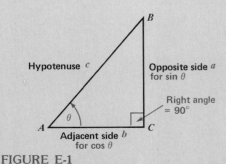

FIGURE E-1
The right triangle.

Appendix E
Trigonometric Functions

TABLE E-1. Trigonometric Functions

ANGLE	SIN	COS	TAN	ANGLE	SIN	COS	TAN
0°	0.0000	1.000	0.0000	45°	0.7071	0.7071	1.0000
1	.0175	.9998	.0175	46	.7193	.6947	1.0355
2	.0349	.9994	.0349	47	.7314	.6820	1.0724
3	.0523	.9986	.0524	48	.7431	.6691	1.1106
4	.0698	.9976	.0699	49	.7547	.6561	1.1504
5	.0872	.9962	.0875	50	.7660	.6428	1.1918
6	.1045	.9945	.1051	51	.7771	.6293	1.2349
7	.1219	.9925	.1228	52	.7880	.6157	1.2799
8	.1392	.9903	.1405	53	.7986	.6018	1.3270
9	.1564	.9877	.1584	54	.8090	.5878	1.3764
10	.1736	.9848	.1763	55	.8192	.5736	1.4281
11	.1908	.9816	.1944	56	.8290	.5592	1.4826
12	.2079	.9781	.2126	57	.8387	.5446	1.5399
13	.2250	.9744	.2309	58	.8480	.5299	1.6003
14	.2419	.9703	.2493	59	.8572	.5150	1.6643
15	.2588	.9659	.2679	60	.8660	.5000	1.7321
16	.2756	.9613	.2867	61	.8746	.4848	1.8040
17	.2924	.9563	.3057	62	.8829	.4695	1.8807
18	.3090	.9511	.3249	63	.8910	.4540	1.9626
19	.3256	.9455	.3443	64	.8988	.4384	2.0503
20	.3420	.9397	.3640	65	.9063	.4226	2.1445
21	.3584	.9336	.3839	66	.9135	.4067	2.2460
22	.3746	.9272	.4040	67	.9205	.3907	2.3559
23	.3907	.9205	.4245	68	.9272	.3746	2.4751
24	.4067	.9135	.4452	69	.9336	.3584	2.6051
25	.4226	.9063	.4663	70	.9397	.3420	2.7475
26	.4384	.8988	.4877	71	.9455	.3256	2.9042
27	.4540	.8910	.5095	72	.9511	.3090	3.0777
28	.4695	.8829	.5317	73	.9563	.2924	3.2709
29	.4848	.8746	.5543	74	.9613	.2756	3.4874
30	.5000	.8660	.5774	75	.9659	.2588	3.7321
31	.5150	.8572	.6009	76	.9703	.2419	4.0108
32	.5299	.8480	.6249	77	.9744	.2250	4.3315
33	.5446	.8387	.6494	78	.9781	.2079	4.7046
34	.5592	.8290	.6745	79	.9816	.1908	5.1446
35	.5736	.8192	.7002	80	.9848	.1736	5.6713
36	.5878	.8090	.7265	81	.9877	.1564	6.3138
37	.6018	.7986	.7536	82	.9903	.1392	7.1154
38	.6157	.7880	.7813	83	.9925	.1219	8.1443
39	.6293	.7771	.8098	84	.9945	.1045	9.5144
40	.6428	.7660	.8391	85	.9962	.0872	11.43
41	.6561	.7547	.8693	86	.9976	.0698	14.30
42	.6691	.7431	.9004	87	.9986	.0523	19.08
43	.6820	.7314	.9325	88	.9994	.0349	28.64
44	.6947	.7193	.9657	89	.9998	.0175	57.29
				90	1.0000	.0000	∞

SINE VALUES
In Table E-1, $\sin \theta$ increases from 0 for $0°$ to 1 for $90°$. The sine increases with θ as the opposite side becomes longer. However, the maximum sine ratio is 1 because no side of the triangle can be larger than the hypotenuse.

COSINE VALUES
The values for $\cos \theta$ start from 1 as its maximum value for $0°$. Then $\cos \theta$ decreases to 0 as θ increases to $90°$. The cosine function becomes smaller as the adjacent side is shorter for larger angles.

TANGENT VALUES
The values for $\tan \theta$ increase with θ as the opposite side becomes larger. However, $\tan \theta$ values should be considered below and above $45°$, as

  $\tan \theta$ from 0 to $45°$ increases from 0 to 1.

  At $45°$, $\tan \theta = 1$ because the opposite and adjacent sides are equal.

  Above $45°$, $\tan \theta$ is more than 1 because the opposite side is larger than the adjacent side.

IMPORTANT VALUES
Some values are worth memorizing, since they summarize the trend of the entire table of trigonometric functions. These are:

ANGLE	$\sin \theta$	$\cos \theta$	$\tan \theta$
$0°$	0	1	0
$30°$	0.5	0.866	0.577
$45°$	0.707	0.707	1
$60°$	0.866	0.5	1.73
$90°$	1.0	0	∞

ANGLES MORE THAN 90°
The angles considered so far are *acute* angles, less than $90°$. The $90°$ angle is a right angle. Angles larger than $90°$ are *obtuse* angles. The complete circle is $360°$. After $360°$ or any multiple of $360°$, the angles just repeat the values from $0°$.

For angles from 90 to $360°$, the full circle is divided into four quadrants, as shown in Fig. E-2. To use the table of trigonometric functions for obtuse angles in quadrants II, III, and IV, convert to equivalent acute angles in quadrant I by the following rules:

In quadrant II, use $180° - \theta$.
In quadrant III, use $\theta - 180°$.
In quadrant IV, use $360° - \theta$.

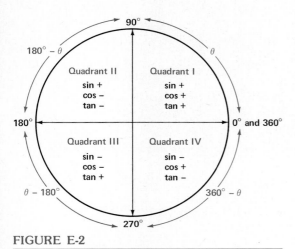

FIGURE E-2

The four quadrants of a circle.

Also, use the appropriate sign or polarity for the trigonometric functions in different quadrants, as shown in Fig. E-2.

Note that the conversions are only with respect to the horizontal axis, using 180 or 360° as the reference. This way the obtuse angle is always subtracted from a larger angle.

All the functions are positive in quadrant I. Notice that the tangent alternates in polarity through quadrants I, II, III, and IV.

In all the quadrants, the sine is + when the vertical ordinate is up or − when the ordinate is down. Similarly, the cosine is + when the horizontal abscissa is to the right or − to the left. The hypotenuse has no polarity.

Finally, the tangent is + when the sine and cosine have the same polarity, either both + in quadrant I or both − in quadrant III. For more details on trigonometry, selected books are listed in the bibliography.

Electrical Symbols and Abbreviations

Appendix F

Table F-1 summarizes the letter symbols used as abbreviations for electrical characteristics with their basic units. All the metric prefixes are in Table F-2, for multiple and fractional values. In addition, Table F-3 shows electronic symbols from the Greek alphabet.

TABLE F-1. Electrical Characteristics

QUANTITY	SYMBOL	BASIC UNIT
Current	I or i	ampere
Charge	Q or q	coulomb
Power	P	watt
Voltage	V or v	volt
Resistance	R	ohm
Reactance	X	ohm
Impedance	Z	ohm
Conductance	G	siemens
Admittance	Y	siemens
Susceptance	B	siemens
Capacitance	C	farad
Inductance	L	henry
Frequency	F or f	hertz
Period	T	second

Capital letter for I, Q, and V generally used for peak, rms, or dc value; small letter for instantaneous values. Small r and g usually for internal values such as r_p and g_m of a tube.

TABLE F-2. Multiples and Submultiples of Units*

VALUE	PREFIX	SYMBOL	EXAMPLE
$1\,000\,000\,000\,000 = 10^{12}$	tera	T	$THz = 10^{12}\,Hz$
$1\,000\,000\,000 = 10^{9}$	giga	G	$GHz = 10^{9}\,Hz$
$1\,000\,000 = 10^{6}$	mega	M	$MHz = 10^{6}\,Hz$
$1\,000 = 10^{3}$	kilo	k	$kV = 10^{3}\,V$
$100 = 10^{2}$	hecto	h	$hm = 10^{2}\,m$
$10 = 10$	deka	da	$dam = 10\,m$
$0.1 = 10^{-1}$	deci	d	$dm = 10^{-1}\,m$
$0.01 = 10^{-2}$	centi	c	$cm = 10^{-2}\,m$
$0.001 = 10^{-3}$	milli	m	$mA = 10^{-3}\,A$
$0.000\,001 = 10^{-6}$	micro	μ	$\mu V = 10^{-6}\,V$
$0.000\,000\,001 = 10^{-9}$	nano	n	$ns = 10^{-9}\,s$
$0.000\,000\,000\,001 = 10^{-12}$	pico	p	$pF = 10^{-12}\,F$

*Additional prefixes are exa $= 10^{18}$, peta $= 10^{15}$, femto $= 10^{-15}$, and atto $= 10^{-18}$.

TABLE F-3. Greek Letter Symbols

NAME	LETTER CAPITAL	LETTER SMALL	USES
Alpha	A	α	α for angles, transistors, and amplifier circuits
Beta	B	β	β for angles, transistors, and amplifier circuits
Gamma	Γ	γ	
Delta	Δ	δ	Small change in value
Epsilon	E	ϵ	ϵ for permittivity, also base of natural logarithms
Zeta	Z	ζ	
Eta	H	η	
Theta	Θ	θ	Phase angle
Iota	I	ι	
Kappa	K	κ	κ for coupling coefficient, dielectric constant
Lambda	Λ	λ	λ for wavelength
Mu	M	μ	μ for prefix micro, permeability, amplification factor
Nu	N	ν	
Xi	Ξ	ξ	
Omicron	O	o	
Pi	Π	π	π is constant of 3.1416 for ratio of circumference to diameter of a circle

Appendix F

Electrical Symbols and Abbreviations

TABLE F-3. Greek Letter Symbols (*Continued*)

NAME	LETTER		USES
	CAPITAL	SMALL	
Rho	P	ρ	ρ for resistivity
Sigma	Σ	σ	Summation
Tau	T	τ	Time constant
Upsilon	Υ	υ	
Phi	Φ	ϕ	Magnetic flux, angles
Chi	X	χ	
Psi	Ψ	ψ	Electric flux
Omega	Ω	ω	Ω for ohms; ω for angular velocity

This table includes the complete Greek alphabet, although some letters are not used for electronic symbols.

Color Codes

Appendix G

Included here are color codes for chassis wiring, carbon resistors, and small fixed capacitors with mica or ceramic dielectrics. Most of these codes are standarized by the Electronic Industries Association (EIA). Members are not required to follow the codes, but it is industry practice to do so where practical.

CHASSIS WIRING

Colors for the wires in electronic circuits generally follow the system in Table G-1. By noting the wiring color code, you can often save time in tracing the connections. The colors may be either solid or helical stripes on white insulation.

In addition, blue is used for high side of antenna connections.

For stereo connections in audio equipment the right channel uses red (high side) with green, and the left channel has white (high side) with blue.

CARBON RESISTORS

For ratings of 2 W or less, carbon resistors are color-coded with either bands or the body-end-dot system as summarized in Table G-2.

The color values summarized in Table G-3 apply to both resistors and capacitors. However, the colors for voltage ratings apply only to capaci-

TABLE G-1. Chassis Wiring Color Code

COLOR	ABBREVIATION	CONNECTED TO
Red	Red or R	High side of voltage source, B+ for tubes
Blue	Blue or B	Amplifier tube plate, transistor collector, FET drain
Green	Grn or G	Tube control grid, transistor base, FET gate, input of diode detector
Yellow	Yel or Y	Tube cathode, transistor emitter, FET source
Orange	Orn or O	Screen grid of tube, second base of transistor
Brown	Brn or N	Heaters or filaments
Black	Blk or K	Chassis ground return
White	Wht or W	Return for control grid (AVC bias) or base of transistor
Gray	Gra or A	Ac power line

Appendix G
Color Codes

TABLE G-2. Color Codes for Carbon Resistors (see Fig. G-1)

AXIAL LEADS	COLOR	RADIAL LEADS
Band A	First significant figure	Body A
Band B	Second significant figure	End B
Band C	Decimal multiplier	Dot C
Band D	Tolerance	End D

Notes: Band A is double width for wirewound resistors with axial leads. Body-end-dot system with radial leads is a discontinued standard. For resistors with color stripes and axial leads, body color is not used for color-coded value.

Film resistors have five stripes; fourth stripe is multiplier, and fifth is tolerance.

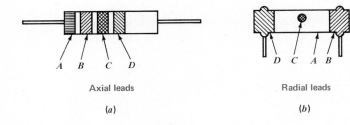

Axial leads
(a)

Radial leads
(b)

FIGURE G-1

Resistor color coding. (a) Color stripes on R with axial leads. (b) Body-end-dot colors on R with radial leads.

TABLE G-3. Color Values for Resistor and Capacitor Codes

COLOR FIGURE	SIGNIFICANT FIGURE	DECIMAL MULTIPLIER	TOLERANCE,* %	VOLTAGE RATING*
Black	0	1	20	
Brown	1	10	1	100
Red	2	10^2	2	200
Orange	3	10^3	3	300
Yellow	4	10^4	4	400
Green	5	10^5	5	500
Blue	6	10^6	6	600
Violet	7	10^7	7	700
Gray	8	10^8	8	800
White	9	10^9	9	900
Gold		0.1	5	1000
Silver		0.01	10	2000
No color			20	500

*Tolerance colors other than gold and silver are used for capacitors only. Also, colors for voltage rating apply just to capacitors.

TABLE G-4. Preferred Values for Resistors and Capacitors

20% TOLERANCE	10% TOLERANCE	5% TOLERANCE
10*	10*	10*
		11
	12	12
		13
15	15	15
		16
	18	18
		20
22	22	22
		24
	27	27
		30
33	33	33
		36
	39	39
		43
47	47	47
		51
	56	56
		62
68	68	68
		75
	82	82
		91
100	100	100

*The columns list numerical values and their decimal multiples for ohms or picofarads.

tors. Also, only gold or silver is used for carbon-composition resistor tolerance, but all the colors apply to tolerances for capacitors or film resistors.

Similarly, the preferred values in Table G-4 are for resistance values in ohms or capacitor values in picofarads. Only the basic value is listed, from which multiple values are derived. As an example, a 1500-Ω R or a 1500-pF C is a standard component value, with a tolerance of either 20, 10, or 5 percent. However, 2000 is only in 5 percent tolerance.

MICA CAPACITORS
These may be coded with old RMA or new EIA methods, military (MIL), Joint Army-Navy (JAN) specifications, or American War Standards (AWS). See Figs. G-2 and G-3.

Appendix G
Color Codes

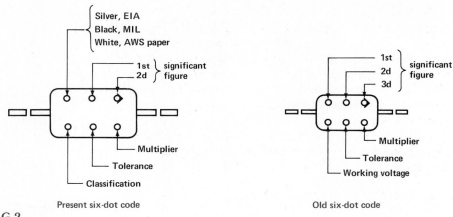

FIGURE G-2 Mica capacitors with six-dot color codes. All rated at 500 WVDC and ±20 percent tolerance.

The new EIA six-dot code starts with a white dot. If this dot is black, it indicates the MIL six-dot code. Or, if this dot is silver, it indicates a paper capacitor in the AWS code. In all three cases, though, the capacitance in picofarad units is read from the next three color dots. However, if the first dot has a color, this indicates the old EIA six-dot code, where the first four dots are used for the capacitor value.

The characteristics indicated by the last dot in the new EIA six-dot code specify five classes from A to E, according to leakage resistance, temperature coefficient, and other factors.

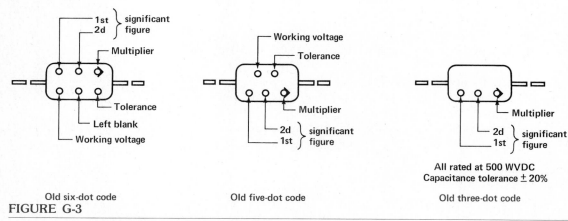

FIGURE G-3 Mica capacitors with obsolete color codes.

CERAMIC CAPACITORS

These have stripes or dots with three colors or five colors. With five stripes or dots, the first and last colors indicate temperature coefficient and tolerance, as listed in Table G-5. The middle three colors give the capacitance in picofarads, with the same color values as for resistors and mica capacitors. The fifth color is for tolerance, in percent for sizes larger than 10 pF but in picofarad units for smaller capacitors. See Table G-5 and Fig. G-4 on the next page.

Most ceramic capacitors are long, in the shape of a resistor, or the disk type. Larger disk capacitors often have the capacitance in picofarads printed on the unit.

TABLE G-5. Color Code for Ceramic Capacitors

COLOR	DECIMAL MULTIPLIER	TOLERANCE ABOVE 10 pF, %	TOLERANCE BELOW 10 pF, pF	TEMPERATURE COEFFICIENT, ppm/°C
Black	1	20	2.0	0
Brown	10	1		−30
Red	100	2		−80
Orange	1000			−150
Yellow				−220
Green		5	0.5	−330
Blue				−470
Violet				−750
Gray	0.01		0.25	30
White	0.1	10	1.0	500

Appendix G
Color Codes

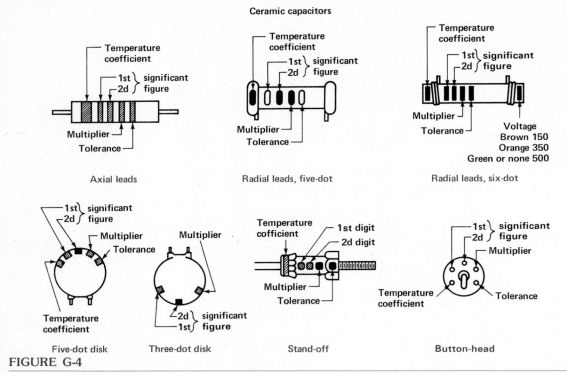

FIGURE G-4

Ceramic capacitors. Color coding for tubular, disk, stand-off, and button-head types.

Soldering and Tools

Appendix H

In addition to the usual workshop tools, working on electronic equipment often requires long-nose or needle-nose pliers to bend the end of a wire, diagonal cutting pliers to cut wire, $\frac{1}{4}$-in socket wrench for machine screws with $\frac{1}{4}$-in hexagonal head, and a Phillips screwdriver. These tools are shown in Fig. H-1.

The cutting pliers can be used when you want to strip about an inch or less of the insulation from the end of stranded wire that is not the push-back type. First crush the insulation by squeezing below the cutting edges of the pliers; then notch the crushed insulation with the cutting edge down to the wire, being careful not to cut the wire; finally pull off the crushed insulation with the cutting edges of the pliers in the notch. A wire stripper is a special tool for this.

With stiff insulation that cannot be crushed, such as the cover on a shielded coaxial cable, slit the insulation with a razor blade along the required length, pull back, and cut it off with the diagonal cutters. The braided shield can be opened by picking it apart with a pointed tool.

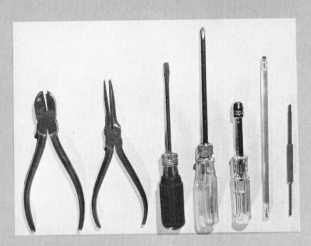

FIGURE H-1

From left to right: cutting pliers, long-nose pliers, slotted-head screwdriver, Phillips-head screwdriver, $\frac{1}{4}$-in socket wrench, slotted-head plastic alignment tool, and hexagonal plastic alignment tool.

Appendix H
Soldering and Tools

SOLDERING
Solder is an alloy of tin and lead used for fusing metals at relatively low temperatures of about 500 to 600°F. The joint where two metal conductors are to be fused is heated, and then solder is applied so that it can melt and cover the connection. The reason for soldering connections is that it makes a good bond between the joined metal, covering the joint completely to prevent oxidation. The coating of solder provides protection for practically an indefinite period of time.

The trick in soldering is to heat the joint, not the solder. When the joint is hot enough to melt the solder, the solder flows smoothly to fill all the cracks, forming a shiny cover without any air spaces. Do not move the joint until the solder has set, which takes only a few seconds.

Either a soldering iron or soldering gun can be used, rated at 25 to 100 W. See Fig. H-2. The gun is convenient for intermittent operation, since it heats almost instantaneously when you press the trigger. The small pencil iron of 25 to 40 W is helpful for soldering small connections where excessive heat can cause damage. This precaution is particularly important when working on PC boards, where too much heat can soften the plastic form and loosen the printed wiring. A soldering iron for FET devices should have the tip grounded to eliminate static charge.

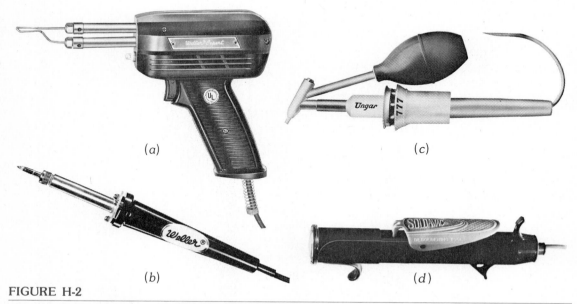

FIGURE H-2
(*a*) Soldering gun. (*b*) Soldering pencil. (*c*) Desoldering iron. (*d*) Soldavac desoldering tool. (*Edysyn Inc.*)

The three grades of solder generally used for electronics work are 40-60, 50-50, and 60-40 solder. The first figure is the percentage of tin, while the other is the percentage of lead. The 60-40 solder costs more, but it melts at the lowest temperature, flows more freely, takes less time to harden, and generally makes it easier to do a good soldering job.

In addition to the solder, there must be flux to remove any oxide film on the metals being joined. Otherwise they cannot fuse. The flux enables the molten solder to wet the metals so that the solder can stick. The two types are acid flux and rosin flux. Acid flux is more active in cleaning metals but is corrosive. Rosin flux is always used for the light soldering work in making wire connections.

Generally, the rosin is in the hollow core of solder intended for electronics work, so that a separate flux is unnecessary. Such rosin-core solder is the type generally used. It should be noted, though, that the flux is not a substitute for cleaning the metals to be fused. They must be shiny clean for the solder to stick.

DESOLDERING

On printed-wiring boards, desoldering to remove a defective component can be more important than the soldering. From the wiring side of the PC board, desolder at the holes where the component leads go through the board to join the printed wiring. Three methods are:

1. Soldering gun with vacuum attachment.
2. Soldavac tool with a separate iron. The vacuum plunger pulls the molten solder out of the connection.
3. Soldawick or similar metal braid with a separate iron. Heat the braid on the connection, and the molten solder runs up into the braid. This method is most convenient where there is no room for a vacuum tool.

Remove enough solder so that you can actually see the hole where the leads come through. Then the component practically falls off the board. If you try to force the component out, the board can be damaged.

For more details on soldering and tools, the following references can be helpful:

Anderson and Tatro: "Shop Theory," McGraw-Hill Book Company, New York.
"Radio Amateur's Handbook," (chapter on construction practices), American Radio Relay League, Newington, Conn.
"Solder," Kester Solder Co., Chicago, Ill.

Schematic Symbols

Appendix I

SCHEMATIC SYMBOLS

Device	Symbol	Device	Symbol
Ac voltage		Coil or inductance, air-core	
		iron-core	
Amplifier		variable	
AND gate		powdered iron or ferrite slug	
Antenna, general		Conductor, general connection	
dipole		no connection	
loop		Current source	
		Cystal, piezoelectric	
Battery, cell or dc voltage Long line positive		Diodes See Table 30-7 for schematic symbols	
Capacitor, general, fixed Curved electrode is outside foil, negative or low-potential side		Fuse	
variable		Ground, earth or metal frame: Chassis or common return connected to one side of voltage source Chassis or common return not connected to voltage source Common return	
ganged			

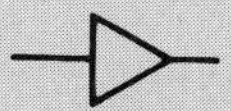

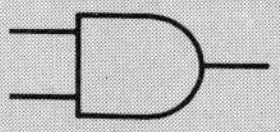

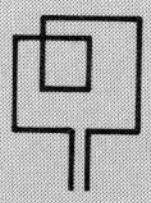

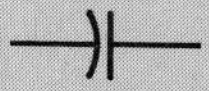

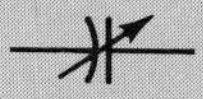

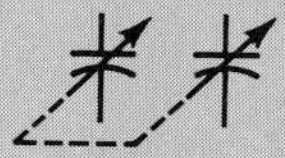

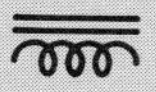

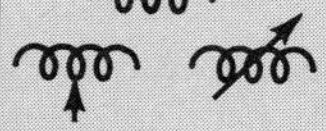

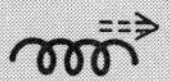

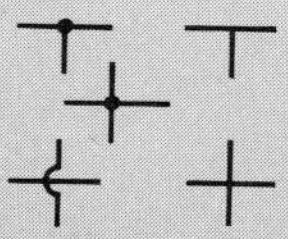

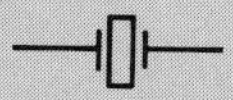

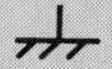

SCHEMATIC SYMBOLS

Device	Symbol	Device	Symbol
Hall potential generator		Spark gap	
Jack		Switch, (SPST)	
plug for jack	Tip / Sleeve	(SPDT)	
		2-pole (DPDT)	
Key, telegraph		3-pole, 3-circuit wafer	
Loudspeaker, general			
phones		Shielding	
		shielded conductor	
Magnet, permanent	PM		
electromagnet		Thermistor, general	
Microphone		Thermocouple	
Meters, letter or symbol to indicate range or function	A mA V	Transformer, air core	
NAND gate		iron-core	
NOR gate		autotransformer	
OR gate		link coupling	
Relay, coil		Transistors and semiconductors See Table 30-7 for schematic symbols	
contacts			
Resistor, fixed		Vacuum tubes See Table 29-3 for schematic symbols	
tapped			
variable			

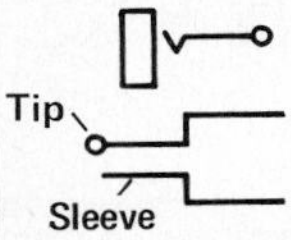

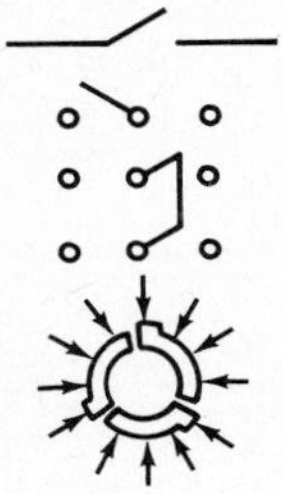

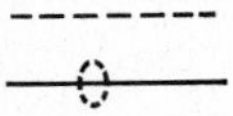

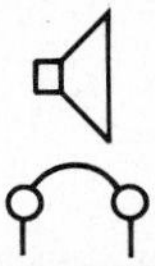

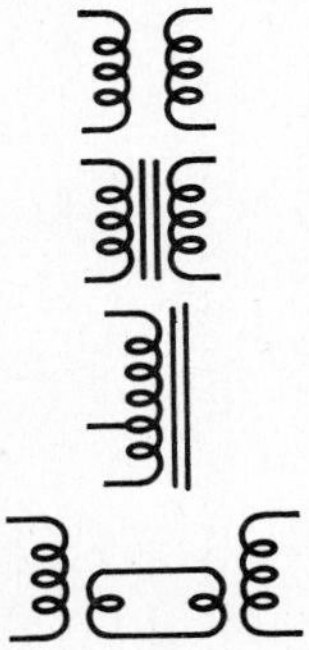

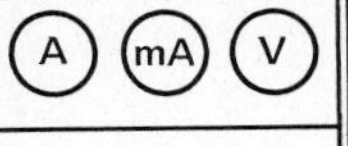

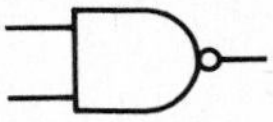

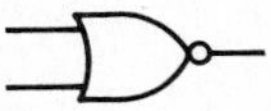

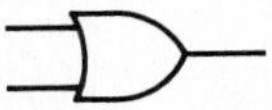

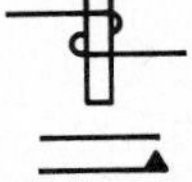

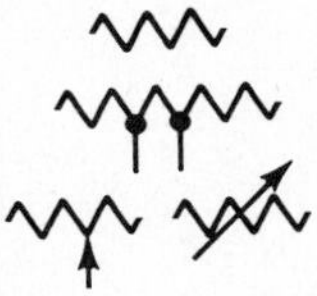

Answers to Self-Examinations

CHAPTER 1
1. T 2. T 3. T 4. T 5. T 6. T 7. T 8. T 9. T 10. T
11. T 12. T 13. T 14. T 15. T 16. T 17. T 18. T 19. F 20. F

CHAPTER 2
1. 2 2. 4 3. 16 4. 0.5 5. 2 6. 25 7. 25 8. 10
9. 0.4 10. 72 11. 8 12. 2 13. 2 14. 4.17 15. 28.8 16. 2
17. 1.2 18. 3 19. 0.2 20. 0.12

CHAPTER 3
1. (d) 2. (c) 3. (d) 4. (b) 5. (c) 6. (d) 7. (c) 8. (b)
9. (b) 10. (d)

CHAPTER 4
1. (b) 2. (a) 3. (a) 4. (d) 5. (a) 6. (c) 7. (c) 8. (b)
9. (c) 10. (b)

CHAPTER 5
1. (c) 2. (c) 3. (c) 4. (c) 5. (d) 6. (b) 7. (d) 8. (a)
9. (d) 10. (d)

REVIEW OF CHAPTERS 1 TO 5
1. (c) 2. (c) 3. (c) 4. (b) 5. (c) 6. (b) 7. (d) 8. (b)
9. (a) 10. (c) 11. (b) 12. (a) 13. (a) 14. (a) 15. (a)

CHAPTER 6
1. T 2. T 3. T 4. F 5. T 6. T 7. T 8. T 9. F 10. T

CHAPTER 7
1. (a) 2. (c) 3. (a) 4. (a) 5. (c) 6. (c) 7. (a) 8. (c)
9. (d) 10. (c)

REVIEW OF CHAPTERS 6 AND 7
1. T 2. T 3. T 4. T 5. F 6. F 7. T 8. T 9. T 10. F

CHAPTER 8
1. T 2. F 3. T 4. T 5. T 6. T 7. F 8. T 9. T 10. T

CHAPTER 9
1. T 2. T 3. T 4. T 5. T 6. T 7. T 8. F 9. T 10. T

REVIEW OF CHAPTERS 8 AND 9
1. T 2. T 3. T 4. T 5. T 6. F 7. F 8. T 9. T 10. T
11. T 12. T 13. T 14. T 15. T

CHAPTER 10
1. (a) 2. (d) 3. (d) 4. (b) 5. (b) 6. (a) 7. (b) 8. (c)
9. (c) 10. (c)

CHAPTER 11
1. (d) 2. (a) 3. (d) 4. (c) 5. (a) 6. (b) 7. (d) 8. (a)
9. (b) 10. (c)

CHAPTER 12
1. (d) 2. (c) 3. (b) 4. (a) 5. (d) 6. (d) 7. (a) 8. (c)
9. (a) 10. (d)

REVIEW OF CHAPTERS 10 TO 12
1. (d) 2. (b) 3. (b) 4. (c) 5. (d) 6. (d) 7. (b) 8. (c)
9. (c) 10. (b)

CHAPTER 13
1. T 2. T 3. T 4. F 5. T 6. T 7. T 8. T 9. T 10. T
11. F 12. T 13. T 14. T 15. F

CHAPTER 14
1. F 2. T 3. T 4. F 5. T 6. T 7. T 8. T 9. T 10. T

CHAPTER 15
1. T 2. T 3. T 4. T 5. T 6. T 7. T 8. T 9. T 10. T
11. T 12. T 13. T 14. T 15. T 16. T 17. T 18. T 19. T 20. T

CHAPTER 16
1. T 2. T 3. T 4. T 5. T 6. T
7. F 8. F 9. T 10. T 11. 28.28 V 12. 1.2 A

13. 70.7 V 14. 3×10^4 cm 15. 0.001 ms 16. 60 Hz 17. 0.01 μs 18. 0.25 MHz
19. 7.07 V 20. 40 V 21. 1000 Hz 22. 180 Hz 23. 11.1 V 24. 120 V

CHAPTER 17
1. 240 Hz 2. Off 3. 120 V 4. 240 V 5. Cathodes 6. Series
7. Dc motor 8. White 9. 208 V 10. On

REVIEW OF CHAPTERS 13 TO 17
1. (b) 2. (a) 3. (c) 4. (d) 5. (b) 6. (d) 7. (a) 8. (d)
9. (c) 10. (a) 11. (d) 12. (a)

CHAPTER 18
1. (b) 2. (c) 3. (c) 4. (d) 5. (c) 6. (d) 7. (b) 8. (d)
9. (b) 10. (a)

CHAPTER 19
1. (a) 2. (c) 3. (c) 4. (d) 5. (a) 6. (c) 7. (c) 8. (b)
9. (d) 10. (c)

CHAPTER 20
1. (c) 2. (c) 3. (c) 4. (c) 5. (b) 6. (b) 7. (c) 8. (b)
9. (c) 10. (d)

REVIEW OF CHAPTERS 18 TO 20
1. (c) 2. (b) 3. (d) 4. (d) 5. (d) 6. (d) 7. (c) 8. (a)
9. (b) 10. (c)

CHAPTER 21
1. (a) 2. (b) 3. (b) 4. (c) 5. (c) 6. (c) 7. (c) 8. (b)
9. (c) 10. (b)

CHAPTER 22
1. (b) 2. (c) 3. (c) 4. (b) 5. (d) 6. (a) 7. (a) 8. (d)
9. (b) 10. (a)

CHAPTER 23
1. (d) 2. (b) 3. (a) 4. (b) 5. (b) 6. (b) 7. (c) 8. (c)
9. (b) 10. (b)

CHAPTER 24
1. (a) 2. (c) 3. (d) 4. (b) 5. (d) 6. (c) 7. (d) 8. (c)
9. (d) 10. (c)

REVIEW OF CHAPTERS 21 TO 24
1. T 2. T 3. T 4. T 5. T 6. T 7. T 8. T 9. F 10. T
11. T 12. T 13. F 14. T 15. T 16. T 17. F 18. T 19. T 20. T
21. T 22. T 23. T 24. T 25. T 26. T 27. T 28. T 29. T 30. T
31. F 32. T 33. T 34. T 35. T 36. T 37. F 38. F 39. T 40. T

CHAPTER 25
1. (b) 2. (c) 3. (a) 4. (c) 5. (c) 6. (a) 7. (b) 8. (c)
9. (c) 10. (a)

CHAPTER 26
1. (d) 2. (m) 3. (n) 4. (j) 5. (h) 6. (l) 7. (o) 8. (k)
9. (e) 10. (a) 11. (b) 12. (f) 13. (g) 14. (i) 15. (c) 16. (p)

CHAPTER 27
1. (c) 2. (b) 3. (d) 4. (c) 5. (d) 6. (a) 7. (d) 8. (d)
9. (a) 10. (b)

CHAPTER 28
1. (d) 2. (a) 3. (b) 4. (c) 5. (d) 6. (b) 7. (a) 8. (b)
9. (a) 10. (d)

REVIEW OF CHAPTERS 25 TO 28
1. 300 2. 300 3. 300 4. 250 5. 250 6. 200 7. 200 8. 14.1
9. 14.1 10. 1 11. 45 12. −45 13. 1 14. 1.41 15. 7.07 16. 600
17. 353.5 18. 8 19. 0.8 20. 1.6 21. 0.8 22. 10 23. 10 24. 10
25. 1 26. 5 27. 50 28. 0.08 29. 40 30. 150 31. T 32. T
33. T 34. T 35. T 36. T 37. T 38. T 39. T 40. T

CHAPTER 29
1. (a) 2. (b) 3. (b) 4. (c) 5. (b) 6. (c) 7. (a) 8. (d)
9. (b) 10. (c)

CHAPTER 30
1. T 2. T 3. T 4. T 5. T 6. T 7. T 8. F 9. T 10. F
11. T 12. T 13. F 14. T 15. F 16. T 17. T 18. F 19. T 20. F
21. T 22. F 23. T 24. T 25. T 26. T 27. T 28. T 29. T 30. T
31. T 32. T

REVIEW OF CHAPTERS 29 AND 30
1. T 2. T 3. T 4. T 5. T 6. T 7. F 8. F 9. T 10. F
11. F 12. T 13. T 14. T 15. F 16. F 17. T 18. T 19. T 20. T
21. T 22. T

CHAPTER 31
1. T 2. F 3. T 4. T 5. T 6. T 7. F 8. T 9. F 10. T

CHAPTER 32
1. 2 2. 1 + 1 3. AND 4. OR 5. NAND 6. 0
7. 0 8. Four 9. MV 10. Flip-flops 11. 1 and 0 12. Four

REVIEW OF CHAPTERS 31 AND 32
1. 20 pF 2. FET 3. 18 4. Linear 5. $(111)_2$ 6. $(10)_2$ 7. AND
8. LOW 9. JK 10. Astable

Answers to Odd-numbered Problems

CHAPTER 1
1. $I = 4$ A
3. See Prob. 4 values
5. 2.2 V
7. 0.2 Ω

CHAPTER 2
1. (a) See Fig. 2-2
   (b) $I = 3$ mA
   (c) $I = 3$ mA
   (d) $I = 1$ mA
3. (b) $R = 21$ Ω
5. (a) $V = 12$ V
   (b) $P = 24$ W
   (c) $P = 24$ W
7. 10,000 V
9. (a) 1496 V
   (b) 108.1 V
   (c) 2.84 V
11. 9.84 V

CHAPTER 3
1. $I = 1$ A, $R_2 = 10$ Ω
3. $V_2 = 0.3$ V
5. $V_T = 30$ V, $I = 1$ A
7. $R_T = 2,552,470$ Ω
9. Each $R = 15$ kΩ
11. $I = 1$ mA
13. $I = 2.5$ mA
15. $R_2 = 25$ Ω
17. $V_2 = 13$ V

CHAPTER 4
1. (b) 45 V
   (c) $I_1 = 3$ A, $I_2 = 1$ A
   (d) $I_T = 4$ A
   (e) $R_T = 11.25$ Ω
3. (b) 20 V
   (c) $I_2 = 2$ A
   $I_3 = 4$ A
5. (a) $I_1 = 1$ A
   (b) $I_T = 1$ A
   (c) $R_T = 10$ Ω
   (d) $P_T = 10$ W
7. (a) 7.14 Ω
   (b) 2 kΩ
   (c) 250 Ω
   (d) 54.6 Ω
   (e) 714 Ω
   (f) 5 kΩ
9. $G_T = 0.038$ S
11. $R_3 = 666.7$ Ω

CHAPTER 5
1. (a) $R_T = 25\ \Omega$
   (b) $I_T = 4\ A$
3. (b) $R_T = 15\ \Omega$
5. (a) $R = 6\ \Omega$
   (b) $R = 24\ \Omega$
7. (a) $V_1 = 2.23\ V$
   $V_2 = 0.74\ V$
   $V_3 = 6.7\ V$
   $V_4 = 22.3\ V$
   (b) $P_1 = 204\ mW$
   $P_2 = 69\ mV$
   $P_3 = 620\ mW$
   $P_4 = 2.08\ mW$
   (c) R_4
   (d) R_4
9. $R_1 = 15\ k\Omega$
   $R_2 = 5\ k\Omega$
13. $R_T = 10.45\ \Omega$
17. (a) $V_1 = V_2 = V_3 = 40\ V$
   $I_1 = I_2 = 2\ mA$
   $I_3 = I_T = 4\ mA$
   (b) $V_{AG} = V_{BG} = +40\ V$
   I is the same as in (a)
11. $V_1 = V_X = 1\ V$
   $V_2 = V_S = 10\ V$
15. (a) $V_2 = 20\ V$
   (b) $V_1 = V_2 = 22.5\ V$

CHAPTER 6
1. $V_1 = 4\ V$
   $V_2 = 8\ V$
   $V_3 = 40\ V$
9. $R_1 = 75\ \Omega$
   $R_2 = 86.4\ \Omega$
   $R_3 = 38.5\ \Omega$
3. $I_B = 0.6\ mA$
5. $I_1 = 3\ mA$
   $I_2 = 6\ mA$
   $I_3 = 30\ mA$
7. $R_D = 555.6\ \Omega$
   $R_E = 740.7\ \Omega$
   $R_F = 500\ \Omega$

CHAPTER 7
1. (a) $R_S = 50\ \Omega$
   (b) $R_S = 5.55\ \Omega$
   (c) $R_S = 0.505\ \Omega$
   (d) 1 mA, 5 mA, and 50 mA
9. (a) $R_1 = 145\ \Omega$
   (b) 150 Ω for 5 mA
3. (a) 300 mA
   (b) 60 V
11. $V_1 = 40\ V$
    $V_2 = 80\ V$
5. (a) 199 kΩ
   599 kΩ
   9999 kΩ
   (b) 20,000 Ω/V
   (c) 10 MΩ
7. $R_1 = 78.3\ \Omega$
   $R_2 = 7.83\ \Omega$
   $R_3 = 0.87\ \Omega$

CHAPTER 8
1. $I_1 = 1.42\ A$
3. $I_1 = 1.42\ A,\ V_{R_1} = 11.36\ V$
   $I_2 = 0.32\ A,\ V_{R_2} = 0.64\ V$
   $I_3 = 1.1\ A,\ V_{R_3} = 4.4\ V$
   $I_4 = 1.1\ A,\ V_{R_4} = 2.2\ V$
5. $I_A = I_1 = 1.1\ A$

CHAPTER 9
1. $V_{Th} = 15\ V$
   $R_{Th} = 3\ \Omega$
   $V_L = 6\ V$
9. $V_{R_2} = 19.2\ V$
3. $I_1 = 5\ A$
   $R_1 = 4\ \Omega$
   $I_L = 3\ A$
11. $V_{R_2} = 19.2\ V$
5. R_L not open
13. $V_{R_3} = 10.6\ V$
7. $V_P = 42\ V$
15. See Fig. 9-28

CHAPTER 10
1. (a) 1024 cmil
   (b) Gage No. 20
   (c) $R = 1.015\ \Omega$
3. (a) 1-A fuse
   (b) 0 V
   (c) 120 V
5. $R = 48\ \Omega$
7. 5000 ft
9. (a) $4.8\ \Omega$
   (b) 4000 ft
11. 2 V
13. $I = 30$ A

CHAPTER 11
1. (a) 1 W
   (b) 2 W
3. (a) 470 kΩ, 10 percent
   (b) 2.2 MΩ, 10 percent
   (c) 33 Ω, 5 percent
   (d) 910 Ω, 5 percent
   (e) 2.2 Ω, 5 percent
   (f) 10 kΩ, 20 percent
5. 50 Ω, 1 W

CHAPTER 12
1. 1.5 mA
3. 600 A
5. 2.88×10^5 C
7. 20 kΩ
9. 1 kΩ

CHAPTER 13
1. 5×10^3 Mx
   5×10^{-5} Wb
3. 0.1 T
5. 24×10^3 Mx
7. 300

CHAPTER 14
1. (a) 200
   (b) 1000
3. (a) 300 G/Oe
   (b) 378×10^{-6} T/(A/m)
   (c) 300
5. (a) 126×10^{-6}
   (b) 88.2×10^{-6}
7. (b) 40 V
   (c) 1000 A/m
   (d) 0.378 T
   (e) 3.02×10^{-4} Wb
   (f) 67×10^4 A/Wb
9. 14.4

CHAPTER 15
1. 9 kV
3. (a) 2 Wb/s
   (b) -2 Wb/s
   (c) 2 Wb/s
5. (a) 0.2 A
   (b) 80 ampere-turns
   (c) 400 ampere-turns/m
   (d) 0.252 T
   (e) 1.512×10^{-4} Wb

CHAPTER 16
1. (a) $I = 12$ A
   (b) $f = 60$ Hz
   (c) $0°$
   (d) 120 V
3. (a) $t = 0.25$ ms
   (b) $t = 0.0625\ \mu$s
5. (a) $f = 20$ Hz
   (b) $f = 200$ Hz
   (c) $f = 0.2$ MHz
   (d) $f = 0.2$ GHz
7. (a) $+10$ and -10 V
   (b) $+10$ and -10 V
   (c) $+10$ and -10 V
   (d) $+15$ and -5 V
9. $I_1 = 40\ \mu$A
   $I_2 = 20\ \mu$A
   $V_1 = V_2 = 200$ V
   $P_1 = 8$ mW
   $P_2 = 4$ mW
11. $I = 2.5$ A
    $f = 60$ Hz
    $0°$

CHAPTER 17
No problems.

CHAPTER 18
1. (a) 2 A/s
   (b) 10,000 A/s
   (c) 10,000 A/s
   (d) −10,000 A/s
3. (a) 10 mV
   (b) 50 V
   (c) 50 V
   (d) −50 mV
5. (a) 60 Hz
   (b) 960 V
   (c) 96 mA
   (d) 0.768 A
7. (a) 300 μH
   (b) 66.7 μH
   (c) 280 μH
   (d) 0.0707
9. (a) $R = 10.52 \, \Omega$
   (b) $R = 0.01 \, \Omega$
11. 1.21×10^{-3} J
13. 82.6 mH

CHAPTER 19
1. At 100 Hz, $X_L = 314 \, \Omega$
   At 200 Hz, $X_L = 628 \, \Omega$
   At 1000 Hz, $X_L = 3140 \, \Omega$
3. (b) $I = 5$ mA
   (c) $V_L = 10$ V
5. $X_L = 1.2$ kΩ
   $L = 3.18$ H
7. $L = 0.159$ H
   $X_L = 10$ kΩ
9. (a) $X_{L_T} = 5$ kΩ
   (b) $I = 2$ mA
   (c) $V_{L_1} = 2$ V
   $V_{L_2} = 8$ V
   (d) $L_1 = 2.65$ H
   $L_2 = 10.6$ H
11. (a) $f = 0.16$ kHz
   (b) $f = 1.28$ kHz
   (c) $f = 0.4$ MHz
   (d) $f = 1.6$ MHz
   (e) $f = 16$ MHz
13. $X_L = 1628.6 \, \Omega$
15. $X_L = 754 \, \Omega$
17. (d) At 500 Hz, $X_L = 785 \, \Omega$
    $I = 12.8$ mA

CHAPTER 20
1. (a) $Z = 100 \, \Omega$
   $I = 1$ A
   $\theta = 0°$
   (b) $Z = 100 \, \Omega$
   $I = 1$ A
   $\theta = 90°$
   (c) $Z = 70.7 \, \Omega$
   $I = 1.41$ A
   $\theta = 45°$
3. (b) $X_L = 377 \, \Omega$
   (c) $Z = 340 \, \Omega$
   (d) $I = 250$ mA
   (e) $I = 40$ mA
5. $Z = 400 \, \Omega$
   $X_L = 400 \, \Omega$
7. $R_e = 94 \, \Omega$
9. $Z = 566 \, \Omega$
   $I = 0.177$ A
   $V_L = 70.7$ V
   $V_R = 70.7$ V
   $\theta_Z = 45°$
11. At 800 Hz,
    $I_R = 0.25$ A
    $I_L = 0.125$ A
    $\theta_I = -26.6°$
13. $X_L = 500 \, \Omega$
    $L = 15.9$ H
15. $Z_T = 583 \, \Omega$
    $I = 0.2$ A
    $\theta_Z = 59°$
17. v_L is a square wave,
    ± 200 V p-p

Answers to Odd-numbered Problems

CHAPTER 21
1. $Q = 400 \ \mu C$
3. $Q = 12 \ \mu C$
   6 V
5. $C = 1062$ pF
7. (a) 200 V
   (b) $Q = 200 \ \mu C$
   (c) $C = 1 \ \mu F$
9. (a) 2.5×10^{-2} J
   (b) 12.5 J
   (c) 3.2 J
11. (a) 0.06 μF
    (b) 74.2 pF

CHAPTER 22
1. 80 pF at 1 MHz
3. (b) $I = 5$ mA
   (c) $f = 1$ kHz
5. (b) $I = 2$ mA
   (c) $V_{C_1} = 2$ V
   (d) $V_{C_2} = 8$ V
7. $f = 1590$ Hz
9. (a) $X_{C_T} = 200 \ \Omega$
   (b) $C = 333.3$ pF
   $C_T = 1000$ pF
11. (a) $X_C = 300 \ \Omega$
    $C = 8.85 \ \mu F$
    (b) $C = 17.7 \ \mu F$
13. $X_C = 169.4 \ \Omega$
15. $I = 0.96$ mA
17. $C = 422$ pF

CHAPTER 23
1. (b) $Z = 50 \ \Omega$
   (c) $I = 2$ A
   (d) $V_R = 80$ V
   $V_C = 60$ V
   (e) $\theta_Z = -37°$
3. $C = 1590$ pF at 100 Hz
   $C = 1.59$ pF at 100 kHz
5. At 60 Hz, $C = 26.59 \ \mu F$
   At 1 kHz, $C = 1.50 \ \mu F$
   At 1 MHz, $C = 1590$ pF
7. $I_C = 15$ mA
   $I_R = 20$ mA
   $I_T = 25$ mA
   $Z_T = 1.2$ kΩ
   $\theta_I = 37°$
   $V_R = V_C = 30$ V
9. For dc or ac,
   $V_1 = 400$ V
   $V_2 = 200$ V
   $V_3 = 100$ V
11. $C = 67 \ \mu F$
    $L = 106$ mH
13. $Z_T = 583 \ \Omega$
    $I = 0.2$ A
    $\theta_Z = -59°$
15. i_C is a square wave, ± 2 mA p-p

CHAPTER 24
1. (a) 0.05 s
   (b) 0.05 μs
   (c) 1 ms
   (d) 20 μs
3. (a) 4 s
   (b) 100 V
5. $v_C = 86$ V
7. $v_C = 150$ V
9. 1.4 ms
11. 0.05×10^6 V/s

CHAPTER 25
1. (a) 100 W
   (b) No reactance
   (c) 1
3. (b) $I = 10$ A, approx.
   (c) $Z = 10 \;\Omega$
   (d) $\theta = 0°$
5. (c) $Z_T = 500 \;\Omega$
   $I = 0.8 \;\Omega$
   $\theta_Z = 53°$
7. (a) $X_L = 0$, approx.
   $X_C = 665 \;\Omega$
   (b) $Z_T = 890 \;\Omega$
   $I = 135$ mA
   $\theta_Z = -47.9°$
9. (a) 180°
11. $R = 24 \;\Omega$
13. $R = 267 \;\Omega$
15. $C = 7.7 \;\mu F$
17. $I_T = 6.9$ mA, $\theta_I = -16.9°$
    $Z_T = 60.9 \;\Omega$, $\theta_Z = 16.9°$
19. $R = 9704 \;\Omega$

CHAPTER 26
1. (a) $4 - j3$
   (b) $4 + j3$
   (c) $3 + j6$
   (d) $3 - j3$
3. (a) $5\underline{/-37°}$
   (b) $5\underline{/37°}$
   (c) $3.18\underline{/18.5°}$
   (d) $4.24\underline{/-45°}$
5. $Z_T = 65.36 + j23.48$
7. (a) $4.5\underline{/14°}$
   (b) $4.5\underline{/34°}$
   (c) $100\underline{/-84°}$
   (d) $100\underline{/-60°}$
9. $Z_T = 12.65\underline{/18.5°}$
11. $Z_T = 5.25\underline{/-14.7°}$
13. $R = 5.08 \;\Omega$
    $X_C = 1.27 \;\Omega$
15. $R = 21.4 \;\Omega$
    $X_L = 10.2 \;\Omega$
17. $Z_T = 50\underline{/-37°} = 40 - j30 \;\Omega$
    $I = 2\underline{/37°} = 1.6 + j1.2$ A
    $V_R = 80\underline{/37°} = 64 + j48$ V
    $V_L = 120\underline{/127°} = -72 + j96$ V
    $V_C = 180\underline{/-53°} = 108 - j144$ V
19. $Z_T = 4.05\underline{/7.4°}$ kΩ
    $I = 19.75\underline{/-7.4°}$ mA
21. $Z_T = 13.29\underline{/46.2°}$

CHAPTER 27
1. $f_r = 12.6$ Hz
3. $f_r = 1.624$ MHz
5. (a) $Q = 100$
   (b) $X_C = X_L = 1500 \;\Omega$
   (c) $I = 1$ mA
   (d) $V_C = 1.5$ V

7. $L = 254\ \mu H$ at 1 MHz
   $L = 15.9\ \mu H$ at 4 MHz

9. $C_{max} = 32.7$ pF
   $C_{min} = 21.7$ pF

11. (a) At f_r of 795 kHz,
   $I_T = 0$
   $Z_T = \infty$
   $\theta = 0°$
   (b) At 895 kHz,
   $I_T = 1.25\underline{/90°}$ mA
   $Z_T = 4\underline{/-90°}$ kΩ
   (c) At 695 kHz,
   $I_T = 1.3\underline{/-90°}$ mA
   $Z_T = 3.85\underline{/90°}$ kΩ

13. (a) $Q = 200$
   $BW = 25$ Hz
   (b) $Z_T = 200$ kΩ

15. (a) At 5 kHz,
   $Z_T = 5\ \Omega$
   $I = 100$ mA
   (b) See Fig. 27-3b.

17. (a) $Q = 120$
   (b) $r_S = 78.4\ \Omega$
   (c) $C = 56.5$ pF

CHAPTER 28

1. (a) $C = 0.01\ \mu F$
   (b) $V_R \cong 0$ V
   $V_C = 20$ V
   (c) $V_R = 7.07$ V rms value
   $V_C \cong 0$ V
7. (b) $L = 1.53$ mH

3. $C = 0.064\ \mu F$

9. (a) $C_1 = 0.001\ 59\ \mu F$
   (b) $C_2 = 0.003\ 18\ \mu F$
   (c) $C_3 = 0.007\ 95\ \mu F$

5. Dc level is -8 V
   Max is -11 V
   Min is -5 V
   Ac voltage is 6 V p-p

CHAPTER 29

1. (a) $V_{C_2} = 150$ V
   (b) $V_b = 190$ V
7. $R_L = 130$ kΩ
13. $V_b = 180$ V
   $V_{c_2} = 96$ V
   $V_k = 3$ V

3. $I_b = 0.8$ mA, constant value

9. $A_V = 40$

5. $g_m = 4000\ \mu S$

11. (a) $A_V = 12$
   (b) $1200\ \mu V$

CHAPTER 30

1. 0.6 V

3. (a) 160 V
   (b) 138 V

5. $V_C = 7.5$ V
   $V_{CE} = 6.7$ V
   $V_B = 1.4$ V
   $V_{BE} = 0.6$ V
   $V_E = 0.8$ V

7. $\beta = 21.4$
   $\alpha = 0.955$

9. 2 W

11. 7500 μS

13. 2.8 mA

15. (a) 50
    (b) 20 V
    (c) 250

17. $I_C = 5.2$ mA p-p $A_I = 52$
    $V_C = 8.32$ V p-p $A_V = 41.6$
                          $A_P = 2163.2$

19. All conducting

CHAPTER 31
No problems.

CHAPTER 32
1. (a) 100111
   (b) 111110

3. 14

5. (a) $x = \overline{AB}$
   (b) $x = \overline{A + B}$
   (c) $x = AB$
   (d) $x = \overline{A} + \overline{B}$

Index

(A)

A batteries, 240, 600, 604
Absolute temperature, 726
Acceptor elements, 628
Admittance, 533
Air gap, magnet, 271–272
α (alpha) temperature, 214–215
  transistor, 647
Alternating current, 31–32, 310–336
  average value, 318
  cycle, 312–313
  effective value, 318–319
  frequency, 319–321
  nonsinusoidal, 329–331
  peak value, 317–318
  period, 321–322
  phase angle, 323–327
  rms value, 318–319
  sine-wave, 314–316
  three-phase, 344–345
Alternating-current circuits, 498–540
  analysis, 498–518
  complex numbers, 519–540
  impedance, 503, 506
  meters, 510–511
  ohms, types of, 511–512
Alternating-current circuits (cont.):
  parallel type, 505–507
  phase angle, 504, 506
  phasors, types of, 512–513
  power factor, 509
  series type, 503–505
Alternating-current power line, 337–351
Alternating voltage, 310–336
Alternation, 313
AM, defined, 3
AM radio band, 4, 321
Amateur radio, 5, 720
Ammeters, 129–135
  shunts for, 131–135
Amp-clamp probe, 149
Ampere, 24
Ampere-turns, 280–281, 291
Amplification factor, 609
Amplifiers:
  class, 649–650
  transistor, 639–654
  tube, 606–608
AND function, 689, 694
Anode, 600
Antiresonance, 562
Apparent power, 509
Armature, 339
ARRL, 5
Atomic number, 15, 721–723

Atomic structure, 12–17
Audio frequencies, 320
Autotransformer, 364–365
Average ac value, 318
Average dc axis, 573–574
Ayrton shunt, 133–135

(B)

B battery, 240–241
B voltage, 600, 604
B-H curve, 284–285
Ballast, lamp, 341
Bandpass filter, 583–586
Band-stop filter, 584–585
Bandwidth, 555–558
 half-power points, 558
 measurement, 558
Bank, parallel, 79, 94–95
Barrier potential, 630–631
Base, transistor, 626, 637–638
Bass tone, 320
Batteries, 22, 32, 240–252
 alkaline, 250–251
 carbon-zinc, 243–244
 charging, 249–250
 Edison, 251
 lead-acid, 247–250
  specific gravity, 248–249
 mercury, 250–251
 nickel-cadmium, 251–252
 parallel, 245–247
 primary, 242, 250
 secondary, 242–243, 251
 series, 245–247
 shelf life, 244
 storage, 242–243
 types, 250–251
Baud unit, 705
Beam power tubes, 615
β (beta), transistor, 646
Bias, base, 642
 grid, 606

Bias stabilization, 651–652
Binary numbers, 685–689
 addition, 688
 multiplication, 688
Bipolar transistors, 657
Bistable circuits, 701
Bleeder current, 121
Bohr atom, 15
Branch currents, 76–78
Bridge circuit, 100–101
Bridge rectifier, 636
Broadcasting, radio, 3–5
Brushes, motor, 340
Brown and Sharpe gage, 204–205
Btu, heat unit, 726
Bulk, FET, 652
BX cable, 342
Bypass capacitor, 577–580
Byte unit, 706

(C)

C battery, 240
C voltage, 600, 604
Calorie, 726
Capacitance, 418–441
 capacitor types, 424–430
 charging, 419–421
 discharging, 421
 distributed, 433
 energy stored, 434
 farad unit, 422
 stray, 431–434
 time constant, 466–467, 476–478
 voltage dividers, 464–465
Capacitive coupling, 575–577
Capacitive diodes, 660
Capacitive reactance, 442–464
 applications, 448, 463, 575–580
 calculations, 444–446

Capacitive reactance (cont.):
 frequency effect, 445
 phase angle, 457–458
Capacitor action, ohmmeter, 435–436
Capacitors, 9, 424–430
 ceramic, 426
 color code, 429–430, 737–740
 coupling, 463–464
 diode, 660
 electrolytic, 426–428
 mica, 425
 Mylar, 426
 paper, 425
 parallel, 430
 series, 430–431
 tantalum, 428
 testing, 435–436
 tolerance, 428
 troubles in, 434–436
 variable, 426
 voltage rating, 428–429
Carbon resistors, 226–227, 236
Cascaded amplifiers, 644–645
Cathode, thermionic, 600–601
Cathode-follower, 640
Cathode-ray tube, 616–617
CB circuit, 639–641
CC circuit, 642
CE circuit, 640–644
Celsius scale, 726
Cermet resistors, 227
Cgs units, 268, 724–727
Channel, FET, 652–653
Charge, electric, 17–26
 coulomb unit, 17–18
 electric field, 20, 289–290
 types of, 25–26
Chassis ground, 101–104
Choke coil, 405–406, 489–490

Circuit, defined, 28–31
Circular mil, 205–206
Clock generator, 701
C/MOS IC chips, 678–679
Coaxial cable, 206–207
Coercive force, 286
Collector, transistor, 626, 637
Collector rings, 339
Color codes, 735–740
   capacitors, 429–430, 736–740
   resistors, 230–231, 735–736
   transformers, 365–366
   wiring, 735
Common-base circuit, 639–641
Common-collector circuit, 642
Common-emitter circuit, 640–652
   bias stabilization, 651–652
   load line, 647–650
Common mode, 680
Commutator, 339
Compensating diodes, 659
Complement, binary, 706
Complementary symmetry, 678–679, 682
Complex numbers, 519–540
   j operator, 520–521
   polar form, 526–528
   rectangular form, 524–526
Compound winding, field, 339
Conductance, 27–28, 82–83, 119–120
   siemens unit, 27
Conductors, wire, 203–218
   circular mils, 205–206
   printed, 207–208
   resistance wire, 214
   temperature coefficient, 214–215

Conductors, wire (cont.):
   wire sizes, 204–206
Constant-current source, 255–256
Constant-k filter, 584
Constant-voltage source, 255
Continuity testing, 153–154
Core, coil, 366–368
Corona effect, 220
Cosine function, 728–731
COS/MOS IC chips, 678–679
Coulomb unit, 17–20
Counting circuits, 703–705
Coupling:
   capacitive, 489, 463–464
   transformer, 574–575
Covalent bonds, 218, 626–627
CRT, 616–617
Cryogenics, 215
Current:
   alternating, 31–32, 310–336
   ampere unit, 24
   conventional, 30–31
   direct, 30–31
   hole, 218–219, 629
   ion, 216–217
   magnetic field, 26–27
Current dividers, 118–120
Cutoff frequency:
   filter, 581
   transistor, 647
Cutoff voltage, grid, 605
Cycle, defined, 319–321

(D)

Damping resistance, 563–564
Darlington pair, 642
D'Arsonval movement, 128
De Morgan's identity, 692

Decade resistance box, 227–228
Decibel scale, 149
Degaussing, 286–287
Delta connections, 193–195, 344
Demagnetization, 286–287
Depletion-type FET, 654
Depletion zone, 631
Desoldering, 207, 743
Detector, diode, 636
Diac, 656
Diamagnetism, 274
Dielectric constant, 423–424
Dielectric materials, 219–220
Differential amplifier, 679–681
Diffusion current, 630
Digital electronics, 685–712
   AND gate, 689, 694
   binary numbers, 685–688
   clock generator, 701
   counters, 703–705
   flip-flops, 701–703
   logic, 689–694
   NAND gate, 685, 695–696
   NOR gate, 686, 696–698
   OR gate, 689–690, 694–695
   symbols, 706
Digital meters, 150–151
Diode gate circuits, 694–695
Diode rectifier circuits, 633–636
Diodes:
   Esaki, 659
   light-emitting, 660
   semiconductor, 633–636, 659–660
   symbols, 659, 663
   testing, 636
   tubes, 600–604
   tunnel, 659
   varactor, 660

Diodes (cont.):
  varistor, 659
  Zener, 659–660
DIP package, 676–677
Dipole magnet, 274
Direct current, defined, 31–32
Direct-current circuits:
  Kirchhoff's laws, 161–174
  meters, 126–160
  networks, 175–202
  Ohm's law, 39–53
  parallel, 74–90
  series, 54–73
  series-parallel, 91–125
  voltage dividers, 116–125
Discrete components, 597
Dissipation, power, 47–49
Donor elements, 628
Doping, semiconductor, 34, 218, 627–629
DPDT switch, 209
Drain, FET, 626, 652
Drift current, 630
DTL circuits, 693–698
Dynamometer, 510
Dyne unit, 725, 727

(E)

Eddy currents, 366
Edison cell, 251
Edison wiring system, 342
EDP, defined, 6
Effective value, 318–319
Efficiency, 365
EHF band, 719
EIA, 716
Electric field, 20, 289–290
Electrolyte, 217
Electrolytic capacitors, 426–428
Electromagnetism, 273, 294–309

Electron, 11–17
Electron gun, CRT, 616
Electron volt, 46
Electronics:
  components, 7–10
  defined, 1
  divisions, 5–7
Elements, chemical, 14, 721–723
Emitter, transistor, 626, 637
Emitter-follower, 642
Energy:
  in capacitance, 434
  in inductance, 372
  units, 46–47, 725
Enhancement-type FET, 654
Epitaxial transistors, 657
Erg unit, 725, 727
Esaki diode, 659
Extrinsic semiconductors, 627

(F)

Fahrenheit scale, 726
Fan-in, 693
Fan-out, 693
Farad unit, 422
Faraday's law, 303–306
FCC, 3, 719–720
Ferrite beads, 275
Ferrite core, 368
Ferrites, 275
Ferromagnetic materials, 274
FET, 652–655
Field-effect transistor amplifier, 654
Field-effect transistors, 652–655
  characteristic curves, 654–655
  g_m, 654
  electrodes, 652
  transconductance, 654

Field winding, 339
Filament, tube, 600–601
Filters, analysis of, 570–592
  ac bypass, 577–580
  ac coupling, 574–577
  bandpass, 583–586
  band-stop, 584–585
  constant-k, 584
  high-pass, 582–583
  interference, 586–587
  L-type, 581–582, 585–586
  low-pass, 581–582
  m-derived, 584
  π-type, 581–582
  resonant, 584–586
  T-type, 581–582
Flip-flop circuits:
  counters, 703–705
  D-type, 702–703
  JK-type, 703
  set-reset, 701–702
Fluctuating direct current, 571–574
Fluorescent lighting, 340–341
Flux:
  electric, 289–290
  magnetic, 267–270
  soldering, 743
FM, defined, 3
FM radio band, 5, 321, 720
Form factor, 319
Forward current, 631–632
Forward voltage, 631–632
Frequency:
  allocations, 720
  audio, 320
  defined, 319–321
  harmonic, 331
  radio, 320
Frequency spectrum, 717–718
Full-wave rectifier, 635–636
Fuses, 210–211
  testing, 152

(G)

Gain, amplifier, 607–608, 644
Galvanometer, 128
Gate:
 FET, 626, 652
 SCR, 655
Gate circuits, 689–695
Gauss unit, 268–269, 277
Gaussmeter, 276
Generator, ac, 311–314
 three-phase, 338–340
Getter, tube, 601
Gilbert unit, 281
g_m:
 FET, 654
 tube, 609
Graphs, plotting, 44–45
Ground, chassis, 101–104

(H)

Half-power frequencies, 558
Half-wave rectifier, 602–603, 633–634
Hall effect, 276
Harmonic frequencies, 331
Heat sink, 658
Henry unit, 354
Hertz unit, 319–320
High-pass filters, 582–583
High side, circuit, 62, 101
High-voltage probe, 150
Hole charge, 25–26, 218, 628–629
Hole current, 629
Horsepower unit, 46
Hot resistance, 215, 617
Hot-wire meter, 510
Hum, 636
Hybrid IC chip, 674
Hysteresis losses, 285–286, 366–367
Hz unit, 319–320

(I)

IC chips, 2–3, 673–684
IEEE, 716
IGFET, 653
Impedance:
 capacitive, 459–462
 complex form, 523–524
 defined, 329
 inductive, 398, 401–402
 matching, 257–258
 parallel, 506
 series, 503–505
Induced voltage, 302–306
 nonsinusoidal, 407–408
 sine-wave, 386–390
Inductance, 352–378
 chokes, 405–406
 defined, 354
 energy stored, 372
 henry unit, 354
 mutual, 358–360, 370–371
 parallel, 369
 self-inductance, 356
 series, 369–371
 stray, 371
 time constant, 408–409, 473–476
 transformers, 360–366
 troubles, 372–373
 variable, 368–369
Induction:
 electrostatic, 419–420
 magnetic, 270–271, 294–309
Induction motor, 339
Inductive reactance, 379–393
 calculations, 381–384
 chokes, 405–406
 frequency effects, 382
 phase angle, 386–390
Inductors, 9–10
Instant-start lamp, 341
Insulators, 219–220

Integrated circuits, 673–684
 differential amplifier, 679–681
 digital, 674, 693–712
 hybrid, 674
 linear, 674, 681
 monolithic, 673
 operational amplifier, 680–681
 packages, 676–677
 production, 675–679
 thick-film, 674
 thin-film, 674
Interference filters, 586–587
Internal resistance:
 coil, 403–404
 generator, 252–258
Intrinsic semiconductor, 627
Ions, 25–26, 33, 216–218
 bonds, 217
 charge, 25–26
 current, 216–217
 in semiconductors, 629
IR voltage drops, 58–62
Iron-vane meter, 510

(J)

j operator, 520–521
JFET, 653
Joule unit, 46, 725, 727
Junction (PN), 630–633
 barrier voltage, 630–631
 forward voltage, 631–632
 reverse voltage, 632
 voltampere curve, 632–633

(K)

K temperature, 726
Keeper, magnet, 272
Kelvin scale, 726
Kilowatthours, 46–47

Kirchhoff's laws, 161–174
  branch-current method, 164–167
  current law, 161–163
  mesh-current method, 169–172
  node-voltage method, 167–169
  voltage law, 163–164
Knife switch, 209

(L)

L, inductance, 354–356
L-type filters, 581–582, 585–586
Laminations, core, 367–368
Lead-acid battery, 247–250
  charging, 249
  chemical action, 247–248
  specific gravity, 248–249
Leakage current, transistor, 630, 647
Leakage flux, 358
LED, 660
Lenz' law, 301–302, 356
Light-emitting diode, 660
Linear IC chips, 674, 681
Linear resistance, 45
Litz wire, 404
Load current, 30, 120–123
Load-line analysis, 647–650
Load resistance, 30
Loading, voltmeter, 140–142
Logic:
  negative, 693
  positive, 693
  symbols, 689–692, 706
Logic circuits, 693
Loose coupling, 359
Low-pass filters, 581
Low-power ohms, 149

L/R time constant, 408–409, 473–476
LSI chips, 674

(M)

m-derived filter, 584
Magnetic field, 26, 289–290, 294–296
Magnetic units, 277, 280–293
Magnetism, 265–279
  air gap, 271–272
  electromagnets, 273
  ferrites, 275
  field lines, 265–266
  flux, 267–270, 277
  Ohm's law, 287–288
  permanent magnets, 273–274
  permeability, 282–283
  poles, 266–267
  shielding, 275–276
  types of, 274
  units, 277, 280–293
Magnetomotive force, 280, 287–288
Maxwell unit, 268, 277
Mercury cell, 250–251
Mesa transistors, 657
Mesh current, 169–172
Metal-film resistors, 227
Meters:
  ac, 510–511
  dc, 126–160
  digital, 150–151
  VOM, 147–150
  VTVM, 147–151
Metric prefixes, 43, 733
Millman's theorem, 191–193
Mks units, 268, 727
Mmf, 280, 287–288
Molecules, 14

Monolithic IC, 673
Motors:
  action, 298
  parts of, 338–340
Multimeters, 147–151
Multiplier, voltmeter, 136–140
Multivibrators, 700–701
MOSFET, 653
Mutual inductance, 358–360, 370–371
  coupling coefficient, 359
  transformers, 360–366
MV circuits, 700–701

(N)

NAND function, 691, 695–696
Nanosecond, 321
Negation, logic, 690
Negative resistance, 659
Network theorems, 175–202
  conversion of sources, 187–191
  delta-wye, 193–195
  Millman, 191–193
  Norton, 184–188
  superposition, 175–177
  T and π, 193–195
  Thevenin, 177–184, 188
Neutron, 17
Nickel-cadmium cell, 251–252
Node, 167–168
Noise, tube, 618
Nonlinear resistance, 45
Nonsinusoidal waveforms, 329–331, 465–466
NOR function, 691, 696–698
Norton's theorem, 184–188
NOT function, 690–691
NPN transistors, 625, 636–639
Nucleus, in atom, 17

(O)

Octave, 331
Oersted unit, 282
Ohm unit, 27
  types of, 511–512
Ohmmeters, 142–147
  back-off scale, 143–145
  low-power, 149
  testing: capacitors, 435–436
    coils, 372–373
    continuity, 153–154
    diodes, 636
    resistors, 234–235
    transformers, 373
    transistors, 661–662
  zero adjustment, 146–147
Ohm's law, 39–53
  magnetic circuits, 287–288
Ohms-per-volt rating, 139–140
Open circuit, 67, 84–85, 104–106
  capacitance of, 434
  voltage tests, 152–153
Operational amplifier, 680–681
OR function, 689–690, 694–695, 699
  exclusive, 690
  inclusive, 690
Orbital electrons, 15–16
Oscillators, 598–599
Oscilloscope, 617

(P)

Parallel circuits, 74–90
  batteries, 245–247
  capacitors, 430
  compared with series, 107

Parallel circuits (cont.):
  conductances, 82–83, 119–120
  current dividers, 118–120
  inductances, 369–370
  open, 84–85
  resistors, 78–82
  shorted, 85–86
Paramagnetic materials, 274
Parameters, tube, 609
Passband, filter, 582
PC board, 207–208
Peak value, 317–318
Pentode tubes, 612–614
Period, defined, 321–322
Permanent magnets, 273–274
Permeability, 282–283, 291
Permeance, magnetic, 290–291
Permittivity, electric, 290–291
Phase angle, 323–327, 398–399, 402, 448–452, 457–458
Phasor:
  defined, 325
  types of, 512–513
Photoelectricity, 33
Phototubes, 615–616
Physics units, 724–727
  cgs system, 724, 727
  mks system, 724, 727
  heat units, 726–727
  SI units, 724, 727
  temperature scales, 726
π-type filter, 581–582
π-type network, 193–195
Picofarad, 422
Pilot lamps, 212
Plate, tube, 600
Plate resistance, 609
PN junction, 630–633

PNP transistors, 625, 636–639
Polarity:
  to ground, 102–104
  of IR drops, 61–63
Poles, magnetic, 266–267
Positive logic, 693
Potential, defined, 20–21
Potential difference, 20–22
Potentiometers, 228–230
Power:
  apparent, 509
  calculations, 48–49
  real, 507–510
  VARs, 509
  watt, 45–47
  wattmeter, 510–511
  units, 726–727
Power factor, 509
Power line, ac, 337–351
  advantages, 338–339
  three-phase, 344–345
  wiring, 341–344
Preferred values, 231, 737
Primary cells, 242, 250
Printed wiring, 207–208
Probes:
  amp-clamp, 149
  high voltage, 150
Protons, 11–17
PUT, 656
Pulsating direct current, 571–574

(Q)

Q, charge, 17–20
Q factor, coil, 403–404
  resonant circuit, 552–555
  bandwidth, 556–558
  damped, 563–564
  measurement of, 553

Q point, amplifier, 649
Quadrature phase, 324

(R)

Radian angle, 314
Radio, 1, 3–5
Radio frequencies, 320, 717–720
Radix, 685–686
RAM, defined, 707
Ramp voltage, 331
Rapid-start lamp, 341
RC coupling, 463–464, 575–577
RC time constant, 466–467, 476–486
Reactance:
  capacitive, 442–456
  inductive, 379–393
Real power, 507–510
Reciprocal formula, 78–80
Rectifier circuits, 599, 633–636
  bridge, 636
  full-wave, 635–636
  half-wave, 602–603, 633–634
  hum, 636
  ripple, 603, 606
  voltage doubler, 636
Reluctance, magnetic, 287–288, 291
Remote cutoff tubes, 613–614
Reset state, 701–702
Residual induction, 286
Resistance, 8, 27, 56–58, 78–82
  parallel, 78–82
  series, 56–58
  specific, 212–214
Resistance wire, 214
Resistors, 225–239

Resistors (cont.):
  carbon, 226–227, 236
  cermet, 227
  color code, 230–231, 735–736
  decade box, 227–228
  metal-film, 227
  parallel, 78–82
  power rating, 232–233
  preferred values, 231, 737
  series, 56–58
  tolerance, 231
  troubles, 234–235
  variable, 227–230
  wirewound, 225–226, 236
Resonance, 541–569
  bandwidth, 555–558
  calculations, 550–552, 564–565
  damping, 563–564
  parallel, 546–549
  Q, 552–555, 561–562
  series, 542–546
  tuning, 558–561
Resonant filters, 584–586
Retentivity, 286
Reverse current, 630, 651
Reverse voltage, 630, 632, 647
Rheostats, 228–230
Ripple, ac, 603, 636
Rms value, 318–319
ROM, defined, 707
Romex cable, 342
Root-mean-square, 318–319
Rosin-core solder, 743
RTL, 693, 698

(S)

Saturation:
  magnetic, 284–285
  transistor, 642
Sawtooth waveform, 330

SCR, 655–656
Screen grid, tube, 610–614
Secondary cell, 242–243, 249
Secondary emission, 611–612
Semiconductors, 8, 34, 218–219, 624–684
  covalent bonds, 626–629
  devices, 625–626
  diac, 656
  diodes, 633–636, 659–660
  doping, 627–629
  FETs, 652–655
  integrated circuits, 673–684
  PN junction, 630–633
  SCR, 655–656
  symbols, 663
  thyristors, 655–656
  transistors, 636–655
  triac, 656
  UJT, 656
Sensitivity, voltmeter, 140
Series circuits, 54–73
  ac source, 503–505
  batteries, 245–247
  capacitances, 430–431
  compared with parallel, 107
  dc source, 54–73
  inductances, 369–371
  open in, 67–68
  resistances, 56–58
  resonant, 542–546
  voltage dividers, 116–118
  voltage drop, 58–62
Series-parallel circuits, 91–115
  analysis, 97–100
  open, 104–106
  short, 104–106
Series voltages, 64, 245–247
Set state, 701–702

Index

SHF band, 719
Shielding, 275–276, 366
Shock, electric, 49
Short circuit, 85–86, 104–106
Shot effect, tube, 618
Shunts, meter, 131–135
SI units, 268, 290–291, 724–727
Siemens unit, 27, 290
Signal, defined, 598
Silicon, 218, 626–627
Silicon-controlled rectifier, 655–656
Sine function, 728–731
Sine wave, 314–316
Single-ended, 680
Sinusoid, 314–316
Skin effect, 404
Slip rings, 339
Slow-blow fuses, 211
Solar battery, 252
Soldering, 207, 741–743
Solenoid, 297
Sonic frequencies, 321
Sound waves, 323
Source, FET, 626, 652
Space charge, tube, 602
Spaghetti for wire, 206
SPDT switch, 209
Specific gravity, 248–249
Specific resistance, 212–214
SPST switch, 209
Square wave, 330
Stabilization, bias, 651–652
Standard broadcast band, 4
Starter, lamp, 341
Static charge, 17, 20, 33
Stop-band filter, 582
Storage cells, 242–243, 251
Stray capacitance, 431–434
Stray inductance, 371, 431–434
String, series, 57, 93–94
  heaters, 59–60

Substrate, FET, 652
Superconductivity, 215
Superposition theorem, 175–177
Supersonic frequencies, 6, 321
Suppressor grid, 612
Susceptance, 533
Switches, 208–210
  three-way, 343–344
Switching circuits, logic, 689–693
Symbols:
  diode, 659, 663
  letter, 606, 650, 732–734
  logic, 706
  schematic, 744–745
  transistors, 663
  tubes, 620

(T)

T networks, 193–195
T-type filters, 581–582
Tangent function, 728–731
Tank circuit, 549
Taper, resistance, 129
Taut-band meters, 129
Television broadcasting, 5
Tesla unit, 269, 277, 291
Testing:
  capacitors, 435–436
  coils, 372–373
  continuity, 153–154
  diodes, 636
  fuses, 211
  resistors, 234–235
  transformers, 373
  transistors, 661–662
Thermal runaway, 651
Thermionic emission, 33, 600–601
Thermistors, 215
Thermocouple, 510

Thevenin's theorem, 177–184, 188
Three-phase power, 344–345
  delta, 344
  four-wire, 345
  wye, 344–345
Thyristors, 656
Time base, 331
Time constants, 472–497
  long, 483–486
  L/R, 408–409, 473–476
  RC, 466–467, 476–486
  short, 483–484
  universal curves, 486–488
Toggle switch, 209
Toroid magnet, 271–272
Transconductance:
  FET, 654
  tube, 609
Transformer coupling, 574–575
Transformers, 360–366, 574–575
  autotransformers, 364–365
  color codes, 365–366
  isolation, 365
  troubles, 373
  turns ratio, 361–363
Transistors, 3, 34, 218–219, 624–672
  alpha (α), 647
  amplifier circuits, 639–645
  analysis of load line, 647–650
  beta (β), 646
  bias stabilization, 651–652
  common-base (CB), 639–641
  common-collector (CC), 642
  common-emitter (CE), 640–652
  curves, 645–646
  emitter-follower, 642

768 Index

Transistors (cont.):
  FET, 652–655
  forward bias, 631–632
  heat sink, 658
  reverse voltage, 632
  symbols, 663
  testing, 661–662
  troubles, 660–662
  type numbers, 658–659
Transmission line, 206–207
Treble tone, 320
Triac, 656
Trigonometric functions, 728–731
  table, 729
Triode tubes, 604–610
Troubles:
  capacitors, 434–436
  coils, 372–373
  diodes, 636
  resistors, 234–235
  transistors, 660–662
  tubes, 617–618
Truth tables, 691–692
TTL, 693, 698–700
Tubes, vacuum, 7, 597–623
  amplifier, 597, 606–608
  beam power, 615
  bias, grid, 606
  curves, 608, 610
  CRT, 616–617
  diodes, 600–604
  parameters, 609
  pentodes, 612–614
  photo, 615–616
  sockets, 614–615
  symbols, 606, 620
  tetrodes, 610–612
  thermionic emission, 600–601
  triodes, 604–610
  troubles, 617–618
Tuning, 558–561
Tunnel diode, 659

Turns ratio, 361–363
Twin-lead cable, 206–207

(U)

UHF band, 719
UJT, 656
Ultrasonics, 6
Unijunction transistor, 656
Unipolar transistor, 657
Units:
  electrical, 732
  magnetic, 277, 280–293
  physics, 724–727
  practical, 42
  SI, 268, 290–291, 724–727
Universal motor, 339–340
Universal shunt, 133–135

(V)

Vacuum tubes, 7, 597–623
Valence, electron, 16
VAR unit, 509
Varactor diodes, 660
Variac, 369
Variometer, 368
Varistor, 659
Vector, 325
VHF band, 719
Volt unit, 20–22
Voltage dividers, 116–125, 464–465
  capacitive, 431, 464–465
  load currents, 122–123
Voltage drops, 58–62, 66
Voltaic cell, 241–242
Voltampere, 509
Voltampere characteristic:
  diode, 632–633
  resistance, 45
Voltmeters, 135–142
  digital, 150–151

Voltmeters (cont.):
  loading effect, 140–142
  multipliers, 136–140
  ohms-per-volt, 139–140
VOM, 147–150
VTVM, 147–151

(W)

Watt unit, 45–47
Wattmeter, 510–511
Wavelength, 322–323
Wavetrap filter, 586
Weber unit, 268, 277, 291
Weston movement, 128
Wheatstone bridge, 100–101
Wire:
  color code, 735
  gage sizes, 204–206
Wirewound resistors, 225–226, 236
Work, units, 46–47, 725, 727
Wye network, 193–195, 344–345

(X)

X_C, capacitive reactance, 442–456
X_L, inductive reactance, 379–393

(Y)

Y, admittance, 533
Y connections, 344–345
Y network, 193–195

(Z)

Z, impedance, 398, 503–506
Zener diode, 659–660
Zero-ohms adjustment, 146–147